D0024505

ATLAS of ANATOMY

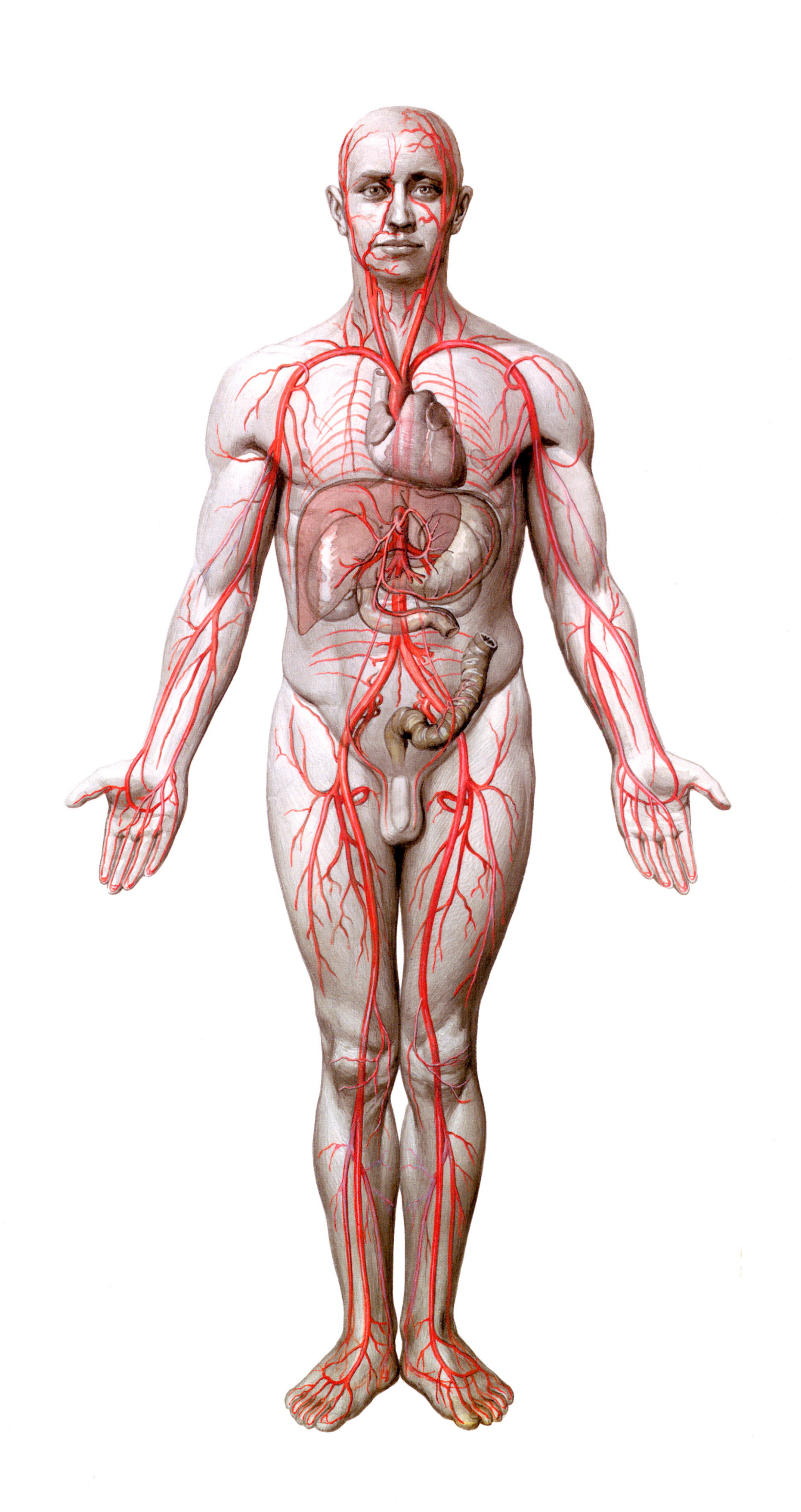

ATLAS of ANATOMY

h.f.ullmann

h.f.ullmann is an imprint of Tandem Verlag GmbH
Original title: Atlas der Anatomie
ISBN for the German edition: 978-3-8331-5468-3

h.f.ullmann is an imprint of Tandem Verlag GmbH

English translation by Mo Croasdale, Ann Drummond, David Hefford, Judith Phillips and Katherine Taylor
in association with First Edition Translations Ltd, Cambridge, UK
Editing by Sue Peter
Typesetting by The WriteIdea
Cover design: Simone Sticker
The images on the cover are all taken from the content.

ISBN 978-3-8331-5192-7

10 9 8 7 6 5 4 3 2 1
X IX VIII VII VI V IV III II I

www.ullmann-publishing.com

Printed in China

Everything starts with anatomy

Anyone wanting to really understand the human body has to start with anatomy. Many issues relating to health and illness, to physical performance capabilities, and functions, are based on anatomy and cannot be understood without a basic knowledge of the human organ system and structures. It is not without reason that all basic healthcare training courses—be they in medicine or nursing, physiotherapy, or alternative healer training—begin with an in-depth study of anatomy. In an age when every human being ought to be taking the greatest possible responsibility for their own body, it is worth following the professional example and always having a basic anatomical reference book to hand for issues regarding your own health. With this Atlas of Anatomy you have just such an intelligible reference book in your hands.

A journey of discovery

The Atlas of Anatomy is an invitation to undertake an adventurous journey of discovery through the human body, starting with external appearance and ending in a single human cell.

The individual stops along the way:

- External appearance
- Musculature, bones, and joints
- Brain and the nervous system
- The eye
- The ear and organ of equilibrium
- Olfactory organs
- Respiratory tract and lungs
- The heart
- Circulatory system and blood vessels
- Lymphatic system
- The spleen
- Digestive system
- Kidneys and urinary tract
- Hormone system
- Sexual organs
- The cell

Perfect tour guaranteed

In the interests of improved comprehension, each of the 16 chapters is preceded by an explanatory text with descriptive semi-schematic images. These serve to ensure a basic understanding of the image plates that follow.

The graphic anatomical material in this atlas has been selected from a pool comprising some of the best material in the world: more than 600 images, the majority of which are classic color drawings, together with a number of explanatory diagrams and X-ray images.

All of the anatomical structures are clearly labeled and specialist Latin terms in general use have also been included. Comprehensive captions explain the images. Specific attention is paid to anatomical structures, which are susceptible to pathological changes or injury. The detailed index ensures quick access to any topic required.

Contents

External appearance

It was Andreas Vesalius (1514-1564), a Belgian physician to Emperor Charles V, who first described anatomy as a science in itself. Anatomy is the study of the construction of the human body, from the muscles, bones, and organs, down to the tiniest cellular structures that can only be seen with an optical or electronic microscope. Anatomy involves studying the body's physical features, and to do this, the body is formally divided into regions, the skin being regarded as the body's outer sheath.

Build

Whether someone is tall or short, broad or slender is determined largely by their build, or physique. Gender can also usually be determined from build: while men tend to have broad shoulders and a narrow pelvis, women have narrower shoulders and a wider pelvis to accommodate a growing baby during pregnancy. Women also have well-developed mammary glands, while in men these are rudimentary. Overall, the female form is softer and more rounded, due largely to the higher proportion of fat under the skin, while the male body is more angular. Men usually have stronger bones and a more developed musculature than women and are generally taller. Typically, men also have a larger larynx and more body hair, especially on the face and chest.
However, physical differences are not only determined by gender: genetic and external factors such as climate and diet also play an important role.

Body regions

For purposes of understanding and communication globally, each body region and anatomical part has been allocated a formal (usually Latin) name. The naming of the areas is related to their anatomical structures. For example, the upper arm is known as the deltoid area, after the deltoid muscle; the region around the navel is the umbilical area (from the Latin umbilicus = navel), and the upper abdominal area between the ribs and the navel, is the epigastric region (from the Greek epi = next to, beside; gastricus = of the stomach).

Skin

The skin covers the entire body surface. It is made up of three layers: the epidermis, dermis, and the subcutaneous layer. The epidermis is further subdivided into five layers: the surface of the epidermis consists of flattened, cornified cells piled one on top of the other (stratum corneum). The cornified layer is thicker in areas, such as the soles of the feet and palms of the hand, which are subject to the most wear and tear. Beneath the superficial layer is the so-called clear layer, and beneath that a granular layer. Beneath the granular layer is the prickle cell layer, which consists of mature living skin cells (keratocytes), while in the basal layer of the epidermis vigorous cell multiplication occurs. Dead skin cells are constantly being shed and these are replaced by new ones from the basal epidermal layer. The basal epidermal layer also contains pigment cells, which give the skin its color. The epidermis contains no blood vessels; its living cells are nourished by diffusion of nutrients from the most superficial layers of the underlying dermis.
The interface between the basal epidermal layer and the dermis is undulating. The dermis consists of elastic tissue that makes it both firm and resilient. Unlike the epidermis, it contains numerous tiny blood and lymph vessels, as well as many tiny nerve branches that originate from so-called lamellar corpuscles. These corpuscles are sensory cells, arranged in layers like an onion; they are found in both the dermis and the subcutaneous layer and transmit sensations of pain, itching, pressure, and information regarding temperature.
The subcutaneous layer consists of fatty and connective tissue, and is flexibly bound to the tissue beneath it (e.g., muscle or bone). It too contains a network of nerve fibers and tiny blood and lymph vessels. The thickness of the subcutaneous layer is determined by the level of physical demand as well as the individual's age and gender.
Hair, nails, and sweat glands are among the skin's so-called appendages. Hair growth starts at the hair root, or bulb, which reaches down into the subcutaneous layer. Slightly above it is a tiny muscle that is connected to the hair and contracts in response to cold, pulling the hair upright. Sebaceous glands around the hair produce sebum, a fatty substance that is discharged at the base of the hair. Sebaceous glands are also found on parts of the body with less hair, such as the nose, ears, and back. In these areas sebum is discharged through pores in the skin.

The bulbous ends of the sweat glands extend into the dermis and also partly into the fatty tissue of the subcutaneous layer. Their secretions reach the surface of the skin via convoluted ducts.
The finger- and toe-nails are thin, rounded, hard plates of a tough protein called keratin. They provide support and protection for the tips of the fingers and toes.

Function of the skin

The primary function of the skin is to protect the body against environmental stresses, such as water, light, heat, as well as mechanical and chemical threats. The fatty substance excreted by the sebaceous glands and the protective acid layer created by the sweat glands, both help to prevent pathogens such as bacteria and molds from entering the body. The pigment cells in the epidermis filter out ultraviolet rays and protect the skin from sunburn when the sun is strong. The skin's other main function is temperature regulation, i.e., keeping the body temperature at a constant level. If the ambient temperature is high, the tiny blood vessels in the skin expand. This increases the diameter of the vessels, which in turn causes the blood to flow more slowly, so that more warmth can be lost from the surface. When it is cold, the blood vessels in the skin contract, with the opposite effect: thus the skin acts as an "insulating layer" protecting the body from environmental influences and minimizing heat loss. Increased sweat production also cools the body down: as it evaporates it also gives off heat.
The body can lose up to two liters of water per day through perspiration, something that may be advantageous in certain conditions, for example, in cases of renal disease.
The skin is also a sensory organ, with a great many tiny nerve endings that perceive touch, pain, and temperature stimuli, thus providing us with information on our environment.

Tension lines of the skin

One characteristic feature of the skin is that it does not possess the same level of elasticity all over. This results in tension lines, which always follow the direction of least elasticity. This is why incisions, such as those required for minor surgical procedures, especially if they involve the hands or feet, should be made in or parallel to these lines.

Fig. 1
Physical differences between the male and female body. The male body is more angular, the muscles more defined, and the shoulders wider in proportion to the pelvis than the female body. The female body is typically an hour-glass shape. There are also clear differences in the breasts and sexual organs.

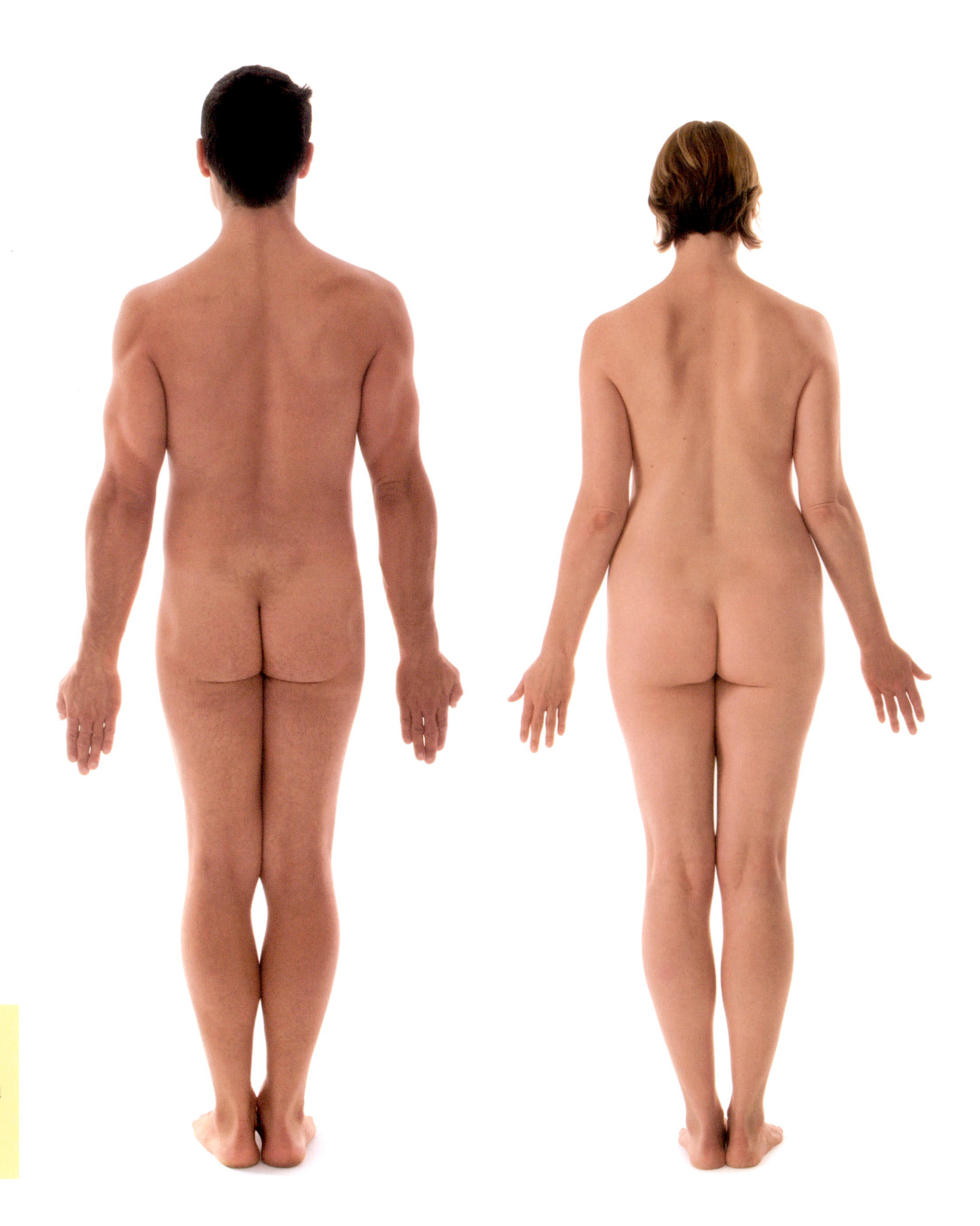

◆
Fig. 2
Physical differences between the male and female body from behind.

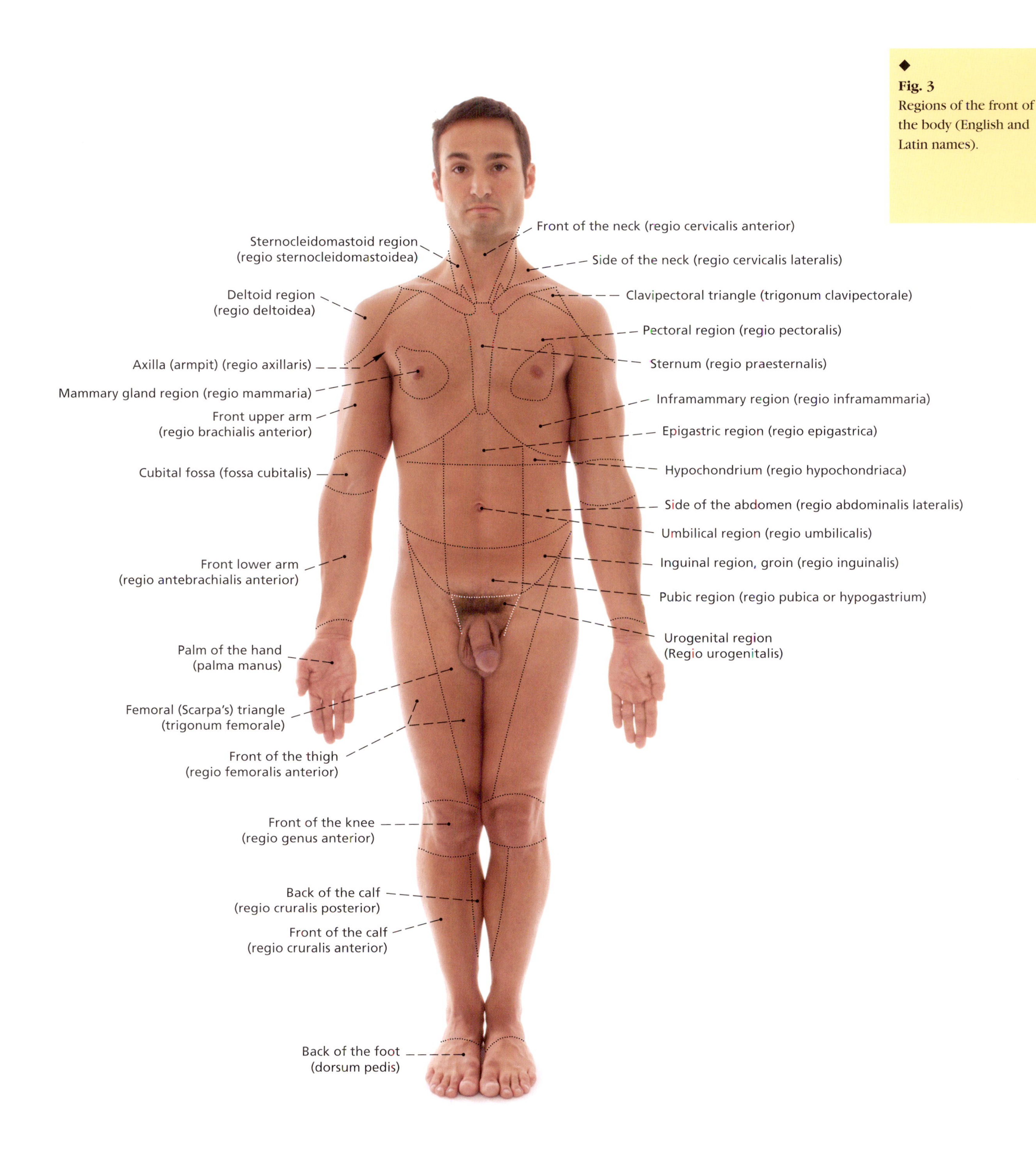

◆
Fig. 3
Regions of the front of the body (English and Latin names).

◆
Fig. 4
Regions of the back of the body.

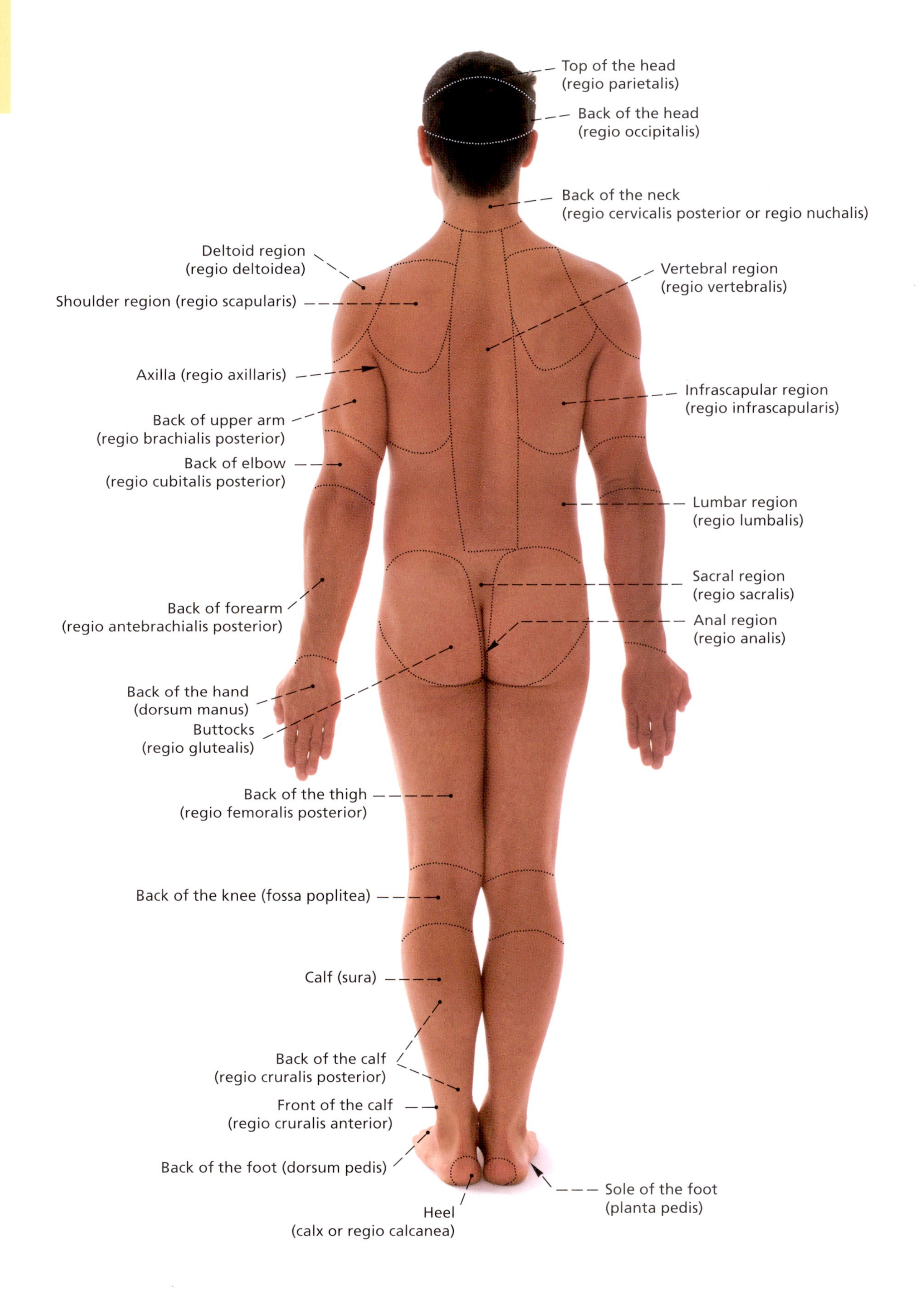

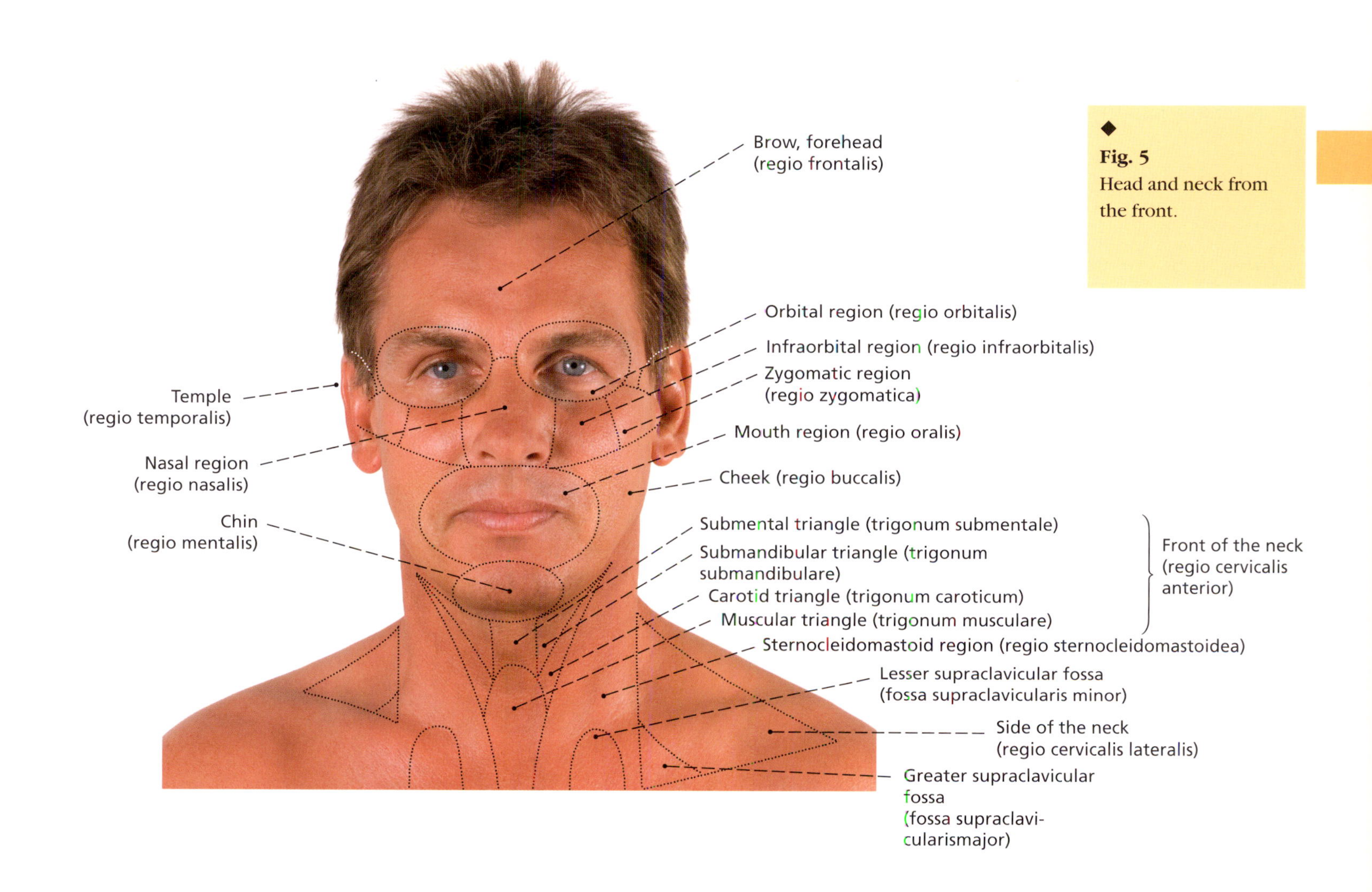

◆ **Fig. 5**
Head and neck from the front.

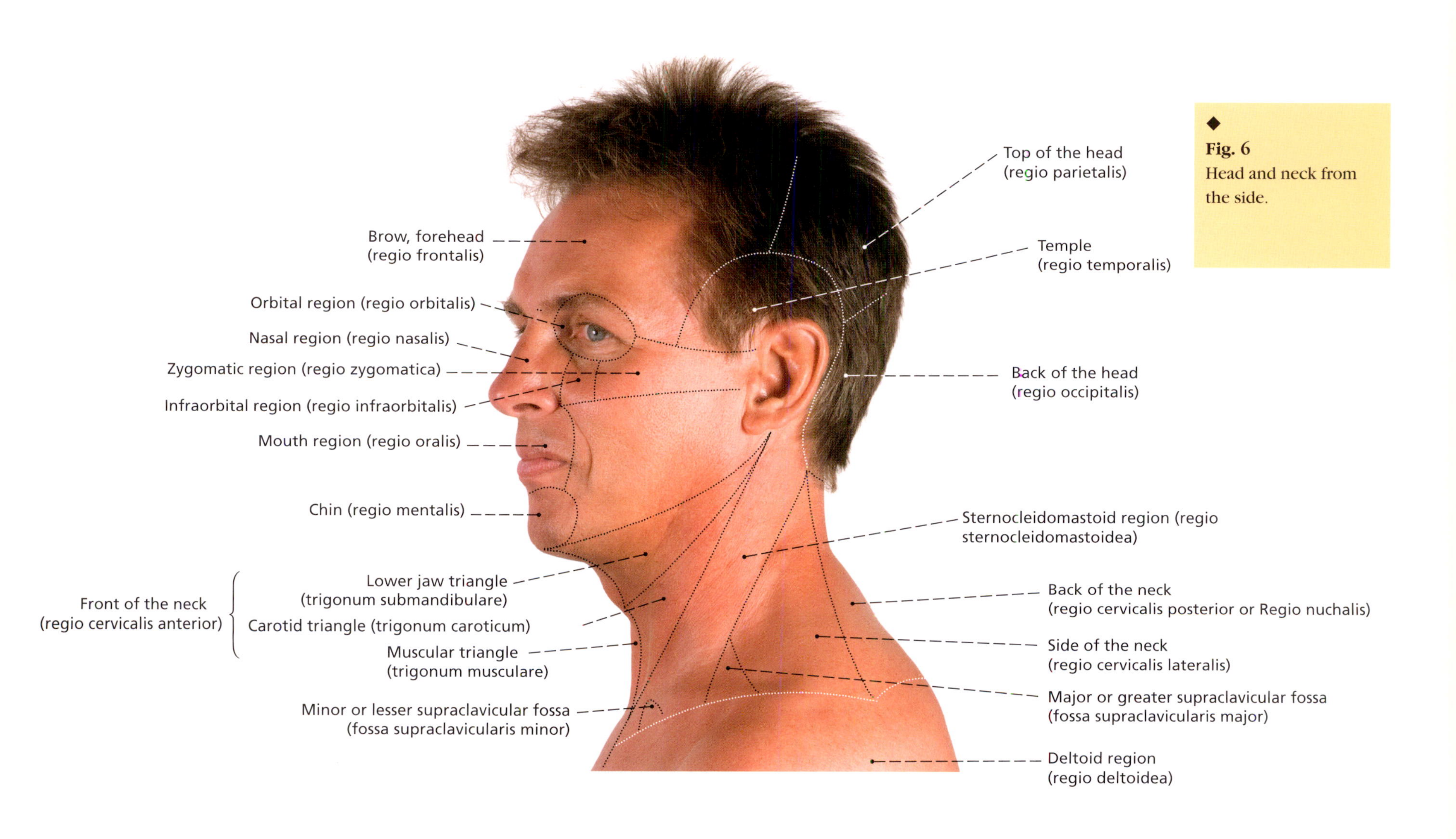

◆ **Fig. 6**
Head and neck from the side.

Fig. 7
Cross-section of the skin. There are three main layers: 1. the epidermis with a horny layer and a basal layer, 2. the dermis and 3. the subcutaneous layer with fatty tissue. The lamellar corpuscles in the subcutaneous layer and adjoining dermis are involved in the perception of touch.

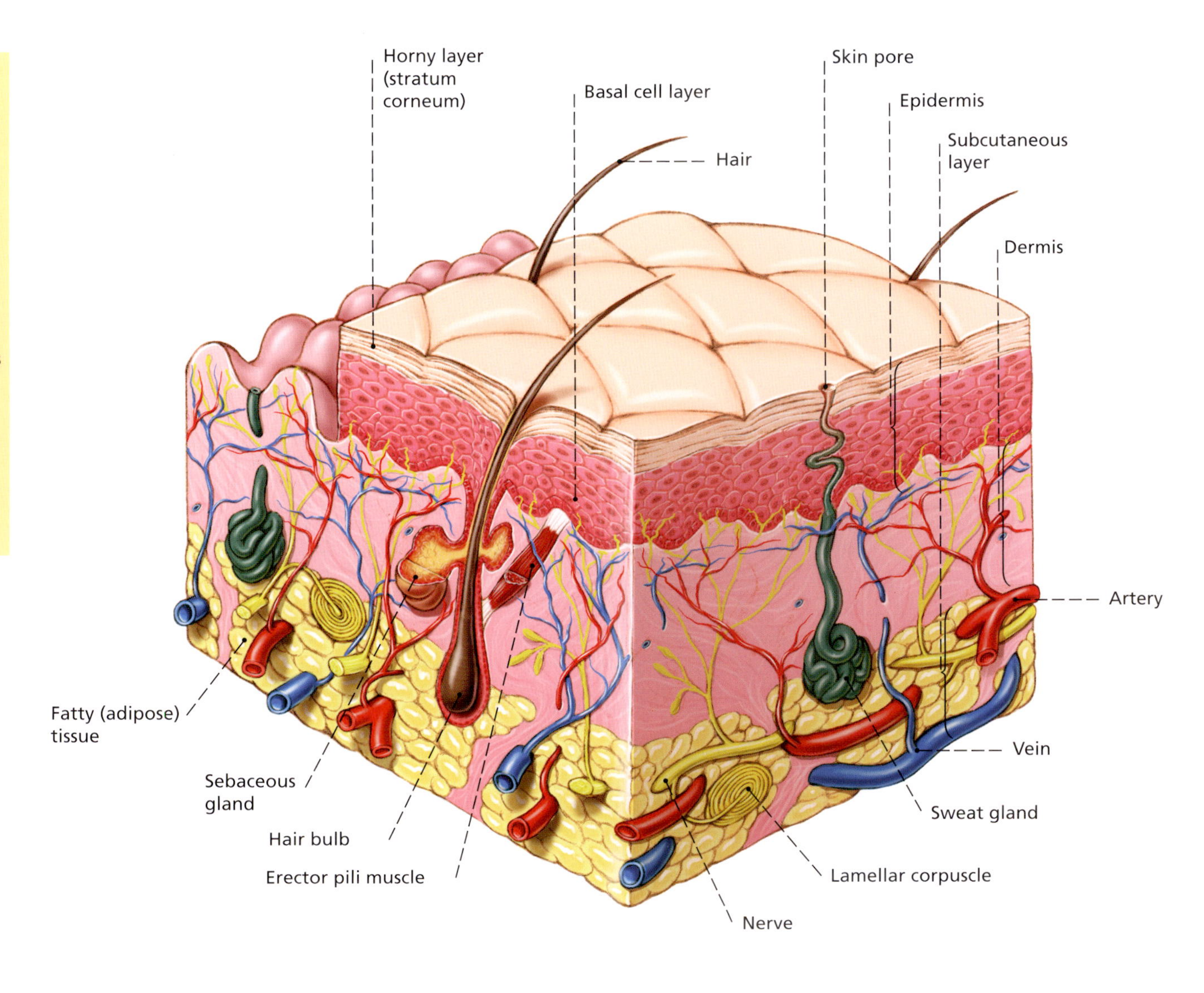

Fig. 8
Magnified cross-section of the epidermis as seen under a microscope. The basal and prickle cell layers form the basal cell layer. Pigmented cells in the epidermis are what give the skin its color.

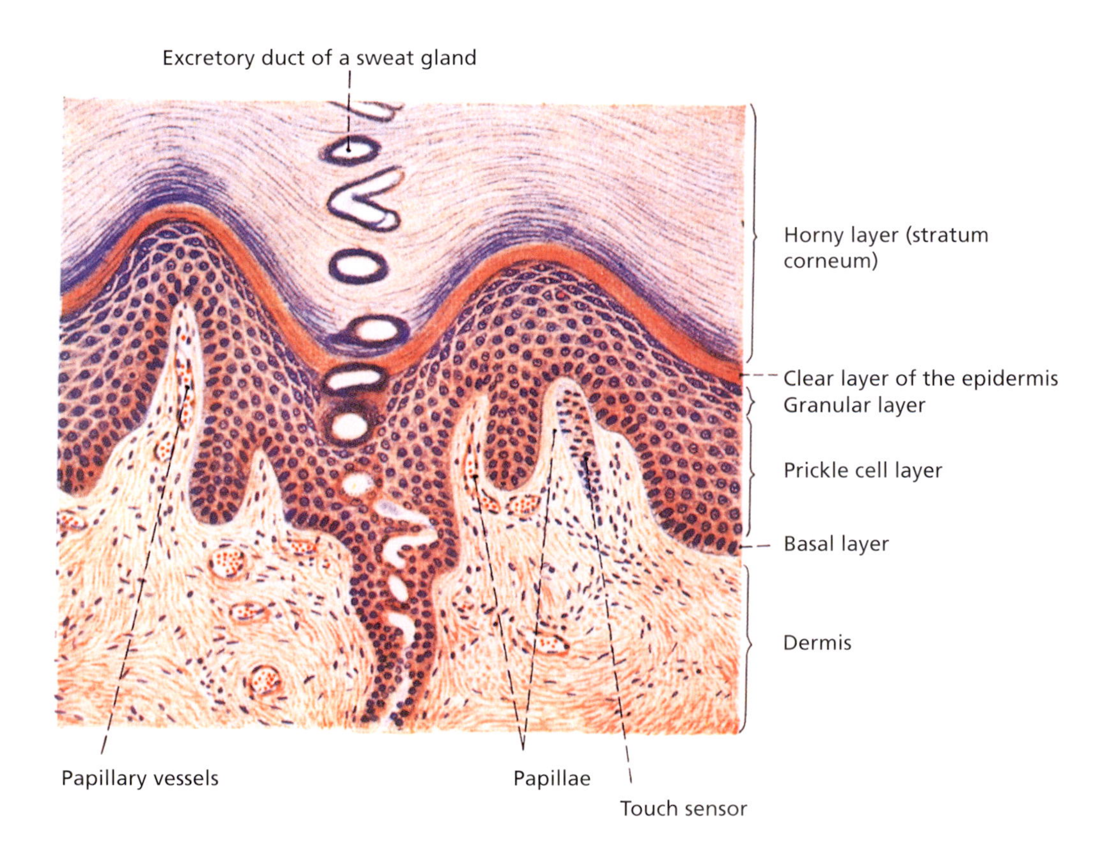

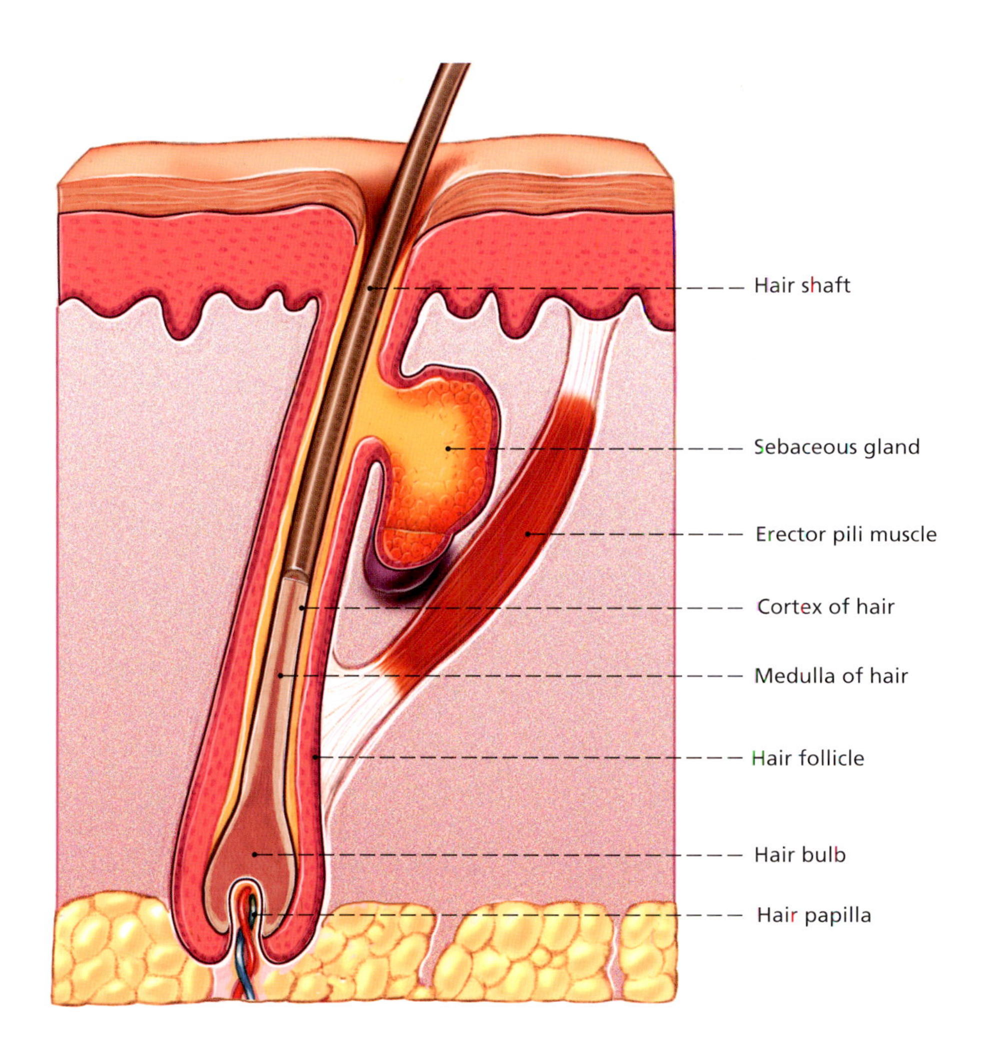

◆
Fig. 9
Parts of a hair (longitudinal section).

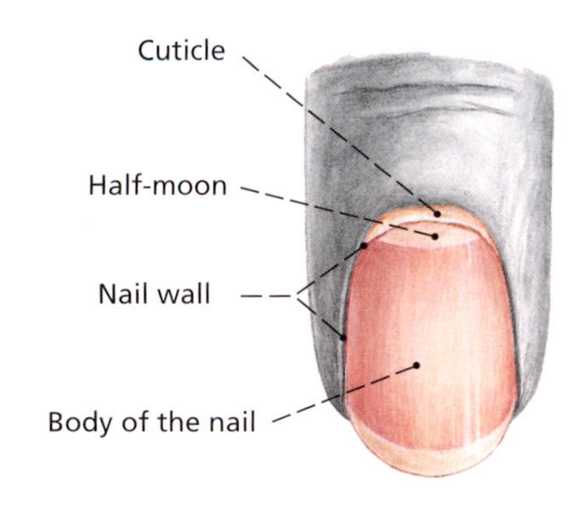

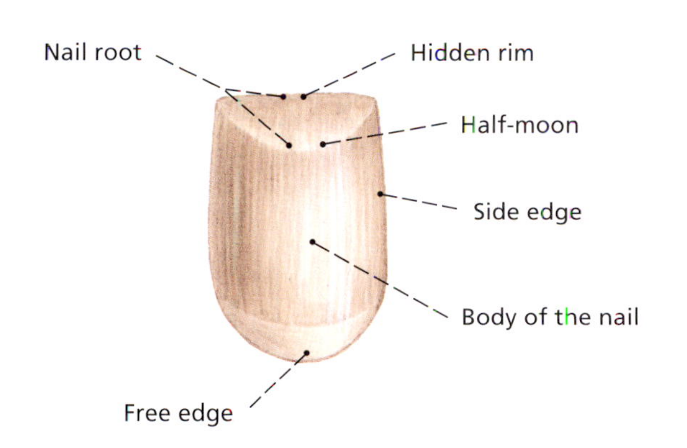

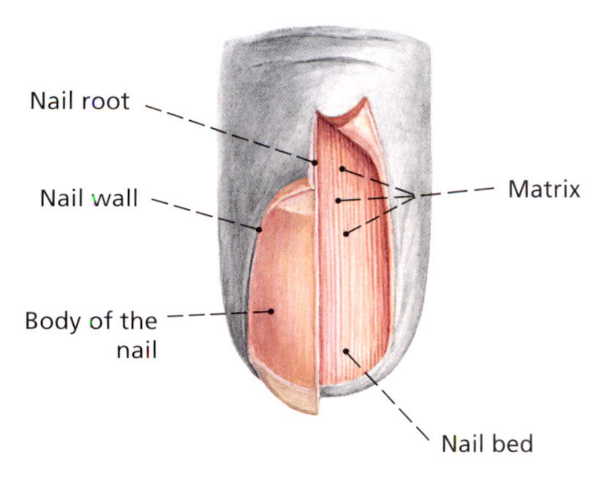

◆
Fig. 10
Structure of the end of a finger (including the nail).

◆
Fig. 11
Parts of the finger nail.

◆
Fig. 12
End of the finger after partial removal of the nail.

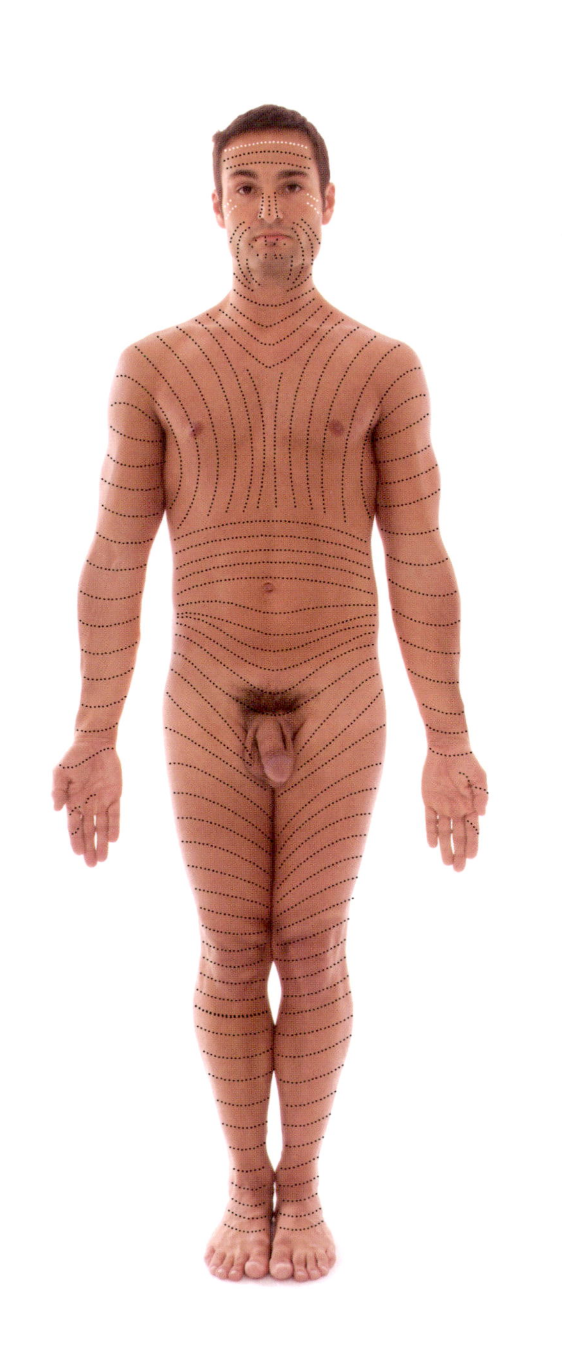

◆
Fig. 13
Tension lines of the skin on the front.

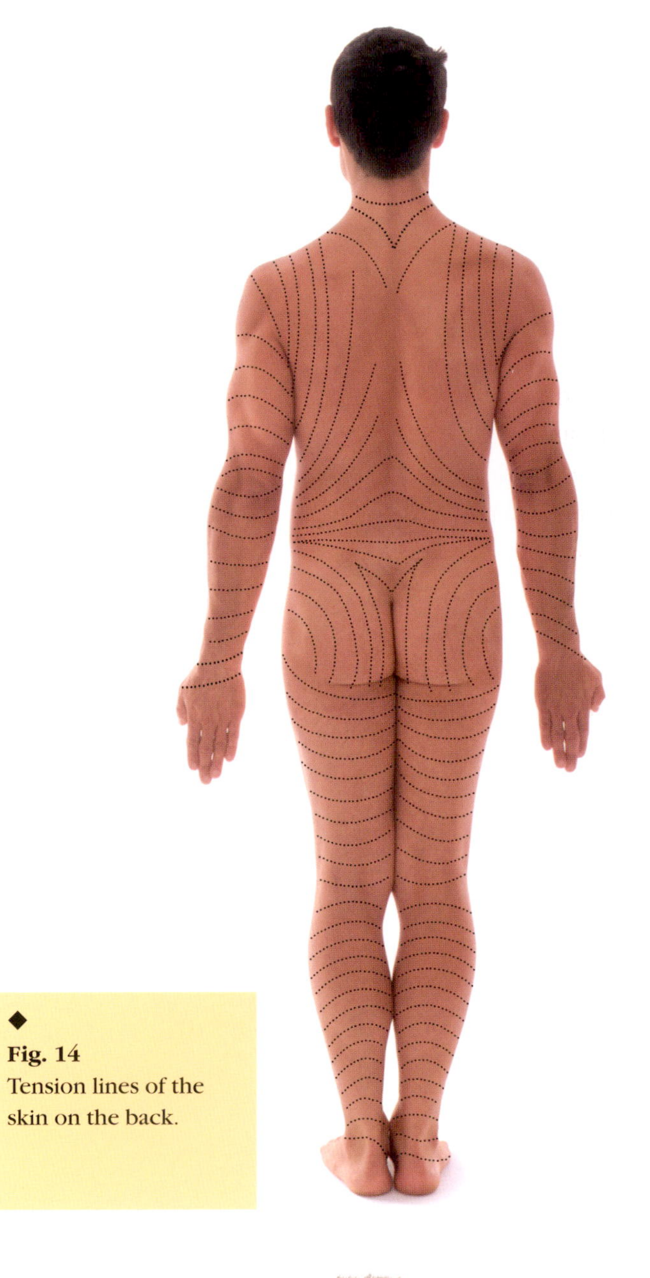

◆
Fig. 14
Tension lines of the skin on the back.

◆
Fig. 15
Tension lines of the skin around the head and neck from the front (a) and from the side (b). The tension lines in the facial area frequently correspond to the formation of lines and wrinkles as we age.

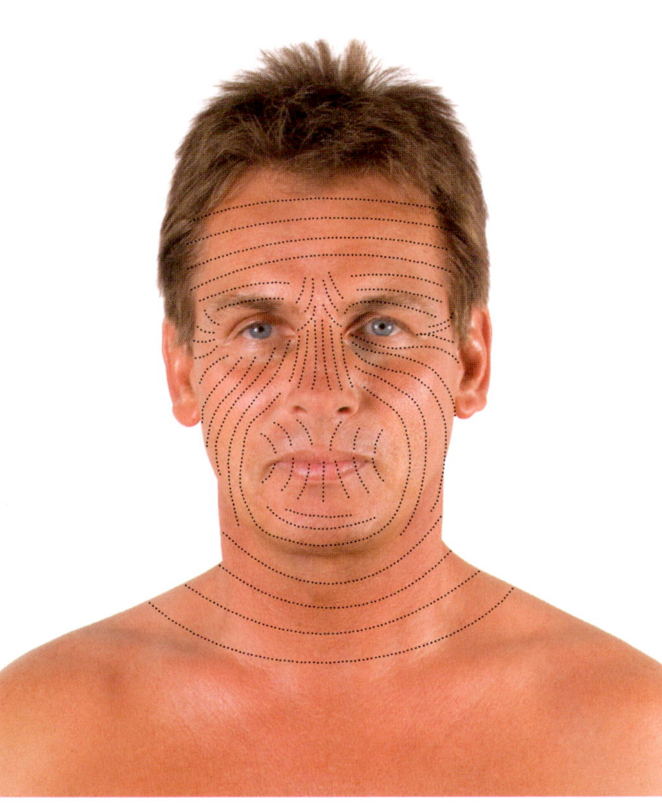

a

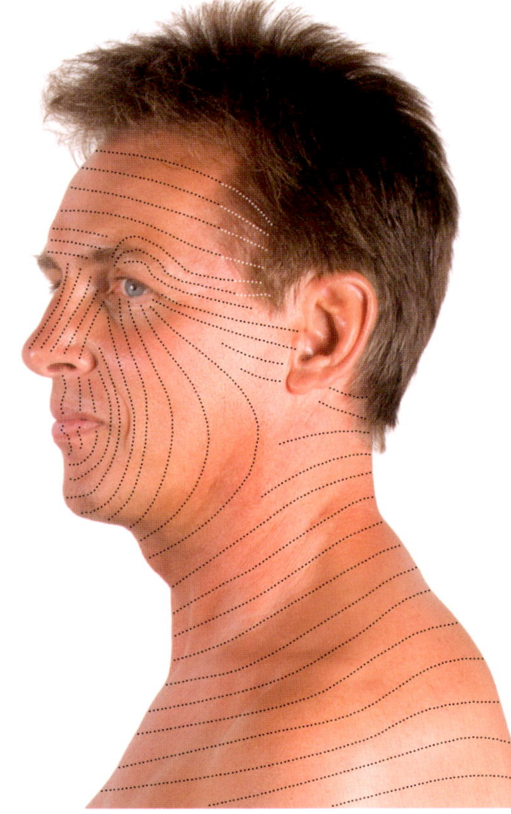

b

Symptoms and diseases

Benign changes in the skin (nevi)

A non-hereditary deformation of the skin or mucous membrane that is usually benign. Most nevi are present at birth and are called birthmarks. These marks, which are usually sharply delineated, can be large or small, flat or raised, hairy or smooth, with a coarse or rough surface, and range in color from white to red, blue, brown or even dark black.
Treatment is not usually necessary unless the nevus becomes a problem, either cosmetically or for other reasons. Care must be taken, however, if a nevus increases in size, changes in color and starts to hurt, itch or bleed. Seek medical advice instantly, as there is a risk of the nevus turning malignant. Some commoner types of nevi are described here.

• *Port wine stain*

Flat, benign growth on the skin; made up of blood vessels. These light red to blue-red marks are usually irregular in shape. Sometimes a baby is born with a port wine stain on the back of its neck; this is also known as a **stork bite**. It disappears during the first few months of the baby's life. A port wine stain around the eye can be an indicator of eye disease.

• *Spider nevi*

Star-shaped changes in the blood vessels of the skin. These are red, pinhead-sized lesions with tiny rays that usually occur on the face; as the name implies, they look like a spider. They are caused by a bulge in a blood vessel. In cases of chronic liver disease, they are frequently observed all over the body.

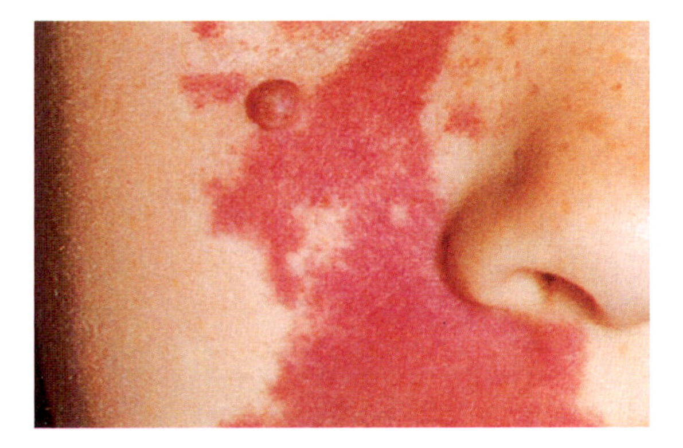

◆
Port wine stain
A port wine stain is a congenital expansion of blood vessels in the skin.

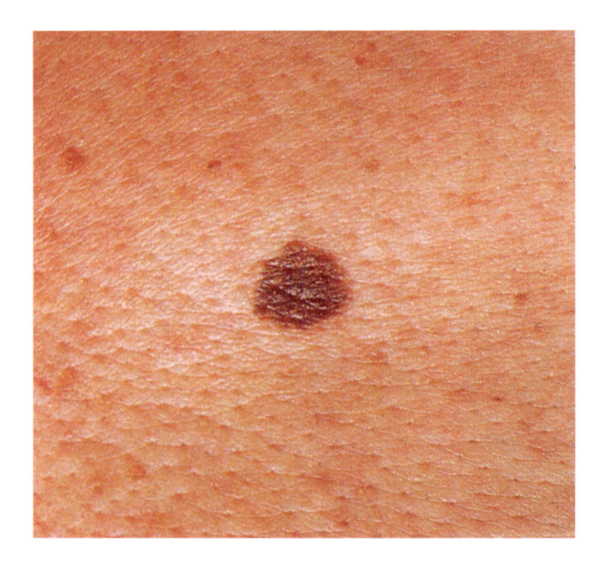

◆
Pigmented moles
Dark marks (moles) on the skin are not usually dangerous, but if they change in shape and/or color, then they need to be checked by a dermatologist, as this can be an indication of skin cancer.

• *Acquired nevus*

A nevus, pigmented mole, or liver spot, that is not visible at birth. They can vary in size; colors range from yellow-brown to black, and they can be flat or raised, smooth or hairy. They usually develop during the first 30 years of life. Acquired nevi are almost always harmless.
In rare cases they can lead to skin cancer, especially if they are often exposed to the sun. They should therefore be checked regularly for signs of change. Suspicious moles can be removed under local anesthesia.

• *Pigmented moles*

Stain-like discoloration of the skin, caused by melanin deposits. They include **age spots**, which usually occur in areas that are exposed to the sun (e.g., the backs of the hands). **Café-au-lait marks** are round or irregularly shaped changes in the skin, usually the color of milky coffee, that occur individually or in groups, and are usually present at birth or appear shortly afterwards. **Liver spots** are small, round or oval dark brown pigmentation marks that can occur in childhood, but do so more frequently in midlife.

• *Keloid scars*

A raised scar that some people develop after an injury, burn, or surgery. When the skin is injured, the body forms scar tissue between the skin and subcutaneous layer to close open wounds. This new scar tissue contains no pores or hairs, nor does it have the tiny divisions seen on normal skin. In some people, the formation of the scar does not stop when the wound is sealed off, but continues growing. This results in coarse, knotty or band-like tissue growths full of tiny blood vessels. Keloids can be treated with injections of cortisone-like substances or by plastic surgery.

Acne

Acne is one of the most common skin diseases. Many adolescents suffer from acne, with spots and inflamed pustules that can leave unsightly scars on the face and trunk. Acne is neither dangerous nor contagious, but it can have a serious effect on emotional well-being.

Skin contains a dense network of sebaceous glands, consisting of the gland body where the sebum is produced, and an excretory duct. Sebum lubricates the hair of the head and face, and impregnates the surface of the skin, protecting it from heat, cold, moisture, and drying out.

Why is acne more common in adolescents?

Sebum production increases considerably at puberty. Cells lining the wall of the excretory duct harden more quickly, then break free and can block the duct. The hardened cells seal the pore, stopping the sebum from flowing. This is the perfect place for bacteria to grow and multiply. The result is a suppurating infection. The face and upper body are most affected, as the skin possesses more sebaceous glands and larger excretory ducts in these areas.

- An **excess of male sex hormones** can increase sebum production and cause the skin to harden.
- In **stress situations**, the adrenal cortex produces androgens. This is why spots and pimples often appear before tests or exams, a date, or an interview.

Skincare can help

It is important to follow an appropriate skincare regime.

- Skin cleansing: The skin should be cleansed thoroughly twice a day. The best product for this is a soap-free cleansing lotion that is pH-neutral and will not destroy the skin's natural protective layer. A mild facial toner is disinfecting, and stimulates the circulation. It should not contain more than 35% alcohol since alcohol, like soap, dries the skin. A face mask applied once or twice a week will loosen dead skin cells and hardened sebum.
- Blackheads (also called comedones) should never be squeezed with the fingers; a proper tool for this purpose is available from pharmacies. Pimples, spots, and sore, inflamed pustules should only ever be treated by a specialist.
- Cosmetic cover-ups (e.g., sulfur-based creams) can be beneficial, both physically and emotionally.

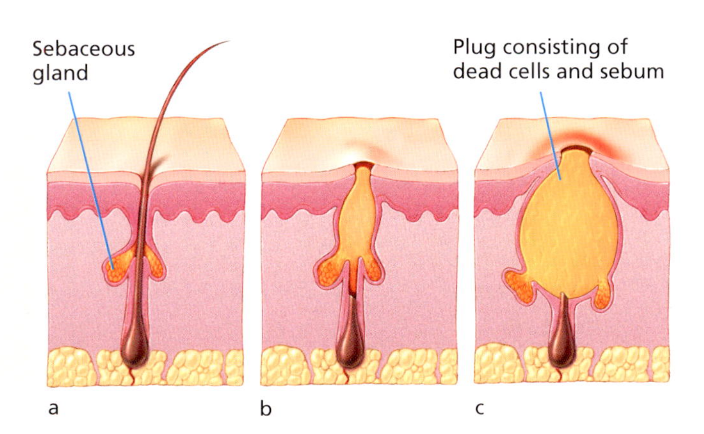

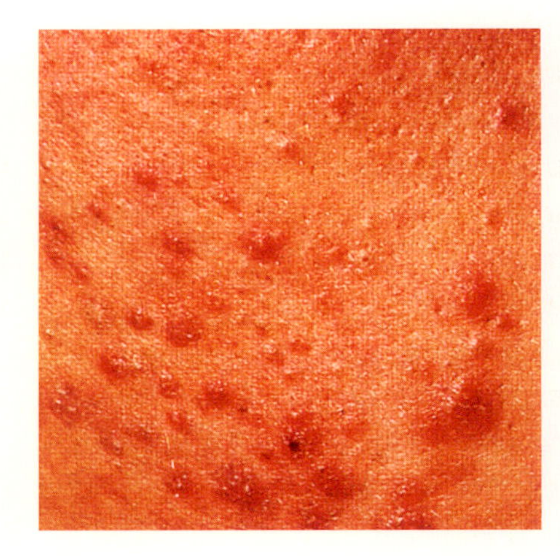

Acne is often most common on the cheeks.

Medical treatment

A medical practitioner may prescribe a variety of medical treatments for acne, depending on the cause:

- Anti-androgens balance hormone levels, which in turn reduces the production of sebum. The contraceptive pill can help young women.
- Antibiotics are both antibacterial and anti-inflammatory. They help reduce the risk of unattractive scars.
- Locally applied **keratolytics/vitamin A products** will loosen hardened skin.
- **UV radiation** in low doses encourages healing.

Supporting measures

Getting plenty of sleep, and exercise in fresh air both encourage healing. So will a balanced diet, ideally one that is low in carbohydrates and animal fats but with sufficient fiber to ensure regular digestion. A number of relaxation methods can be learnt to help with handling stressful situations.

Viral diseases

• *Herpes*

There are two types of herpes virus. Type 1 mainly affects the skin of the face, around the nose and lips, the cornea of the eye, and the mucous membranes in the mouth. Type 2 affects the genital area, and is passed on during sexual intercourse or at birth.
The initial infection usually occurs in childhood and can often be missed. Typical symptoms are itching, burning, and a feeling of tightness, which are followed by tiny blisters that burst open and spread quickly. After about a week, the blisters dry out and start to form a crust. However, the virus remains in the tissue, and a new infection may re-occur later in the same place.
The virus can become active when resistance is low, perhaps causing a cold sore in the event of a cold or flu, but can also be stimulated by too much sun. Left alone, the itchy blisters will heal without scarring.
A herpes infection of the **cornea** is particularly painful, since this area contains a large number of nerve endings. The eye becomes red and weeps, and the patient is extremely sensitive to light. The infection is treated with eye cream or drops.

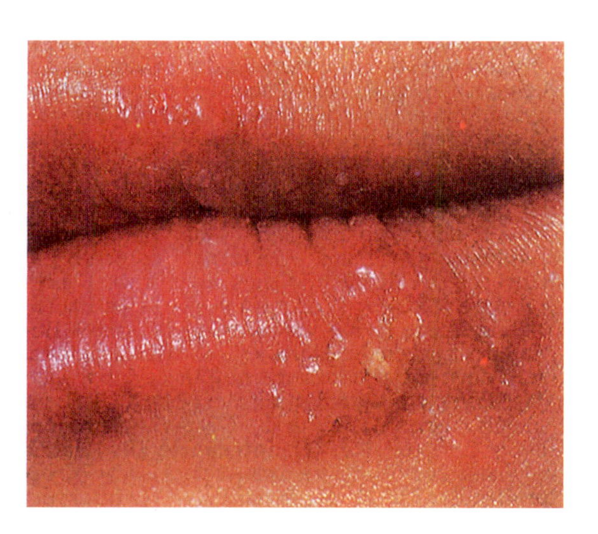

◆
Herpes
Herpes blisters are most common on the lips.

• *Shingles (herpes zoster)*

A skin disease affecting the nerves of the skin that is caused by the same virus as chicken pox. After an outbreak of chicken pox, the virus remains in the body and can become active again when the immune system is weakened or in the event of serious illness. Shingles is more common in older people, whose natural resistance has started to decline. Tiny red blisters form on a belt- or band-shaped area of skin, most often on the side of the chest, and usually only on one side. The affected area reflects the affected nerve's area of distribution. Severe, burning pain and itching are typical symptoms. The blisters form crusts and heal after about two weeks, although the neuralgia can continue for longer. Usually only the pain is treated, with painkillers.

Changes in the nails

These are usually symptoms of skin or metabolic diseases. Brittleness can be a sign of a nutritional deficiency (lack of calcium and iron), while changes in color can be a sign of poor circulation, with the tissue under the nail having a white or blue shimmer to it.
A number of toxins can cause white lines across the nails, although these must not be confused with the random white spots that are caused by air penetrating the lamellae of the skin and which are harmless. Watch-crystal nails are rounded towards the fingertip; the fingertips rounded upwards. They are a sign of a chronic oxygen shortage in people with heart/lung disease.

• *Inflammation of the nail bed (onychitis)*

This is usually due to incorrect nail care, but can also be caused by injury or an in-growing nail. The signs are inflamed redness and swelling of the nail bed, accompanied by often severe, throbbing pain. If pus accumulates under the nail, the body of the nail turns yellow-green. It is important that treatment be given as early as possible to prevent the infection spreading to the whole finger, or even the hand or arm.
Treatment consists of hot soapy baths, alcohol compresses, and antibiotic creams; if necessary the finger may be splinted. If these measures are not successful, a small surgical procedure can be carried out under local anesthetic.

• *Infection of the nail fold*

This is usually the result of minor injury, an in-growing nail or impaired resistance, and is also frequently seen in diabetics. In cases of acute infection, usually caused by bacteria, viruses, or fungi, the nail wall becomes thickened and red. Another common symptom is pus. Chronic infections of the nail fold are often caused by a yeast infection and frequently result in wart-like growths on the nail bed (granulation).
An infection of the nail fold is treated with hot soapy baths, alcohol compresses, antibiotic creams and, if these measures are not sufficient, a small surgical procedure can be carried out under local anesthesia.

• *Fungal nail infection*

Caused in most cases by fungal infections (ringworm or yeast-like fungi), or more rarely by molds. The nail turns yellow and thickens; it may split or splinter, and typically becomes grooved. Depending on the particular cause, the nail is treated with anti-fungal solutions or creams. In severe cases the medication is given in tablet form.

Neurodermatitis

Agonizing itching and obvious changes in the skin are just two of the problems that those suffering from neurodermatitis have to live with. Although the causes are not yet known and there is no known cure, there is much that can be done to alleviate the symptoms.

What causes the disease?

Although neurodermatitis is now one of the most common skin diseases, little is known of the cause. We know only that it is caused by the interplay between a number of factors. These include:

- hereditary predisposition,
- allergies (e.g., to certain foodstuffs),
- psychological factors (especially stress),
- environmental factors: suspected causes may include chemical residues in air and water, and chemically treated foods.

The symptoms occur most frequently on the elbows, the backs of the knees, the face, and behind the ears.

Agonizing symptoms

Although neurodermatitis can occur at any age, it usually first appears before the age of 12 months. Babies may develop tiny lumps of skin, blisters, and crusts primarily on the face (cheeks and forehead), and in flex lines on the elbows and backs of the knees, neck, wrists, and shins. However, the dry, reddened, thickened, flaking areas can occasionally spread over other parts of the body.

For sufferers the biggest problem is the unbearable itching. Scratching damages the dry skin, and the open areas become infected. It is also typical that periods without symptoms alternate with periods during which sufferers are unable to rest by day or night. Many patients are affected emotionally by the appearance of disfiguring changes in their skin.

One of the most effective measures is the avoidance of certain foods. Fish, eggs, nuts, and sugar are thought to be among the possible causes; close observation will indicate which ones affect a particular sufferer.

Treatment

In the acute stage, antihistamines can alleviate the itching. Because of their side effects, cortisone-based creams should only be used for the shortest possible time. It is important to treat infections of raw areas without delay.

It is important to check for any precipitating factors in the diet. Some of the foods that are believed to encourage or increase neurodermatitis are milk and dairy products, fish, pork, nuts, citrus fruits, sugar, and confectionery.

The correct skincare is also important. A quick, cold shower (approx. 68°F or 20°C) in the morning is recommended, using unperfumed, alkali-free soap. In the acute stage, the skin should have as little contact as possible with water.

As stress can also encourage the disease, psychotherapeutic procedures and relaxation methods such as autogenic training are particularly important. Climate cures in the mountains, by the sea, or in the desert are also extremely beneficial. The sun (in moderation) and fresh air are recommended. However, sufferers should avoid spending long periods of time sunbathing.

Skin cancer

The frequency of skin cancer is increasing from year to year. However, there are some effective tools that can be used in the battle against this highly visible and therefore easily recognized cancer: information, prevention, and early detection.

Skin cancer almost always occurs in the upper cell layers of the skin. Depending on the type of cell affected, it occurs as basal cell (basalioma), prickle cell (spinalioma), or pigmented cell cancer (malignant melanoma), Whatever the type of skin cancer, it is important for the sufferer to take advantage of every option for prevention, early detection, and early treatment.

Risk factor: the sun

While not every case of sunburn will necessarily result in skin cancer, an excessive amount of ultraviolet radiation will permanently damage the structure of the skin cells, leading in time to cell deformation and skin cancer. The same applies for the effects of X-rays and prolonged exposure to irritants such as arsenic and tar. Children who spend a lot of time outside and whose skin is particularly sensitive to the sun are especially at risk. People with light skin and hair, who have lots of moles, or who have already had skin cancer in the past, need to be especially careful in the sun.

Early detection

Skin cancer and its early stages are visible on the skin, and can sometimes even be felt. The chances of detecting and treating it in its early stages are very good, so the skin should be checked regularly and carefully. A medical practitioner should be consulted at the first sign of any suspicious skin changes. The earlier skin cancer is detected, the better the chances of treating it.

Basal cell cancer (basalioma)

Basal cell cancer is the most benign form of skin cancer since it does not metastasize. It almost always occurs on areas of the skin that are directly exposed to the sun: the face, the ears, and other parts of the head that are not covered by hair. Initially, it looks like a white spot, the surface of which contains a network of microscopic blood vessels. Later, an indentation forms in the middle, surrounded by a wall. The best treatment for a basalioma is surgery.

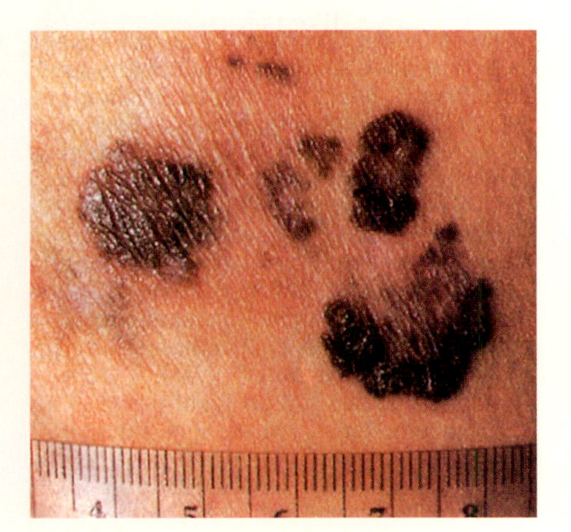

◆ Irregularly shaped moles, moles that have started to change, and black-blue moles, must always be checked by a dermatologist.

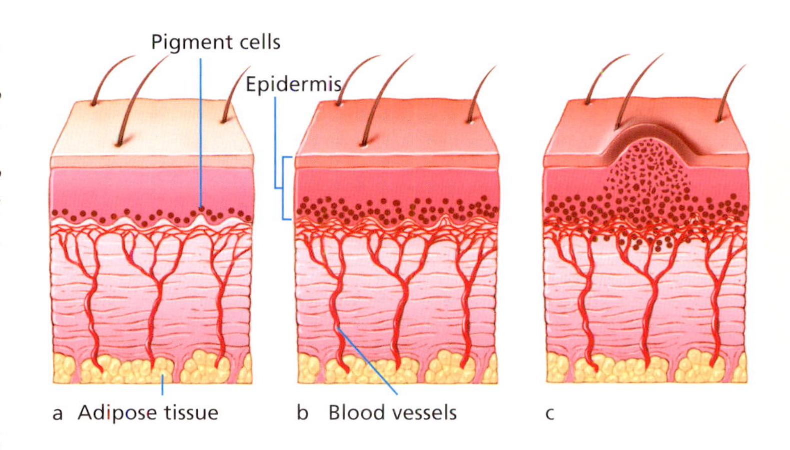

◆ Healthy skin contains only a few moles (a). If cancer is suspected, the blood vessels will expand, the pigment cells of the mole increase, and the dermis will appear swollen and red (b). Cancerous changes will have already occurred if irregularly shaped moles continue to increase, spreading to other skin layers and causing discoloration and swelling on the surface of the skin (c).

Prickle cell cancer (spinalioma)

Prickle cell cancer also occurs on the parts of the body that receive the most of the sun's rays, such as skin on the head that is not covered by hair, the backs of the hands, and the lower arms. A spinalioma will often develop from a benign early stage, a reddish, sharply delineated growth on the skin with a hard surface.
The best treatment method is surgery; in advanced stages the adjoining lymph nodes will also be removed. Chemotherapy is only necessary if the tumor has metastasized to other organs.

Pigmented cell cancer (malignant melanoma)

"Black cancer" is the most dangerous type of skin cancer. It often occurs on parts of the body that are not exposed to the sun, such as the back. Cancer must always be suspected at the appearance of a new mole or if an existing one suddenly changes in size, color, or shape, and/or starts to burn, itch, weep, or bleed.
Treatment depends on the size of the tumor. If the tumor measures less than 0.75 millimeters, only the melanoma and a strip of adjoining healthy skin will be removed. With larger tumors that often will have metastasized, the adjoining lymph nodes will also be removed.

Eczema

Eczema is due to abnormal changes in the upper layers of the skin. It usually occurs as an itchy rash with tiny lumps, blisters, or flaking skin. The cause is usually an oversensitivity to an external (contact eczema) or internal (endogenic or atopic eczema) stimulus. **Atopic eczema** is the best known form of endogenic eczema. Contact eczemas can be caused by an allergy to specific materials, such as cosmetic ingredients or dyes in clothing. Repeated contact with certain substances is often what over-sensitizes the skin to them (e.g., aggressive detergents).

- ***Hand eczema***

Eczema may be caused by regular contact (for professional reasons) with irritant materials, for example, the lime or cement used by builders. This type of eczema is different from the allergic hand eczemas that are caused by contact with materials to which an individual is hypersensitive. The third major group consists of hand eczemas caused by bacterial or fungal infections. The treatment varies accordingly: avoiding the materials that cause it (including retraining if necessary), anti-allergenic hyposensitization, or antibiotics, respectively.

- ***Cradle cap (infantile seborrheic eczema)***

A gray/yellow crust that forms on the scalp of babies due to overproduction of sebum. Mild forms in adults are known as dandruff. The typical sign of cradle cap is a flaking, itching skin rash on the top of the head, cheeks, and neck. It can also spread to the backs of the knees, elbows, and other folds of the body. The baby or child will often scratch until the skin is raw. The causes are not yet fully known, however, it is assumed that allergic reactions play a part. The symptoms can be relieved with medical products, while damp compresses, short periods in the sun, moisturizing creams, lotions, and oil baths will all help. The illness usually clears up by itself by the time the child is at school, but it can continue as atopic eczema into adulthood.

Diagnostic methods

Most cases of eczema present a typical clinical picture that can instantly be identified by a medical practitioner. If this is not possible, the practitioner may express suspicion and may take a sample of skin for further **histological examination**. In this case, a thin layer of tissue is stained to reveal the cell and tissue structures. This examination will be carried out if there is any suspicion of skin cancer.

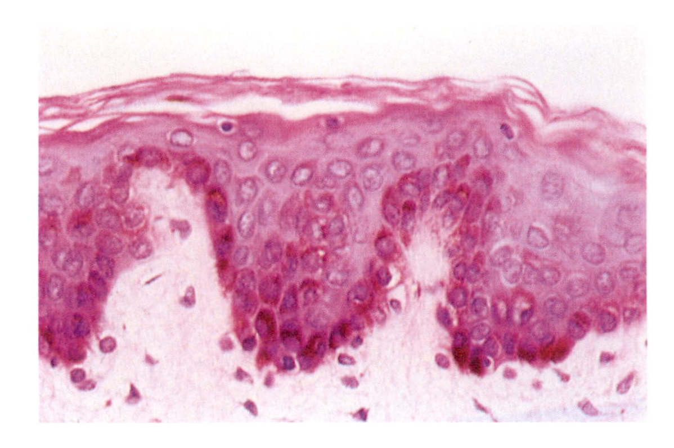

◆
Histological examination
A section of the skin is placed under a microscope, where the structure can clearly be seen.

Treatment methods

Most cases are treated externally. In diseases of the skin, nails, and hair that are caused by other illnesses (e.g., metabolic), the particular cause must be treated.
Antibiotic creams and lotions are available for inflammatory processes. Cortisone is used mostly to treat allergic or inflammatory conditions, such as eczema.
Cortisone is a hormone that is produced by the body. It is also known as a stress hormone, since the body increases its production when under stress. It raises the blood sugar level, is anti-inflammatory, and can suppress the formation of white blood corpuscles, thereby reducing the body's defense responses. Prolonged application of cortisone creams and lotions will thin the skin and make it prone to injury. If a cortisone-based product is to be used for a long period of time, this must be under close medical supervision.
Antimycotics (antifungals) work by either killing the fungus (fungicides) or inhibiting its growth (fungistatics). They are available as drops or tablets, gels, creams, tinctures, and nail lacquer. The choice of product is determined by the type of infection.
Phytotherapy (treatment with plants) may also be used to prevent and treat some skin diseases. Plant-based products, known as phytopharmaceuticals, use specific plants, parts or extracts of plants, usually have fewer side-effects and are milder than synthetically produced medicines. However, as they are not entirely free from side-effects, they should not be used indiscriminately or without medical advice. Arnica and chamomile are two examples of useful phytopharmaceuticals.
Light therapy is extremely helpful in treating some skin diseases. Also known as phototherapy, this treatment is given with intensive light. For instance, ultraviolet light may be used to relieve flaking and other skin diseases. Before treatment, the patient may be given a product that makes the skin more sensitive to light.

Muscles, bones, and joints

Bones, muscles, joints, and tendons are known collectively as the musculoskeletal system. While the bones form a solid framework (the skeletal system), and provide the body with necessary stability, the joints and musculature are what enable us to move.

Muscles

The external shape of a person is not only determined by the size and length of his or her bones, but also by the muscles that move them. Muscles make up about 40% of our total body weight. There are two types of muscle: the striated muscles of the skeletal system, and the smooth muscles of the internal organs. The movements of the striated muscles (the striations are visible under a microscope) are voluntary, and we control them via the nerves of the central nervous system. The movements of the smooth muscles (which are not striated under a microscope), on the other hand, are not subject to voluntary control. Smooth muscles are found in the walls of the internal hollow organs (e.g., stomach, intestine, bladder, gall bladder), blood vessels, and bronchi. Their movements are initiated by stretch stimuli and controlled by the autonomic nervous system.

Each skeletal muscle consists of countless cylindrically shaped muscle fibers, and these in turn are made up of tiny contracting threads, the muscle fibrils.

A number of muscle fibers are bundled together by connective tissue to form a muscle; each muscle therefore consists of numerous bundles of fibers. On most muscles, the external sheath forms a tough skin, the fascia. This skin separates the various muscle groups from each other and enables them to move separately.

Almost all of the muscles, especially in the arms and legs, are connected to tendons rather than directly to a bone. Tendons are joined directly to the bones, and transfer movements from muscles to bones. On the flexing sides of limbs in particular, the tendons (which are relatively long there) are contained inside special sliding tubes, the tendon sheaths.

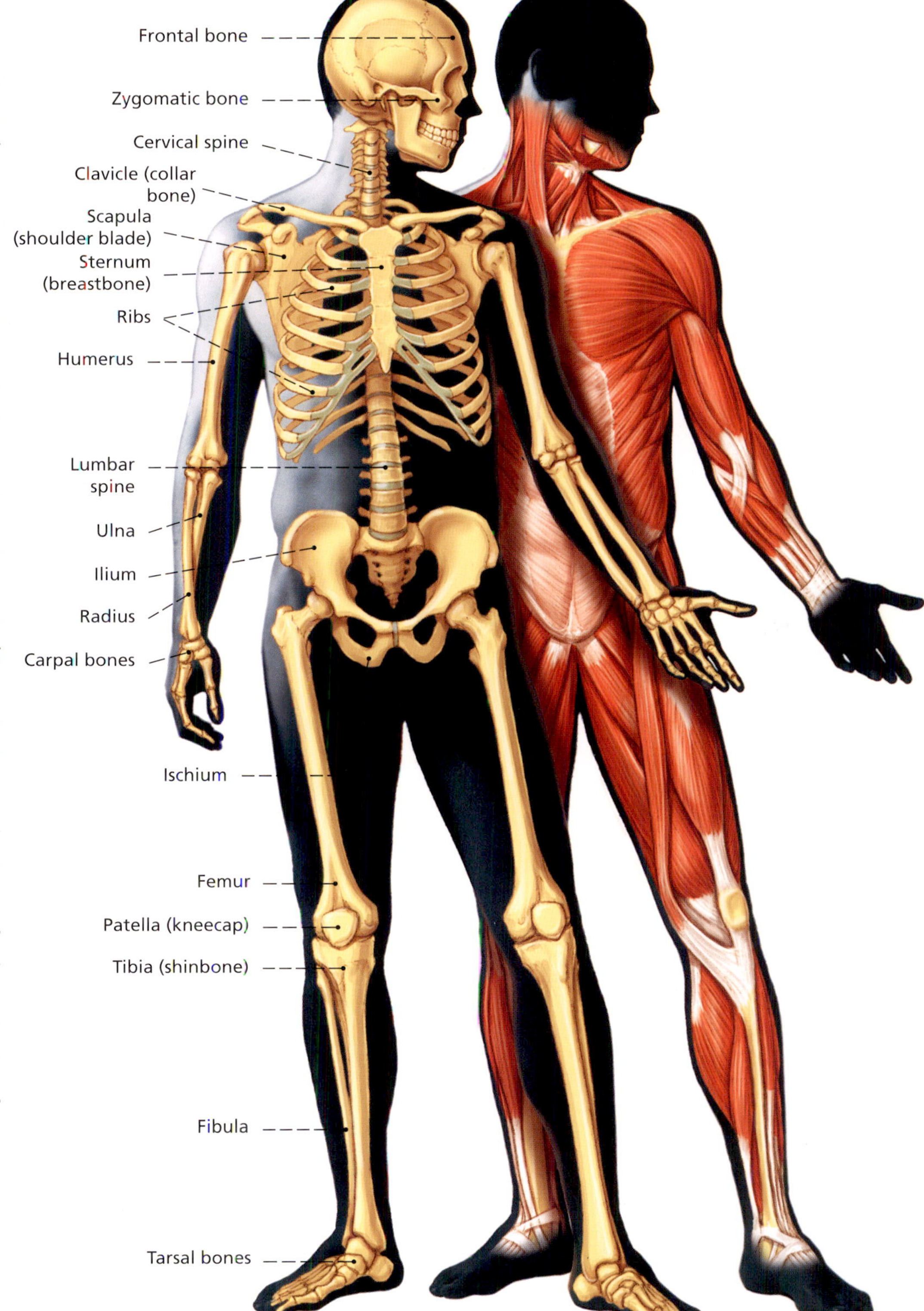

◆ **Fig. 16**
The human musculoskeletal system consists of the skeleton plus musculature, with tendons, bones, joints, and ligaments.

Muscle movement

Each muscle can contract or increase its tension (known as muscle tone) without contracting. In striated (skeletal) muscles, the impulse for this starts in the central nervous system, while in smooth muscles it starts in the autonomic nervous system (see also chapter on brain and nervous system). For example, a number of muscles that connect the bones of the upper and lower arm act together to bend or stretch the arm at the elbow. Bending is achieved when the biceps (two-headed muscles) contracts on the front of the upper arm, which shortens the distance between its attachment points on the upper and lower arm. Conversely, the arm is extended by the muscles on the back of it contracting.

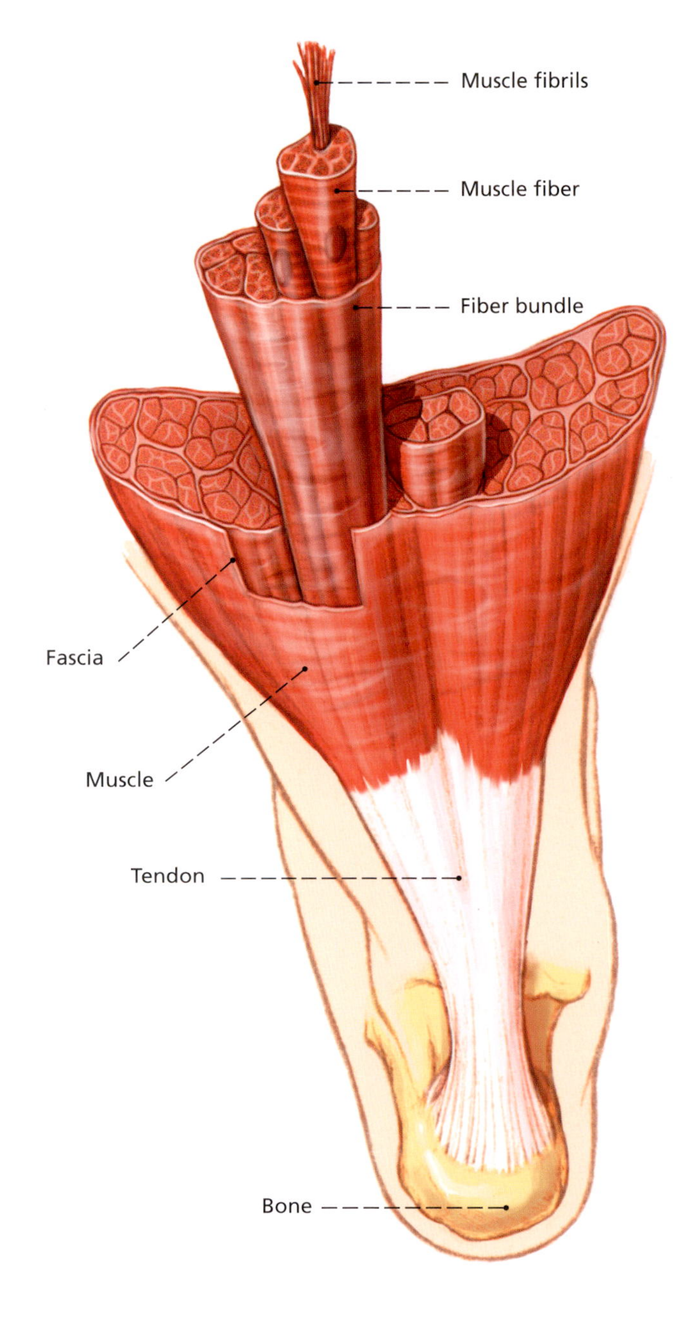

◆
Fig. 17
A skeletal muscle is constructed of muscle fibrils, muscle fibers, and fiber bundles. The whole muscle is covered by a sinewy skin, the muscle fascia.

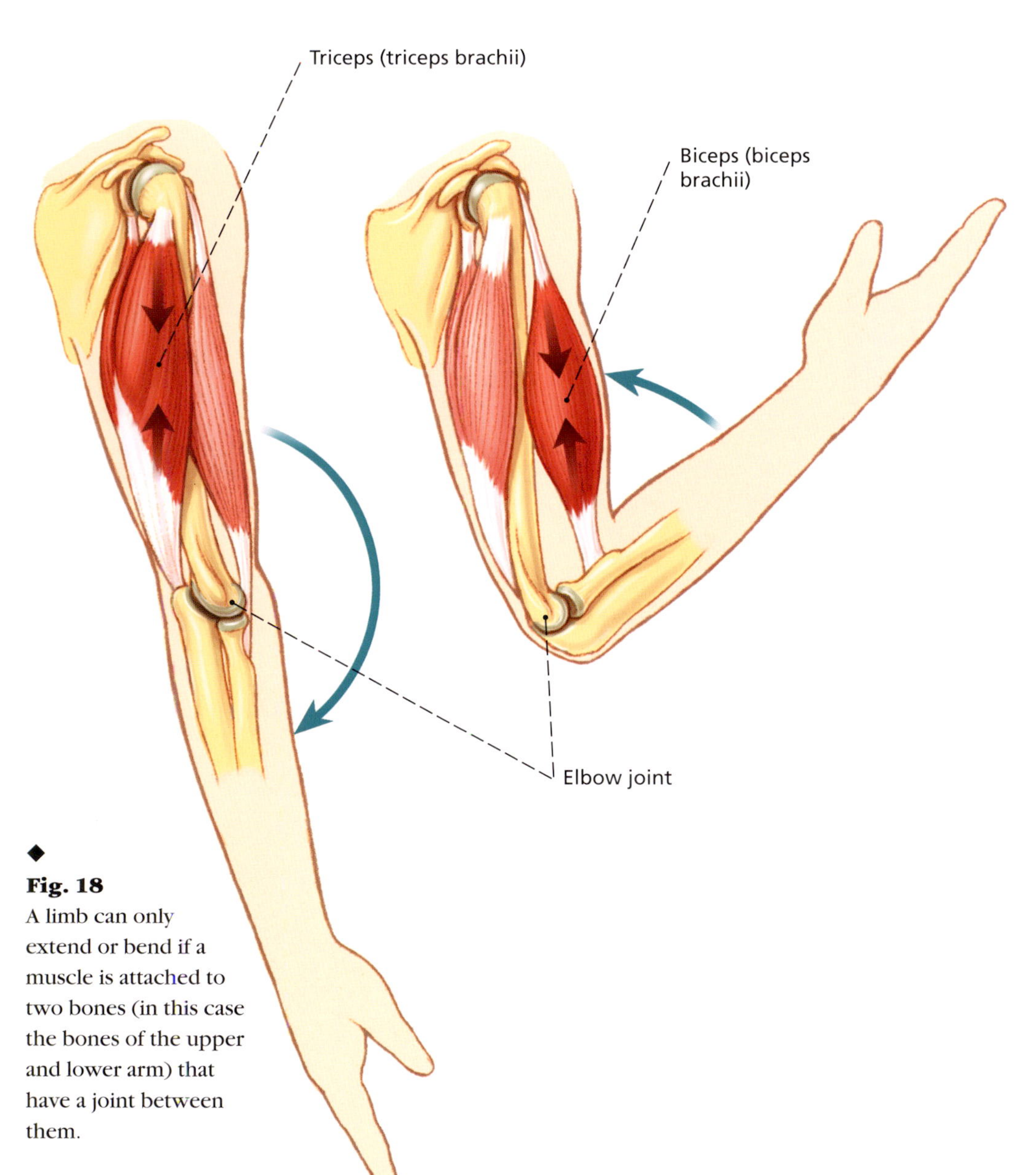

◆
Fig. 18
A limb can only extend or bend if a muscle is attached to two bones (in this case the bones of the upper and lower arm) that have a joint between them.

Joints and ligaments

Body movements are effected at the connecting points between the individual bones, the joints. This is the point where the ends of two bones (occasionally three), covered by tough cartilage, meet. The ends of the bones are surrounded by a capsule of tough connective tissue that is attached to the bone, and often further reinforced by firm elastic ligaments. So joints are made up of three parts: the cartilage-covered surface of the joint, the joint capsule, and the hollow area between the surfaces of the joints, the joint cavity. This cavity contains a viscous liquid, the synovial fluid, which is produced by the synovium (inner skin of the joint), and ensures that the joint surfaces rub together smoothly.

What movements can be carried out by a joint are determined primarily by its shape. For example, while a hinged joint such as the elbow can only perform movements around a single axis, a ball joint such as the shoulder or hip, can carry out a wider range of movements on several planes.

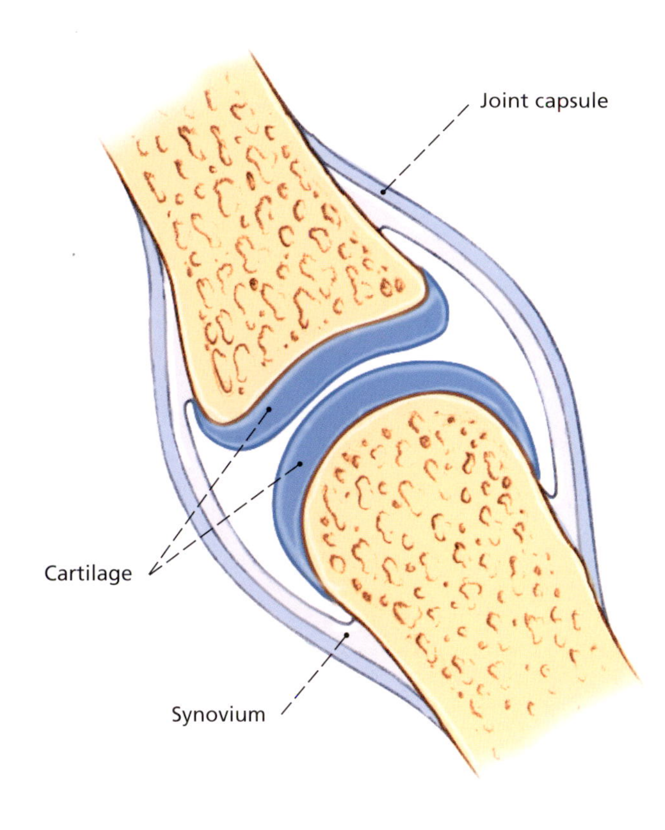

◆
Fig. 19
Example: components of a joint.

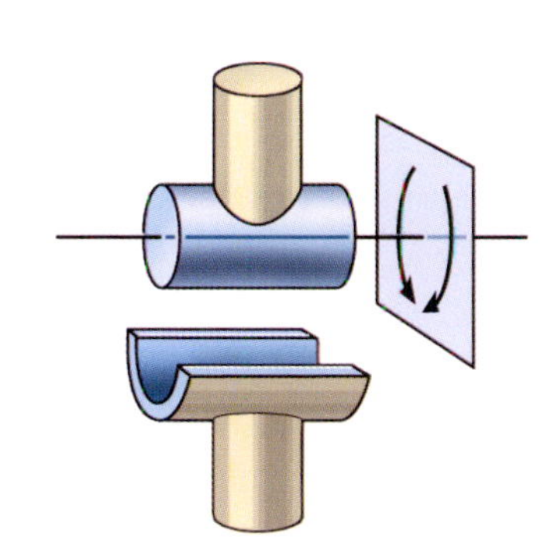

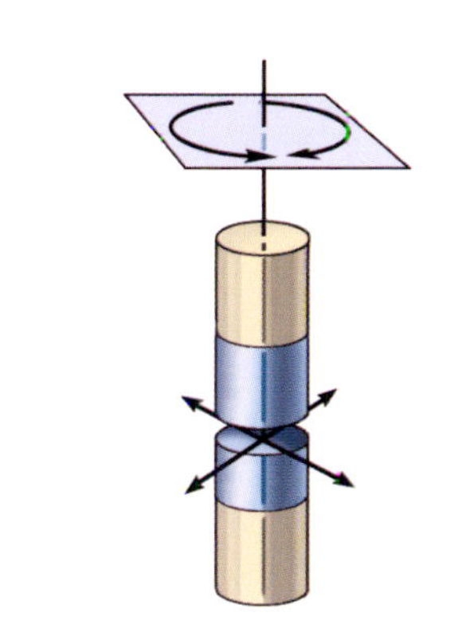

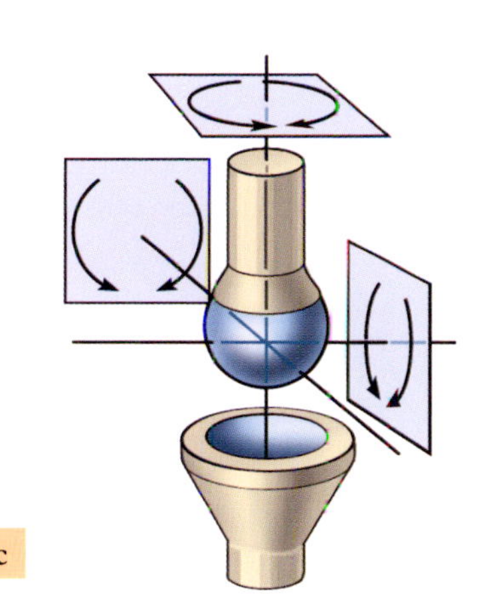

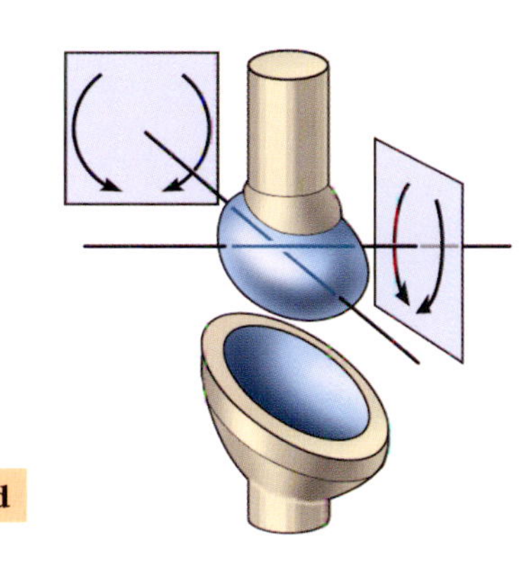

◆
Fig. 20
The various joint shapes.
(a) Hinge joint: bends and flexes two bones against each other in one axis. Example: elbow.
(b) Level joint: rotating movements around the longitudinal axis. Example: joints between the vertebral bodies.
(c) Ball joint: movements on all planes. Example: shoulder and hip joints.
(d) Ellipsoid joint: movements on two planes. Example: thumb joint.

Bone

There are some 200 bones in the human body, which together comprise the skeleton. The skeleton not only gives the body the necessary stability, but also determines physique. It is possible to differentiate between men and women from the shape of the skeleton alone, as men have stronger bones and a narrower pelvis than women, while women have a wider pelvis in order to provide room for a growing baby in the event of pregnancy.

The skeleton is also the storage facility for minerals such as calcium and phosphate. These mineral salts are taken from the bones when the requirement for them is greatest. This occurs, for example, during pregnancy and lactation, the formation of bones and teeth in babies, or during growth stages in children and adolescents, when the mineral salts are passed into the blood. Many bones also perform another function: they are the production site for blood cells in the bone marrow.

Bone is composed primarily of protein (collagen) fibers, commonly organised into sheets of parallel fibers, and calcium salts. Both of these enable the bone to cope with considerable mechanical loads, and make it resistant to pressure and bending. The bone structure is subject to constant reconstruction, which is the task of special cells: bone-forming cells (osteoblasts), and cells that break down bone tissue (osteoclasts). These procedures are controlled by hormones from the parathyroid and thyroid glands. Vitamin D is particularly important in bone metabolism, as it enables calcium to be absorbed from the intestine and passed into the bone substance. We obtain vitamin D from a healthy diet (green vegetables, butter, fish, and eggs), and from sunlight.

Construction of the bones

The bones are all covered by a tough skin, the periosteum, through which blood vessels penetrate the interior of the bone. Just underneath this skin is a hard cortical layer (compact bone), while the interior of the bone is made up of a sponge-like tissue with tiny hollow areas, the spongiosa, which contains the bone marrow; in long tubular bones this is also contained in the medullar cavity. Bones are divided into various types according to their form and function:

- Tubular bones (example: bones of the arms and legs), consist of a tubular shaft, which makes them particularly stable, and two thicker ends.
- Short bones (example: carpal bones) are usually cuboid or block-shaped, and have a thin outer layer.
- Flat bones (example: skull, ribs, sternum, the iliac wings). Blood cells are produced in the marrow of these bones.
- Bones with air-filled sinuses (example: upper jaw bone). The sinuses have a membrane on the inside that secretes mucus.

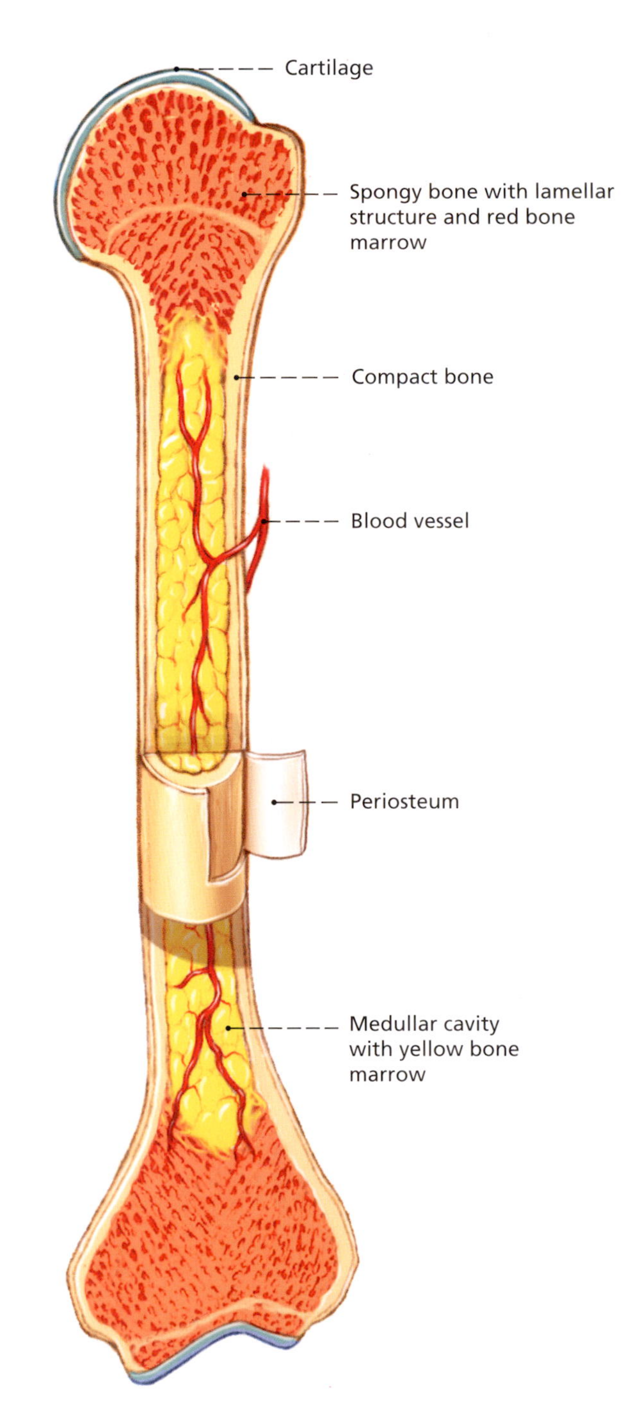

Fig. 21
Construction of a bone (humerus). In the sections near a joint (seen here at the top and bottom), the bone substance is arranged in lamellae, and is spongy in appearance (from the latin spongiosa meaning sponge). This structure guarantees a high level of resistance to pressure and tensile stress.

◆
Fig. 22
Human skeleton musculature from the front (a) and from behind (b).

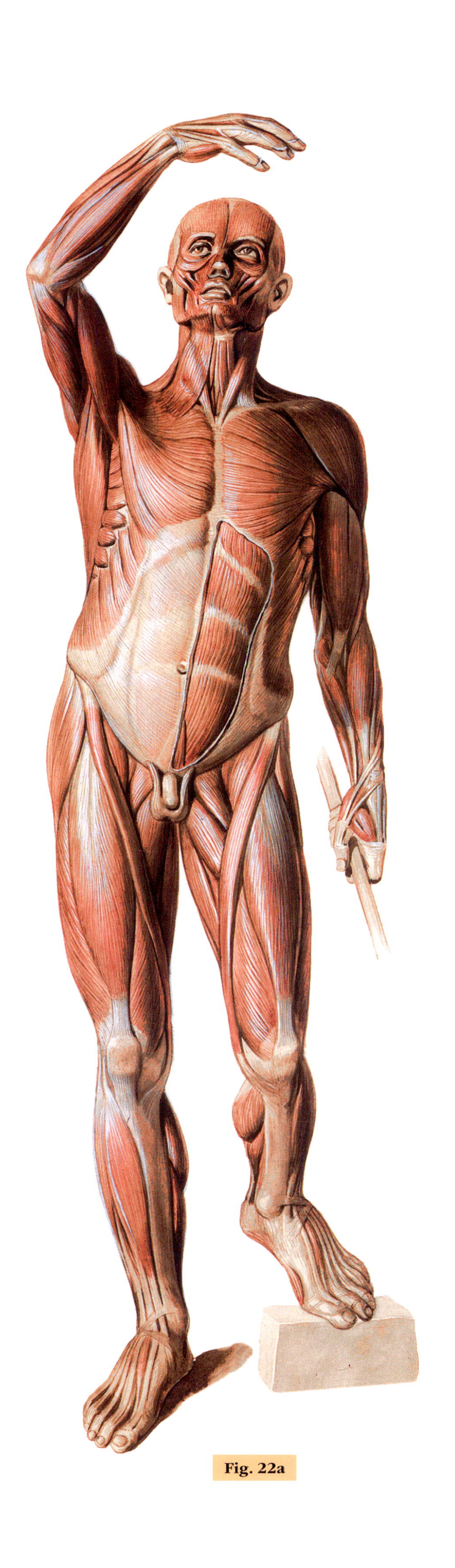

Fig. 22a

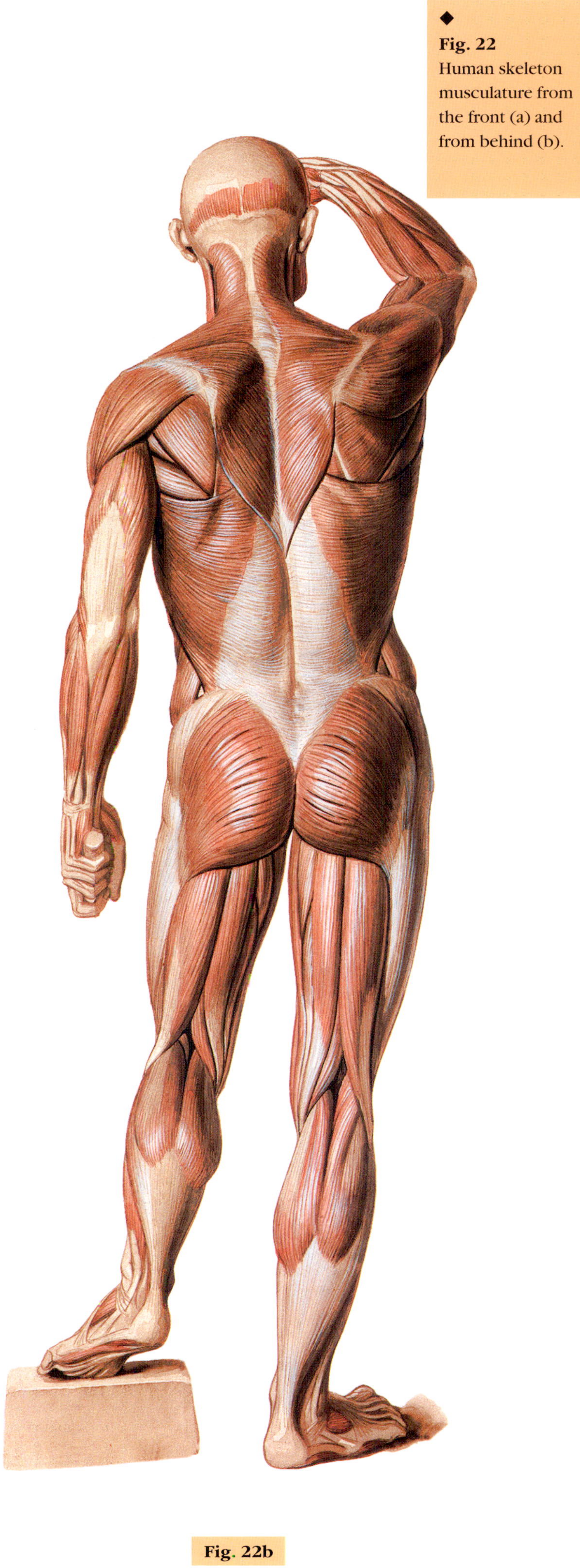

Fig. 22b

Fig. 23
Facial muscles from the side.

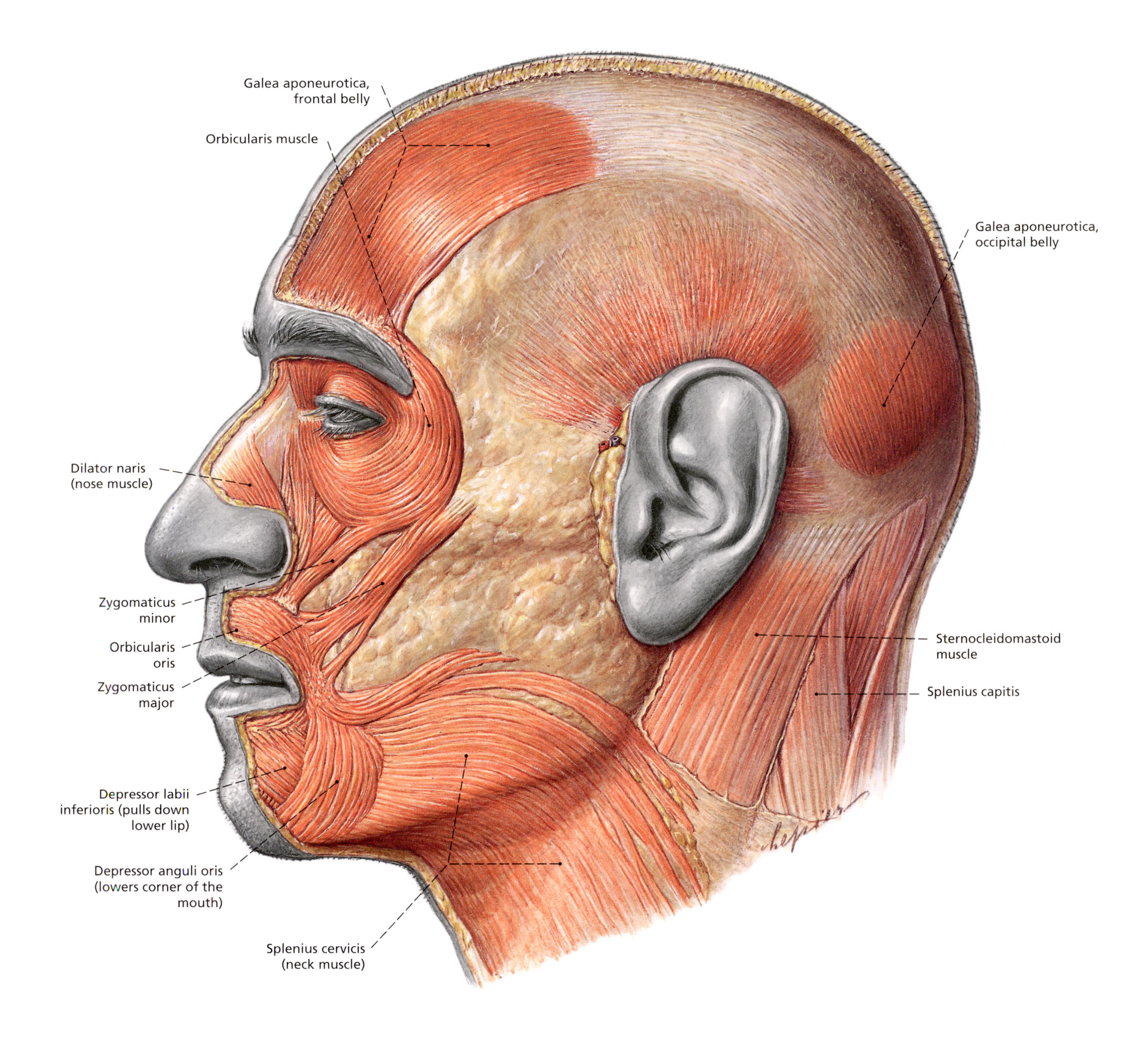

◆
Fig. 24
Head and chewing muscles from the side. Some of the muscle skin and smaller muscles have been removed.

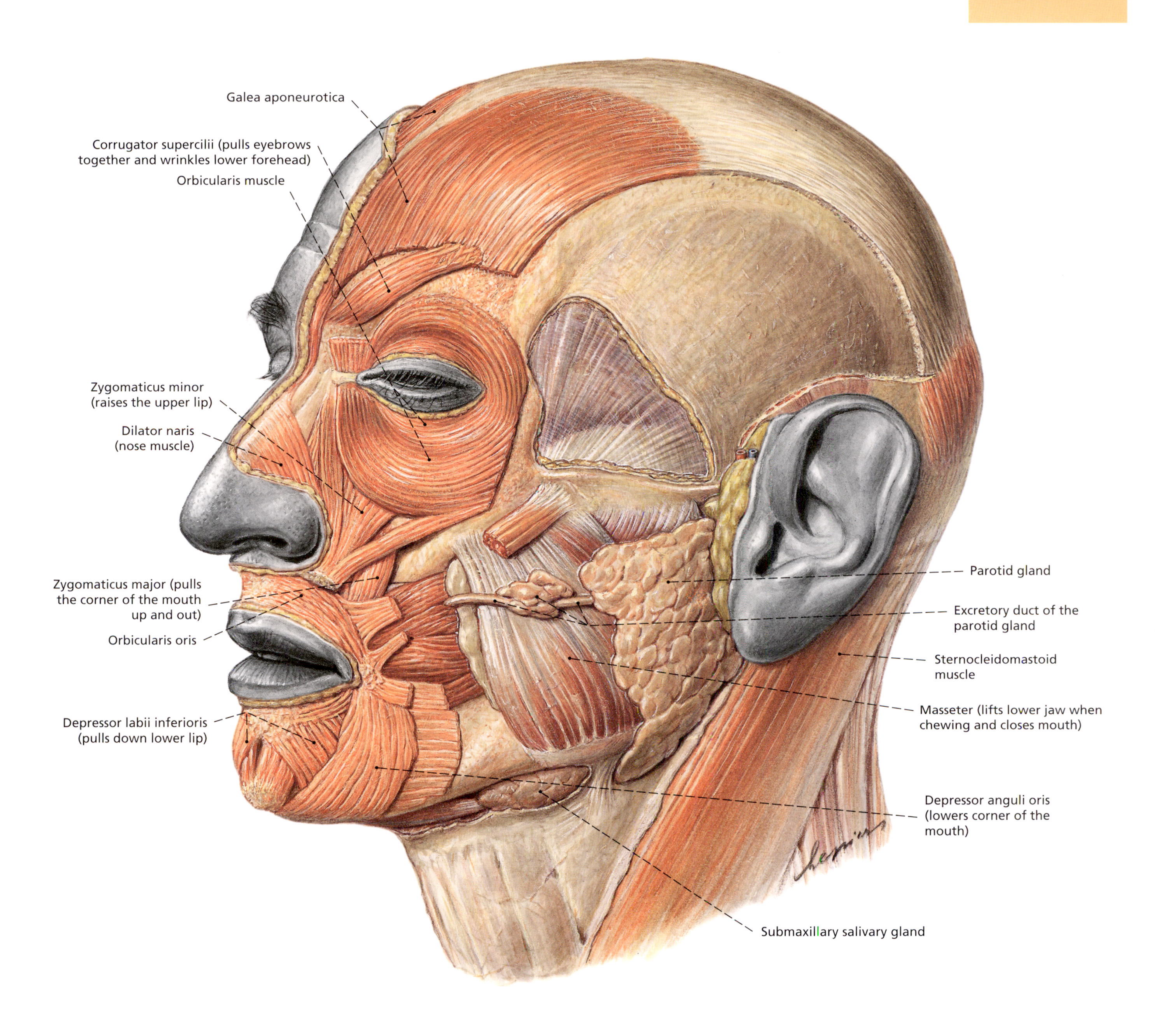

◆
Fig. 25
Superficial (left side of the body) and deep neck muscles (right side of the body) from the front.

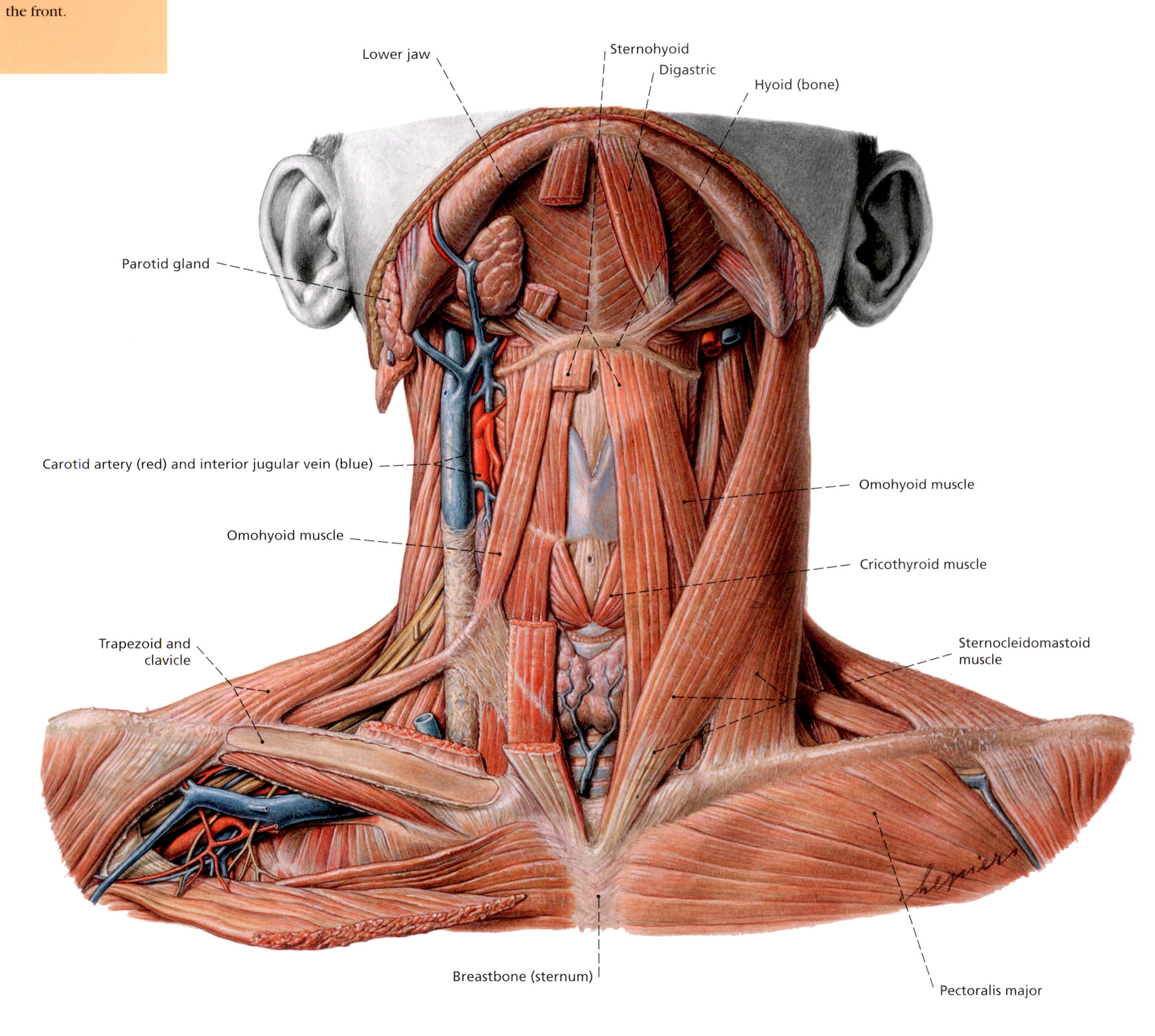

◆
Fig. 26
Deep neck muscles and upper part of the back muscles.

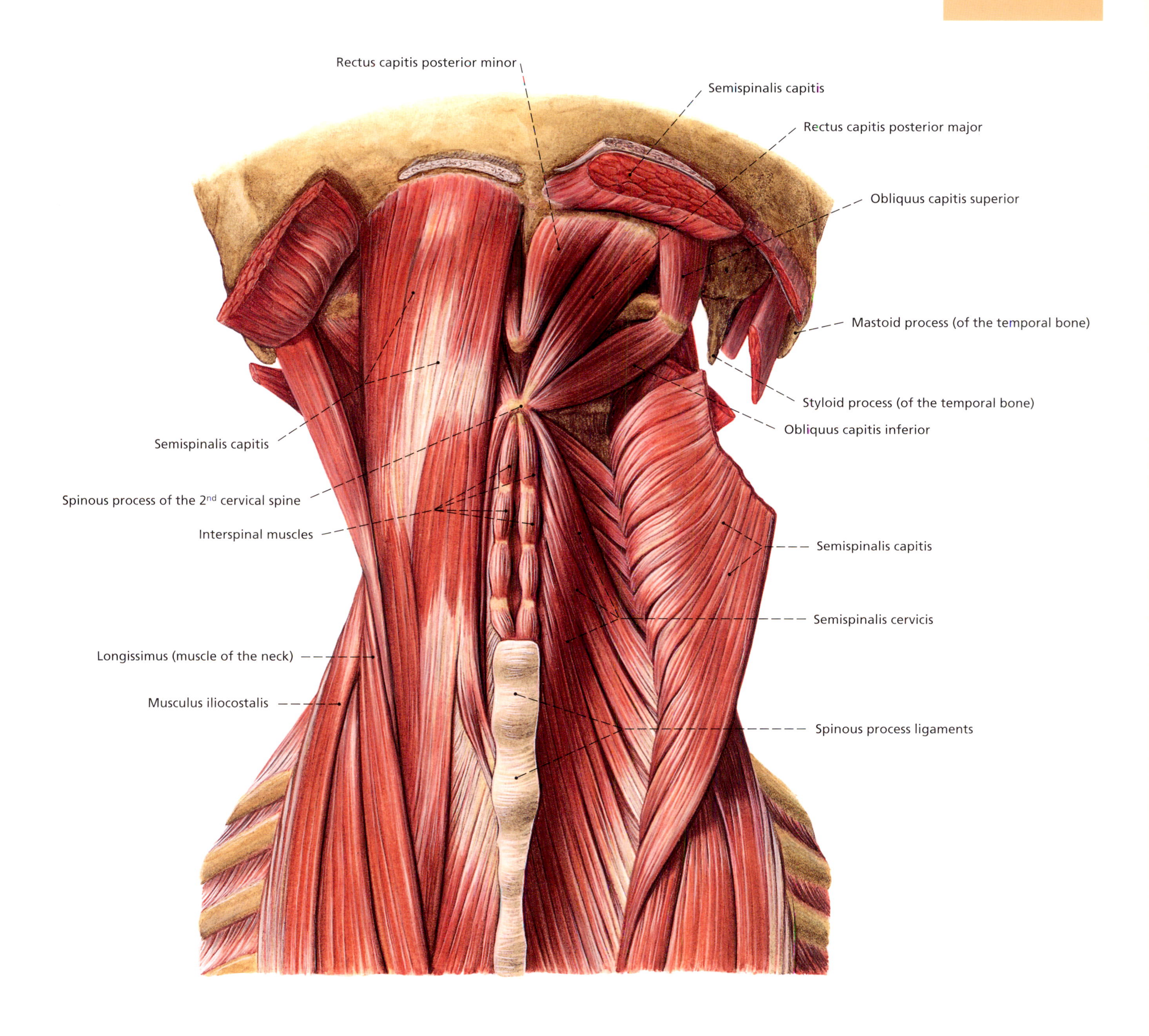

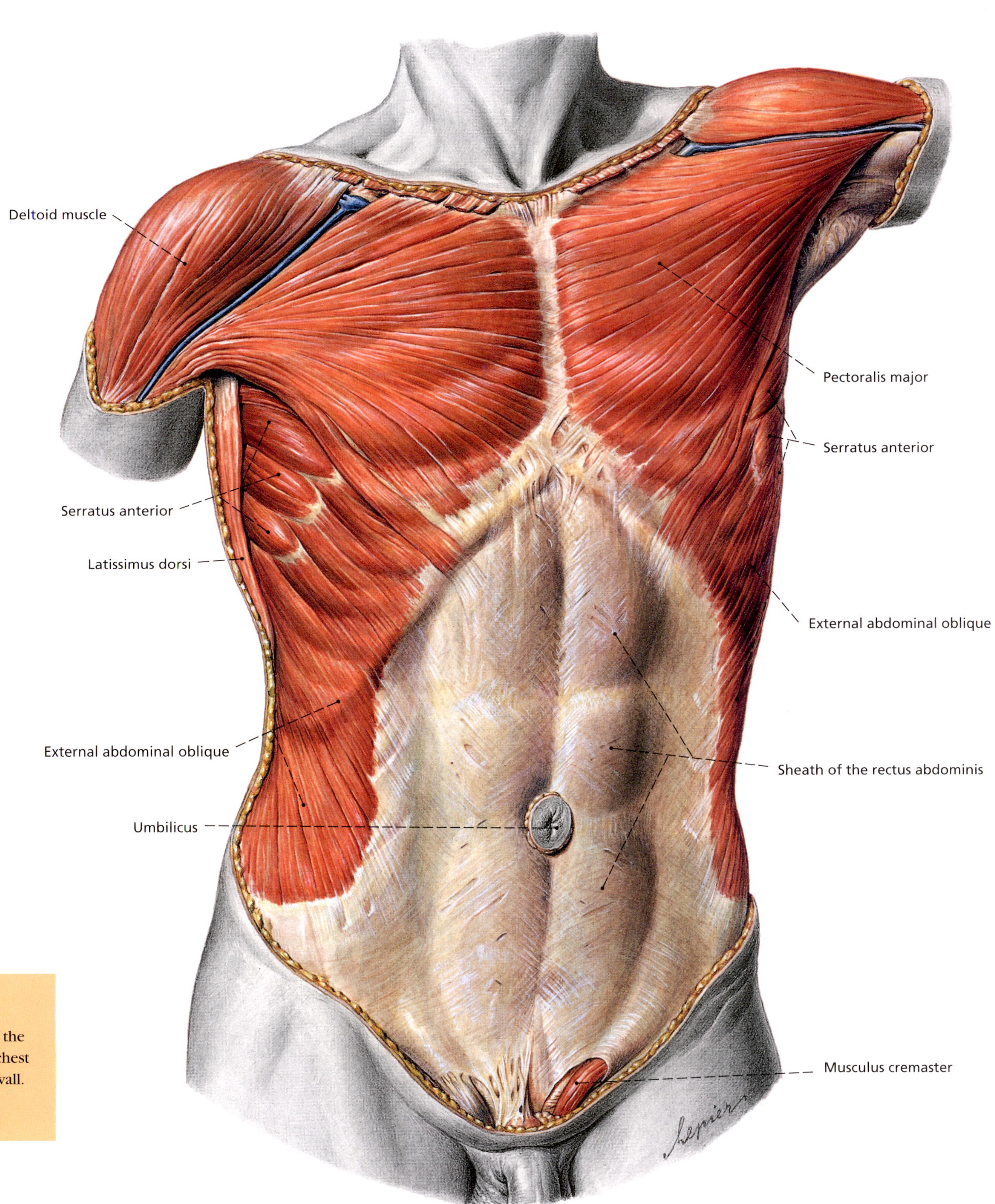

◆
Fig. 27
External layer of the musculature of chest and abdominal wall.

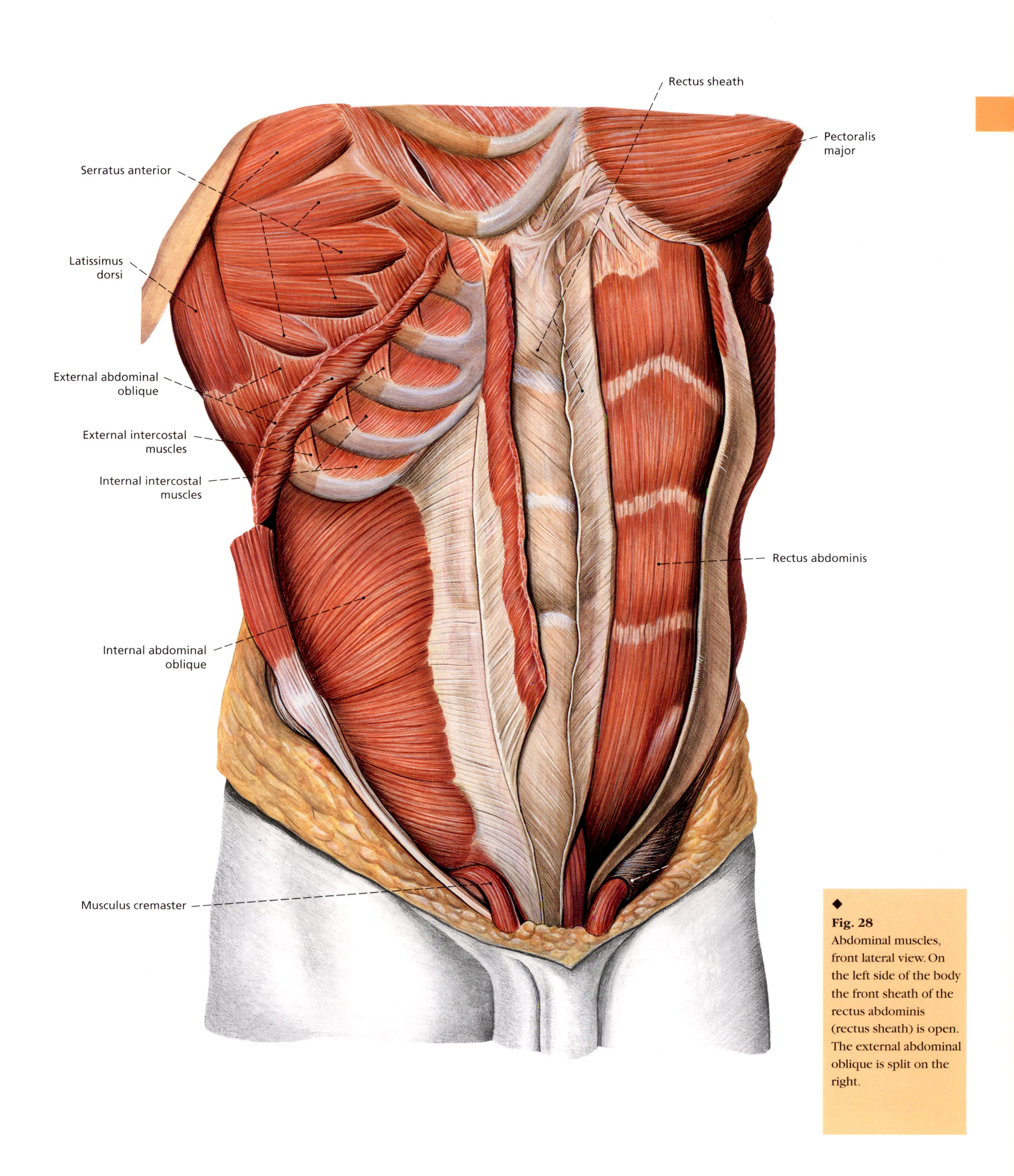

◆
Fig. 28
Abdominal muscles, front lateral view. On the left side of the body the front sheath of the rectus abdominis (rectus sheath) is open. The external abdominal oblique is split on the right.

◆
Fig. 29
Dome of the diaphragm and its various sections on extreme exhalation. The picture shows the exit points of blood vessels and the esophagus. Some of the ribs and part of the sternum have been removed at the front.

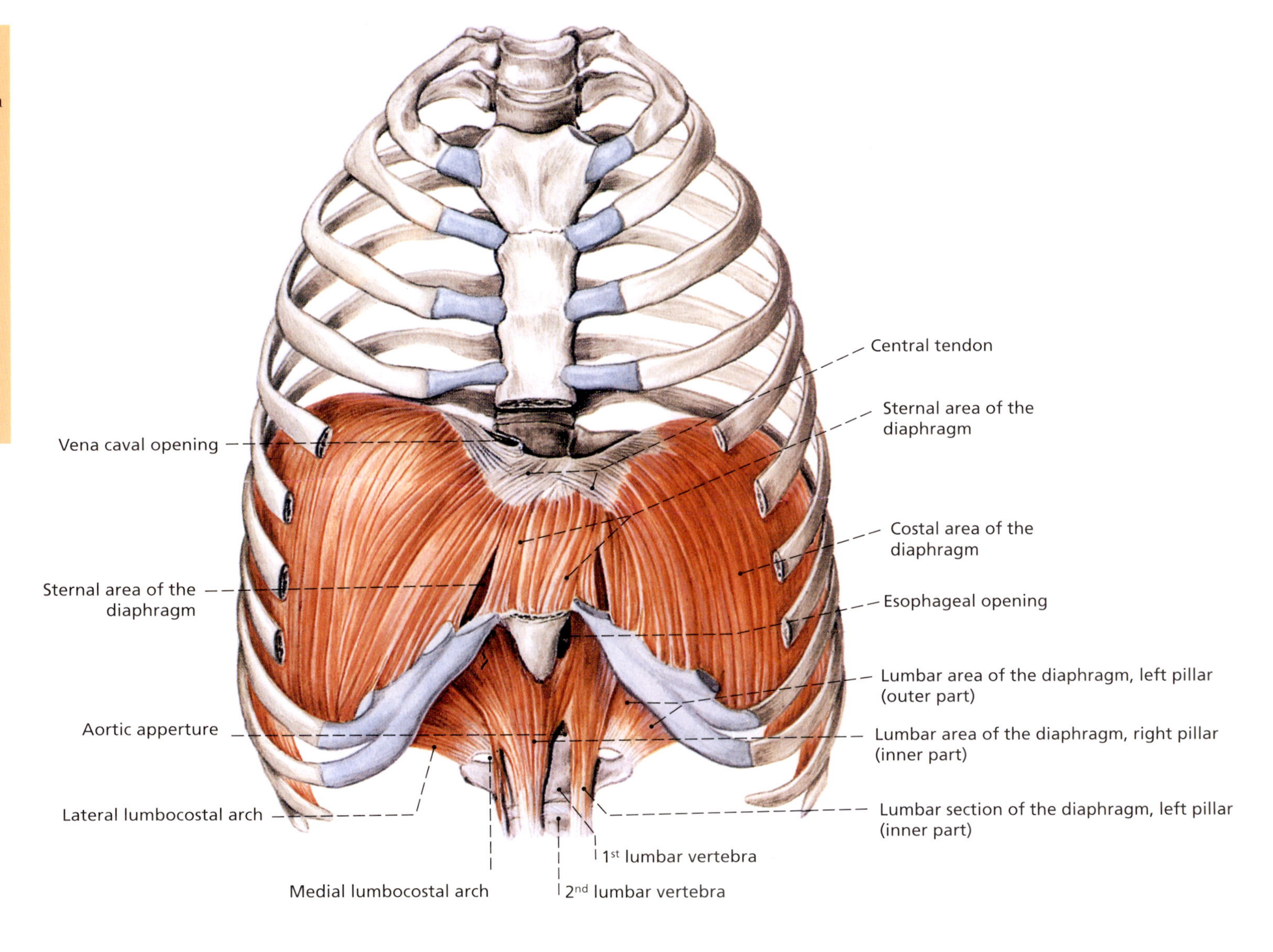

◆
Fig. 30
Behavior of the dome of the diaphragm and the abdominal wall on inhalation and exhalation.

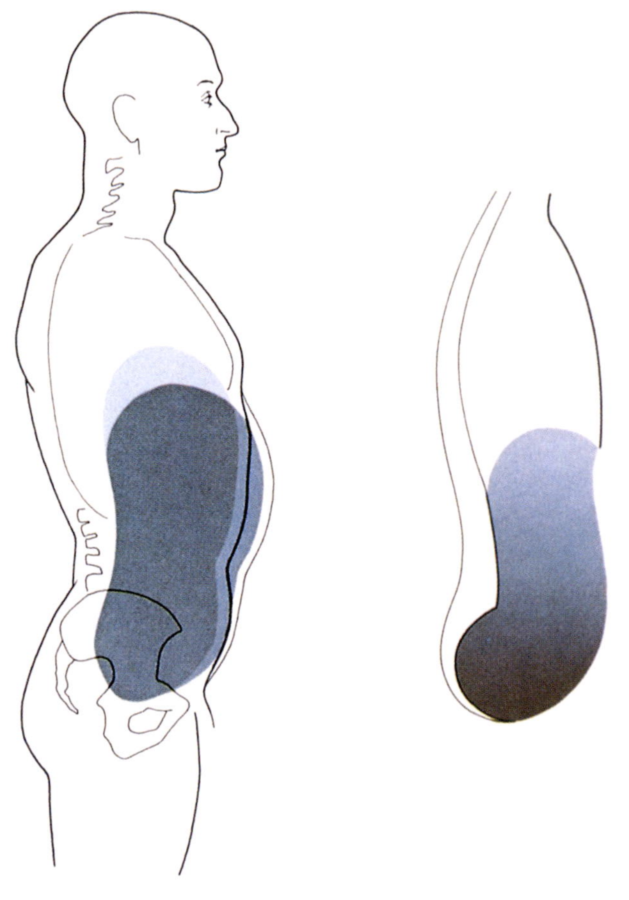

◆
Fig. 31
Diaphragm and inner abdominal muscles seen from the front. The abdominal viscera have been removed.

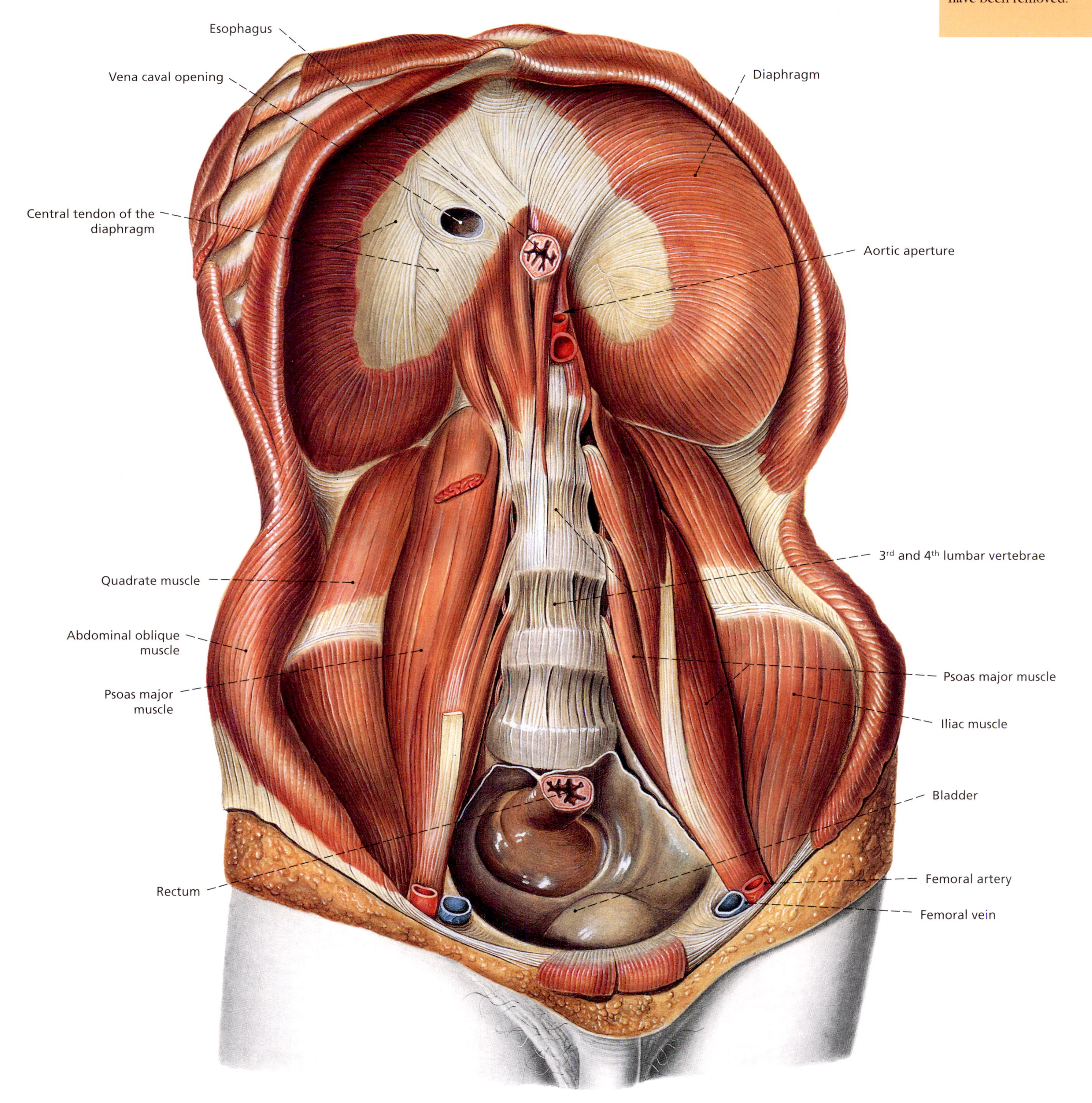

◆
Fig. 32
Back musculature, superficial layer.

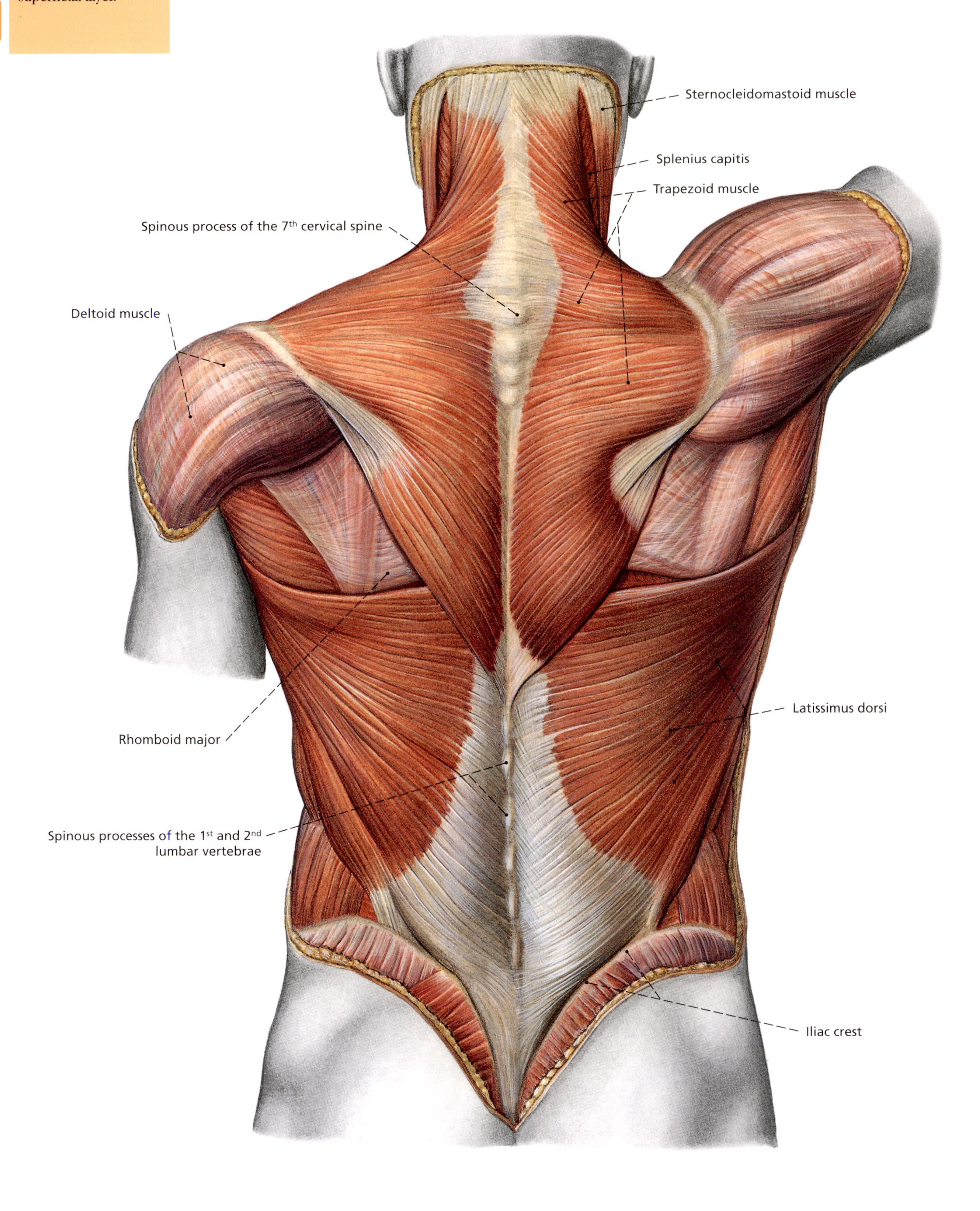

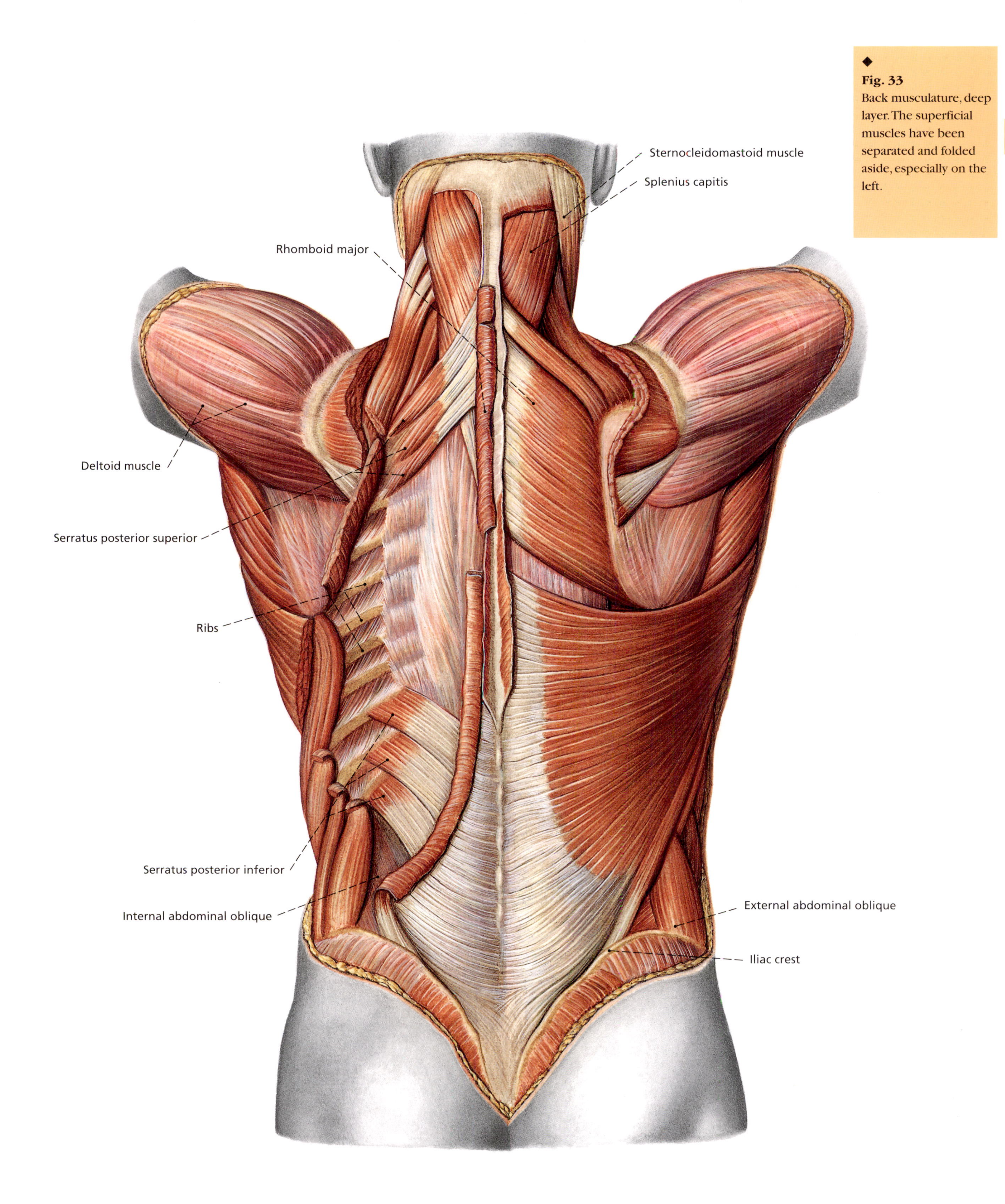

◆
Fig. 33
Back musculature, deep layer. The superficial muscles have been separated and folded aside, especially on the left.

◆
Fig. 34
Deep layer of the back muscles in the area of the spinous processes of the vertebrae and the base of the ribs.

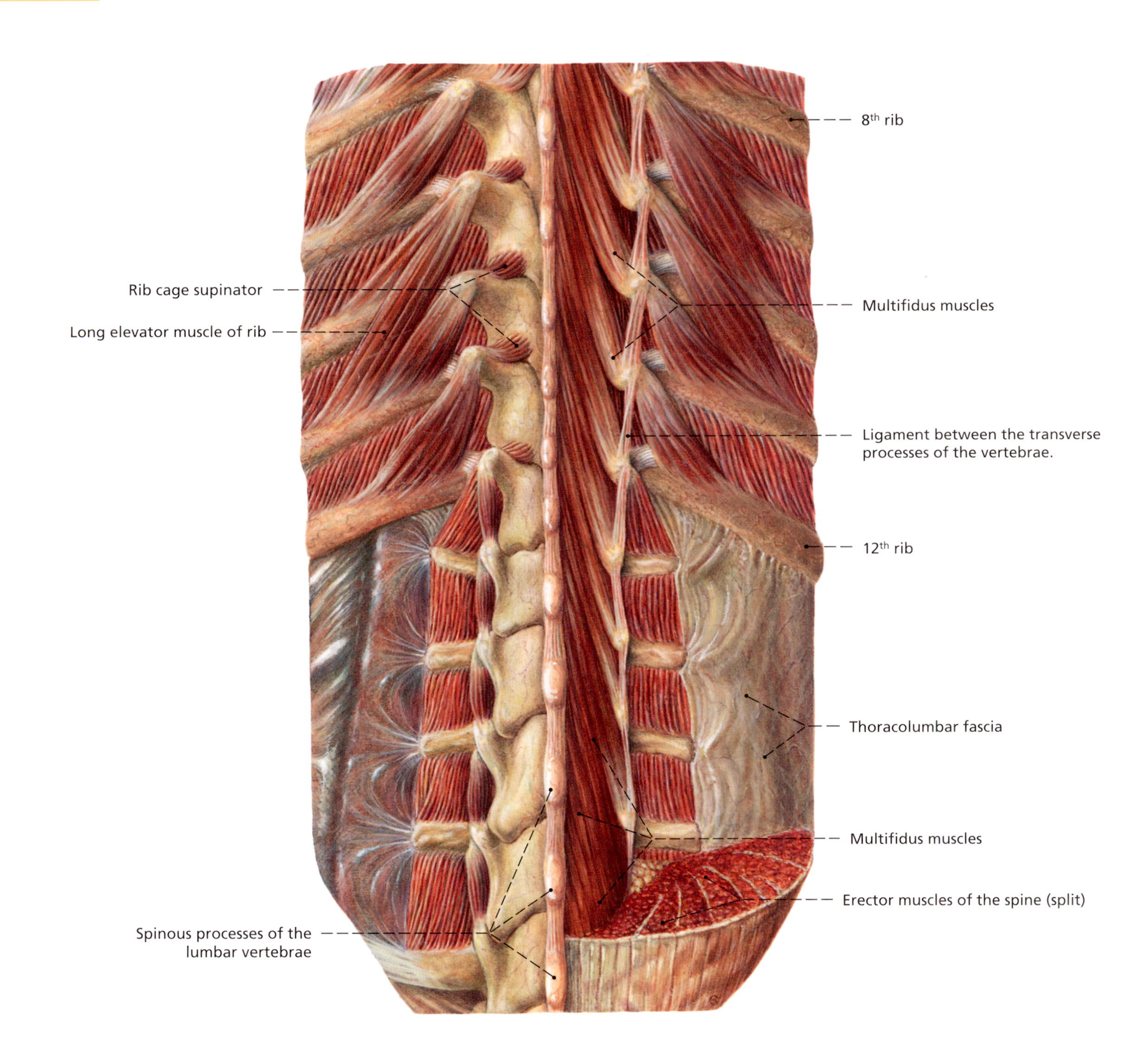

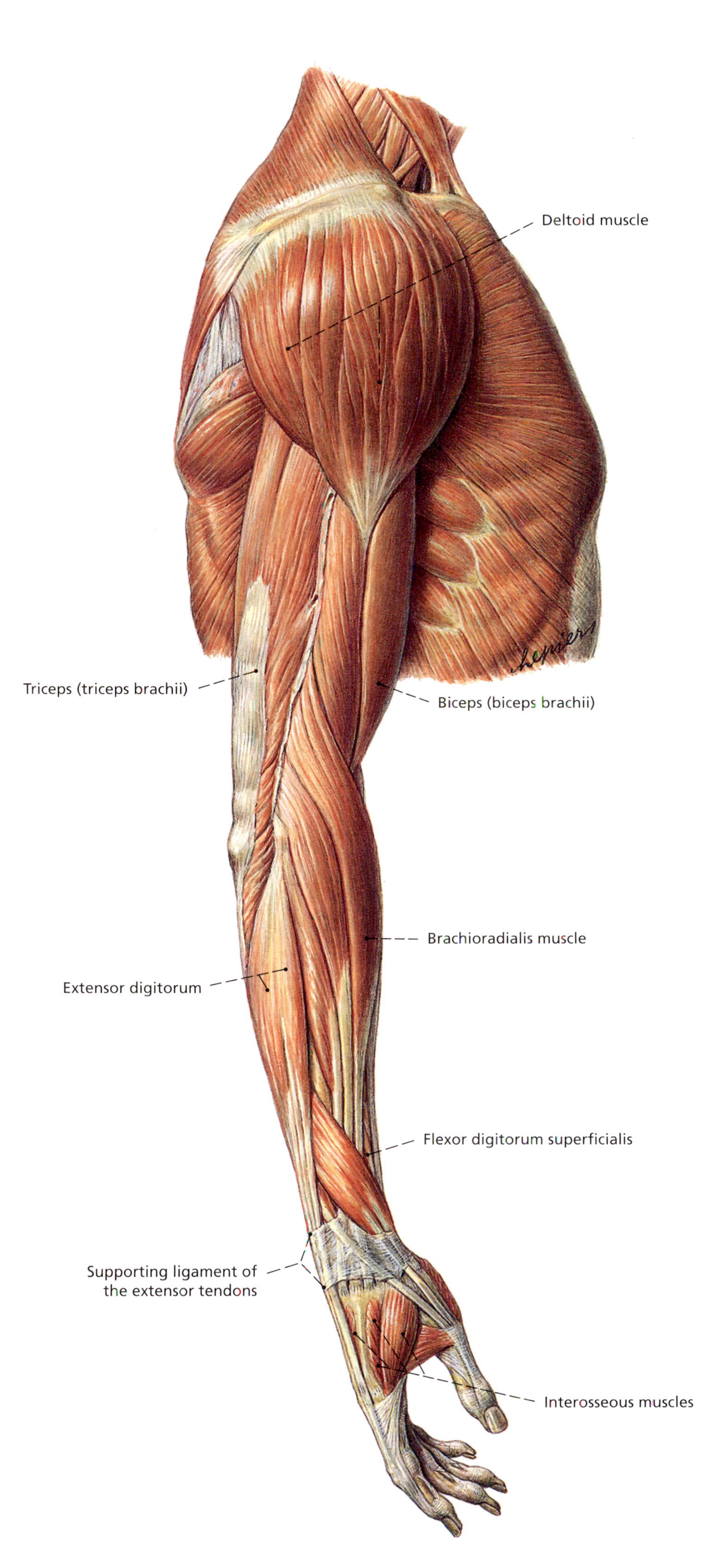

◆
Fig. 35
Musculature of the right arm from the side.

◆
Fig. 36
Musculature of the left upper arm from the front.

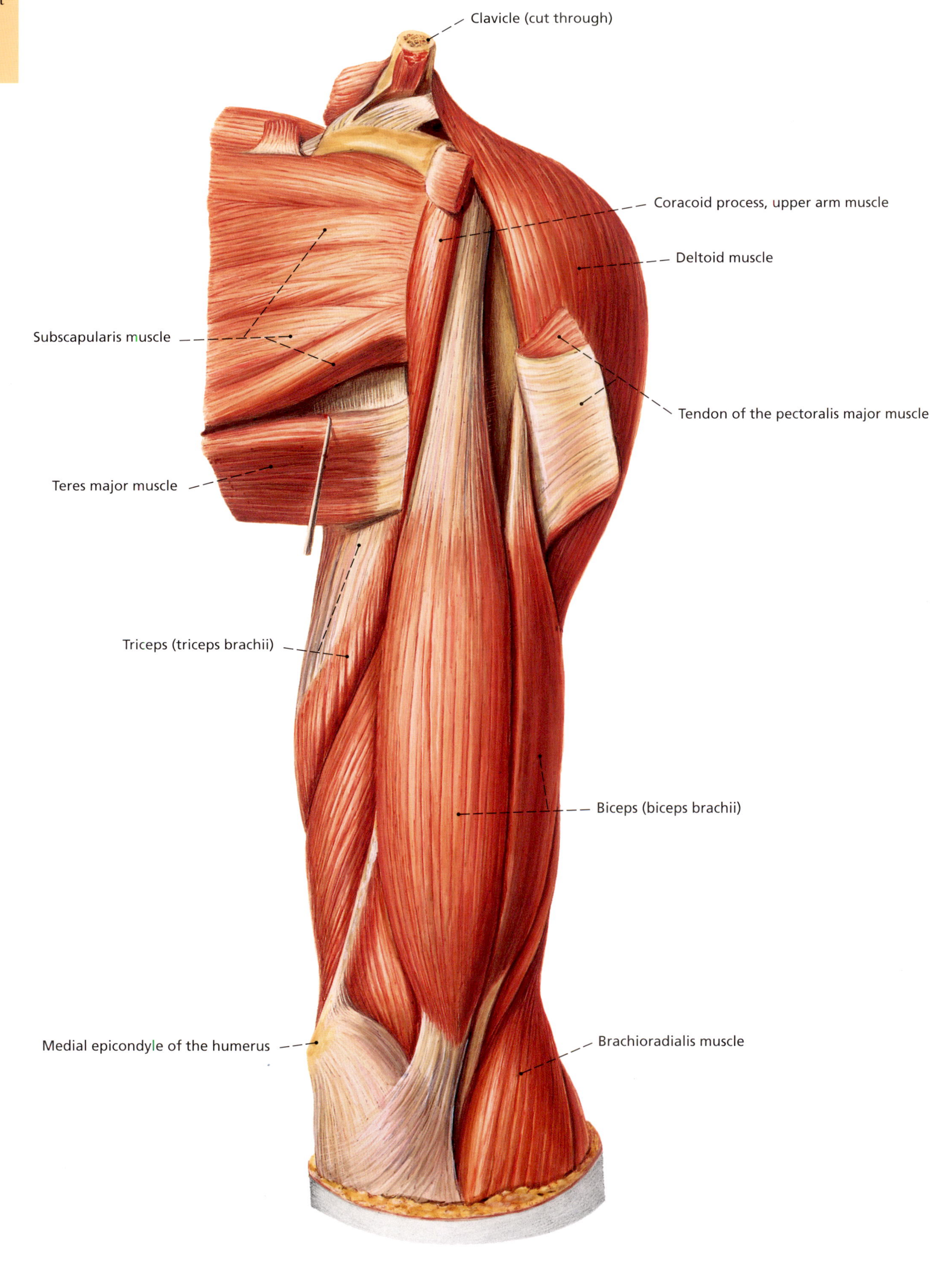

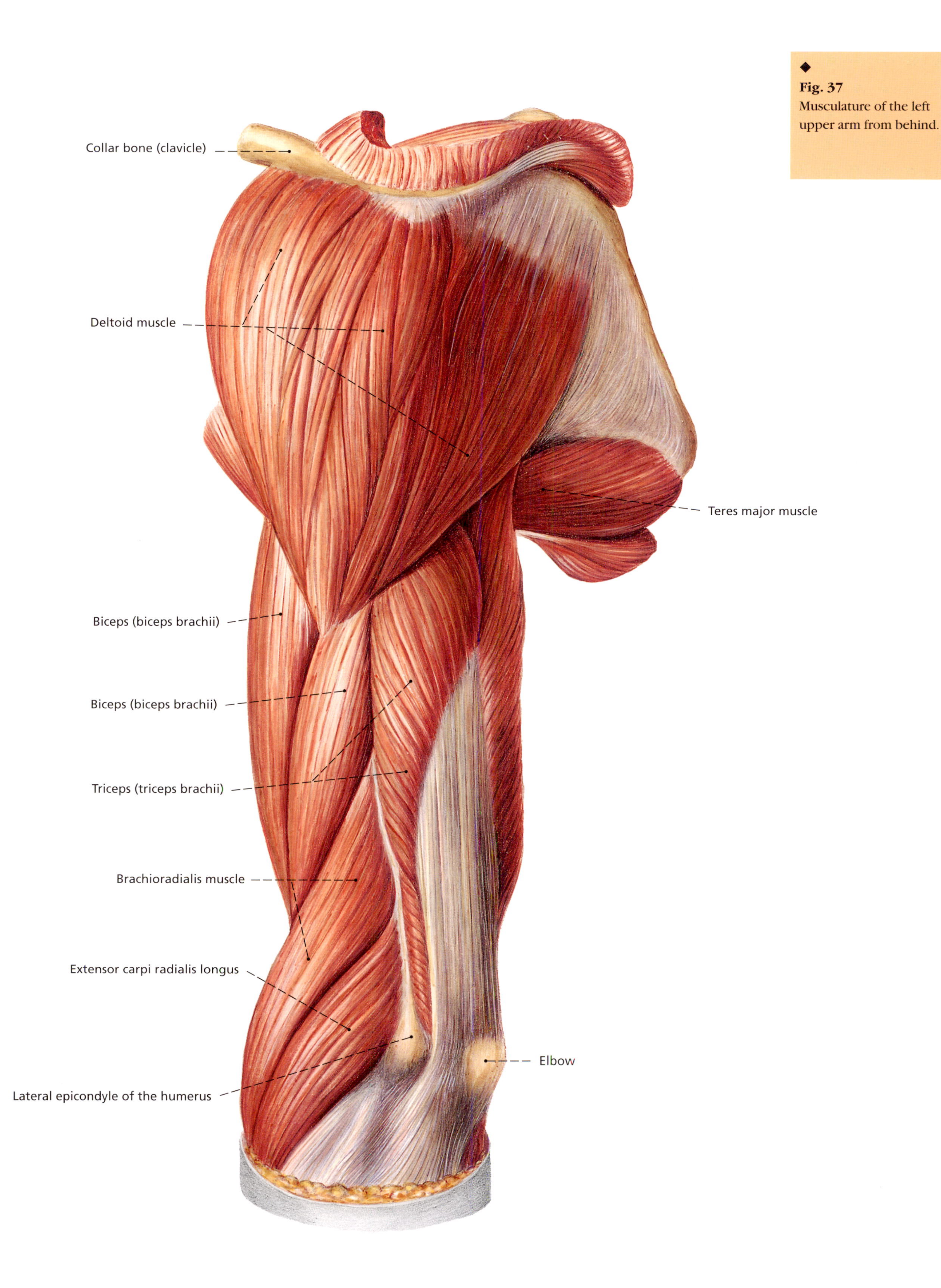

◆
Fig. 37
Musculature of the left upper arm from behind.

Fig. 38
Musculature of the left lower arm from the front.

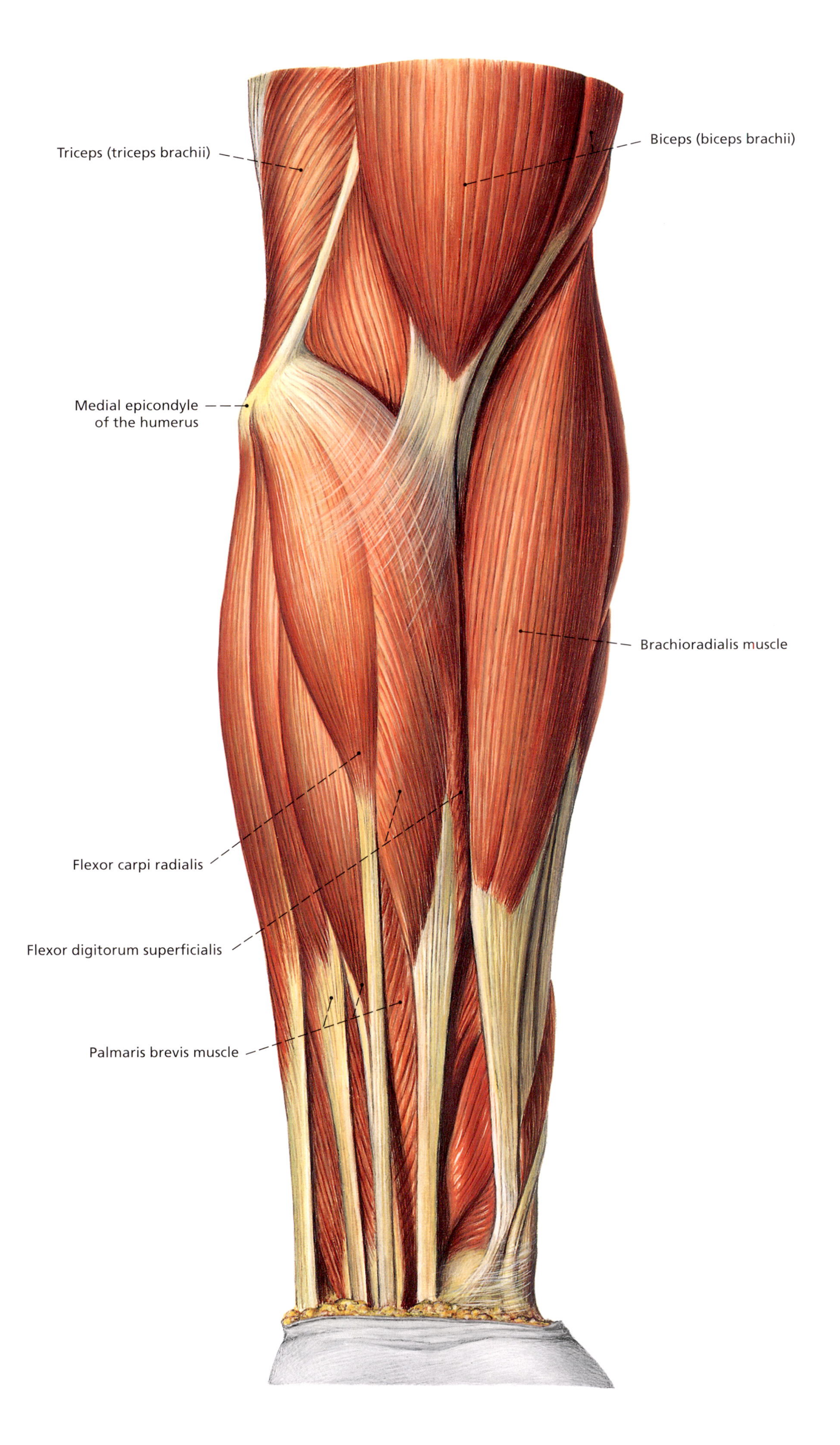

◆
Fig. 39
Musculature of the left lower arm from the side

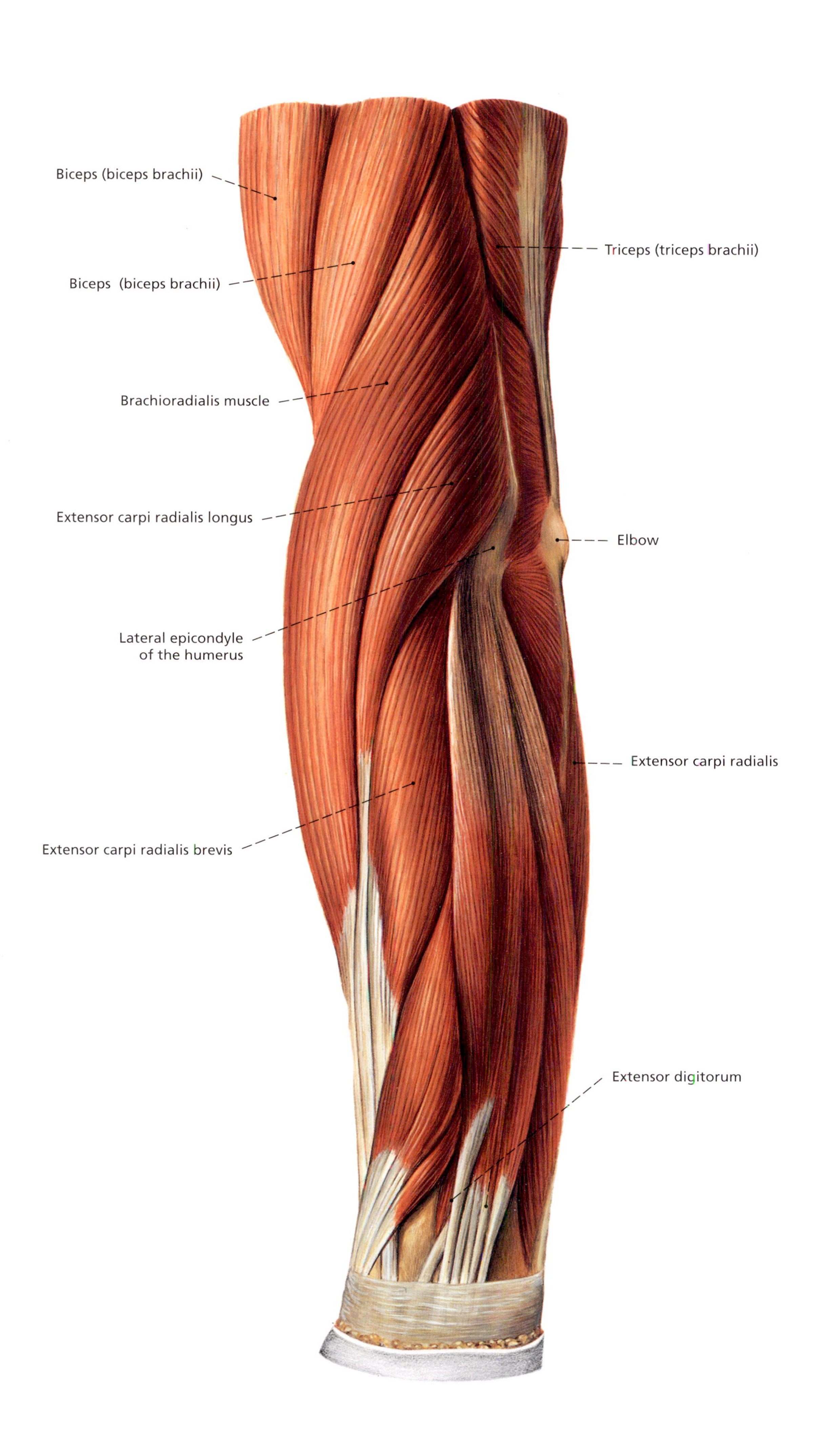

◆
Fig. 40
Musculature of the left lower arm from behind.

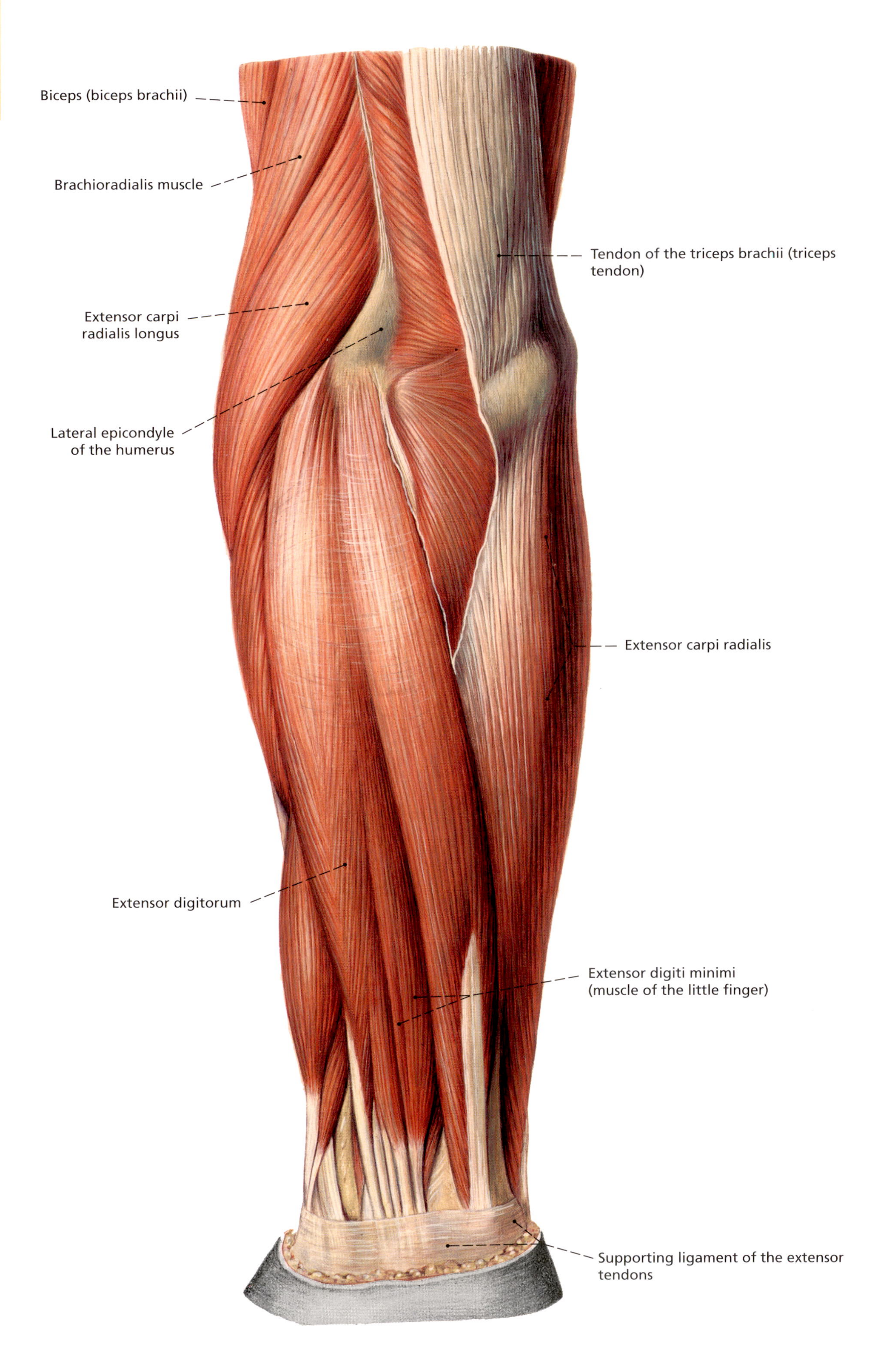

◆
Fig. 41
Superficial muscles of the palm of the hand. The tendon of the palmaris brevis muscle widens to become a tendon plate.

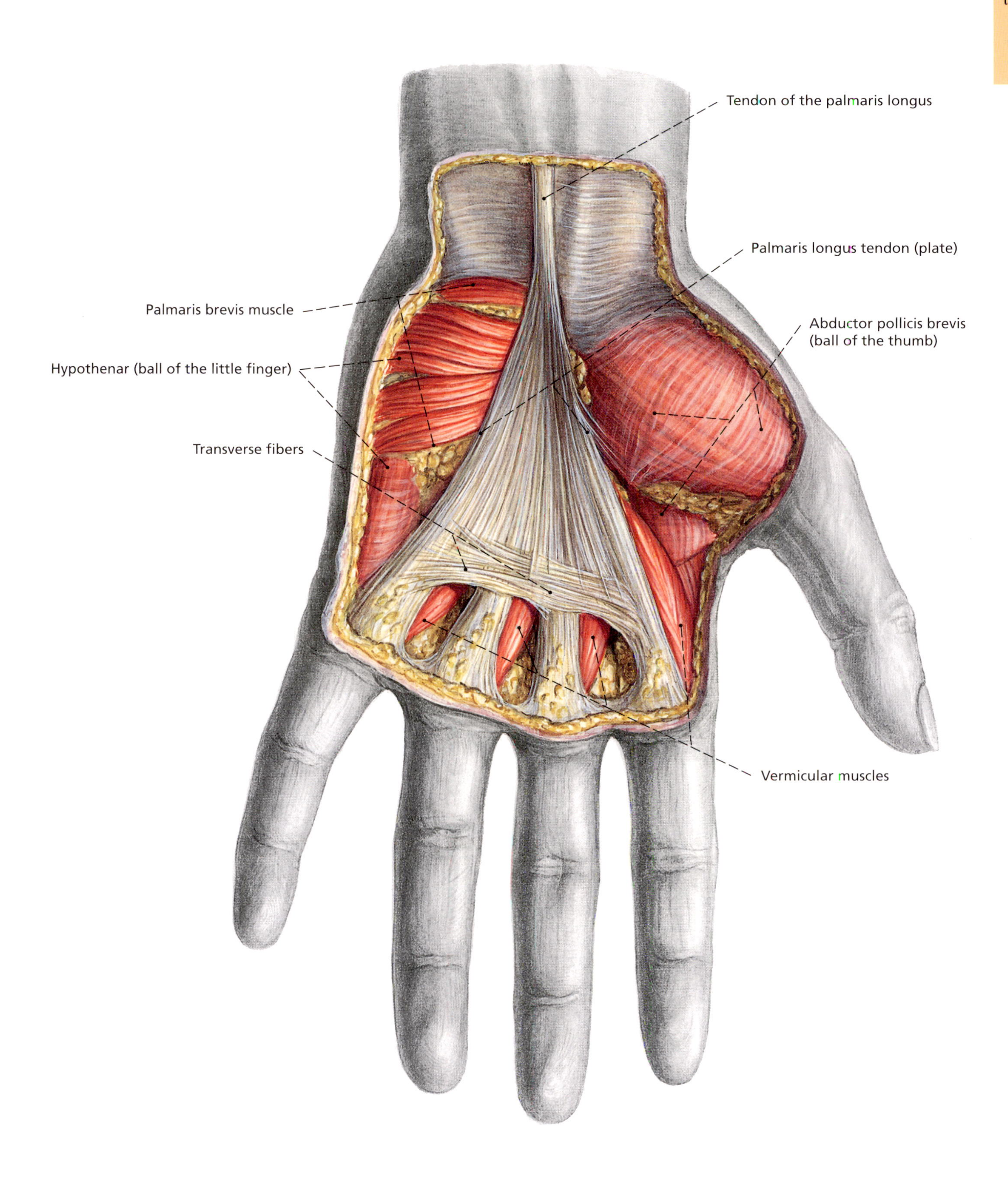

◆
Fig. 42
Tendon sheaths of the palmaris longus.

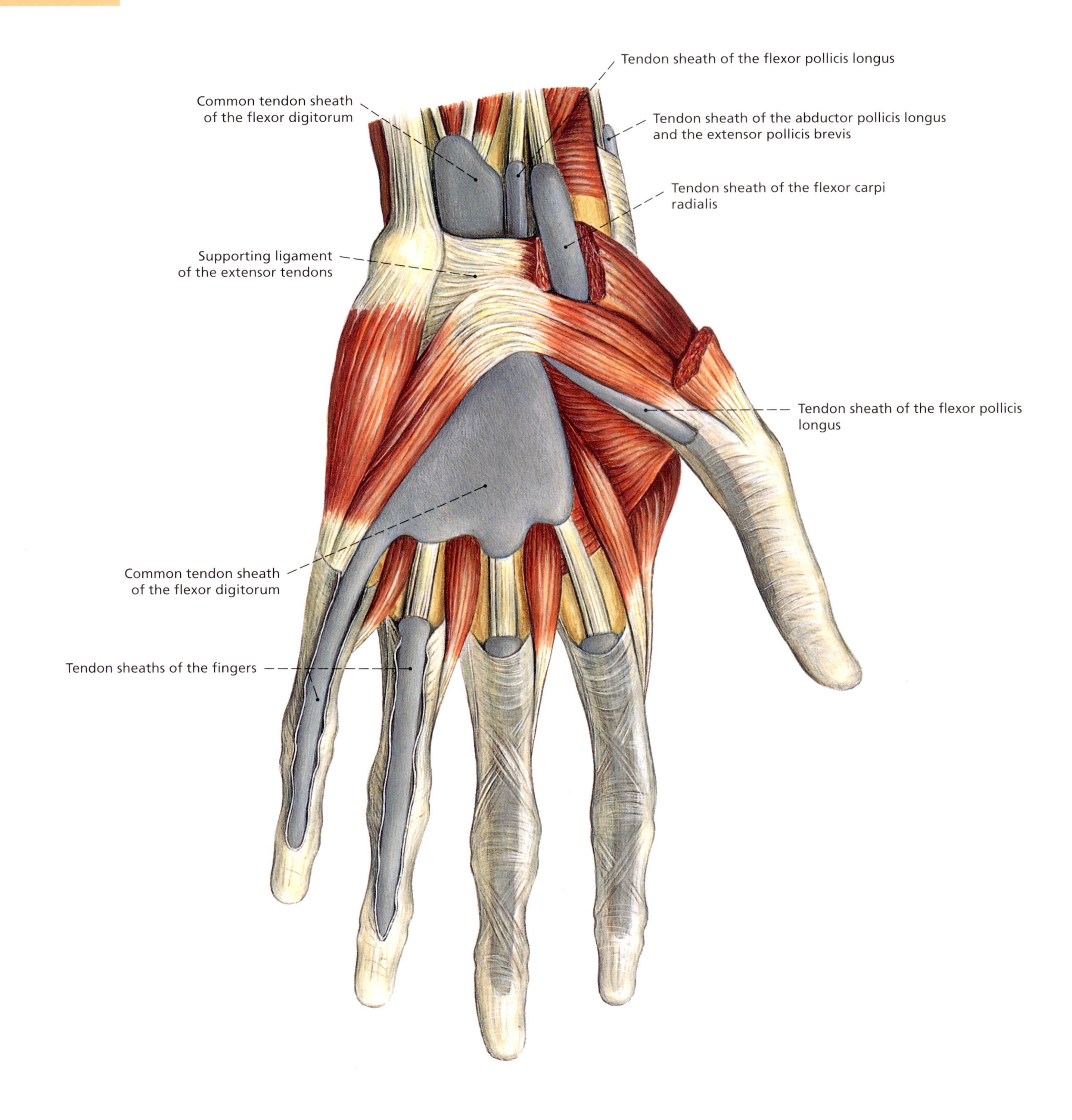

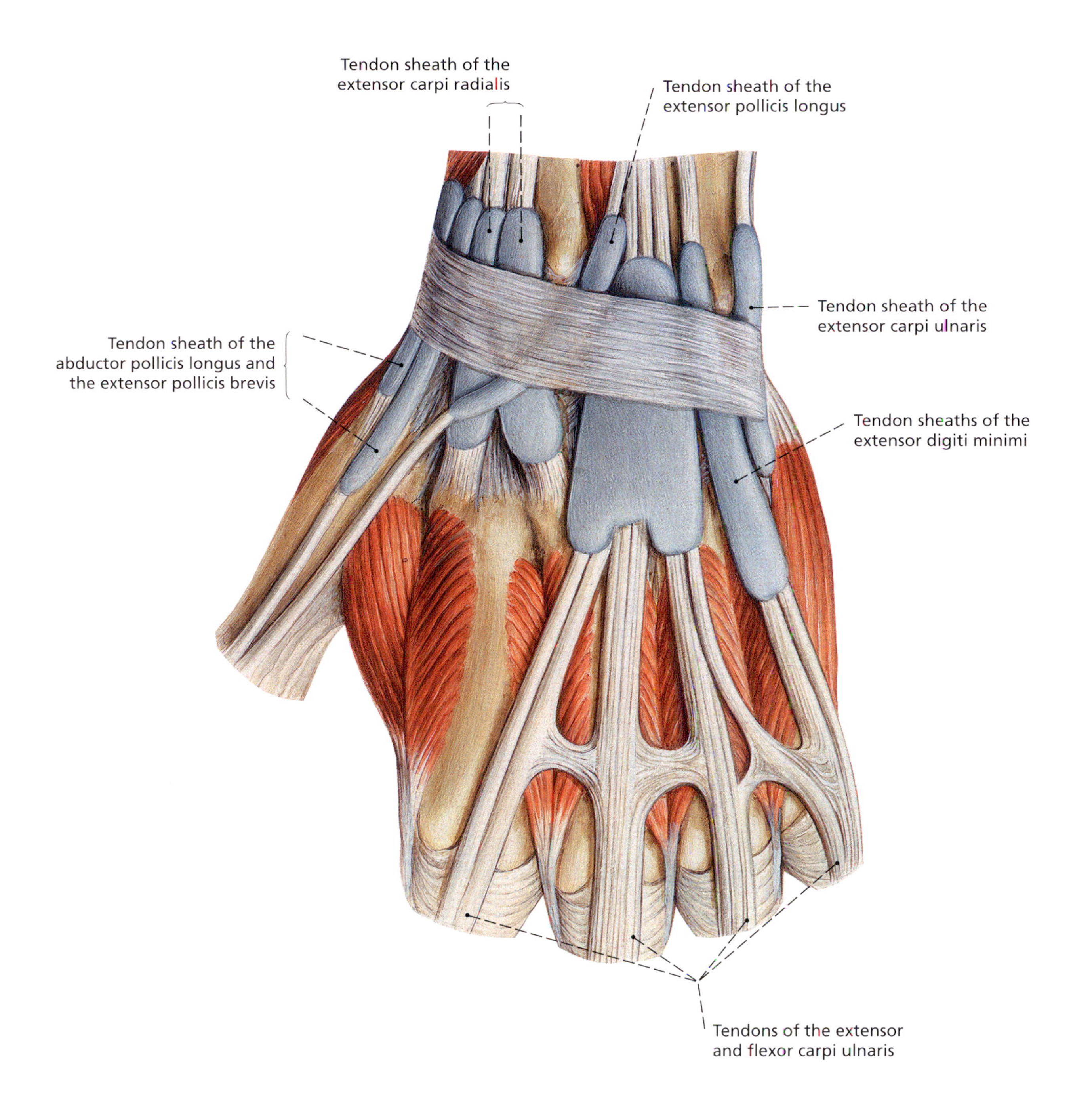

Fig. 43
Tendon sheaths of the back of the hand.

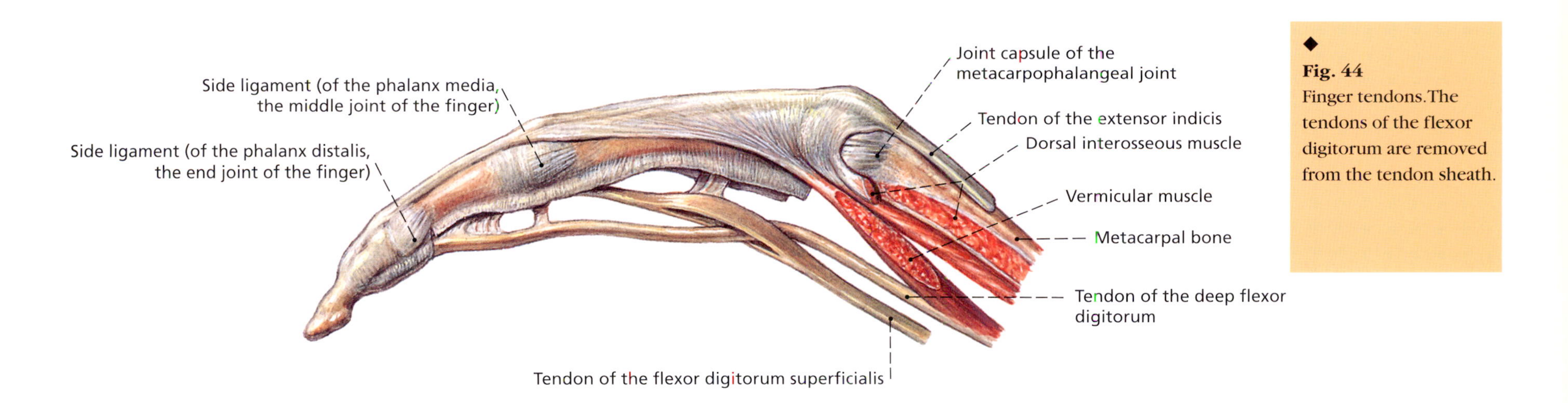

Fig. 44
Finger tendons. The tendons of the flexor digitorum are removed from the tendon sheath.

◆
Fig. 45
Musculature of the back of the hand with tendons.

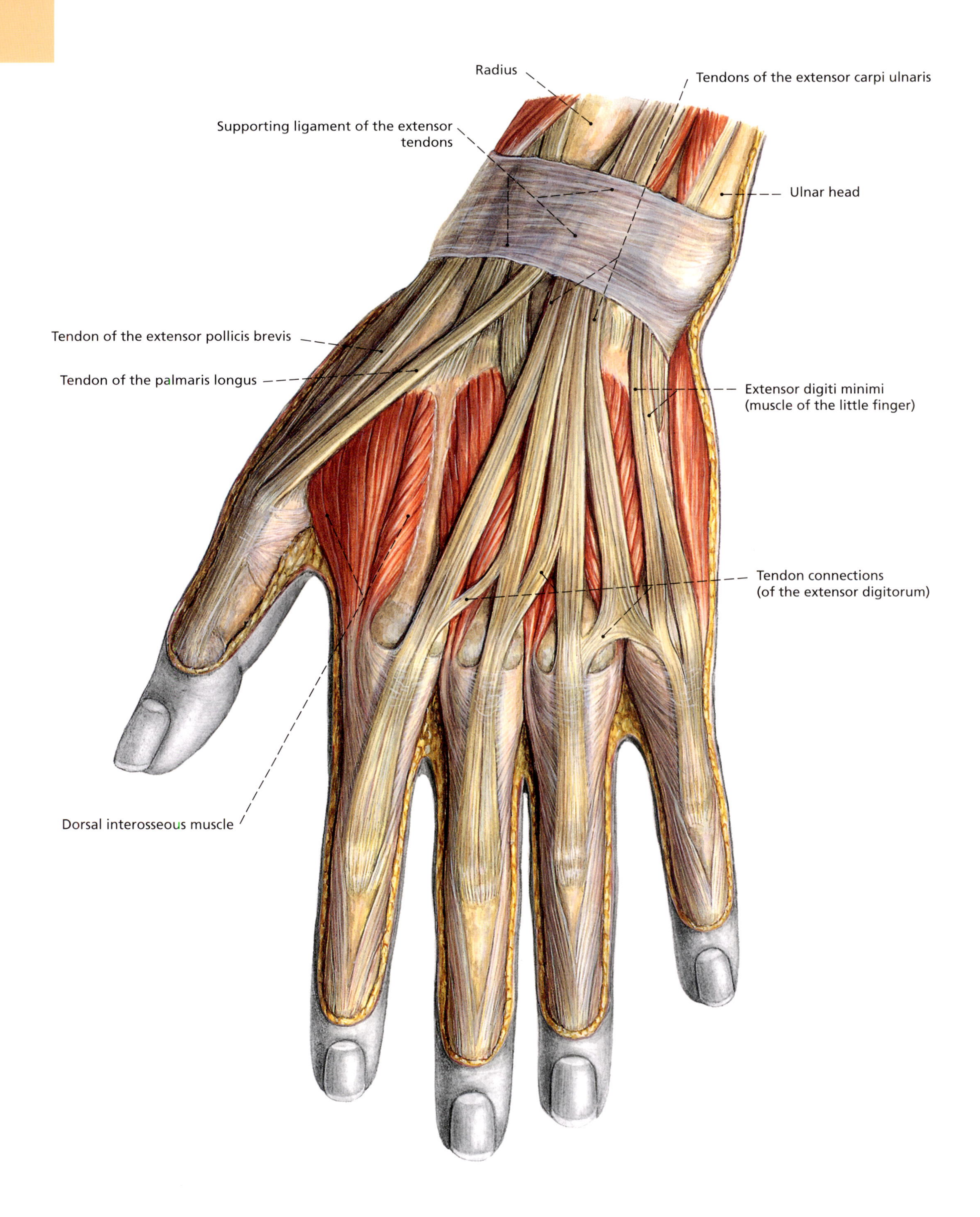

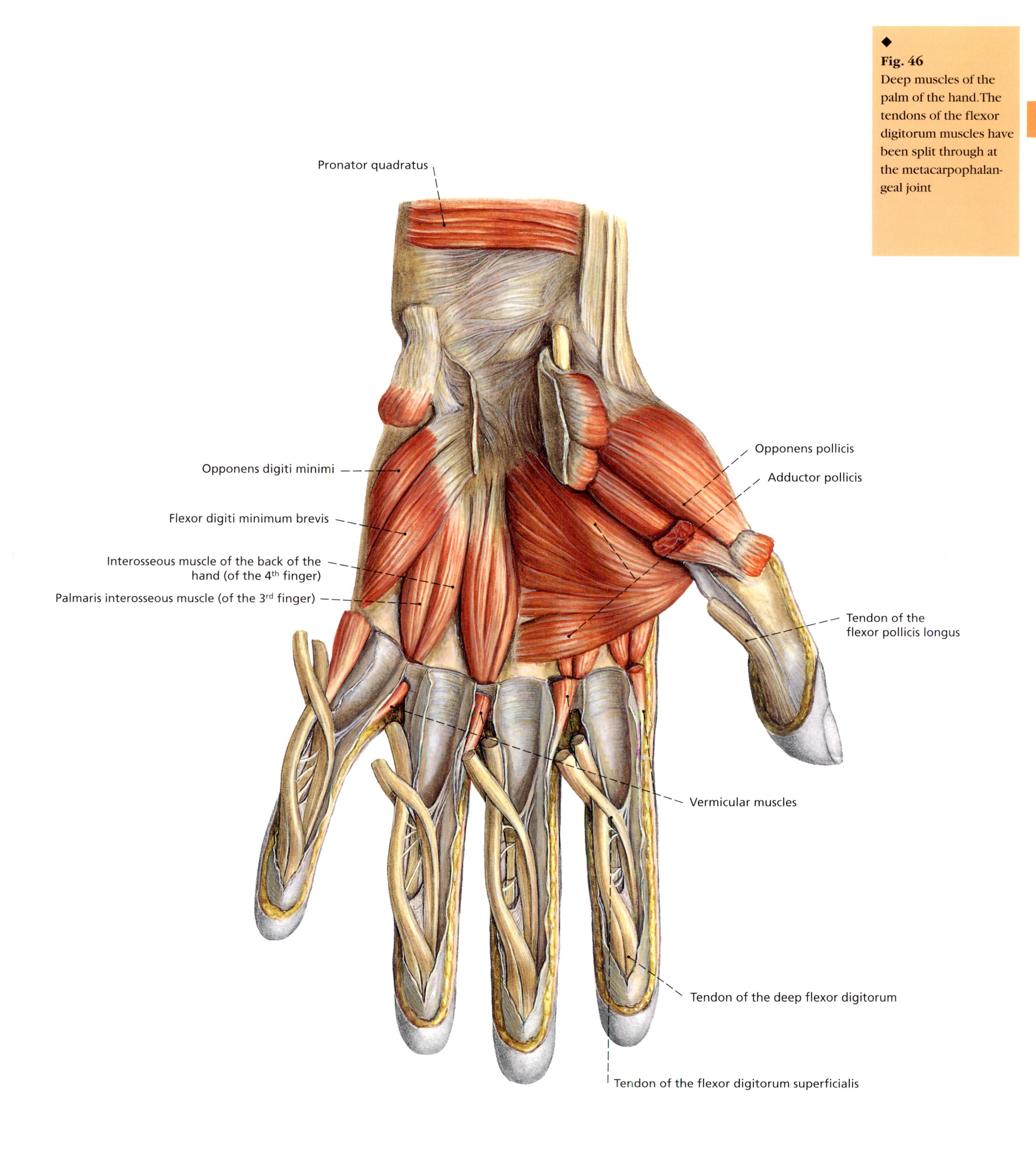

◆
Fig. 46
Deep muscles of the palm of the hand. The tendons of the flexor digitorum muscles have been split through at the metacarpophalangeal joint

◆
Fig. 47
Upper thigh musculature from the side (right).

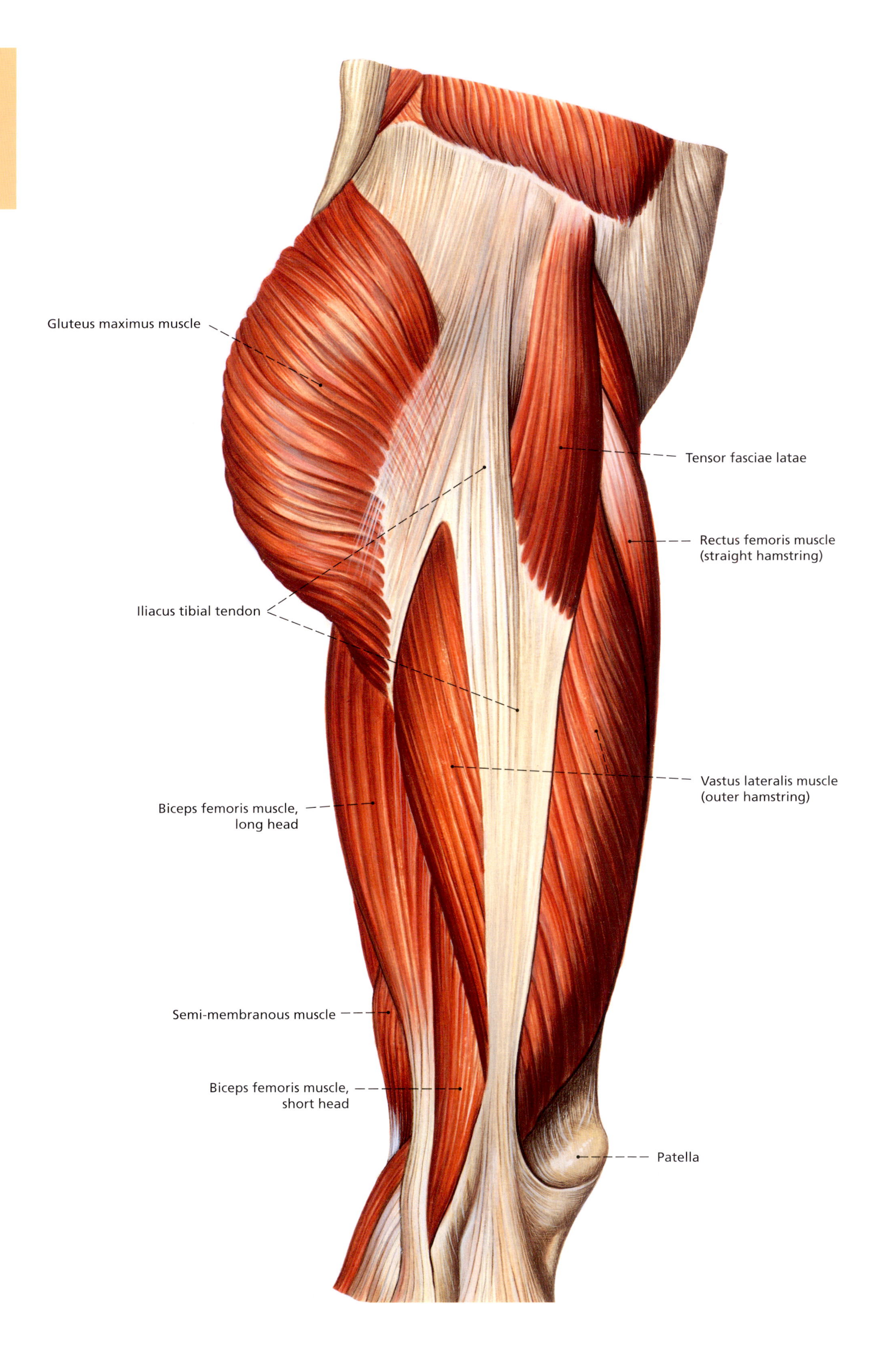

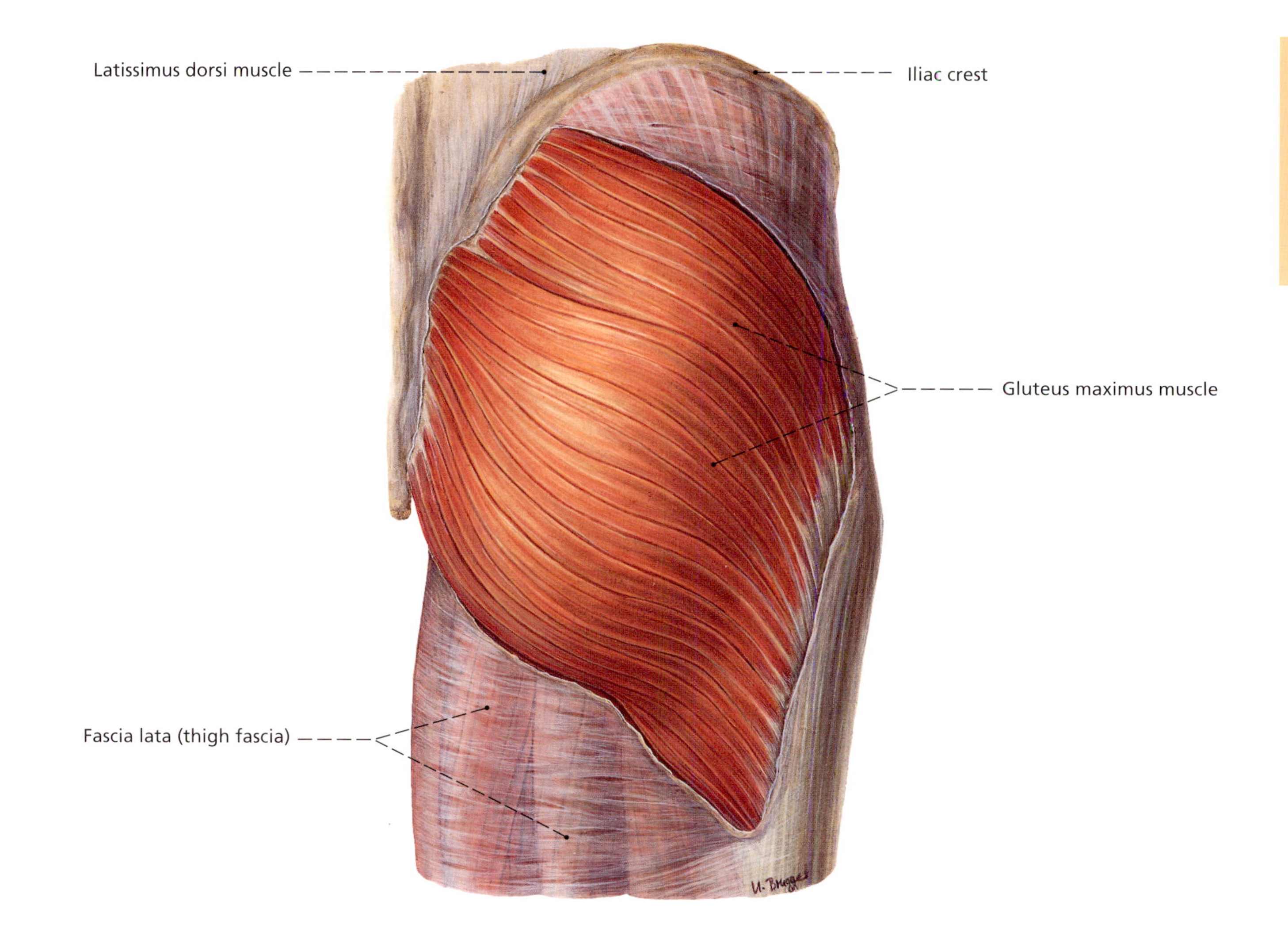

◆
Fig. 48
Gluteus maximus from behind (right) after removal of the fascia.

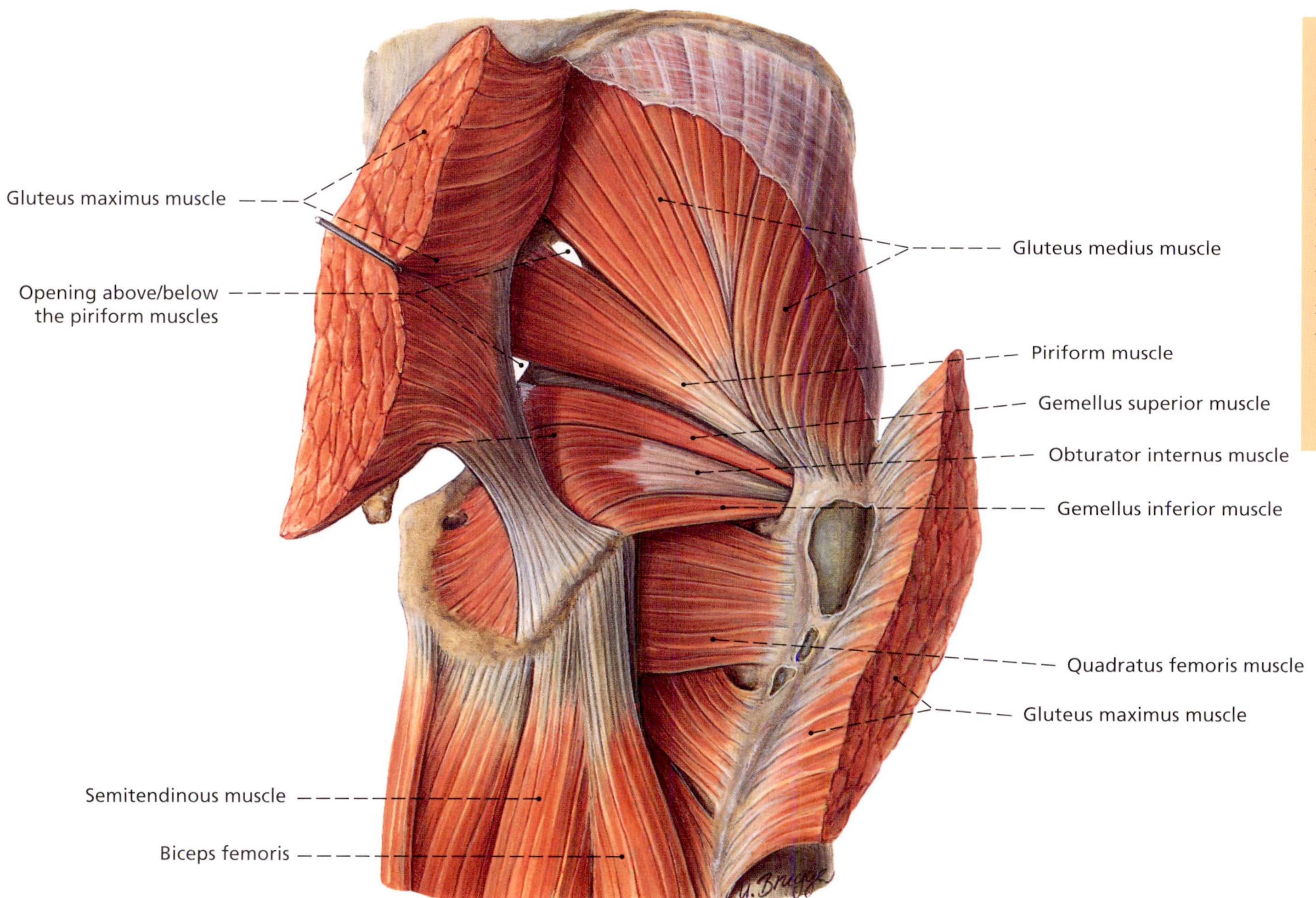

◆
Fig. 49
Deep layer of the gluteus muscles. The gluteus maximus is split and folded aside. The sciatic nerve passes through the opening beneath the piriform muscles into the upper thigh (right).

Fig. 50
Thigh muscles from the front (right leg).

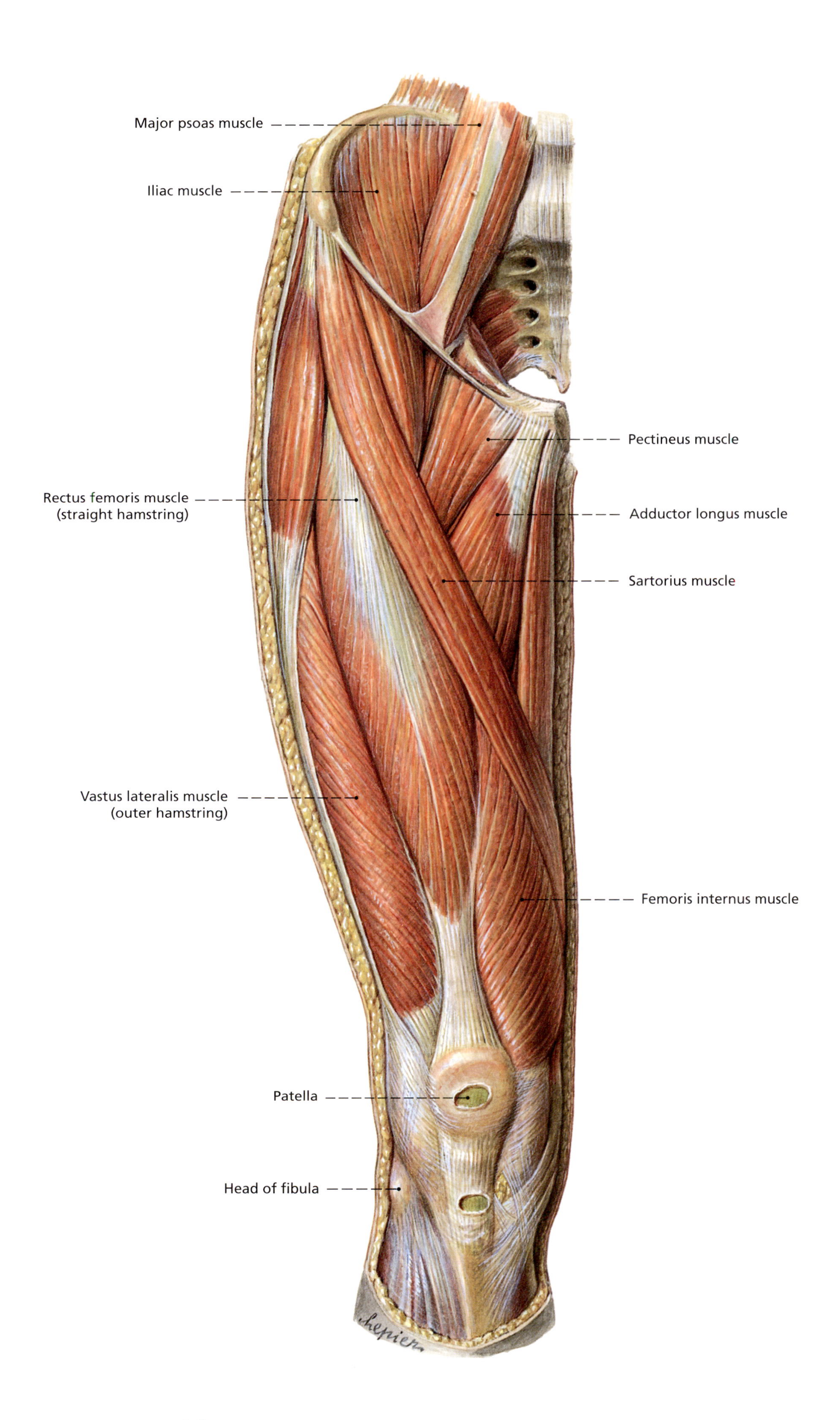

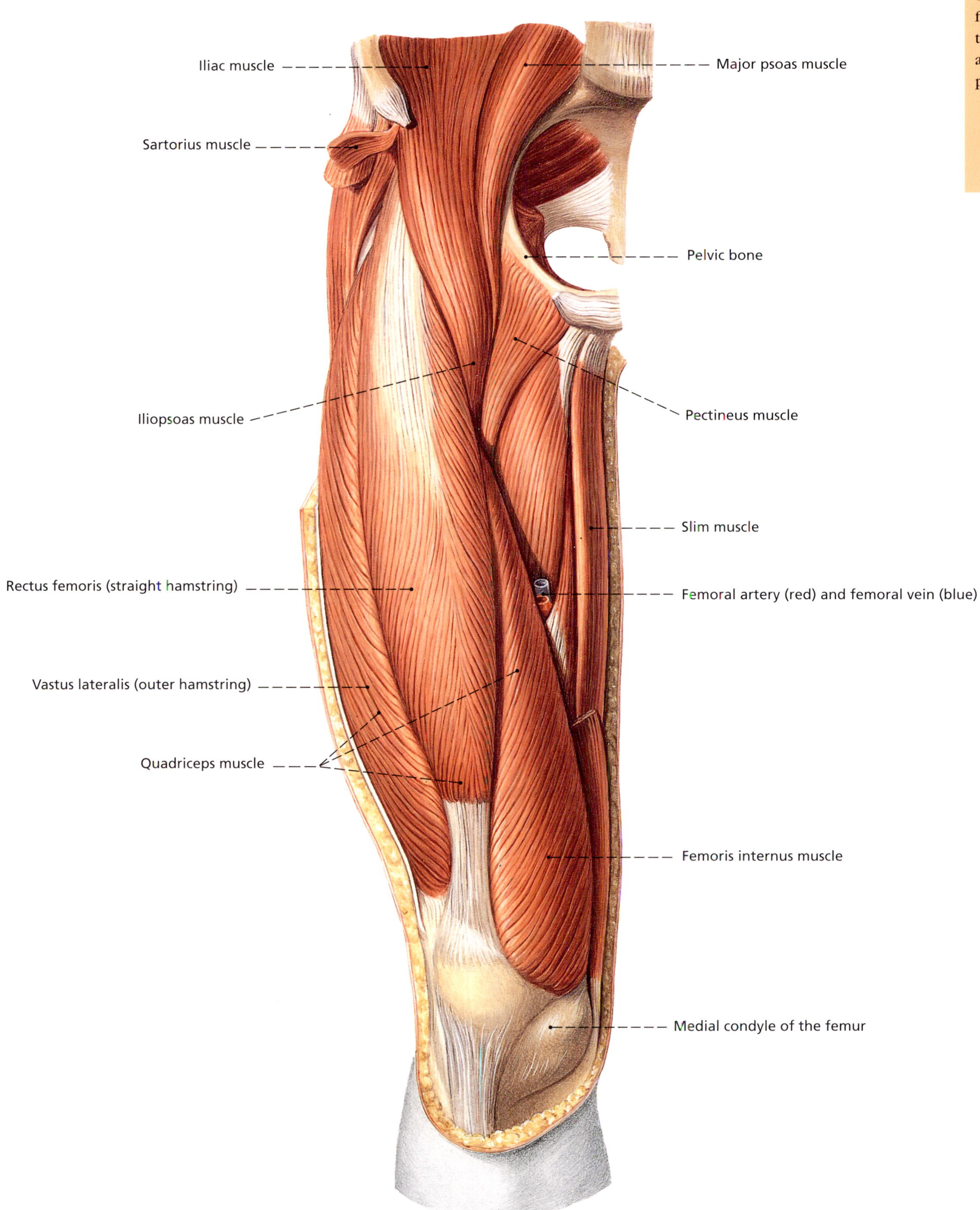

◆
Fig. 51
Thigh muscles from the front after removal of the sartorius muscle and the thigh tendon plate (right leg).

◆
Fig. 52
Musculature of the femur and hip from behind (right leg). Gluteus maximus and gluteus medius have been partly removed.

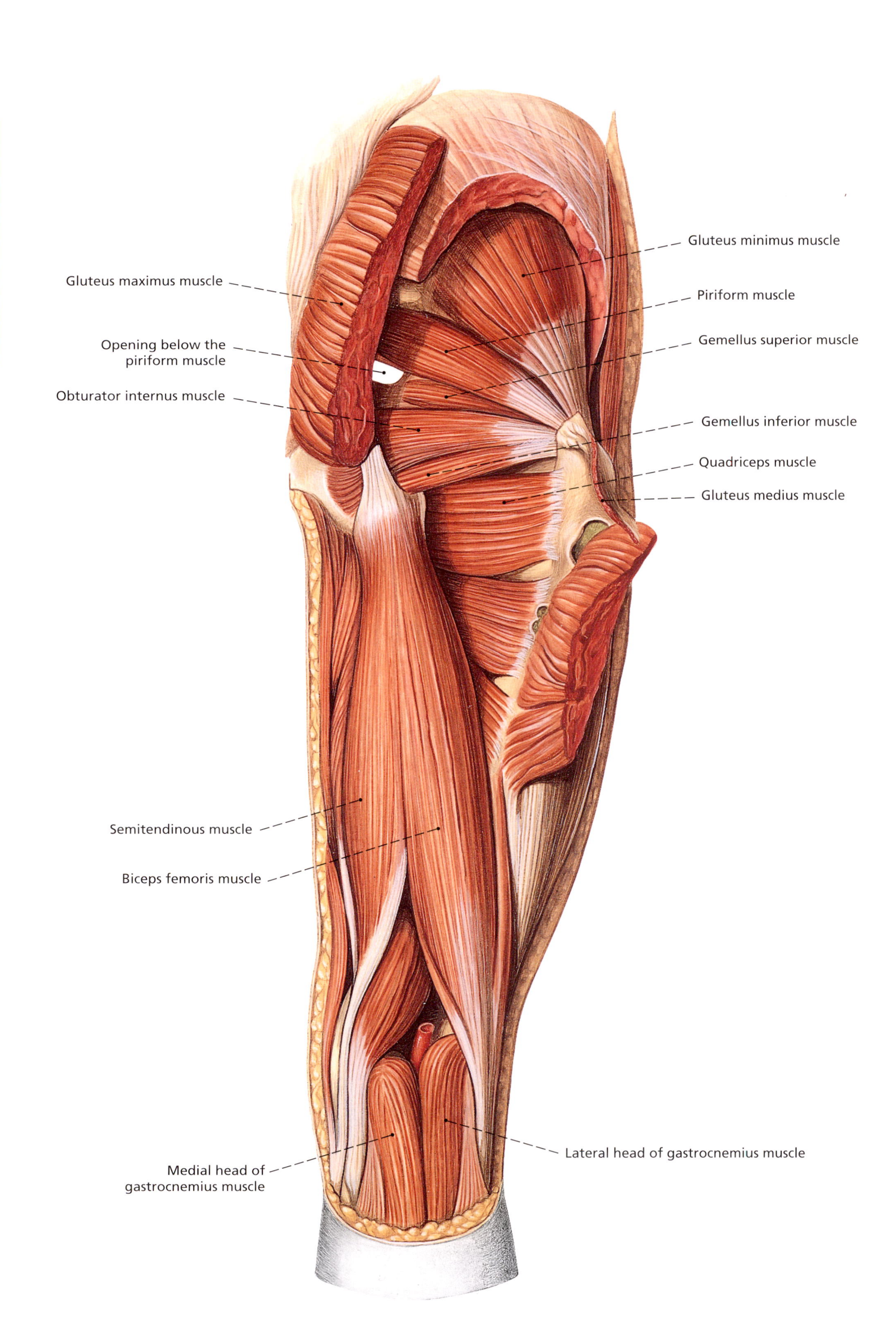

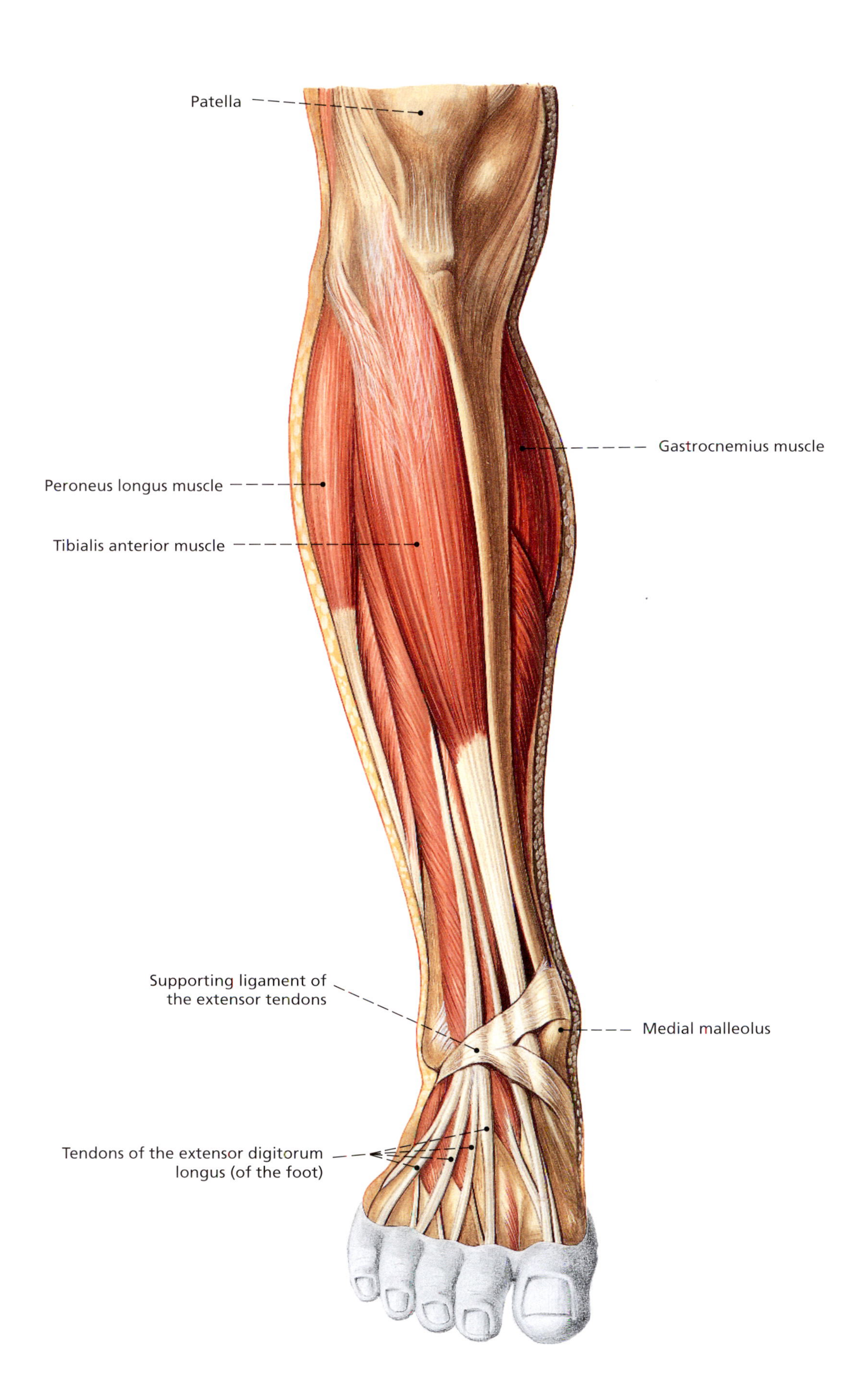

◆
Fig. 53
Musculature of the right lower leg from the front.

◆
Fig. 54
Musculature of the right lower leg and foot from the outer side.

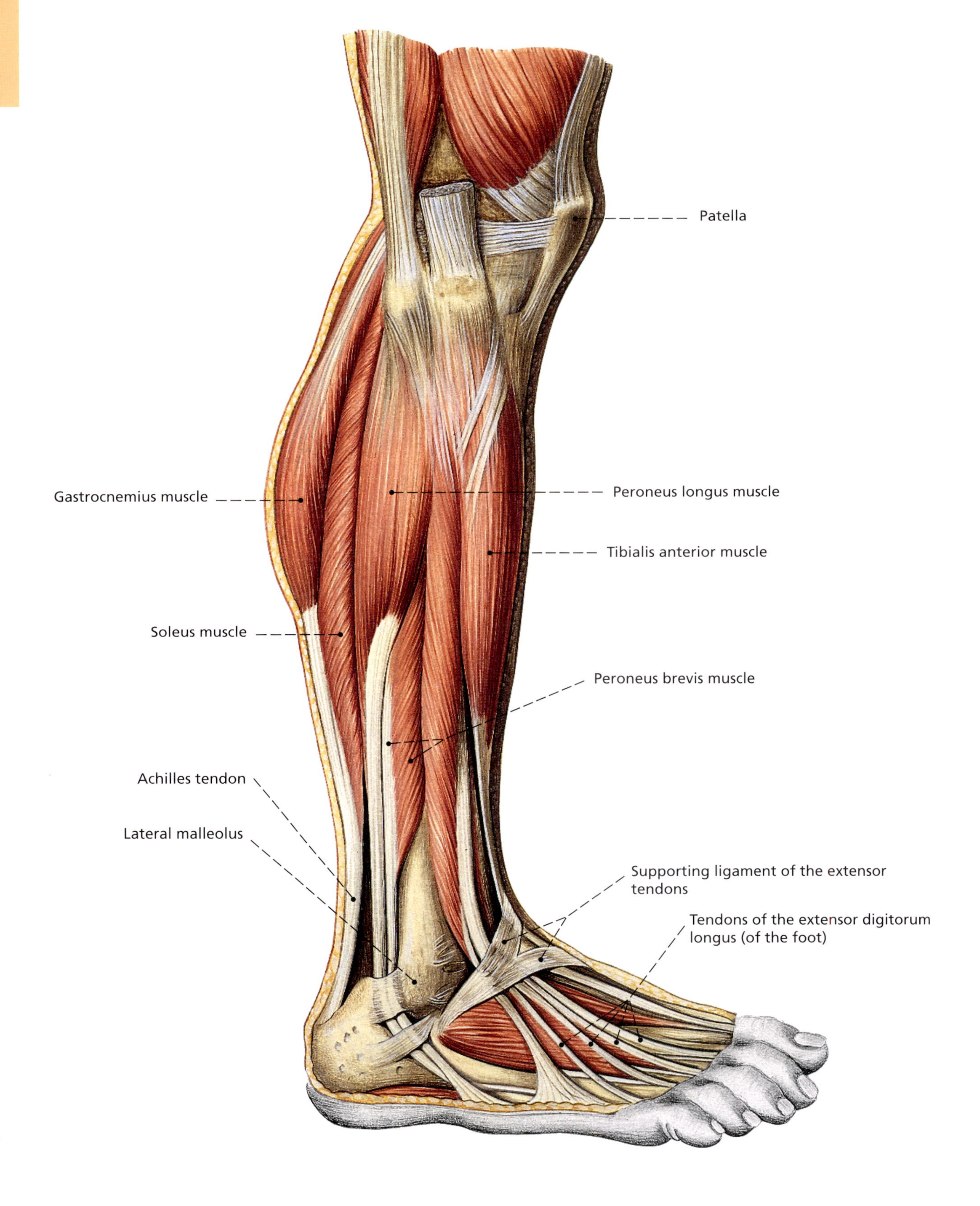

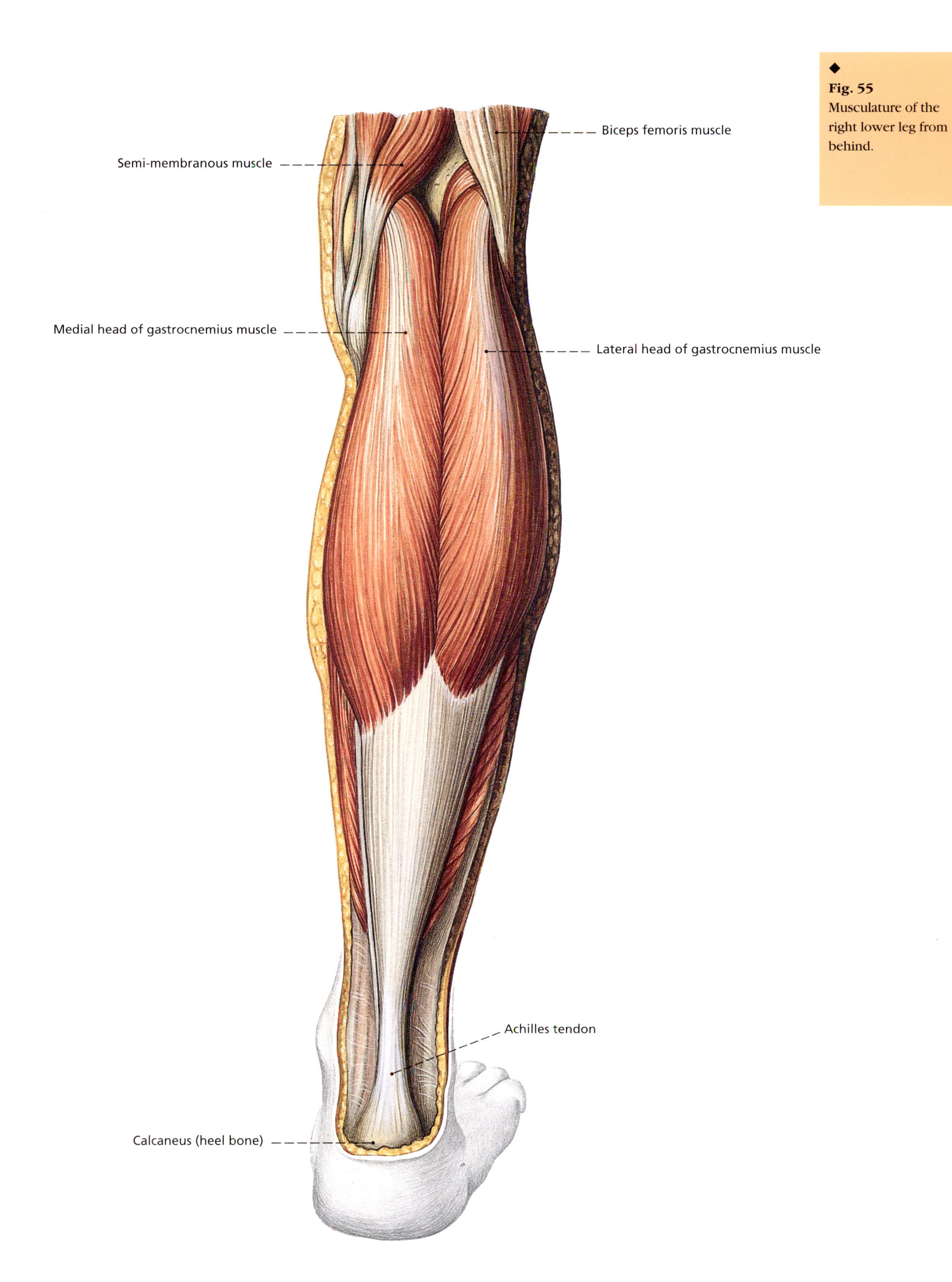

◆
Fig. 55
Musculature of the right lower leg from behind.

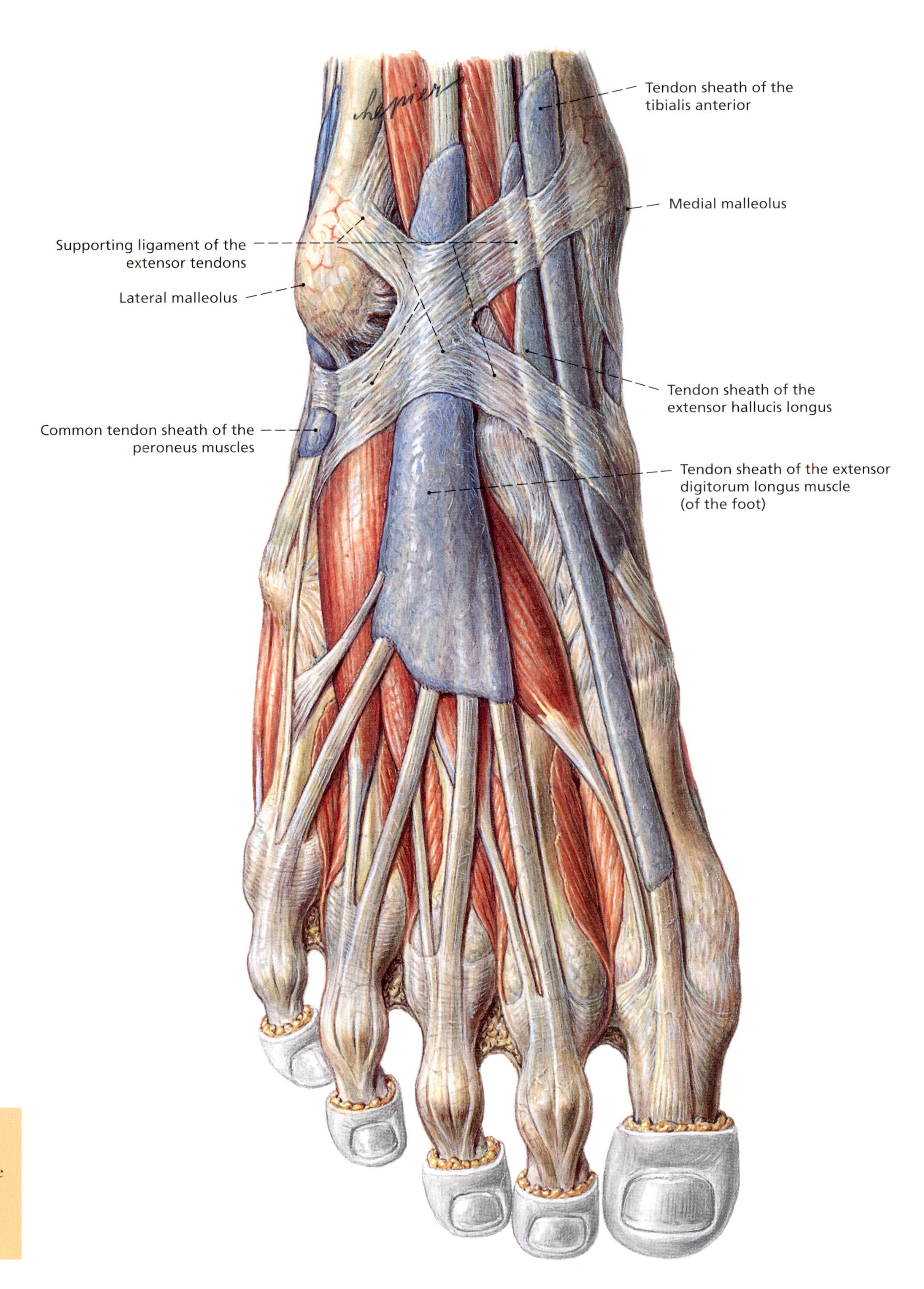

◆
Fig. 56
Tendon sheaths of the back of the right foot.

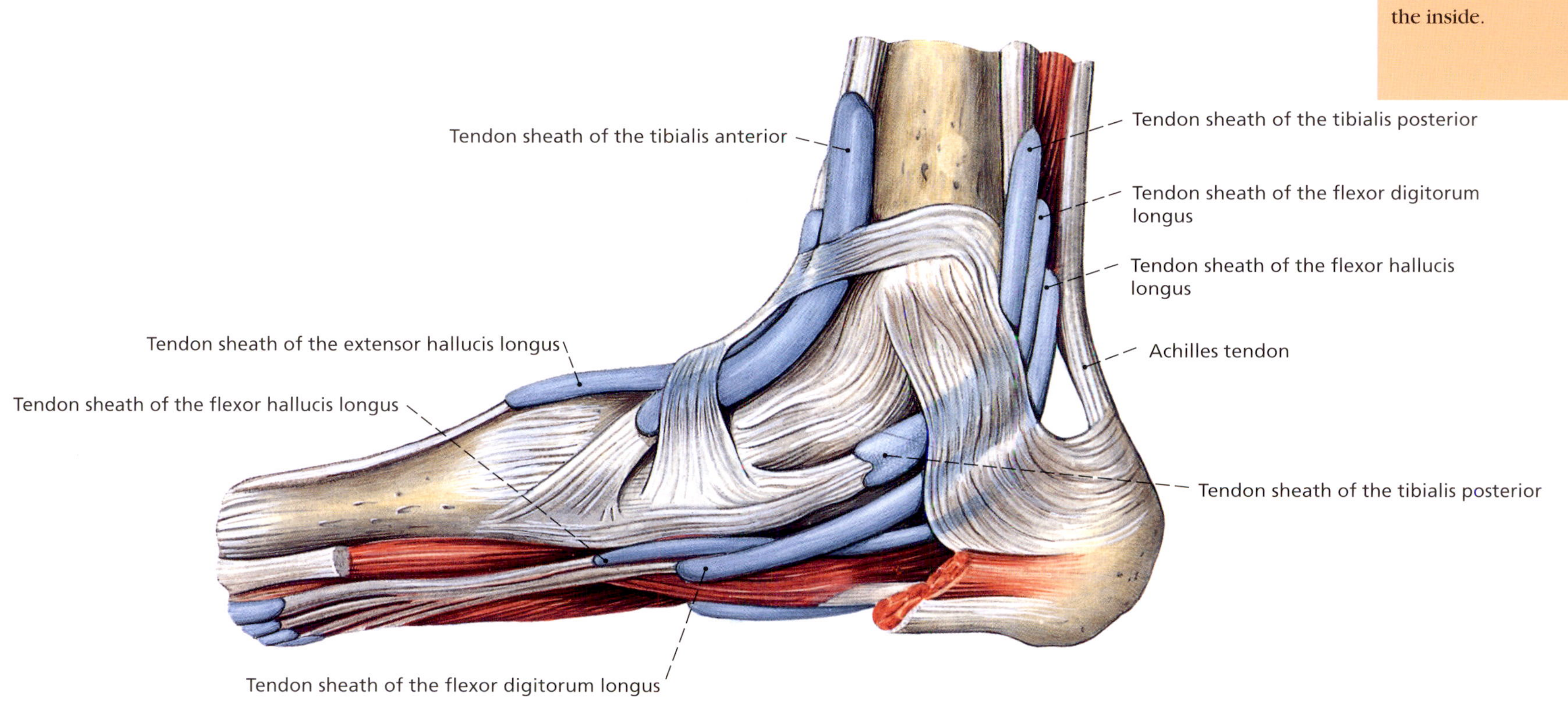

◆
Fig. 57
Tendon sheaths of the right foot, seen from the inside.

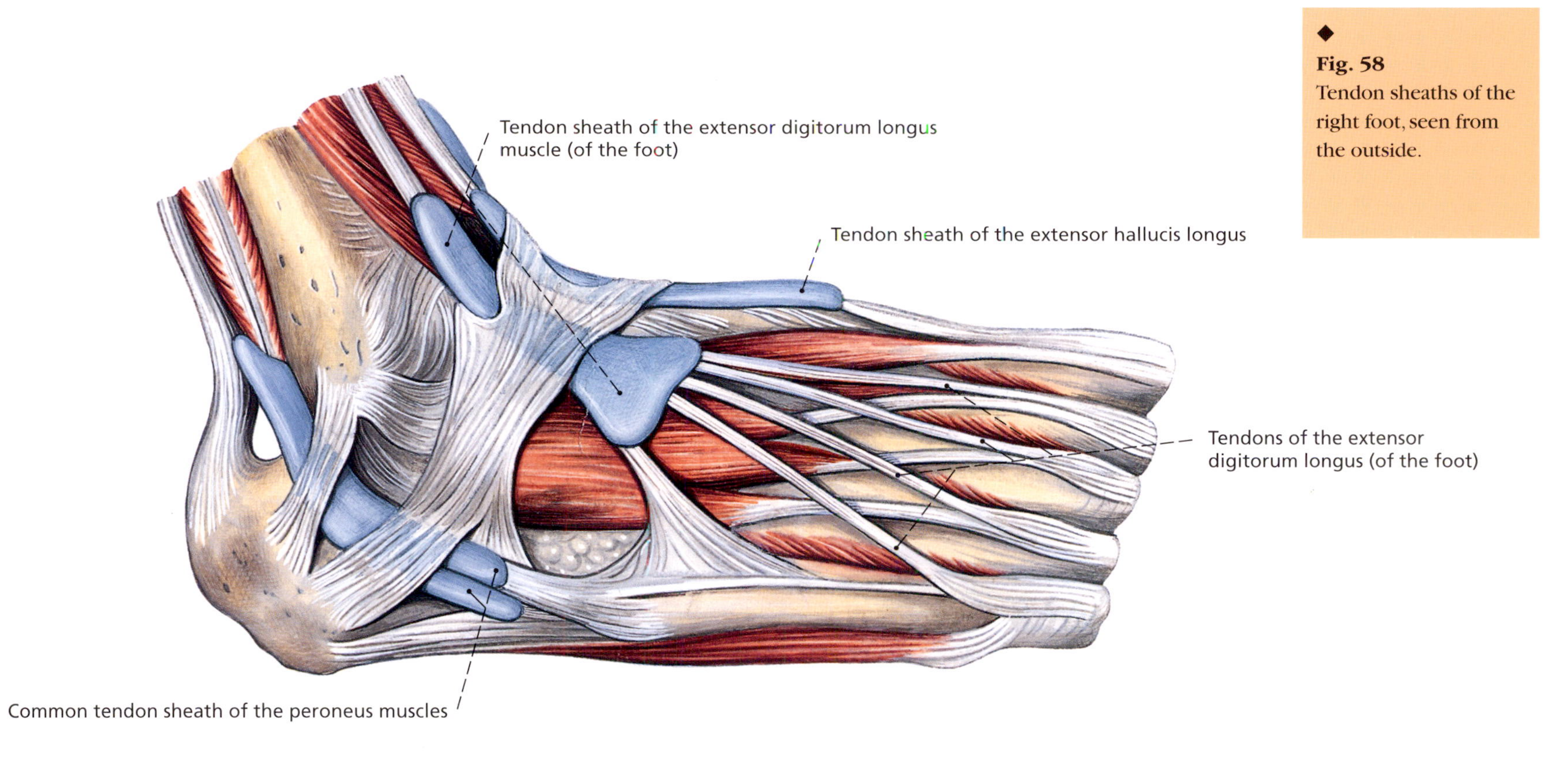

◆
Fig. 58
Tendon sheaths of the right foot, seen from the outside.

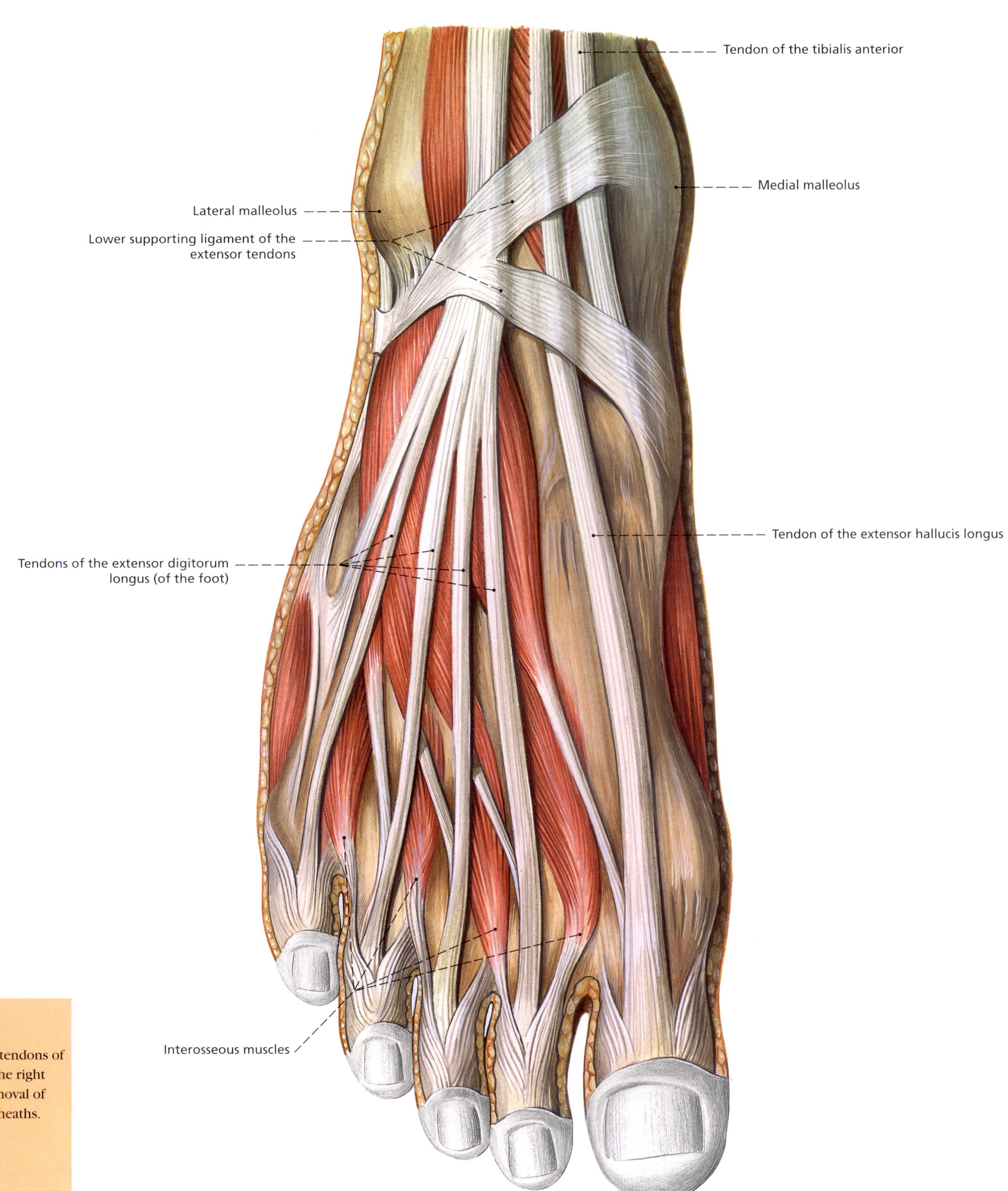

◆
Fig. 59
Muscles and tendons of the back of the right foot after removal of the tendon sheaths.

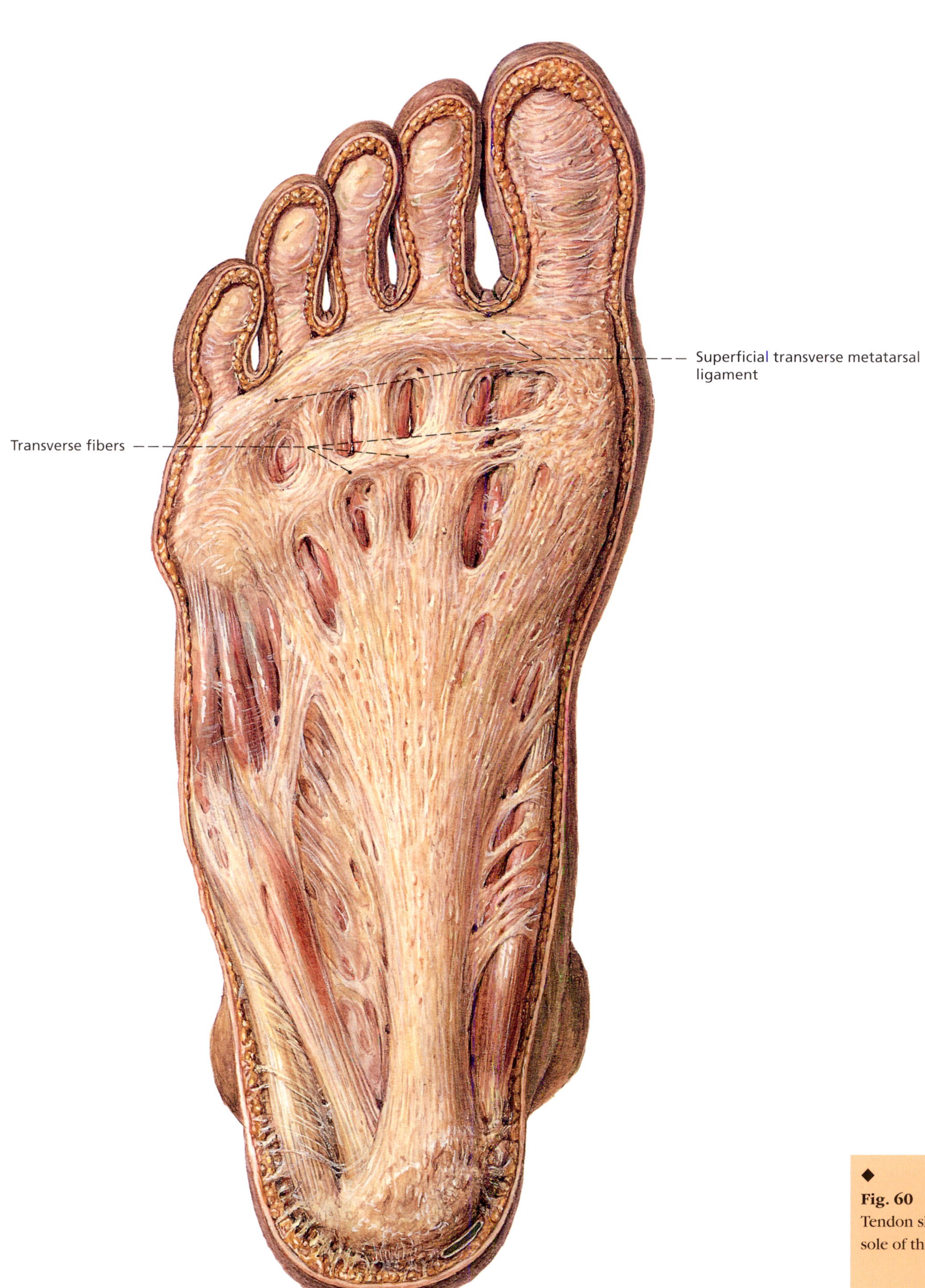

◆
Fig. 60
Tendon sheaths of the sole of the right foot.

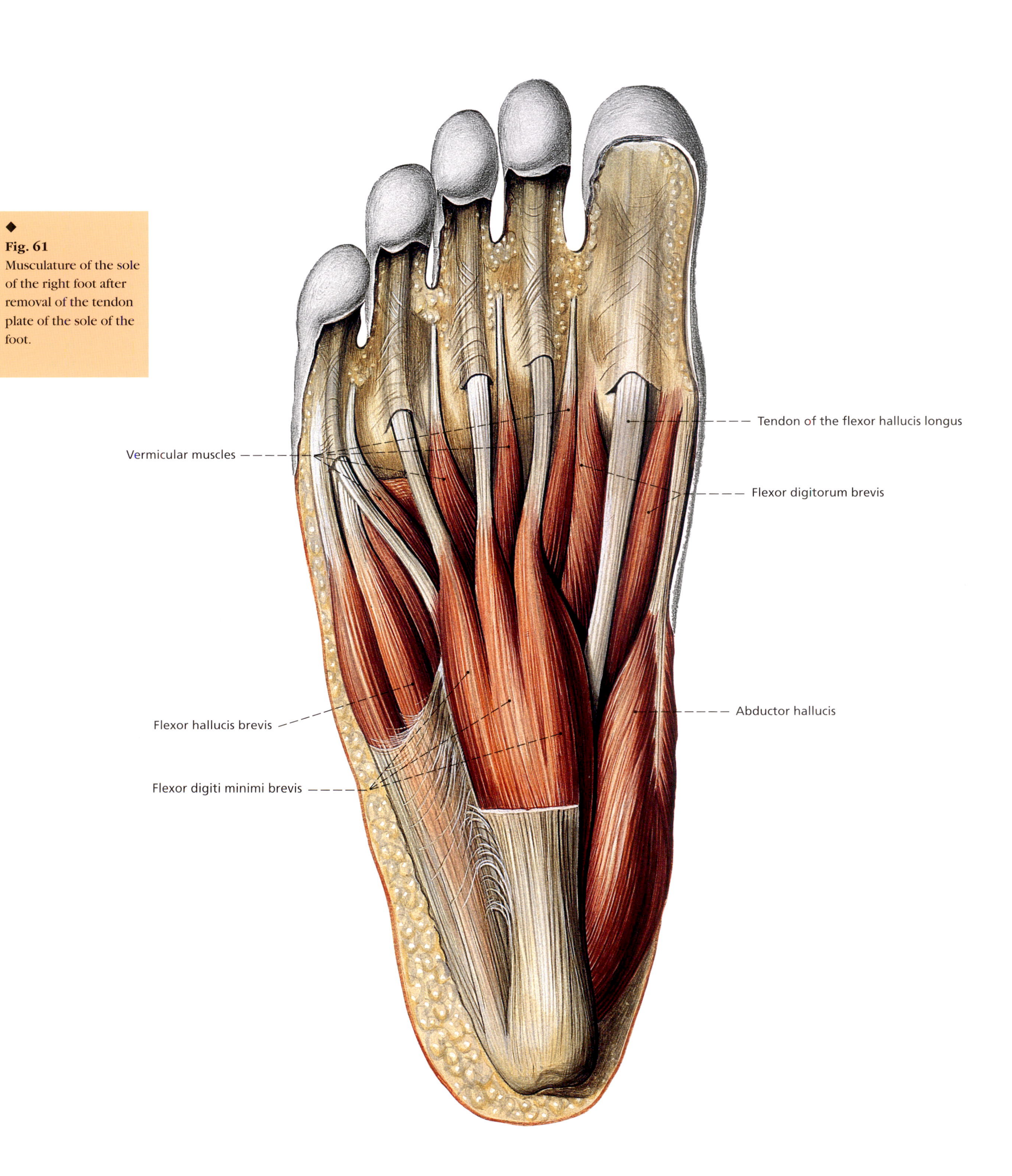

◆
Fig. 61
Musculature of the sole of the right foot after removal of the tendon plate of the sole of the foot.

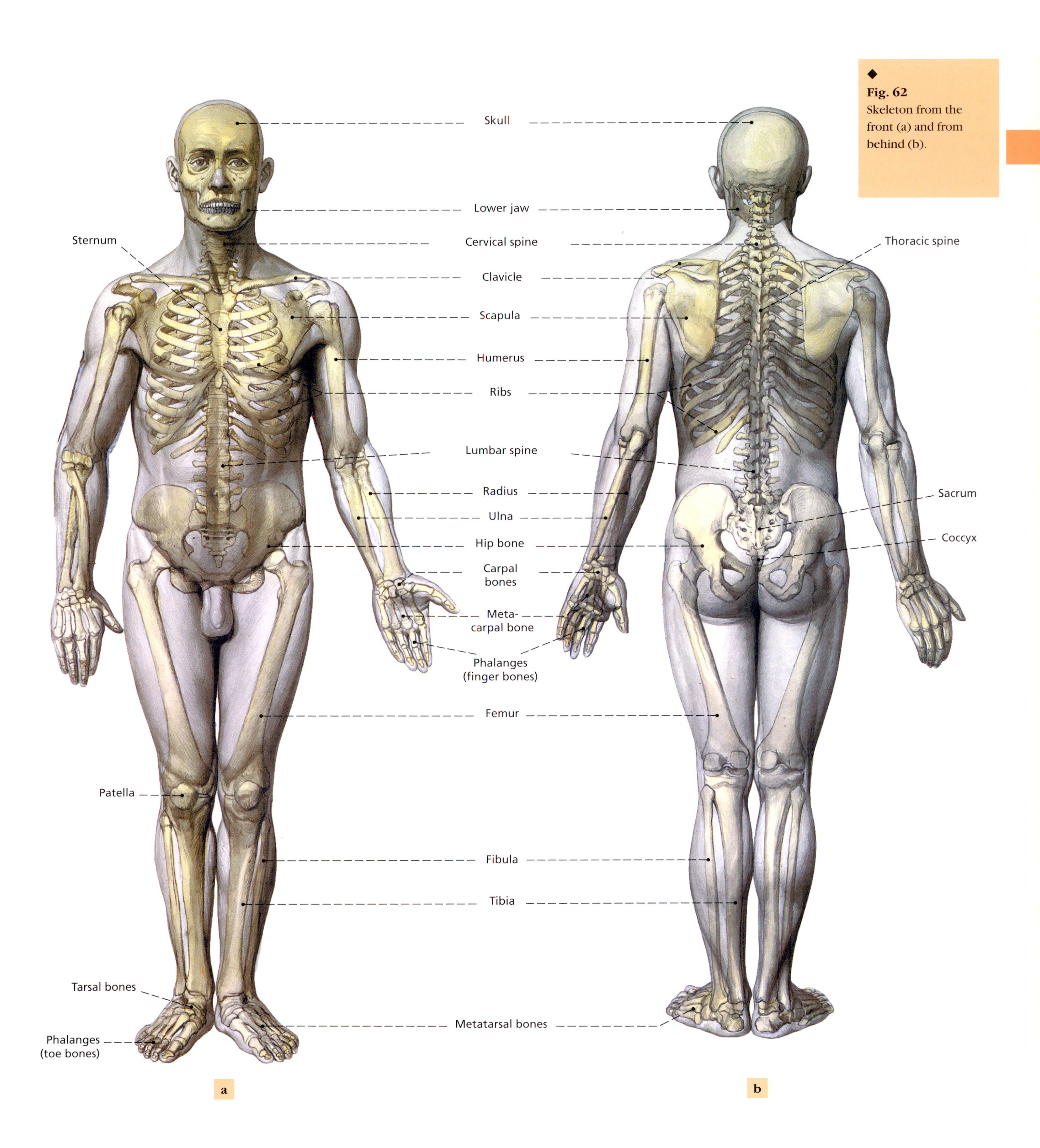

◆ **Fig. 62** Skeleton from the front (a) and from behind (b).

◆
Fig. 63
Skull from the front. The individual skull bones are marked in color.

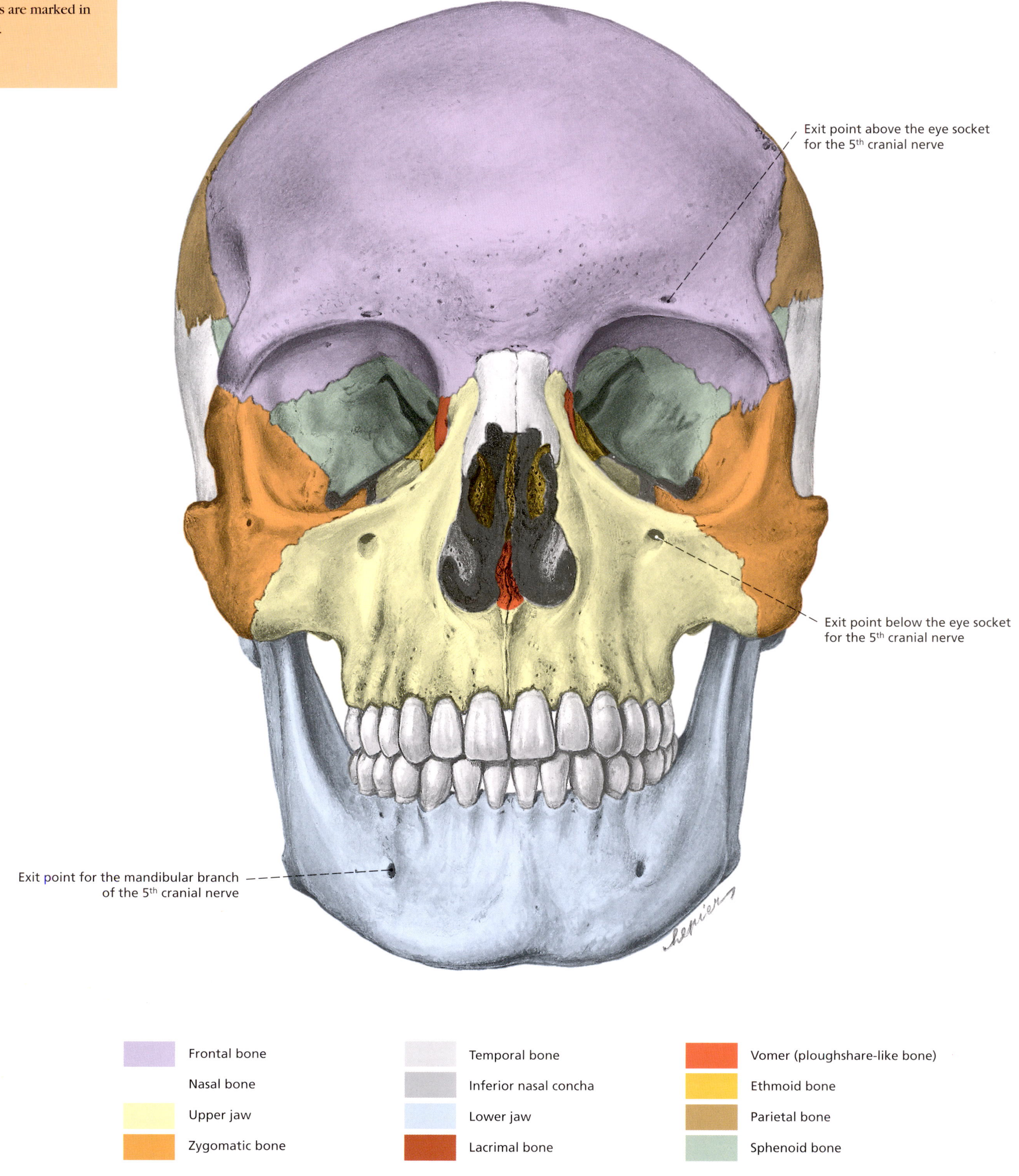

Frontal bone
Nasal bone
Upper jaw
Zygomatic bone
Temporal bone
Inferior nasal concha
Lower jaw
Lacrimal bone
Vomer (ploughshare-like bone)
Ethmoid bone
Parietal bone
Sphenoid bone

◆ **Fig. 64**
Skull from the side. The individual skull bones are marked in color.

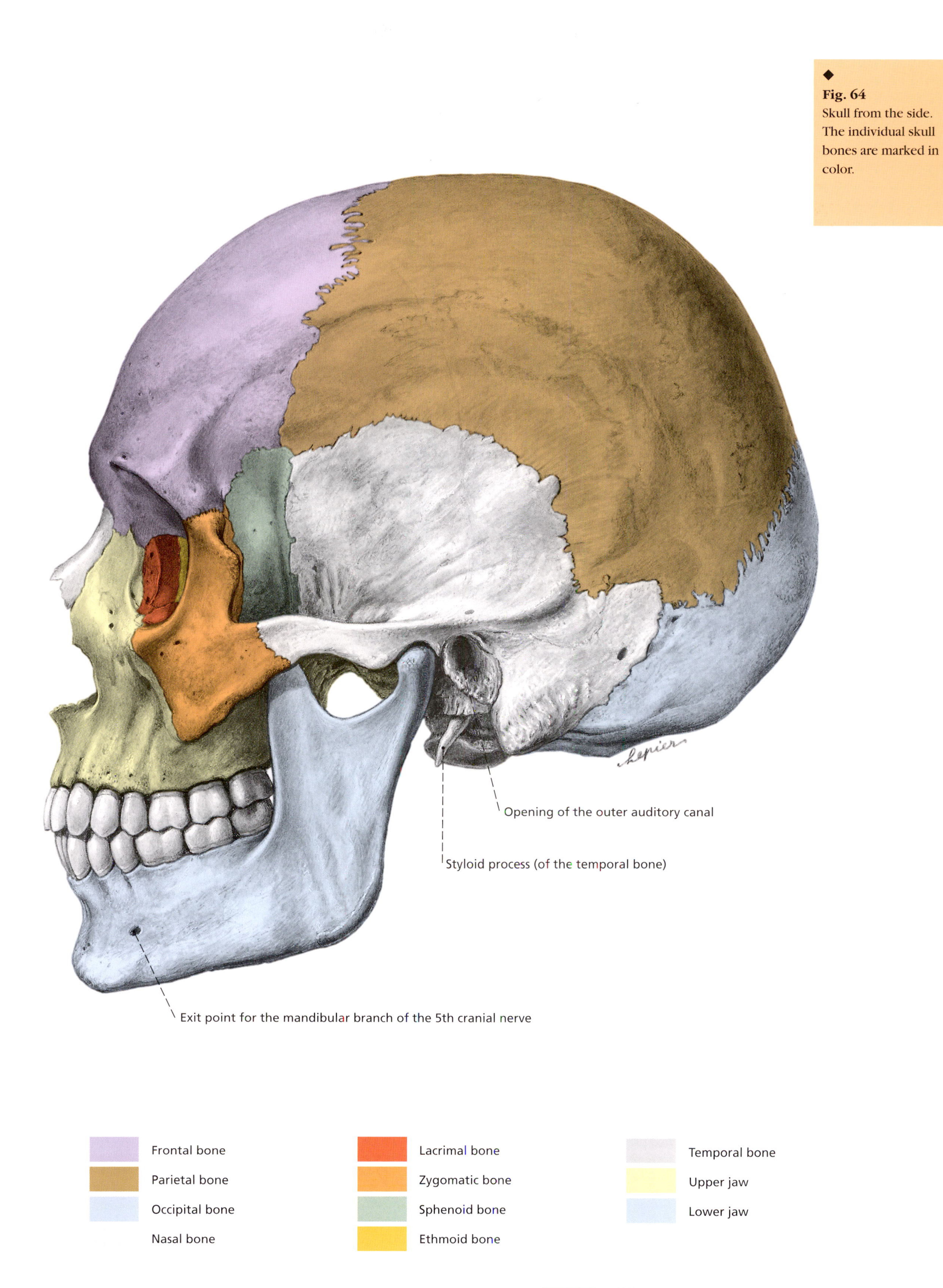

◆
Fig. 65
Skull bones from behind.

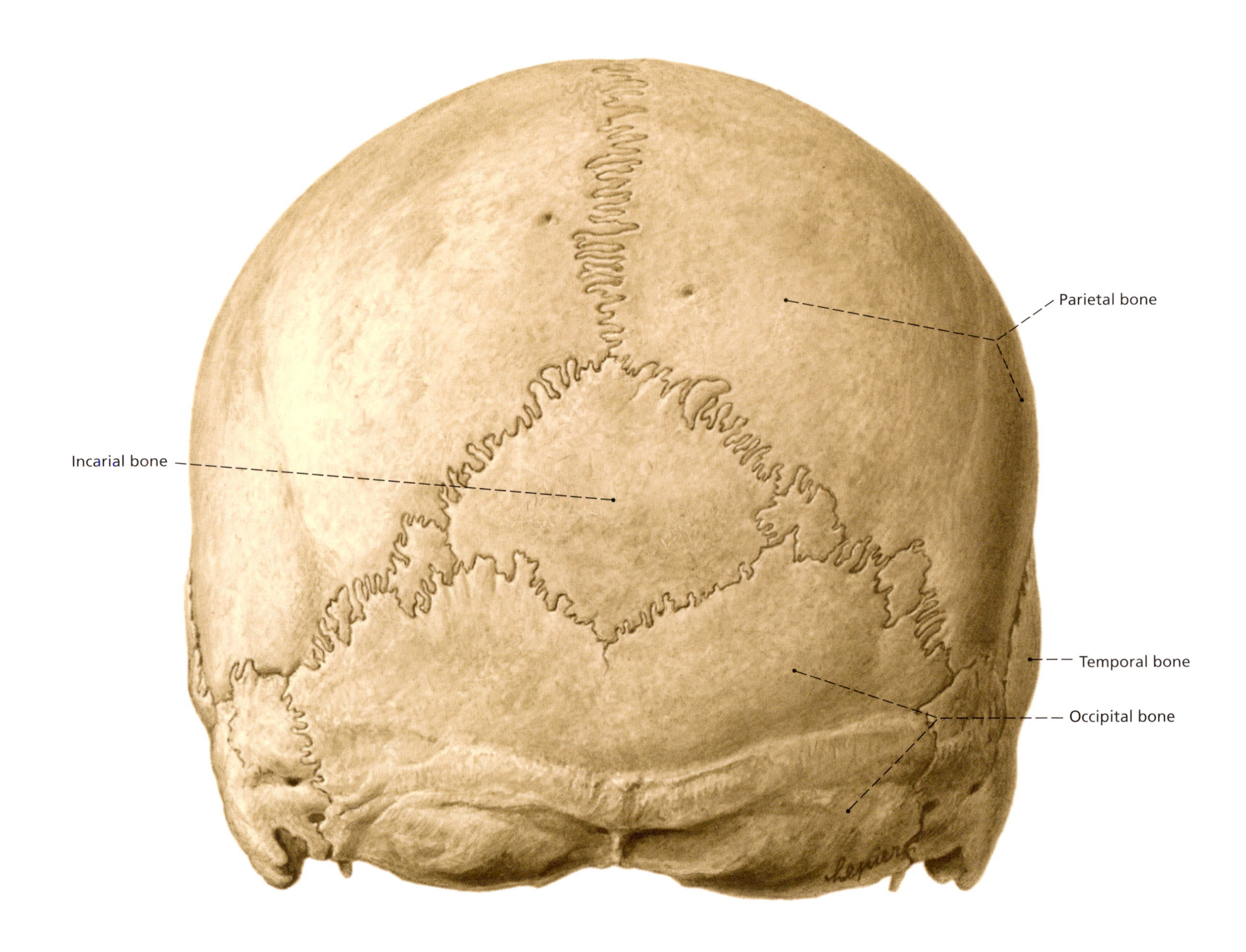

◆
Fig. 66
Cross-section through the skull, side view. The individual skull bones are marked in color.

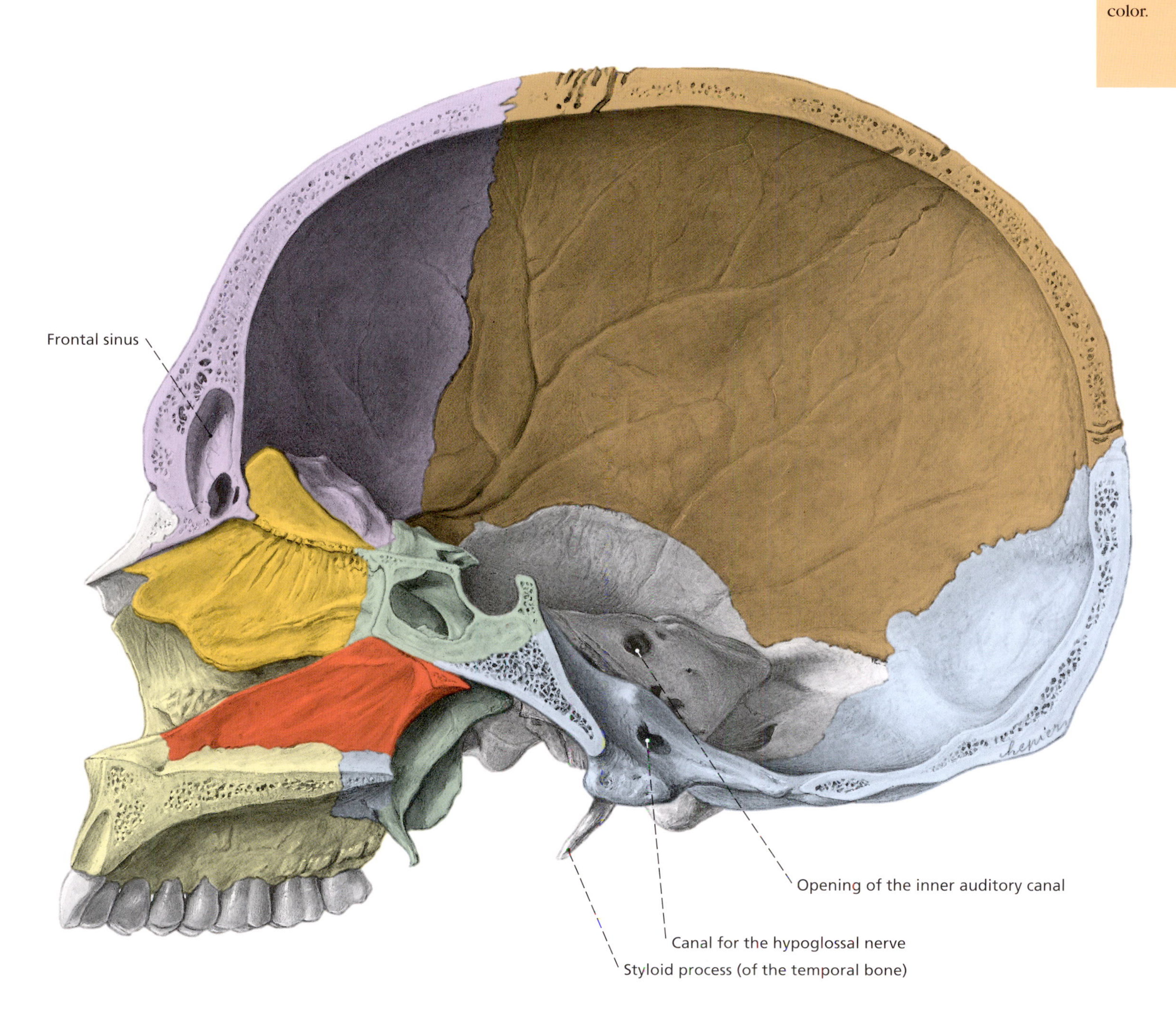

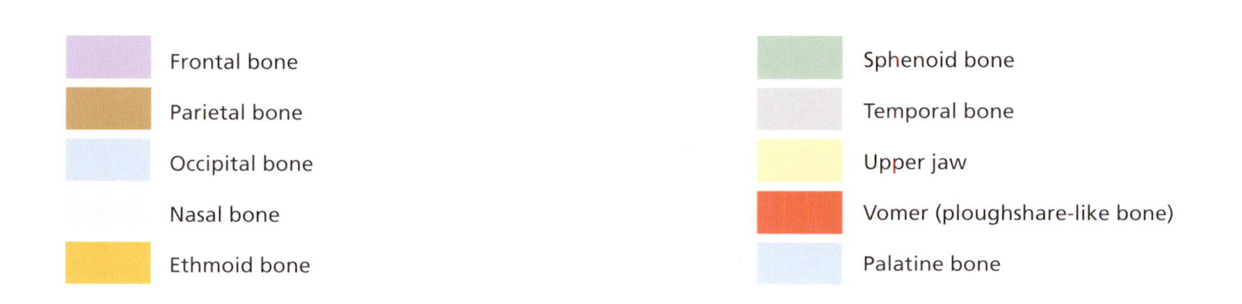

◆
Fig. 67
Base of the skull (seen from above and inside) with openings for vessels and nerves. The individual skull bones are marked in color.

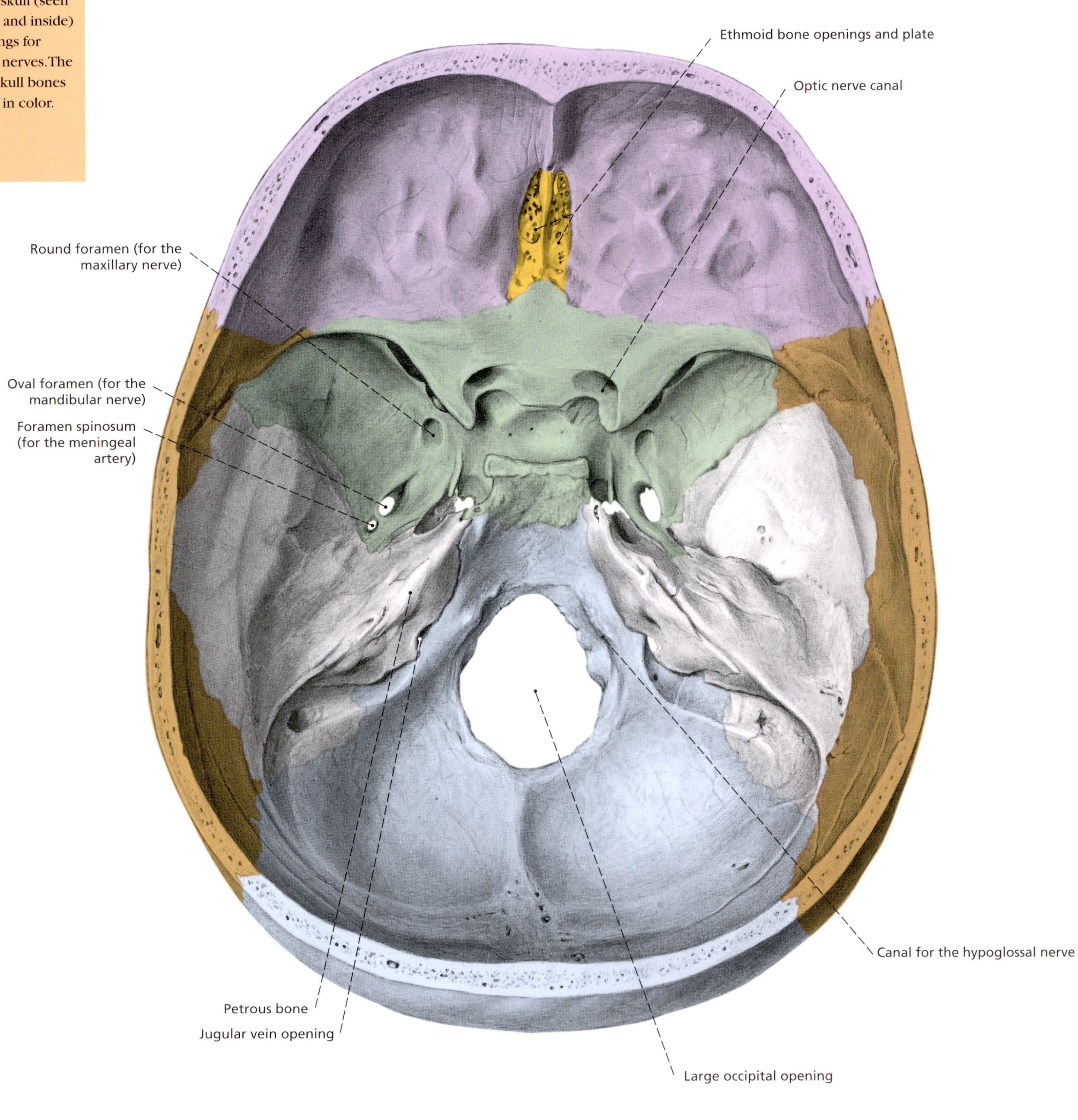

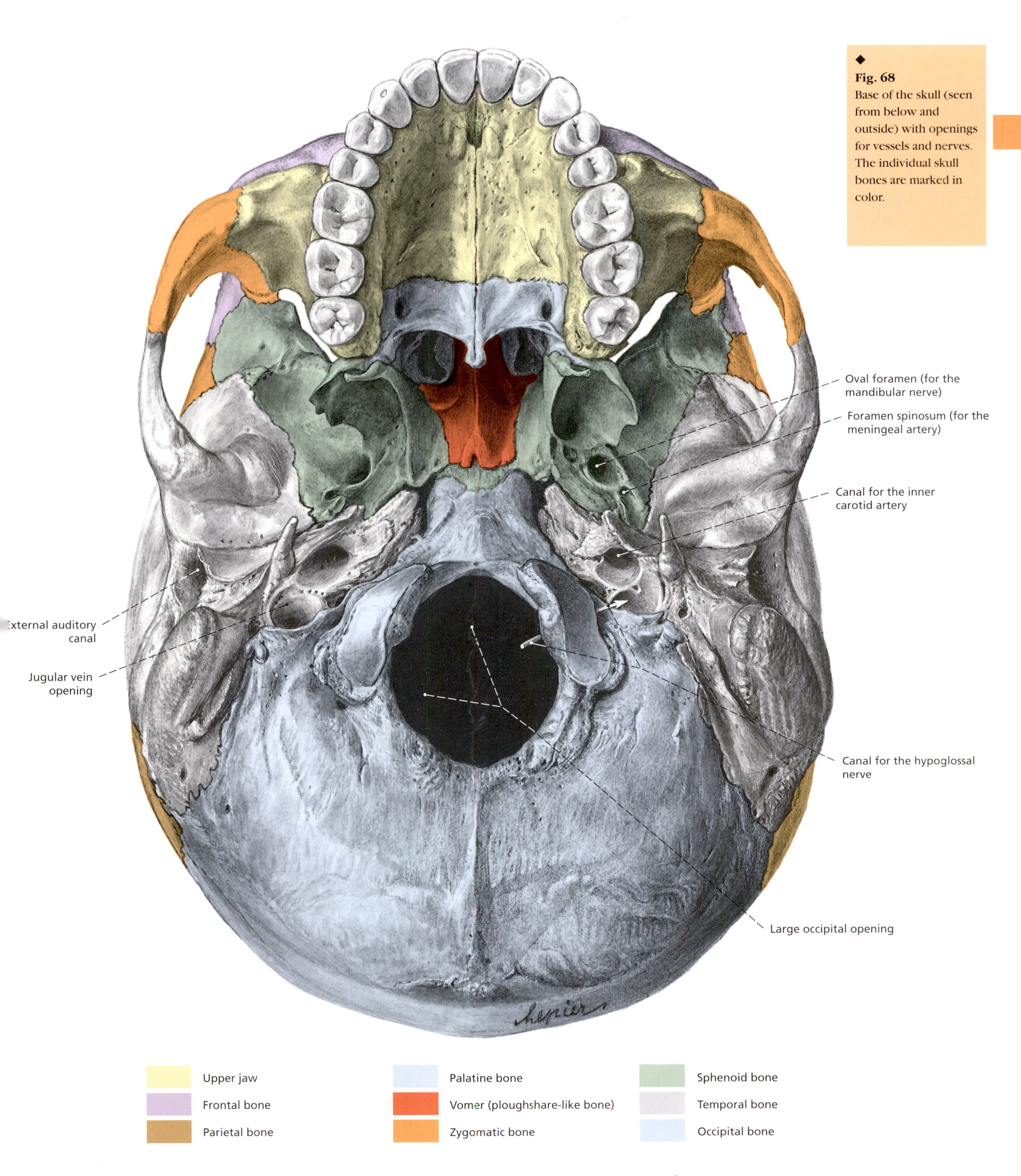

◆
Fig. 68
Base of the skull (seen from below and outside) with openings for vessels and nerves. The individual skull bones are marked in color.

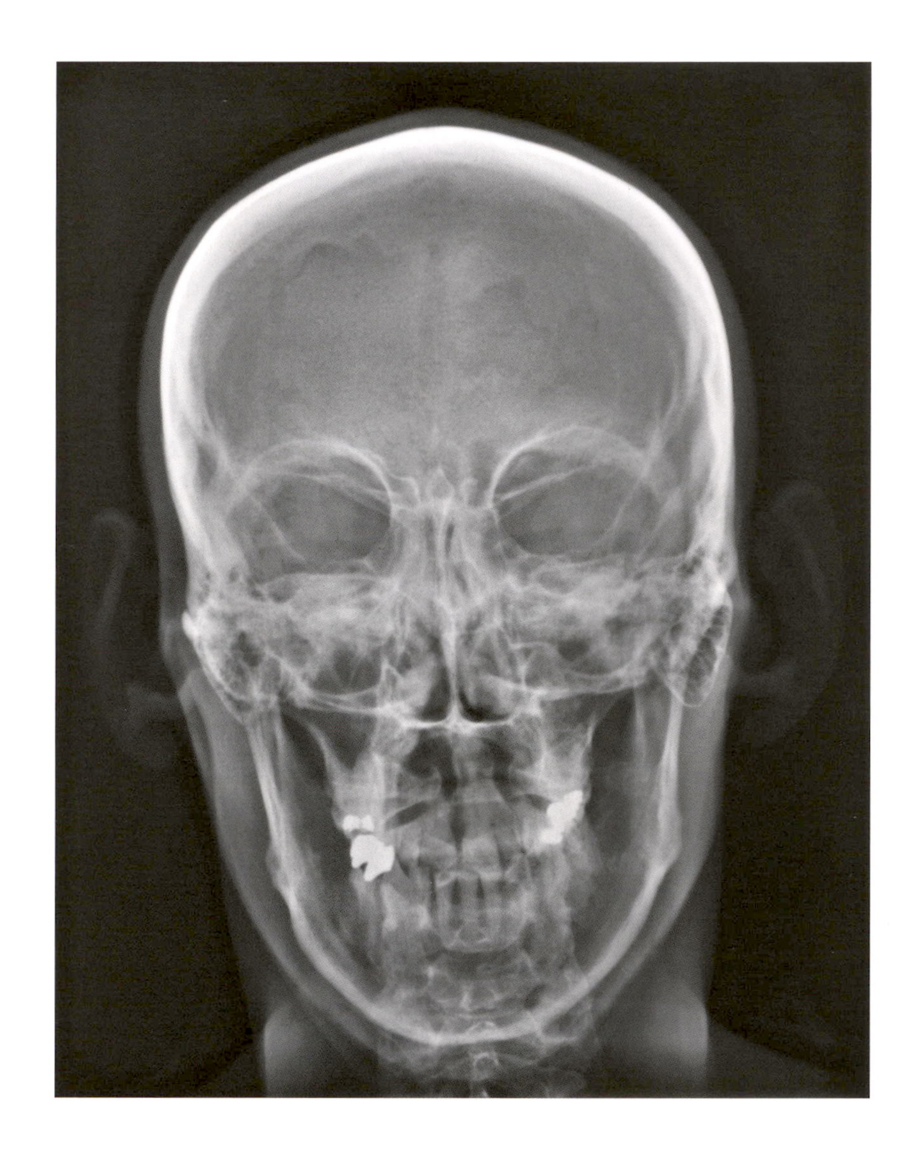

◆
Fig. 69
X-ray of the skull from the front. Fillings in the teeth of the upper and lower jaws can be seen in white.

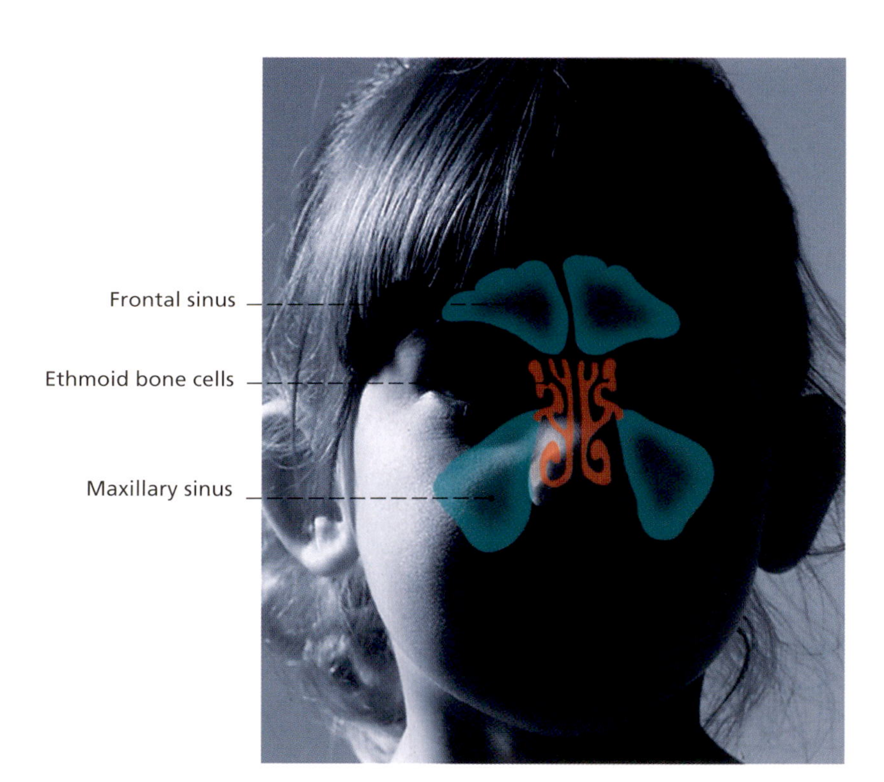

◆
Fig. 70
The paranasal sinuses have been drawn onto the skin of the face.

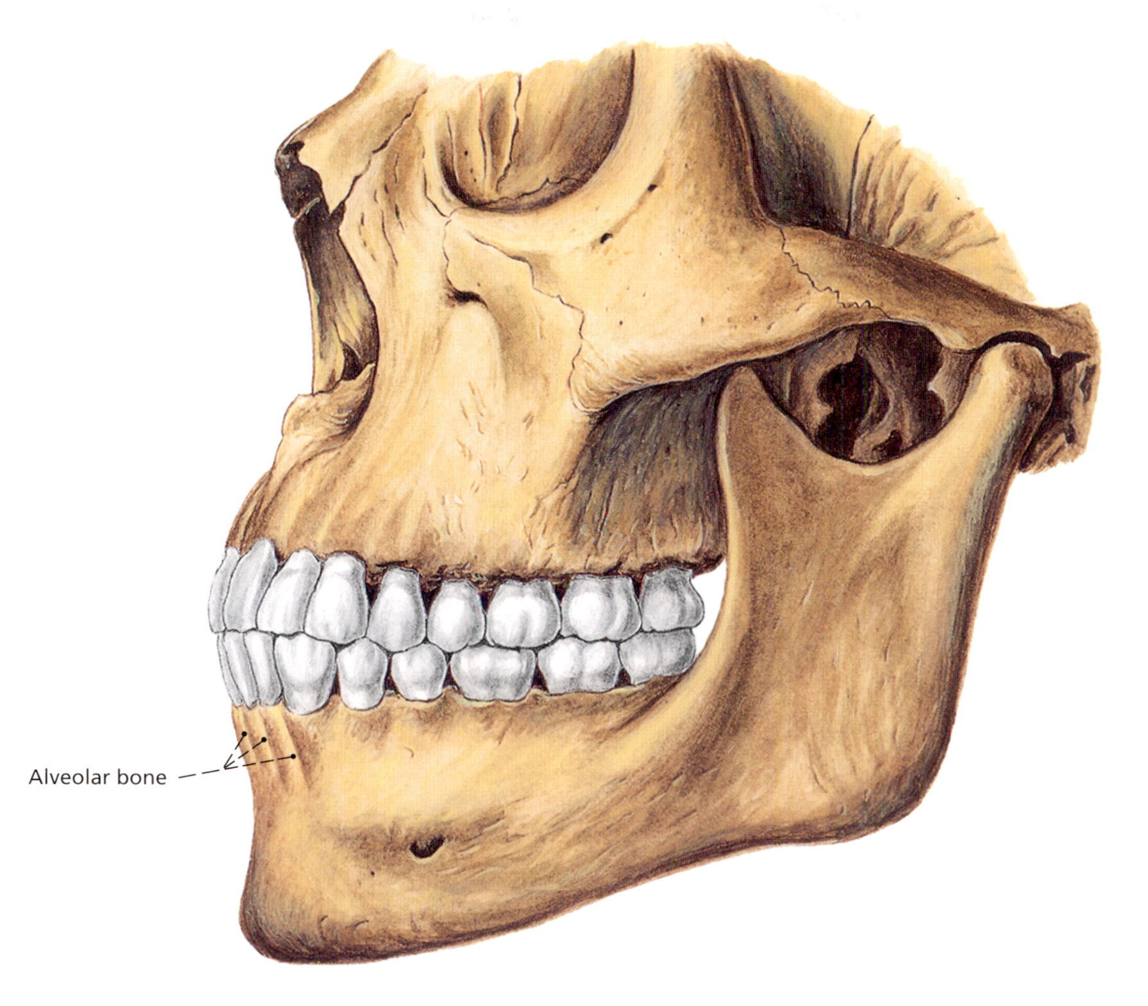

◆
Fig. 71
Upper and lower jaws from the side.

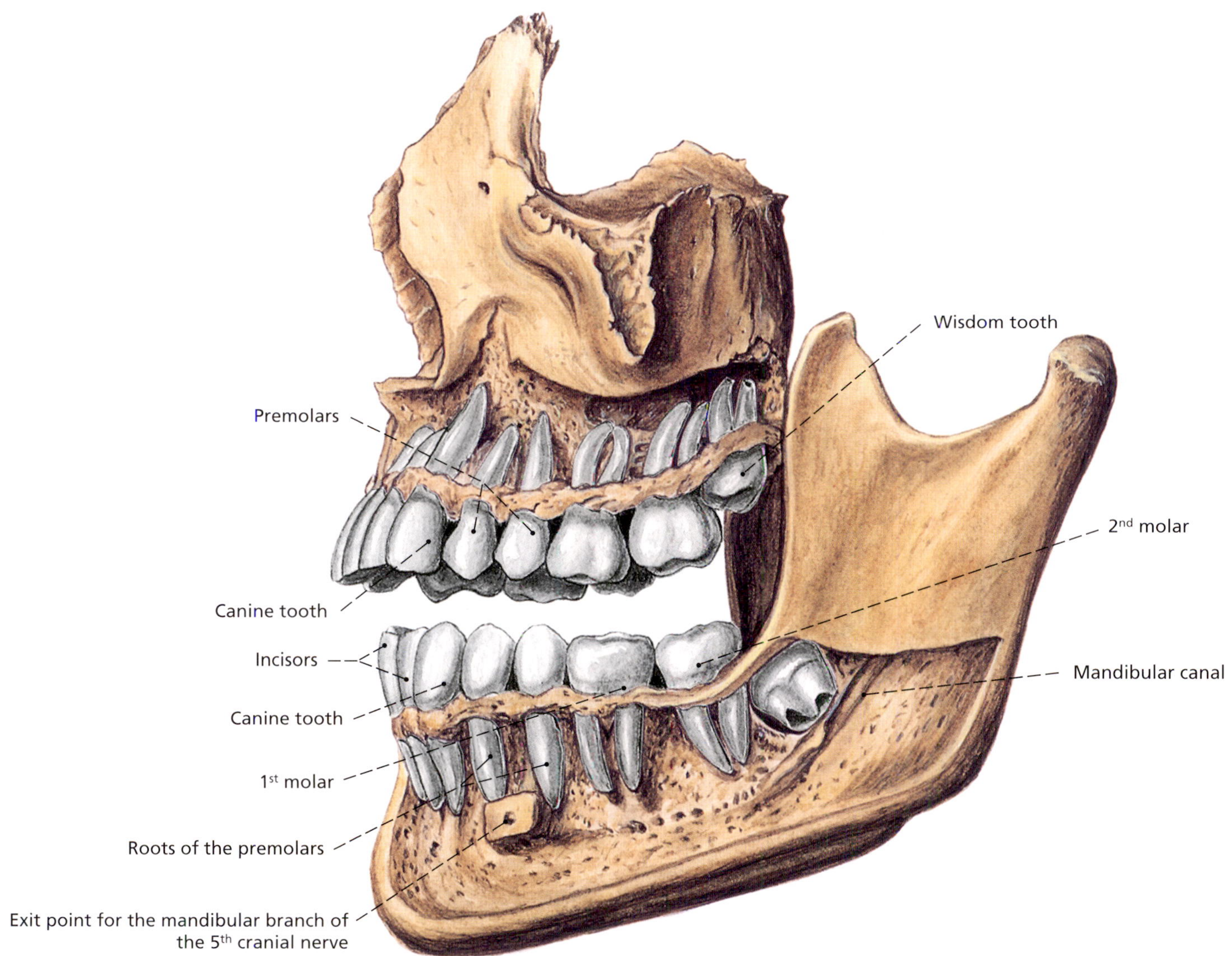

◆
Fig. 72
Upper and lower jaws with teeth after removal of the outer bone layer, exposing the roots of the teeth.

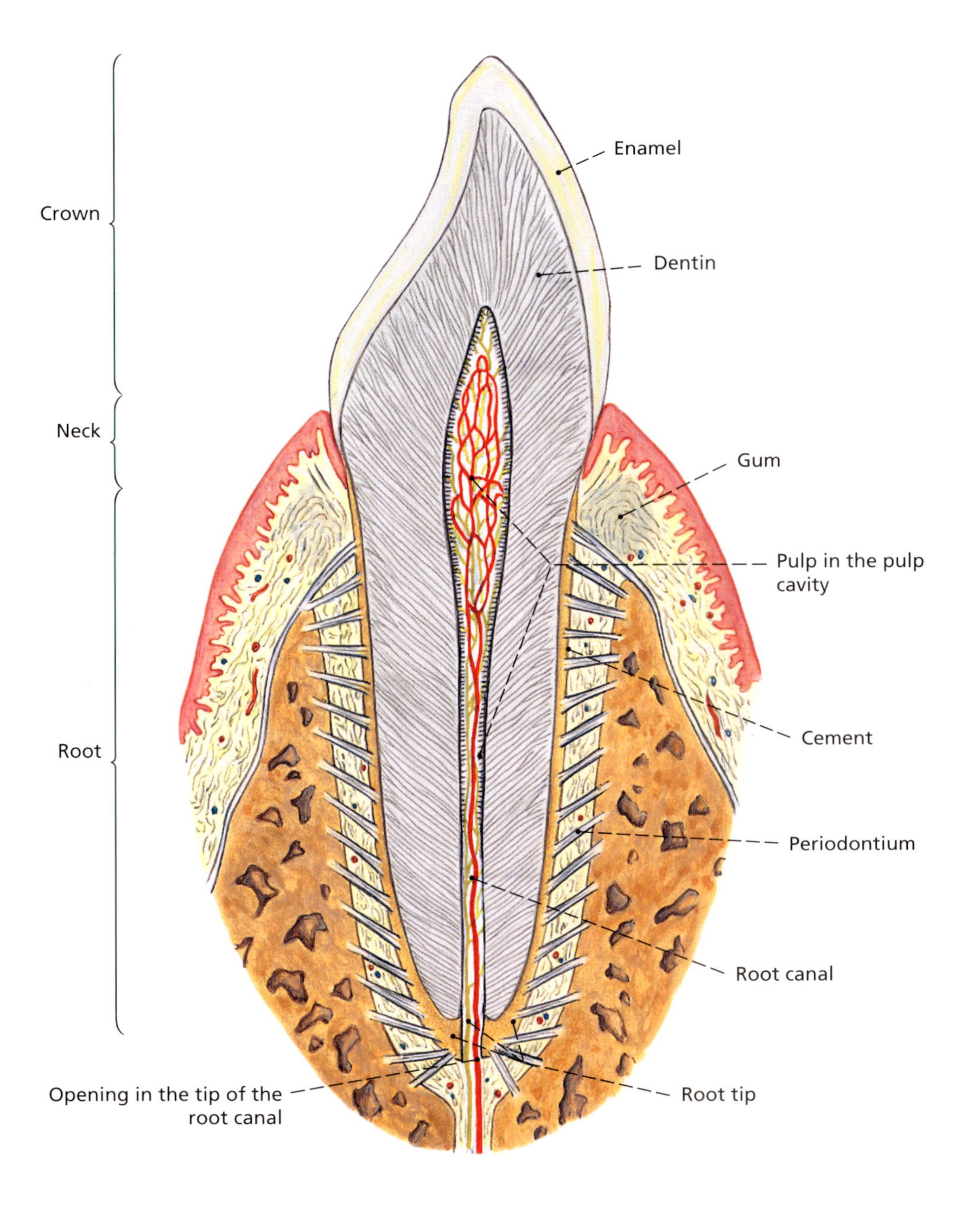

◆
Fig. 73
Longitudinal section of an incisor. Nerves and vessels pass into the root canal through an opening in the tip of the root.

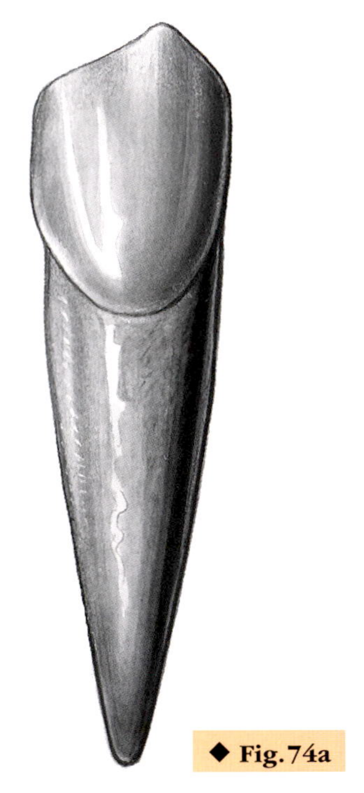

◆ **Fig.74a**

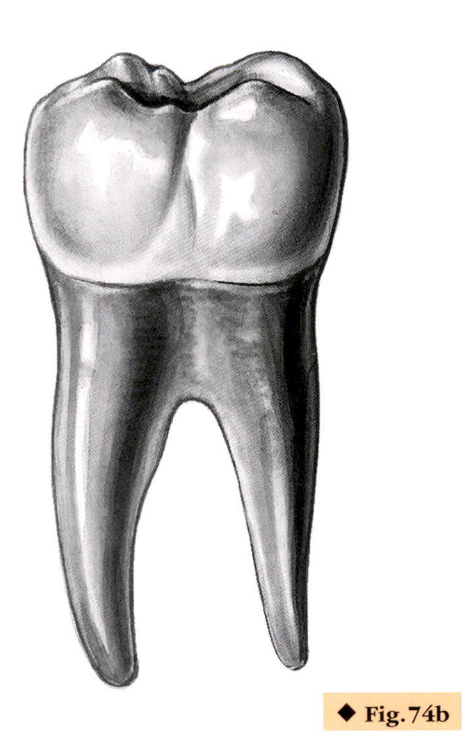

◆ **Fig.74b**

◆
Fig. 74
Permanent canine tooth with one root (a) and second baby molar (b) with two roots.

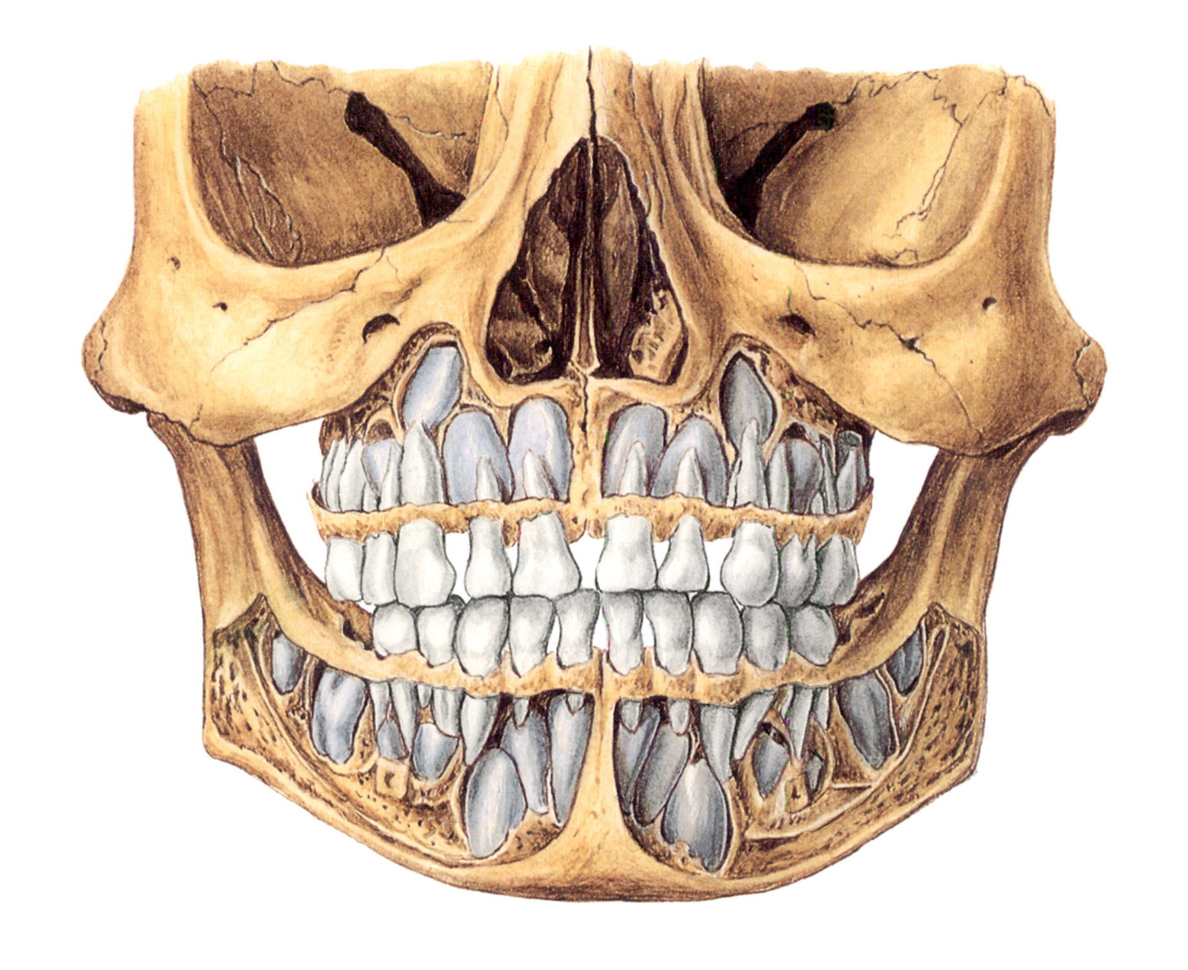

◆
Fig. 75
Viscerocranium (facial bones). The outer layer of upper and lower jaw bones has been removed in order to expose the dental germs of the permanent teeth (blue).

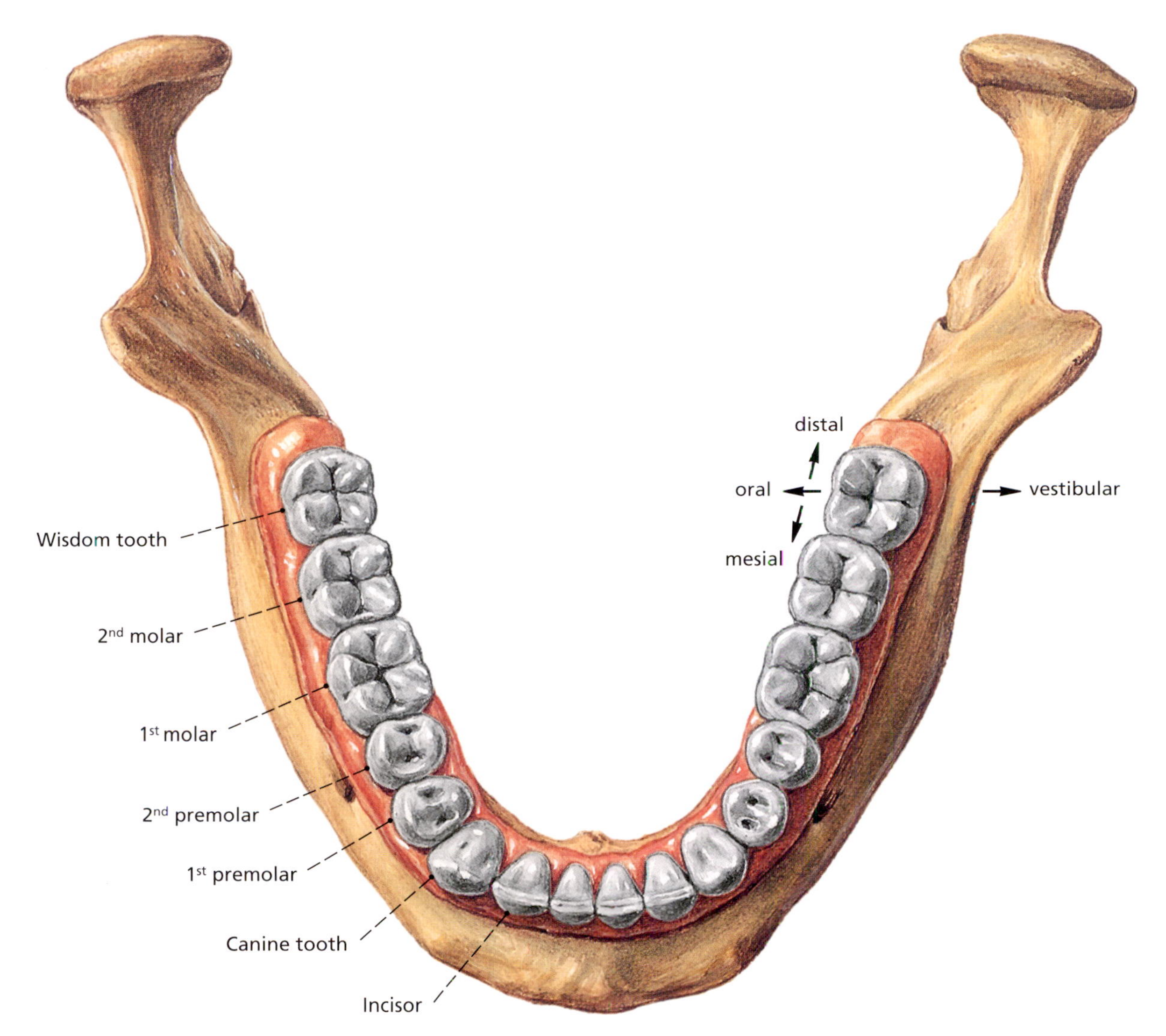

◆
Fig. 76
Adult lower jaw seen from above. The upper and lower jaws each have 16 teeth. The terms distal, oral, mesial and vestibular are used for orientation, e.g., for local definition of damage to tooth (distal = toward the back, oral = towards the oral cavity, mesial = towards the middle of the jaw, vestibular = on the side of the outer jaw surface).

◆
Fig. 77
Spine with thoracic cage (chest) and pelvis split lengthwise down the middle; (a) from the right, (b) from the left.

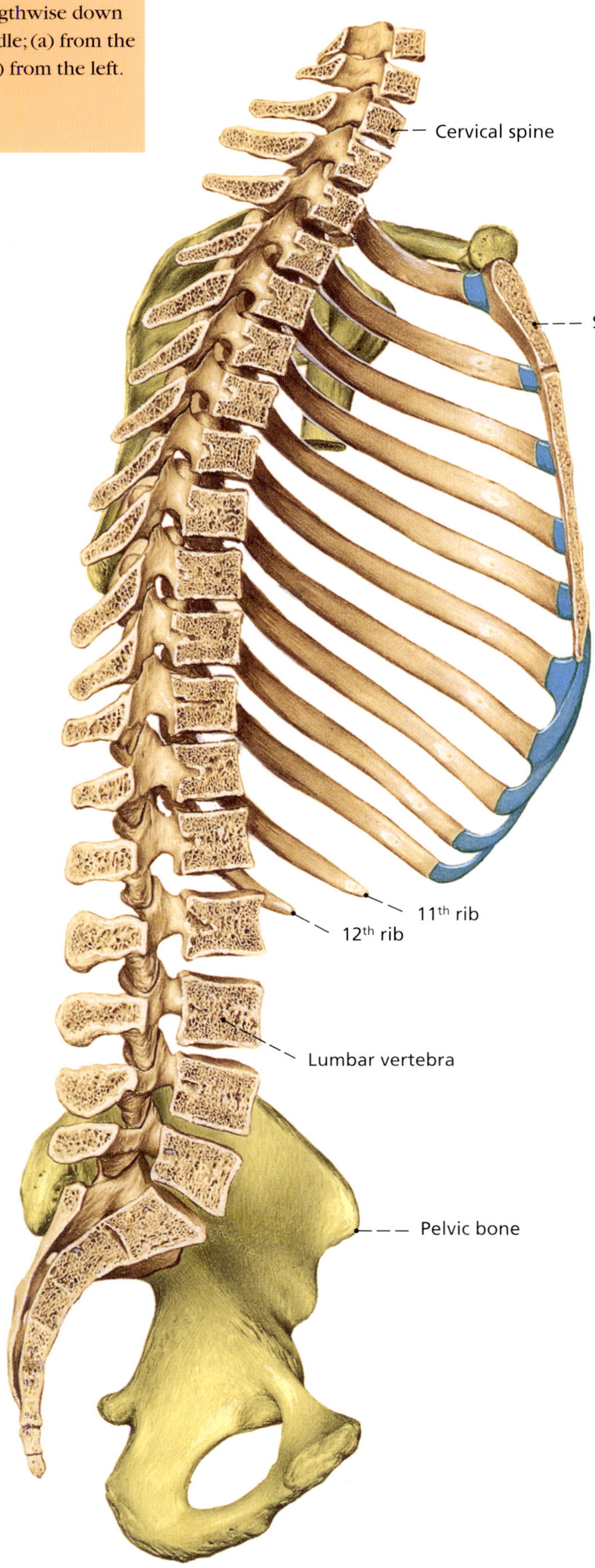

Cervical spine
Sternum
11th rib
12th rib
Lumbar vertebra
Pelvic bone

a

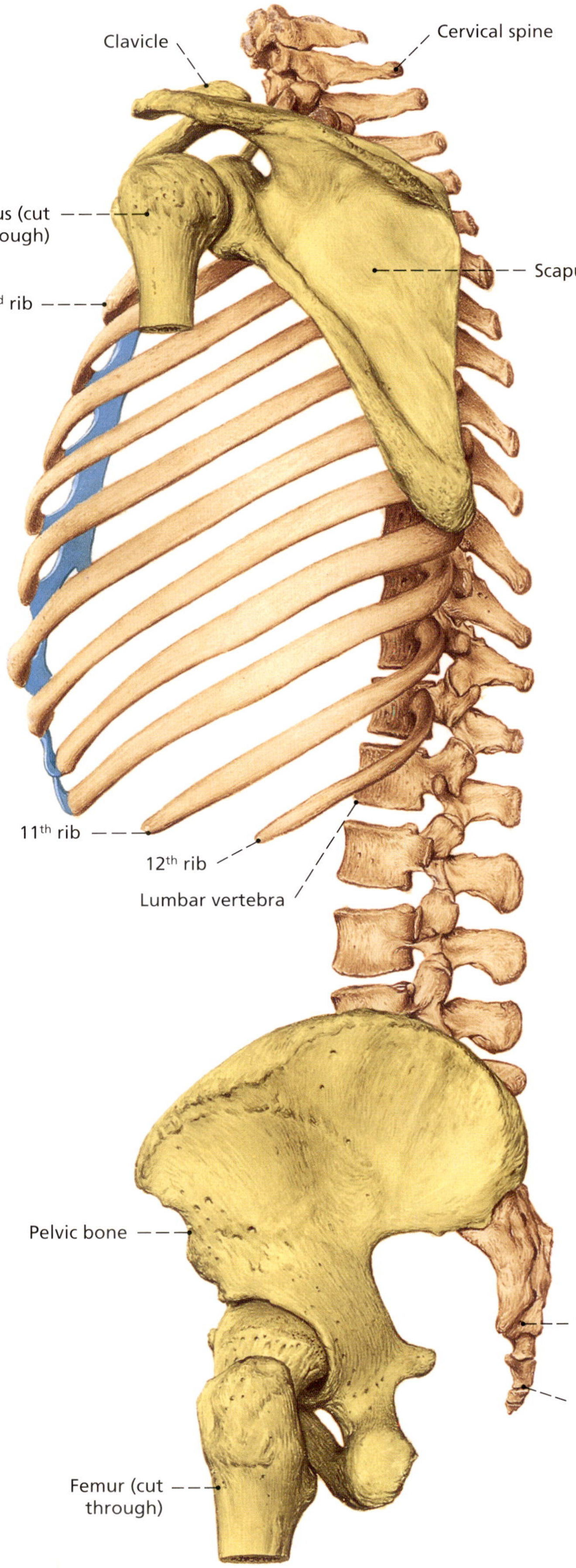

Clavicle
Cervical spine
Humerus (cut through)
Scapular
2nd rib
11th rib
12th rib
Lumbar vertebra
Pelvic bone
Sacrum
Coccyx
Femur (cut through)

b

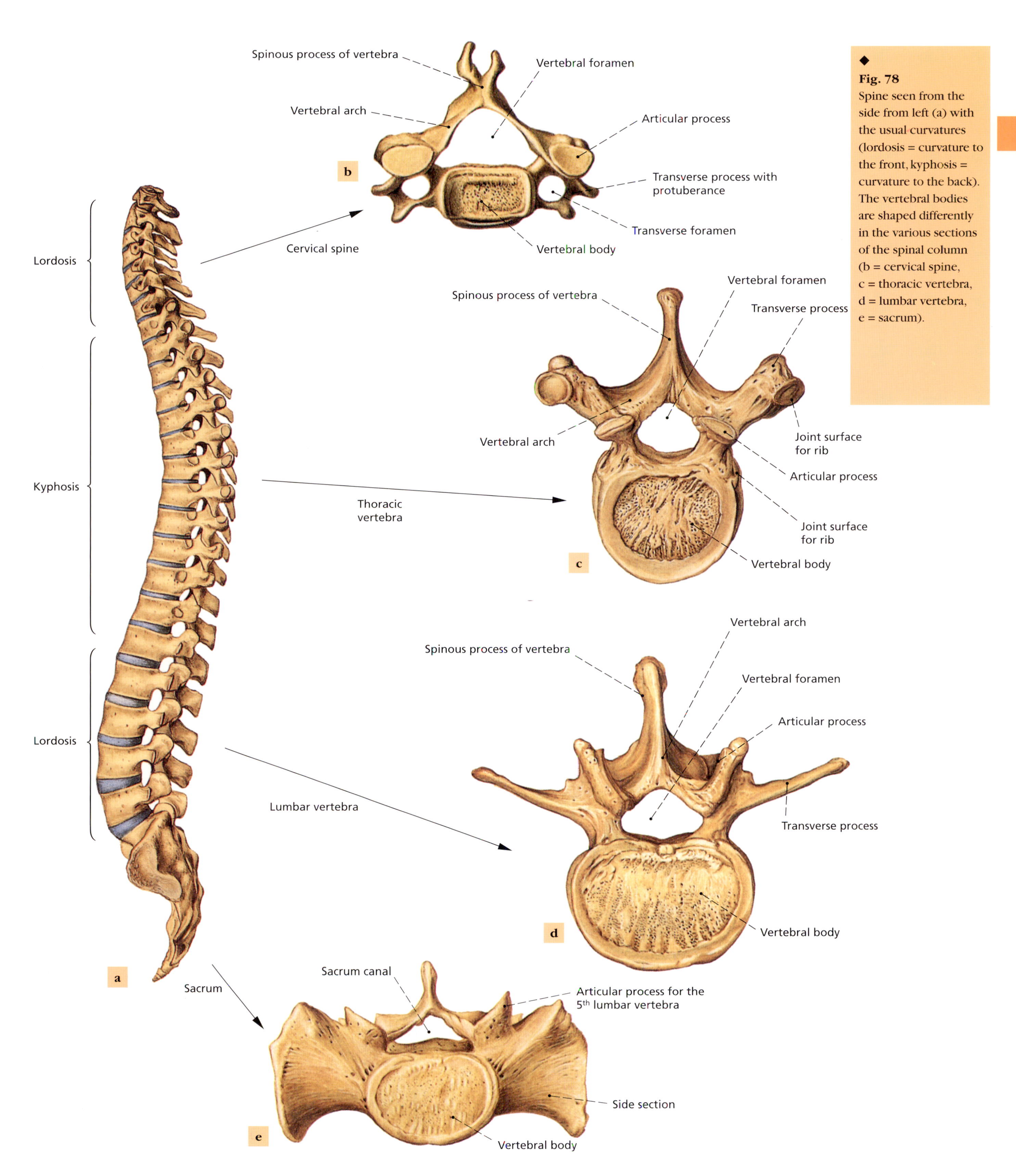

◆
Fig. 78
Spine seen from the side from left (a) with the usual curvatures (lordosis = curvature to the front, kyphosis = curvature to the back). The vertebral bodies are shaped differently in the various sections of the spinal column (b = cervical spine, c = thoracic vertebra, d = lumbar vertebra, e = sacrum).

◆
Fig. 79
Ligaments of the spinal column and joints between the vertebrae and ribs. Side sections of the anterior longitudinal ligament.

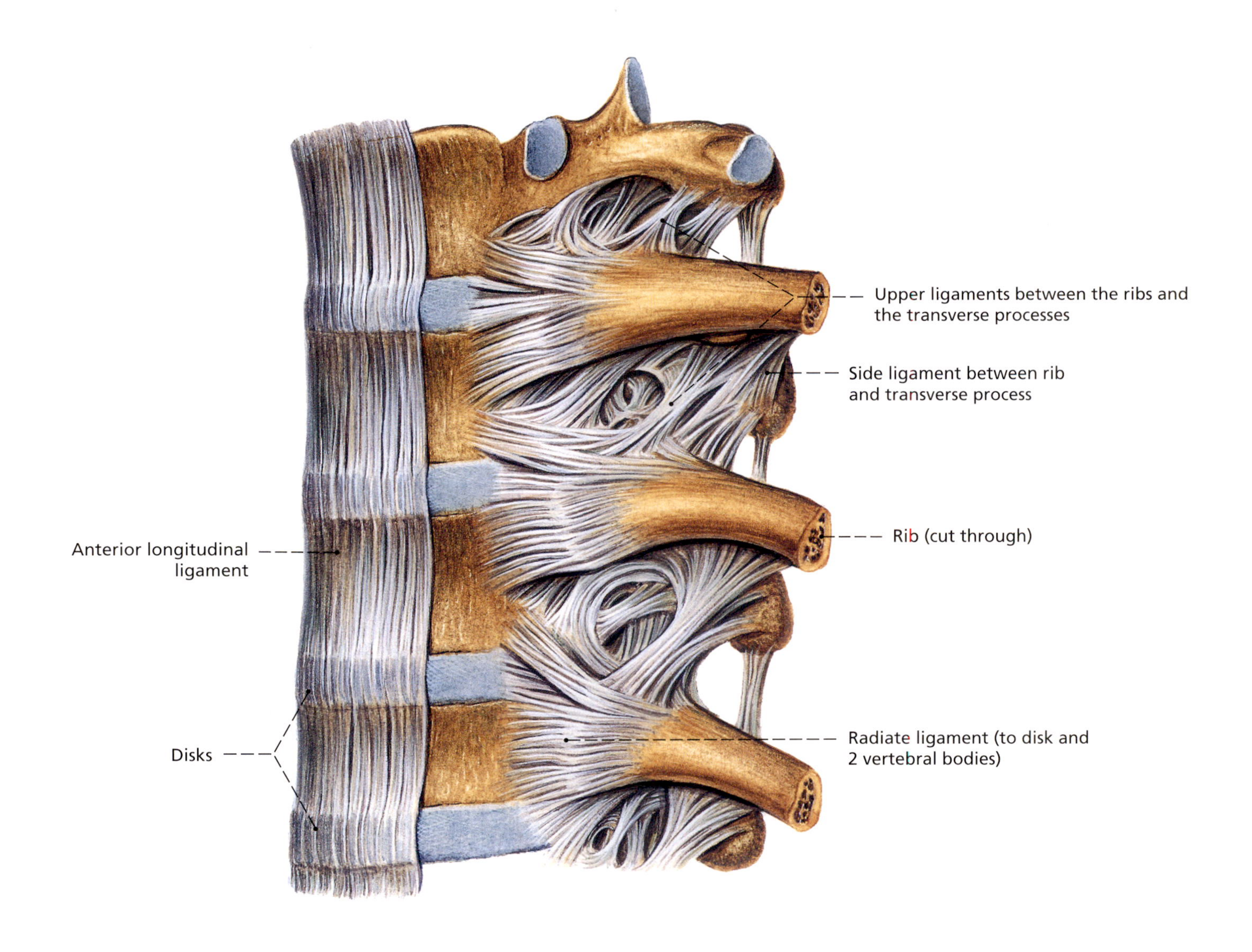

◆
Fig. 80
Illustration of a vertical section in the region of the lumbar vertebrae.

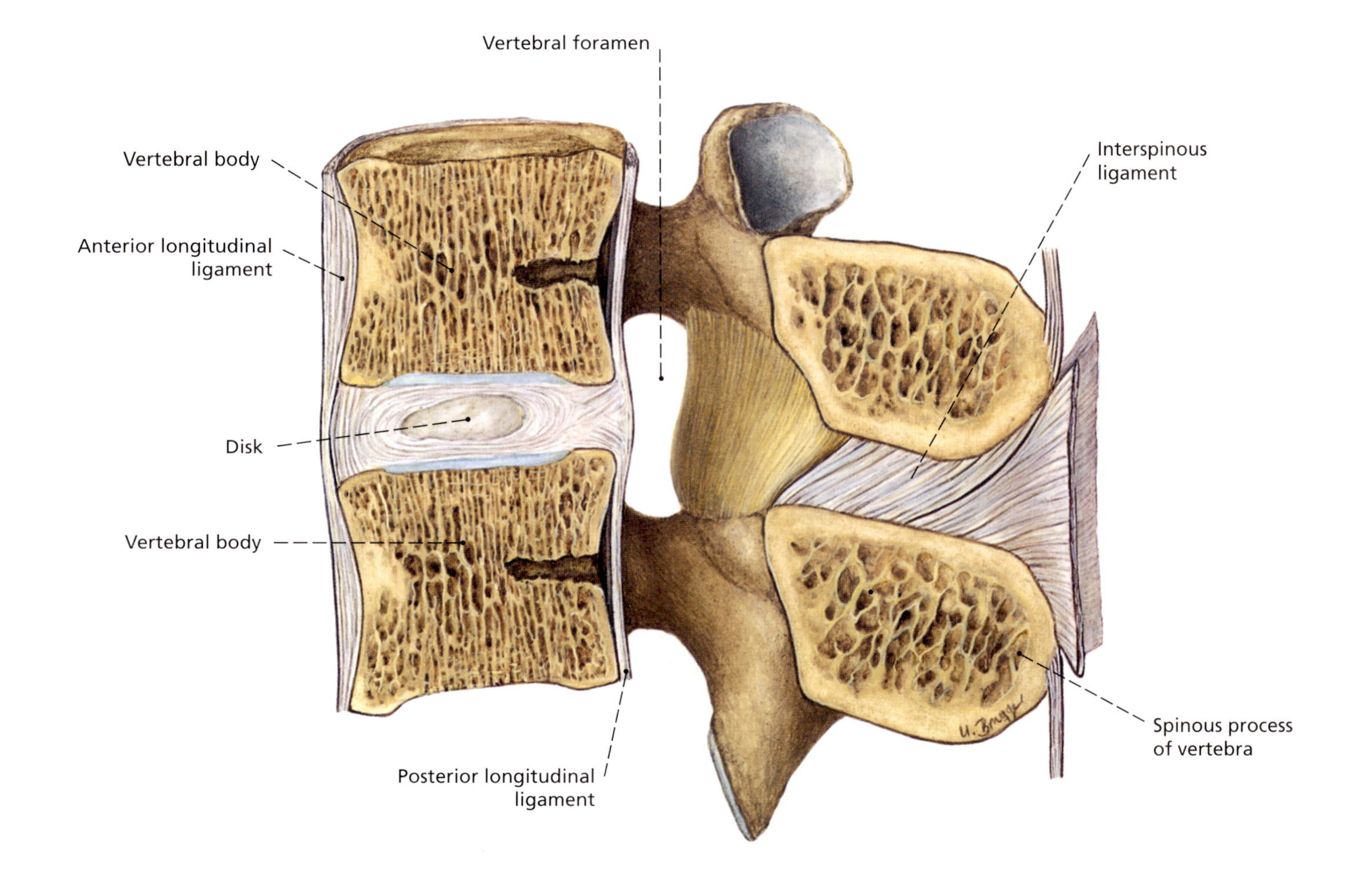

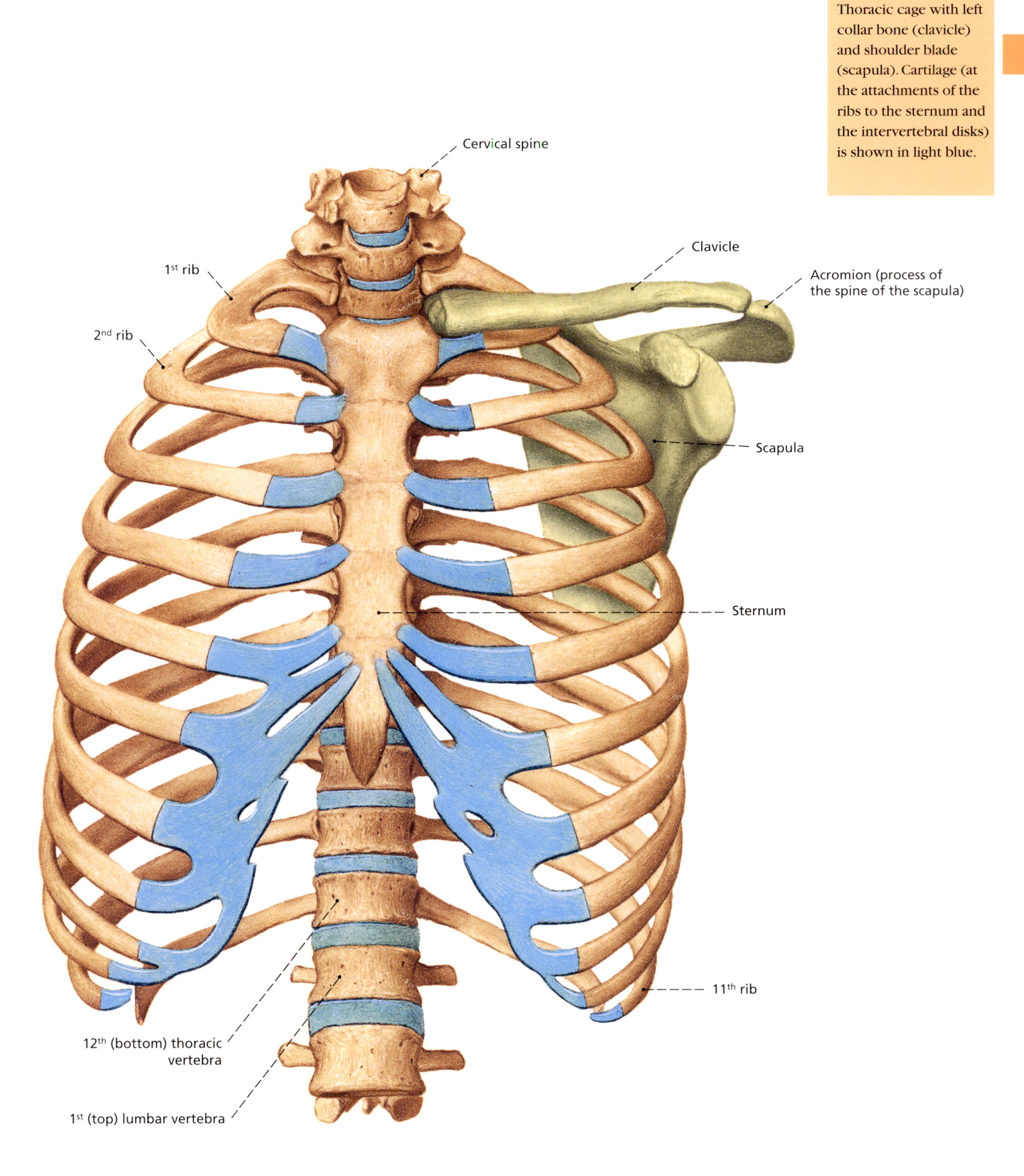

◆
Fig. 81
Thoracic cage with left collar bone (clavicle) and shoulder blade (scapula). Cartilage (at the attachments of the ribs to the sternum and the intervertebral disks) is shown in light blue.

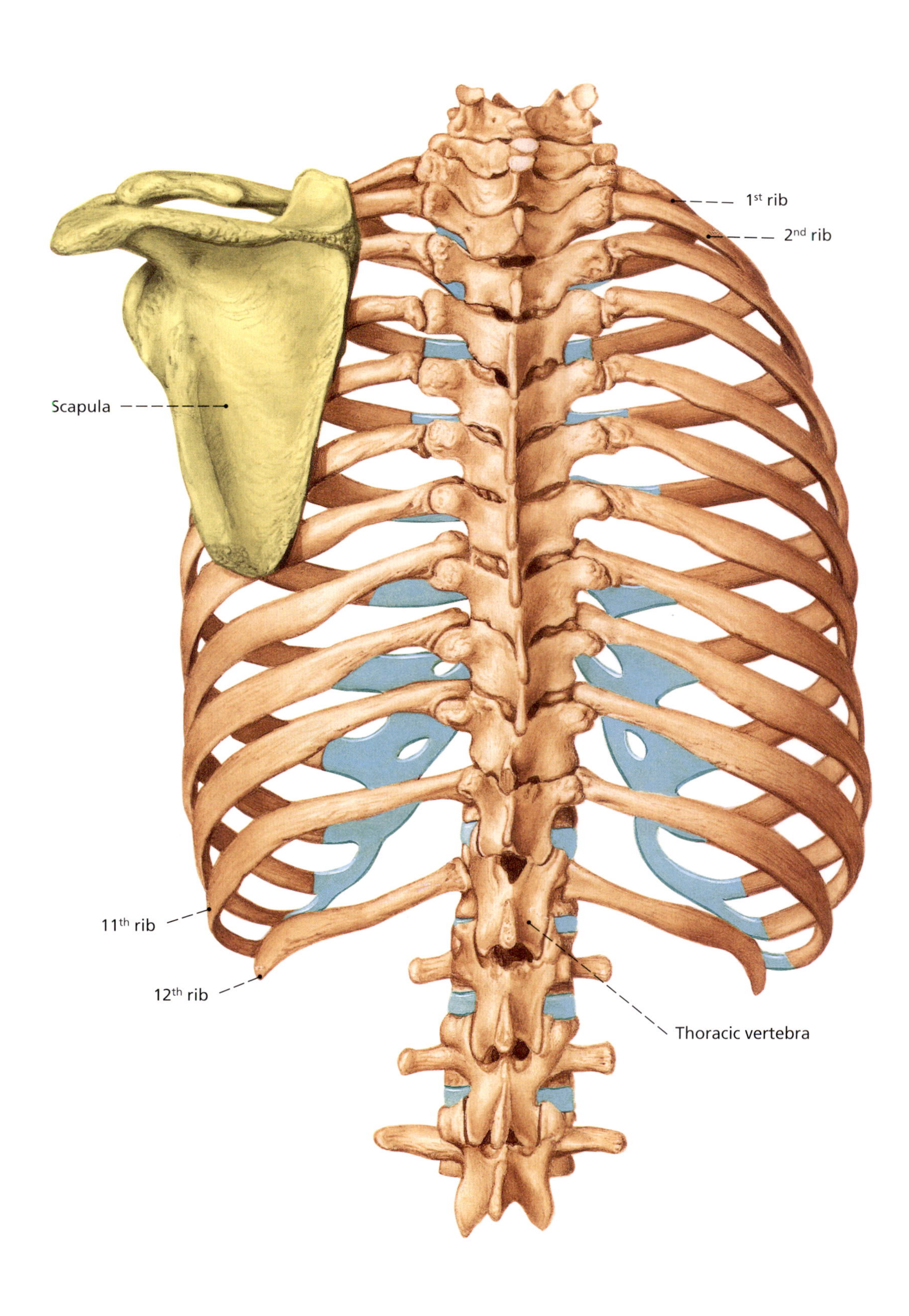

◆
Fig. 82
Thoracic cage (chest) with left shoulder blade (scapula) from behind. The cartilaginous areas (at the attachments of the ribs to the sternum and the intervertebral disks) are shown in light blue.

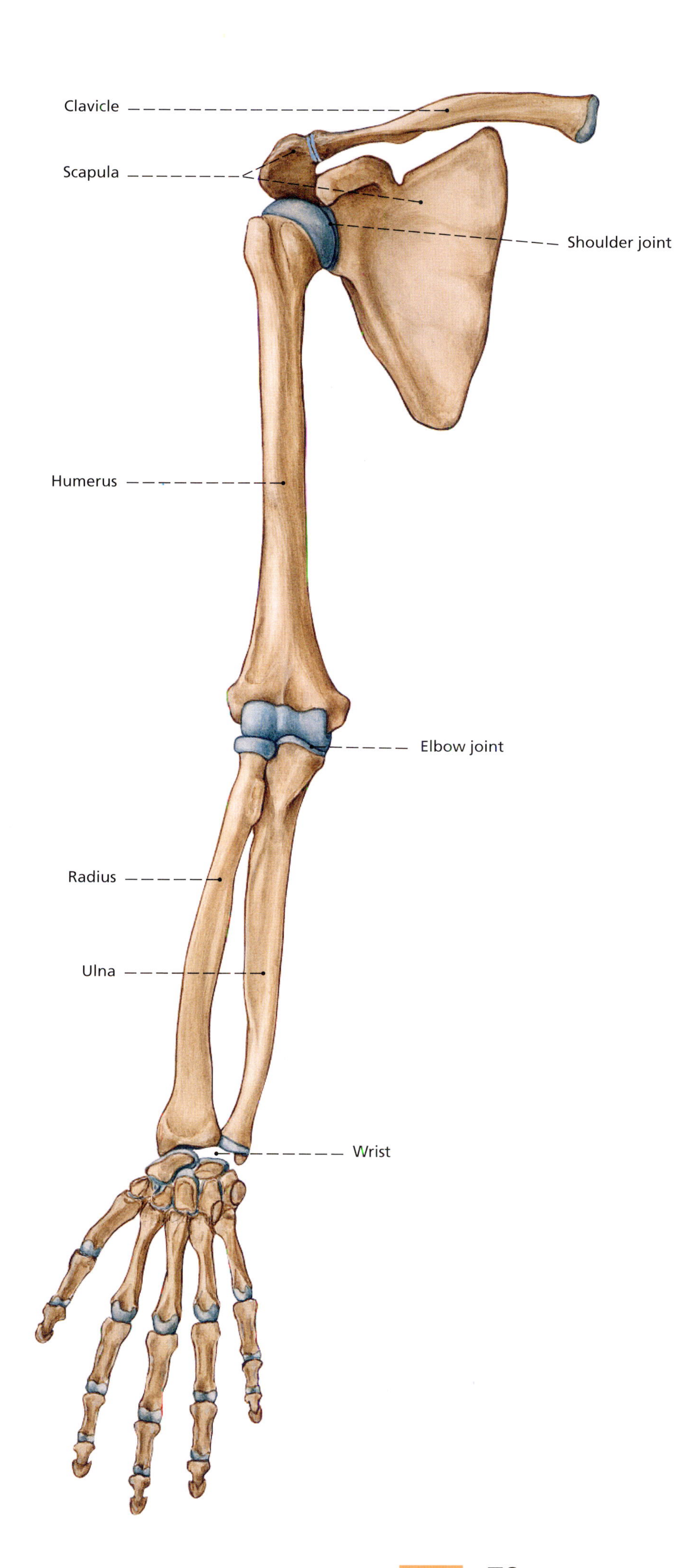

Fig. 83
Bones and joints of the right arm.

◆
Fig. 84
Left humerus from the front (a) and behind (b).

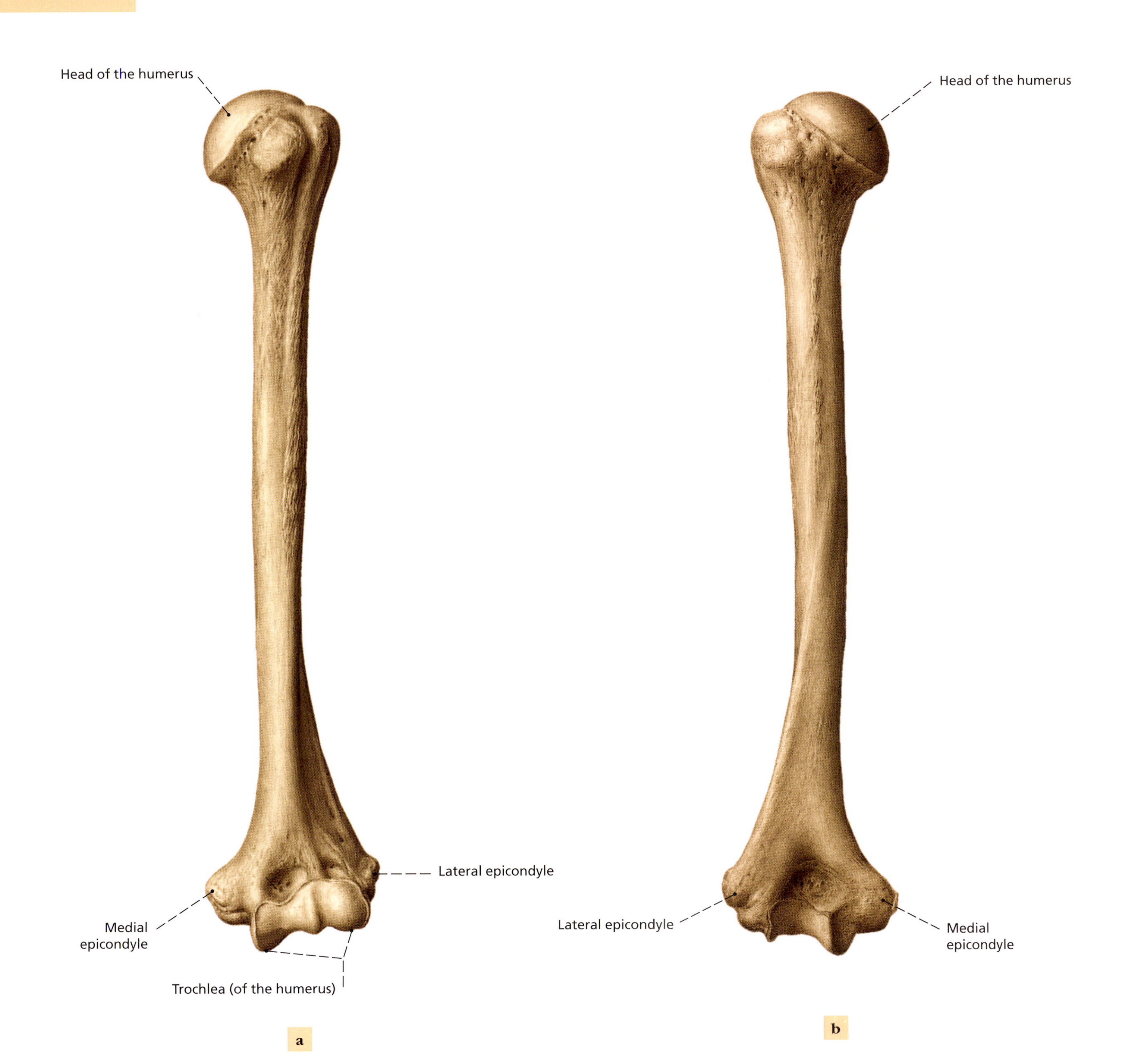

Fig. 85
Bones of the lower arm: left ulna from the front (a), from behind (b) and from the left (c).

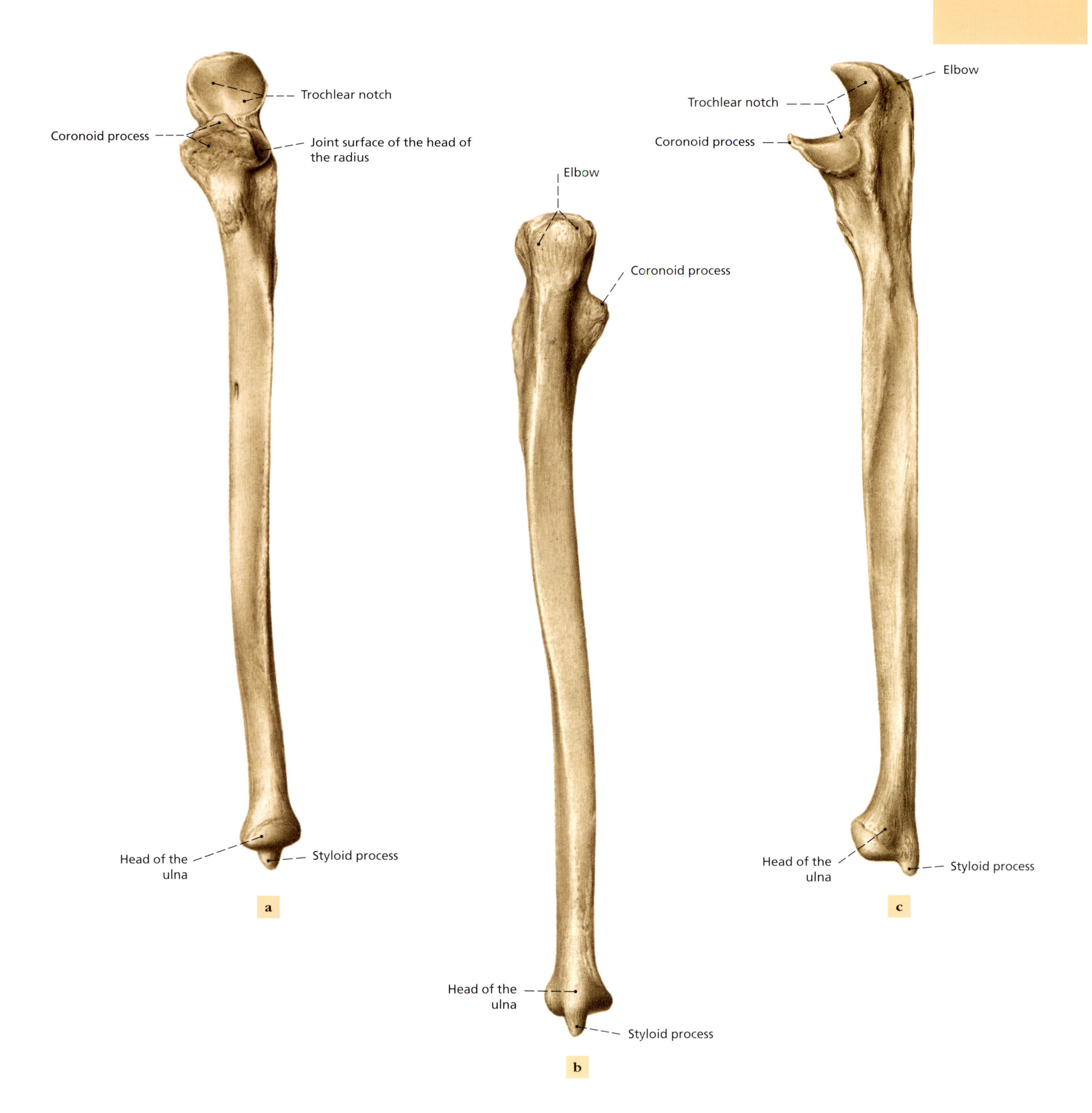

◆
Fig. 86
Bones of the lower arm: left radius from the front (a), from behind (b), and from the right (c).

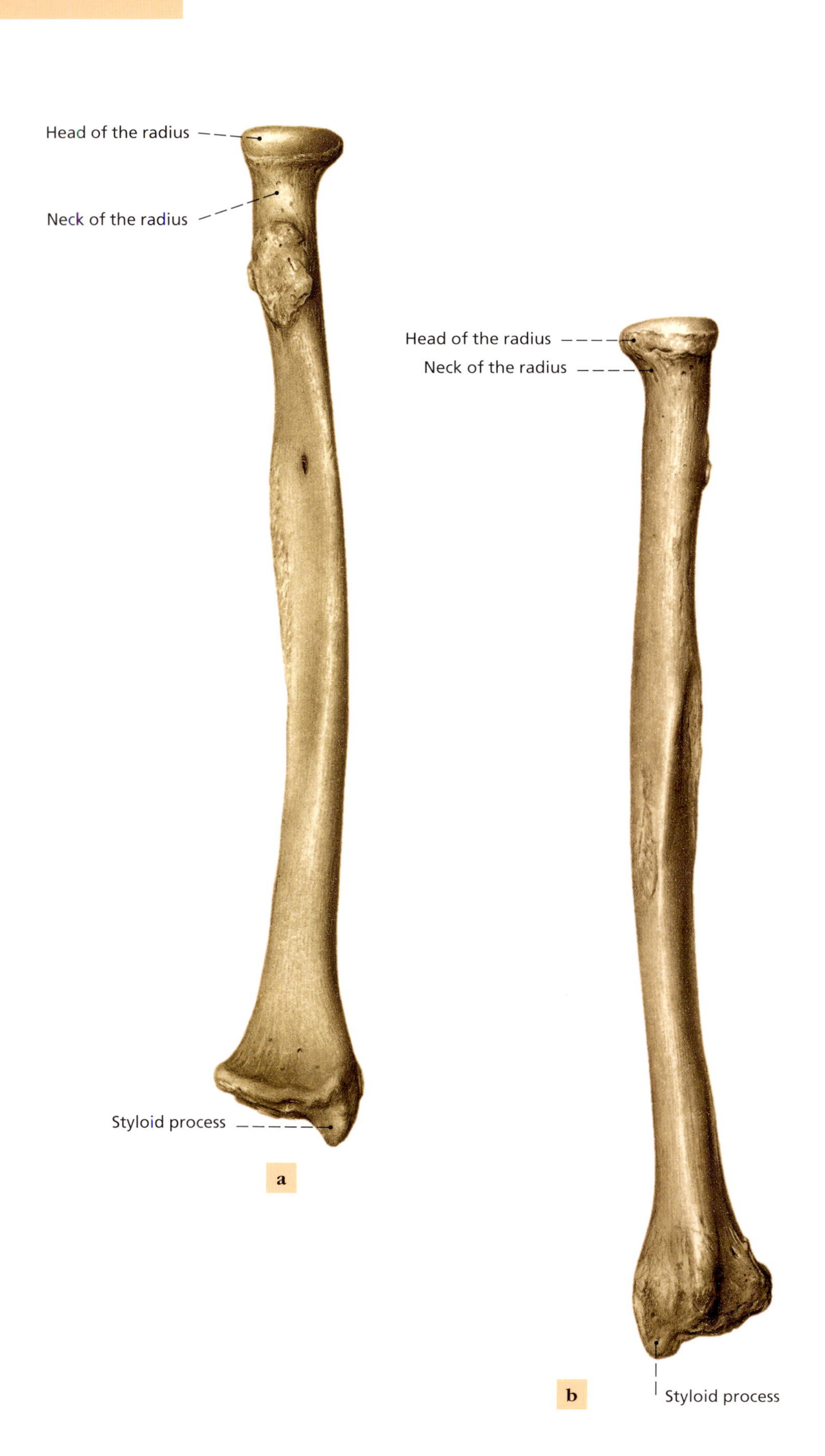

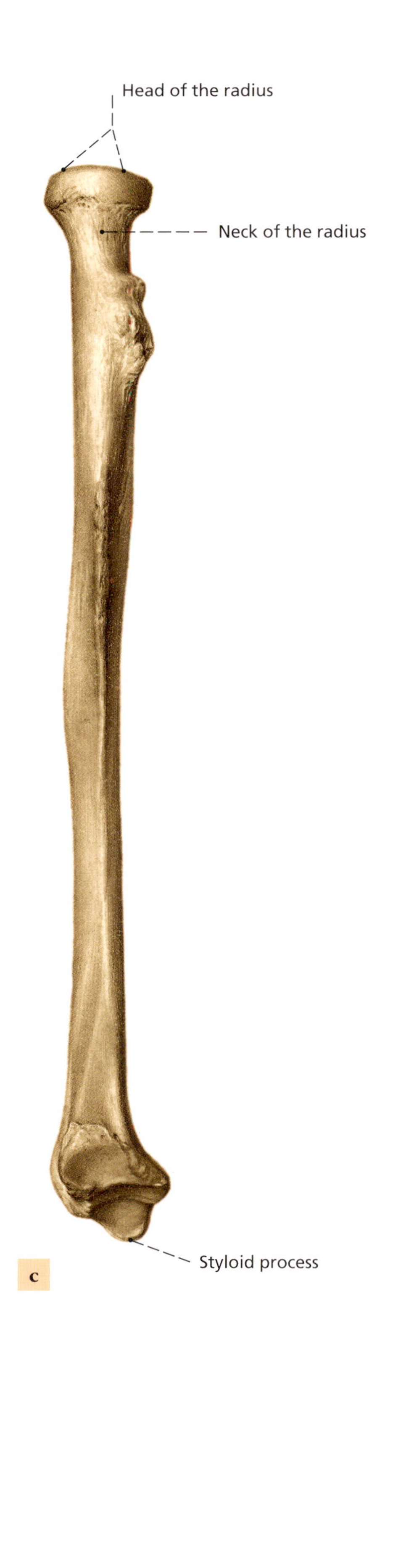

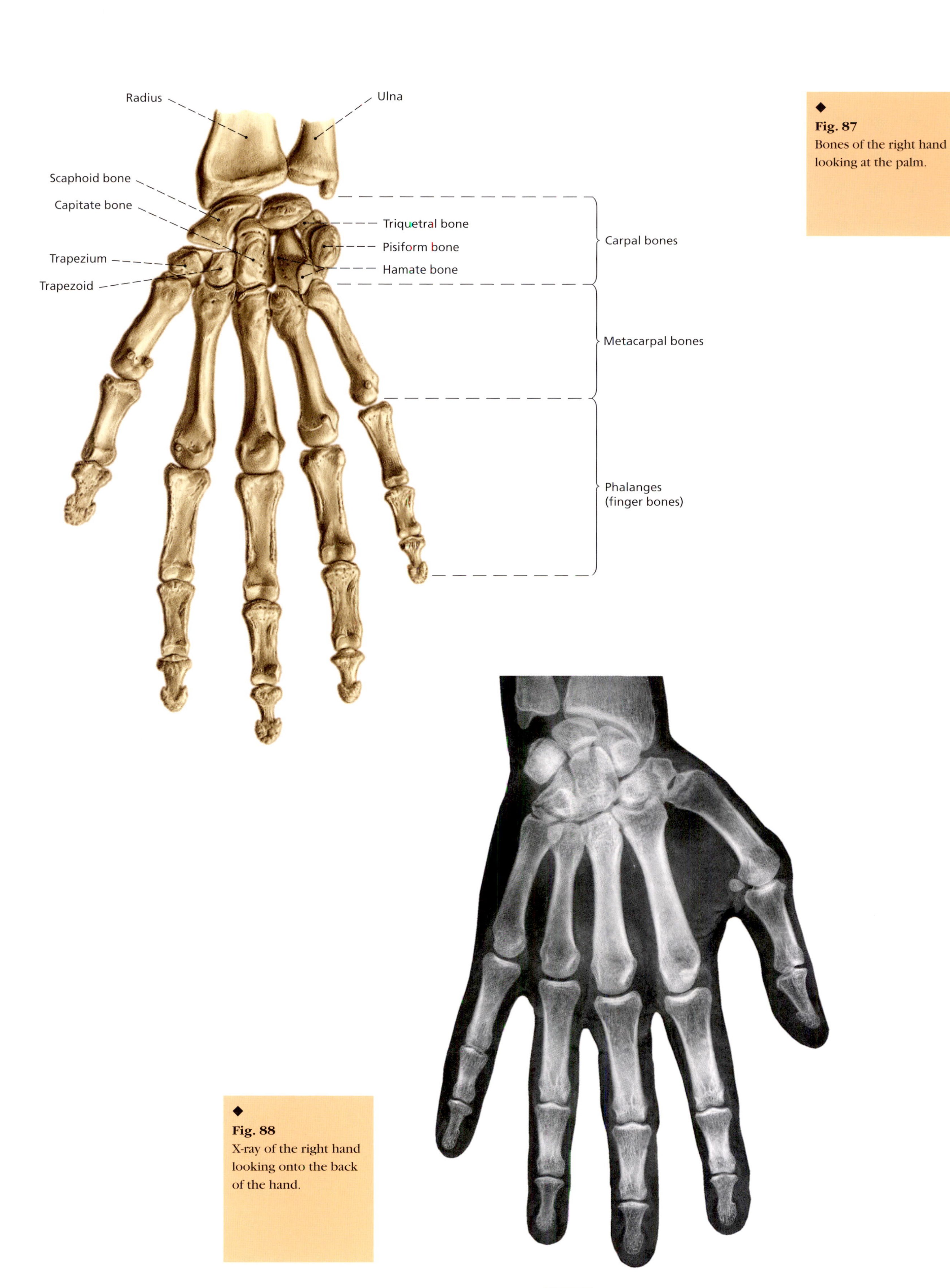

◆
Fig. 87
Bones of the right hand looking at the palm.

◆
Fig. 88
X-ray of the right hand looking onto the back of the hand.

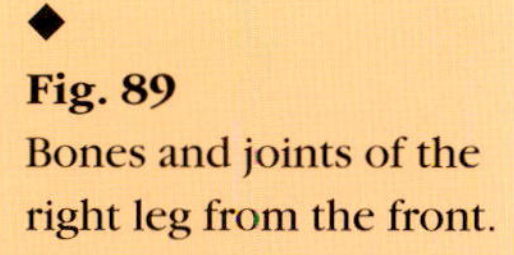

Fig. 89
Bones and joints of the right leg from the front.

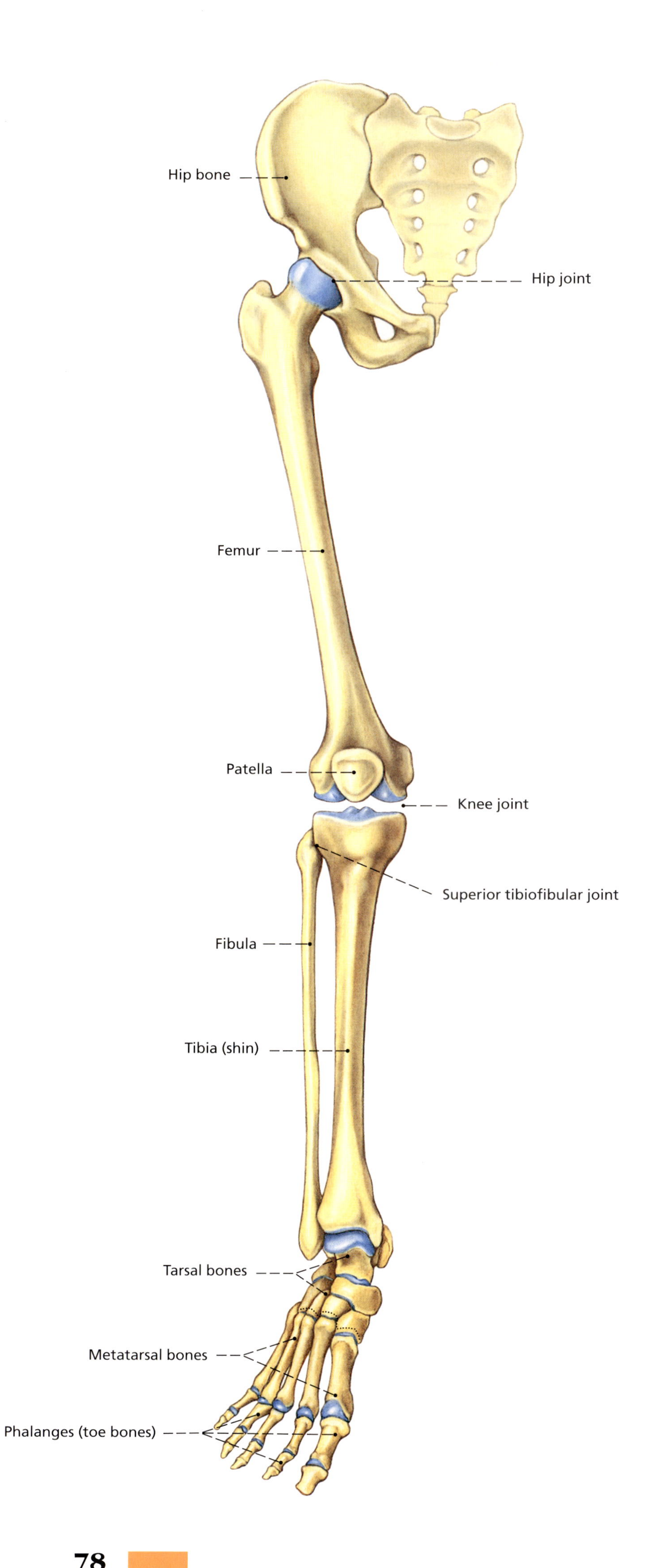

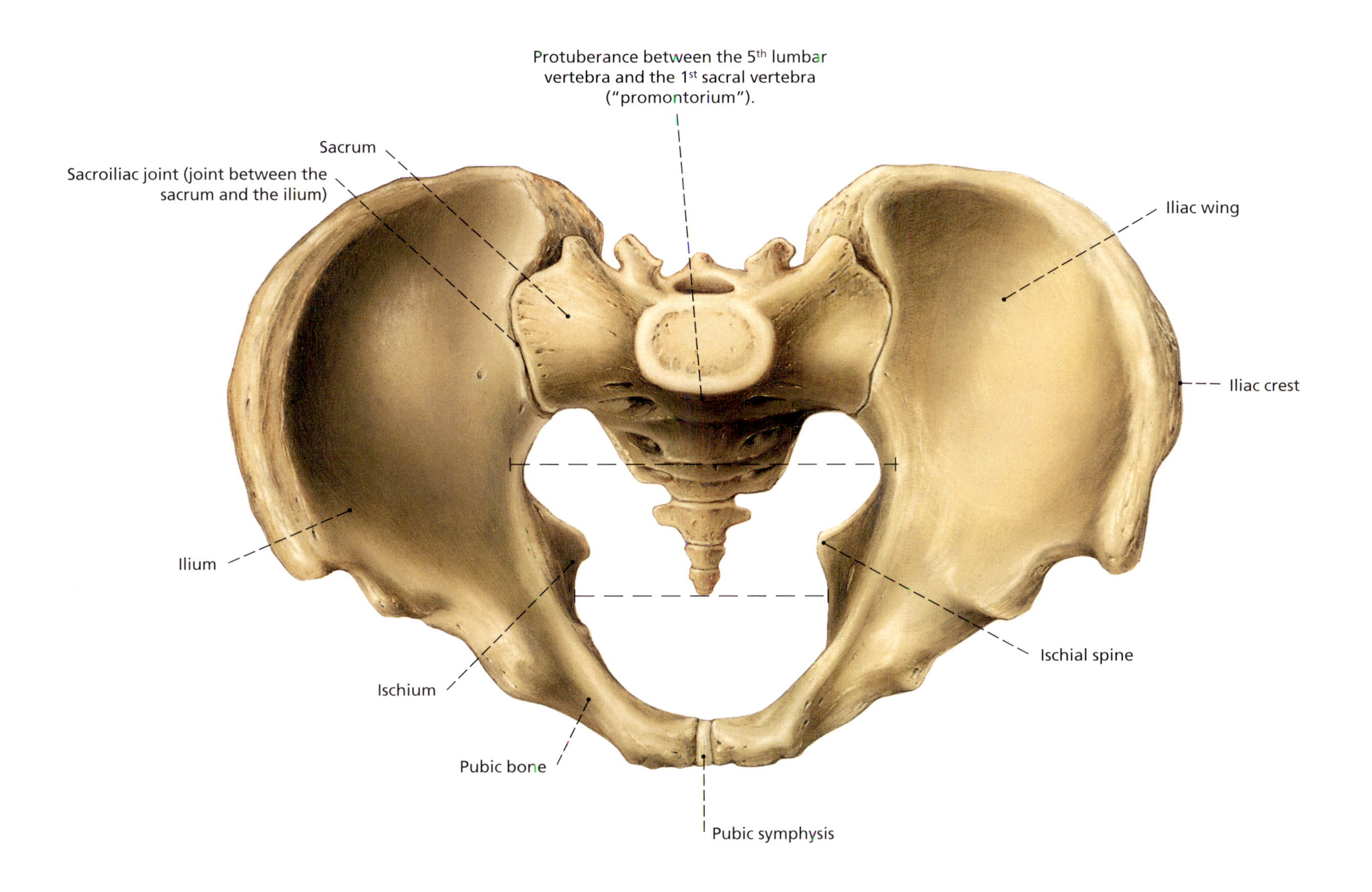

◆
Fig. 90
Pelvic girdle from above. The broken line above the ischial spine indicates the diameter of the large pelvis, the broken line below the ischial spine indicates the diameter of the small pelvis.

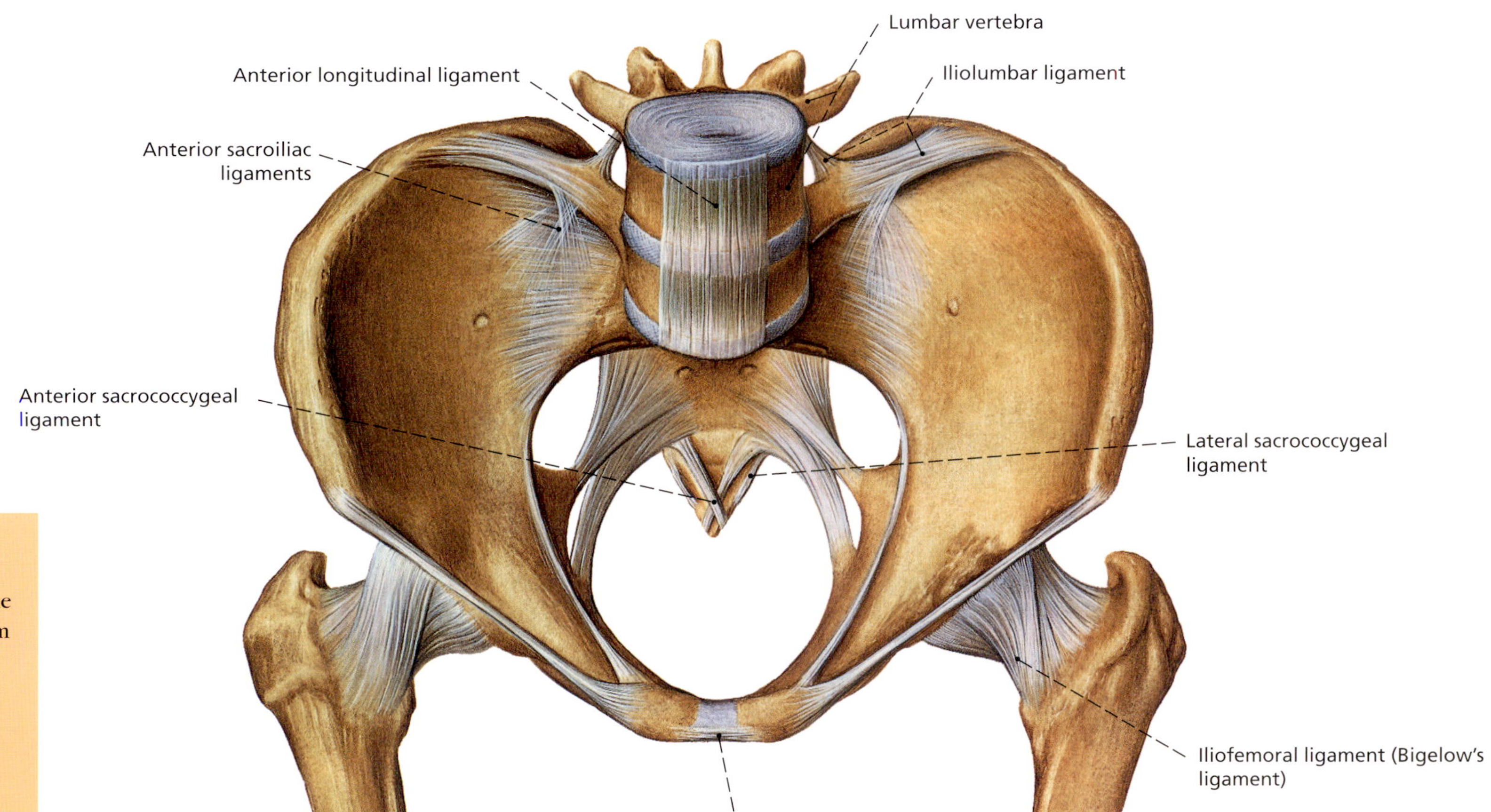

◆
Fig. 91
Bone formation in the male pelvis seen from the front. The pelvic opening is much smaller than in the female (see below).

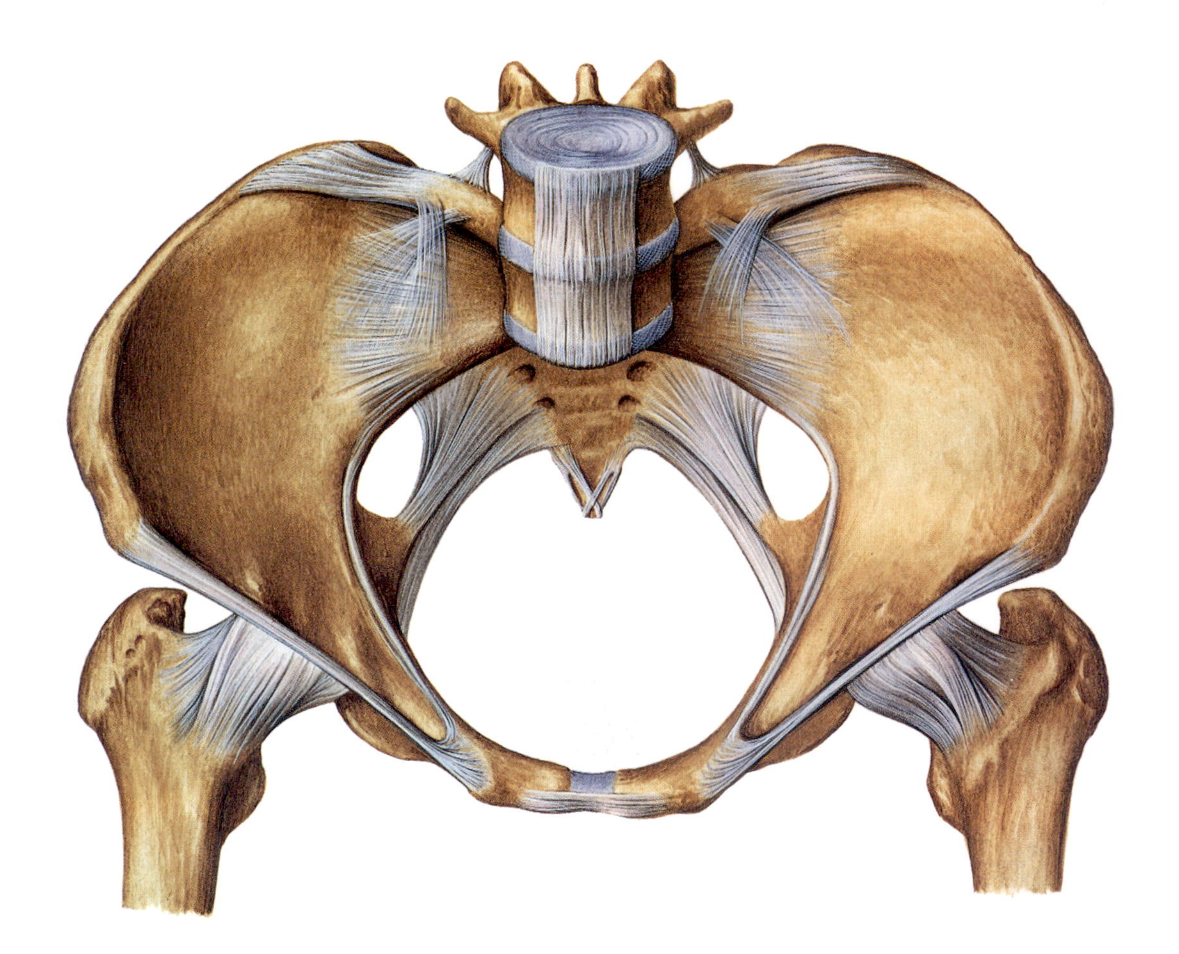

◆
Fig. 92
Bone formation in the female pelvis seen from the front.

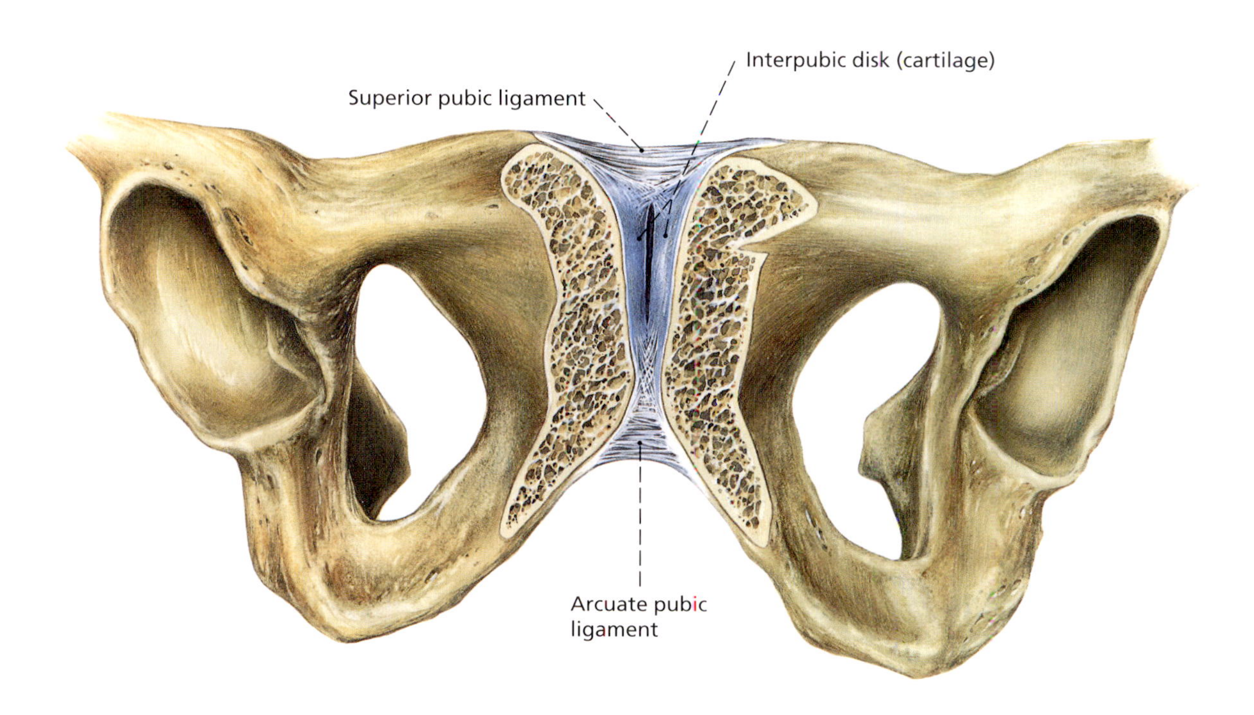

◆
Fig. 93
Pubic symphysis with both hip joints seen from the front and below. Diagonal section towards the longitudinal axis of the pubic symphysis. Cartilage is shown in light blue.

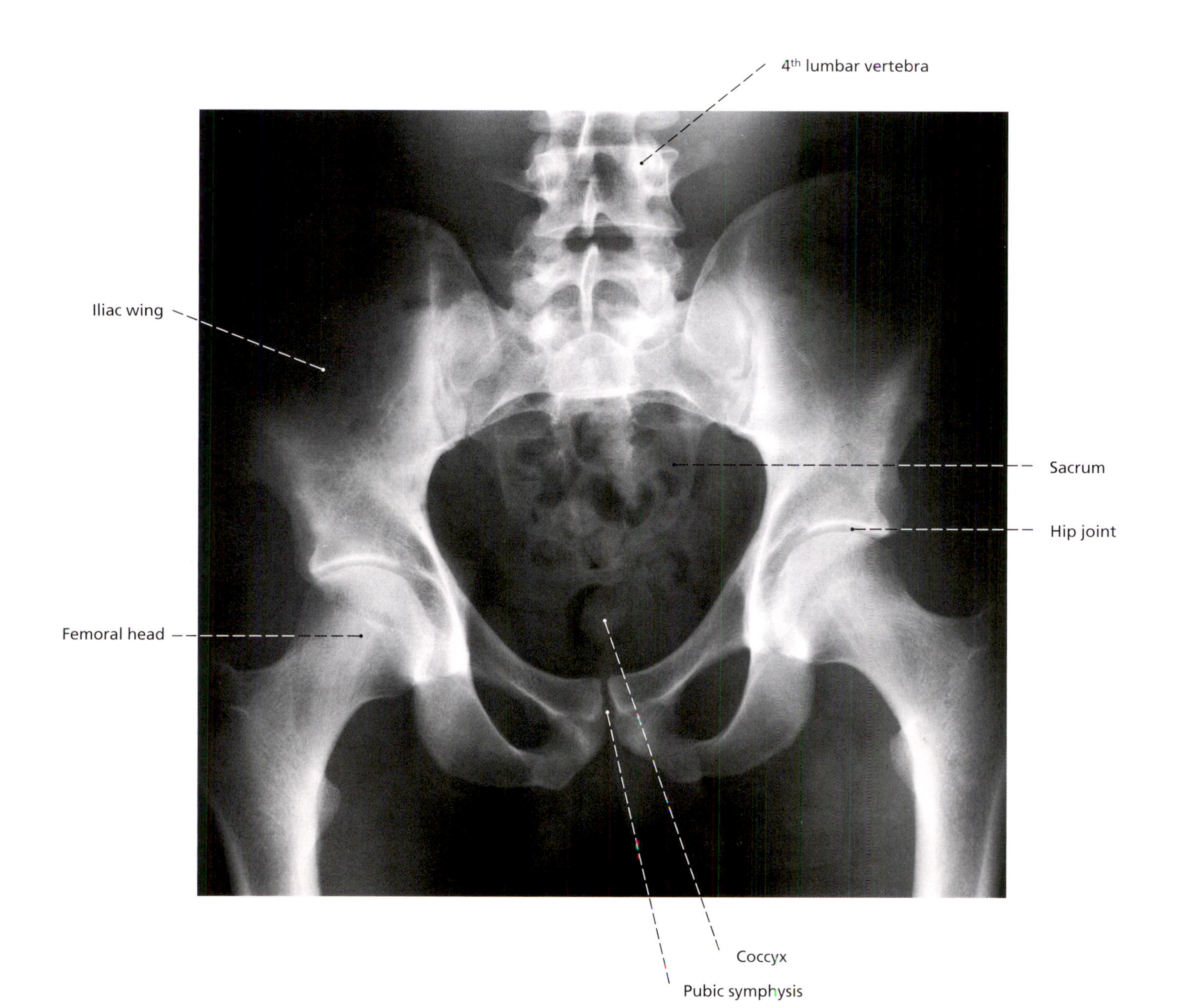

◆
Fig. 94
X-ray of the pelvis from the front.

◆
Fig. 95
Femur of the right leg from the front (a) and behind (b).

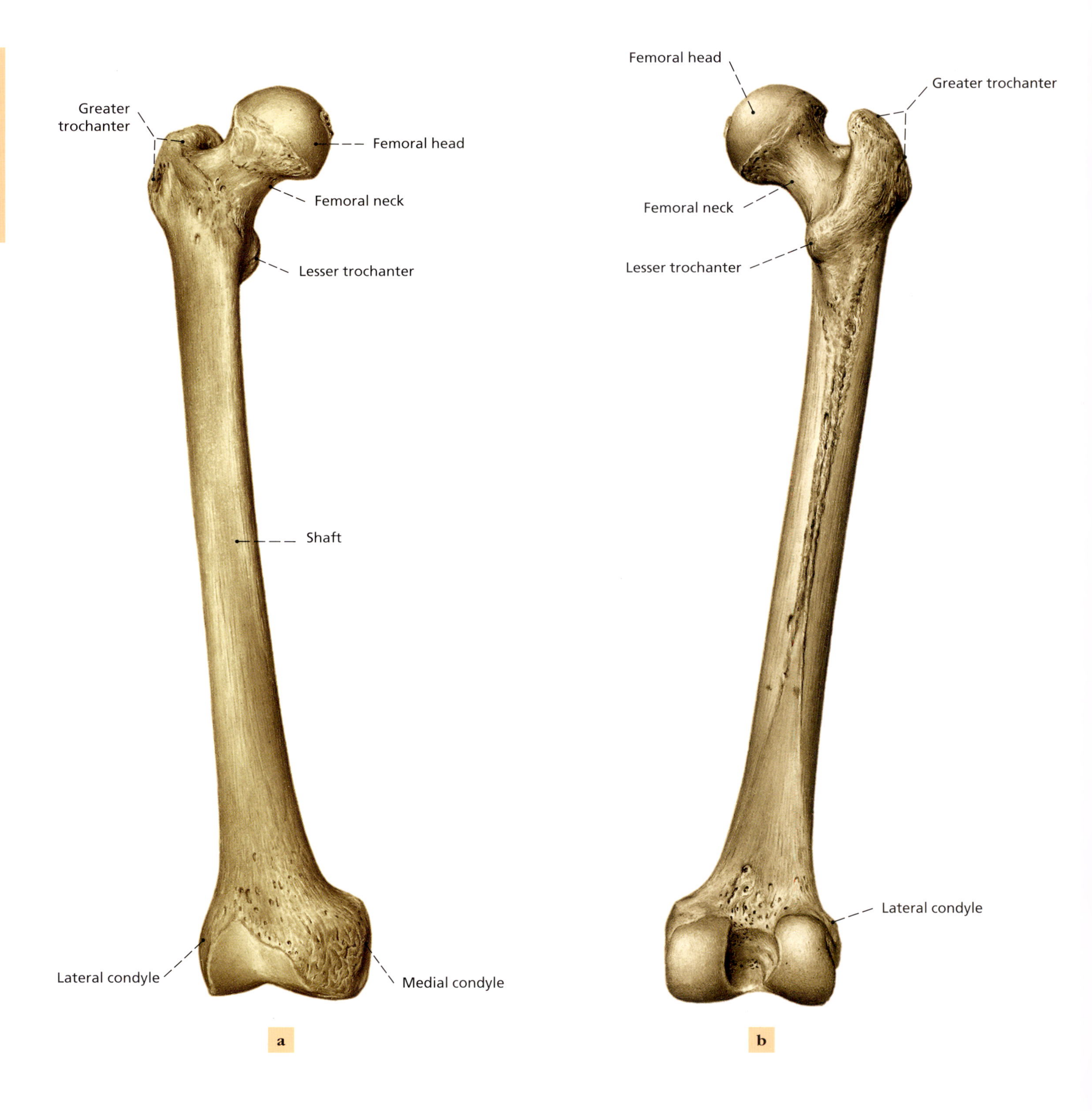

◆
Fig. 96
Cross-section through the right femur in the middle of the shaft.

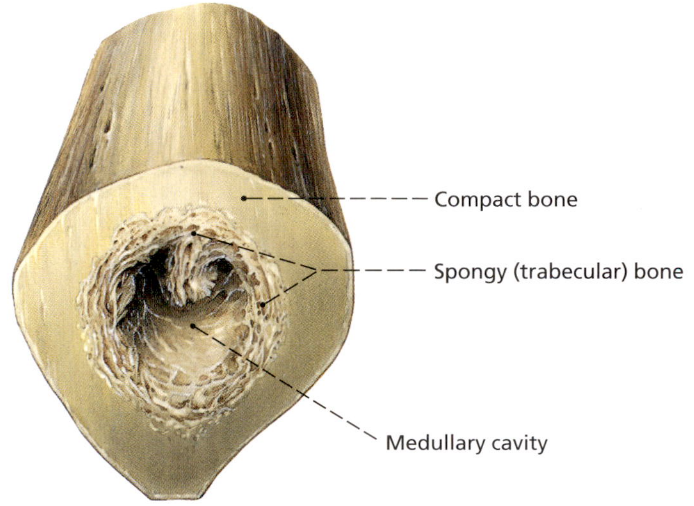

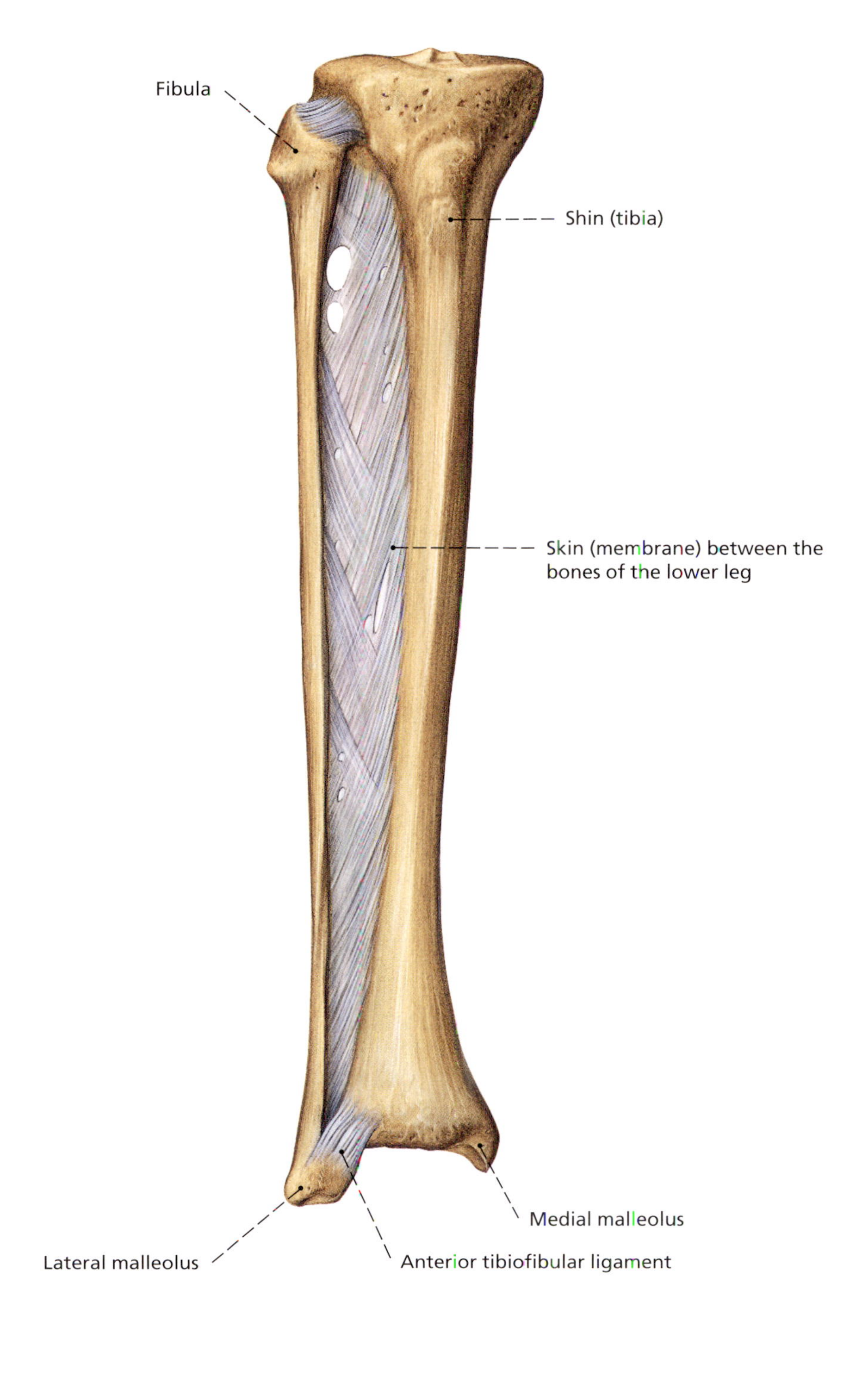

◆
Fig. 97
Bone formation of the right lower leg seen from the front.

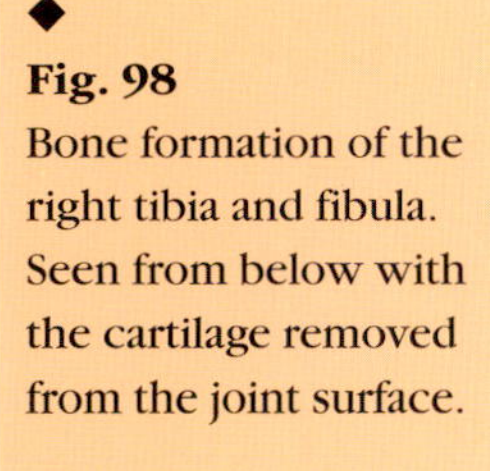

◆
Fig. 98
Bone formation of the right tibia and fibula. Seen from below with the cartilage removed from the joint surface.

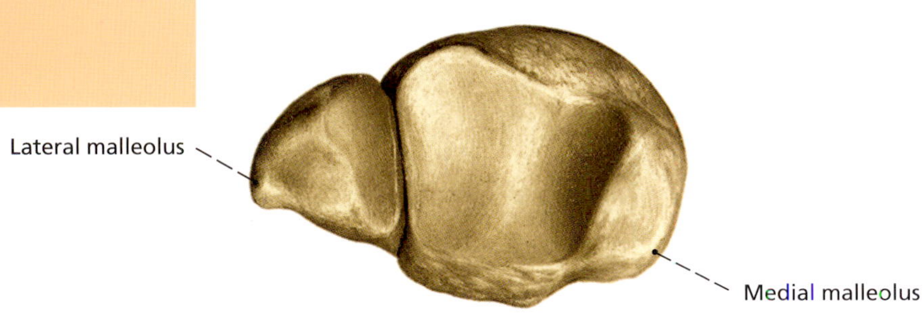

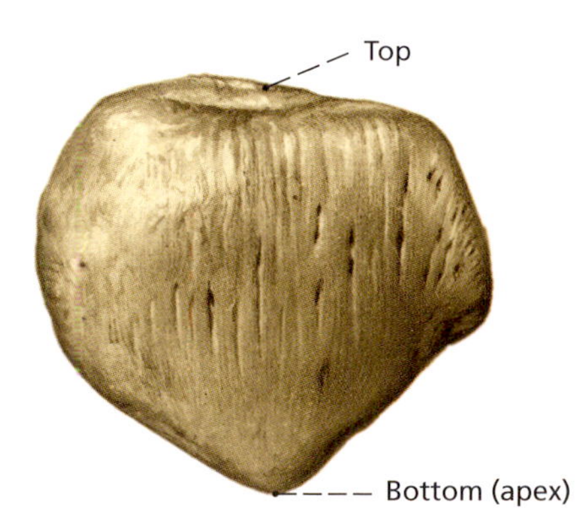

◆
Fig. 99
Right kneecap (patella) seen from the front.

◆ **Fig. 100**
Skeleton of the right foot seen from above.

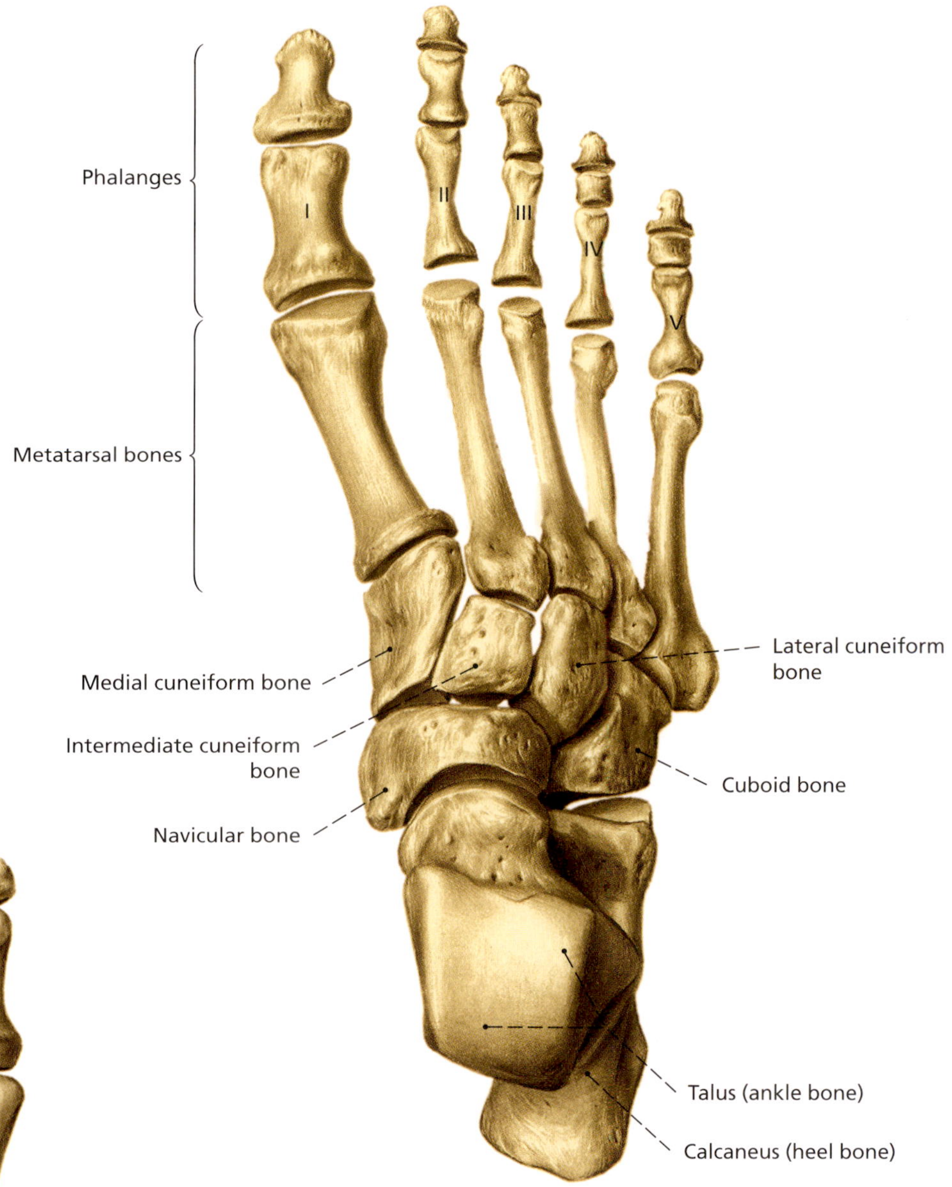

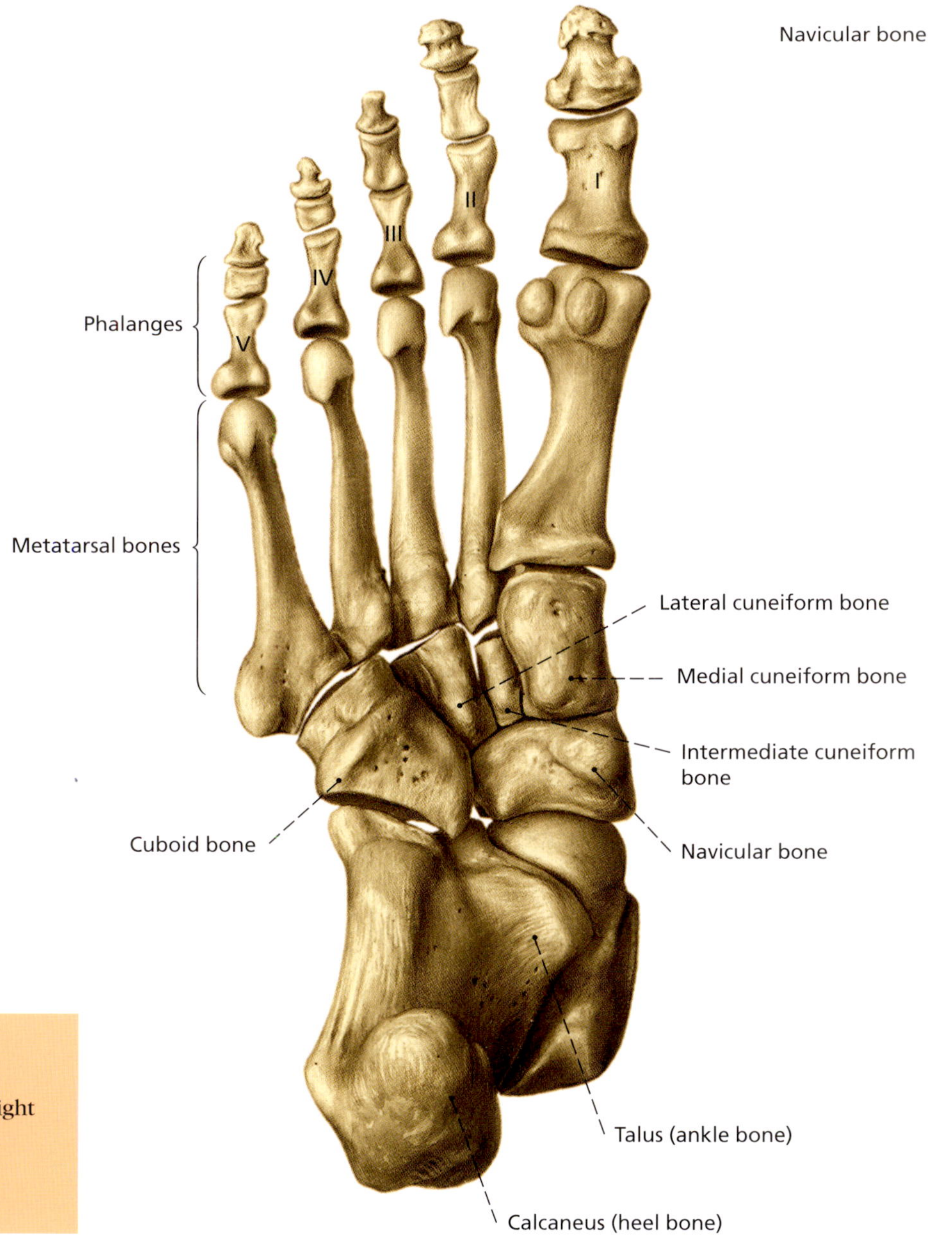

Phalanges
Metatarsal bones
I
II
III
IV
V
Medial cuneiform bone
Intermediate cuneiform bone
Navicular bone
Lateral cuneiform bone
Cuboid bone
Talus (ankle bone)
Calcaneus (heel bone)

◆ **Fig. 101**
Skeleton of the right foot from below.

Phalanges
Metatarsal bones
I
II
III
IV
V
Lateral cuneiform bone
Medial cuneiform bone
Intermediate cuneiform bone
Navicular bone
Cuboid bone
Talus (ankle bone)
Calcaneus (heel bone)

◆
Fig. 102
Skeleton of the right foot. Interior aspect.

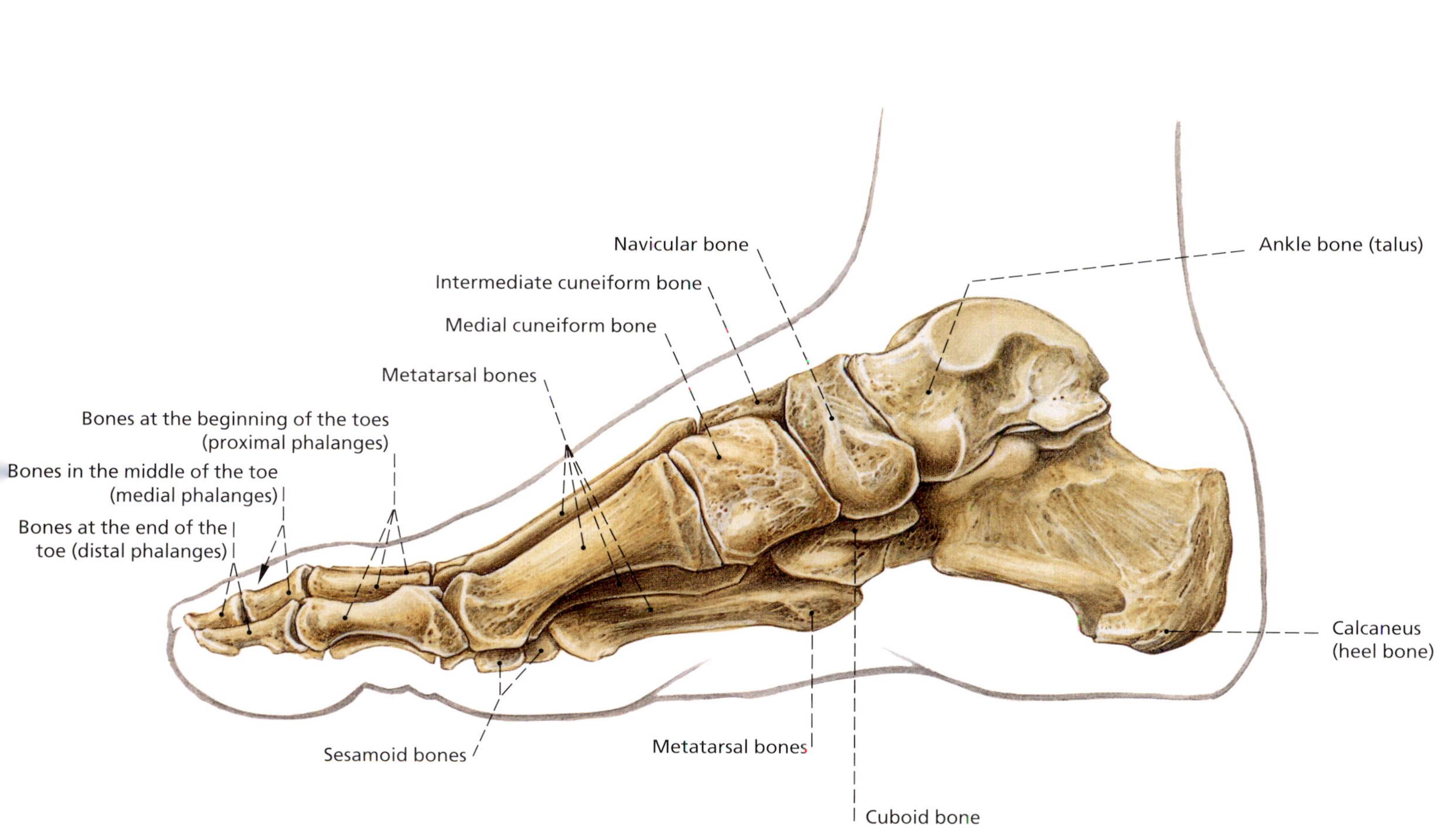

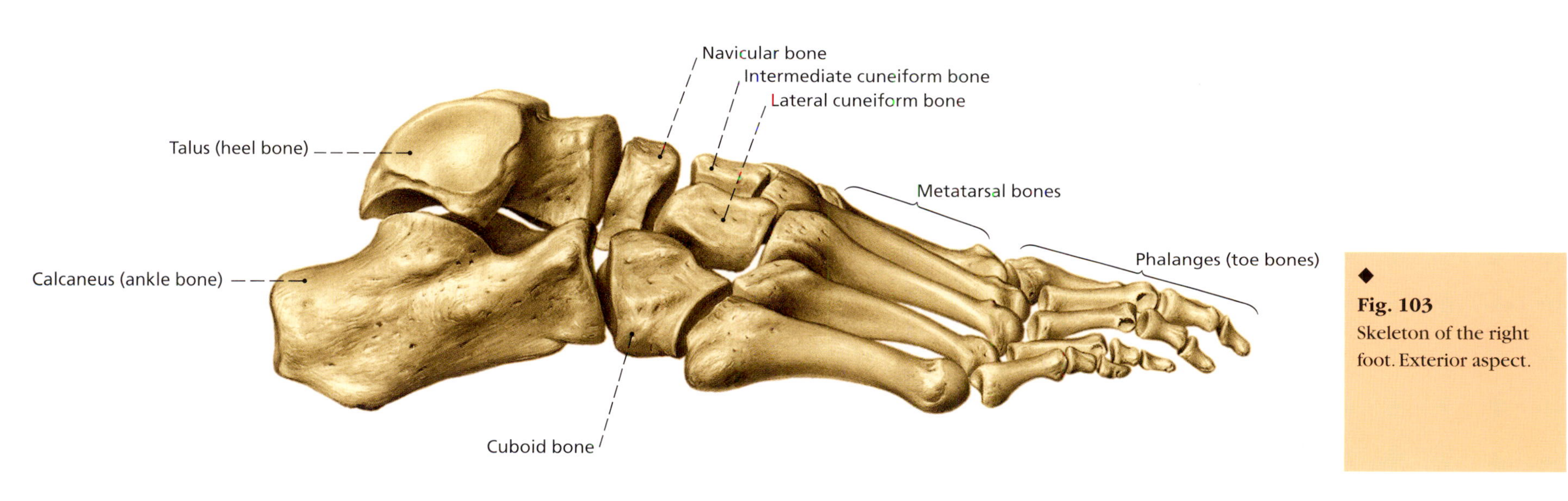

◆
Fig. 103
Skeleton of the right foot. Exterior aspect.

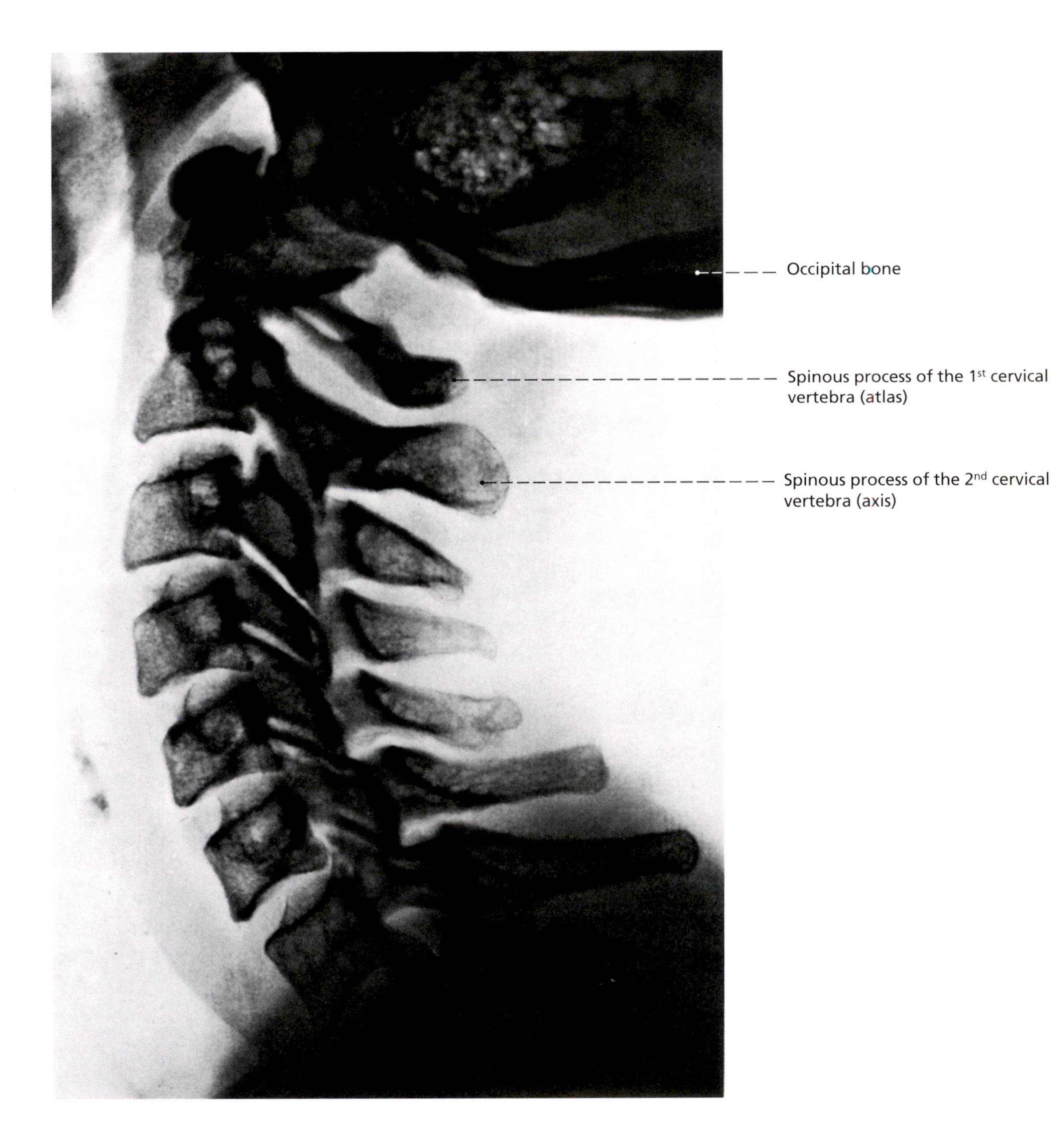

◆
Fig. 104
X-ray of the cervical spine. Seen from the left side. The first two cervical vertebrae are called the atlas and the axis. The convex curvature (lordosis) of the cervical spine is easy to see.

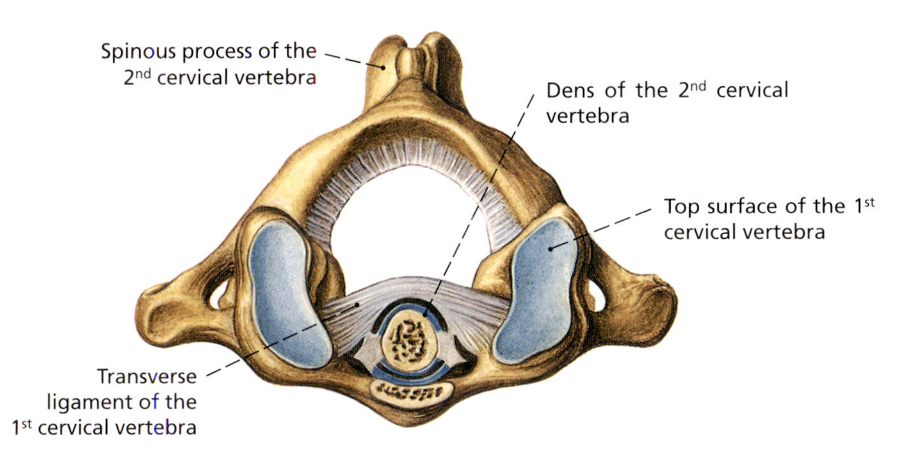

◆
Fig. 105
Atlanto-occipital joints seen from above. The joint surfaces of the 1st cervical vertebra (atlas) can be seen, as well as a cross-section of the dens (odontoid process) of the 2nd cervical vertebra.

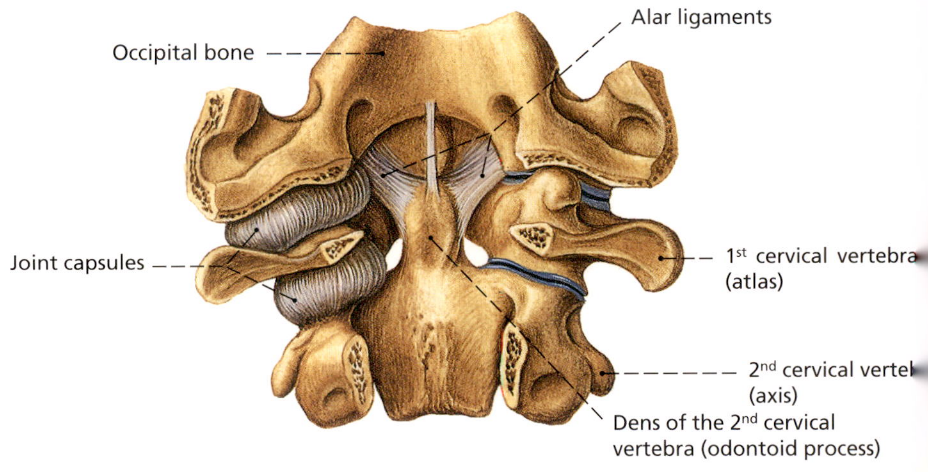

◆
Fig. 106
Atlanto-occipital joints seen from behind. Vertical section through the occipital bone, of which only a small section is seen; the vertebral arches of the 1st and 2nd cervical vertebrae have been removed. The ligaments of the joint capsules are on the left.

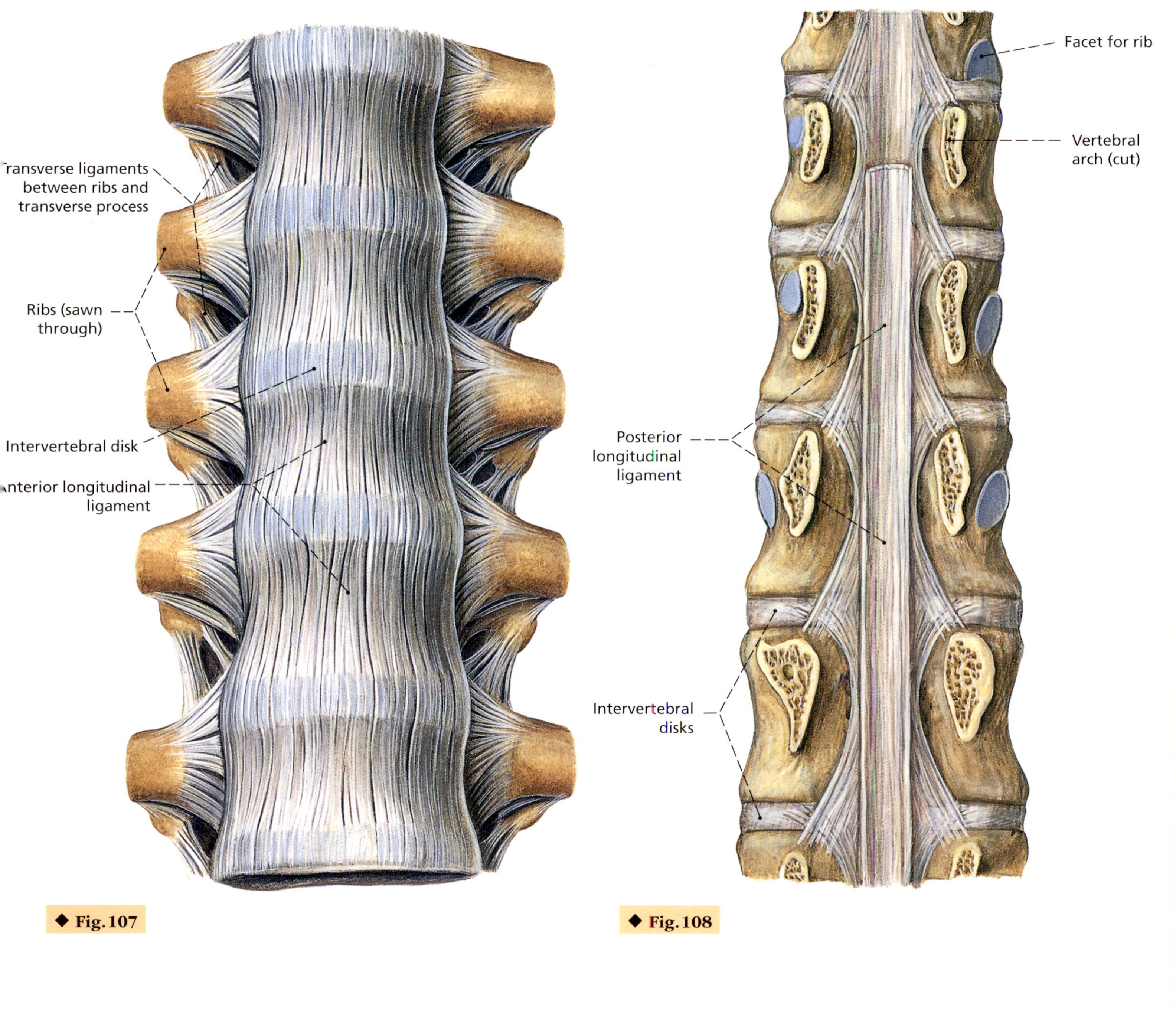

◆ Fig. 107

◆ Fig. 108

◆
Fig. 107
Ligament structure of spine as seen in the lower thoracic vertebral region (anterior view). The anterior longitudinal ligament, which covers the vertebral bodies down the whole length of the spinal column, is most striking.

◆
Fig. 108
Ligament structure of spine (posterior view, lower thoracic and upper lumbar regions): the vertebral canal has been opened by removing the vertebral arches. The posterior longitudinal ligament runs through the vertebral canal from the neck to the lumbar region, but is distinctly narrower than the anterior longitudinal ligament.

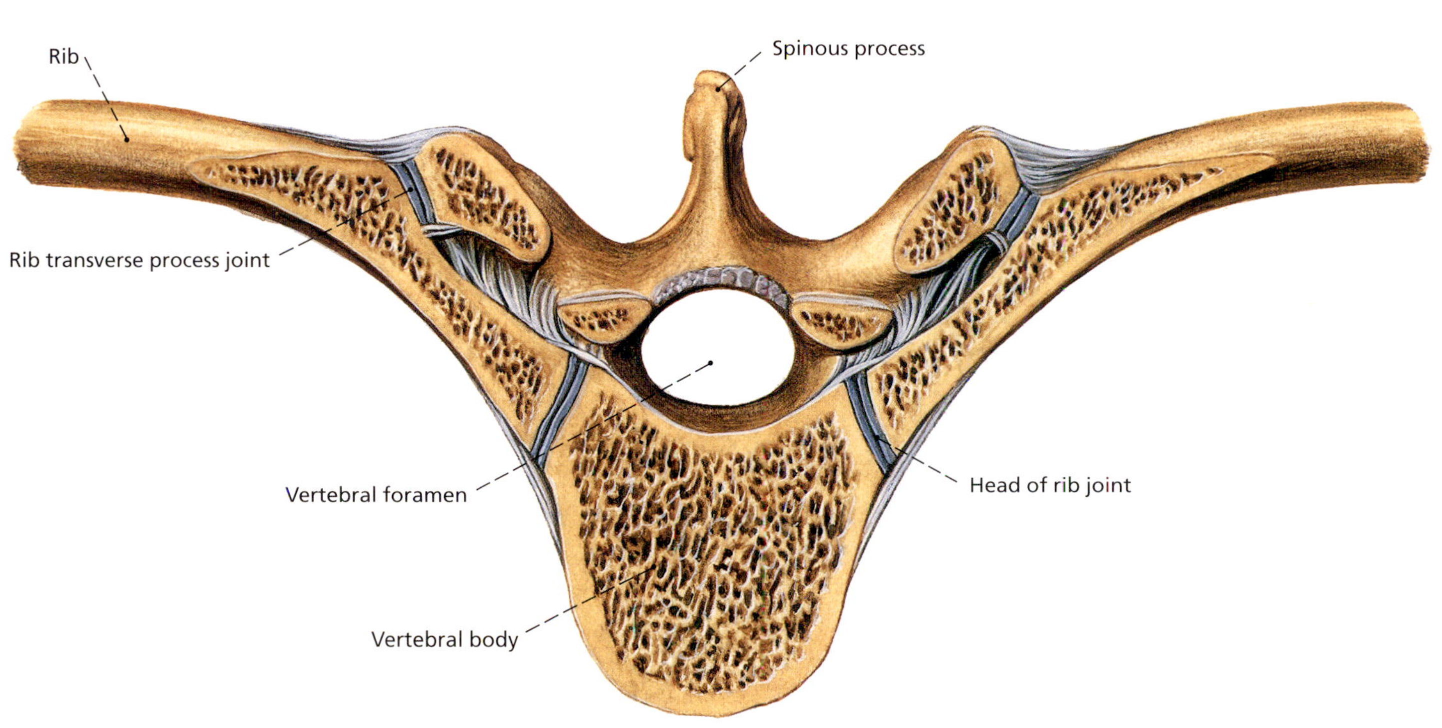

◆
Fig. 109
Costovertebral joints. Cross-section seen from above the lower section of a head of rib joint.

Muscles, bones, and joints

◆
Fig. 114
Shoulder joint (right). Vertical cross-section seen from the front. The joint cartilage is shown in light blue.

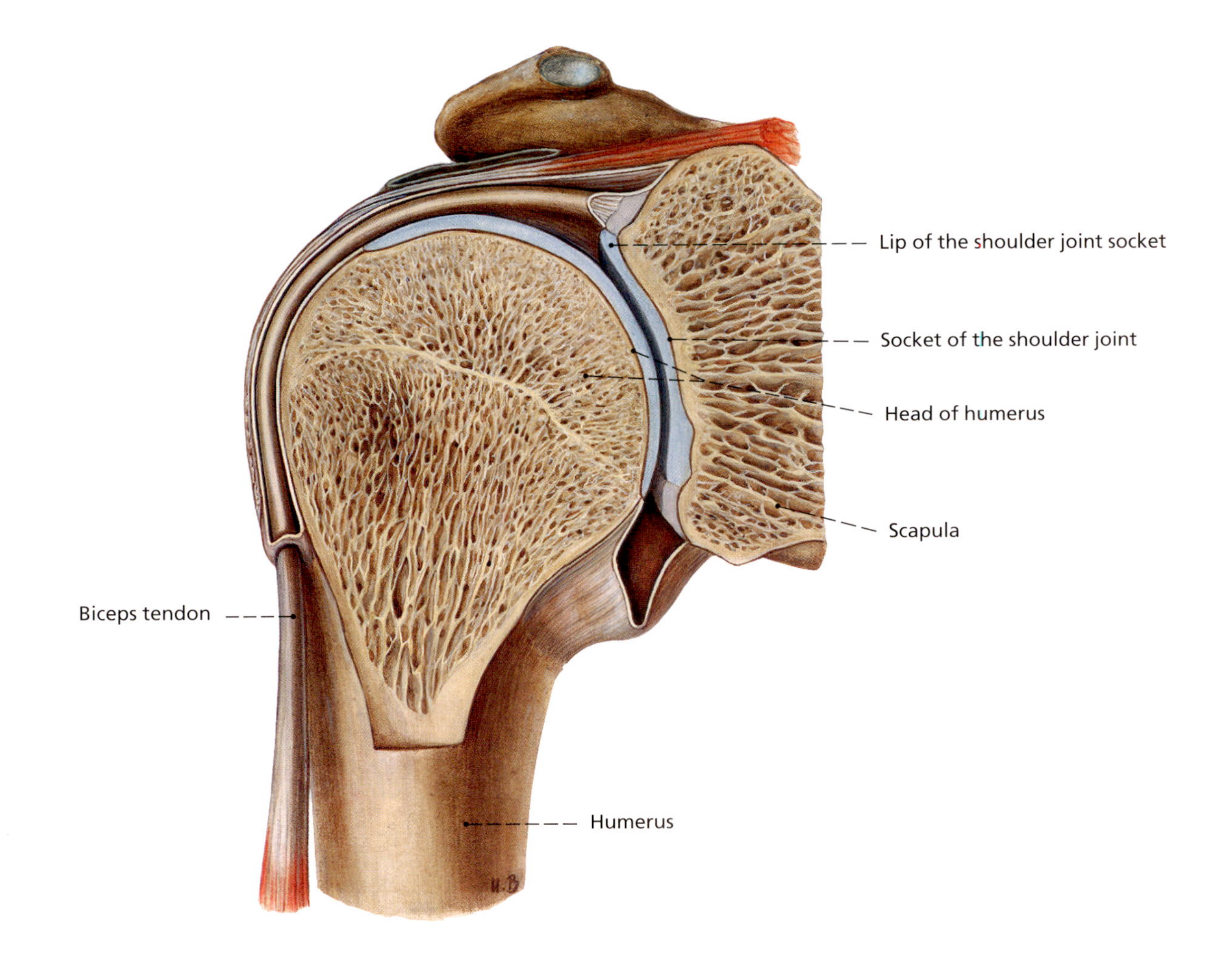

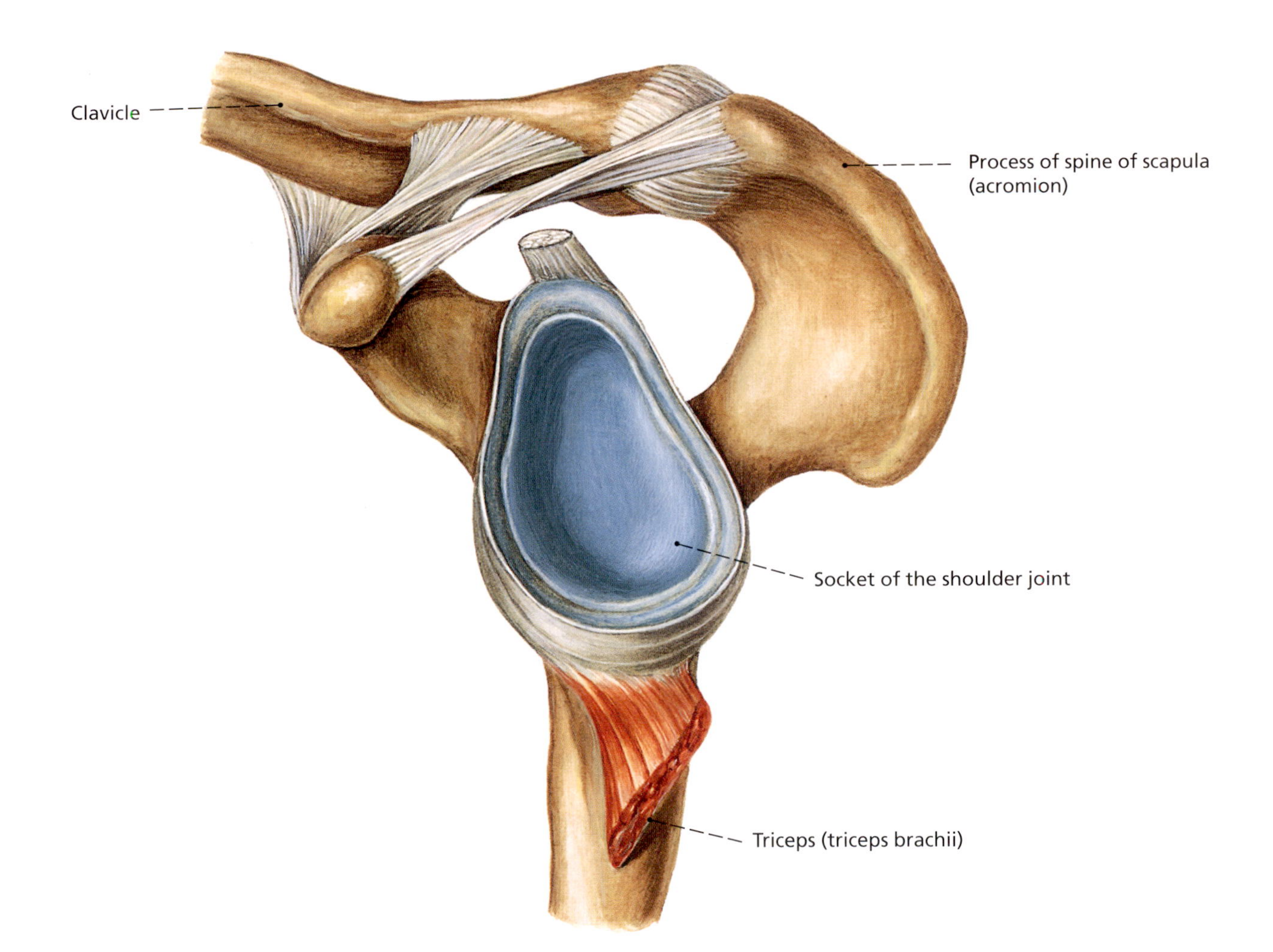

◆
Fig. 115
Shoulder joint (left). Joint cavity of scapula, lateral view, with sectioned joint capsule and humerus head removed. The joint cartilage is shown in light blue.

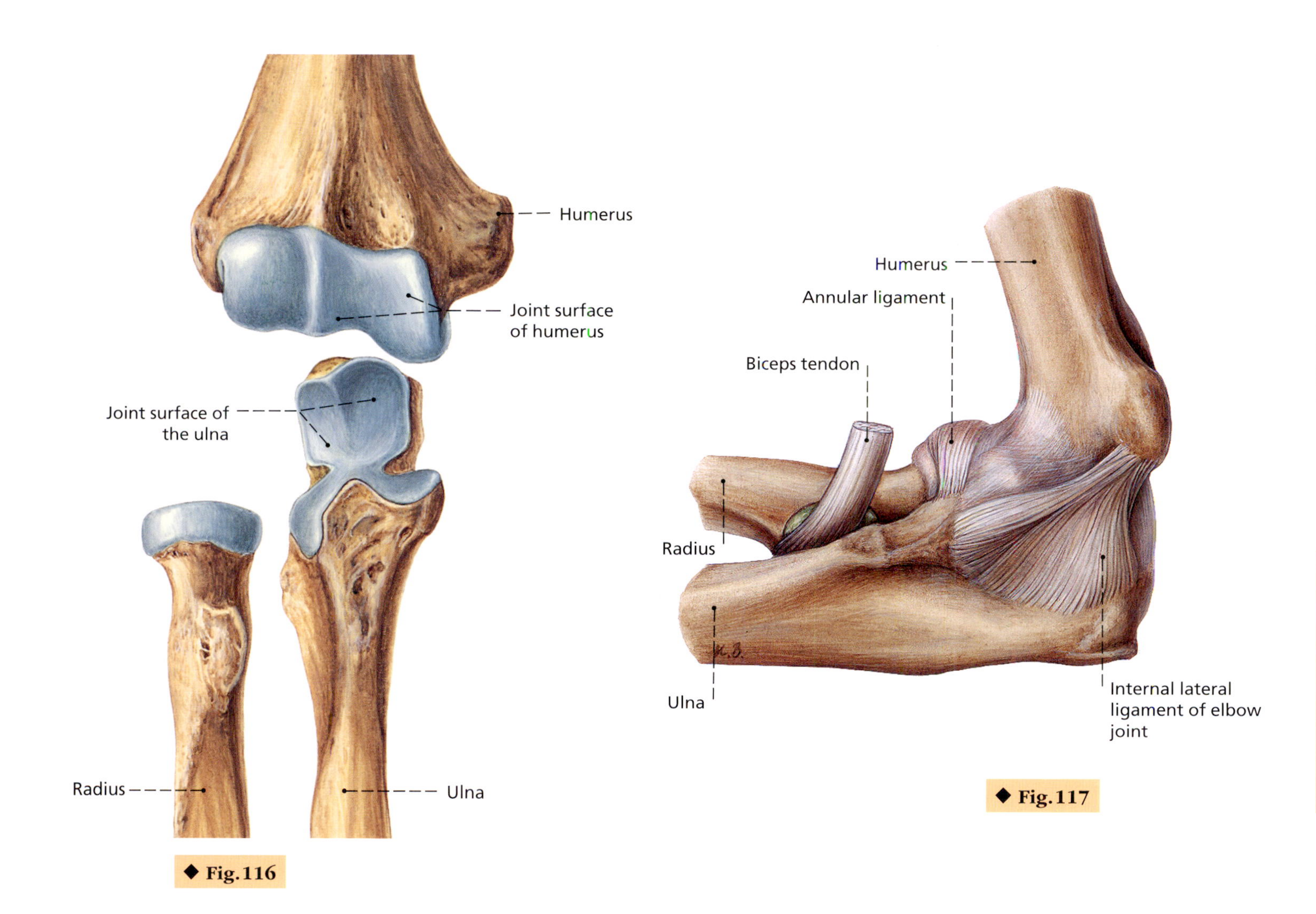

◆ Fig. 116

◆ Fig. 117

◆ **Fig. 116**
Bones of right elbow joint. Cartilage is shown in light blue. The gaps between the joint heads are enlarged for the purposes of illustration.

◆ **Fig. 117**
Right elbow joint (from inside) in flexed position with ligament of the joint tightened.

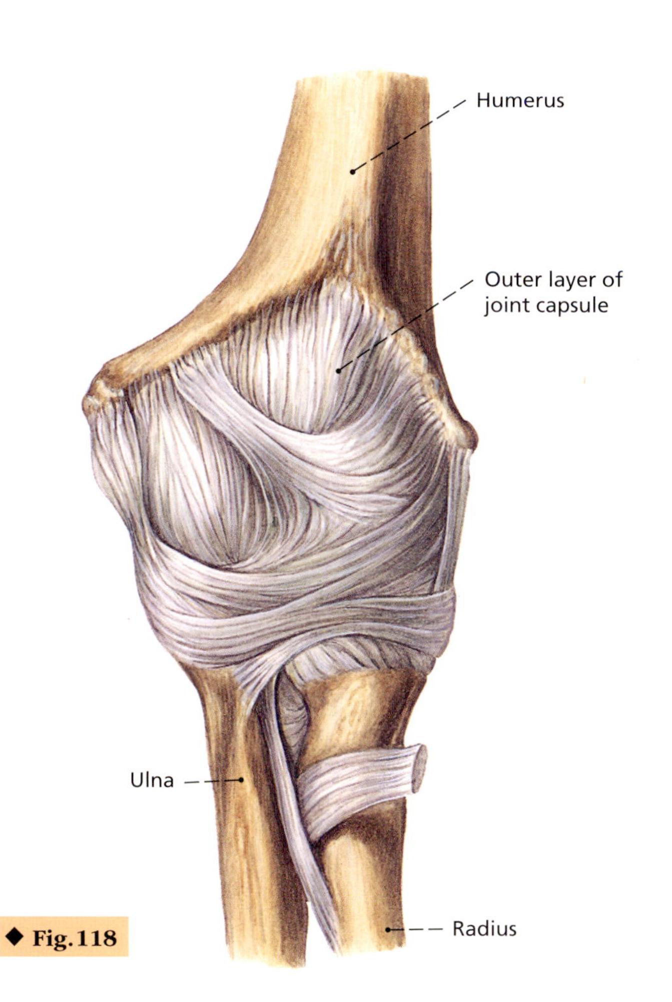

◆ Fig. 118

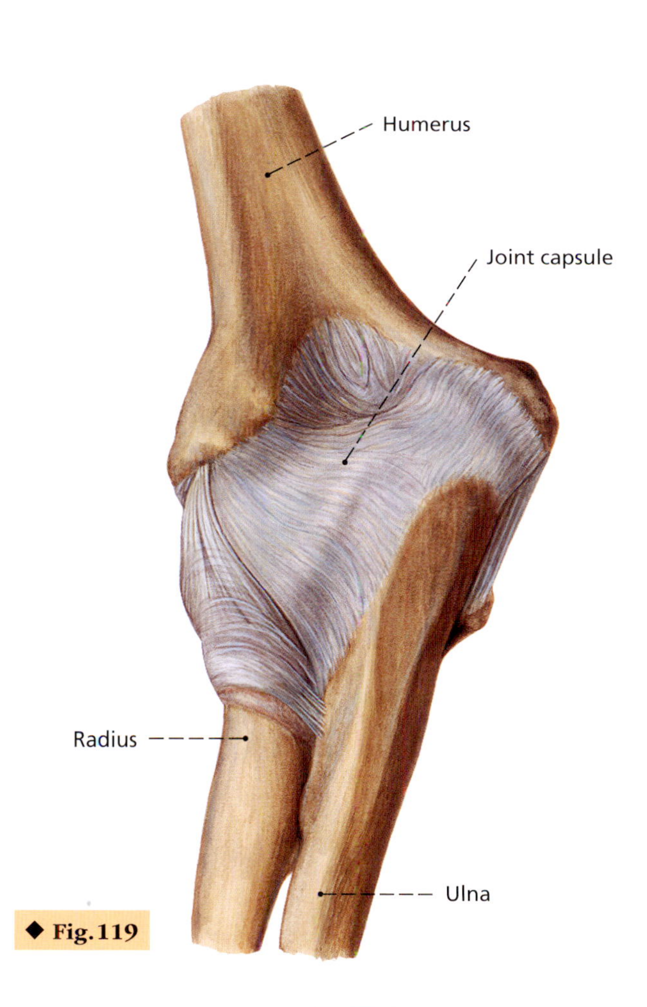

◆ Fig. 119

◆ **Fig. 118**
Left elbow joint (from front) in extended position with lateral ligaments and anterior surface of the joint capsule.

◆ **Fig. 119**
Left elbow joint (from behind) in extended position with lateral ligaments and posterior surface of the joint capsule.

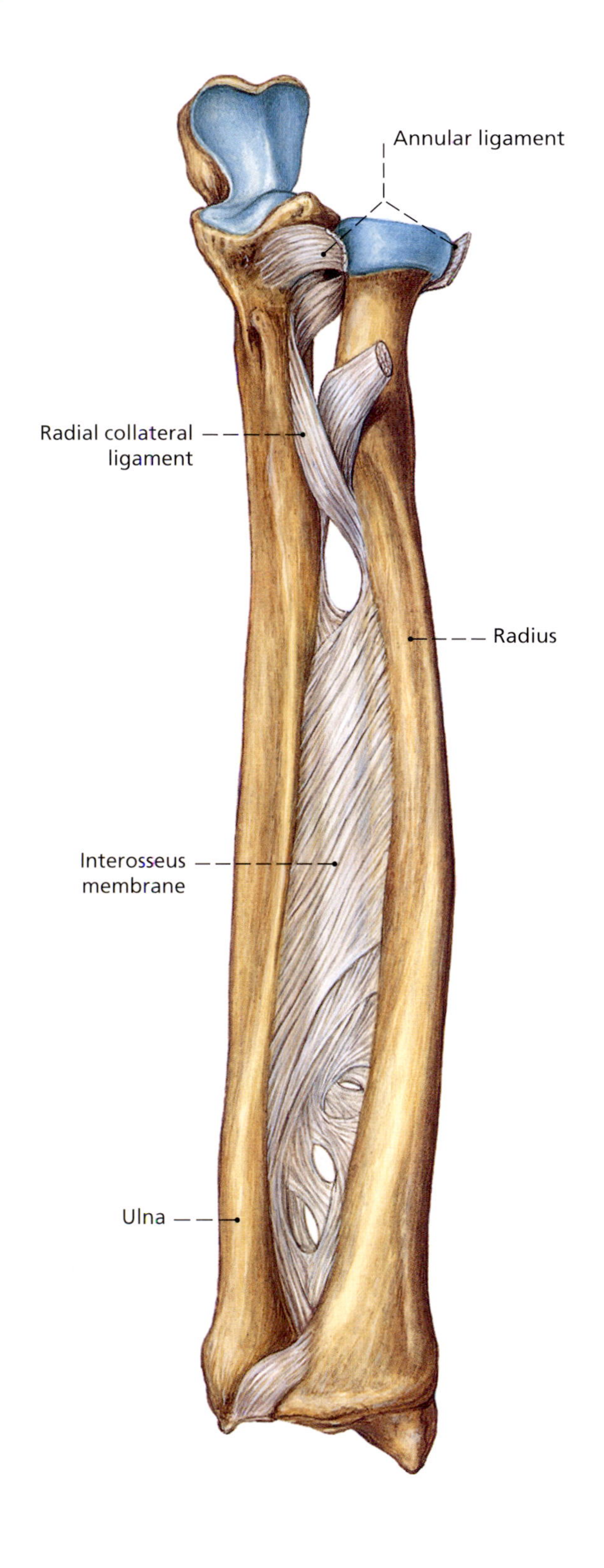

◆
Fig. 120
Articulation of the bones of the left forearm. Joint cartilage shown in light blue.

Ulna
Radius
Annular ligament
Proximal radio-ulnar joint
Elbow

◆
Fig. 121
Proximal elbow joint between left radius and ulna, anterior view from above. The joint is strengthened by the annular ligament of the radius. Cartilage is shown in light blue.

◆
Fig. 122
Cross-section through both bones of the left forearm from below.

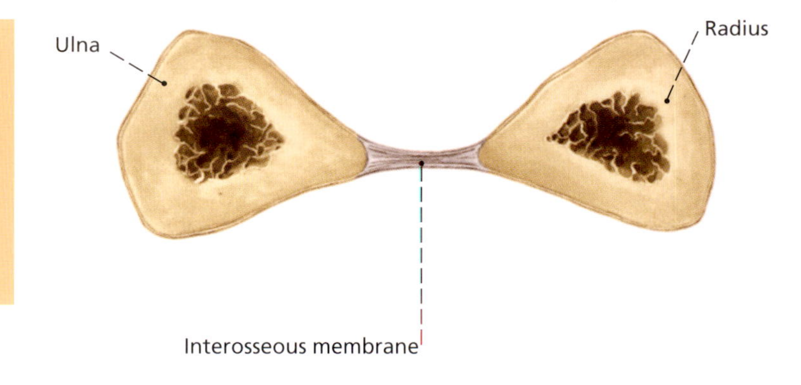

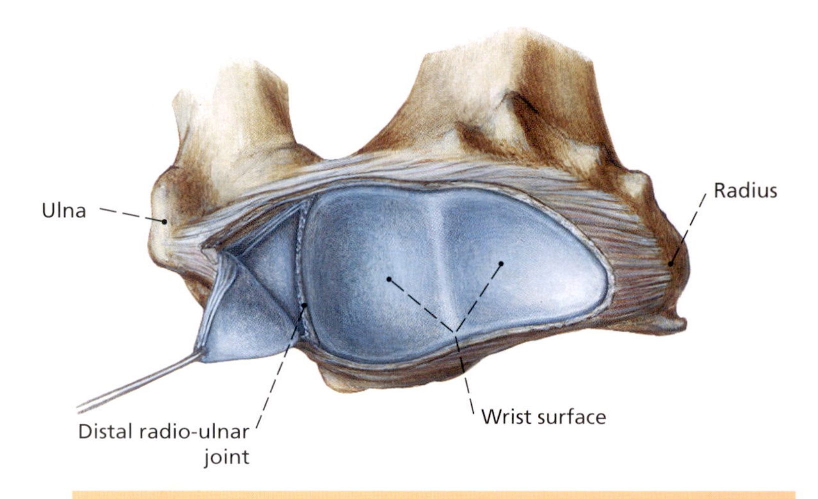

◆
Fig. 123
Joint connection between ulna and radius in left wrist area. The cartilage surface of the radial joint cavity is shown in light blue.

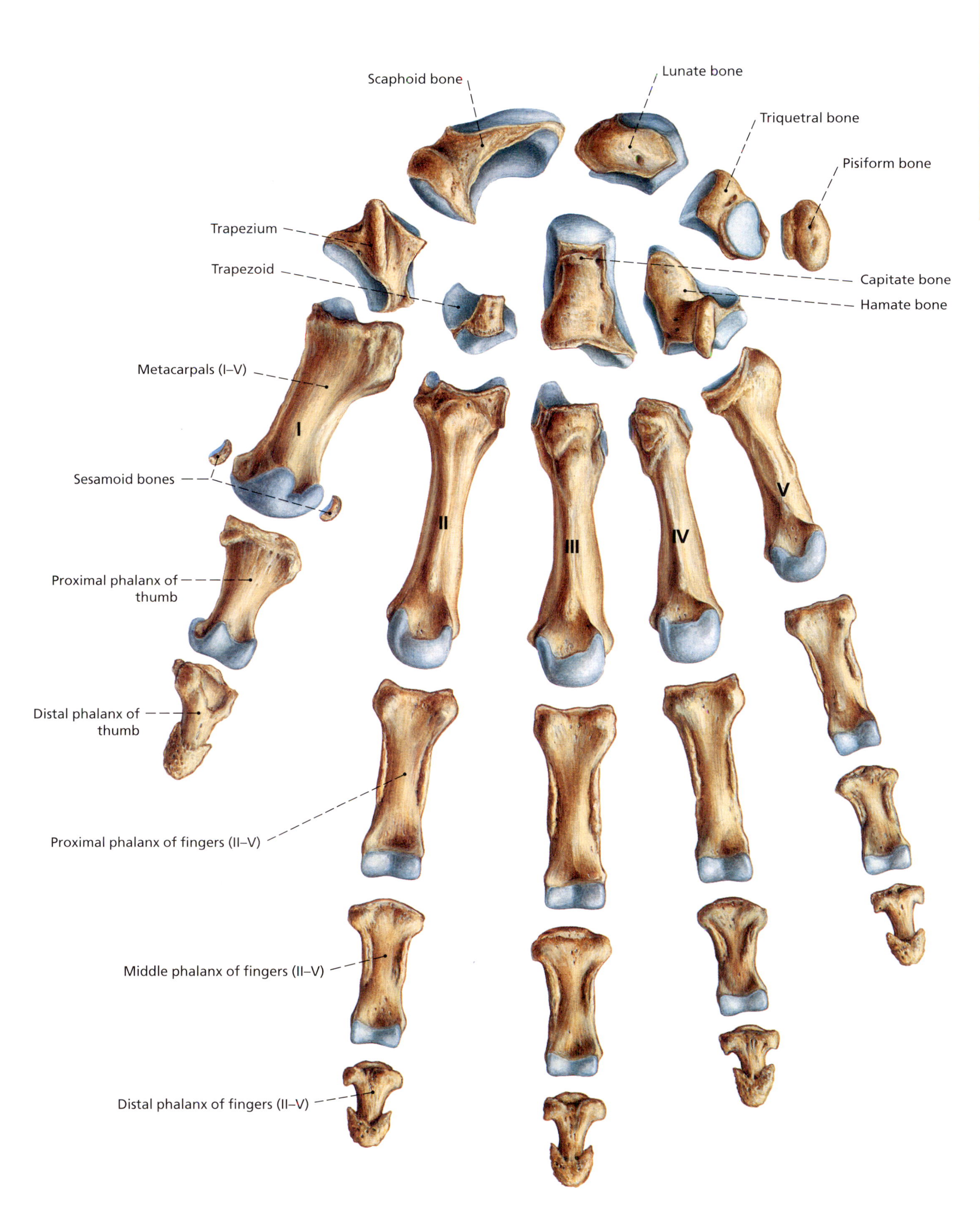

◆
Fig. 124
Bones of the right hand. View of inner surface. Gaps between the individual bones are enlarged for the purposes of illustration (I–V = metacarpals of hand from thumb to small finger).

◆
Fig. 125
Front view of the bones of the female pelvis. Vertical cross-section from above.

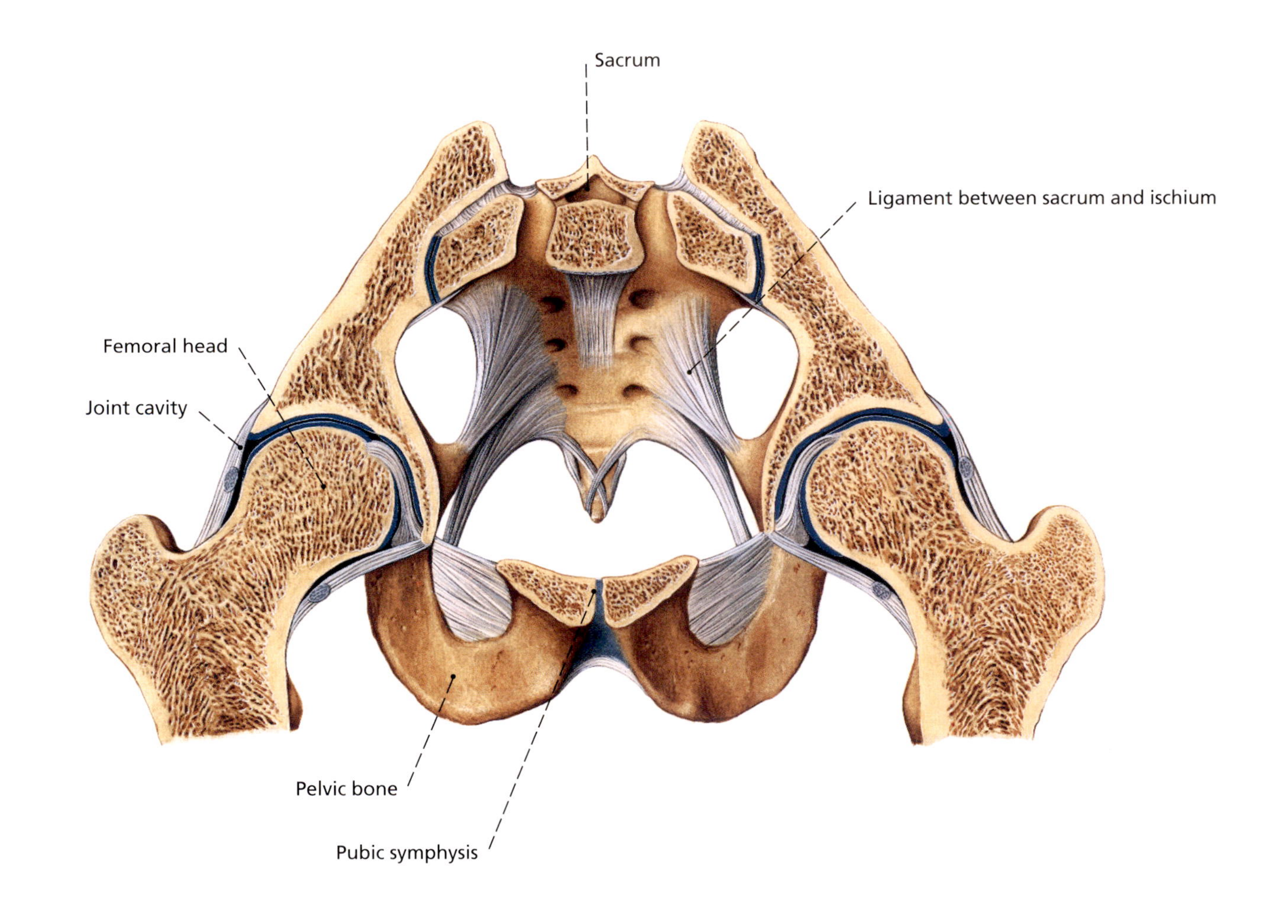

◆
Fig. 126
Bone connections of the female pelvis, lateral view. Longitudinal section through the pubic symphysis and center of vertebral body.

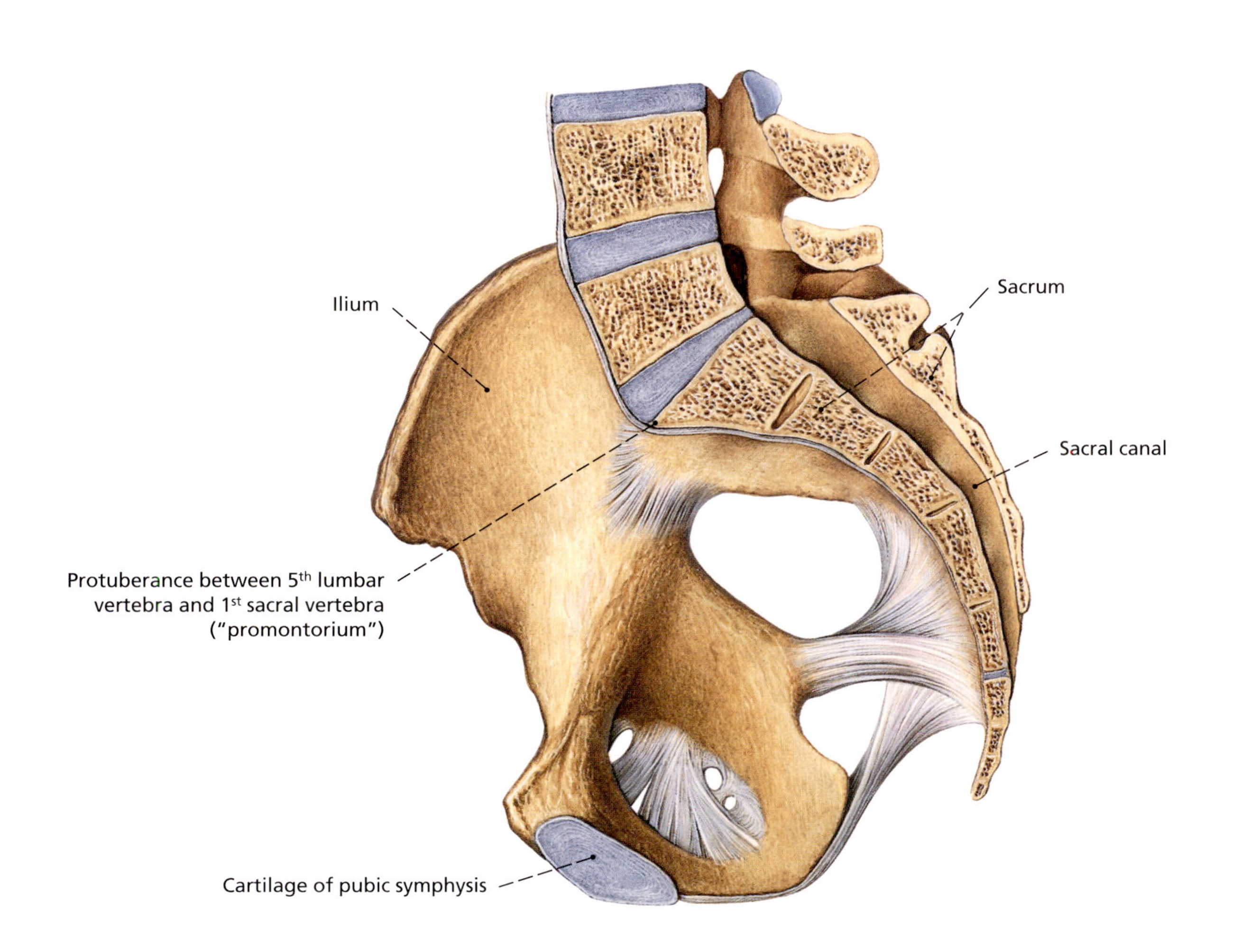

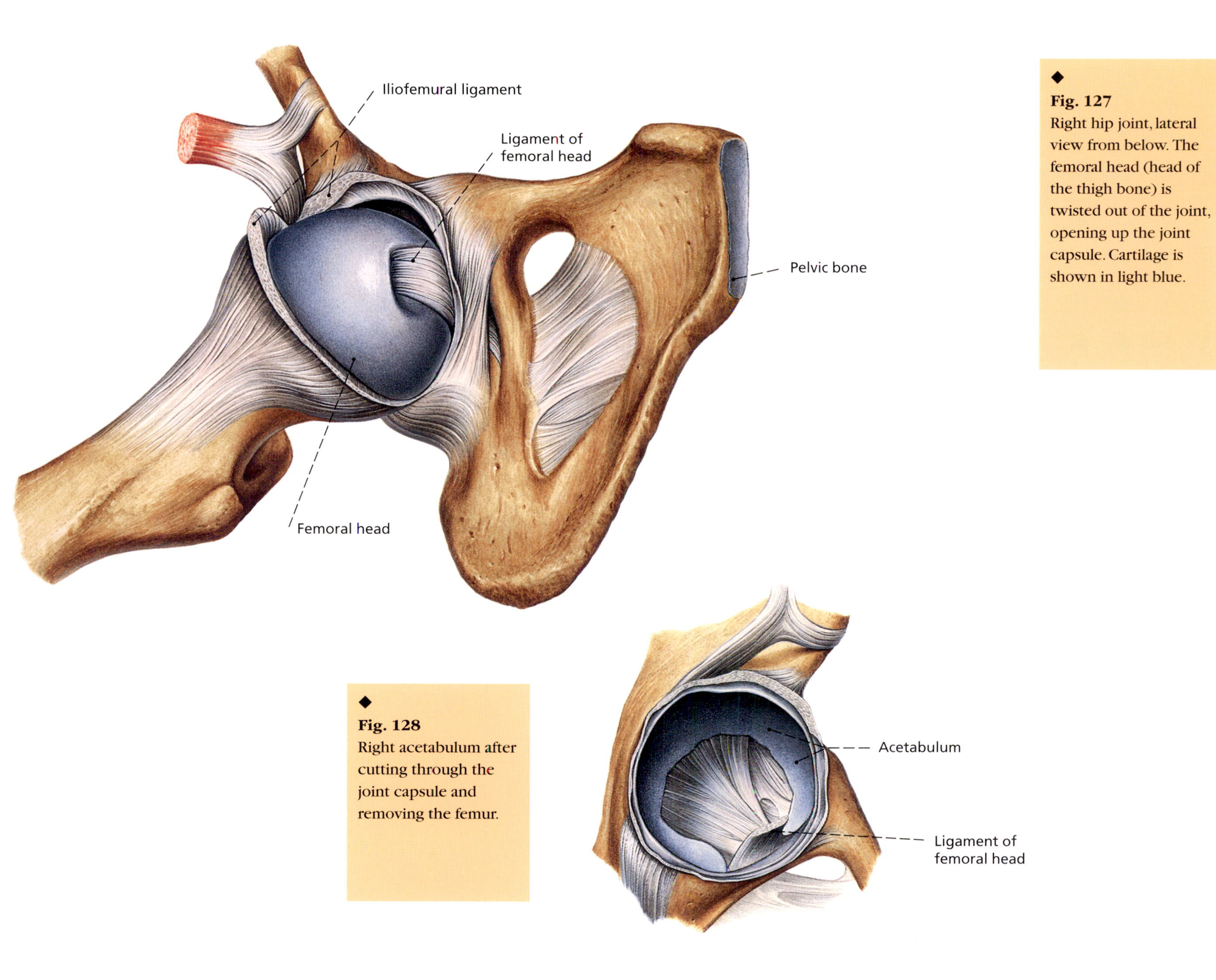

◆ **Fig. 127**
Right hip joint, lateral view from below. The femoral head (head of the thigh bone) is twisted out of the joint, opening up the joint capsule. Cartilage is shown in light blue.

◆ **Fig. 128**
Right acetabulum after cutting through the joint capsule and removing the femur.

Tensor fasciae latae
Acetabulum
Femoral head
Ligament of femoral head
Femoral neck
Obturator externus muscle
Vastus lateralis muscle

◆ **Fig. 129**
Vertical section through the right hip joint with musculature, anterior view.

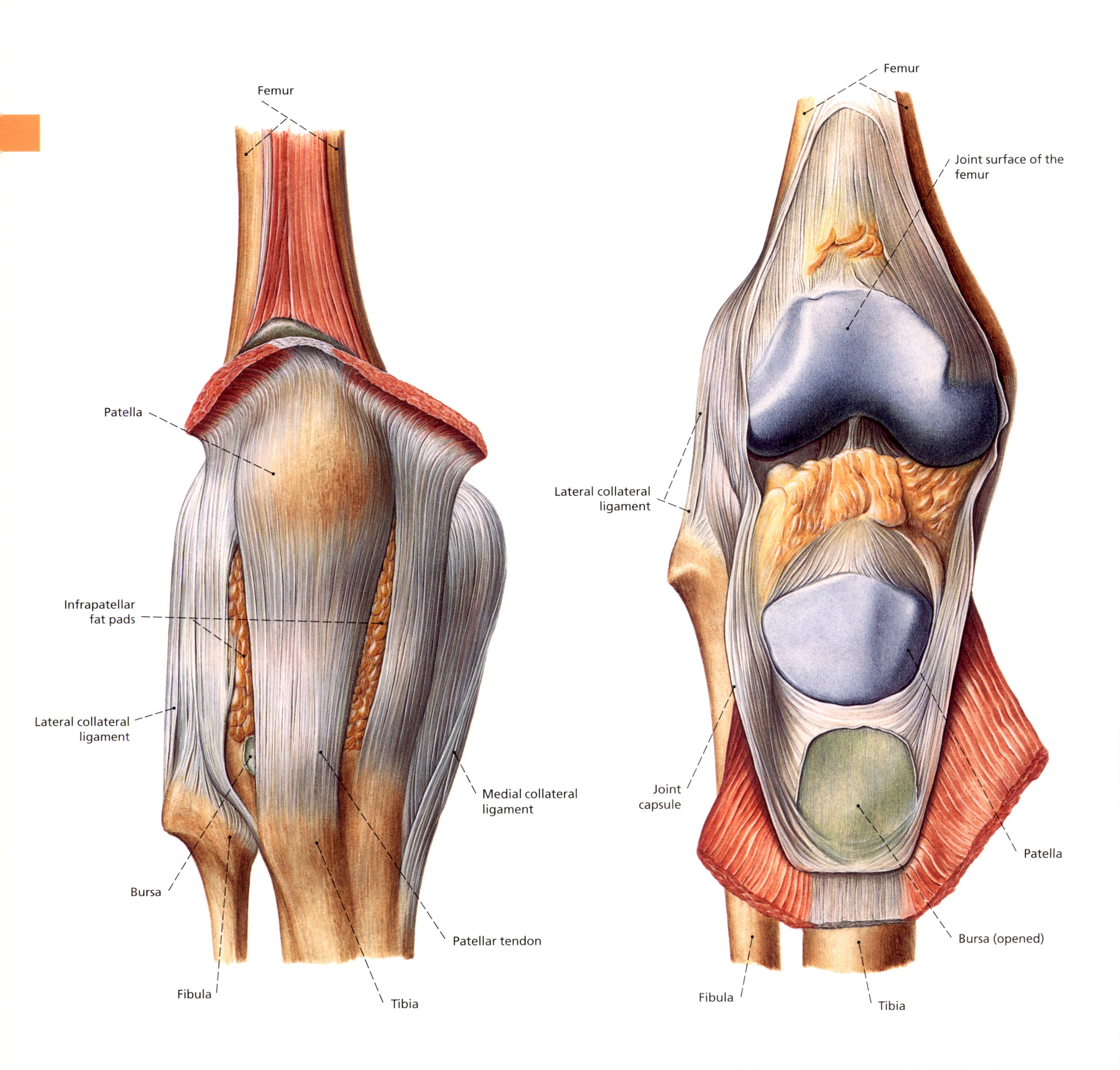

◆
Fig. 130
Right knee joint, anterior view, with closed joint capsule.

◆
Fig. 131
Right knee joint, anterior view. The front section of the capsule has been dissected at the femoral muscles and folded down. Cartilage is shown in light blue.

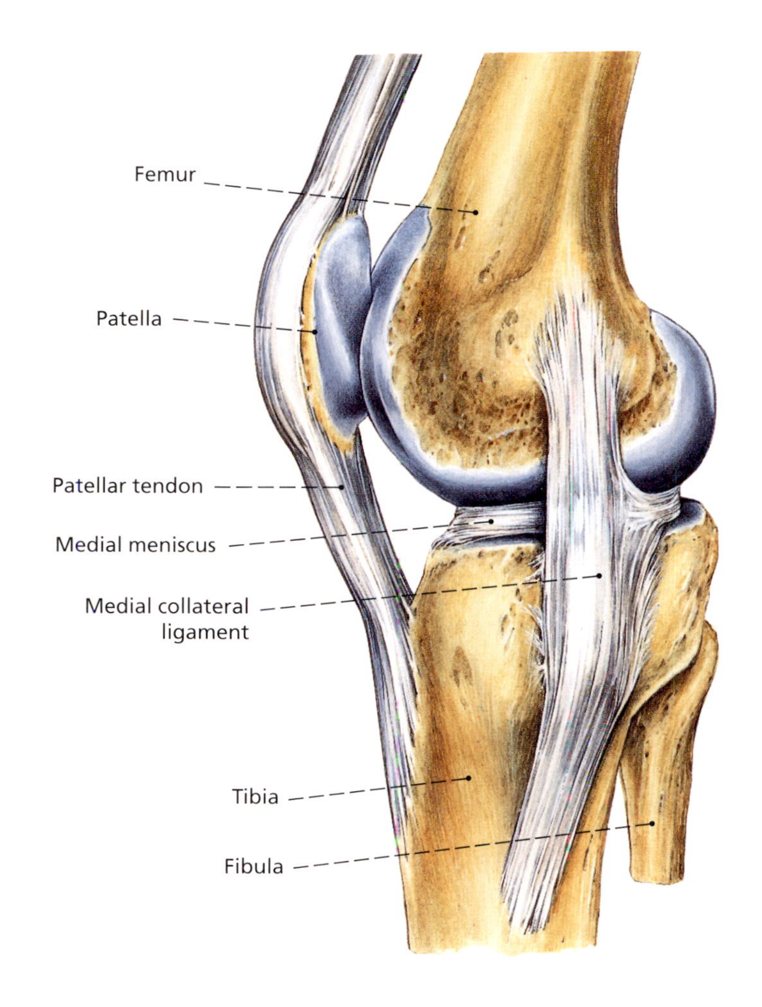

◆
Fig. 132
Right knee joint extended with medial collateral ligament, seen from the middle of the body. The posterior fibers of the medial collateral ligament coalesce with the medial meniscus. Cartilage is shown in light blue, ligaments in gray.

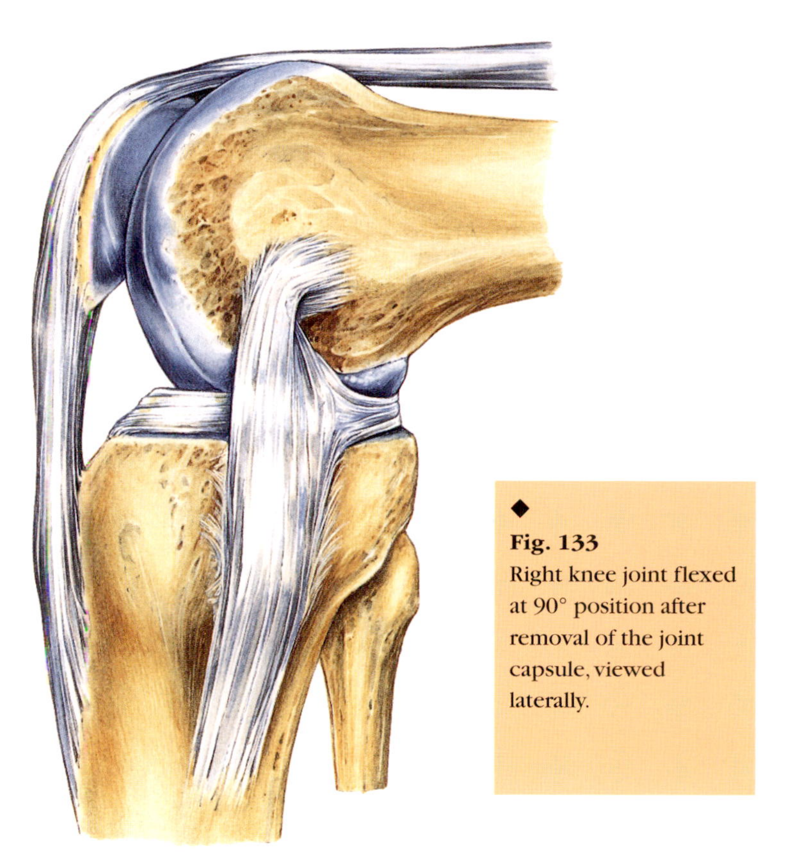

◆
Fig. 133
Right knee joint flexed at 90° position after removal of the joint capsule, viewed laterally.

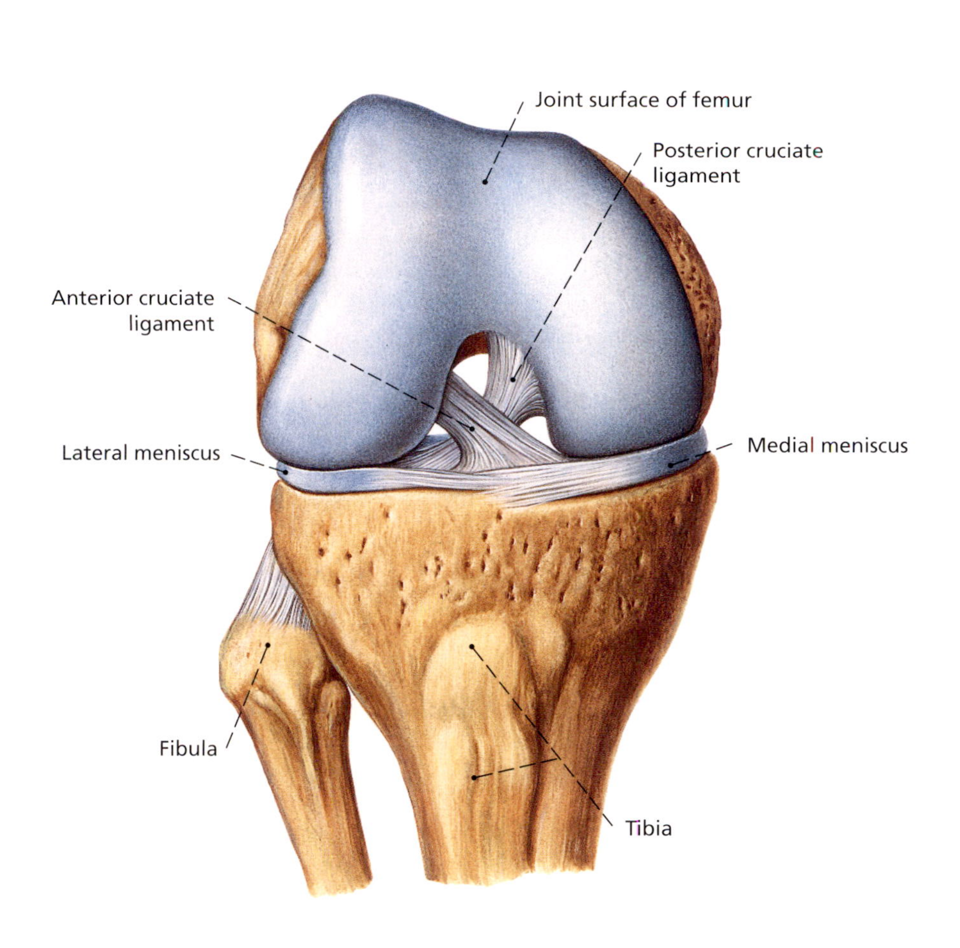

◆
Fig. 134
Right knee joint flexed at 90° position, anterior view, after removal of the joint capsule and collateral ligaments. The anterior and posterior cruciate ligaments are clearly visible.

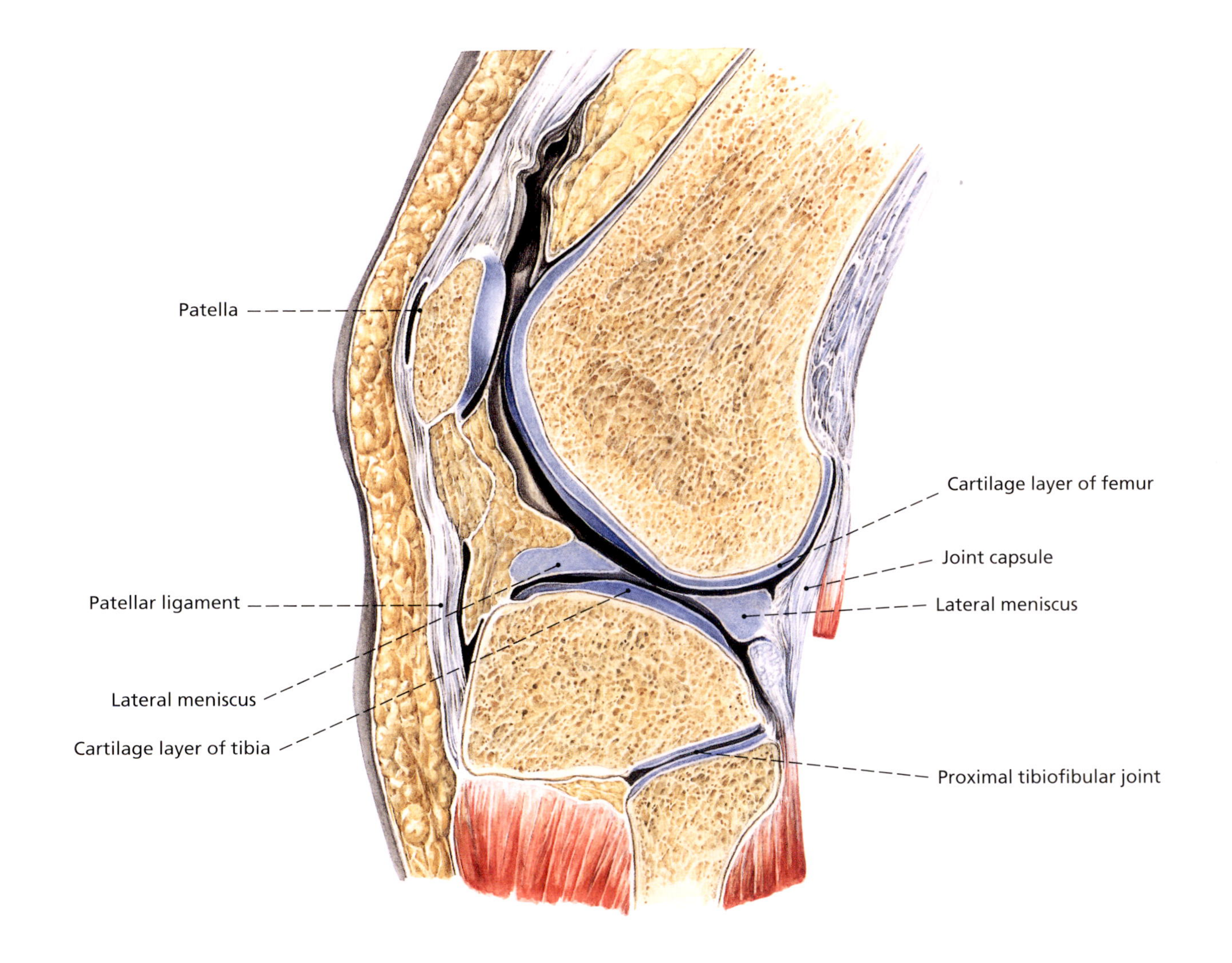

◆
Fig. 135
Right knee joint. Longitudinal section through the lateral part of the joint (viewed from the side). Joint surfaces and lateral meniscus shown in light blue.

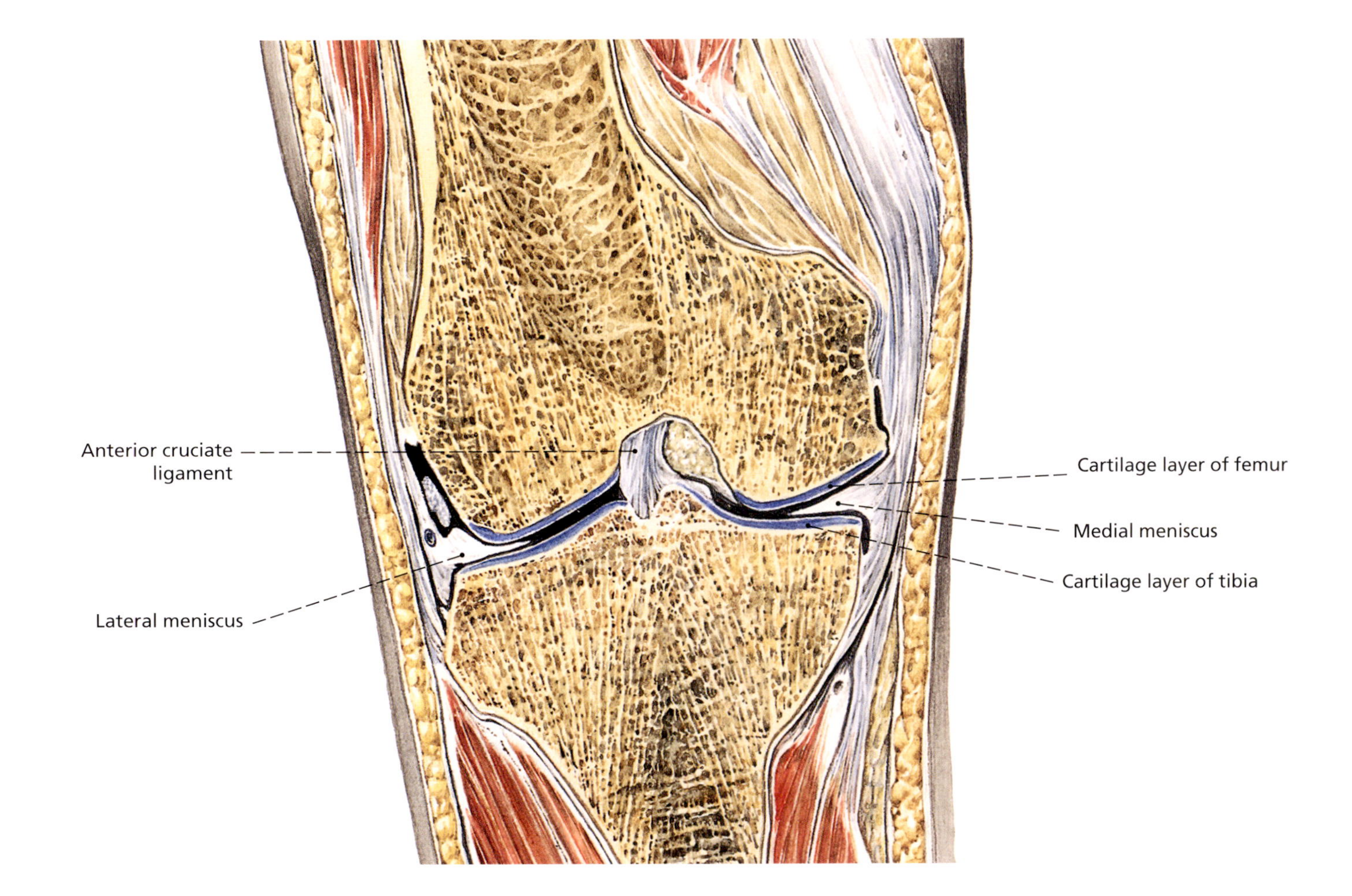

◆
Fig. 136
Longitudinal section through the middle of right knee joint (viewed from front).

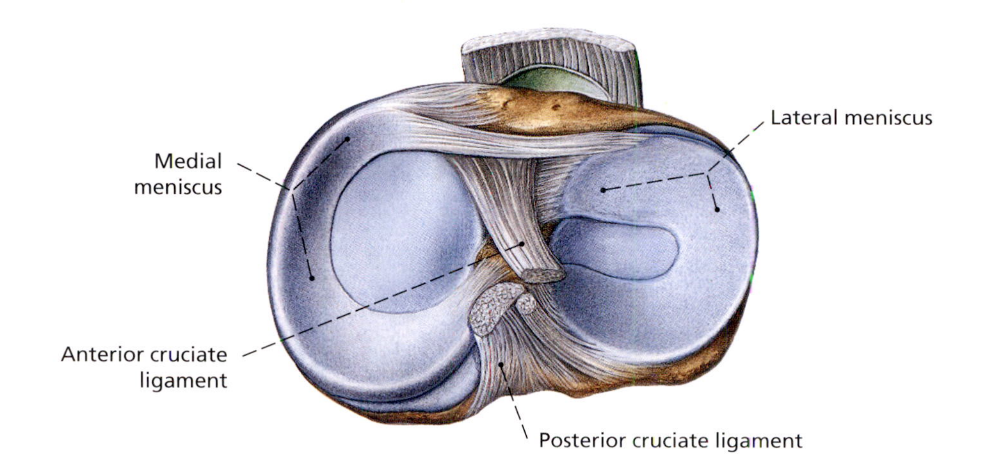

◆
Fig. 137
Right knee joint from above after cutting laterally through the joint capsule, and the cruciate and collateral ligaments. The lateral and medial menisci are visible.

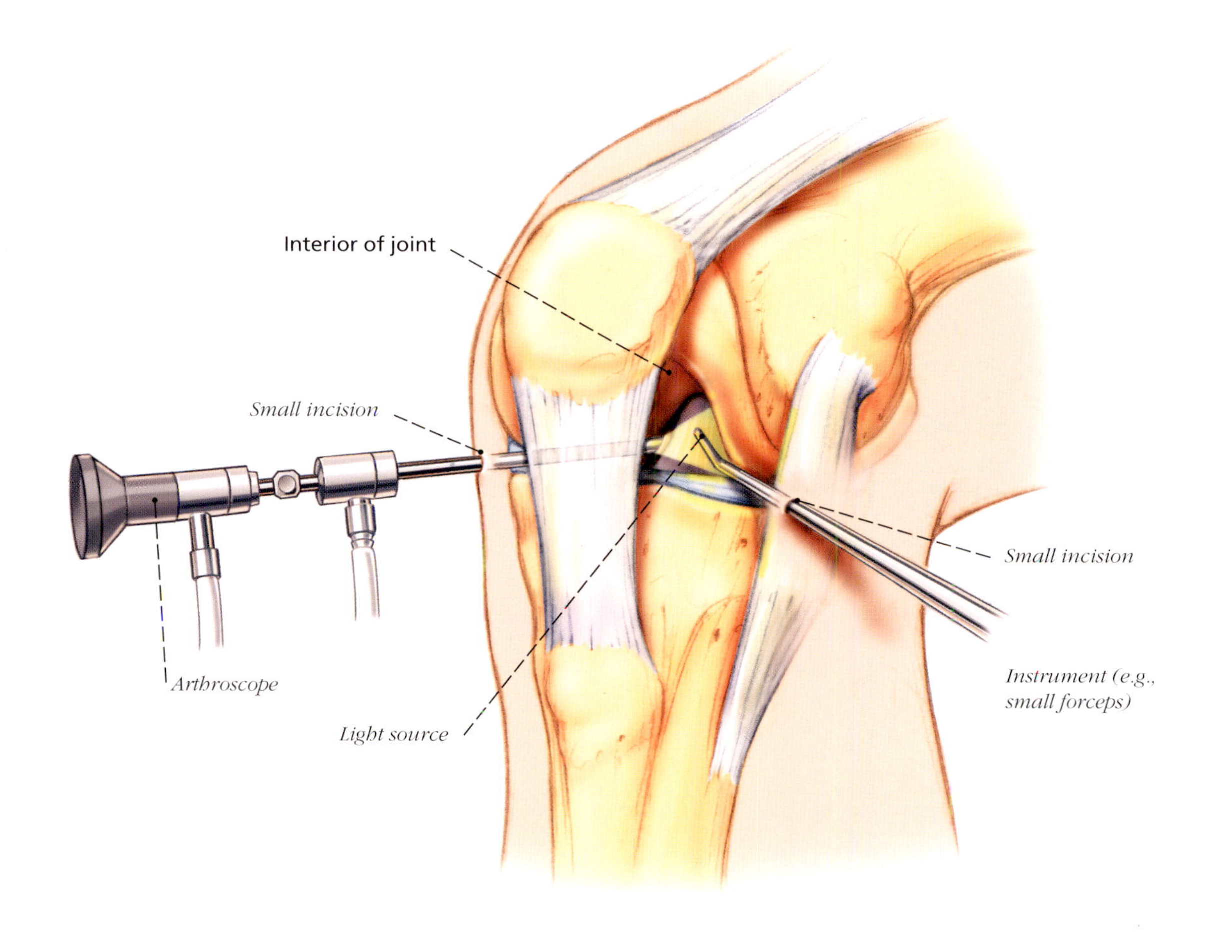

◆
Fig. 138
Endoscopic removal of a damaged meniscus using an arthroscope. In the past, the knee joint was opened up completely, but now this procedure only involves small incisions in the skin.

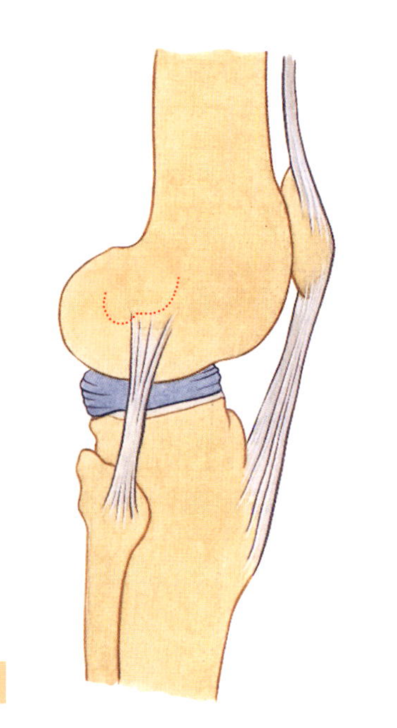

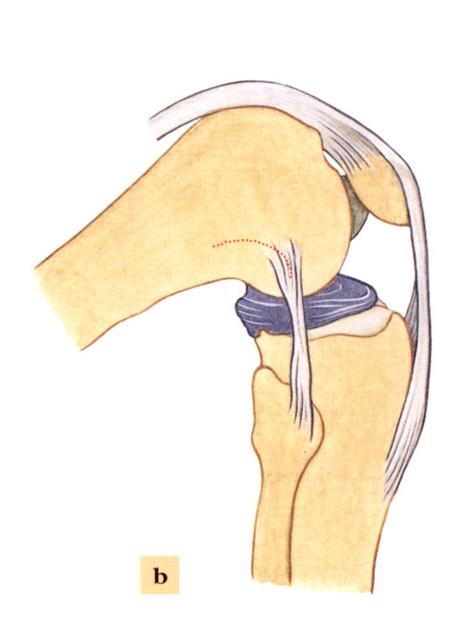

◆
Fig. 139
Displacement of medial and lateral menisci during movement of the right knee joint, lateral view:
(a) in extended position
(b) in flexed position.
The shift in the flexion/stretching axis during the course of movement is marked in red.

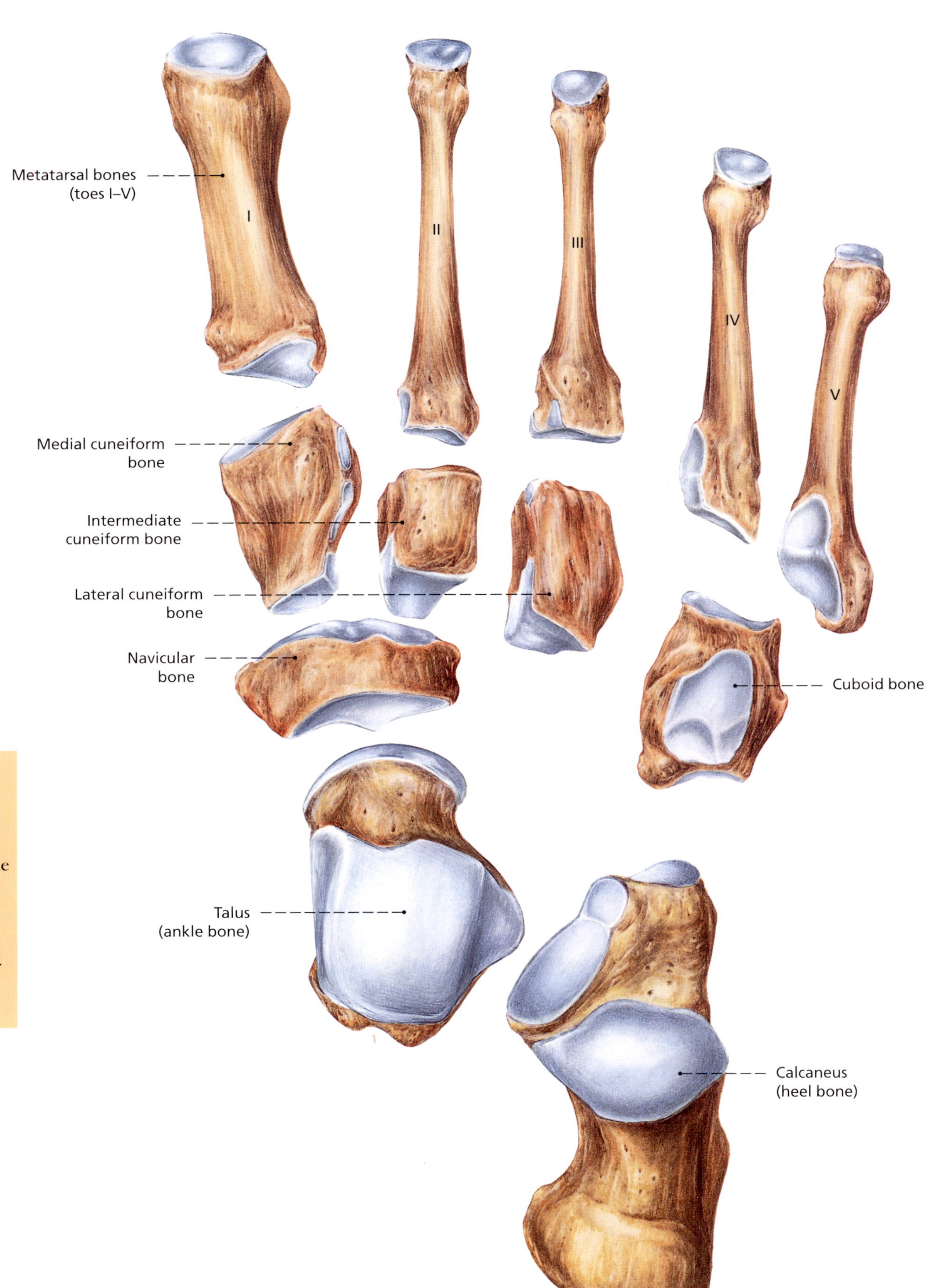

◆
Fig. 140
Tarsal and metatarsal bones (right foot), viewed from above. The gaps between the bones have been increased for the purpose of illustration.

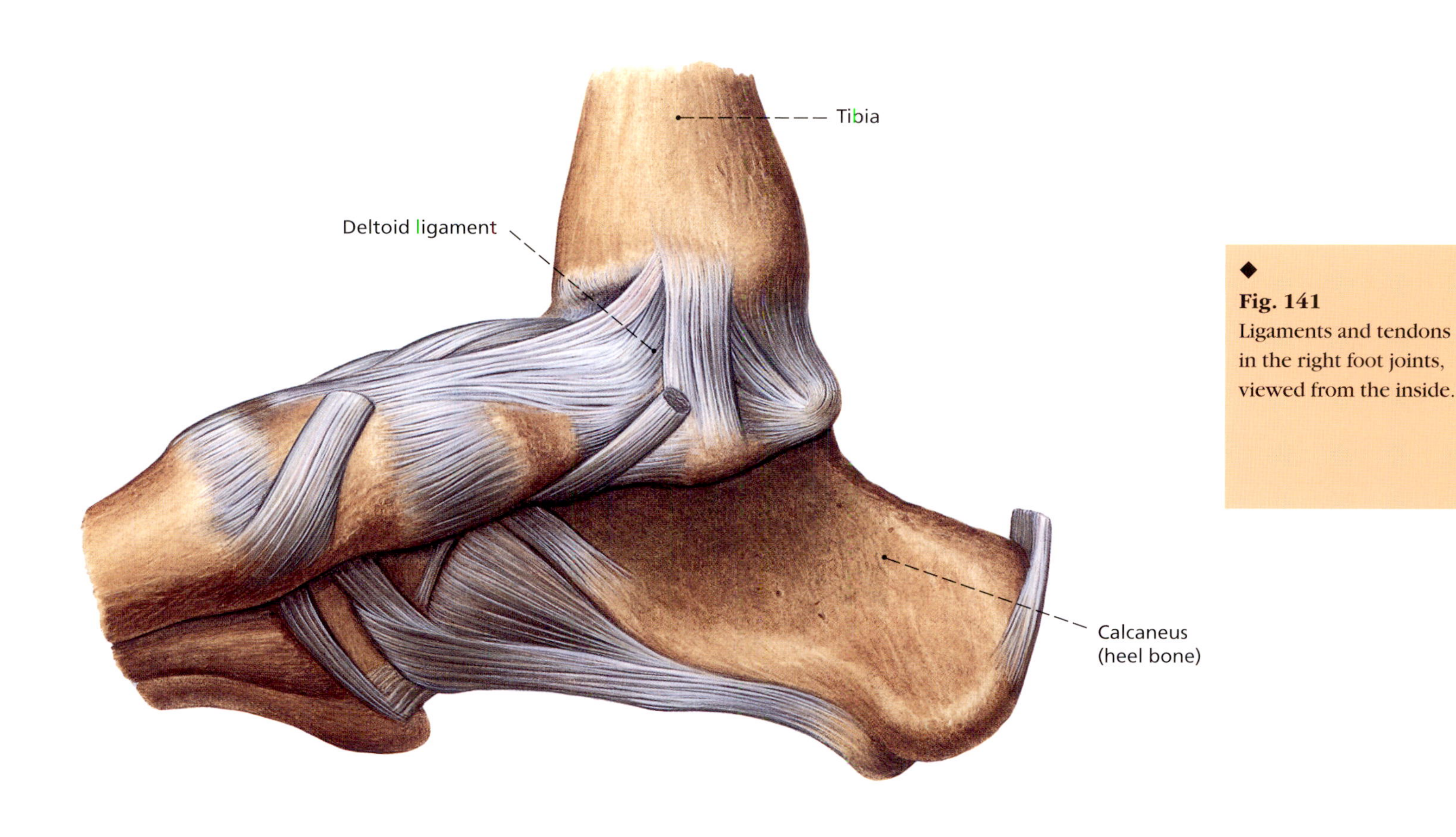

◆
Fig. 141
Ligaments and tendons in the right foot joints, viewed from the inside.

Tibia
Fibula
Anterior talofibular ligament
Calcaneofibular ligament
Calcaneus
(heel bone)

◆
Fig. 142
Ligaments and tendons in right foot joints, seen from outside. Common sports-related injuries include ruptures of the talofibular and calcaneofibular ligaments.

Symptoms and diseases

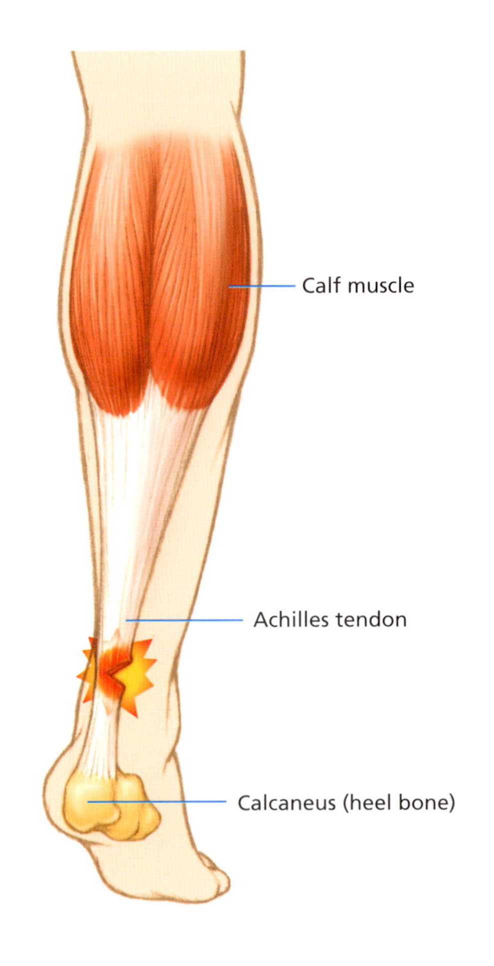

◆ **Ruptured Achilles tendon**
In the area around the heel bone, the Achilles tendon runs just below the skin. It is subjected to major forces, especially during exercise, and is therefore liable to rupture.

Tendons and ligaments

• *Ruptured tendon*

Painful laceration or rupture of a previously damaged tendon, usually caused by a fall, blow, or kick, by sudden excessive stress, or jerking of a muscle. A typical example is an **Achilles tendon** rupture as a result of a sporting injury (e.g., in football, sprinting, or jumping). The rupture point is usually at the point where the tendon meets the calf muscle, just above the ankle. The classic sign of a ruptured Achilles tendon is an inability to stand on your toes. Surgery is required in nearly all cases.

• *Tendonitis*

Inflammation of the tendon sheath which may result from a joint disorder, inflammation of the surrounding soft tissue, excessive or constant unilateral strain on muscles, tendons, and ligaments, e.g., while involved in sport, or an incorrect seating position at the workplace. The inner wall of the tendon sheath does not produce enough synovial fluid and becomes raw, producing slight inflammation and pain.
Physical therapy (ultrasound, electro-therapy, and physiotherapy), as well as anti-inflammatory and analgesic treatments can relieve the symptoms and promote healing.

• *Ruptured ligament*

Complete or partial tear of one or more ligaments in a joint as a result of trauma. Ligaments in the foot and knee joints are often injured through simultaneous stress and rotation, e.g., when playing football and tennis, or skiing. A torn ligament can also be caused by overusing or overstretching a joint, especially in older people.
In the case of a complete rupture, the joint is either immobilized in plaster to allow it to heal or the ligament is reconstructed surgically. Nowadays a torn ligament is frequently treated by arthroscopy: specialist instruments are "fed" into the joint through an endoscope under local anesthetic.

• *Ligament sprain*

The ligaments which stabilize the joint become strained through excessive twisting. A ligament sprain occurs particularly when the bone is twisted out of its usual position during an accident. People with naturally weak or stressed ligaments are especially susceptible to this type of injury. The most common type of ligament sprain is probably a "twisted" ankle. Sprained ligaments tend to swell up and be very painful; they can take a long time to heal. Less serious injuries are treated by applying ice, wrapping the affected joint in elastic bandages, and physiotherapy; in severe cases the joint is kept immobile in a plaster cast.

Joints

• *Arthritis*

Acute or chronic inflammation of the joint. Acute arthritis can develop after a joint injury, if germs penetrate the joint through a wound. Bacteria from an infection elswhere in the body can also get into the joints through the blood system, producing inflammation. Acute arthritis is often an accompanying symptom of a viral infection, such as herpes, mumps, rubella, or hepatitis, or metabolic disorders such as gout, diabetes, or thyroid hyperfunction. It is primarily treated with anti-inflammatory and pain-relieving drugs. Corticosteroid-based preparations are also used, as well as physiotherapy and heat treatments. Rheumatoid arthritis, on the other hand, is a chronic systemic disease which, if untreated, can lead to severely limited mobility in all joints. It is discussed in more detail on page 106.

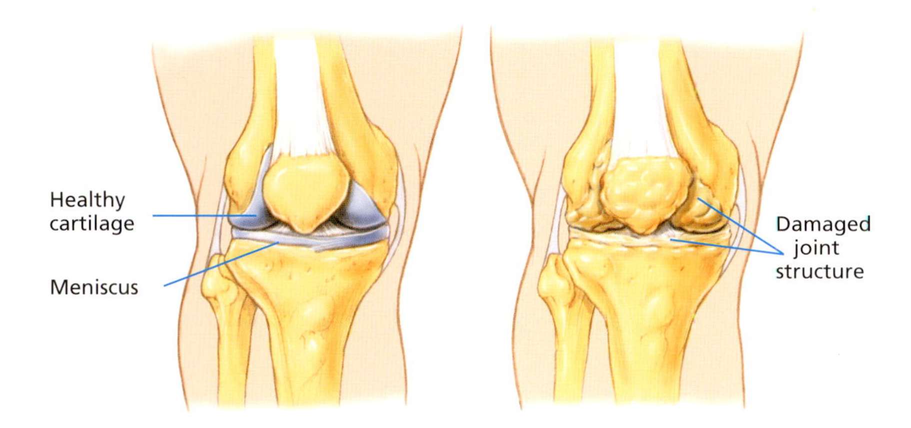

◆
Arthritis
The cartilage layer of a healthy knee joint (left) is gradually destroyed by inflammation of the joint, even resulting in bone deformity (right); mobility can become severely restricted.

• ***Joint effusion***

Fluid discharges into the joint. The fluid produced by the synovial membrane can be clear, sanguineous, or pus-filled. Effusion may build up after an accident, as a result of inflammation, or following an already existing joint disorder.
It produces swelling of the joint and pain when touched or moved. The knee joint is most frequently affected. Treatment involves removing the fluid by tapping, immobilization, cold treatment, and administering drugs. A suppurating joint effusion is caused by an open wound which becomes infected with bacteria, or can be connected with a general infection in the body, the pathogens of which penetrate the joint.

• ***Inflammation of the bursae (bursitis)***

Acute or chronic inflammation of a bursa which is filled with synovial fluid. It is mainly caused by damage to the joint surface, excessive movement, or more rarely by infection. The bursae in the elbow, shoulder, kneecap, and Achilles tendon are particularly susceptible. The inflammation presents with swelling and redness, pain, and restricted movement. Usually the symptoms subside quickly if the joint is kept immobilized.
If the inflammation persists, the fluid build-up can be aspirated and anti-inflammatory medication applied. It is rarely necessary to remove the bursa surgically.

• ***Joint debris (arthroliths)***

Small fragments which move freely within the joint, formed as a result of injuries, when parts of the cartilage or bone break off. Arthroliths are frequently found in knee, elbow and ankle joints and can remain in place without showing any symptoms for a long time; they can, however, lead to painful loss of movement when impacted.

• ***Joint swelling***

A symptom of many joint disorders, often accompanied by redness and pain in the joint. Swelling of the joint occurs in acute inflammation (arthritis), chronic wear and tear (arthrosis), and joint effusion, as well as after accidents, injuries, infections, or as a result of constant overuse.

• ***Ankylosis***

Pronounced and long-term stiffening or fixation of a joint, as a result of disease processes, with fibrous or bony union accross the joint. Stiffening can also be congenital, or develop as the result of an injury and inactivity. Artificial ankylosis (arthrodesis) can be carried out surgically as a treatment for chronic arthrosis, in order to alleviate pain.

Muscles

• ***Muscle inflammation (myositis)***

Painful disease and wasting of a muscle group. It can manifest itself in the form of sore, aching and weak muscles, and signs of paralysis. Myositis is mainly caused by an impaired immune system which attacks, and gradually destroys, the body's own musculature. Sometimes it can also be caused by bacteria, viruses, or fungi. Inflammation of the muscles can become steadily worse over a period of years. For this reason it should be treated with medication at the earliest possible stage.

• ***Myosclerosis***

Hardening of individual muscles or a whole muscle group. Myosclerosis often occurs in the back or shoulder areas. Sometimes knots can be felt in the affected muscles. The hardening is usually caused by excessive or unilateral strain on a muscle group, e.g., through repetitive desk work.
Treatment involves electro-therapy (treatment with low electric currents), massage, and heat treatment. To avoid recurrence of the disease, it may be necessary to make permanent lifestyle changes. Occupational repetitive strain of the muscles can certainly be alleviated by taking short breaks and specific, corrective exercises.

• *Muscle cramp*

Painful, involuntary contractions of a muscle in the form of spasms, caused by insufficient blood supply, exhaustion, or nervous or metabolic disorders. Imbalances in the body's mineral and water levels can often result in muscle cramps, such as a calcium or magnesium deficiency. In an acute attack, you can try to stretch the muscle against the direction of the cramp (if you have a cramp in the calf muscle, pull the toes towards the knee with your hand).

• *Torn muscle*

Partial, or in rare cases, complete severing of a muscle. Usually caused by a bump or kick against the tightened muscle. A torn muscle produces sudden, searing pain. A dent is sometimes visible at the point of injury, which later swells up, often forming a bruise. Smaller tears usually heal when immobilized and followed by physiotherapy, while larger tears may require surgery.

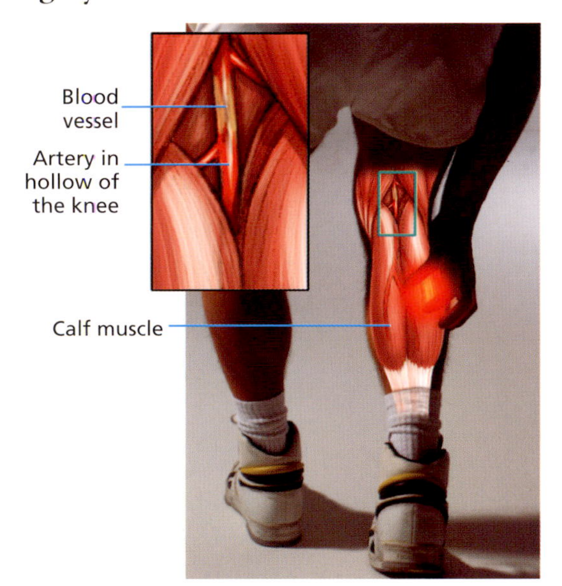

◆
Cramp in the calf
One possible cause of calf cramps is impaired circulation in the blood vessels that supply the muscles.

• *Sore muscles*

Aching muscles a day or two after unaccustomed strain. As the muscles work, metabolic agents formed during activity, such as lactic acid, affect the nerve ends in the muscles causing pain. The best way to avoid this is regular physical activity. A hot bath, gentle massage, and loosening and stretching exercises can bring relief.

• *Bruised muscles*

Injury to the muscle tissue and its supporting blood vessels as a result of kicks or blows. Blood seeps from the damaged blood vessels under the skin, producing a bruise. To promote healing, the affected limbs should be raised and ice applied to the injured spot.

• *Muscle atrophy*

Wasting of the muscle mass. It is normal for muscles to gradually waste away if they are not used regularly. Sudden, morbid atrophy of the muscles is usually hereditary. Muscle strength and mobility decrease disproportionately with age. Muscle atrophy can either come from the muscles themselves (muscular dystrophy) or from the nerves controlling them (neurogenic atrophy). Progressive atrophy in the muscles can often be slowed down with physiotherapy.

• *Muscle strain*

Small lacerations in individual muscle fibers, caused by overstretching and cramping in muscles unused to exercise. The affected muscle is painful to move or touch and feels hard. To help it heal, the part of the body in question must be protected and supported by a special bandage.

• *Myasthenia gravis*

Pathological weakness of the muscles. To begin with, it is mainly the muscles of the eyes, palate, and throat that are affected. It results in difficulties in swallowing, nasal-sounding speech, and drooping eyelids. Sometimes mobility in the arms, legs, or torso muscles can become restricted at a later stage. The disease is caused by a disorder of the immune system, in which the body forms auto-antibodies which block the transmission of nerve signals to the muscles.

Bones

• *Broken bones (fractures)*

A break in a bone caused by pressure, tension, or impact. Symptoms of a fracture include pain, whether immobile or in use, swelling, difficulty in moving, and deformity. Depending on the type of break, this is called either a linear, diagonal, or comminuted fracture. In the case of open fractures, soft body parts (e.g., skin, tissue) are also damaged, and there is an open wound. All fractures must be immobilized immediately. The aim of treatment is to join the bone parts together in such a way that the bone regains complete functionality.
This can be achieved using a plaster cast or by surgically stabilizing the broken ends with metal pins or plates.

• *Bone fragility*

To a certain degree, loss of bone mass is a normal feature of aging. It is exacerbated by calcium deficiency in the diet, insufficient physical activity, and heavy smoking over decades. In women undergoing the menopause bone fragility can also be

◆
A broken bone
If the injured person has to be moved before medical help is available, the break should be put in a makeshift splint. Sticks, umbrellas, boards, poles, and pieces of cloth are all useful items for this purpose.

the result of hormone-induced osteoporosis. In addition, it can be a side effect of certain medicines (e.g., corticosteroids), and of cancer.

• ***Bone inflammation***

Infection of the bone, which is usually suppurative and which can also affect the periosteum and bone marrow. It is caused by pyogenic organisms, which penetrate open wounds. Most commonly, the organisms travel through the bloodstream from inflammed areas elsewhere in the body. The outer skin becomes red and inflamed and the bone is very painful; fever and shivering can develop. It is usually treated with antibiotics. Sometimes the infected bone lesion has to be removed surgically and the resulting cavity flushed out.

• ***Bone softening***

Skeletal change in which the bones become unusually pliable and soft. The most common form of bone softening is rachitis (rickets) which is caused by vitamin D deficiency and insufficient calcium absorption in growing children.

• ***Inflammation of the periosteum***

Inflammation of the membrane covering the bone, either bacterial in origin or caused by external trauma. Pathogens may travel through the bloodstream from elsewhere in the body and get into the tiny vessels of the periosteum. The inflammation can damage the underlying bone by limiting its nourishment. Rest and antibiotics are the common treatments.

• ***Bone cancer***

Rare, malignant tumor produced by the cells in the bone tissue, especially affecting children and adolescents. The tumor can be associated with progressive degeneration of the bone tissue, but also with the growth of new, brittle bone tissue. It quickly results in secondary tumors (metastases), usually in the lungs. Initial treatment involves the administration of substances to inhibit the growth of the cancer cells. Surgery is almost always required.

• ***Inflammation of the bone marrow (osteomyelitis)***

Infection of the bone marrow is normally caused by germs carried in the bloodstream and can therefore spread through the whole body. As a rule, whole bones are affected, along with their periosteum. As bone tissue destruction is common, rapid surgical intervention is essential. Antibiotics may be applied directly to the focus of inflammation during surgery.

• ***Brittle bone disease***

Hereditary disorder of bone formation with an increased level of bone fragility. The bone structure and periosteum are so unstable that even a small amount of pressure can cause the bone to bend or break, or even lead to serious deformities. Although the symptoms of this condition are treatable, there is no therapy for the underlying disease.

• ***Greenstick fracture***

Fracture in children in which the bone is only cracked. The periosteum is undamaged. It results in a typical radiological picture, with pain when the area is pressed or bent. Long bones such as the thigh or upper arm are commonly affected. As children grow rapidly, the fracture usually heals quickly.

• ***Scheuermann's disease***

Disease of the bone and cartilage in the thoracic and lumbar spine. The end plates of individual vertebral bodies become inflamed through increased stress, resulting in lack of circulation to the bone and ultimately in growth disturbances. The disease begins mainly at around 14 years of age, and is associated with pain and sometimes loss of movement. As the disease progresses, it can lead to softening of the bone and spinal deformity.
After puberty the process usually comes to an end. The bones normally regenerate within three years and harden again. Treatment mainly consists of physiotherapy. A support corset must be worn to correct the rounded back. In serious cases an operation is needed to stiffen the spine.

Arthritis (rheumatism)

Rheumatism is a non-specific term applied to a wide range of conditions that cause pain in the muscles, joints, and fibrous tissues, including minor aches as well as serious diseases. However, the medical profession uses specific terms to describe rheumatological conditions, such as rheumatoid arthritis. Chronic arthritis is the general name given to an inflammation of the joints, which though not life-threatening, is nonetheless insidious.

More women than men suffer from arthritis across the globe. The illness appears mainly between the ages of 30 and 50. However, its basic symptoms (pain and loss of mobility) occur in a whole range of disorders. In the first instance it has to be established whether the condition is due to age-related wear and tear of the joints (i.e., osteoarthritis) or whether there is an inflammatory process at work. Only the latter condition is covered by the term rheumatoid arthritis, which may also be referred to by physicians as chronic polyarthritis.

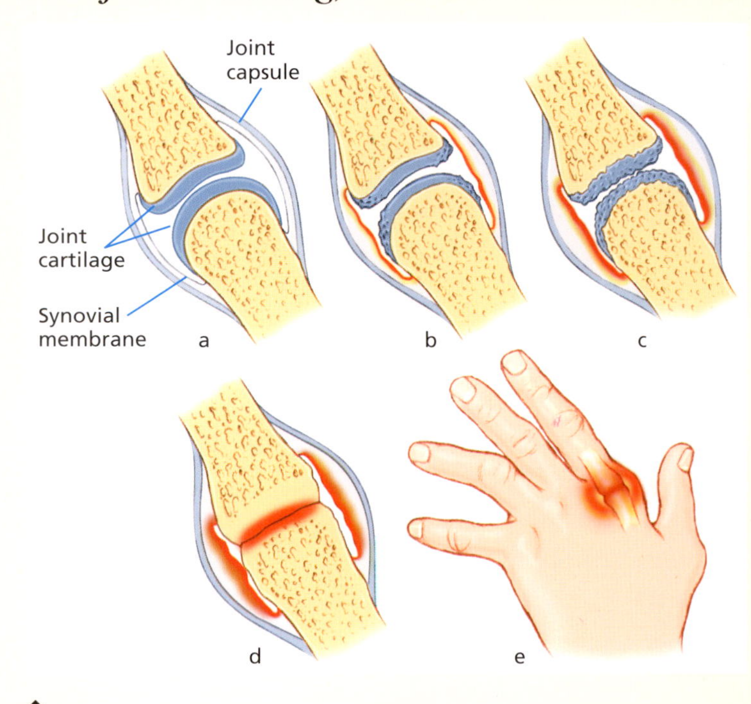

◆
Changes in a finger joint caused by rheumatoid arthritis: in a healthy joint (a) the layer of cartilage prevents the bones from rubbing against each other. In the finger of a person affected by rheumatoid arthritis, the synovial membrane becomes inflamed, and the cartilage is gradually damaged (b and c), until the bones rub directly against each other (d) and joint deformity occurs (e).

Symptoms of the disease and their course

Rheumatoid arthritis begins insidiously, usually with pains in the finger and toe joints. The joints of both sides of the body are affected symmetrically. In the morning, sufferers may have difficulty in tying their shoelaces because their fingers are so stiff.

As the disease progresses, usually spasmodically, the joints swell up, fingers become weak, the hand can no longer make a fist, and sometimes the skin around the affected areas becomes warm and red. If left untreated, joints can become deformed, with misalignments and stiffening sometimes causing serious dysfunction: fingers become crooked, with the palm turned outwards, and toes become claw-shaped; the last of these symptoms results in the affected foot being unable to straighten, making walking very difficult. At the locus of the diseased joints, the skin forms soft, bulging lumps known as rheumatic nodules. Muscles, tendons, and ligaments are frequently also affected. In the progressive course of the disease, the joint cartilage and eventually the underlying bone is damaged. This inflammatory process causes extreme pain.

Causes

As clusters within families have been detected, it seems that susceptibility to rheumatoid arthritis may be hereditary. Specialists include rheumatoid arthritis within the category of autoimmune diseases. The immune system, vital in the defense against disease agents, can no longer distinguish between the body's own and exogenous substances in these diseases, so it directs itself against the body's own tissue. We still do not know which substances trigger these reactions.

Diagnostic methods

A blood test may reveal the rheumatoid factor (specific blood markers), whose presence confirms the autoimmune disorder, as well as other factors indicative of inflammatory changes. In addition to the presenting symptoms, X-rays and magnetic resonance imaging scans may show changes in the joint structure. As rheumatoid arthritis does not always present with an acute, standard range of symptoms, the rheumatologist has to exclude other possible disorders through a series of tests. Arthrography, a procedure in which a radioactive substance is injected to highlight the affected sections of the joint, can reveal an inflammatory process at a very early stage. Ultrasound can be used to show inflammatory changes only in larger joints such as the knee or shoulder. In arthroscopy, an endoscope is inserted into the joint through a small incision, allowing every part of the joint, including cartilage and ligaments, to be examined internally.

Treatment

Until the triggers for this autoimmune disease are known, treatment is limited to alleviation of the symptoms, which can be achieved in different ways:

- Pain relief and control of joint inflammation: joint inflammation can now be controlled by the use of a number of disease-modifying anti-rheumatic drugs (DMARDs), such as methotrexate, sulfasalazine, oral corticosteroids, and other anti-inflammatory agents with analgesic properties. So-called biological drugs, which decrease the levels of inflammatory molecules such as cytokines, may also be effective. Although these medications often have side-effects, such as fatigue, water retention or stomach problems, they can significantly delay the destruction of the joint and threat of invalidity.
- Certain medicines (e.g., corticosteroids) can also be injected into the affected joint. They alleviate symptoms quickly and effect substantial improvement.
- Physical therapy at an early stage can help to prevent or delay joint stiffening and deformities. This includes first-line measures such as hot and cold applications, electro- and ultrasound therapies, massages, physiotherapy, as well as systematic and sustained movement exercises. Therapeutic use of mud baths, carbon dioxide or hydrogen sulphide-based substances (balneotherapy) can be a useful complement.
- Specially designed shoes or splints can facilitate mobility.
- The effectiveness of dietary measures is still under debate. A diet containing sufficient calcium and iron-rich food, vitamin E, and fish oil, is often recommended.
- In the case of seriously limited functionality and movement, the inflamed synovium can be washed out, or an artificial joint fitted.

The best hope of success lies in long term therapy involving careful control of the progress of the disease.

During advanced stages of the condition, patients are sometimes unable to carry out everyday tasks themselves due to the pain and joint stiffening. When this happens it may be of benefit to undertake rehabilitation and occupational retraining.

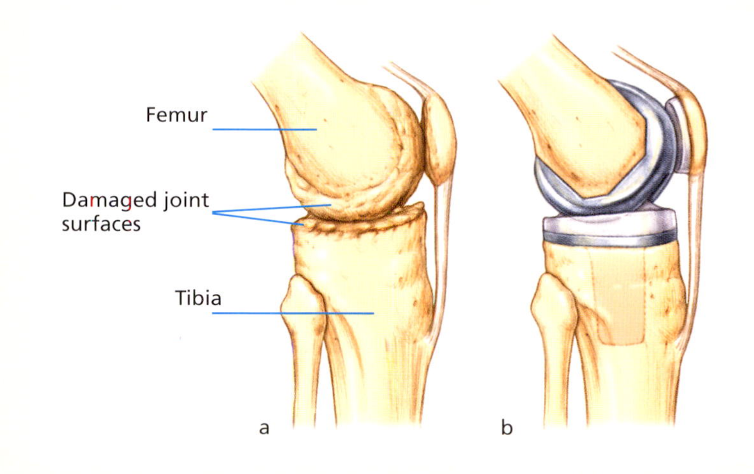

◆
When joints are seriously damaged (a), a prosthesis, in this case for the knee joint, may be required in order to regain mobility (b).

◆
The most important way of maintaining movement in the joints is by doing appropriate exercises, either under expert supervision or on your own.

Osteoarthritis

Osteoarthritis is also a form of rheumatic disease. Unlike inflammatory arthritis, however, osteoarthritis is a degenerative process. By the age of 60 everyone has some wear and tear in their joints, as osteoarthritis is a normal feature of aging. Many people may show the first signs of joint damage at only 30 years of age, without actually noticing them.

All joints consist of a joint head and a joint cavity, both of which are covered by transparent, white cartilage. When we are young, this cartilage is as smooth as glass, so friction is minimal. The joint is lubricated by a fluid called synovial fluid, which is formed in the synovial membrane. The cartilage itself does not contain any blood vessels; this gives it the advantage of being able to withstand considerable stress. The disadvantage, however, is that it has to extract all the nutrients it needs from the synovial fluid, making it less capable of regeneration. As we get older, the cartilage becomes rough, more tissue is broken down than is formed, the cartilage layer becomes thinner or is completely destroyed in places, and its elastic, cushioning properties are severely restricted. As it does not contain any calcium carbonate, cartilage itself is not visible on an X-ray, but it is possible to see clearly the decreased gap between the bone head and the cavity, which represents the thinned-down layer of cartilage in the intra-articular space. This narrowing of intra-articular space is the classic sign of osteoarthritis. The most common attendant symptoms are morning stiffness, noises when moving, and pain at the start of every movement.

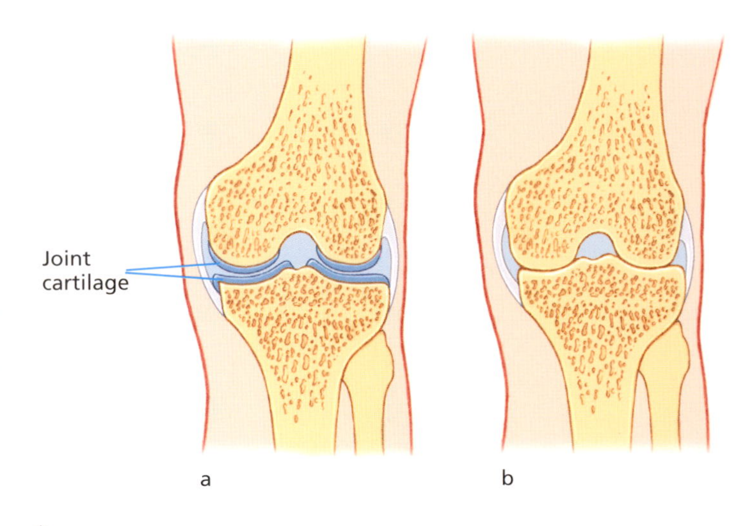

◆ In a healthy knee (a) the cartilage layer is evenly thick. In a diseased knee (b) the cartilage has degenerated completely and the rub directly against each other.

Causes

One of the reasons why humans often develop osteoarthritis over the years is the fact that they walk upright. The main burden of the body's weight is thus borne by the hip joints and knees, although the spine also bears a great deal of stress as a result of the body's constant balancing act.

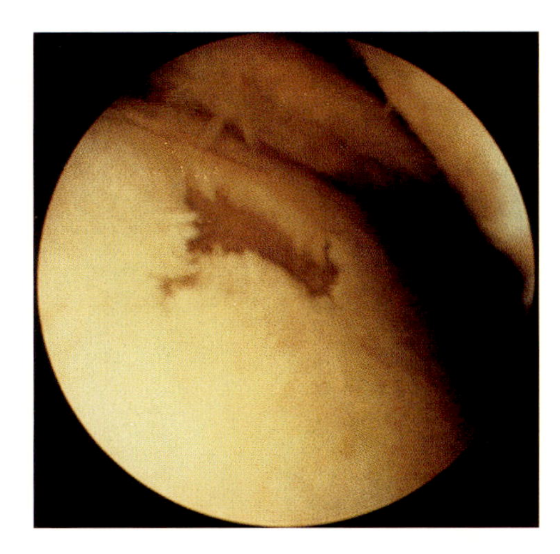

◆ Cartilage damage (centre of picture) can be clearly identified by means of a knee endoscopy.

Walking upright is only one of the reasons for cartilage wear and tear, however. Aggravating factors include:

- being overweight,
- incorrect or excessive weight-bearing during serious sports or certain occupations,
- malposition of joints causing unilateral stress.

As a consequence of all these factors, heavy mechanical strain and wear is put on a joint. Damage can also occur a considerable time after an injury to cartilage tissue caused by a sprain, bruise, or a broken bone. Basically any joint can be affected, though knees, hips, ankles, and spine are particularly susceptible, as well as the fingers, especially in the dominant hand.

Dormant or active osteoarthritis?

Two people with exactly the same X-ray results can experience quite different levels of pain. A patient with osteoarthritis can go for years or even decades without any discomfort, before the classic symptoms appear, often triggered by incidents such as climatic changes, physical over-exertion, or a flu-like bug. This stage of osteoarthritis is accompanied by painful inflammation of the joint and often swelling with sharp pains, which occur even when resting. In the case of activated osteoarthritis of the shoulder joint, for example, the surrounding soft areas (muscles and tendons) are also affected. If this stiffness in the shoulder causes the arm to be held close to the body for too a long period, this can lead to serious loss of mobility and stiffness that needs months of treatment.

Treatment for osteoarthritis

There is no cure for the causes of osteoarthritis. The aim of treatment is to contain the inflammation, alleviate pain, and improve functionality.

- ***Fighting inflammation***

An elastic bandage or splint may be used to rest the joint until the pain subsides. This brings it back from an inflamed to a resting stage. Cold or alcohol-based compresses, may help in the treatment of inflammation, as may natural mud or paste packs in less serious cases. Inflammation can lead to tissue fluid penetrating the intra-articular space, resulting in an effusion that can be removed (aspirated) using a syringe. This is usually done under local anesthetic. The pain and pressure then ease off. It is essential to move the joint again carefully and with increasing intensity once the inflammation has subsided.

- ***Relieving pain***

One-off injections of corticosteroids into an affected joint give rapid relief, and avoid the unpleasant side effects associated with the long-term use of corticosteroids (e.g., water retention in the tissue). Anti-inflammatories and pain-killers may be prescribed. Acupressure, acupuncture, and neural therapy may also be used to treat the pain.

- ***Surgical intervention***

If there is no improvement, surgical intervention may be unavoidable. An endoscopic procedure (arthroscopy) allows the cartilage surface to be smoothed and bits of cartilage and bone which have rubbed off to be removed. If the damage is more extensive, a partial or complete artificial joint, made from metal, plastic or ceramic, may be fitted. Very good results have been achieved in the last few decades, especially with hip joints.

◆
A healthy diet with plenty of fruit and fibre-rich food is particularly important in osteoarthritis.

More movement and less stress

When the condition is stable, and also in the prevention and early treatment stages, it is good to keep the joints moving, but without putting too great a load on them, e.g., swimming or gentle cycling. Therapeutic exercises under the supervision of a physiotherapist can be usefully supplemented by gentle exercise at home.
Some examples: bending or stretching the leg at the knee. This can be done with or without an elastic bandage held in both hands and placed around the sole of the foot like reins. This provides slight counter-resistance when stretching the leg. In osteoarthritis of the hand joints, practicing gripping the hand around a soft ball is very effective. All movements are performed better in warm water as it makes them less painful. Exercises involving weight being pushed away, even knee bends, are not advisable. To prevent initial pain it is a good idea to move the joint first of all without any stress, e.g. by swinging the knee a few times before standing up. Warm baths, infrared light therapy, and mud packs promote the circulation, ease cramps and may alleviate pain. Overweight patients are advised to lose weight. Simple devices like a footstool under the desk rest the knees when sitting. Shopping bags on wheels or walking sticks are also useful for taking the load off joints.
As there is no cure for osteoarthritis, it is absolutely essential for the affected patient to find a balance between necessary strain and the maximum load-bearing capacity.

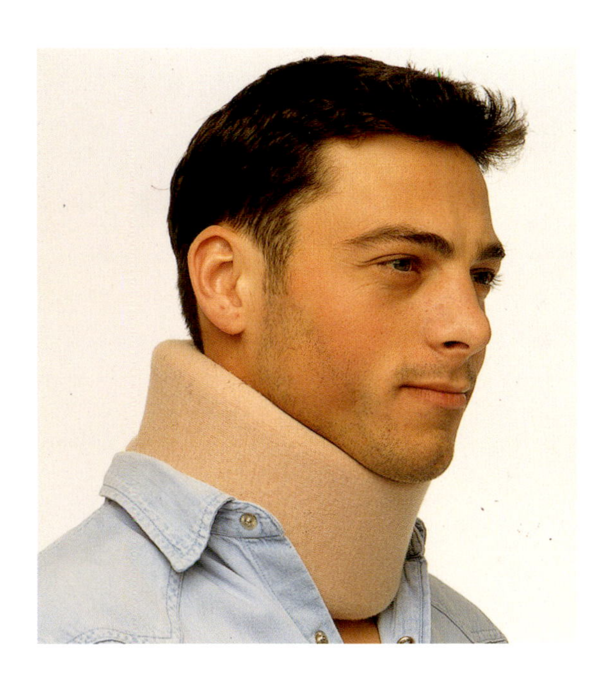

◆
Neck collar
The cervical spine must be supported and pressure taken off it after excessive stress, such as whiplash.

Head and neck

• *Whiplash*

Injury to the cervical spine, usually caused by a rear-end collision. The heads of those affected have usually not been supported at the neck. The torso and head are suddenly propelled forward and then back again. Depending on how bad the impact is, ligaments can be strained, wrenched or torn. The impact can also cause sprains, dislocations and fractures in the intervertebral disks of the neck, or even to craniocerebral injuries. Usually stiffness in the neck and pain in the back of the head only develops hours or even days after the accident. The head may be immobilized by a neck collar; drugs are effective in relieving the pain as well as relaxing the muscles, and physiotherapy aids the healing process. It normally takes a few weeks until the neck can be moved again without pain. This is why it is particularly important that the height of the headrest of the car seat is adjusted to the appropriate level: its top edge should be about eye level.

• *Neck pain*

Pain caused by tension in the neck muscles which can be caused by awkward movements, or by cold, wet or draughty conditions. Very often there is an organic reason for the neck pain, such as a slipped disk in the cervical spine region, an acute condition of the vertebral joints, muscular rheumatism, or a sprain in the vertebral bodies. Treatments designed to stimulate circulation relieve the pain, such as heat application, massage, and fango packs. Analgesics are also required as pain often leads to malpositions, which in turn exacerbates pain. In the case of a strain, special manipulation by an expert hand can also help (see Chiropractice).

• *Wry neck (torticollis)*

Congenital or acquired malposition of the head involving a painful, stiff neck. It is caused by malformations of the cervical spine as a result of damage to the neck muscles, serious scarring or skin shrinkage, muscle tension, or paralysis. A wry neck can also be an adverse side-effect of treatments involving antipsychotic drugs (e.g., for schizophrenia). Congenital wry neck is caused by a shortened, wasted, or torn sternocleidomastoid muscle. A wry neck can be corrected by means of gentle stretching of the neck muscles or surgery. In the case of muscle tension, wearing a neck collar, ultrasound and heat treatment, and physiotherapy are all helpful.

• *Fractured skull*

Bone fracture in the region of the cranial bone. It takes the form of either closed linear fractures (caused by injury from a blow, fall, or impact) in which the cranial roof is directly injured, or comminuted fractures where the whole skull is deformed and pieces of bone can be displaced.
A basilar skull fracture occurs as a result of extreme external force being applied, especially in road accidents; classic signs are raccoon eyes, bleeding from the nose or ears, concussion, and frequently paralysis of the cerebral nerves, especially the auditory and facial nerve. There are different forms of treatment depending on the particular nature of the injury and the symptoms. After simple fractures without signs of concussion, the patient can often leave hospital after a few days. If the cerebral membrane is ruptured and bone penetrates the brain matter, surgery and antibiotics are essential.

◆
Wry neck
If untreated, holding the head twisted to one side can lead to permanent posture damage.

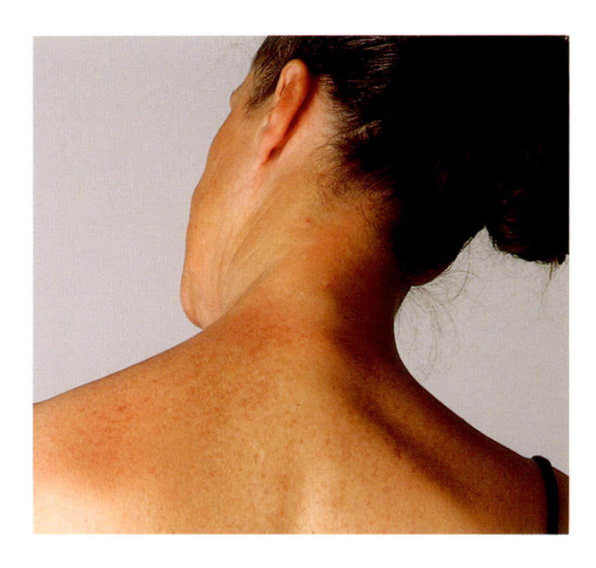

• *Craniocerebral injury*

Injury to the scalp, bony skull and brain as a result of external trauma. In the case of less serious injuries, the deeper regions of the brain remain mostly intact. Open injuries to the skull can cause cerebral hemorrhages and a rupture of the cerebral membrane, with the result that cerebrospinal fluid can sometimes actually leak from the ears and mouth. Brain injuries are divided into four degrees of severity:

- Level I – no loss of consciousness from mild trauma (e.g., concussion with symptoms of dizziness, confusion, visual distortions).
- Level II – loss of consciousness for up to 30 minutes.
- Level III – loss of consciousness for up to 2 hours.
- Level IV – loss of consciousness for over 4 hours, up to potentially lethal coma.

Treatment is based on the severity of the injury, ranging from simple dressing of the head wound to surgical removal of a hematoma in the brain by a neurosurgeon.

The spine

• *Incorrect posture*

Malposition of the spinal column (e.g., curvature of the spine), a condition that is usually acquired rather than congenital. Acquired poor posture is sometimes the result of an injury to the spinal column or having legs of different lengths. In most instances, however, it can be attributed to sitting in the wrong position and weak stomach and back muscles. Bad posture can be prevented by varied sporting activity, regular exercise, and sitting upright. Special physiotherapy exercises can correct bad posture, but it is important to make sure that posture is correct, even in childhood.

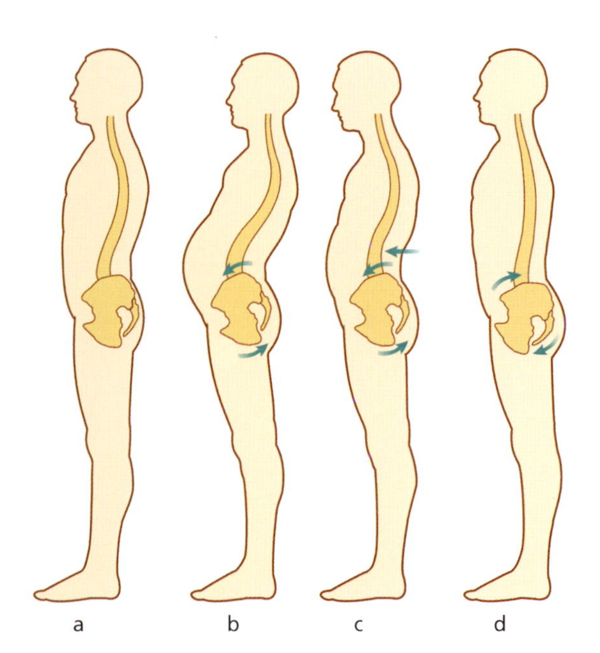

◆

Bad posture
Over a long period, poor posture can lead to deformities in the spine: compared to normal posture (a) the lumbar spine is pushed forward when the back is concave (b); when the back is rounded (c) the thoracic spine curves backwards, while in the flattened back (d), the classic S-shape of the spine is absent.

• *Lordosis*

The spine curves forward in the cervical and lumbar regions. When only slightly pronounced, this curvature is normal.

• *Roundback*

Back curves forward. It is often the result of bad posture at work, but also manifests itself in Scheuermann's disease, a disorder of the spine common in adolescents. In terms of treatment, physiotherapy and a suitable sporting activity like swimming are recommended.

• *Hunchback (kyphosis)*

Spine curves backwards. A hunchback often develops in later years (old age stoop) when the dorsal vertebrae are affected by osteoporosis. It also occurs in adolescents and children, however, mainly as a result of vitamin D deficiency (rickets), when bones are inadequately calcified. In addition, a hunchback can develop from diseases of the skeleton or the surrounding support muscles.

• *Curvature of the spine (scoliosis)*

The spine curves to the side and individual vertebral bodies become twisted. Slight curvatures of the spine are relatively common. They can even be detected prior to puberty in many cases, and girls are three to five times more frequently affected than are boys. The reasons for scoliosis developing are mostly unknown. Symptoms can be alleviated with the aid of physiotherapy. If the scoliosis progresses, a support corset may be recommended, and in chronic cases surgical intervention is the only remedy.

• *Low back pain*

Problems in the lower section of the spine are mainly caused by incorrect and excessive strain on the spine and intervertebral disks through carrying and sitting the wrong way, imbalanced posture, high heels, and osteoporosis; but they can also result from psychological tension leading to muscular cramp. Back pain may also be an attendant symptom of menstruation or during pregnancy. Low back pain is treated with exercises to correct posture, and targeted relaxation, if necessary through the administration of pain- and tension-relieving drugs.

• ***Sciatica***

Pain along the sciatic nerve which can be longstanding or occur in separate attacks. The sciatic nerve is the longest nerve in the body, running from the buttocks to the tip of the toe. The pain usually begins in the hip and can extend into the leg. Pain in the sciatic nerve is caused by irritation of or pressure on the nerve and its roots. The most common cause of sciatica is a herniated disk. Other causes include tumors in the spinal cord or pelvis, disorders of the spine (particularly its lumbar section), as well as fractures and strains. In many cases it is difficult to determine the cause of sciatica. Treatment usually takes the form of analgesics and physiotherapy to strengthen the muscles in the stomach and back.

• ***Lumbago***

A sharp pain in the lumbar region of the back which usually happens suddenly, often after a twisting or bending movement. Possible causes include muscle tension, "locking up" of a vertebral joint, or a "slipped" disk. Acute care involves immobilization, analgesics, muscle relaxants, and hot or cold packs; surgery is sometimes required in the case of a herniated disk. Muscle-strengthening exercises (physiotherapy, back training) are good ways of preventing it.

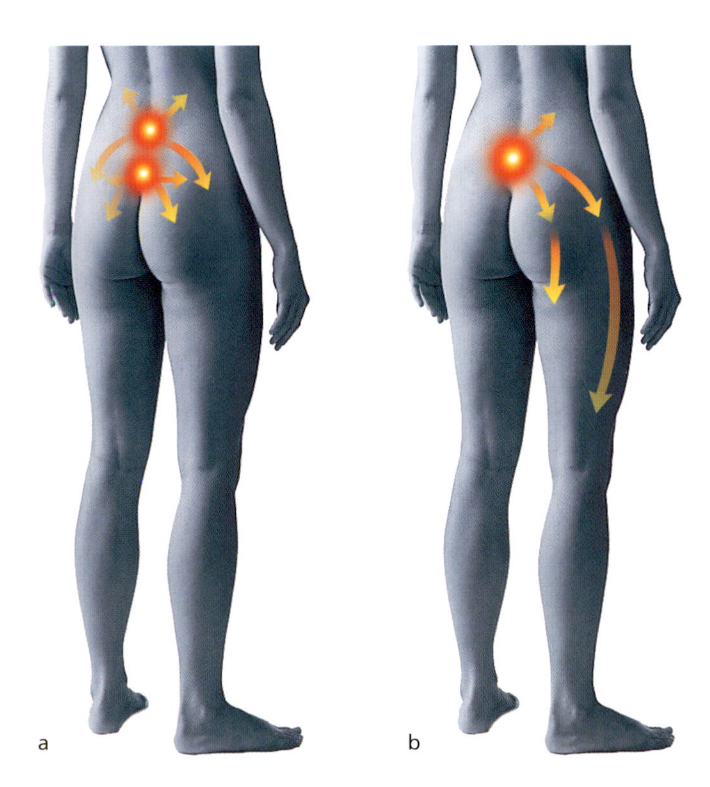

◆
Lumbago and sciatica
While the pain in lumbago (a) can emanate from various areas of the lumbar spine and is concentrated in this area, in the case of sciatica the pain radiates down into the leg (b).

• ***Vertebral facture***

Fracture of a vertebral body, vertebral arch, or the spinous processes. The spinal cord is often damaged by a vertebral fracture. Typical signs are paralysis and hypesthesia. If this type of damage is suspected, the person affected should lie still and should not be moved. An emergency physician must be called immediately.
Fractures of the lumbar vertebrae are extremely painful and are usually the result of trauma (e.g., a traffic accident or a fall from a great height). Movement or tactile disorders, or reduced power in the legs can indicate a lesion in the spinal cord. When giving emergency first aid to those injured in this way it is crucial to ensure that paraplegia does not occur as a result of moving them incorrectly. Often the vertebrae need to be stabilized, which in certain cases can only be achieved by inserting metal plates, wires or pins. Physiotherapy can help counter any long-term adverse effects.

• ***Spondylolisthesis***

When a vertebral body slips forward as a result of a congenital malformation of the spine. It most commonly presents in the lumbar spine and is diagnosed by X-ray. The slipped vertebra can cause inflammation of individual nerve roots in the spinal cord. Typical symptoms are pain and skin sensitivity. Discomfort can be alleviated by exercises for the spine (under the guidance of a physiotherapist) or by wearing a support corset. In severe cases surgery to fuse the affected section of the spine is carried out.

• ***Ankylosing spondylitis***

Chronic inflammation of the vertebral joints, intervertebral disks, and joints adjacent to the spine, resulting in ossification and fusion of the spine (ankylosis), leading to spinal rigidity. The condition begins with stiffening in the back area, with sensitivity in the spine on percussion, and pain, predominantly in the lumbar spine. It is more common in men. The cause of this disease, which occurs in around 1% of the population, remains unclear. Medication and regular physiotherapy is necessary to alleviate the symptoms. There is no cure.

Hand

• *Dupuytren's contracture*

The finger bends and stiffens as a result of connective tissue disease in which the flexible tendon fibers in the palm develop into a tight, gnarled tissue.
This flexion contracture is associated with considerable loss of movement in the fingers, and is especially prevalent in men in the 30–50 age group; it can only be treated by surgery.

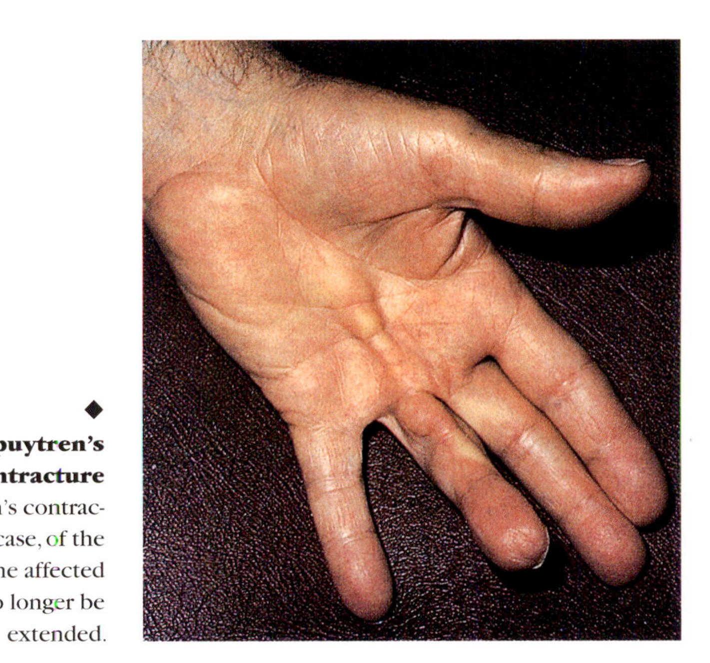

Dupuytren's contracture
In Dupuytren's contracture (in this case, of the ring finger) the affected finger can no longer be extended.

• *Carpal tunnel syndrome*

The carpal tunnel is an osseofibrous band of tissue running across the base of the hand through which tendons and nerves from the lower arm are conducted into the hand. Carpal tunnel syndrome means a thickening of this passageway, often caused by rheumatic inflammation. This thickening makes the tunnel narrower, so the nerves have less space and over time are damaged by the friction and pressure on them. The result is pain on moving the fingers (especially the thumb), numb sensations, and increasing stiffening of the fingers.
In the initial stage, rest and cold treatment can bring relief. In advanced stages, the narrowed tunnel can be widened by a surgical procedure in which adhesions and bumps are removed.

• *Ganglion*

Gelatinous lump. A ganglion usually develops on the wrist, but can also appear on the fingers and feet. Around a tendon sheath or joint capsule, a capsule forms which is tightly packed with tough mucus and can grow to the size of a dove's egg. Such a ganglion often disappears of its own accord, but treatment may be necessary if it is painful or disfiguring. The fluid can be aspirated with a syringe and surgically removed to prevent it reappearing.

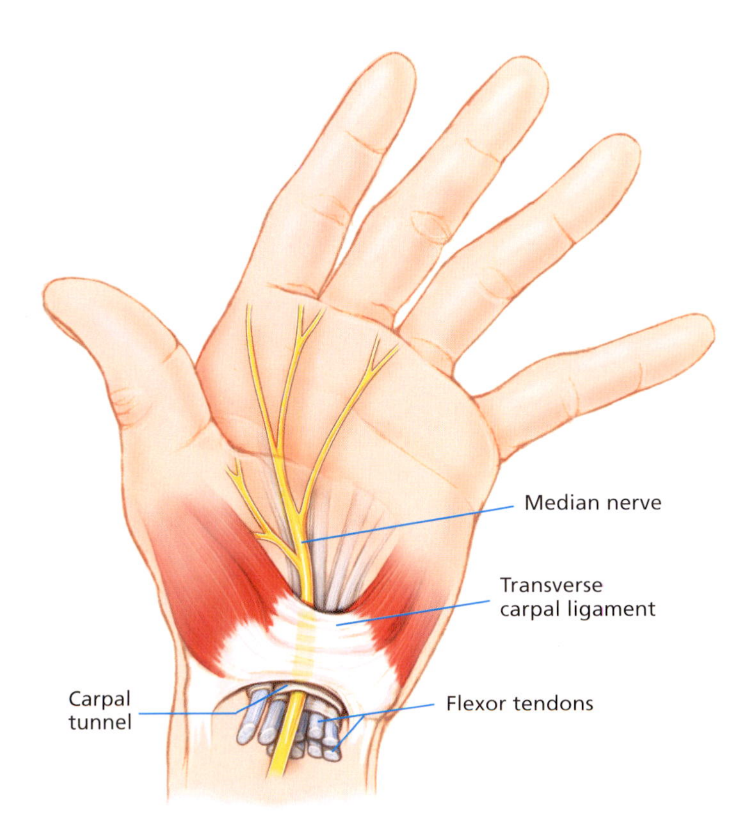

Carpal tunnel syndrome
Swelling in the transverse carpal ligament causes excessive pressure on the nerve which supplies the hand, thus restricting mobility in the tendons of the fingers.

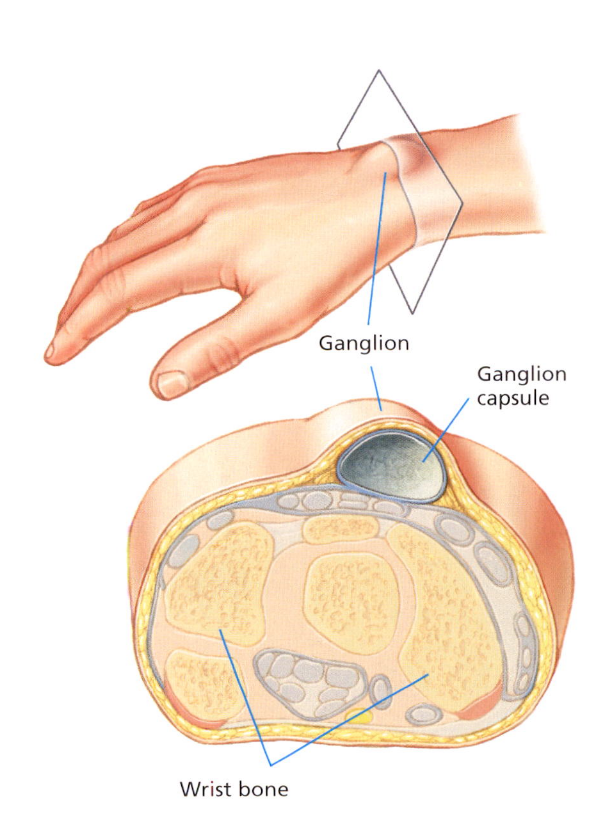

Ganglion
An enclosed capsule forms between the layers of the skin. The ganglion is benign.

Prolapsed (slipped) disk

A slipped disk is not something that happens to a specific age group; it can happen to a 20-year-old athlete just as easily as a 65-year-old pensioner. It occurs most frequently in 40–50-year-olds, and in people whose jobs require them to spend a lot of time sitting down, or who carry heavy loads on a daily basis.

What we call disks are in fact pieces of cartilage between the 25 vertebral bodies that form the spine. They act as shock absorbers and flexible support in the spine. Without them, the spine would not be able to move, and the individual vertebral bodies would rub against each other.

Each disk consists of an external fibrous ring with a central gel-like core.

A slipped disk occurs when the fibrous ring becomes loose or tears; parts of the core are pushed out into the vertebral canal and press against the spinal cord and nerve roots.

The disks of the fourth and fifth lumbar vertebrae are the most common ones to slip, as they experience the most pressure during the day.

A slipped disk is usually diagnosed by a scan (CT or MRI) or myelogram.

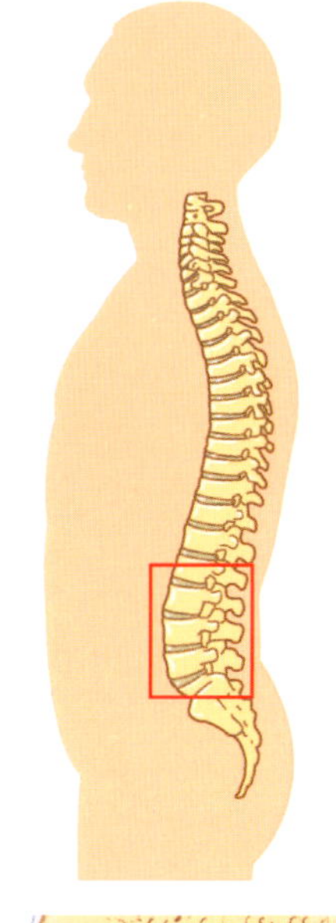

◆ A slipped disk occurs most commonly in the region of the lower lumbar vertebrae. The gel-like core of the disk slips through the vertebrae and presses against the nerves of the spinal cord.

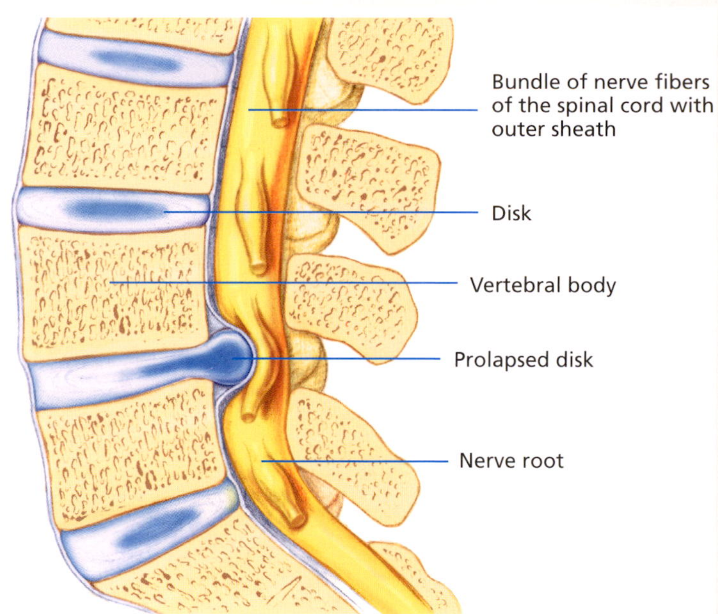

Symptoms

A slipped disk can often be mistaken for a pulled muscle, sciatica, or lumbago. It is extremely painful when a slipped disk presses against the nerve roots. The pain radiates from the back into the buttocks, the leg or even down to the foot. Sneezing, coughing, or straining make the pain worse.

The body tries to find a position that causes the least discomfort to the affected disk. The result is often a bent position with severe muscle tension or cramps on one side.

Irritation of the nerve often causes pins and needles in the extremities, either around the neck and chest or in the lumbar spine, depending on the location of the slipped disk. As the spinal nerves also innervate organs such as the bowel and bladder, damage can cause loss of control or function. If a slipped disk is left untreated, the nerve roots will be damaged. This can result in numbness and paralysis of the arms or legs.

Causes

The disks are also affected by the aging process; they lose elasticity, and shrink. If the fibrous ring of the disk becomes porous and cracks, there is a major risk of the pressure causing the core of the disk to squeeze through. This can happen when lifting a heavy weight or with sudden jerking movements of the spine.

A congenital weakness of the disk structure, the tissue, or curvature of the spine can cause even young people to suffer from a slipped disk. Constantly holding the body crookedly places an uneven burden on the spine, and distributes pressure irregularly on the disks. The core of the disk will always move towards the extended side of the spine.

Treatment

In the first instance, the patient is usually advised to rest for a couple of weeks. It is usually possible to find some relief from acute discomfort of the lumbar spine by lying flat on the back with the legs supported at right angles (e.g., on a chair), a position good for relieving the burden on the disks. Applying heat in the form of a heated pad or mud pack can help to release muscle tension.

A slipped disk does not automatically mean surgery. If the gel-like core has not yet slipped through to the vertebral canal, then significant improvement can be obtained through physiotherapy after a period of rest and once the acute symptoms have subsided; this can also help to prevent a recurrence. The muscles of the back and abdomen are connected to the spine; strengthening these muscles will prevent uneven stress on the spine and further pain.

If the symptoms do not improve, or if the patient experiences sensory dissociation and paralysis, then surgery may be the only solution. This may be performed as an out-patient procedure. A small incision made above the affected disk gives the surgeon access to the spine once the muscles of the back have been pushed aside. Using the absolute precision offered by surgical microscopy, the surgeon will then open the spinal cord and, using a special pair of forceps, remove the tissue that has slipped between the vertebral bodies. Alternatively, the gel-like core is extracted, either partly or completely, using a probe.

The muscles of the back, and the ligaments and disks of the spine adapt to the new situation and compensate for it. This does take a little time, however, which is why the patient, although now free from pain, should take great

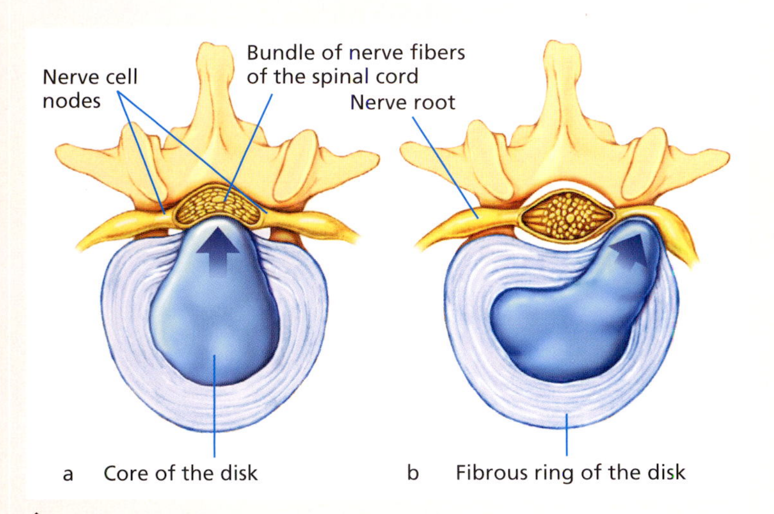

◆
A prolapsed (slipped) disk presses directly against the vertical nerves of the spinal cord (a) or on the nerves (b) on the side of the vertebral body.

◆
In an acute prolapse, the patient is placed flat on their back with their legs at a right angle.

care for several weeks and be very slow to resume normal activities. The patient may start swimming (in warm water) two weeks after surgery; backstroke is gentler on the spine than breaststroke. Walking is also recommended, ideally on soft ground. Sports such as tennis, judo and skiing that put a tremendous strain on the spine should be avoided for the first six months, as should any activities that require sudden and rotating movements. Physical exertion and heavy lifting can also cause a relapse.

The chances of recovery after physiotherapy or surgery are very good; only 5–8% of patients will need further surgery.

Prevention

People with an increased risk of a slipped disk are not the only ones who should avoid incorrect stress on their back. Sitting for long periods in a hunched position, a mattress that is too soft, lifting heavy weights, jerky rotating movements, and poor posture, all put extreme stress on the disks, and may lead to a prolapse. This can be avoided, for instance, by:

- making sure that the back is held straight and upright when lifting heavy objects or bending down, and squatting whenever possible,
- taking regular breaks to walk around or do some loosening exercises when spending long periods sitting at a desk or in the car,
- making sure that the weight in shopping bags is distributed evenly and carrying the same weight in both hands,
- exercising the muscles of the back and abdomen under professional guidance.

Lots of physiotherapists, health centers and sports studios offer back training. In some cases the costs may even be paid by an insurance company.

Backache

According to experts, in industrial countries backache has taken on almost epidemic proportions. What is often referred to, out of ignorance, as a damaged disk, frequently has various causes, and may even be the result of constant emotional stress and strain.

In the US, nine out of ten adults experience back pain at some time in their lives, and five out of ten working adults experience back pain every year. In some senses, this is not surprising as the back contains a large number of pain-transmitting nerve fibers and numerous sensors for all kinds of stimuli such as cold or pressure.

Backache is rarely a symptom of anything serious. In most cases, it is a reminder to be more relaxed and kinder to one's body. A physically and mentally stressed posture puts a tremendous strain on the muscles of the back. The back is then unable to support the spine evenly and without tension on one or the other side.

Muscle tension (cramps)

Sitting hunched over a desk for hours on end; sleeping on an old, lumpy mattress on unsuitable slats or with the head in a poor position (e.g., on a pillow that is too thick or too flat); wearing high heels; being overweight; a lack of exercise; and constant emotional stress, may all lead to excess tension of the back muscles. The back is thrown out of its usual natural balance, which in turn increases the tension in the muscles.

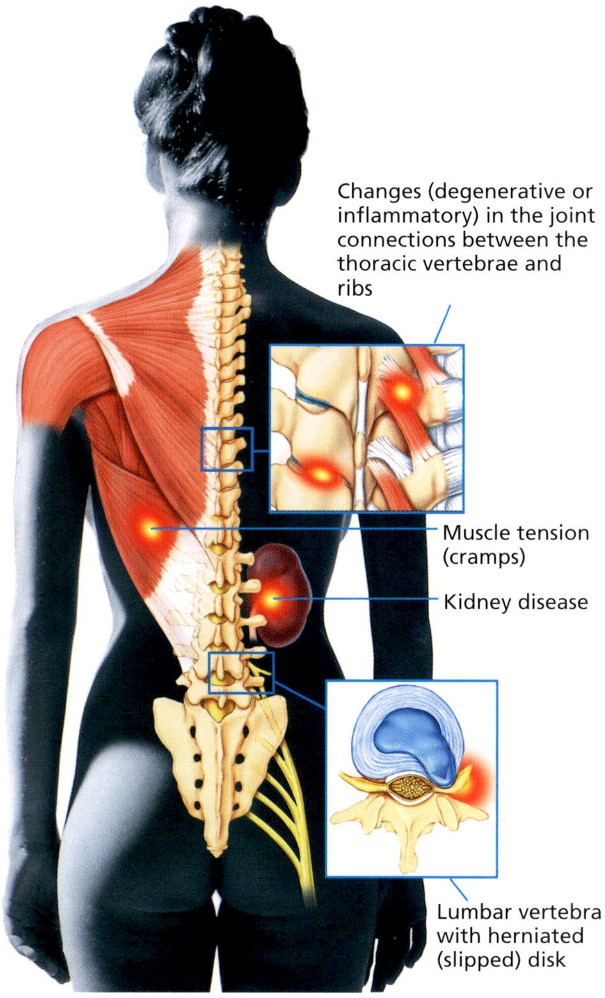

◆ The causes of backache are usually in the area of the spine and the muscles of the back. However, pain due to diseases of the internal organs, such as the kidneys, can also radiate into the back.

Faulty posture: lordosis

What the experts call a "hollow back," or lordosis, is a postural anomaly, usually diagnosed in children, that can be rectified with appropriate exercises. It is not necessarily congenital. It is more common in men than in women.

Ligament laxity

In half of all women over the age of 40, the ligaments in the region of the lumbar vertebrae become looser as hormone levels decline. This is often the reason for the dragging backache that many women experience.

Strains of the spine

Years of overburdening the spine can be every bit as painful as whiplash.

The cervical spine is subjected to all sorts of stresses over the course of the day. Although it is extremely flexible, it has to work hard, together with the surrounding muscles and ligaments, to hold the head stable. Cervical spine syndrome covers all sorts of complaints that emanate from the cervical spine, but headache and stiff neck are the main symptoms.

The thoracic spine is less affected by wear, since the disks in this area are more than capable of absorbing the vibrations to which the spine is subjected when walking. However, anyone who spends hours slumped in front of the computer or TV, or behind the steering wheel, will soon start to feel pain in this area.

The disks of the lumbar spine are the most severely affected, as they suffer most when we spend hours standing, or bend down incorrectly, or lift and carry heavy weights. Furthermore, the exit point of the sciatic nerve is on the bottom lumbar vertebra.

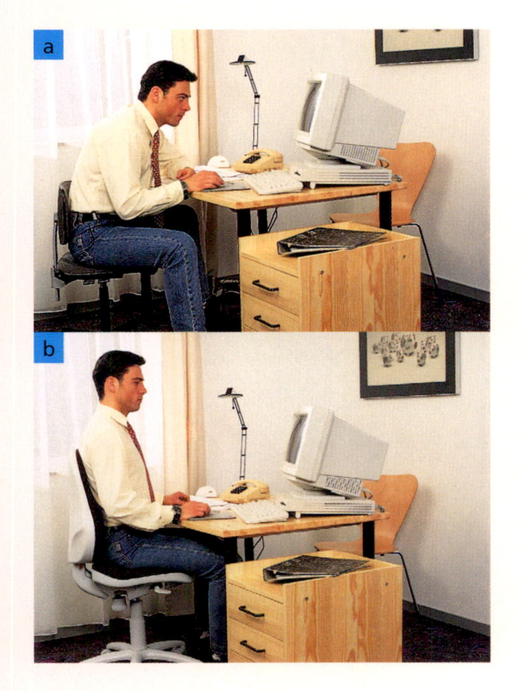

◆ The most common causes of backache are inappropriate seating positions and poorly adjusted office chairs (a); an upright position and a high backrest will support the spine and the muscles of the back (b).

Lumbago and sciatica

Both lumbago and sciatica are symptoms of a damaged disk. In lumbago, the pain radiates suddenly from the lumbar region, but can occasionally radiate out towards the chest. The patient may be wholly incapable of moving; the back muscles will be rigid with spasm, and the spinous processes of the vertebrae extremely sensitive to pressure.

Pain along the course of the sciatic nerve is called sciatica. If the disk moves, it will irritate and possibly even squash this nerve. The main symptoms are also experienced in the lumbar region, but, unlike lumbago, radiate into the buttocks or down one leg as far as the ankle. The outer edge of the affected leg may be numb. Sneezing, coughing, or straining makes the pain worse.

Bed rest with special supports, hot baths, and specific relaxation exercises may help alleviate the pain.

Medical advice must be sought without delay if:

- there is paralysis or numbness,
- the pain is severe and continues for several hours,
- the pain keeps recurring.

Disturbed life balance

Anyone who fails to allow themselves the time to relax is neglecting his or her spine. People who "put on a brave face" even when under pressure are asking a lot of their spines. The same applies to people who think they must always have everything under control. Many of those affected reject the idea that their emotional state may contribute towards their symptoms and that only a change of circumstances can bring about a permanent solution to their problems.

Be responsible for yourself

It is often difficult to be responsible for oneself. It is much easier to go for an injection if your back "is playing up," and get a prescription for pills to make your muscles relax. Of course, such measures do indeed provide immediate relief, but they are not a permanent solution. With a little initiative, one can protect oneself against backache and later damage. A few points to observe:

- sit up straight, at least at work, with the upper and lower leg at a right angle,
- spend at least ten minutes a day doing exercises to strengthen the muscles of the back and abdomen,
- avoid wearing heels over 3 centimeters high for long periods of time,
- carry heavy loads close to the body, and ideally balanced evenly on both sides,
- lose weight if you need to,
- buy a slatted base and mattress from a specialist shop, where your own lying profile (based on your weight and the shape of your spine) can be established,
- go to exercise classes to learn (under guidance) how to strengthen your back muscles,
- swim (alternating between breaststroke, backstroke, and crawl) and walk (on soft ground) regularly, and
- make sure you are emotionally balanced by keeping stress at bay, and getting enough rest and relaxation.

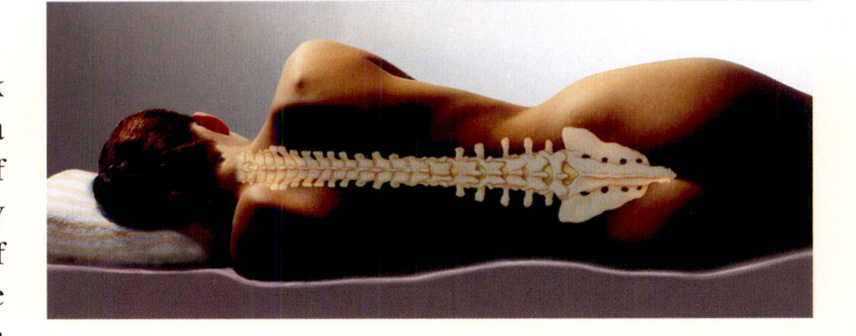

◆ Your spine will remain straight even when you are lying on your side on a special "disk-friendly" mattress.

Shoulder and arm

• *Shoulder–arm syndrome*

The collective term for irritations of the nerve roots around the neck, caused by changes to the cervical spine in the event of disk damage, by inflammation, or, more rarely, by tumors. The irritation causes usually severe pain that can radiate from the back of the neck down the arms to the fingers. It can result in stiffness, "pins and needles," and weakness in the arm and hand with sensory dissociation, cause muscle wasting, and interrupt the blood supply to the fingers. As well as reducing pain, the underlying cause of the problems in the cervical spine needs to be treated.

• *Inflammation of the shoulder joint*

Inflammation of the shoulder joint, caused by an infection of the joint cartilage or capsule, by straining, or as the consequence of a rheumatic disease. Streptococcus and staphylococcus are the main causes of bacterial infections. Antibiotics are used to treat bacterial infections, while anti-inflammatories are given when the cause is not a bacterial infection.

• *Shoulder luxation*

Severe and painful dislocation caused by falling on the shoulder or outstretched arm, as well as other arm movements that are outside the normal range of use. In most cases, the joint head of the upper arm dislocates from the socket, moving outwards and down. There is also a strong possibility of fracture of the upper arm and damage to the ligaments and muscles. If the shoulder muscles are damaged so badly by a violent dislocation that even minor injuries cause a recurrence of the luxation, then the supporting muscles may have to be shortened surgically. In all other cases, the patient is anesthetized while the joint is put back into the socket, and the shoulder is rested.

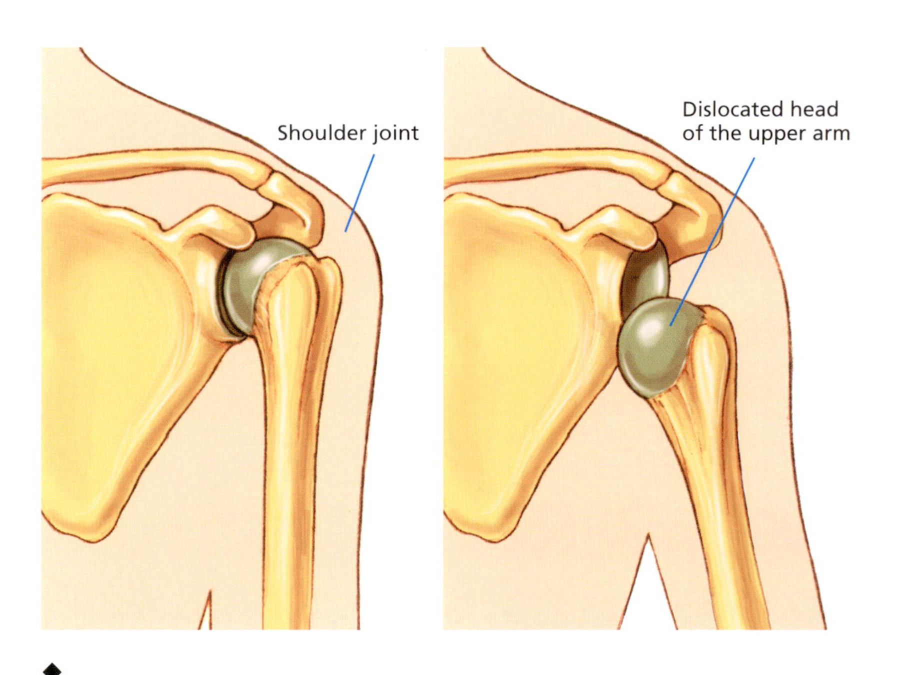

◆
Dislocation
If the shoulder is dislocated, the bone of the upper arm slides out of its socket.

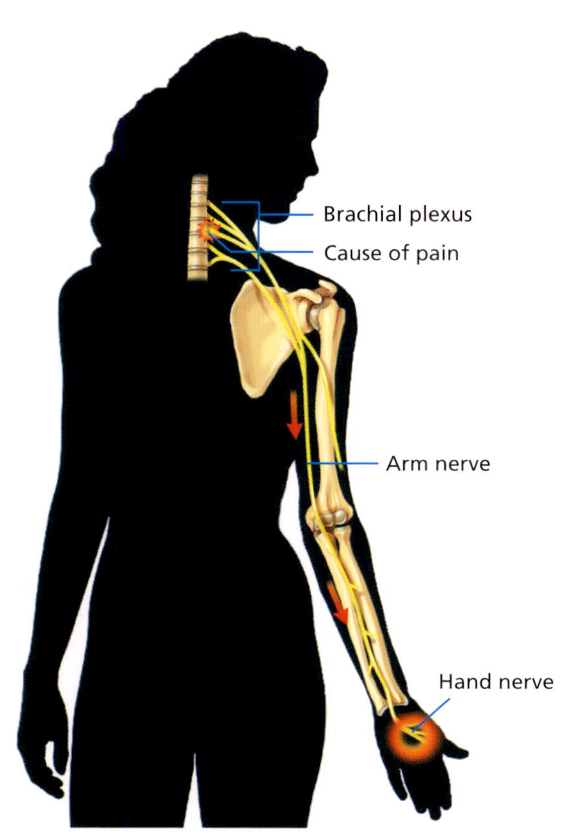

◆
Shoulder-arm syndrome
Injury to the exit points of the nerves of the cervical spine can cause pain and sensory dissociation that radiate down to the hand.

• *Frozen shoulder*

Painful restriction of movement in the shoulder resulting from inflammation or thickening of the membrane lining the joint capsule. There may also be an emotional component, such as constant stress and tension. Treatment with painkillers and heat applications can be long-term, and in stubborn cases surgery will be necessary.

• *Tennis elbow*

A disease is caused by overuse of the elbow. The outer sides of the elbow and the lower arm are painful and sensitive to pressure. The cause is inflammation of the tendon of the lower arm muscle which is responsible for extending the fingers and wrist. Tennis elbow can also be caused by holding a tennis racket incorrectly. Painkillers help to relieve the discomfort, and the inflammation is treated by resting the arm, applying hot, moist compresses, ultrasound, and medication. In severe cases, corticosteroids injected into the base of the tendon can help. Surgery may be necessary if chronic inflammation causes the tendon to adhere to the surrounding tissue.

Hips

• *Hip joint endoprosthesis*

An artificial hip joint is a replacement for a worn or stiff natural one. The head of the femur (thigh bone) is removed surgically and a titanium prosthetic shaft with a ball on the top end is anchored in the remaining bone. The pelvic bone is hollowed out around the site of the natural joint socket to make room for an artificial socket in titanium or plastic. This, together with the movable ball of the prosthetic shaft, forms the new hip joint. The procedure is now a routine operation for older patients following fracture of the femoral neck.

• *Hip luxation*

Dislocation of the hip (dysplasia or "clicky hip"). Usually a congenital defect of the hip joint in which the femoral head is partly or completely outside the socket. Congenital hip dislocation is almost always detected during a postnatal check-up. A temporary splint is applied to the femur using a special diaper (a pillow splint) to align the femoral head and socket correctly. The correction usually stabilizes after a few months, and the defect is rectified. Acquired hip dislocations that arise due to extreme pressure on the hip joint (e.g., from a fall or road accident) are comparatively rare.

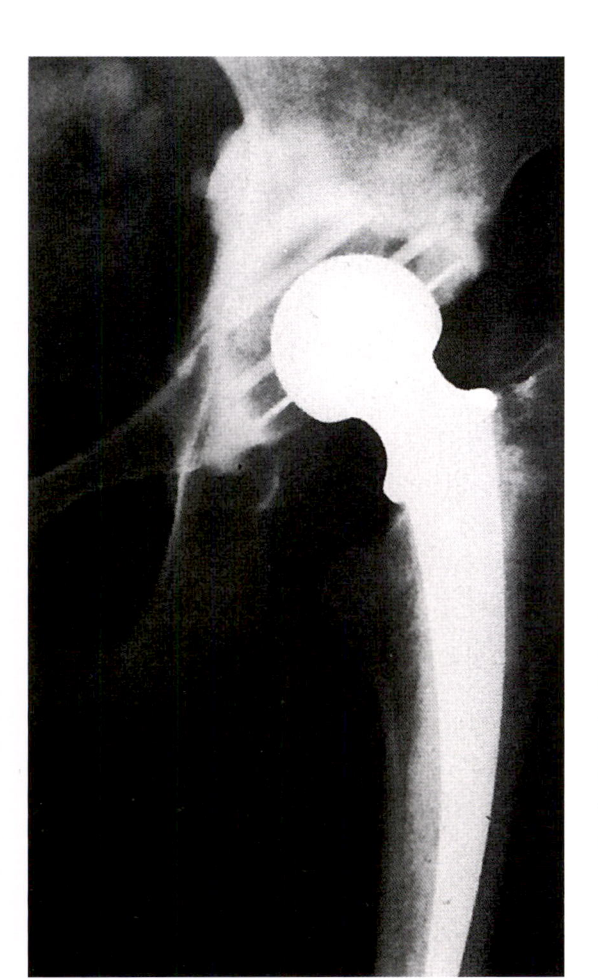

◆
Hip joint endoprosthesis
An artificial hip joint, seen here on X-ray, can help people with severe wear of the joint to live a normal life again.

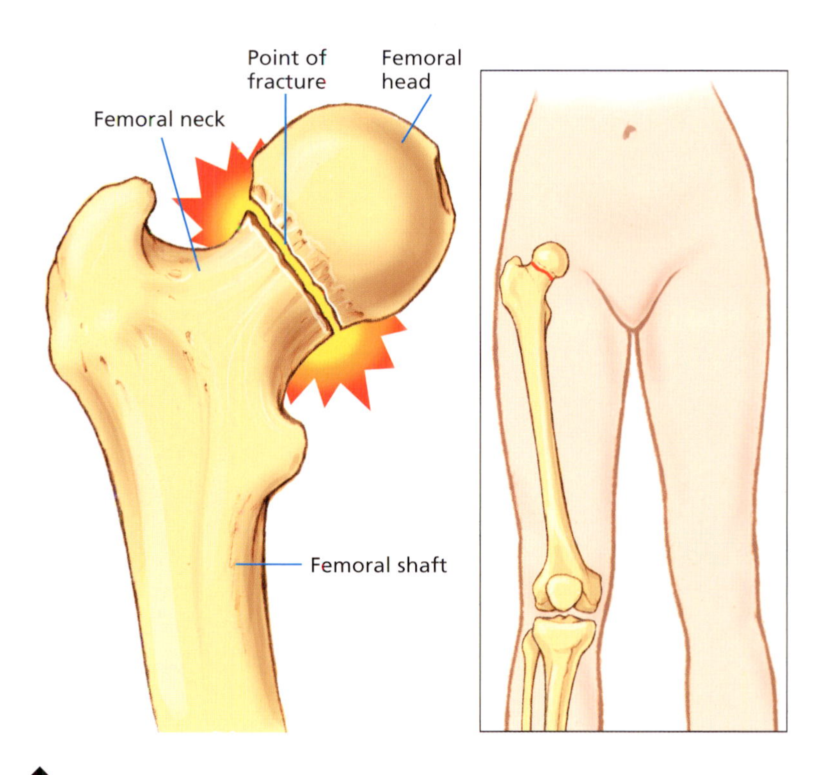

◆
Fracture of the neck of the femur
In a fracture of the femoral neck, the bone breaks just above the femoral head in the socket of the hip joint.

Femur (thigh)

• *Fracture of the femoral neck*

Fracture of the bone at the connection between the head and the shaft of the femur. A fracture of the femoral neck is usually suffered by older people, as the aging process increases the brittleness of their bones (known as senile osteoporosis).
Symptoms of a fracture of the femoral neck are pain, limited mobility of the hip, and a shortening or rotation of the leg to the outside.
If the patient's health permits, surgery is performed to nail the femoral neck or insert a hip joint prosthetic. This will reduce the period that the patient is confined to bed, with its associated risks (pneumonia, thrombosis, embolism), and get the patient back on his or her feet as quickly as possible.

Knee joint

• *Inflammation of the knee joint*

Short- or longer-term inflammation of the knee that often occurs following injury, infection, or infectious venereal disease, but can also be associated with rheumatic complaints. The inflammation may be accompanied by a joint effusion, and can cause an extremely painful suppuration of the joint. This may cause the knee joint to stiffen. Treatment ranges from splinting to surgery involving opening the knee joint.

• ***Dislocation of the knee cap (patellar luxation)***

Severe rotation of the knee cap, usually the result of trauma. The knee cap is thrown out of its normal position, usually to the side, but also upwards, downwards, or even into the joint cavity. Every attempt to move the knee joint causes extreme pain. There is also frequently damage to the tendons and ligaments. Conversely, weak ligaments can cause the knee cap to dislocate. It is not usually possible to manipulate it back into position; surgery is generally required.

• ***Rupture of the cruciate ligament***

Rupture of the anterior or posterior cruciate ligament in the knee. Occasionally the menisci are also affected. The injury is usually caused by a blow to the extended or bent lower leg, or by suddenly turning it inwards. It is a particularly common injury of footballers and skiers. After a complete rupture, the lower leg can be moved to the front and back far more than the thigh can. An effusion usually forms around and inside the knee joint. The ruptured ligament has to be surgically reconstructed.

• ***Meniscal locking***

A knee injury in which the cartilage ring in the joint gets crushed between the femur and the shin. The meniscus slips out of its normal position. The cause is usually extreme rotation of the knee. The symptoms are severe pain in the knee joint and limited mobility. It is usually impossible to straighten the knee. Depending on the severity of the injury, it may be possible to manipulate the meniscus back into its original position; otherwise surgery is required.

• ***Rupture of the meniscus***

Lesion, tear or rupture of the cartilage disk in the knee joint. This can be caused by a sudden turning movement with the knee bent. If the meniscus has already been damaged, then often only minor stress will suffice. A rupture of the meniscus is indicated by sudden severe pain and swelling of the joint. The ruptured cartilage is usually removed surgically.

Foot

• ***Hammer toe***

Hammer-shaped deformation of a toe in which the end joint bends downwards. Usually the second toe is affected. The bent joint presses against the shoe, often causing a corn. A protective felt pad worn over the affected toe can ease the pressure and associated pain. If symptoms persist, minor corrective surgery on the tendon of the toe may be considered.

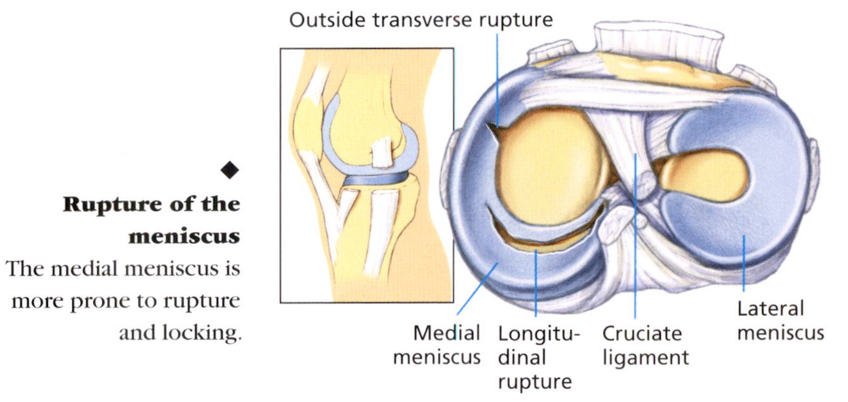

Rupture of the meniscus
The medial meniscus is more prone to rupture and locking.

• ***Clubfoot (talipes equinovarus)***

Deformation of the foot. Both feet turn inwards, with the front of the foot pointing upwards. Afflicted persons walk on the outer edge of the foot. Treatment needs to be given promptly to avoid permanent deformity. The feet are usually put in plaster casts or splints, but in severe cases, corrective surgery can be carried out.

• ***Flat (pronated) foot***

In people with flat feet, the transverse or lengthwise arch of the foot has sunk or is flat. The causes are weakness of the ligaments and tendons, or faulty development of the bones of the foot. Typical symptoms are pain in the arch of the foot and on the heel. Children with the problem soon tire when walking. Corrective measures can be undertaken at an early age using shoe inserts and foot exercises. In adults, the symptoms can be relieved by inserts or orthopedic shoes.

• ***Talipes equinus***

Mispositioning of the foot that causes "tip-toeing." Only the toes and balls of the feet touch the ground. Congenital talipes equinus is often also accompanied by a club foot.
Talipes equinus usually occurs later as the result of paralysis of the extensors. It sometimes develops in bedridden patients as the result of incorrect positioning of the legs.

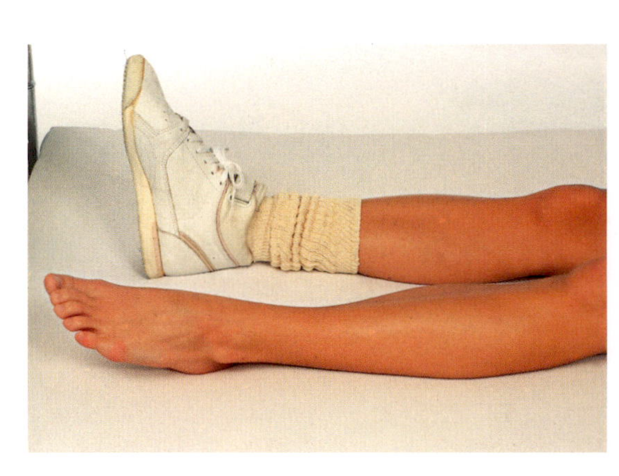

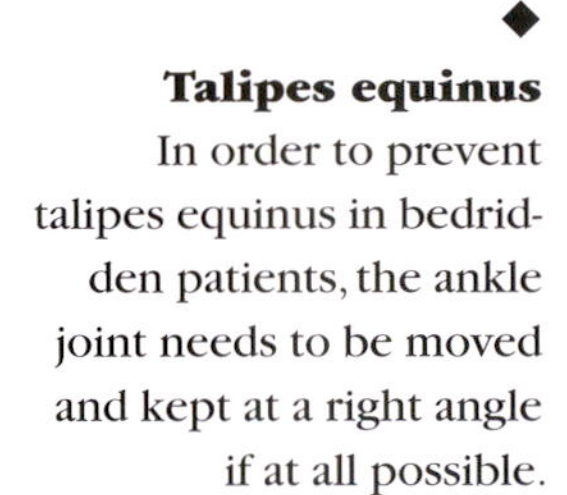

Talipes equinus
In order to prevent talipes equinus in bedridden patients, the ankle joint needs to be moved and kept at a right angle if at all possible.

Movement disorders (ataxia)

Disturbances of the interplay between movements of the body. Trembling movements and an uneven gait are included in ataxia, and indicate a disease of the nerves or brain. Poisoning can also be a cause. The cerebellum is responsible for movement sequences, so diseases, tumors, and age-related changes in the cerebellum will result in movement disorders. If the cause is in the cerebellum, there will often also be balance disturbances. Diseases in other parts of the nervous system, such as the spinal cord or in multiple sclerosis, may also seriously inhibit movements, and cause sensory dissociation in the arms or legs.

Paralysis

• ***Drop hand***

Also known as wrist-drop. In the event of paralysis of a nerve of the lower arm, it is no longer possible to extend the hand. It drops down, and the thumb can no longer be angled out. Perception on the back of the hand may also be disturbed. Drop hand can occur when the arm is broken, and it is also a typical symptom of lead poisoning. The paralysis can be partly overcome with massage and electro-physiotherapy.

• ***Paraplegia***

Paralysis caused by interruption of the nerve transmission fibers in the spinal cord. Accidents, hemorrhage, tumors, and certain diseases of the spinal cord can damage it so much that paraplegia results. The higher up the level of damage in the spinal cord, the more parts of the body will be affected by it.
With very high injuries to the cervical spine, respiration may also be impaired. Without immediate help, respiratory paralysis will lead to death. With complete paraplegia, all mobility and sensory perception below the point of injury will fail; the limbs will be flaccid, and the functions of the excretory and sexual organs will be impaired. If the spinal cord is only partly damaged, there is a lower degree of functional impairment.

Diagnostic methods

X-ray remains the most important method for examination of bones and joints. They show up well on the X-ray image, since the rays penetrate soft and other tissue more easily than bone. The impact of the rays can largely be reduced with modern technology. However, a lead apron must be used to protect the sex glands, and X-rays should be avoided if at all possible during pregnancy.

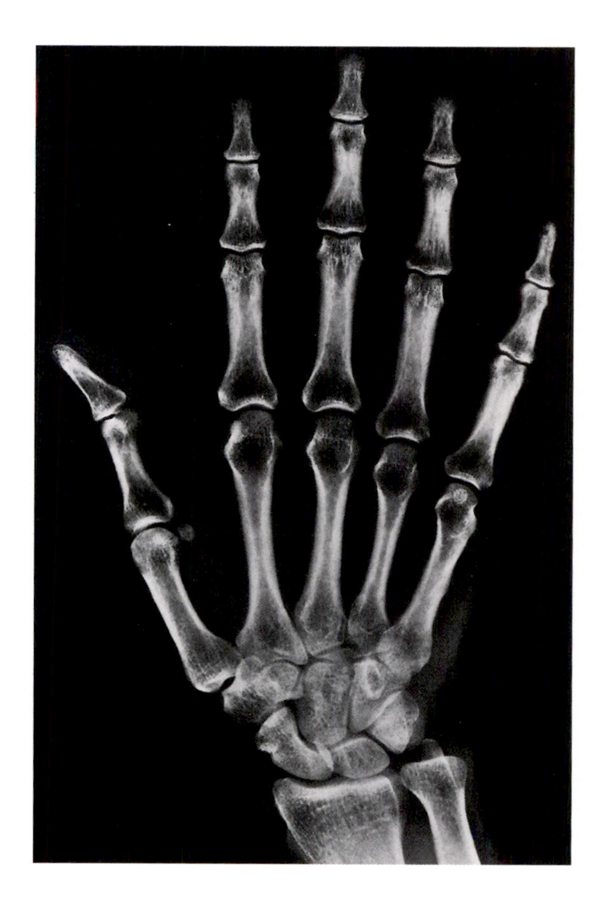

◆
X-ray
Thanks to the discovery and development of X-rays, bones and joints can be visualised with ease.

Computed tomography (CT) is a further development of X-irradiation, and can reveal internal structures of the body, including soft tissue, in sections. The exposure dose is significantly higher than with X-rays, which is why computed tomography is used less often.

Ultrasound (sonography) is only suitable for the evaluation of soft tissue, i.e. muscles. The removal of a tissue sample from a muscle is called a **muscle biopsy**. This procedure is usually carried out under local anesthesia. Individual muscle cells can then be examined under a microscope in order to establish whether the cause is a disease of the actual muscle or of the nerves that supply it.

• ***Arthroscopy***

Procedure to examine the inside of a joint. The endoscope, a tube-like instrument with a light source, offers insight into the joint. The joint is injected with a special liquid for greater clarity. Fine instruments can be guided into the joint through an additional tube.
Arthroscopy may be carried out under local or general anesthetic, depending on the joint involved. Arthroscopies are carried out on the shoulder, elbow, and ankle, and most frequently on the knee. They are not only used to diagnose unclear joint problems, but can also be used to treat ruptures of the meniscus or ligaments, and damage to cartilage. The patient is left with two or three tiny scars, and is usually able to use the joint normally again quite soon.

• *Arthrocentesis (joint tap)*

A needle is inserted into the joint capsule to draw off synovial fluid. In cases of joint inflammation or joint effusion, the fluid may be examined for bacteria or signs of infection.

Treatment methods

• *Physiotherapy*

The use of physical exercise to maintain mobility and physical strength, or to restore them after illness and injury. A physiotherapist may use passive exercises such as bending and stretching on seriously ill or paralyzed people to maintain the mobility of the joints and stimulate the circulation. Active exercises, requiring the patient's co-operation, may be used to correct misalignments and incorrect weight bearing, reduce pain in joints and muscles, relieve stress and cramps, strengthen and build up ligaments, tendons and muscles, and overcome or compensate for paralysis and stiffness.

• *Chiropractice*

A branch of medicine in which pain and impaired movement are treated by specific manipulation, particularly in the spine, arms and legs. These techniques should only be carried out by appropriately trained and qualified specialists.

Weak muscles, possibly resulting from tension or exhaustion, permit the vertebrae to shift against each other. This can constrict the exit points of nerve roots, and blood or lymphatic vessels. Headache, dizziness and pain are often the result, as are lumbago, discomfort in the area of the sciatic nerve, circulatory problems, and high blood pressure.

In what is called the "soft" technique, the chiropractor presses and stretches the tense muscles, rolling over them with gentle, rhythmic movements and light pressure. In the "hard" technique, the joint is restored to its original position with a quick, sharp movement. The patient will hear and feel cracking noises. Chiropractice should not be used if the patient has any infections, tumors or damaged disks. In order to ensure the permanent success of this treatment, the patient must continue to strengthen their muscles with regular physiotherapy and exercise.

◆ **Chiropractice**
If the patient is experiencing discomfort in the cervical vertebrae, the misaligned vertebrae can be restored to their original positions with a sharp movement.

• *Back exercises*

A program of exercises to strengthen the muscles of the back, and prevent and treat postural damage. Health insurers may support and finance special back exercise programs, usually provided by physiotherapists. They consist of exercises to relax the muscles, and teach participants how to use their backs properly, how to avoid incorrect positions, and how to lift and carry correctly. For people whose work involves remaining seated for long periods in the day, it is especially advisable to do regular back exercises.

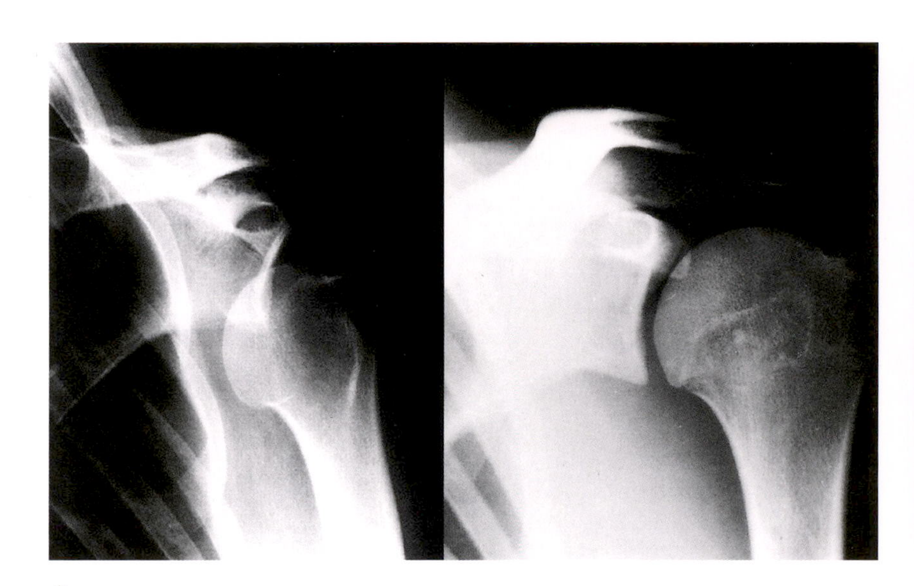

◆ **Manipulation**

• ***Massage***

Mechanical treatment of the skin and underlying tissue. Massage can have a positive benefit on the mind as well as the body. It can relax tense muscles, improve circulation, alleviate pain, and stimulate the internal organs. Classic massage consists of stroking, kneading, tapping, and rubbing movements.

• ***Mudpack***

Paste consisting of warm water and mud that is applied to diseased or painful areas of the body. Mudpacks are used in particular on strained muscles, to treat rheumatic complaints or signs of wear in the joints (arthritis), chronic pain, or after injury.

• ***Mud bath***

Full or partial bath with the addition of peat. Peat is the result of decomposition of plant materials in air-tight conditions. It contains a number of medically effective constituents which, in combination with the warmth of the bath, can help to relieve rheumatic complaints and chronic inflammatory diseases.

• ***Manipulation***

Returning bones that have been pushed apart following fracture or twisting, or bones that have slipped out of their socket, to their normal position. The bones that form a joint are held together by muscles, tendons, and ligaments. External forces, for example, in an accident, can throw them out of their normal positions. Dislocation (luxation) causes severe pain; the joint swells up, and it is not usually possible to move it. Medical help is required as quickly as possible to return it to its normal position. Sometimes this requires anesthesia. Following manipulation it can take several months for the joint to function normally again. Physiotherapy may help.

• ***Arthrocentesis (joint tap)***

This is carried out on a joint effusion to remove excess fluid. The procedure is performed under local anesthesia.

• ***Osteosynthesis***

A surgical procedure to realign bones following a fracture using metal plates, screws, nails, or wires. Once the bones have healed, the metal items are usually removed.

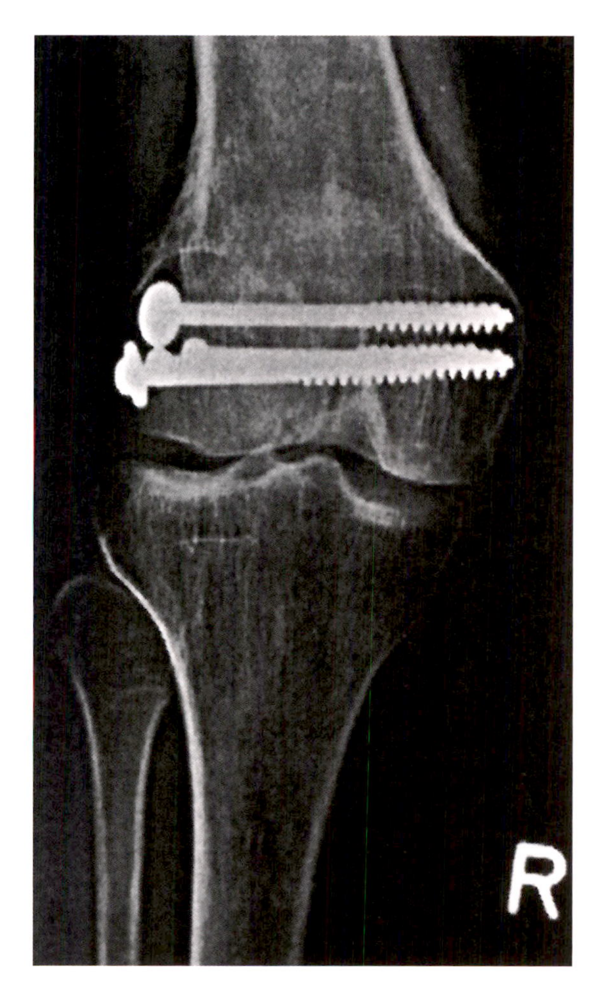

◆
Osteosynthesis
In fractures close to a joint such as the knee, as seen here, it is often necessary to stabilize the bones using screws.

• ***Wire extension***

Method for the treatment of bone fractures where the fractured edges have shifted against each other, and strong permanent tension (traction) is required to return them to their original positions and hold them there until the fracture has healed. In this case, a fixed wire is drilled through the bones and attached to a horseshoe-shaped bracket. This bracket is attached to a pulley device with a weight on one end.

• ***Artificial joints***

Replacement for a damaged or destroyed joint. Artificial joints are made of plastic, ceramic, or metal. There are artificial joints for hips, knees, feet, elbows, and fingers. Hip joints are the ones that are most commonly replaced, since they often fracture, together with the femur, when an older person falls, or as the result of osteoporosis (loss of calcium from the bones). Replacement may also be the most appropriate course of action in advanced arthritis of the hip joint, in the face of severe pain and greatly restricted mobility.
Depending on the level of damage, either the entire joint or just part of it is replaced. Intensive physiotherapy is necessary after surgery to make sure that the patient can move the joint properly.

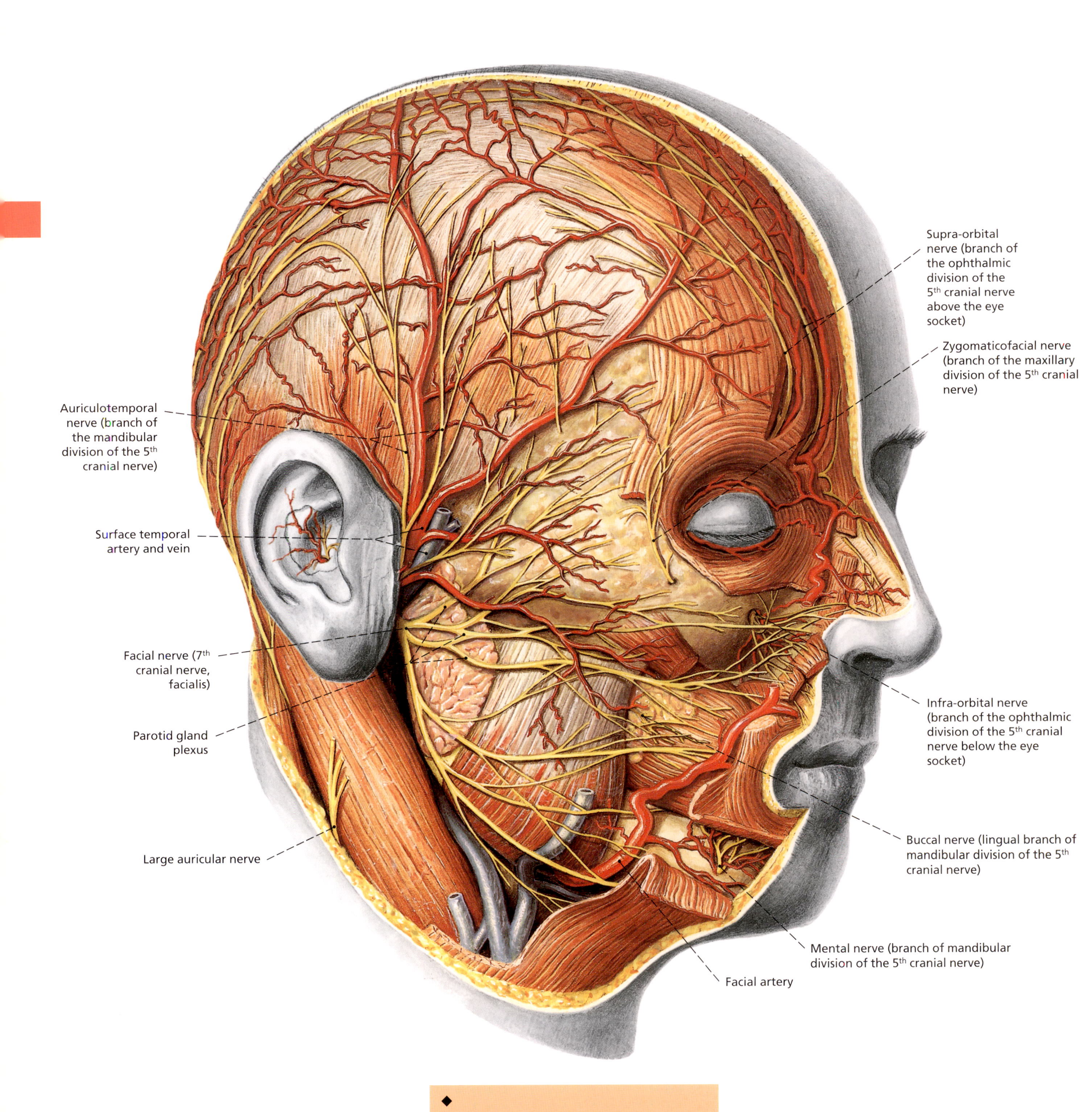

◆
Fig. 145
The head nerves. The neuroplexus in the area of the parotid gland is clearly visible. It contains fibers of the facial nerve (facialis).

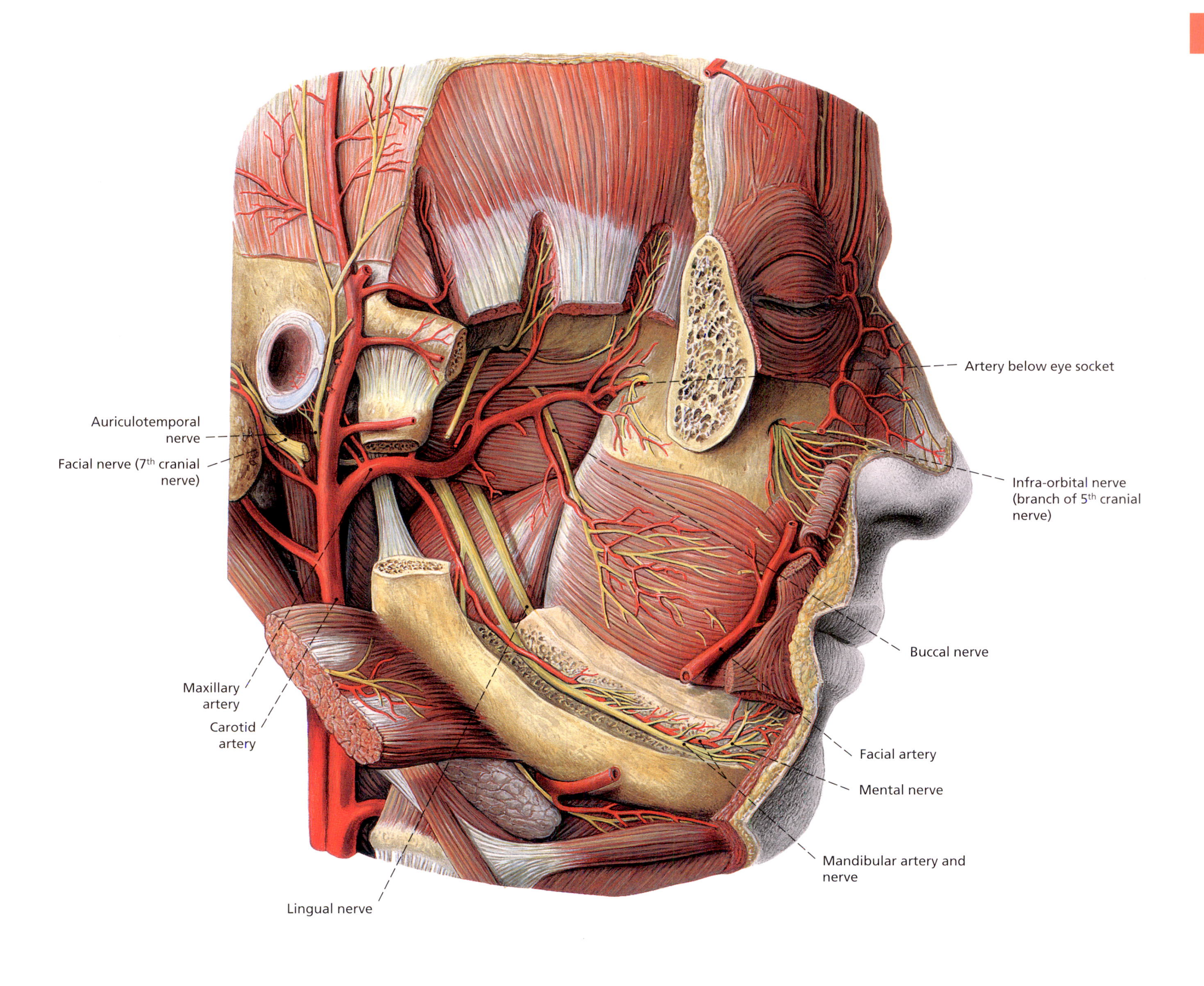

◆
Fig. 146
Nerves in the jaw and cheek area after partial removal of the cheek bone (zygomatic bone) and lower jaw bone (mandible).

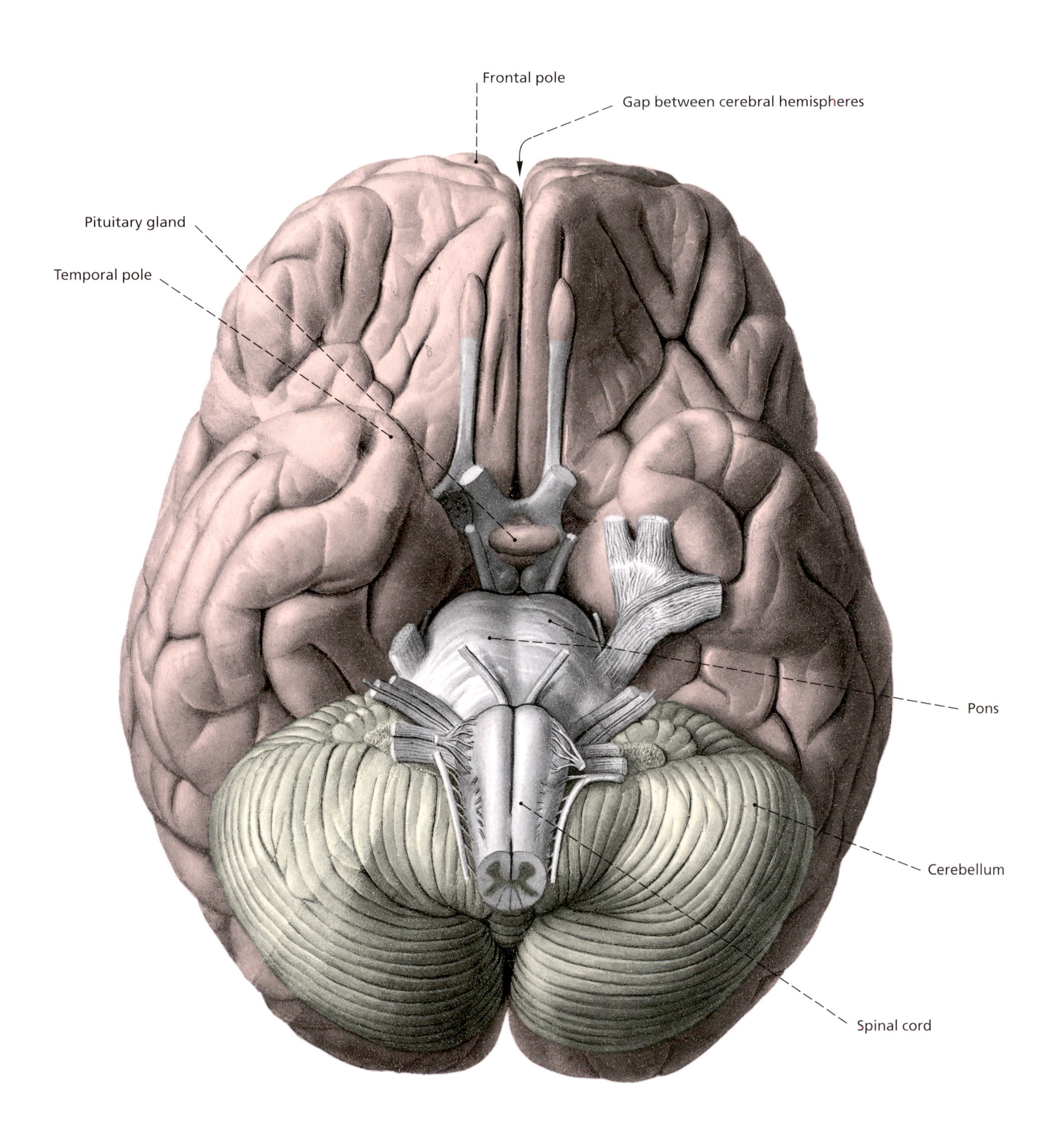

◆
Fig. 150
Brain with cerebellum and upper section of spinal cord, seen from below.

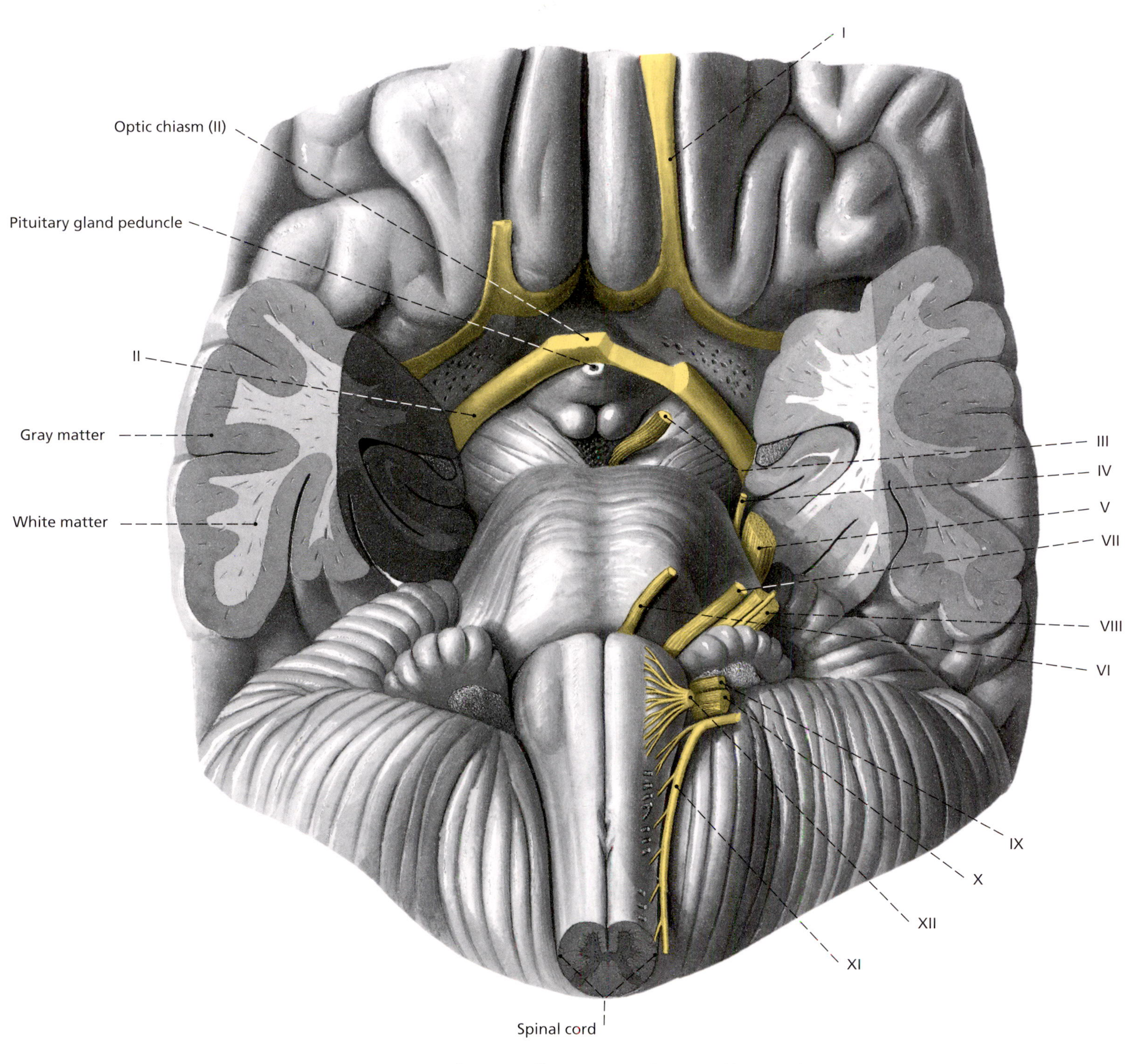

I = 1st cranial nerve, olfactory nerve (nervus olfactorius)
II = 2nd cranial nerve, optic nerve (nervus opticus)
III = 3rd cranial nerve, oculomotor nerve (nervus oculomotorius)
IV = 4th cranial nerve, trochlear nerve (nervus trochlearis)
V = 5th cranial nerve, trigeminal nerve (nervus trigeminus)
VI = 6th cranial nerve, abducens nerve (nervus abducens)
VII = 7th cranial nerve, facial nerve (nervus facialis)
VIII = 8th cranial nerve, vestibulocochlear nerve (nervus vestibulocochlearis)
IX = 9th cranial nerve, glossopharyngeal nerve (nervus glossopharyngeus)
X = 10th cranial nerve, vagus nerve (nervus vagus)
XI = 11th cranial nerve, accessory nerve (nervus accessorius)
XII = 12th cranial nerve, hypoglossal nerve (nervus hypoglossus)

Fig. 151
Exit points of the cranial nerves (I–XII) from the base of the brain.

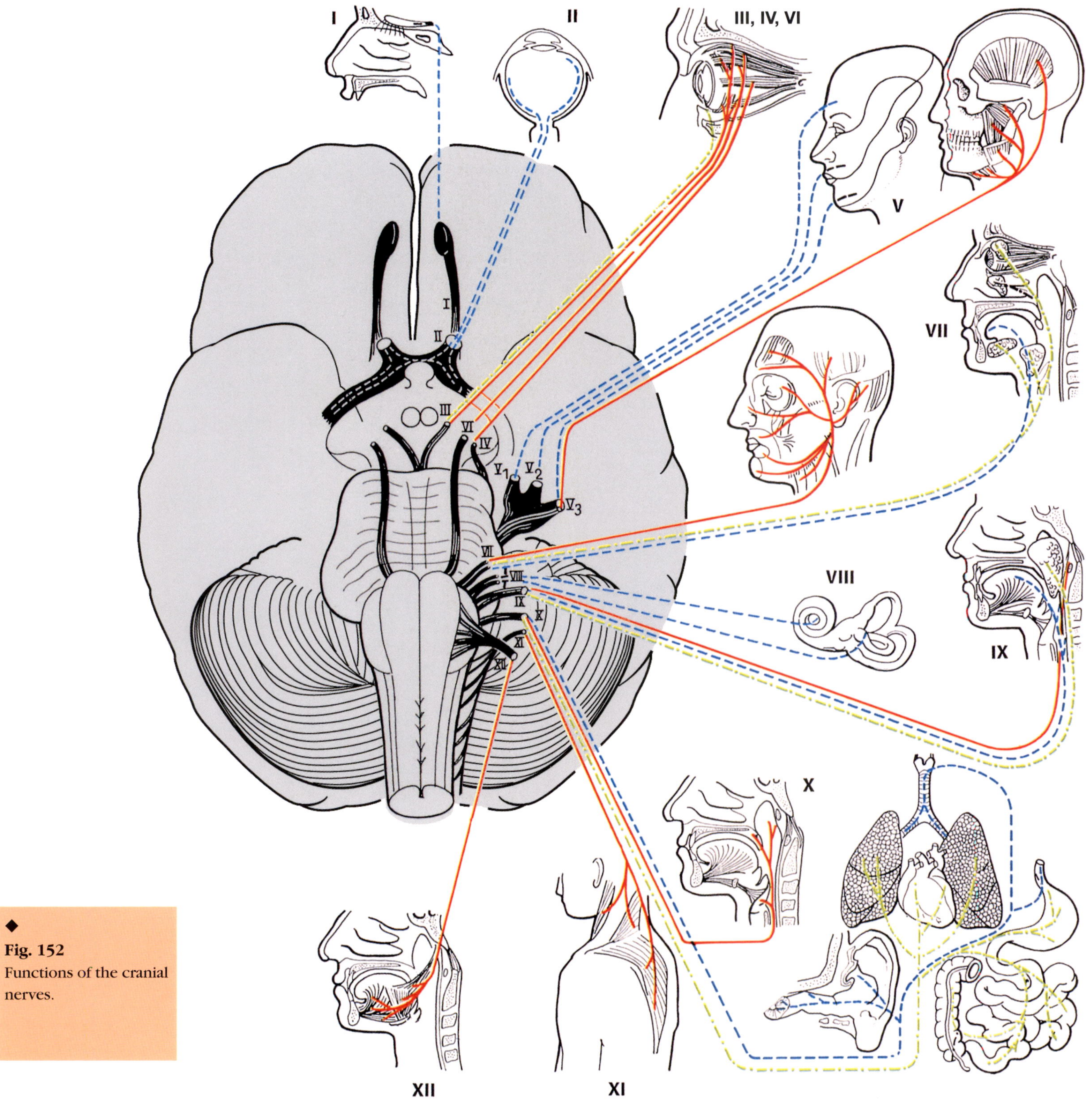

◆
Fig. 152
Functions of the cranial nerves.

I: Olfaction
II: Vision
III: Ocular movements
IV: Ocular rolling movements
V: Motor and sensory innervation of facial skin, eye socket, teeth, musculature of upper and lower jaw
VI: Ocular movements
VII: Facial movements (facial expression), secretion from tear-gland, sublingual gland and mandibular gland, lesser palatine and nasal gland, taste in front two-thirds of tongue
VIII: Hearing, balance
IX: Pharyngeal muscle movements (act of swallowing), mucous membrane sensitivity of pharynx and tympanic cavity, taste in rear third of tongue
X: Laryngeal movements (also vocal cords), motor control of pharyngeal musculature, heart, lungs, stomach, and bowel; sensitivity of external ear and dura mater
XI: Head turning; voluntary muscles of larynx, pharynx, and soft palate, and sternocleidomastoid and trapezius muscles
XII: Tongue muscles and dura mater

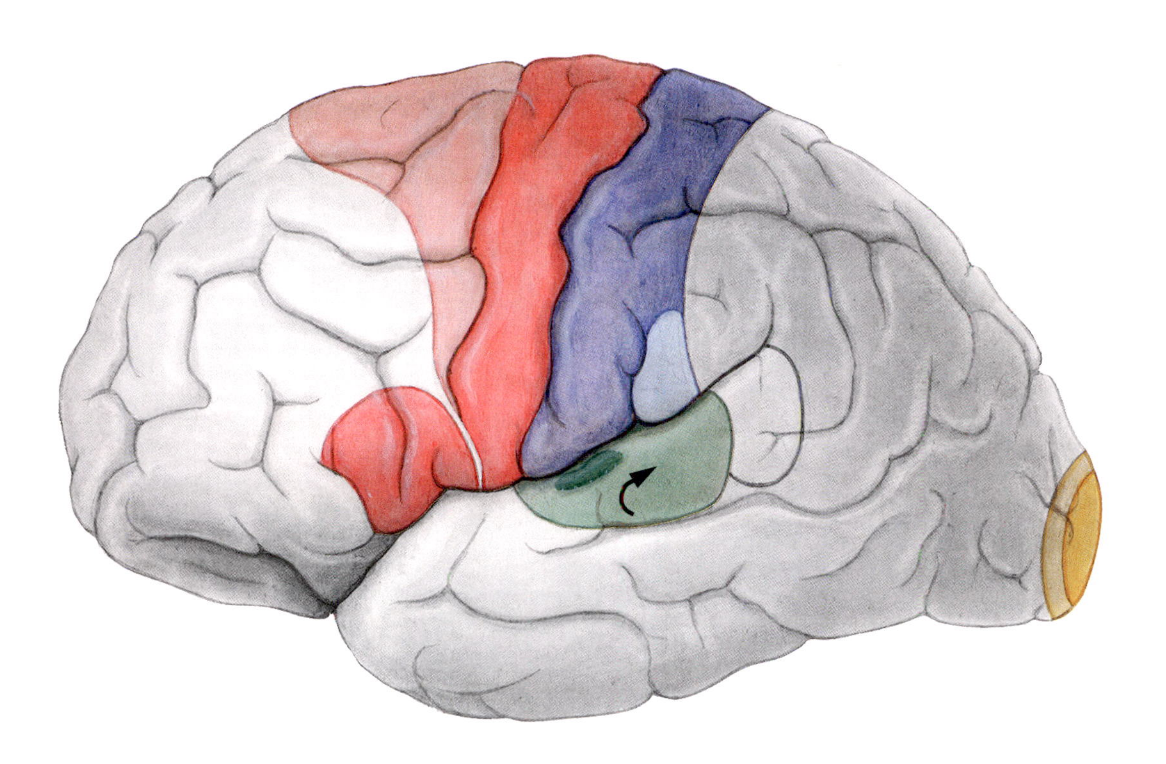

◆
Fig. 153
Cortical fields of the cerebrum viewed from the left-hand side. Projection fields are the demarcated brain areas to which the nerve fibers of an organ of sensation (e.g., inner ear, visual nerve, muscle, skin) lead directly. Association fields, on the other hand, are brain areas from which connections exist to other brain areas. Such association fields serve higher mental functions, e.g., the conscious processing of an image or an auditory sensation. The projection field of hearing (arrow) extends over the upper edge of the temporal lobe onto its inner surface.

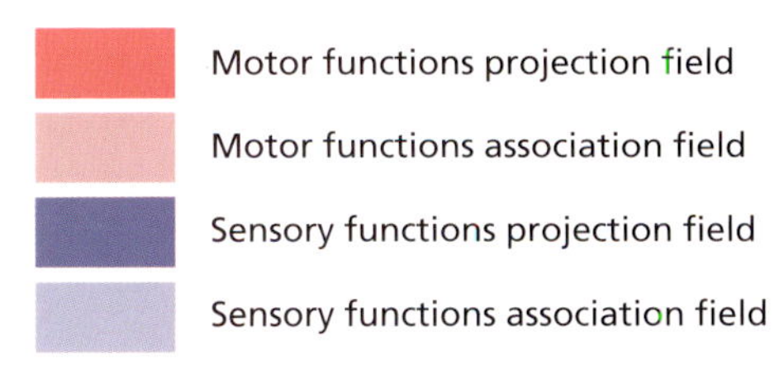

Motor functions = voluntarily controlled movements
Sensory functions = perception of stimuli

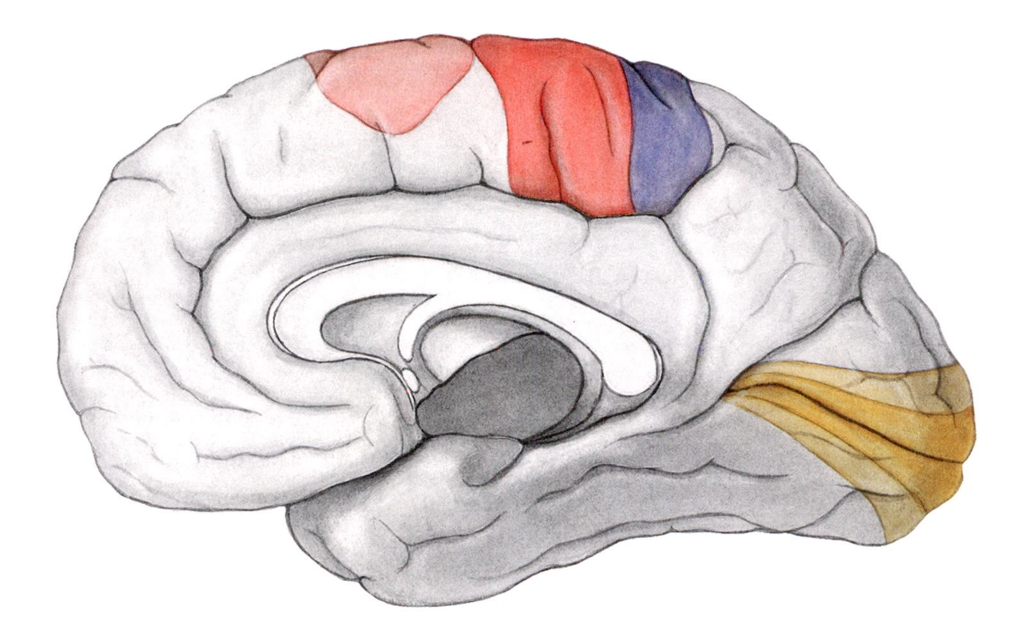

◆
Fig. 154
Cortical fields of the cerebrum, seen from the middle (sagittal section).

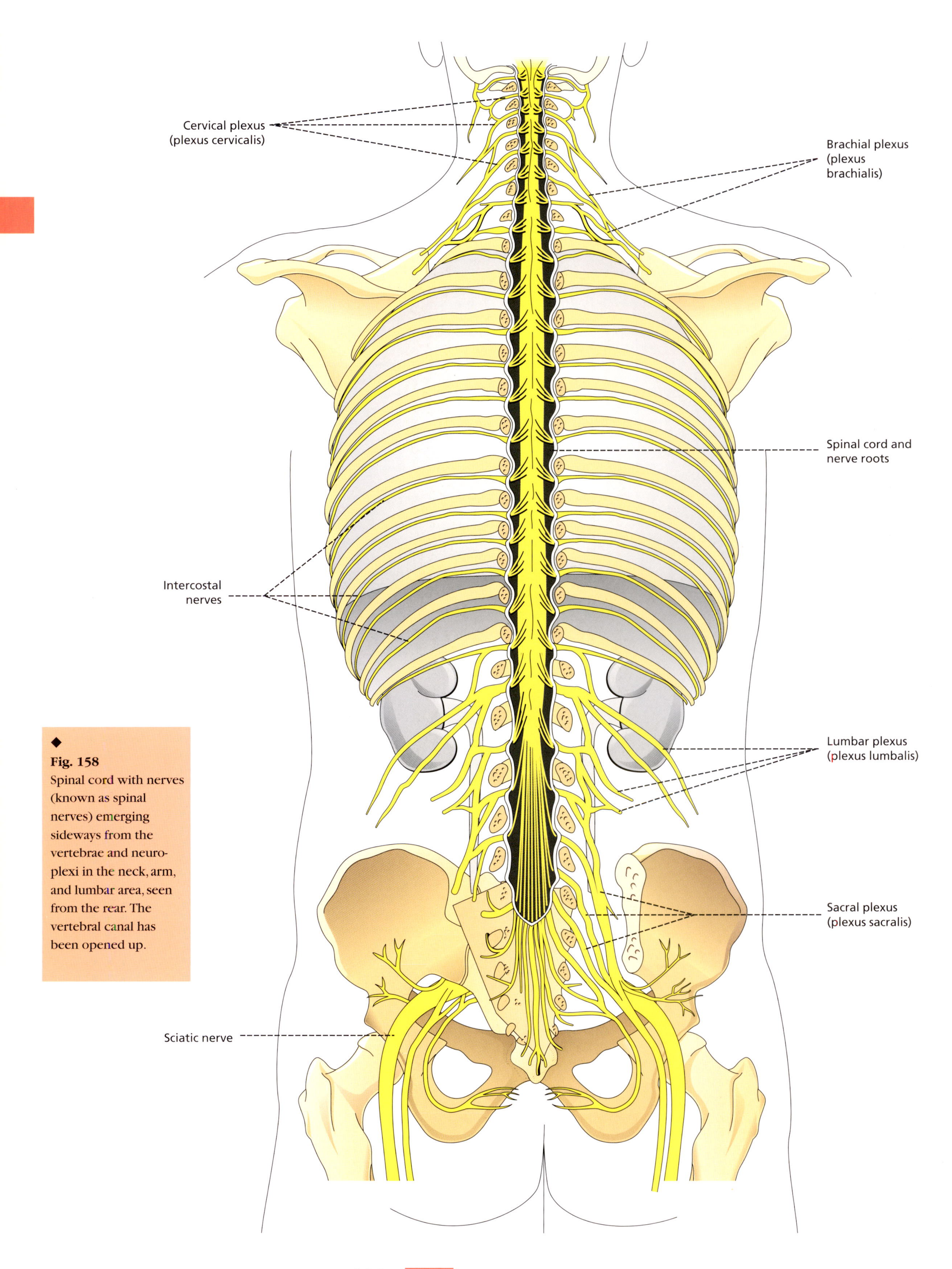

◆
Fig. 158
Spinal cord with nerves (known as spinal nerves) emerging sideways from the vertebrae and neuroplexi in the neck, arm, and lumbar area, seen from the rear. The vertebral canal has been opened up.

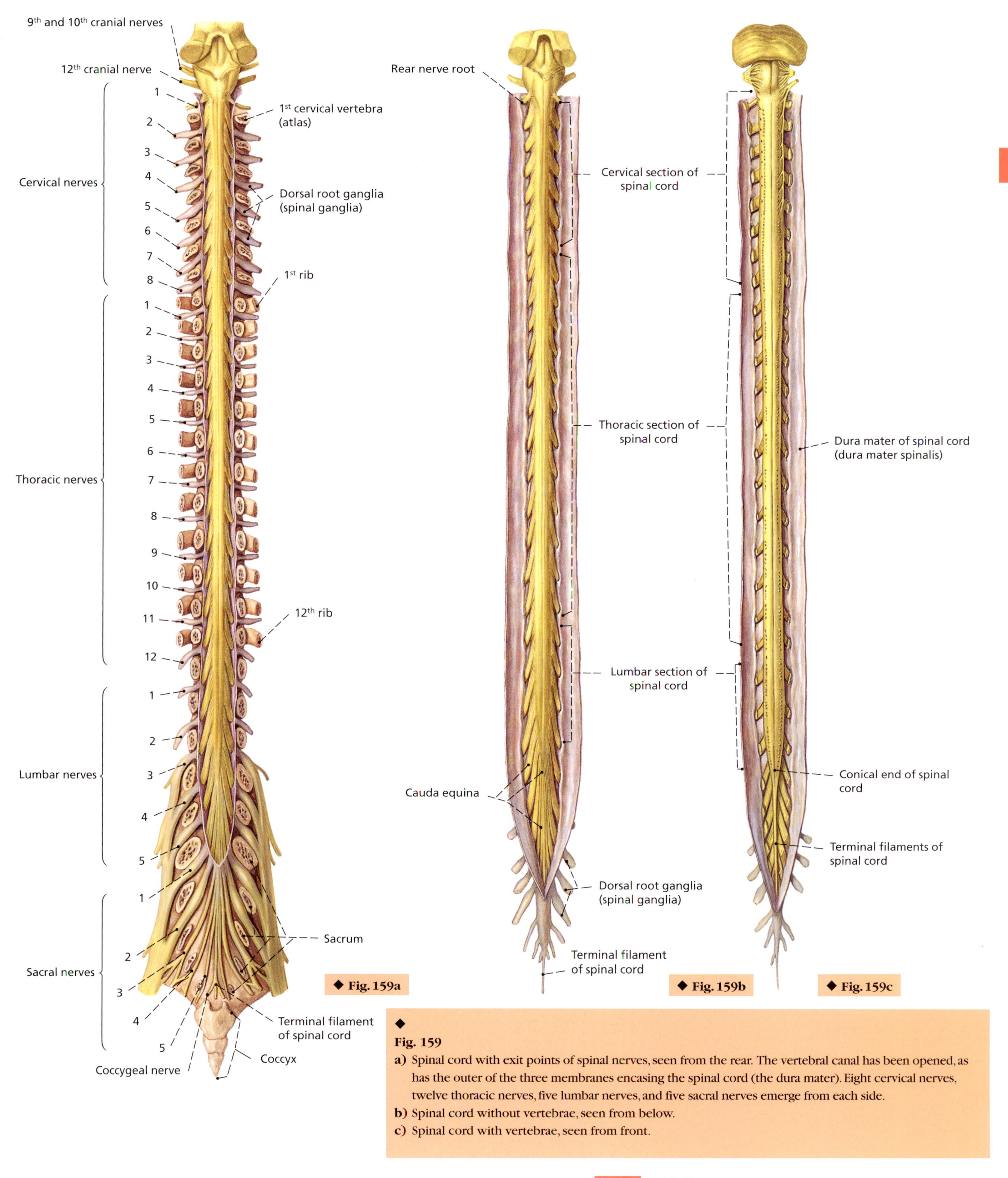

◆

Fig. 159

a) Spinal cord with exit points of spinal nerves, seen from the rear. The vertebral canal has been opened, as has the outer of the three membranes encasing the spinal cord (the dura mater). Eight cervical nerves, twelve thoracic nerves, five lumbar nerves, and five sacral nerves emerge from each side.

b) Spinal cord without vertebrae, seen from below.

c) Spinal cord with vertebrae, seen from front.

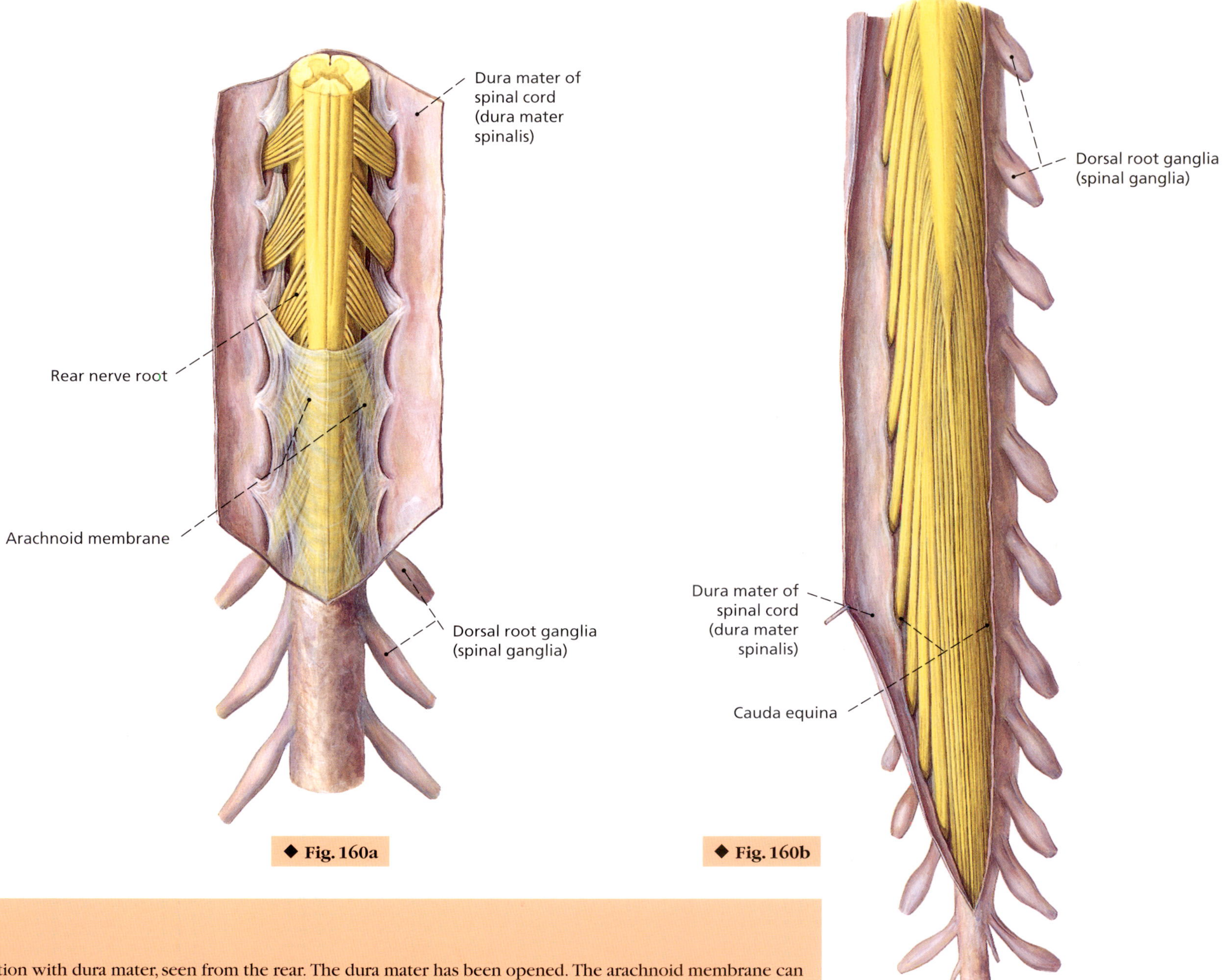

◆ Fig. 160a

◆ Fig. 160b

◆
Fig. 160
a) Spinal cord section with dura mater, seen from the rear. The dura mater has been opened. The arachnoid membrane can also be seen.
b) Lowest section of spinal cord after opening of the dura mater, seen from the front. The spinal cord ganglia (spinal ganglia) can be seen on the right-hand side.

◆
Fig. 161
Spinal nerve roots in the cervical (a), thoracic (b), and lumbar (c) regions. A distinction can be made between the front root and the rear root. The front root carries motor nerve fibers, the rear root sensory nerve fibers.

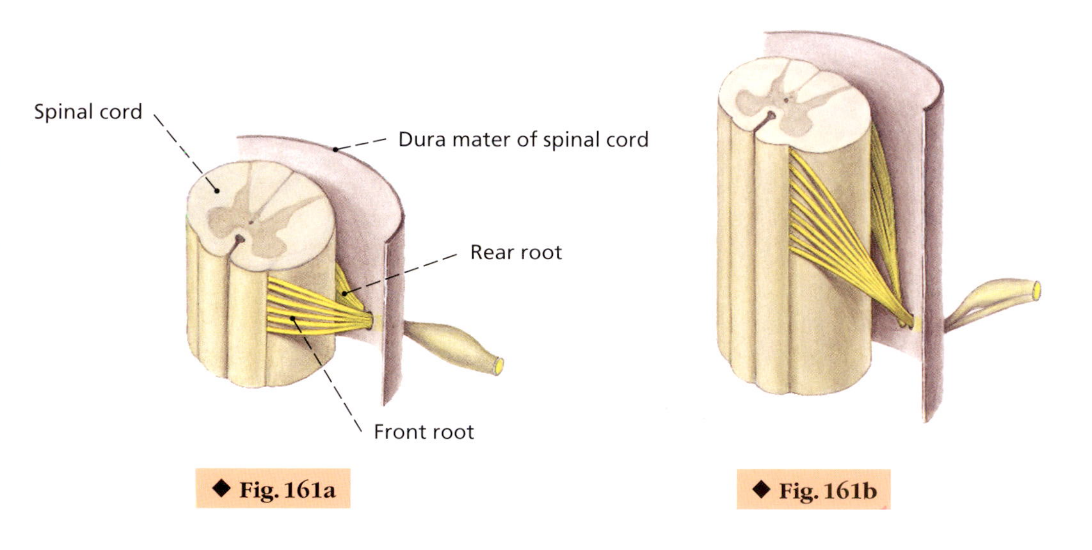

◆ Fig. 161a

◆ Fig. 161b

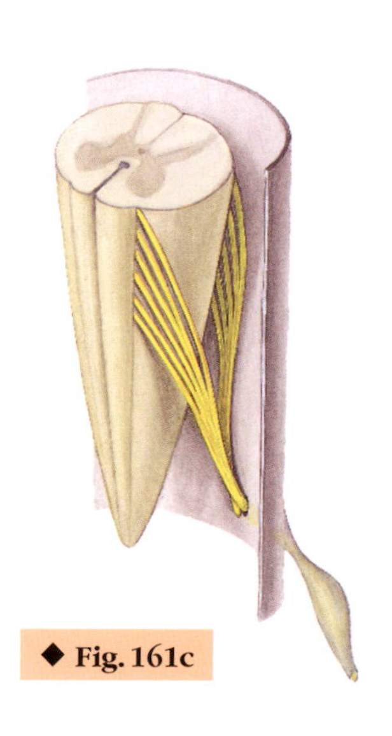

◆ Fig. 161c

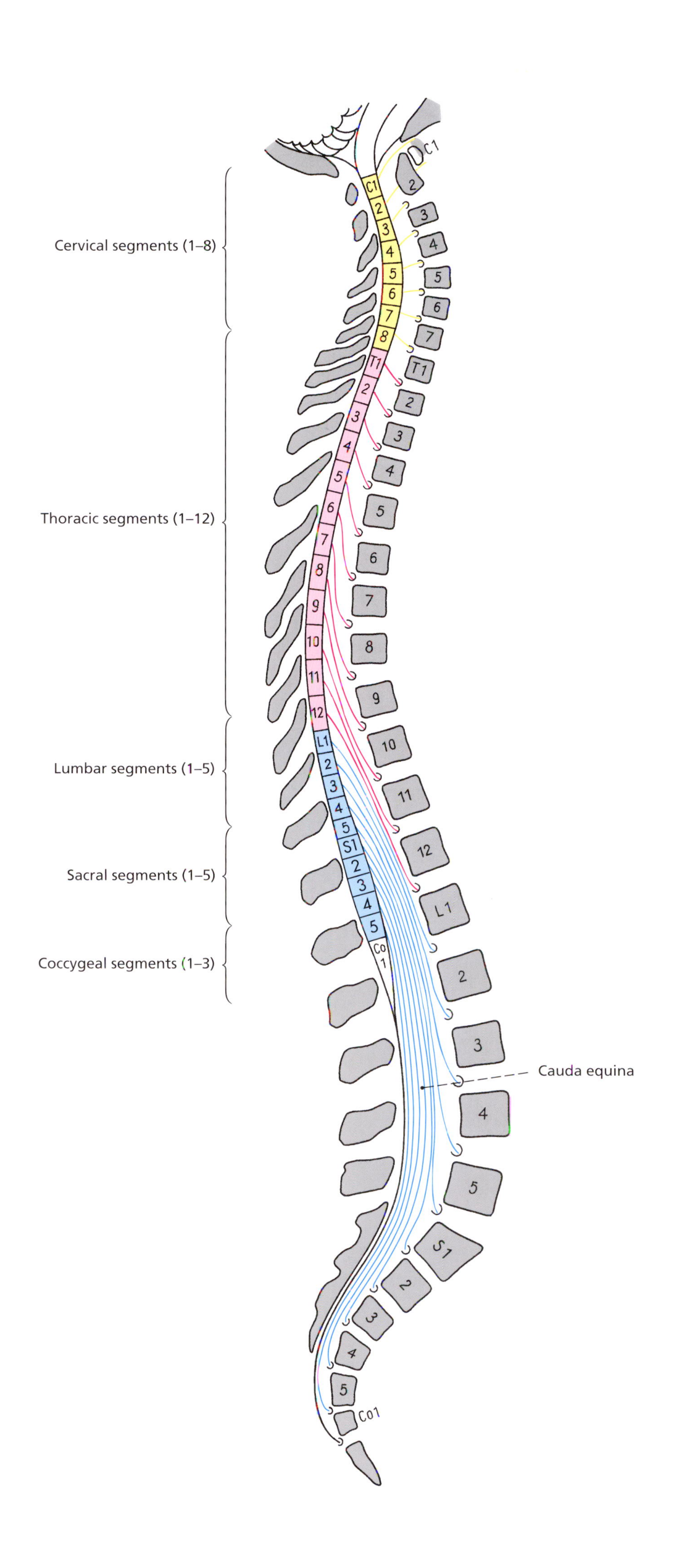

◆
Fig. 162
Schematic representation of spinal cord sections (segments), seen from the side (the head is turned to the right in the image). The cervical, thoracic, and lumbar segments are color-coded. The structures shown in gray represent the vertebrae. The lower end of the spinal cord extends to about the height of the second lumbar vertebra.

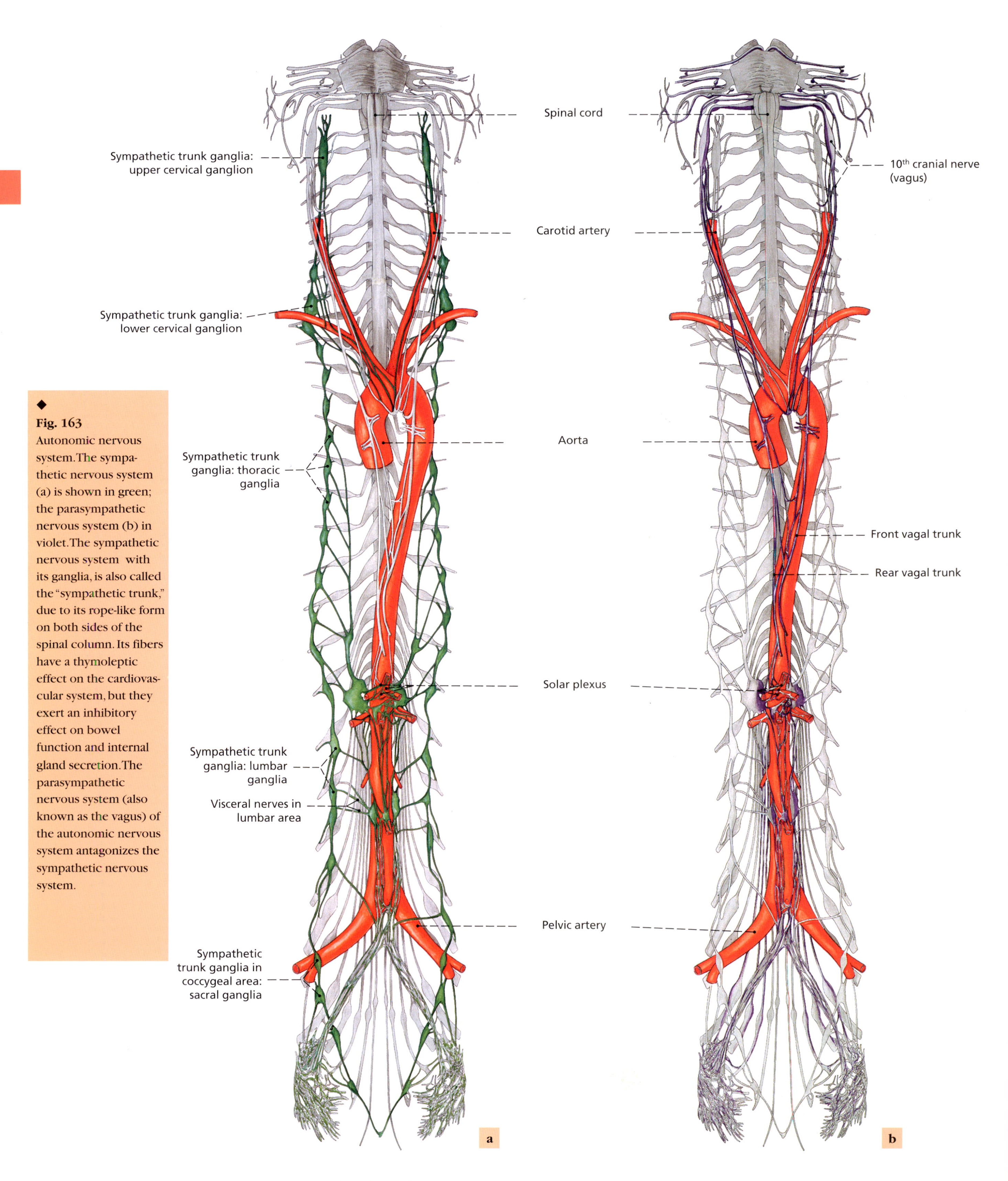

◆
Fig. 163
Autonomic nervous system. The sympathetic nervous system (a) is shown in green; the parasympathetic nervous system (b) in violet. The sympathetic nervous system with its ganglia, is also called the "sympathetic trunk," due to its rope-like form on both sides of the spinal column. Its fibers have a thymoleptic effect on the cardiovascular system, but they exert an inhibitory effect on bowel function and internal gland secretion. The parasympathetic nervous system (also known as the vagus) of the autonomic nervous system antagonizes the sympathetic nervous system.

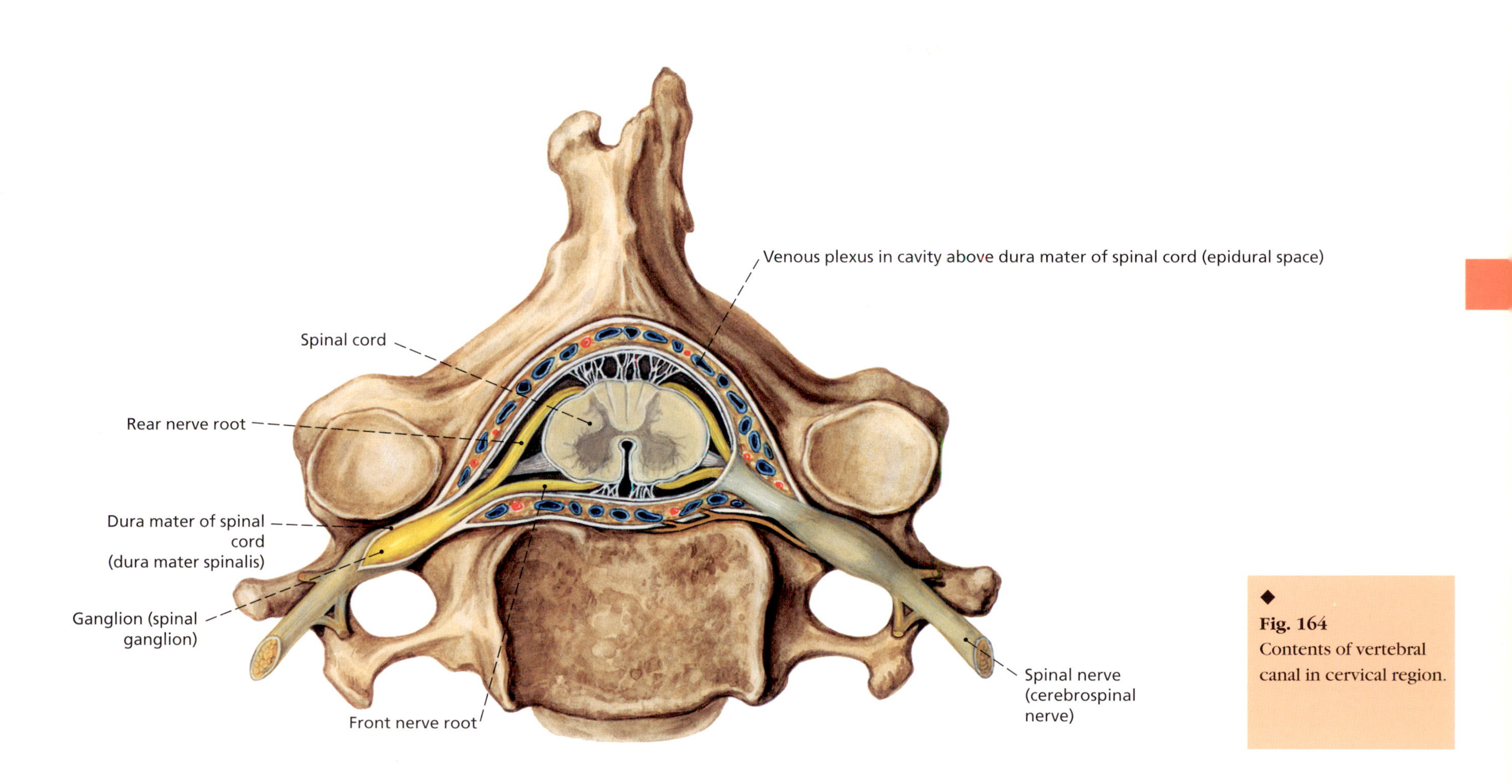

◆
Fig. 164
Contents of vertebral canal in cervical region.

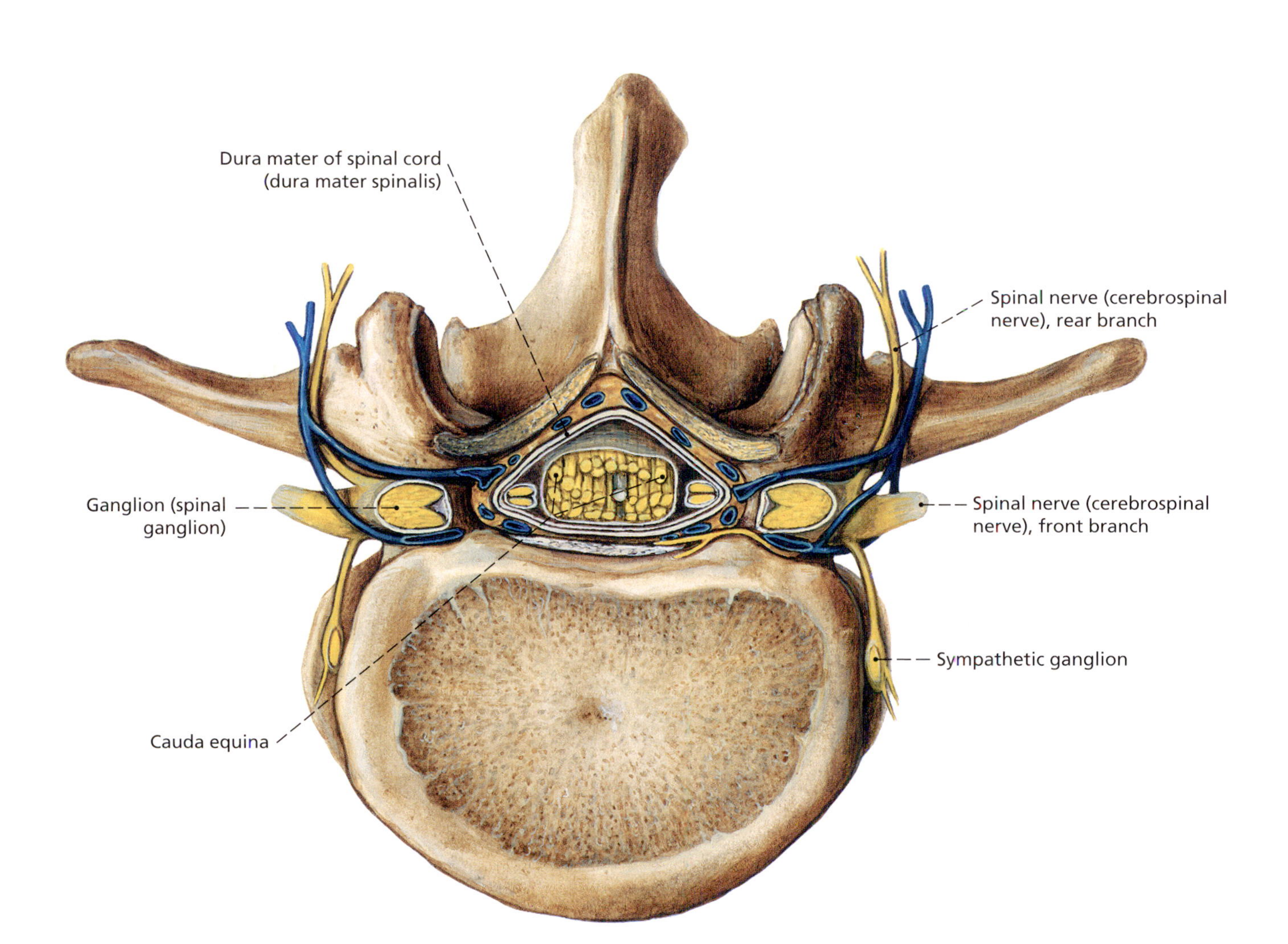

◆
Fig. 165
Contents of vertebral canal at level of third lumbar vertebra.

◆
Fig. 166
Schematic representation of a cerebrospinal nerve (yellow) in the thoracic area.

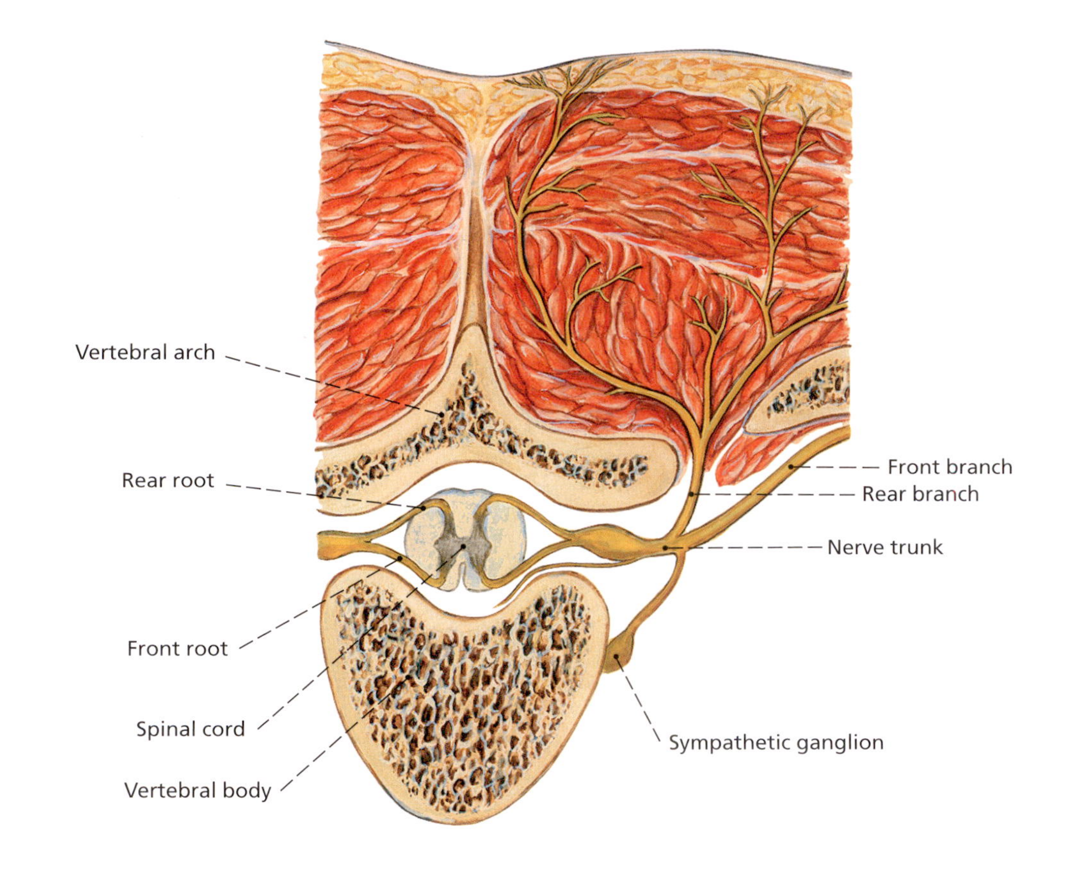

◆
Fig. 167
Contents of vertebral canal at the transition from the lumbar vertebra to the sacral area. The vertebral arches have been removed. Near the venous plexus in the epidural space (upper section of illustration), the opened-up dura mater and pia mater of the spinal cord (as well as the arachnoid membrane) can be clearly distinguished in the vertebral canal.

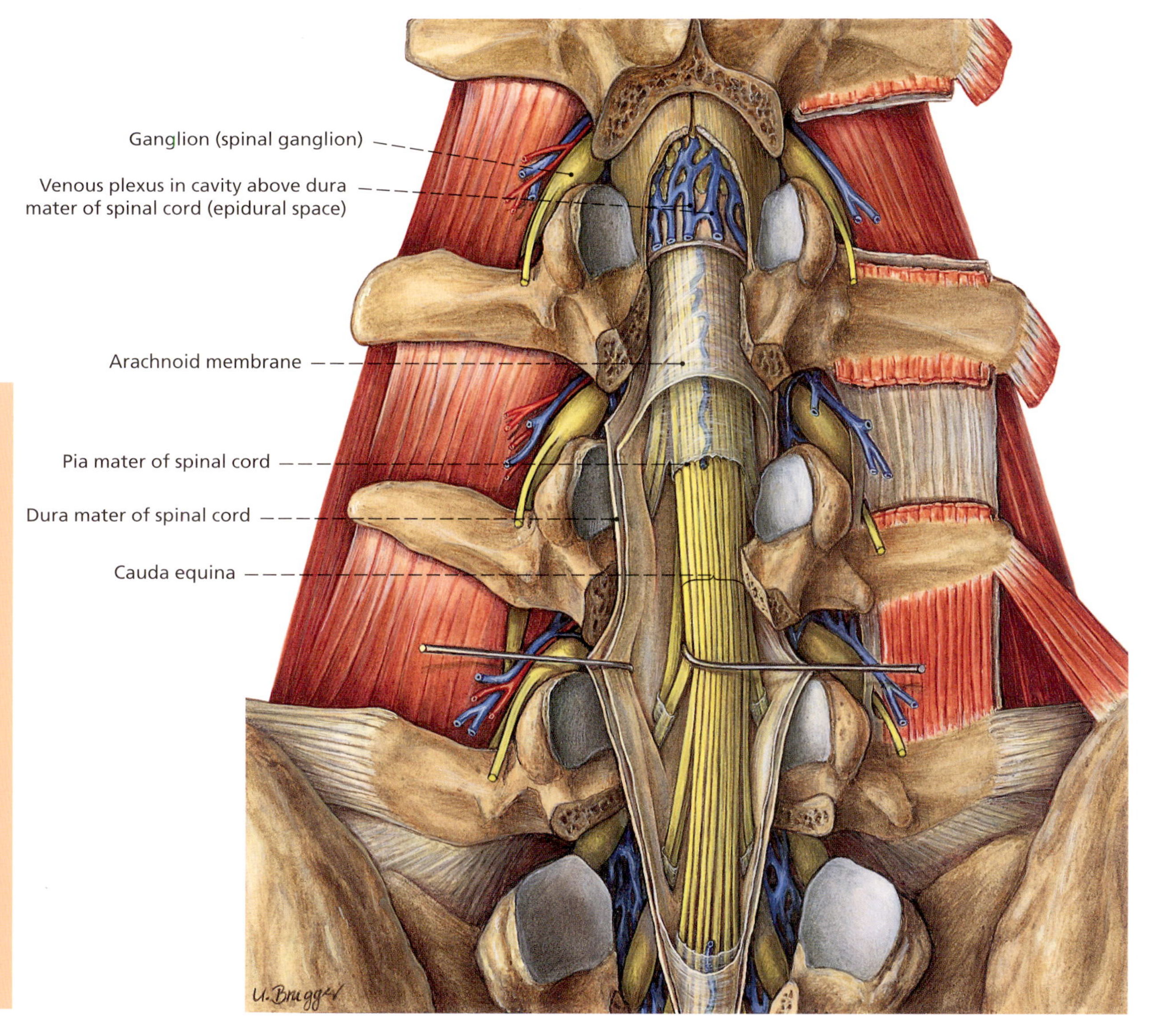

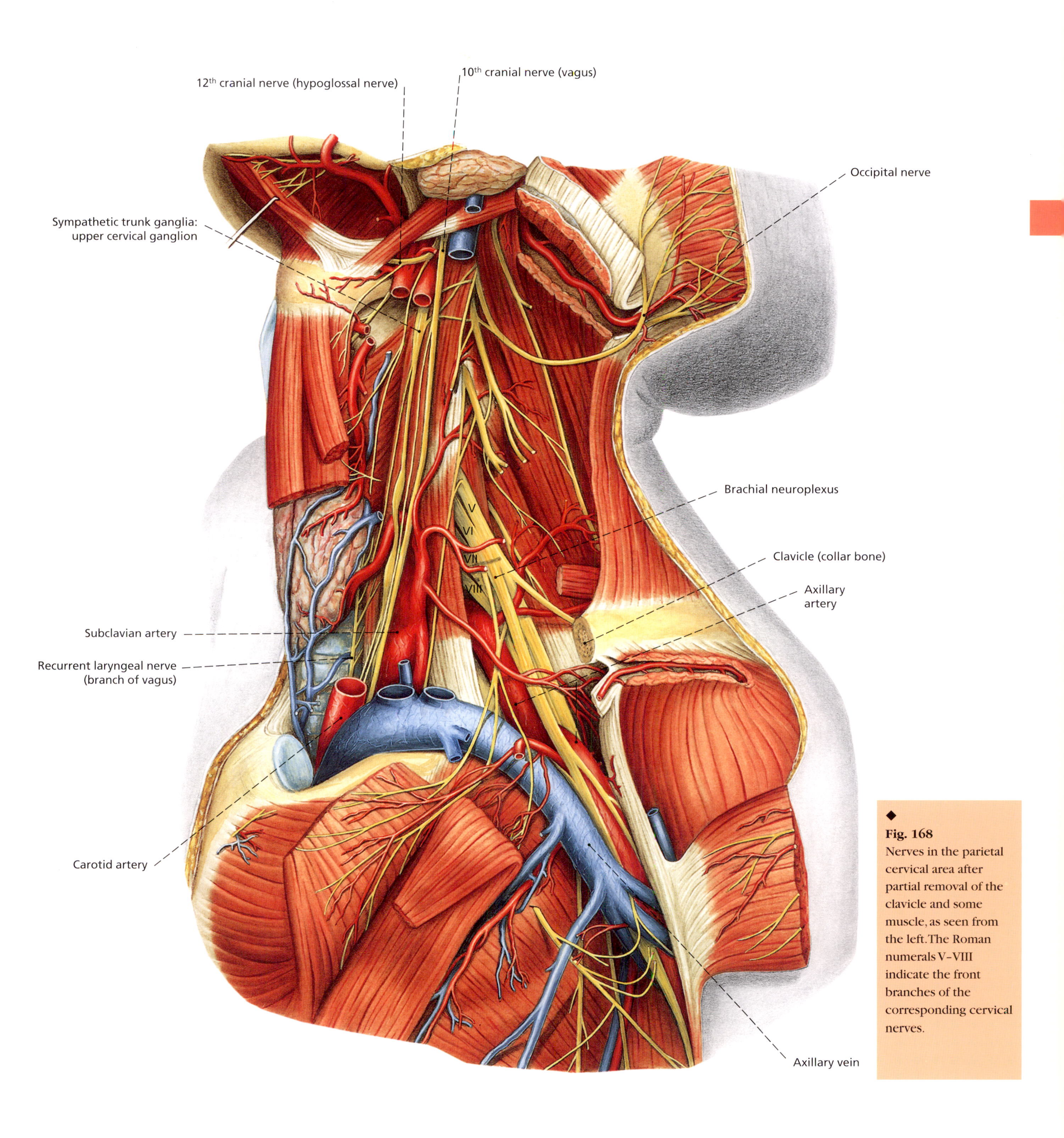

◆
Fig. 168
Nerves in the parietal cervical area after partial removal of the clavicle and some muscle, as seen from the left. The Roman numerals V–VIII indicate the front branches of the corresponding cervical nerves.

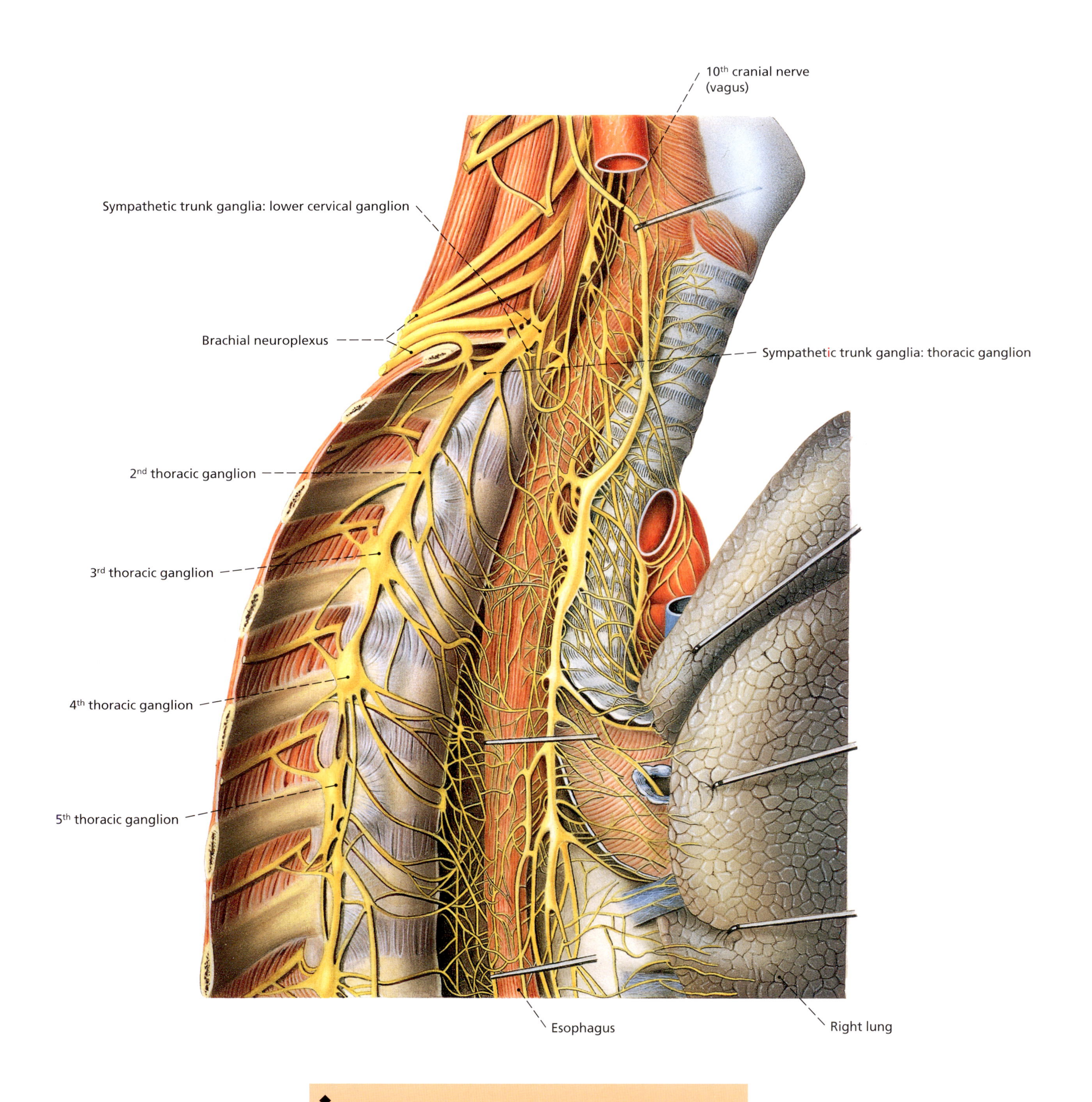

◆
Fig. 169
Nerves in the lower cervical and upper thoracic areas, with sections of the autonomic nervous system. Viewed from the right. The vagus nerve and the right lung have been moved forward to show the esophagus.

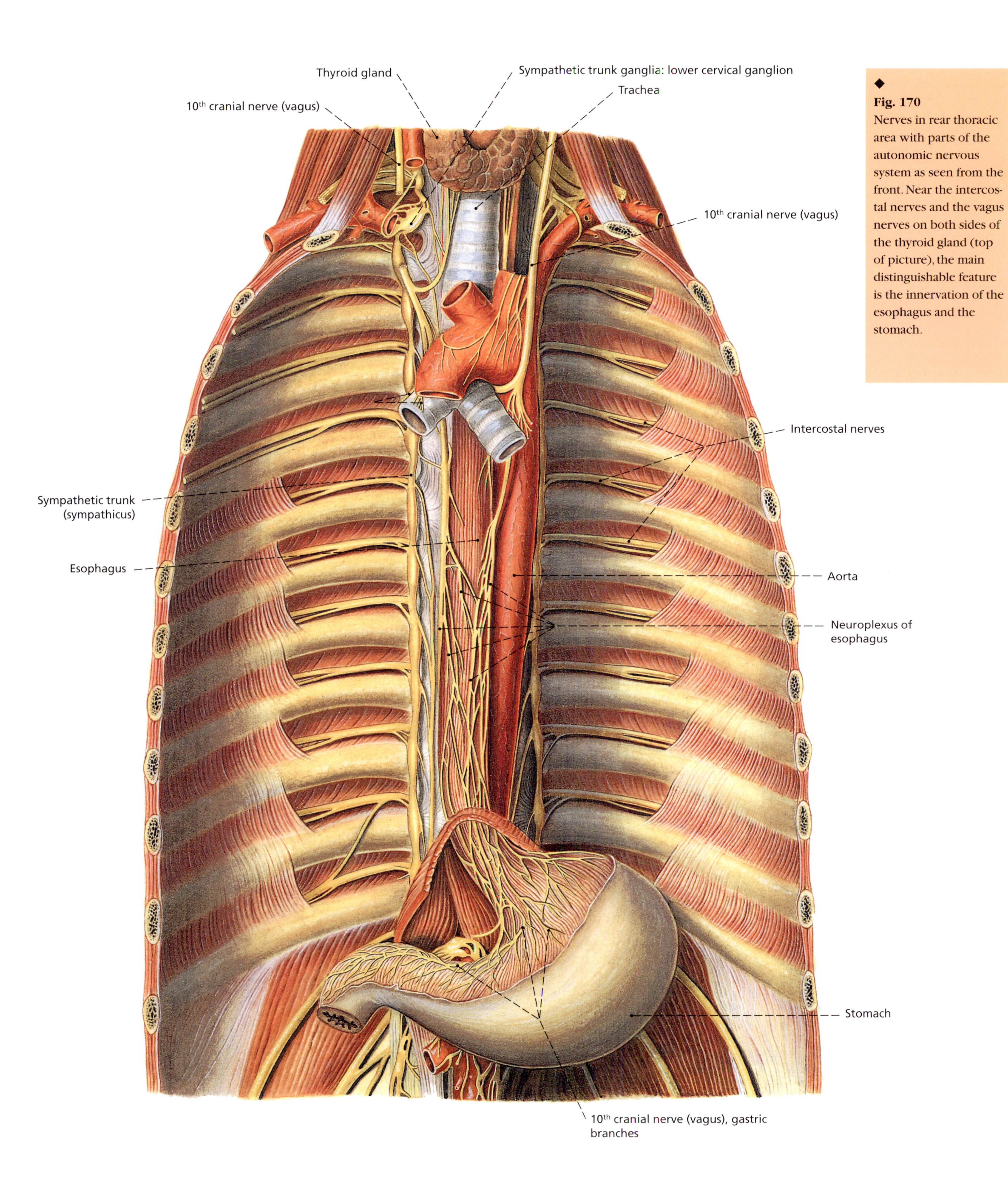

◆
Fig. 170
Nerves in rear thoracic area with parts of the autonomic nervous system as seen from the front. Near the intercostal nerves and the vagus nerves on both sides of the thyroid gland (top of picture), the main distinguishable feature is the innervation of the esophagus and the stomach.

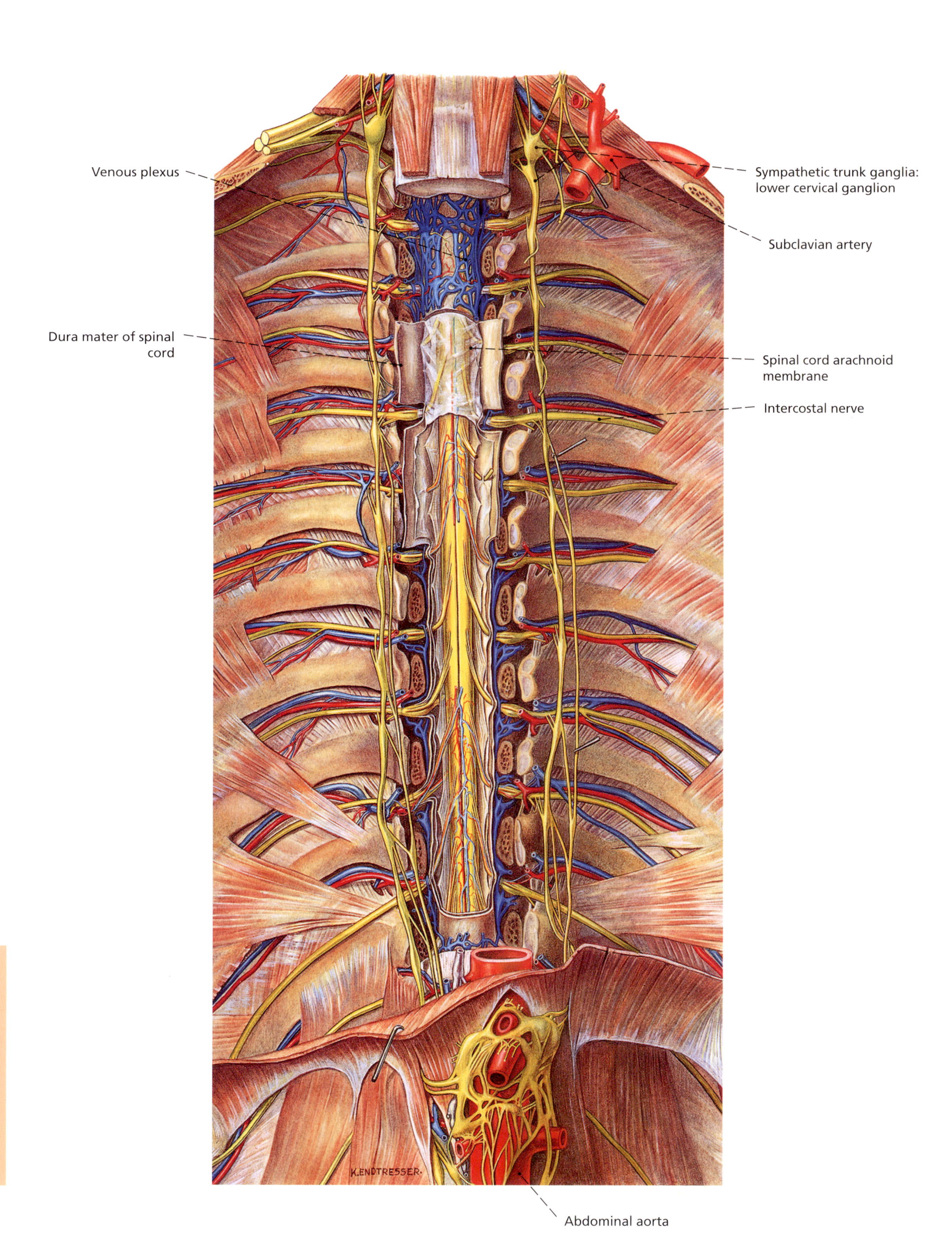

◆
Fig. 171
Spinal cord in open vertebral canal seen from the front (thoracic area). At the bottom of the picture, the neuroplexus of the abdominal artery (abdominal aorta) can be distinguished.

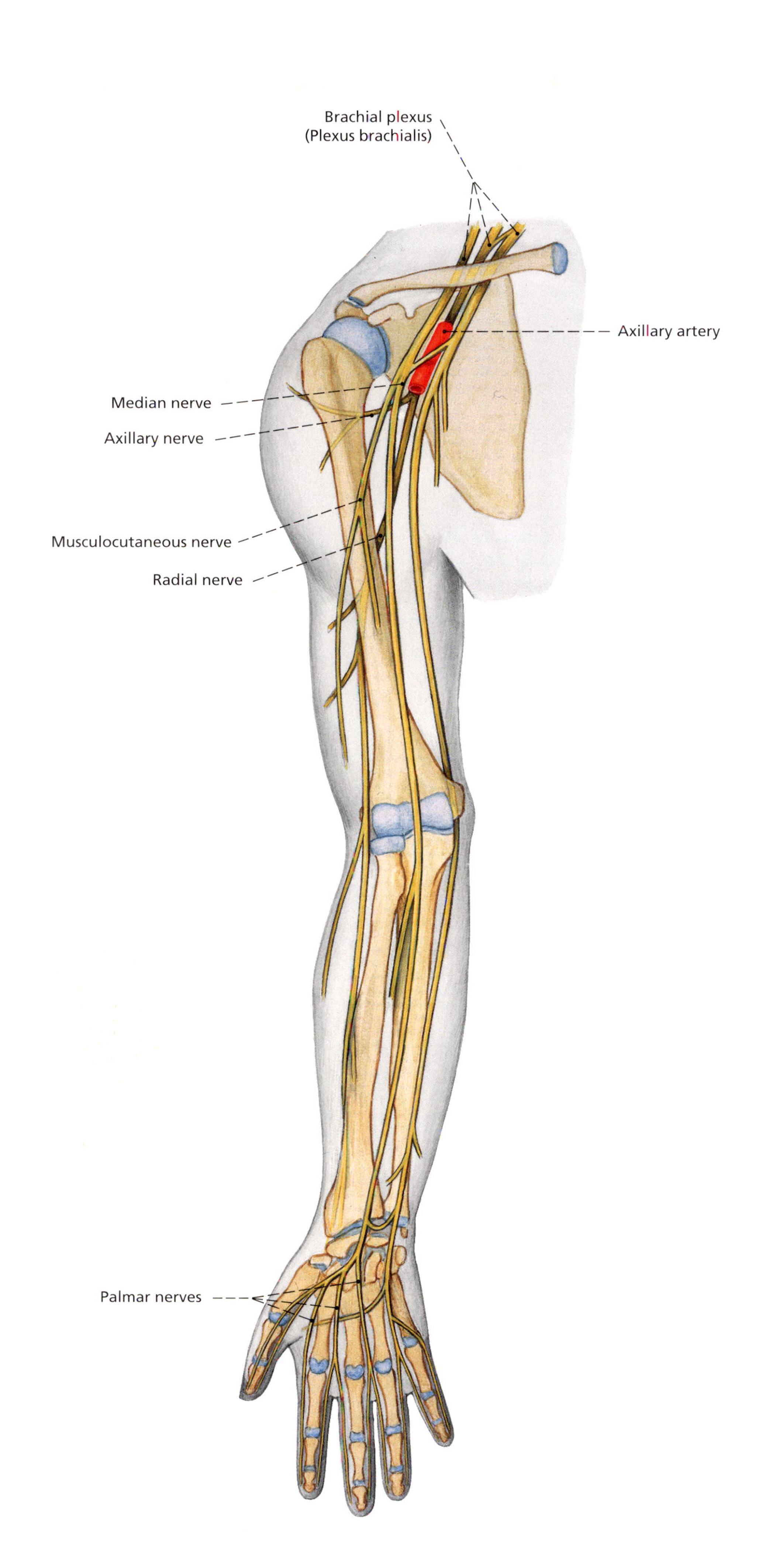

◆
Fig. 172
Brachial nerves seen from the front. The course of the axillary artery (short section marked in red), which enters the upper arm between the brachial nerves, is outlined.

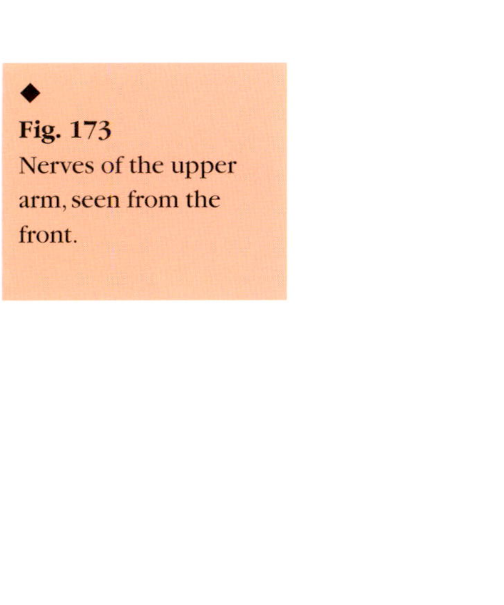
◆
Fig. 173
Nerves of the upper arm, seen from the front.

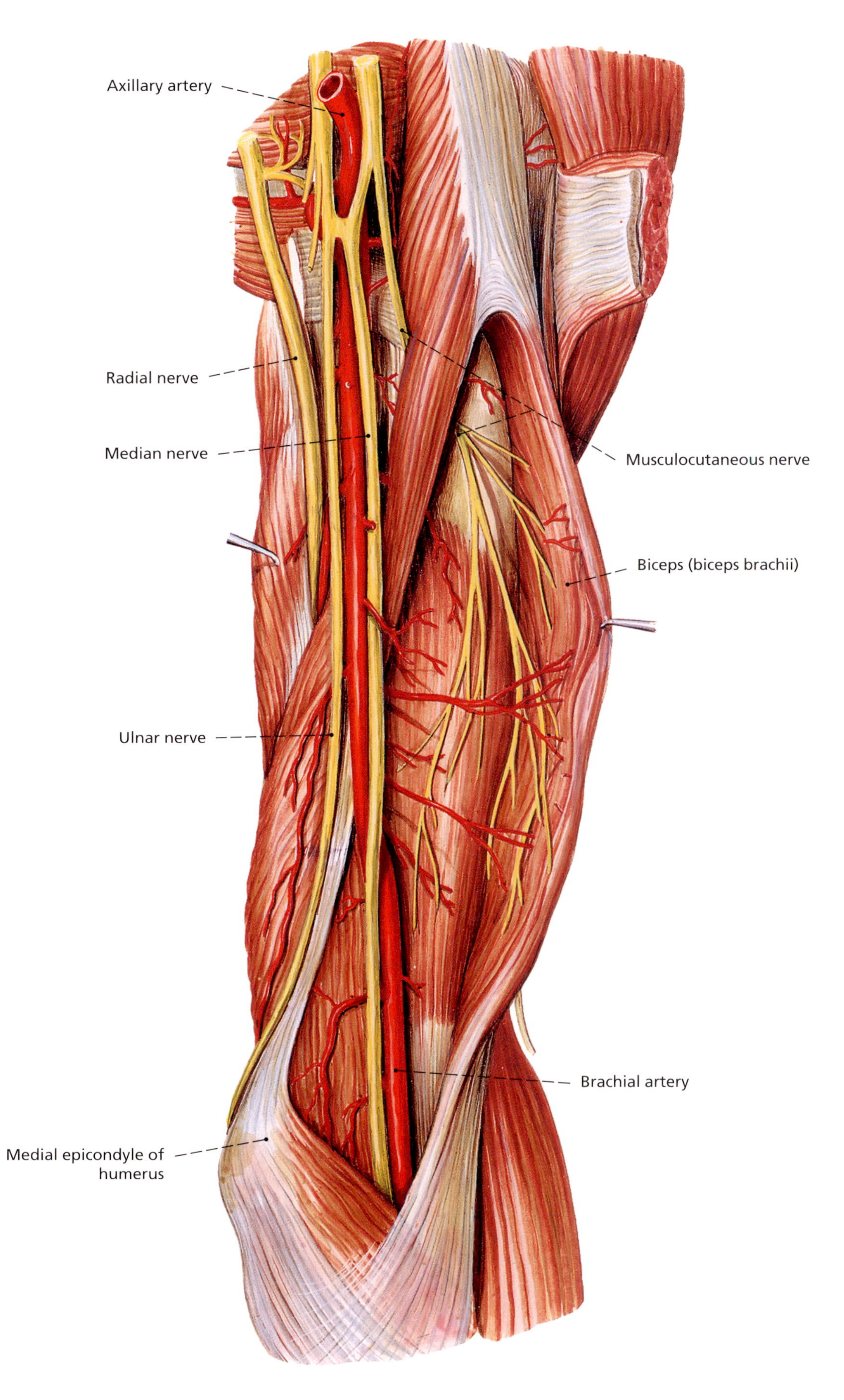

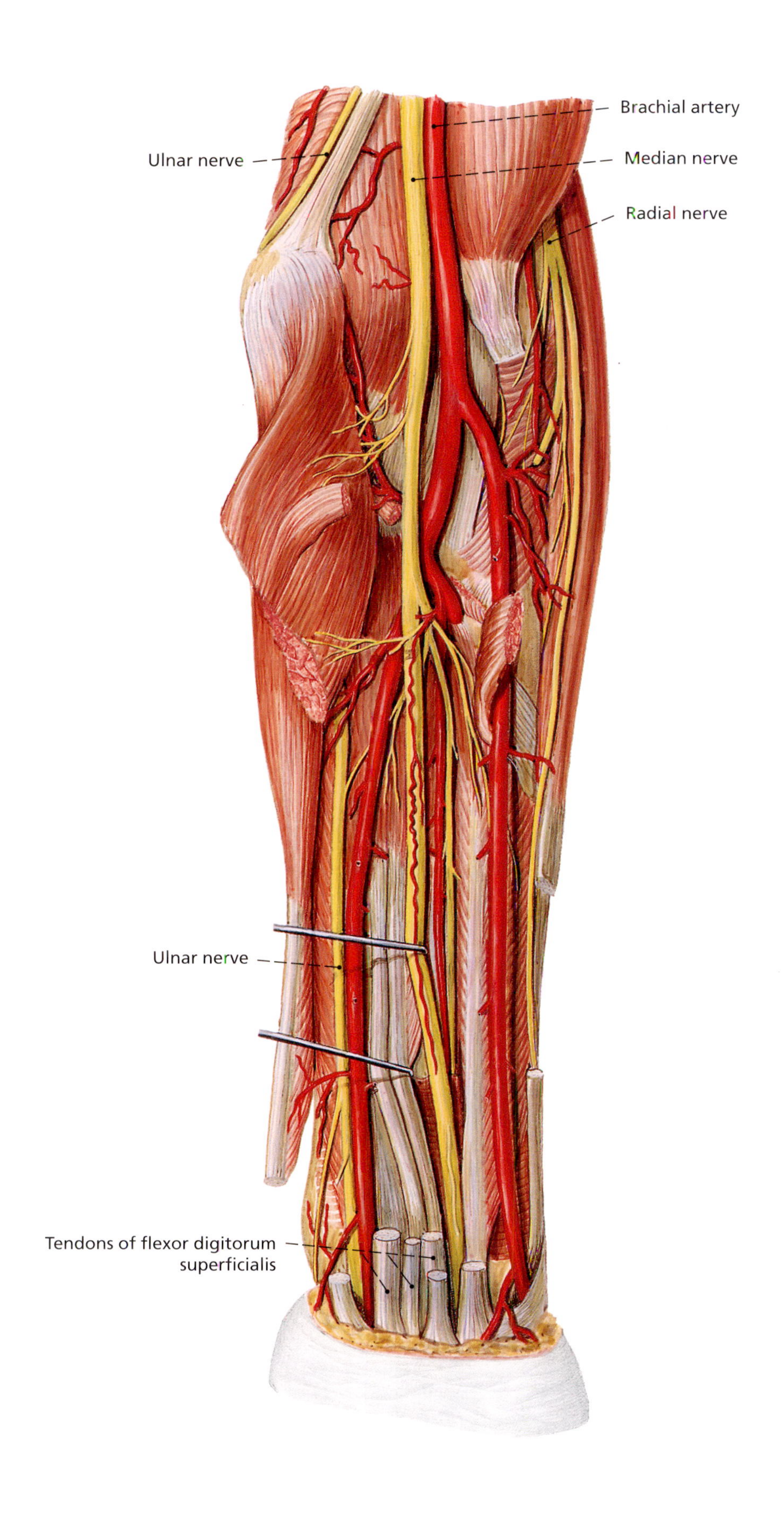

◆
Fig. 174
Nerves of the left forearm, seen from the front. The flexor digitorum superficialis has been partially removed.

◆
Fig. 175
Palmar nerves of the left-hand.

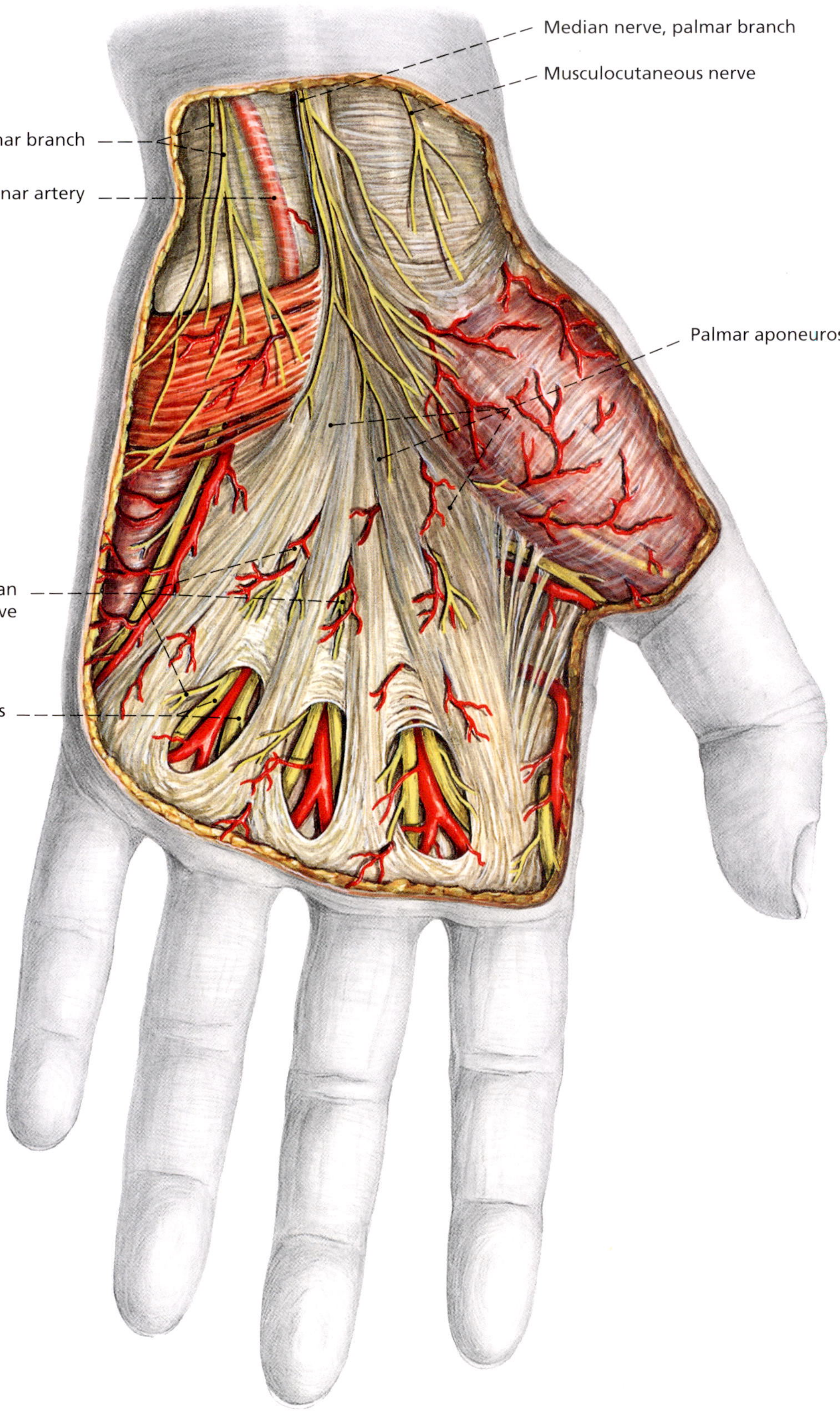

Median nerve, palmar branch
Musculocutaneous nerve
Ulnar nerve, palmar branch
Ulnar artery
Palmar aponeurosis
Palmar branches of median nerve and ulnar nerve
Palm-side finger nerves

◆
Fig. 176
Nerves of index finger, seen from thumb side.

Dorsal digital nerve
Dorsal digital artery
Palm-side finger nerve
Palm-side digital artery

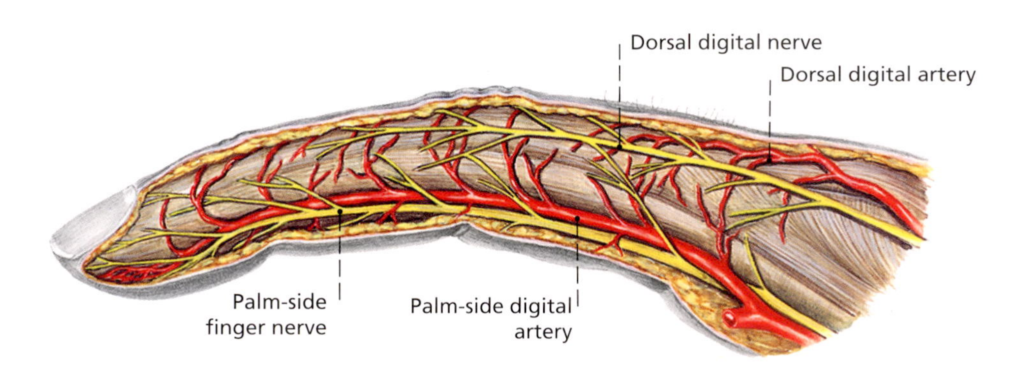

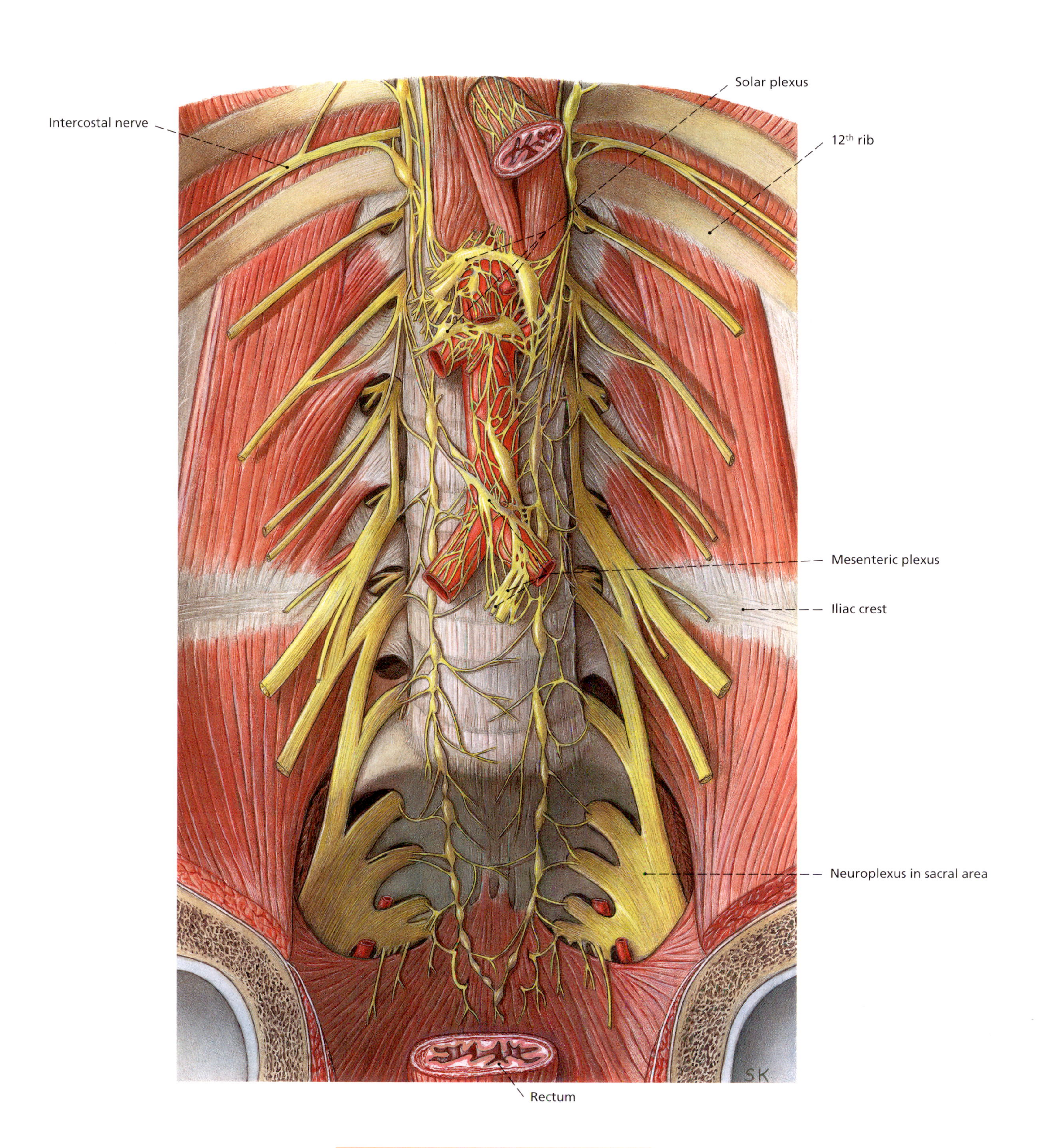

◆
Fig. 177
Neuroplexi of rear abdominal wall, seen from the front.

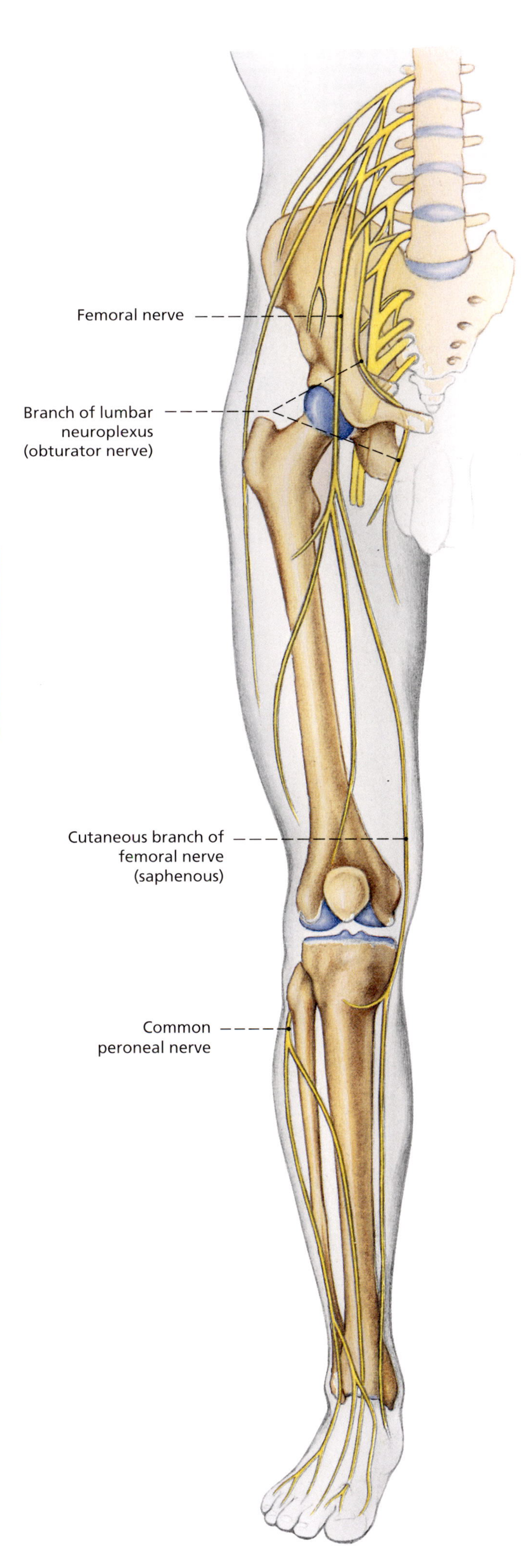

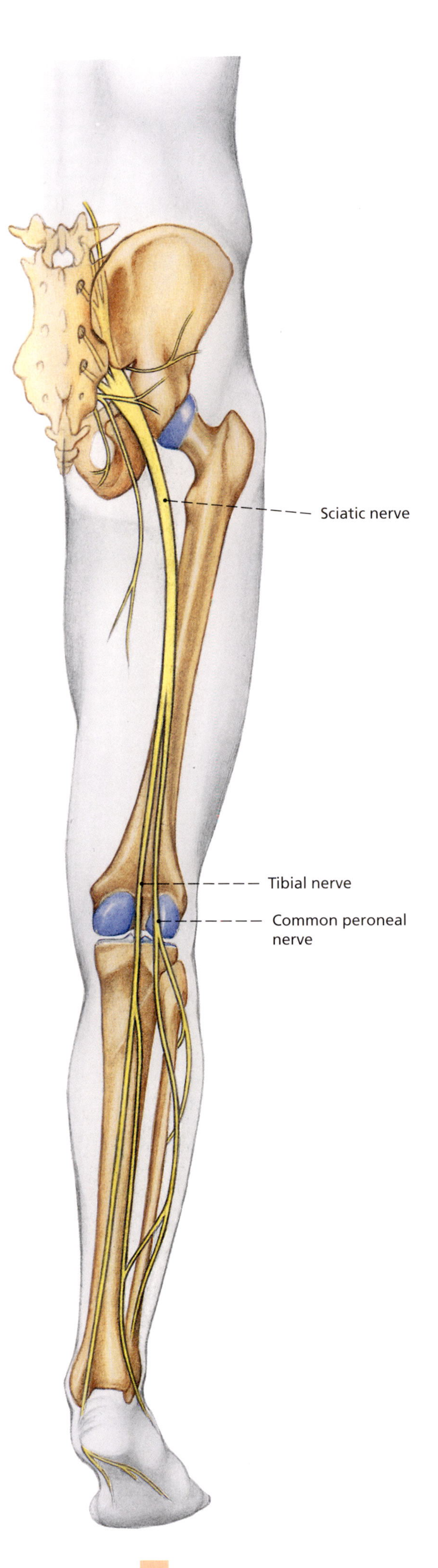

◆
Fig. 178
Nerves of the right leg, seen from the front (a) and rear (b).

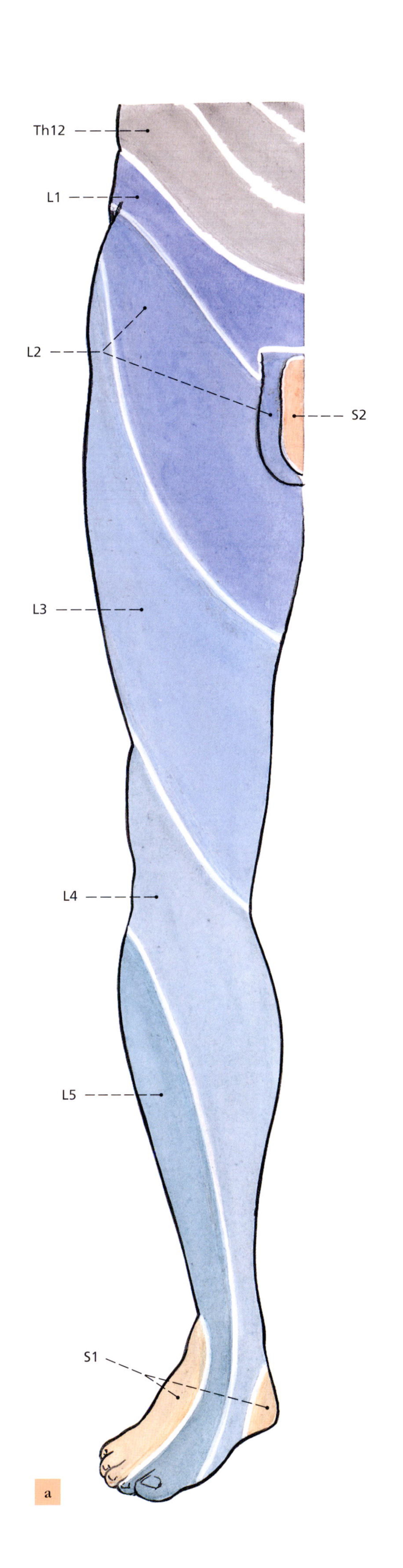

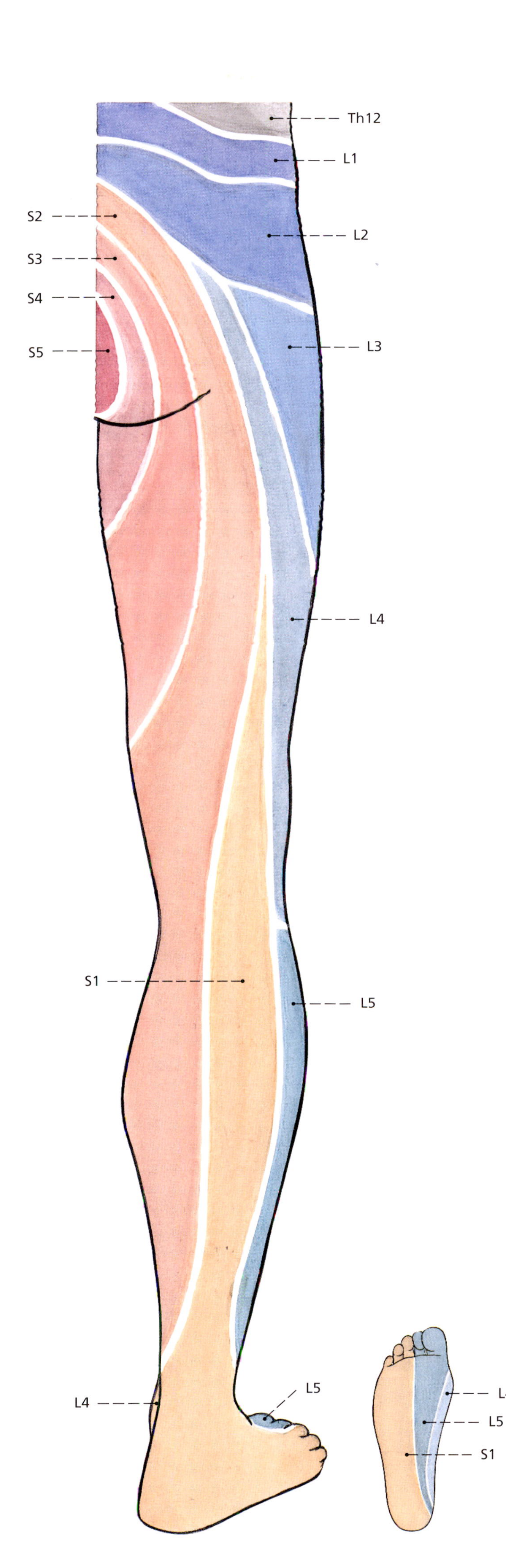

◆
Fig. 179
Innervation of dermatomes of skin of the right leg seen from the front (a) and rear (b). The dermatomes correspond to the sections (segments) of the spinal cord.

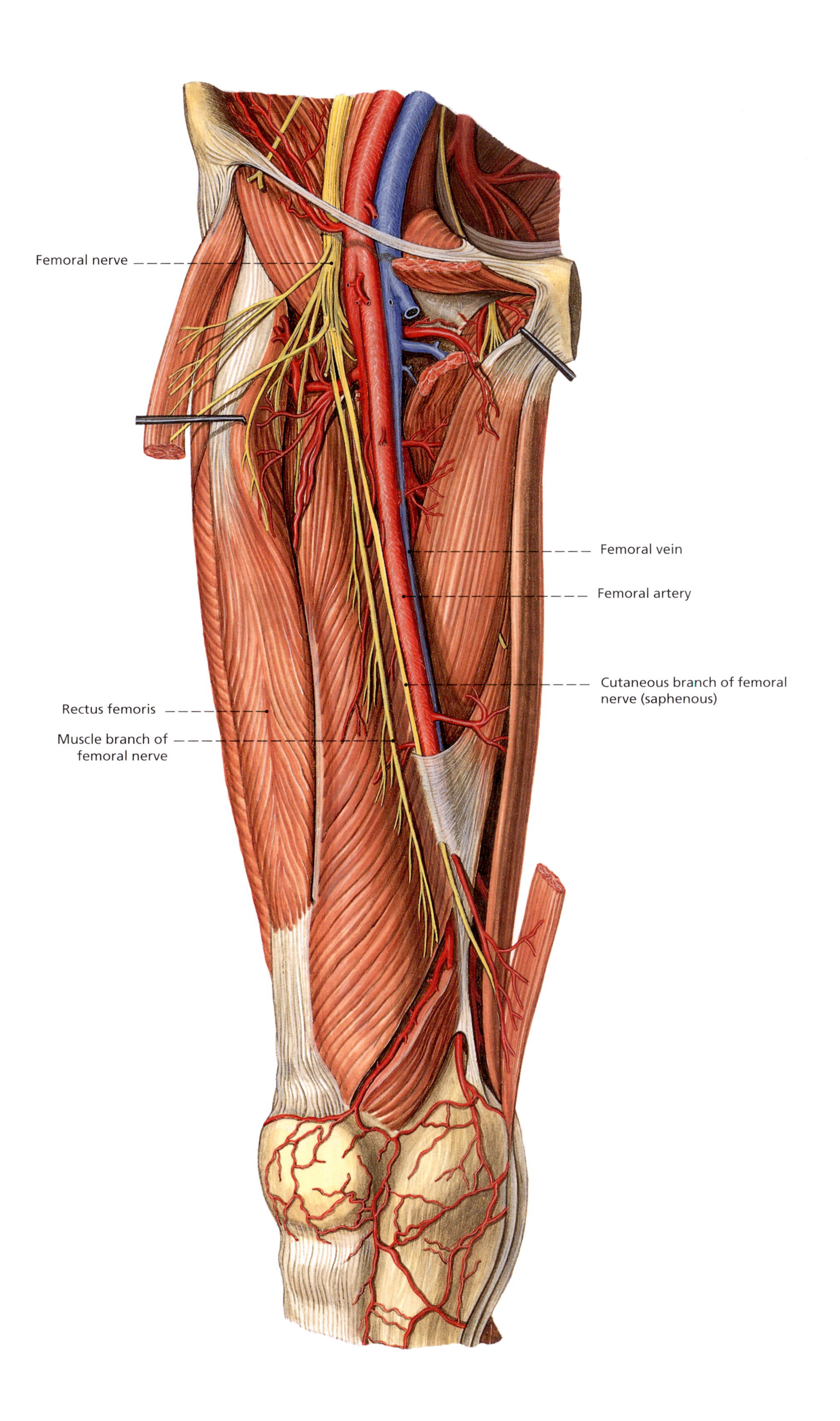

◆
Fig. 180
Nerves of the thigh after partial removal of surface muscles.

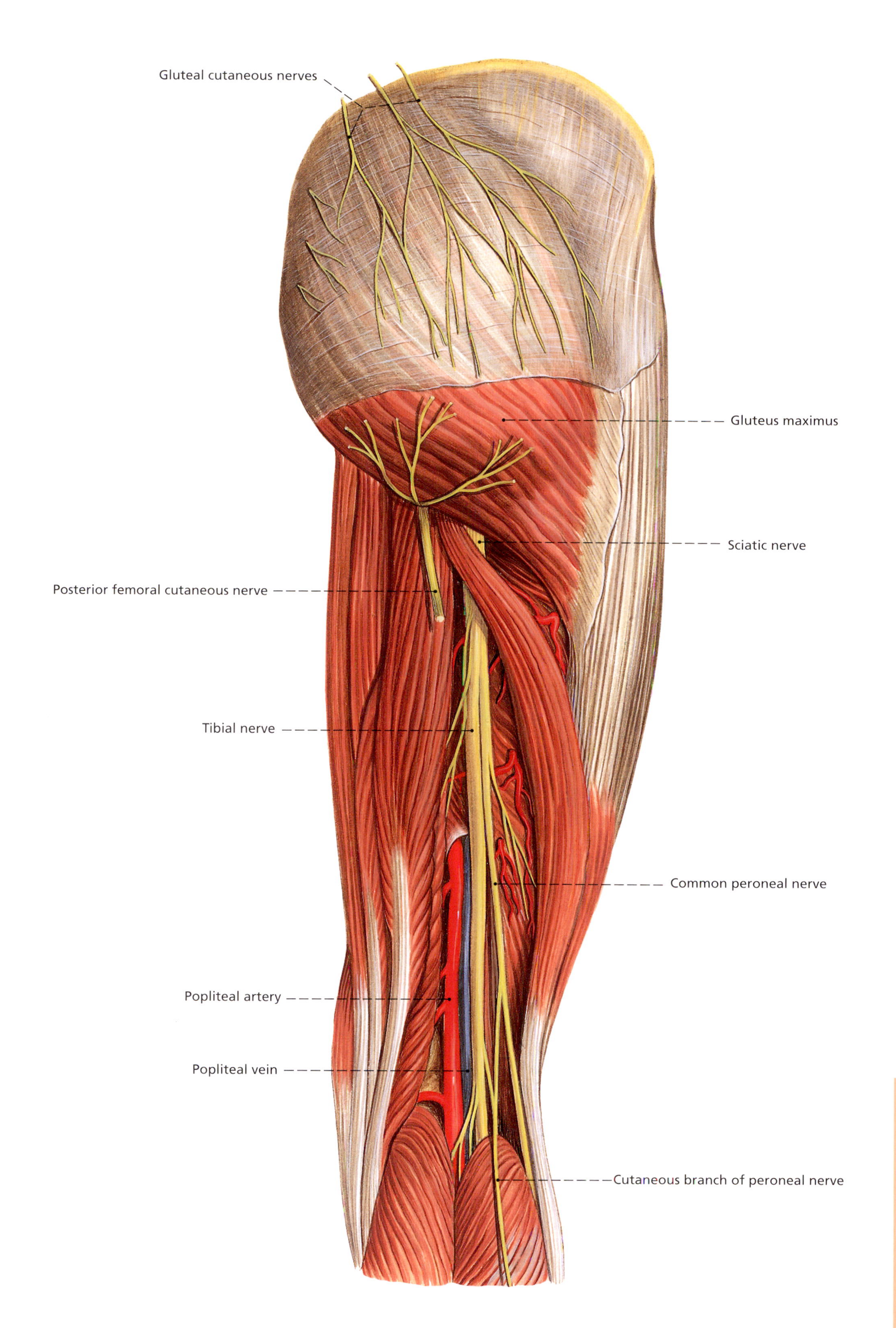

◆
Fig. 181
Nerves of the buttocks area and femur from the rear (right). The sciatic nerve, the fibers of which travel to the sole of the foot, is particularly striking.

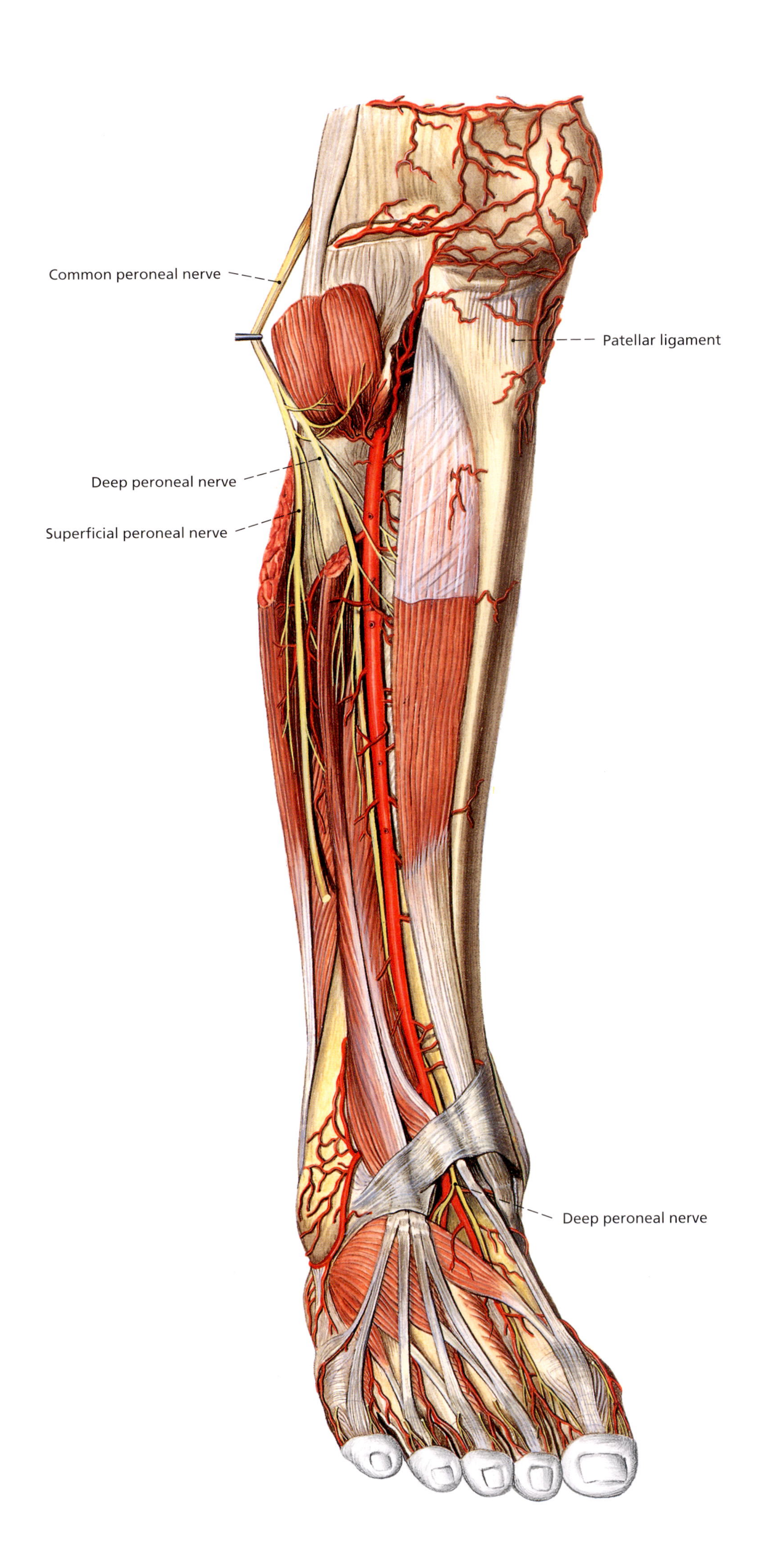

◆
Fig. 182
Nerves of the right lower leg seen from the front after partial removal of musculature from the lower leg.

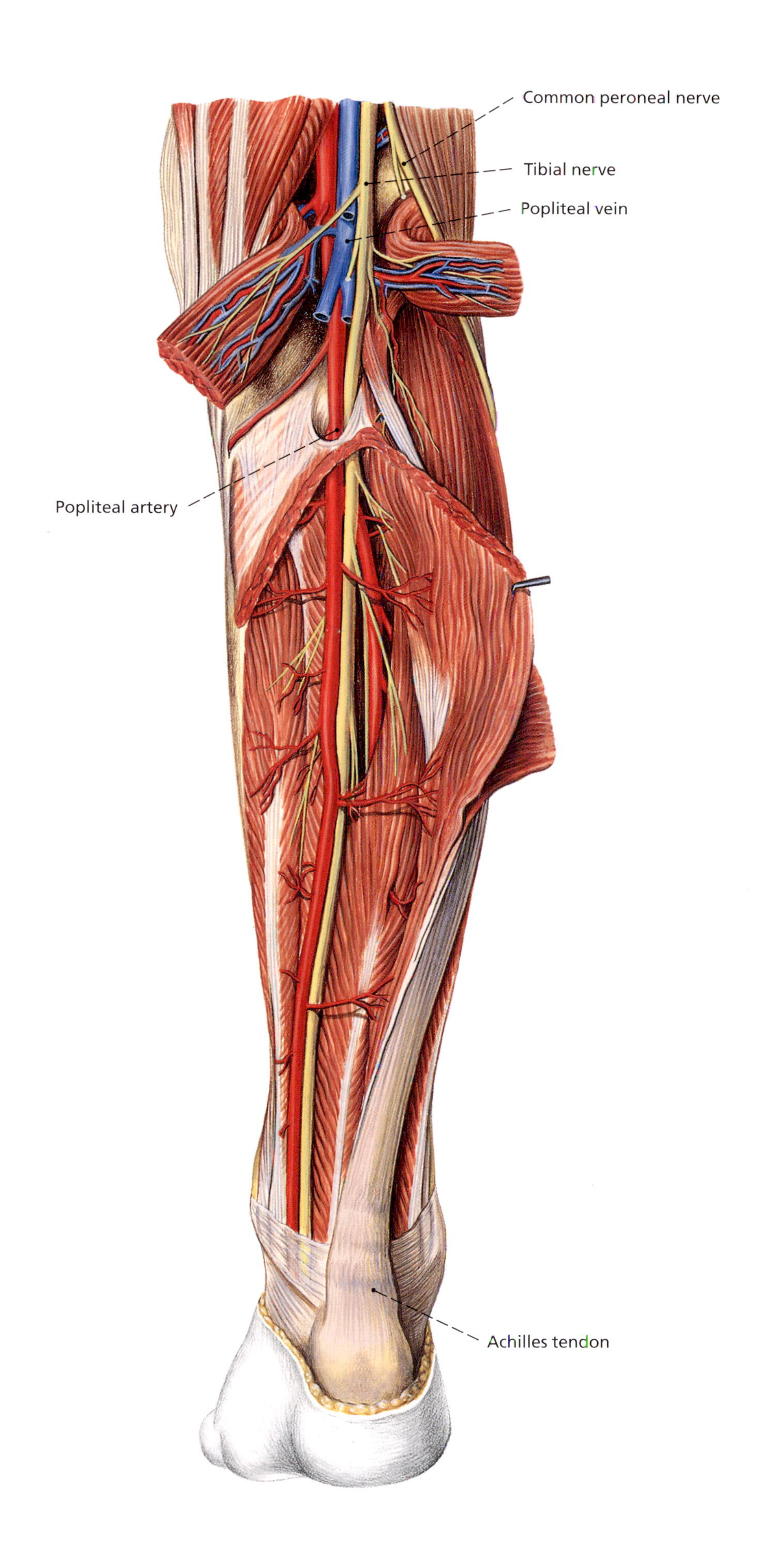

◆
Fig. 183
Nerves of the hollow of the knee and lower leg seen from rear (right). The superficial peroneal muscles have been removed.

◆
Fig. 184
Nerves of back of foot after partial removal of extensor digitorum muscles.

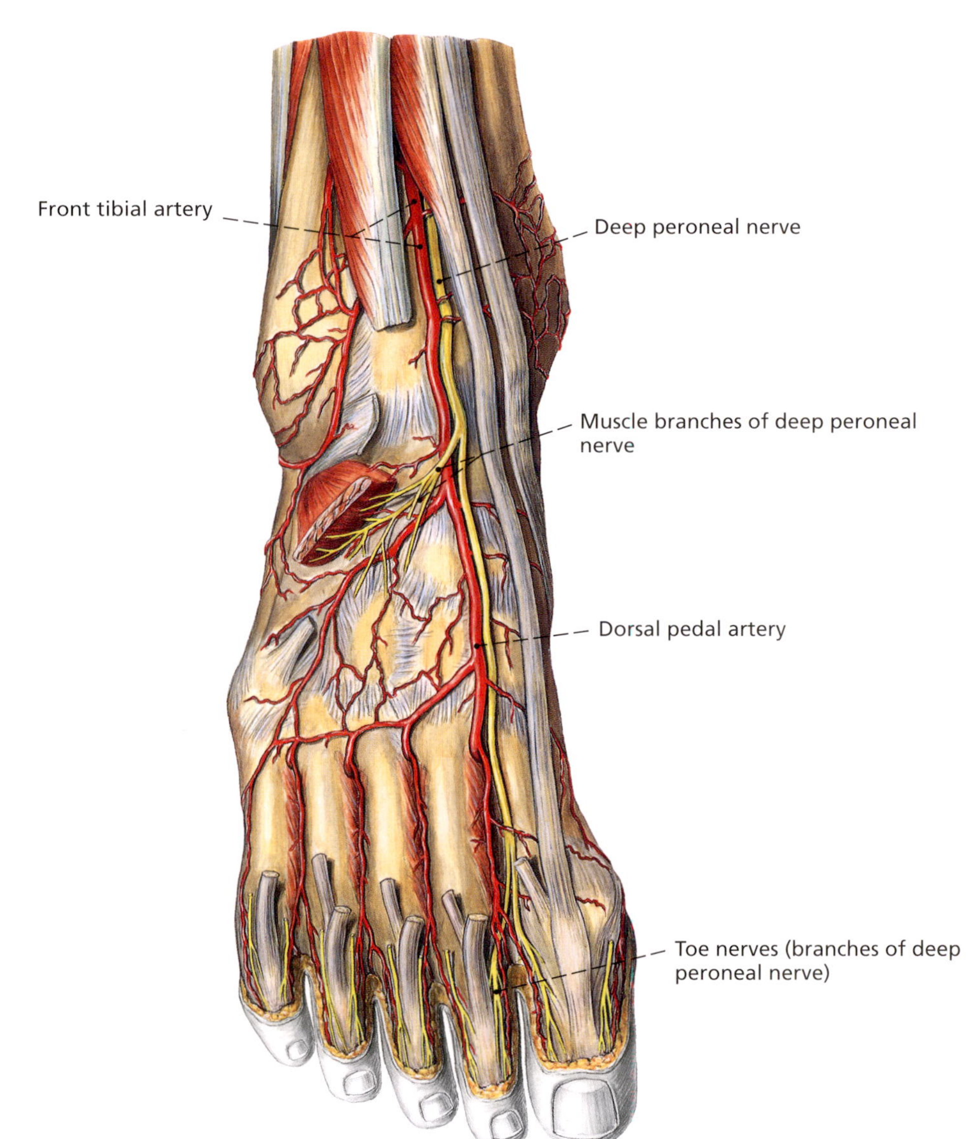

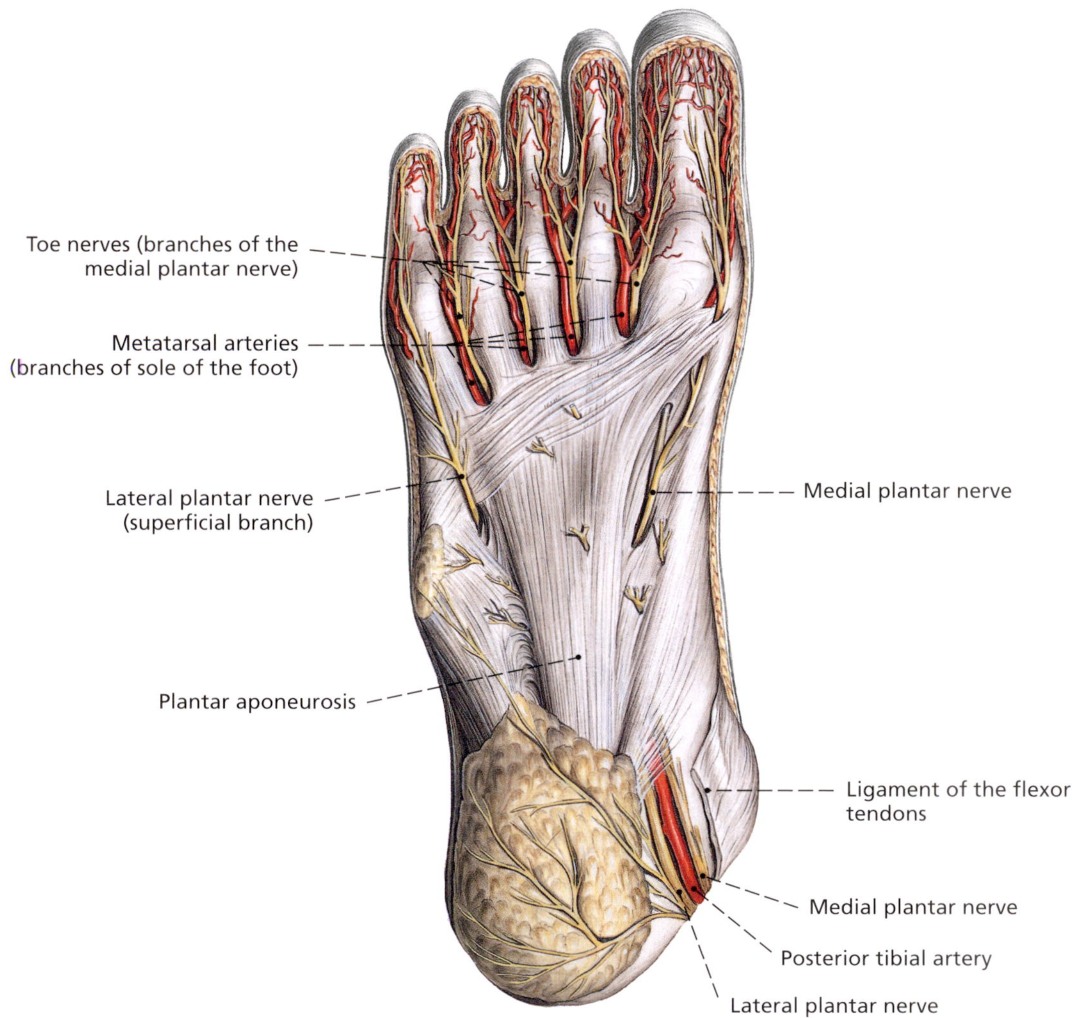

◆
Fig. 185
Nerves of the sole of the foot after partial removal of the supporting ligament of the flexor tendons (right).

Symptoms and diseases

Brain

• *Unconsciousness*

Loss of consciousness is a result of a disturbance in the cerebrum. Those affected are no longer responsive; in severe cases, they also no longer react to pain stimuli.
Unconsciousness can have numerous origins (e.g., head injury, epileptic seizure, sugar deficiency in diabetics). It can rapidly lead to a life-threatening condition, and therefore calls for immediate treatment, irrespective of the cause. The unconscious person should be placed in the recovery position (lateral recumbent position) and the airways should be kept clear.

• *Coma*

A coma is a state of deep unconsciousness as a result of brain damage. Causes can be trauma, hemorrhage, or the result of the brain being deprived of oxygen following a stroke or a cardiac arrest. Even with severe metabolic imbalances (e.g., in the case of diabetes), poisoning due to liver and kidney failure, externally administered poisons (e.g. alcohol), or tumors (which exert increasing pressure on the brain as they grow), there is a risk that the affected person will fall into a coma. Coma is an acutely life-threatening condition, the causes of which must be treated without delay. Secondary damage cannot be ruled out.

• *Brain death*

With the complete extinction of all brain functions, brain death is considered to be the irreversible end of human life, even if heart and lung functions can be maintained by mechanical support.

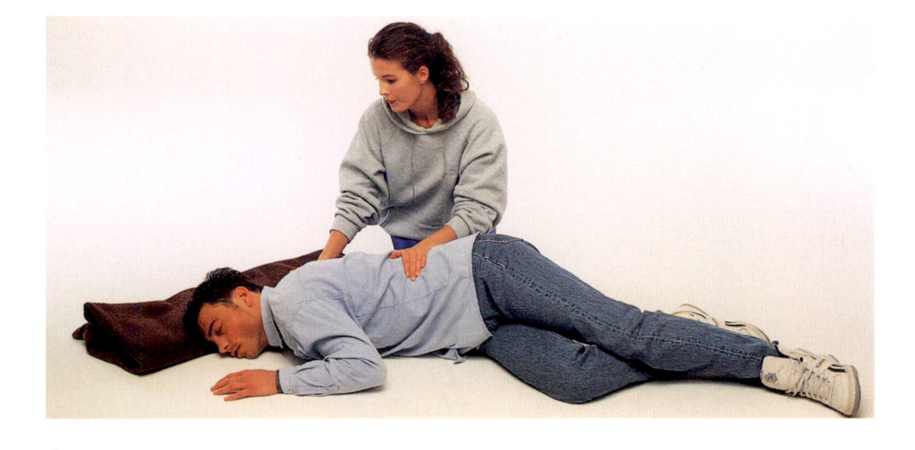

◆
Recovery position

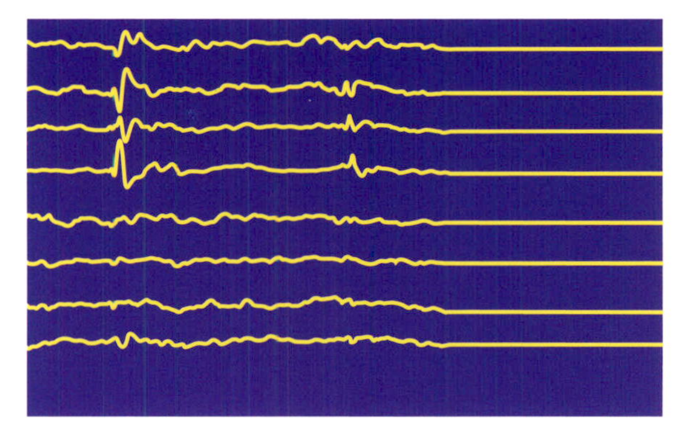

◆
Brain death
Brain death can be established with certainty on the basis of an electroencephalogram (EEG). The lack of activity known as the flatline indicates brain death.

• *Brain swelling (cerebral edema)*

Brain swelling is due to excessive fluid in the tissue gaps of the brain and is accompanied by an increase in intracranial pressure. Depending on the severity, extent, and duration of the swelling, there is a threat of permanent damage. Brain swellings are common accompanying symptoms of brain tumors or abscesses, brain injuries due to accidents, strokes, and diseases or disturbances accompanied by a breach in the blood-brain barrier, i.e., a toxin entering the brain from the blood. Treatment consists of reducing the pressure within the skull using dehydrating medications, the creation of a drainage aperture, and/or treatment of the specific cause, if possible.

• *Brain hemorrhage*

Hemorrhage of cerebral blood vessels into the brain tissue, usually arises from damage due to high blood pressure, injuries, or aneurisms, which destroy brain tissue and, depending on the extent and the location of the bleeding, can lead to a breakdown in specific brain functions. The hemorrhage may lead to the sudden onset of head and neck pains, vertigo, vomiting, pupils of dissimilar sizes, confusion, and signs of paralysis.

• *Disturbances in cerebral blood flow*

Disturbances in cerebral blood flow may be due to a defective blood supply to the brain caused by narrowed or closed blood vessels, or a sudden severe drop in blood pressure. Brief disturbances to cerebral blood flow are usually the result of a sudden fall in blood pressure, and are accompanied by other symptoms, such as vertigo when the victim stands up suddenly. If blood vessels are already abnormally narrowed, this defective blood supply can lead to an **apoplectic stroke**. Depending on the position, frequency, duration, and severity of disturbances to cerebral blood flow, there is a threat of permanent organic brain damage with far-reaching effects (e.g., transient or permanent restrictions of movement, and also loss of intellectual capacity).

• ***Brain damage due to intracranial trauma***

Brain damage, due to increased pressure or as a delayed consequence of bruising of the brain, is not always caused by the direct application of a force from the outside, but more frequently by an increase in the pressure within the skull, triggered by a bleed after a stroke, brain swelling, or tumor. The increase in intracranial pressure causes brain contusion, which can be life-threatening.

• ***Inflammation of the brain (encephalitis)***

Inflammation of brain tissue is due to an infection involving viruses, bacteria, fungi or parasites. It can arise as a complication following infectious diseases such as measles, German measles, and mumps, or as a result of an inflammation in the head area which spreads further. Brain tissue, cerebral membranes, and the spinal cord are frequently simultaneously affected. The symptoms range from the sudden onset of a high fever to headaches and neck stiffness, personality changes, disorientation, paralysis, and epileptic seizures.

• ***Meningitis***

Meningitis is a result of inflammation of the membranes that surround and protect the brain and spinal cord, and is caused by bacteria and viruses. Babies and small children are especially at risk. Meningitis can arise as a consequence of inflammation in the ears, nose, and sinuses, which spreads to the cerebral membranes. However, it is more frequently a late complication of infectious diseases. The symptoms are fever, headaches, and neck stiffness, with pain when the head is bent forwards, disturbances of consciousness, cramps, avoidance of bright lights, and a rash or spots that do not fade under pressure. Children and babies become apathetic and refuse to drink. In the case of a bacterial infection, antibiotics have to be administered at once. Untreated meningitis can lead to complications such as impairment of vision and hearing, or paralysis.

• ***Consequences of accidents – brain injuries***

Concussion is the result of a blunt force trauma to the brain (e.g., due to an impact, a blow, or a collision). The skull bones remain uninjured. The main symptom is immediate loss of consciousness. Memory loss frequently occurs, particularly for recent experiences, such as the occurrence of the accident itself. Additional symptoms are vertigo, headaches, nausea, and retching. With a simple concussion of the brain, the symptoms disappear completely after a few days' bed rest. A full medical examination is needed, however, to exclude internal injuries or brain bleeds.

Cerebral contusion may result from a brain injury due to a collision, blow, or impact, in which the skull bones are intact, but the cerebral membrane and small blood vessels can tear at the location of the trauma or at the opposite point on the brain.

◆ **Concussion**
Most frequent causes are heavy blows to the head or falling backwards onto the head.

Symptoms such as disturbances of consciousness (coma), cramps, pupillary rigidity and impairment of speech may occur, but usually disappear on suitable treatment.

A **skull fracture** is a fracture of one or more of the skull bones, usually due to a collision, blow, or impact. This may be either a closed fracture, in which the roof of the skull is injured but the skin is unbroken, or an open or comminuted fracture, in which the skin is broken and bone fragments can be displaced.

A **skull base fracture** arises due to a severe external trauma (most frequently a road traffic accident); typical indications are monocle hematoma, bleeding from nose or ears, concussion (of the brain), and frequently also paralysis of the cranial nerves (in particular, the auditory and optic nerves). In uncomplicated fractures, the patient can leave hospital after a few days. If the cerebral membrane is opened, however, and bone has invaded the brain substance, surgery and antibiotic treatment are required.

A **craniocerebral trauma** is an injury to the scalp, osteocranium and brain due to the effect of external force. For slight injuries, the deeper-lying brain regions remain mostly intact. Open injuries to the roof of the skull can trigger cerebral hemorrhages and tearing of the cerebral membrane, so that cerebrospinal fluid (CSF) may emerge from the ears and mouth. The most commonly used scale to assess the severity of a brain injury is the Glasgow Coma Scale. Brain injuries can be categorized in the following way:

- ◆ Level I—no loss of consciousness from mild trauma (e. g., concussion with symptoms of dizziness, confusion, and visual distortions).
- ◆ Level II—loss of consciousness for up to 30 minutes.
- ◆ Level III—loss of consciousness for up to 2 hours.
- ◆ Level IV—loss of consciousness for over 4 hours, up to potentially lethal coma.

The treatment depends on the degree of severity, and extends from simple dressing of the wound on the head, to surgery. Many factors affect the prognosis after brain injury, including previous medical history, age, type and location of injury, depth and duration of coma, oxygen levels after the injury, etc.

Alzheimer's Disease

Intensive research is underway to try and disentangle the knot surrounding the possible causes of Alzheimer's disease and to make a cure possible. Current treatments concentrate on using the remaining intellectual capacities, through intensive memory training and care, in such a way as to alleviate the symptoms and slow down the progress of the disease.

At the beginning of this century, the German neurologist, Alois Alzheimer (1864–1915), described a "strange, severe, progressive disease of the cerebral cortex" with irregular atrophy of the brain. The problem, rare at that time, occurred in those over the age of 65. Life expectancy in industrialized countries has risen steeply since then, and as the population is getting older, this illness, named after its discoverer, is expected to display rapid growth in the coming decades.

◆ Sudden loss of short-term memory and the associated loss of spatial and temporal orientation can be the first indications of Alzheimer's disease.

The bridge to memory crumbles

In the early stages, the symptoms (forgetfulness, loss of awareness of space and time) are vague, and resemble the normal signs of aging. But as the disease progresses, the initial normal weakness of memory turns into pronounced amnesia. Long-term memory often remains largely unaffected. Short-term memory, by contrast, fails more and more frequently and for longer periods. Simple everyday tasks, such as the ability to use cutlery or lace-up shoes, are beyond the sufferers' capabilities. It becomes increasingly difficult for them even to find the right words to describe familiar objects in speech. Emotional life usually continues relatively undamaged for longer, although, in the later stages, reactions often do not correspond to the previous behavior patterns of sufferers or to the circumstances, e.g., smiling or crying for no reason, or sudden aggressiveness.

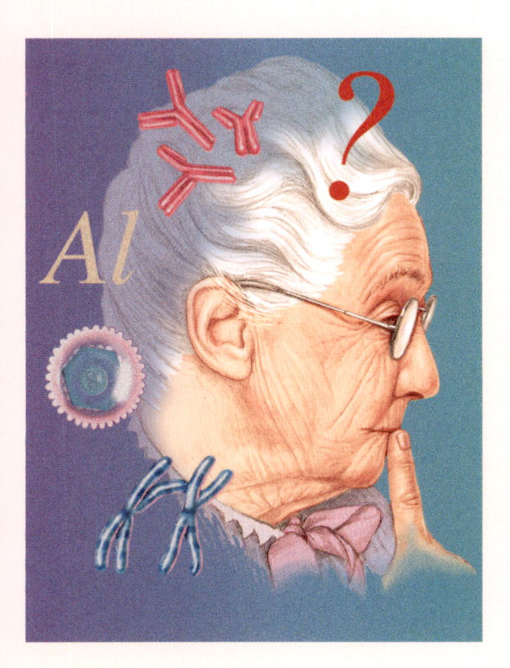

◆ A combination of different factors is assumed to be the cause of Alzheimer's disease.

The further the illness progresses, the more impoverished the speech becomes, until it disappears completely. Disturbances to the gait also occur. Patients can no longer control their bladders and bowels, and in the end they need intensive care.

Treatment of symptoms

As there is currently no medication which can effect a cure, treatment is aimed at alleviating the symptoms and delaying the progression of the disease.
In the initial stages, deterioration of intellectual capacities can be postponed by memory training: talking to patients about the events of their day, reading to them from newspapers, and encouraging their previous interests, without pressurizing them. Special training programs, in which acting techniques are mirrored, using speech, and pictorial and numerical material, may be of assistance. In addition to this, patients remain mobile for longer if they participate in light physical activities and receive regular physiotherapy.
An understanding attitude on the part of carers is extremely important, and is still appreciated, even by seriously ill people. In the later stages of the disease, communication may often only continue at this level. Music therapy and rhythm therapy also frequently have favorable effects in this phase.

• *Brain tumors*

A tumor starting in brain tissue or cerebral membranes, or frequently a metastasis of a malignant tumor outside the brain, especially lung and breast cancers. Brain tumors themselves do not normally form metastases outside the brain and spinal cord. Even benign brain tumors are dangerous, however, because their growth requires space, and leads to an increase in intracranial pressure, compressing and damaging the sensitive brain tissue. Depending on the location and size of the tumor, there can be varying symptoms, such as headaches, vertigo, vomiting, impaired vision, epileptic fits, disturbances of consciousness, behavioral changes, and paralysis. The diagnosis is arrived at by means of a computed tomography (CT) or magnetic resonance imaging (MRI) scan. In most cases, the tumor can be removed surgically.
A **pituitary tumor** is a normally a benign tumoral enlargement of the pituitary gland. Pituitary tumors can cause either an increase or a reduction in the release of hypophyseal hormones. This can result in the absence of menstruation, the cessation of sperm production, failure to develop fully, or gigantism (excessive growth, especially of the feet and hands). Moreover, larger pituitary tumors often press on the optic nerves, and can thus cause impaired vision, with visual field loss. The tumor is diagnosed by means of hormone level readings and X-rays of the skull. It usually has to be removed surgically, as radiation treatment and hormone-regulating medications are rarely sufficient.

• *Epilepsy*

Epilepsy is a convulsive disorder caused by disturbances in the brain. The manifestation of such attacks ranges from short-term absent-mindedness all the way to convulsive seizures. Epilepsy is often congenital but may arise as a result of brain damage, blood vessel disorders, or tumors.
A particularly serious form of such an attack, a major spasmodic seizure, is also known as a **tonic-clonic seizure**. The affected person loses consciousness, usually without any advance warning, falls over, and twitches their arms and legs. Affected persons can do themselves serious injury when they fall. Such an attack lasts anything from one to three minutes. The patient is usually fully responsive again following a restorative sleep.
Long-lasting or recurrent epileptic convulsive seizures, where the sufferer does not regain consciousness between seizures, are described as **status epilepticus**. It can go on for hours or days, and it can lead to death if not tackled in good time. Immediate treatment is needed with high doses of anti-convulsant medication.
The electro-encephalogram (EEG) is used to determine whether changes are present which can lead us to conclude that an epileptic convulsive disorder is present. A CT scan of the head can reveal whether or not a tumor is triggering the attacks.
The most important objective of treatment is to bring the attacks under control. Nowadays there is a whole range of effective medications to reduce the excitability of the brain cells being interfered with and thus prevent the convulsive seizures (anti-epileptics).

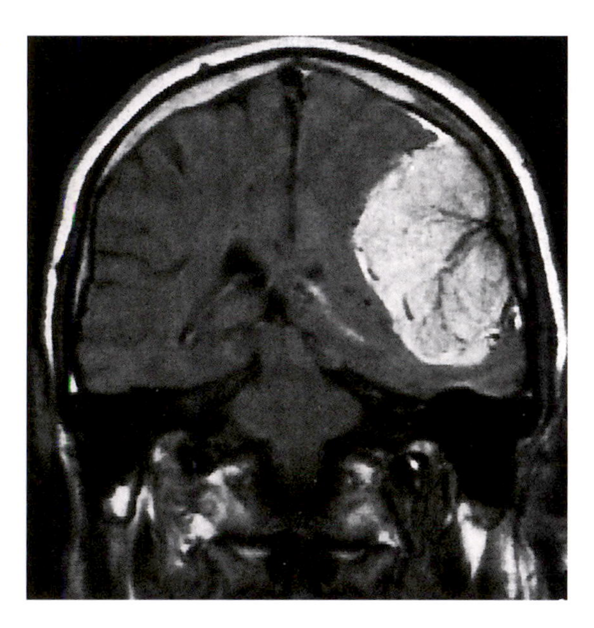

◆
Brain tumor
It can clearly be seen from the MRI scan how the tumor is spreading within the brain.

• *Parkinson's disease*

Parkinson's disease, also described as morbus Parkinson or shaking palsy, is a brain disease in which there is slow, progressive, death of cells (necrobiosis) in the midbrain. The trigger is usually unknown. There may be a hereditary predisposition and/or the stimulus may be a chemical toxin. In rare cases, the origin is a tumor in the midbrain, or a sudden decrease in blood flow to this region. Parkinson's disease can also occur as a result of inflammation of the brain.
Whatever the cause, the necrobiosis of nerve cells in the midbrain leads to a deficiency of dopamine, a chemical neurotransmitter necessary for the transfer of nerve stimuli, and thus ultimately for voluntary active muscle movements, including gestures and facial expression. This results in muscular rigidity: movements slow down, and the face becomes fixed and expressionless. The most common symptom is uncontrollable trembling, which also makes itself noticeable in shaking movements, and is made stronger by agitation. Patients tire quickly, and in subsequent stages there is also impairment of attention and concentration. As a result of their handicap, which means they need help to cope with everyday life, Parkinson's patients frequently suffer from depressive conditions.
Although there is as yet no cure, the symptoms can be alleviated by medication, and the progress of the disease slowed down. In many cases, an operation is also possible, in which certain nerve centers in the brain are deliberately switched off, and the enhanced muscle tension and pronounced trembling are thereby improved.

Headaches

Pain is often traced back to a physical origin alone, but in addition to organic origins, social and emotional factors, such as stress, psychological tensions, or anxiety, may also play a role. Continuous or regularly recurring headaches are an alarm signal from the body, which must not simply be suppressed with tablets.

Migraine

It is estimated that one in nine Americans suffers from migraine, with women being three times more commonly affected than men. A single migraine attack can last for between four and 72 hours. The pulsating, almost unbearable pain which this involves is usually restricted to one side of the head. There is frequently nausea, an urge to vomit and actual vomiting, sensitivity to noise and light, plus misty or flickering vision. The pains can change from one side to the other from attack to attack, or even during an episode, and can become more or less intensive. Migraine pains arise in the cerebral membranes surrounding the brain, and originate in stimulated blood vessels. When the latter dilate, signals are passed on to the cerebral cortex, and a pain signal is triggered there.

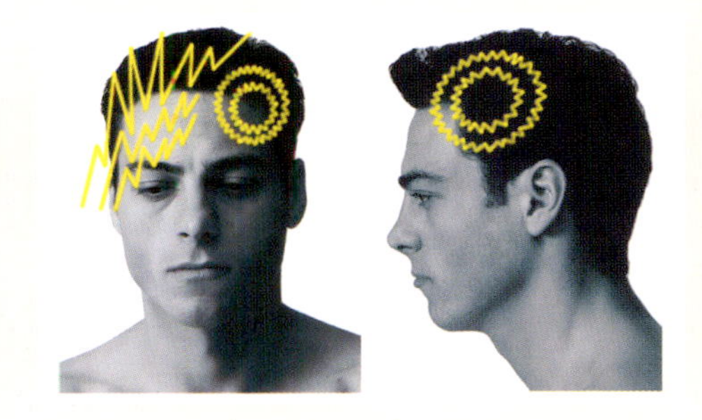

Migraine pains usually originate in one half of the head, but can also transfer to the other side.

Tension headache

The pains are often propagated forward from the neck into the forehead. They are usually perceived as a dull pressure, as though a heavy weight were pressing on the roof of the skull, or as though the head were clamped in a vice. The pains always occur on both sides, and are not made worse by normal physical activity.
With occasional (episodic) tension headaches, the pains usually last for a few hours or days. We refer to them as chronic if the headaches occur for more than 15 days in the month. Nausea and vomiting do not occur, but sensitivity to light and noise are present throughout. There is still no precise explanation as to what the trigger is. Exhaustion, tiredness, overwork, continuous problems in the workplace or at home, can all trigger a tension headache. Psychological problems may exacerbate this type of pain.

Cluster headache

With cluster headaches, several headaches occur one after the other for a few weeks or months, only to vanish again for months or years. Men are about four times more likely to be affected as women. The pains are very intense or shooting. They always occur on one side only, and last for between 30 minutes and two hours. They are perceived most intensely around and behind the eye, which becomes red and weeps. The eyelid also often droops.

Prevention and treatment

Rest, relaxation, and sleep are the best cures for tension headaches or migraine. Cold or hot compresses, ice packs, and massaging of tense back and neck muscles, can all help to drive off the headache. Acupuncture and acupressure reduce the pain for many sufferers. Combinations of analgesics or specific migraine remedies (e.g., triptans) can be prescribed. Behavioral therapy may do more good than tablets. The most successfully tried and tested method is progressive muscle relaxation (PMR). It makes people conscious of stresses by training them to deliberately tense individual muscle groups and then relax them again. Training in overcoming stress can also be of assistance.

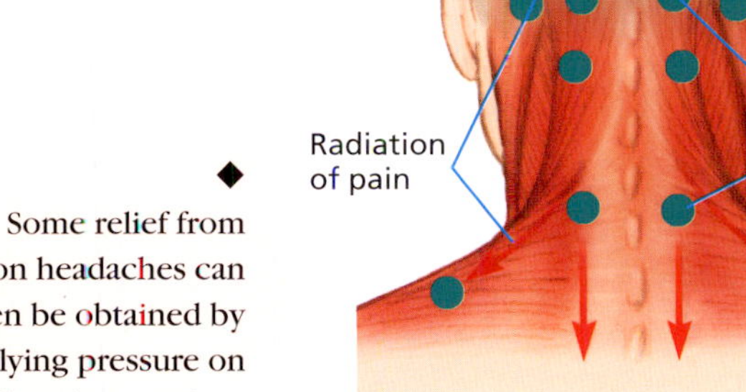

Some relief from tension headaches can often be obtained by applying pressure on the trigger points.

Nerves and nervous system

• *Polyneuropathy*

Polyneuropathy is a non-inflammatory disease of the peripheral nerves, with more and more bundles of nerve fibers gradually being affected. It is most frequently the result of metabolic disorders, predominantly diabetes, but may also be due to chronic misuse of alcohol.

The symptoms usually arise first in the lower extremities. Typical symptoms are sensations of numbness, tingling (formication), muscle weakness and lack of muscular tone, together with a loss of sensitivity in the vicinity of the affected nerves, and gait disturbances. The loss of sensitivity in the feet and hands entails a high risk of injury for those afflicted, as pain can no longer be perceived.

• *Facial nerve paralysis (facial palsy)*

In paralysis of the facial nerve, which serves the facial musculature, the eyelid cannot be completely closed and the lower lid and angle of the mouth droop. There may also be disorders of perception and taste, together with disturbances in the production of saliva and tears. The origins are viral infections, accidents, and inflammations in the brain and face areas. Facial palsy can also arise in the event of circulatory disturbances in the brain (where the facial nerves begin), following a stroke or a brain tumor. In most cases, the paralysis corrects itself without any permanent damage once the cause is treated.

• *Trigeminal neuralgia*

Trigeminal neuralgia is a painful condition of the facial nerves. The causes are not exactly clear. Typical symptoms are paroxysmal pain attacks, e.g., in the forehead, eye, cheek, mouth, nose, teeth, and lower jaw. They last only seconds or minutes, but are so severe that the patient is unable to act. The attack is often accompanied by twitching of the facial muscles. Those over 50 are the most common victims. The attacks can be suppressed with the help of medication, or by injections of local anesthetic.

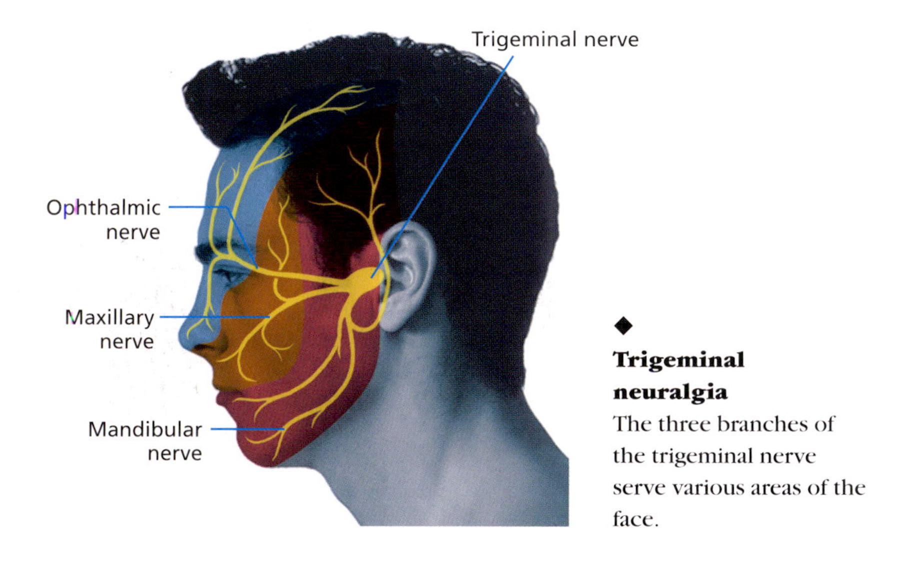

◆ **Trigeminal neuralgia**
The three branches of the trigeminal nerve serve various areas of the face.

• *Creutzfeldt–Jakob disease*

Creutzfeldt–Jakob disease is a rare and, so far, incurable disease of the central nervous system with severe damage to the nerve cells, mainly in the area of the cerebrum, cerebellum, and spinal cord. The disease starts with marked irritability and severely depressed moods. Intellectual disorders (dementia) occur, along with convulsive muscle contractions (spasms) and muscular tremors. Impaired vision and speech are the rule. Further stages are characterized by a wide-ranging loss of all brain functions. A link with infection by the pathogen that causes mad cow disease (bovine spongiform encephalitis) is suspected.

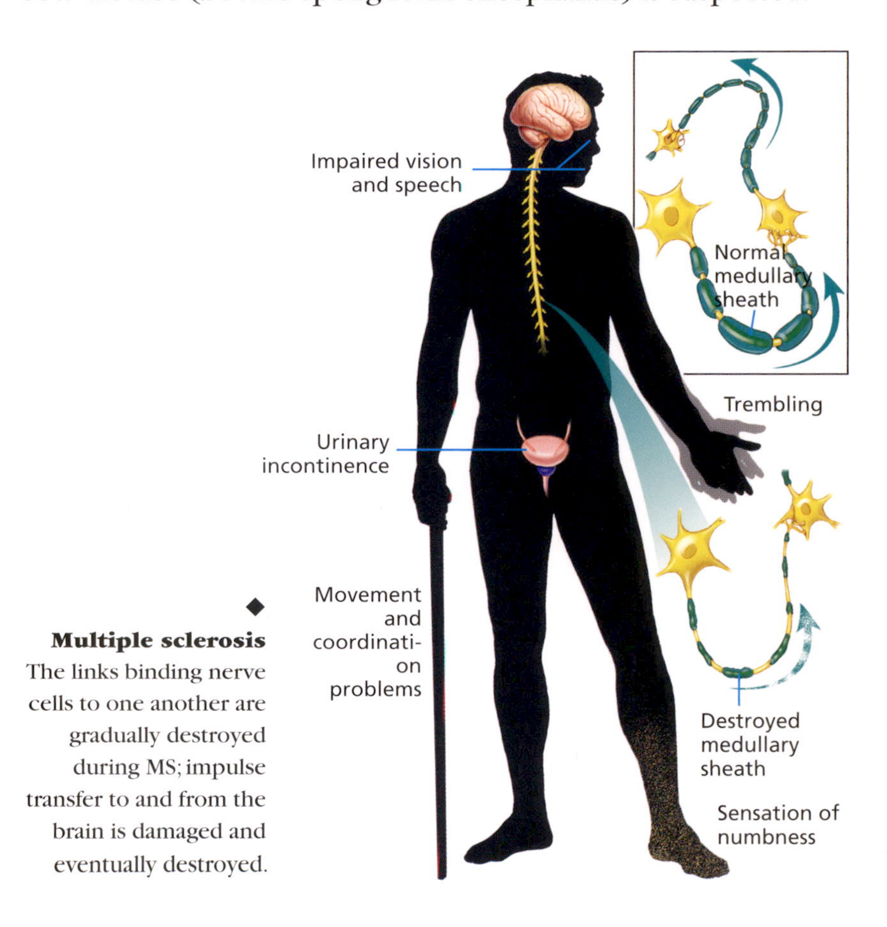

◆ **Multiple sclerosis**
The links binding nerve cells to one another are gradually destroyed during MS; impulse transfer to and from the brain is damaged and eventually destroyed.

• *Multiple sclerosis (MS)*

MS is a chronic episodic progressive disease of the general nervous system. The origin lies in the increasing destruction of the medullary sheaths of nerve fibers in the brain and spinal cord. An error in the immune system, triggered by an earlier viral infection, may also be responsible. It usually manifests itself when sufferers are aged between 20 and 40. Symptoms are varied and, initially, usually only of brief duration: weakness or feelings of numbness in arms and legs, trembling, and indications of paralysis, impaired speech or vision. After some months or years, the disease moves into the next stage, and, as it progresses, the symptoms are no longer intermittent, and disabilities arise. MS is incurable, but the symptoms can be alleviated by means of various types of medication and physical therapies.

Diagnostic methods

The normal **X-ray** can depict the bony skull and spinal column. **Computed tomography (CT)** offers a different type of scan which can also display soft tissue areas. It plays an important role in early recognition of tumors or internal bleeding, because concealed changes can easily be recognized. However, the radiation exposure is comparatively high.

In **magnetic resonance imaging (MRI)**, the body is scanned not by X-rays but by magnetic waves. The result is a layer-by-layer picture of the interior of the body, put together by a computer, and offering a good and reliable representation of the various tissues and cavities. It is used mainly to look for tissue changes, such as tumors, foci of inflammation, cysts and injuries to the ligaments.

Brain waves are measured using an **electro-encephalogram (EEG)**. The nerve cells of the brain are continuously generating low-level electrical currents. These currents can be plotted using small metal plates (electrodes), which are fastened to the scalp for the investigation. Plotting the impulses creates wave-form or jagged lines. The form of these brain waves depends on people's age, their state of consciousness (e.g., sleep, dream phases), and is altered by any pathological changes to the brain. Thus, any increased tendency to convulsions of the brain due to epilepsy or a brain tumor can be detected on an EEG.

Reflex testing is the triggering of an involuntary movement or process due to a stimulus. The best known reflexes are the knee jerk reflex, which is triggered by a blow just under the kneecap, and the pupil reflex, which causes the pupil to dilate or contract, depending on the light intensity. Reflexes enable the body to react instinctively and quickly to external stimuli. Many of the body's protective mechanisms take the form of reflexes. Absent or damaged reflexes may indicate neurological disease or damage to the spinal cord.

The cerebrospinal fluid (CSF) is a clear, colorless fluid surrounding the brain and the spinal cord. It serves to protect these sensitive tissues, and plays a part in the metabolism of the brain and the spinal cord. If it is suspected that certain diseases are present, such as meningitis, CSF is removed and examined. With a **spinal tap** procedure, CSF is removed from the subarachnoid space under local anesthetic, using a hollow needle. The number of white corpuscles and the protein content are determined. **Measuring the pressure in the subarachnoid space** can also provide a lot of information. Any alteration in values can provide indications as to the type of disturbance.

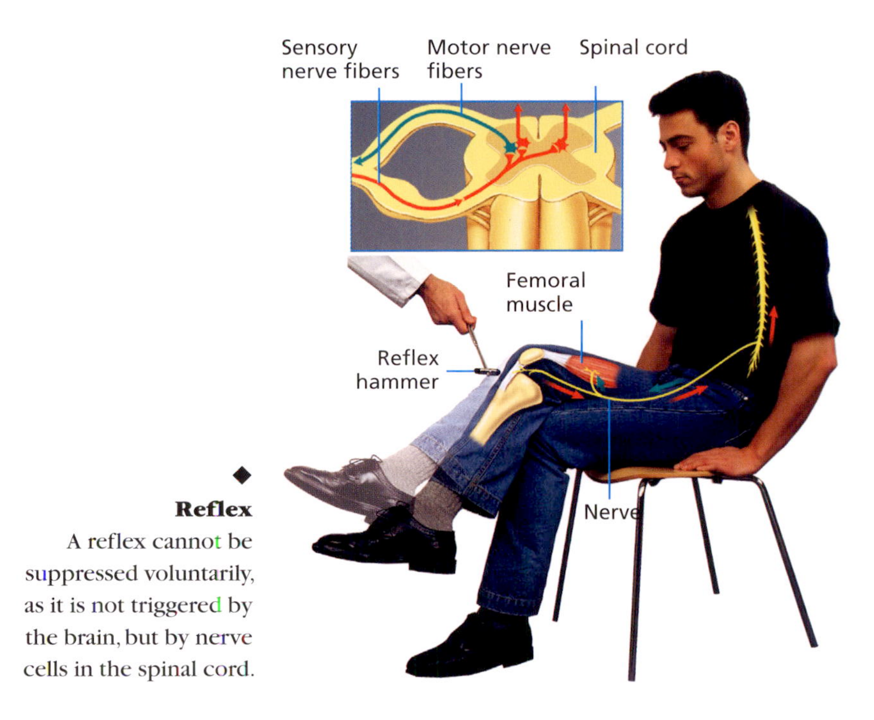

Reflex
A reflex cannot be suppressed voluntarily, as it is not triggered by the brain, but by nerve cells in the spinal cord.

Treatment methods

The therapy is dependent on the injury. While most brain tumors can be operated on today, for injuries to the bone, immobilization is usually the only option.

Medication plays a very important role. Thus, for example, in epileptics the administration of medications known as **anti-epileptics**, makes it possible for most of those affected to live an almost normal life. Even for diseases which are not yet curable, such as Alzheimer's disease or MS, there are medications which can delay the progression of the disease and/or alleviate the symptoms.

However, it remains a fact that the destruction of any part of the brain or nerves is usually irreversible, or is at least linked to severe lifelong limitations. In certain cases a **nerve transplant** may be possible, i.e., the surgical graft of a nerve section.

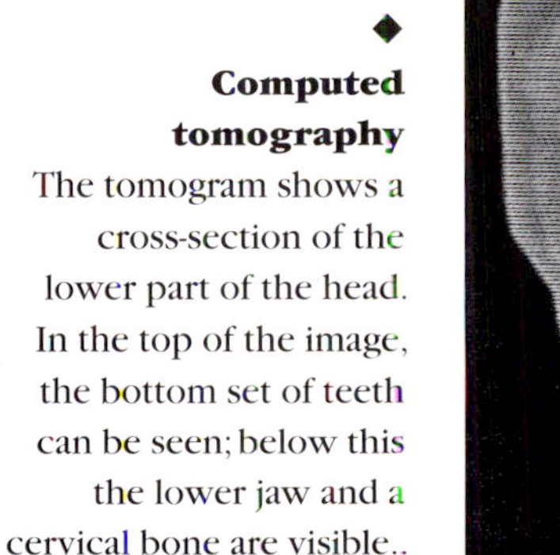

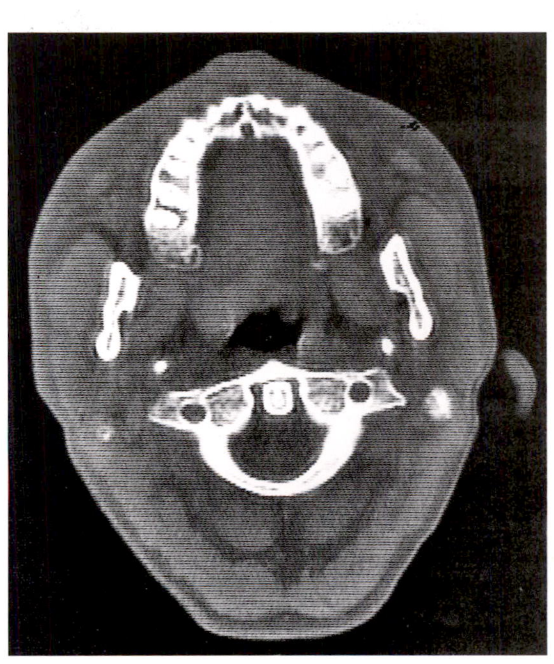

Computed tomography
The tomogram shows a cross-section of the lower part of the head. In the top of the image, the bottom set of teeth can be seen; below this the lower jaw and a cervical bone are visible..

The eye

Light rays enter the eye through the pupil and pass through the lens. Photoreceptors (rods and cones) in the retina convert the light stimuli into an electrical impulse. This impulse is relayed to the visual cortex via the optic nerve. Muscles attached directly to the eyeball enable it to move in a range of directions.

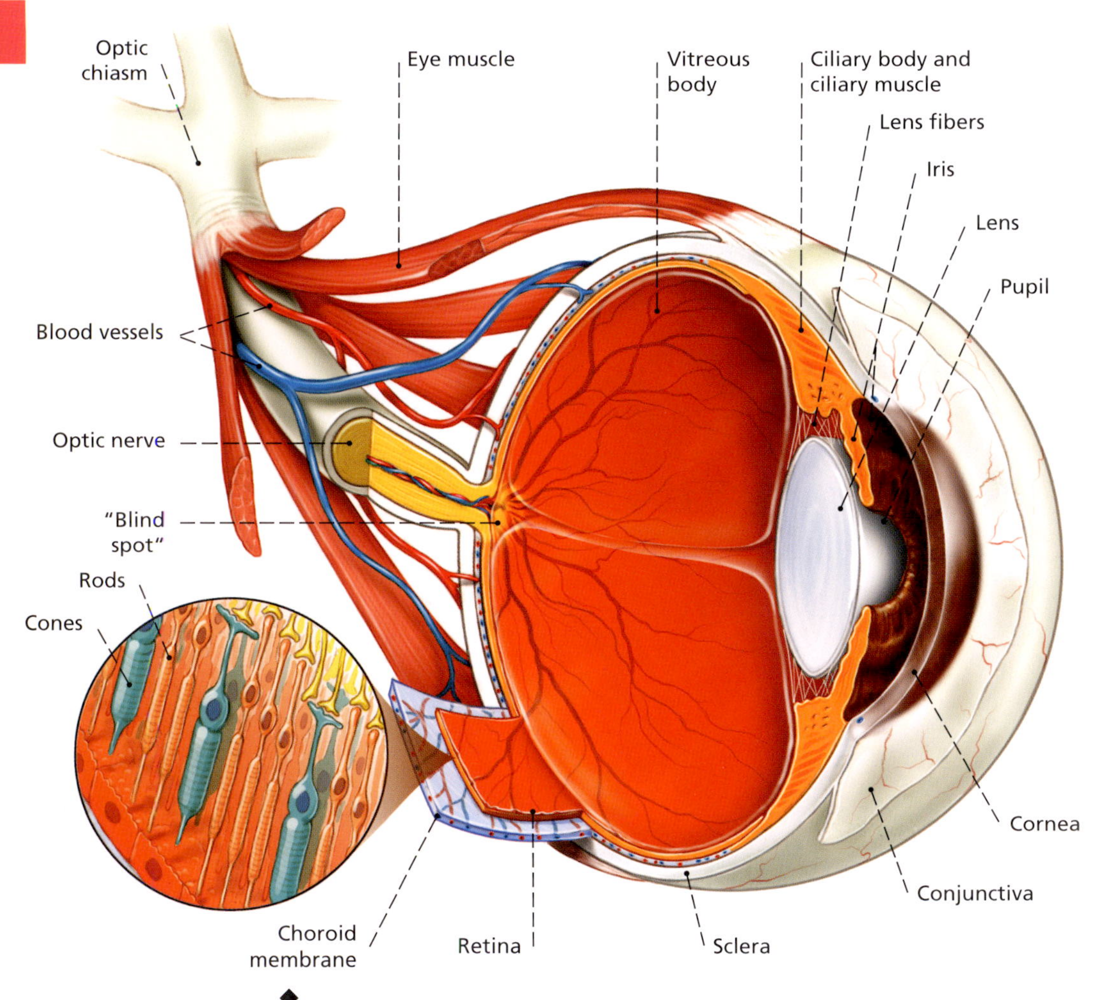

Fig. 186
Light rays enter the eye through the pupil and pass through the lens. Photoreceptors (rods and cones) in the retina convert the light stimuli into an electrical impulse. This impulse is relayed to the visual cortex via the optic nerve. Muscles attached directly to the eyeball enable it to move in a range of directions.

Construction of the eye

When we look at our eyes in the mirror it is the circular iris (which determines the color of the eye) with the pupil in the middle, which we notice first of all. The largest section is formed by the "white" of the eye, the vitreous body, which is comprised of a transparent, gel-like mass, kept in shape by a taut connective-tissue casing, the sclera. This tissue merges with the transparent, strongly curved cornea at the front of the eye. The visible front section of the sclera, as well as the cornea and the inside of the eyelids, are covered with a thin mucous membrane, the conjunctiva, and kept moist by the lacrimal fluid. This fluid comes from the lacrimal glands above the outer corner of the eye and is distributed over the front surface of the eye on blinking. Any dust particles in the eye are washed out by the lacrimal fluid.

The iris is the colored part of the eye. It is made up of ring-shaped muscle fibers and is located directly in front of the lens. It has a hole in the middle known as the pupil. Like the aperture of a camera, the pupil has the flexibility to adjust to prevailing lighting conditions by constricting and expanding.

The transparent lens is located between the iris and the vitreous body. Its task is to concentrate the rays of light entering the eye in such a way that a clear image is formed on the retina. The lens is kept in position by the lens fibers. A ring-shaped group of muscles, the ciliary muscles, can change the curvature of the lens, thus regulating the light refraction. The aqueous humor is formed in the ciliary body, a connective tissue appendage of the ciliary muscles. Aqueous humor drains into the front section of the eye via small ducts. The production and draining of the aqueous humor normally balance each other out so that a constant intraocular pressure prevails.

The retina forms the innermost layer of the eye. It absorbs the light stimuli with its sensory cells and relays them to the brain via the nervous system. The nerve cells are bundled together to form the optic nerve, which extends from the back of the eyeball (the so-called blind spot) to the cerebral cortex.

The visual apparatus

In order for an object to be displayed and perceived clearly on the retina, the light rays entering the eye have first to be refracted and concentrated, a task performed by the cornea and the lens. The lens is flexible: the closer the focused object is to the eye, the more it needs to curve in order to concentrate the rays entering the eye. If the object is further away, it flattens out. This process, which is controlled by the ciliary muscles, is known as accommodation.

Two types of special sensory cells convert the light rays landing on the retina into images: rods and cones. While the light-sensitive rods facilitate sight in dim light and darkness, the cones are responsible for seeing colors.

Rods and cones are distributed across the retina at unequal densities. There are especially high numbers of cones in the area of most acute vision (known as the yellow spot), which is why, for acute vision, the eye always adjusts itself so that the focal point of the rays entering the eye coincides precisely with this spot. Outside of this spot, the rods, which are more sensitive to differences in brightness, predominate. This is why we are best able to seeing moving people or objects in the dark "out of the corner of the eye," i.e., without focusing on them directly. The eye is also able to adapt to different lighting conditions (adaptation) and this process is controlled both by changes in pupil diameter and by sensory receptors in the retina. While the adjustment from light to dark takes several minutes, the eye is able to react to the change from dark to light much more quickly.

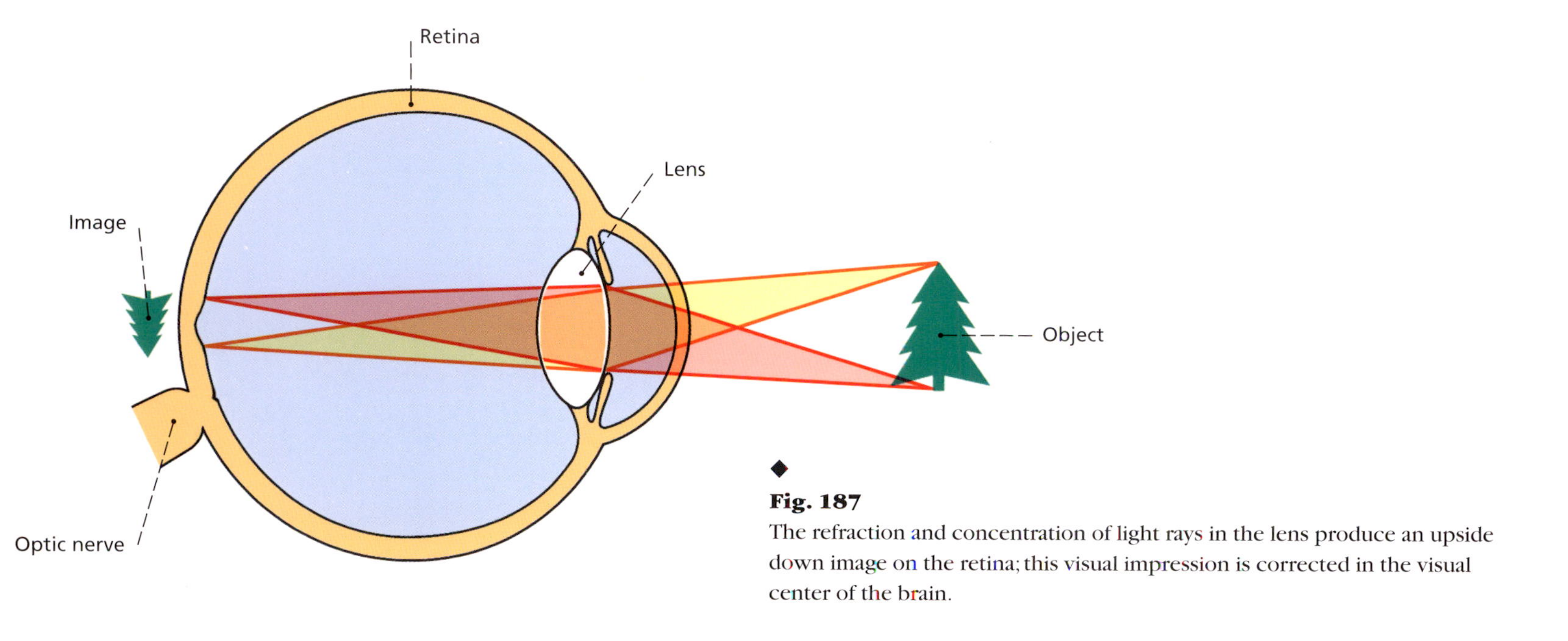

◆
Fig. 187
The refraction and concentration of light rays in the lens produce an upside down image on the retina; this visual impression is corrected in the visual center of the brain.

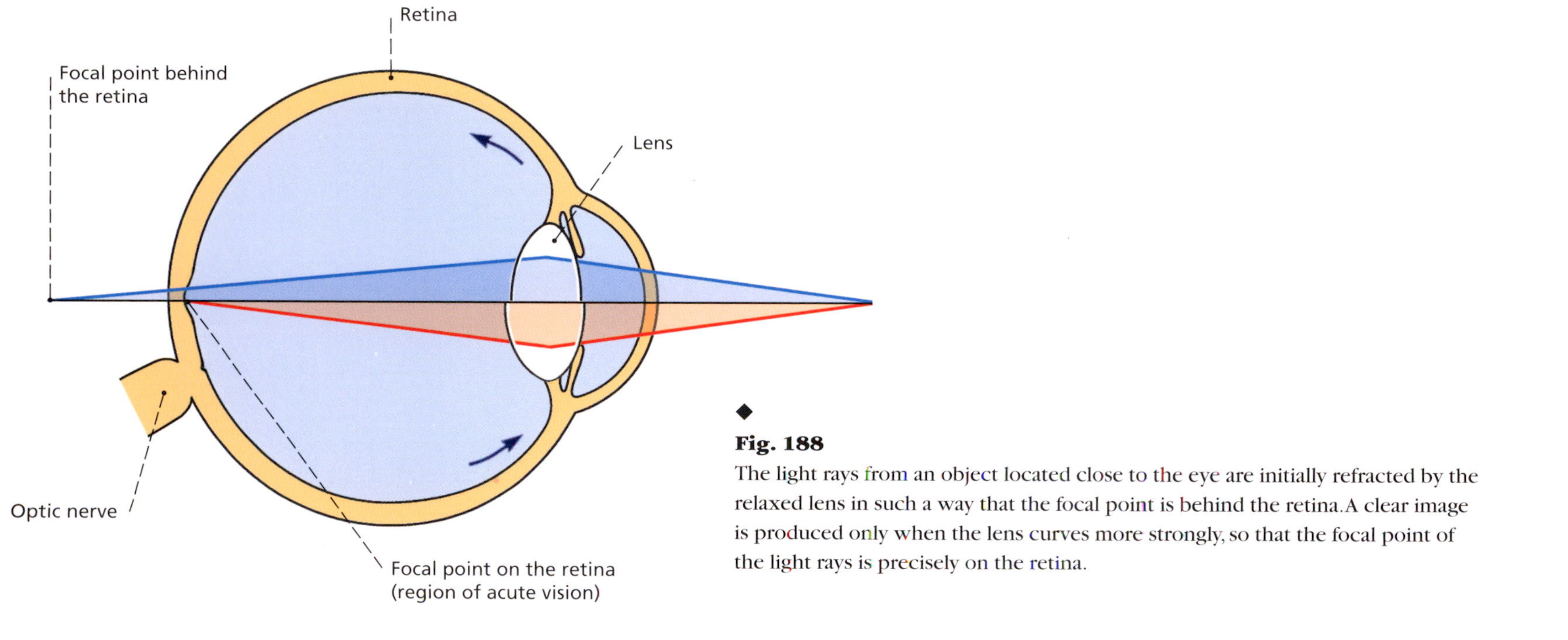

◆
Fig. 188
The light rays from an object located close to the eye are initially refracted by the relaxed lens in such a way that the focal point is behind the retina. A clear image is produced only when the lens curves more strongly, so that the focal point of the light rays is precisely on the retina.

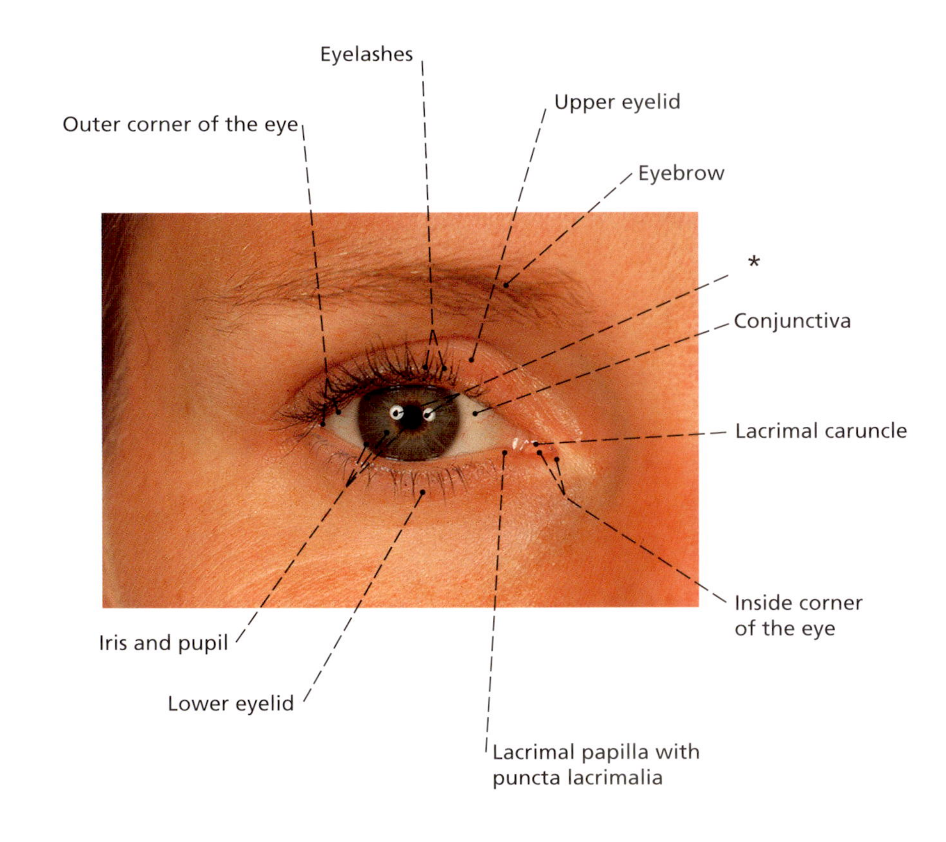

◆
Fig. 189
The structures of the eye when open (right).
* Light reflexes.

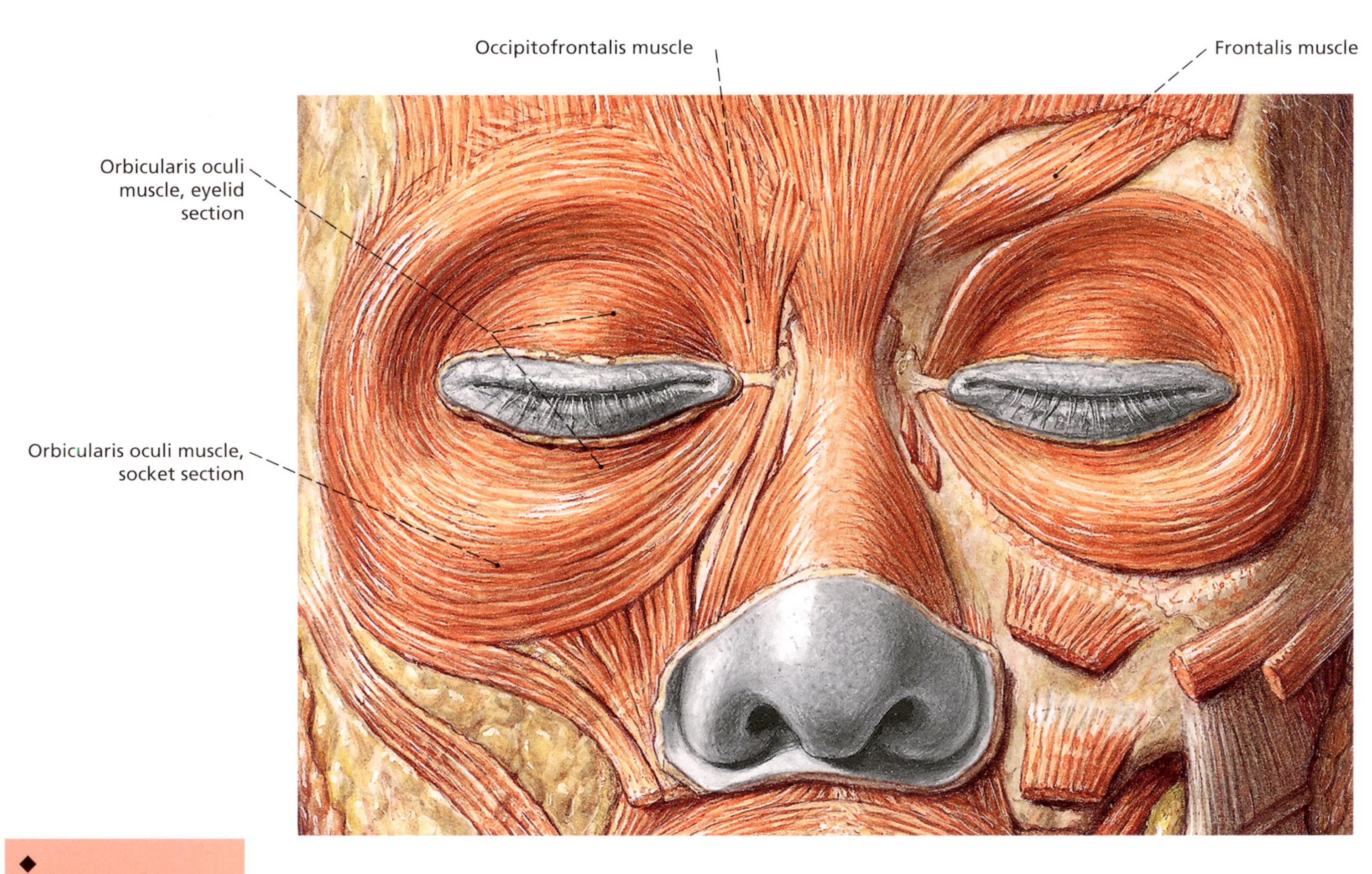

◆
Fig. 190
The muscles surrounding the eye. The orbicularis oculi muscle closes the eye.

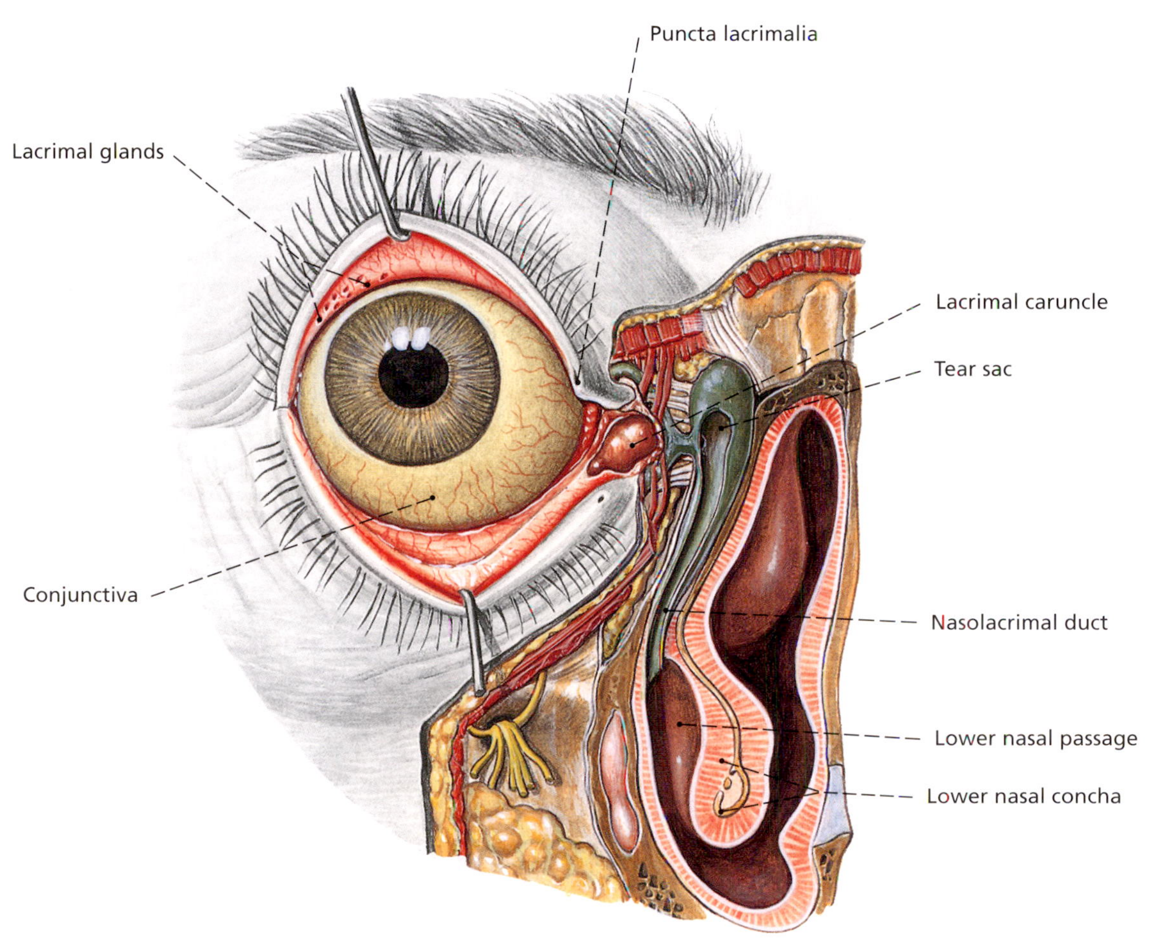

◆
Fig. 191
The lacrimal apparatus of the eye. The nasolacrimal duct is opened up as far as its aperture permits, beneath the lower nasal concha.

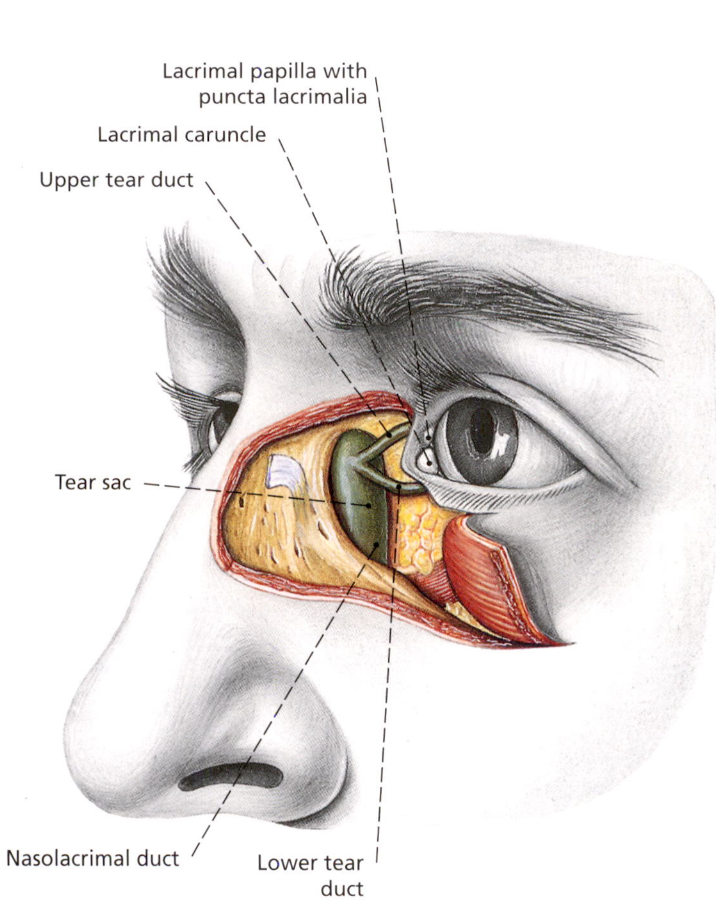

◆
Fig. 192
The lacrimal apparatus after the removal of the orbicularis oculi muscle from the bone.

Upper tear duct
Tear sac
Lacrimal caruncle
Lower tear duct
Nasolacrimal duct

◆
Fig. 193
The lacrimal apparatus after the opening of the nasolacrimal duct and the two tear ducts.

◆ **Fig. 194**
The eye muscles following removal of the side socket wall (left eye).

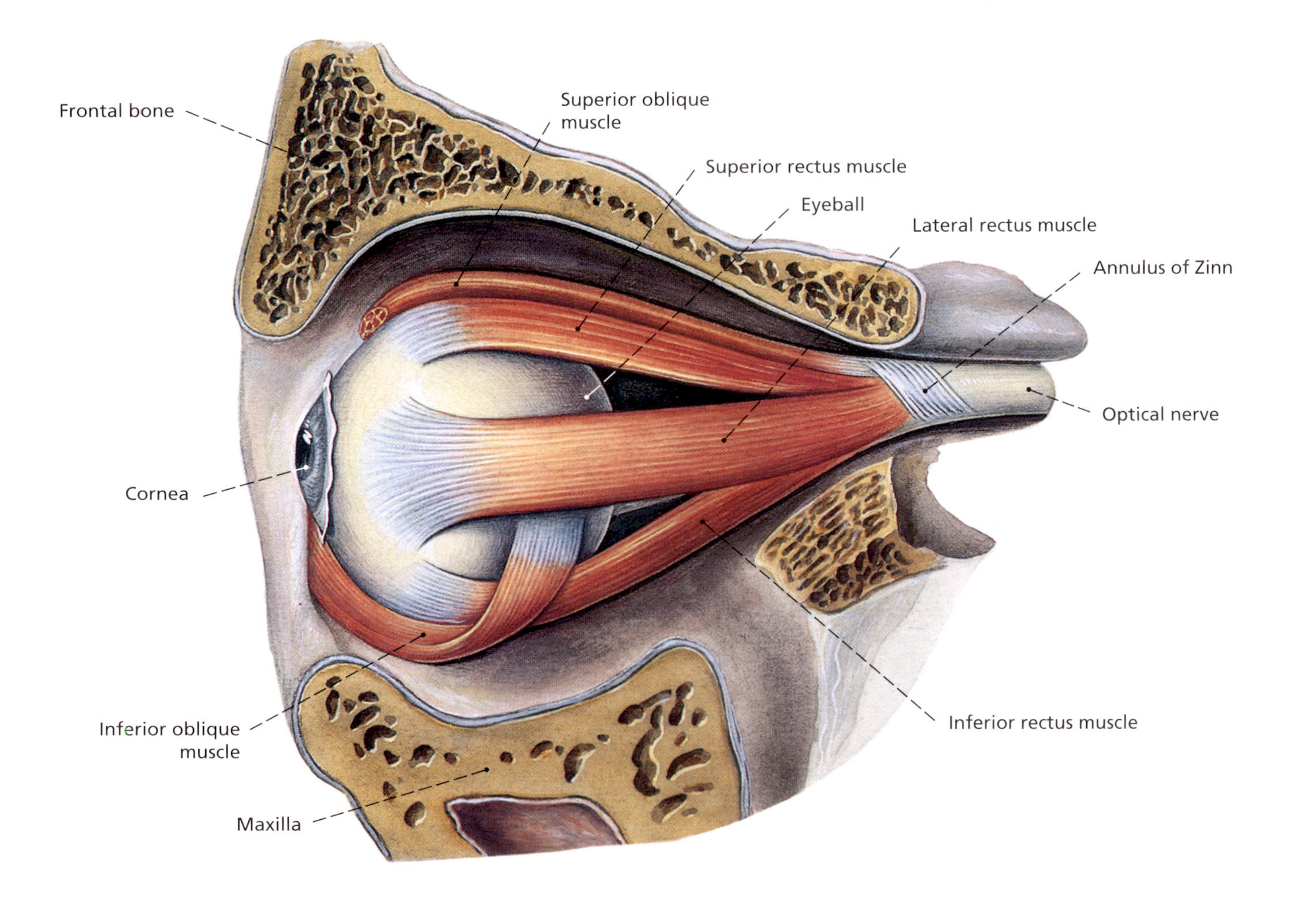

◆ **Fig. 195**
View of the left eye socket from the front. The majority of the eye muscles are connected to the Annulus of Zinn around the optic nerve.

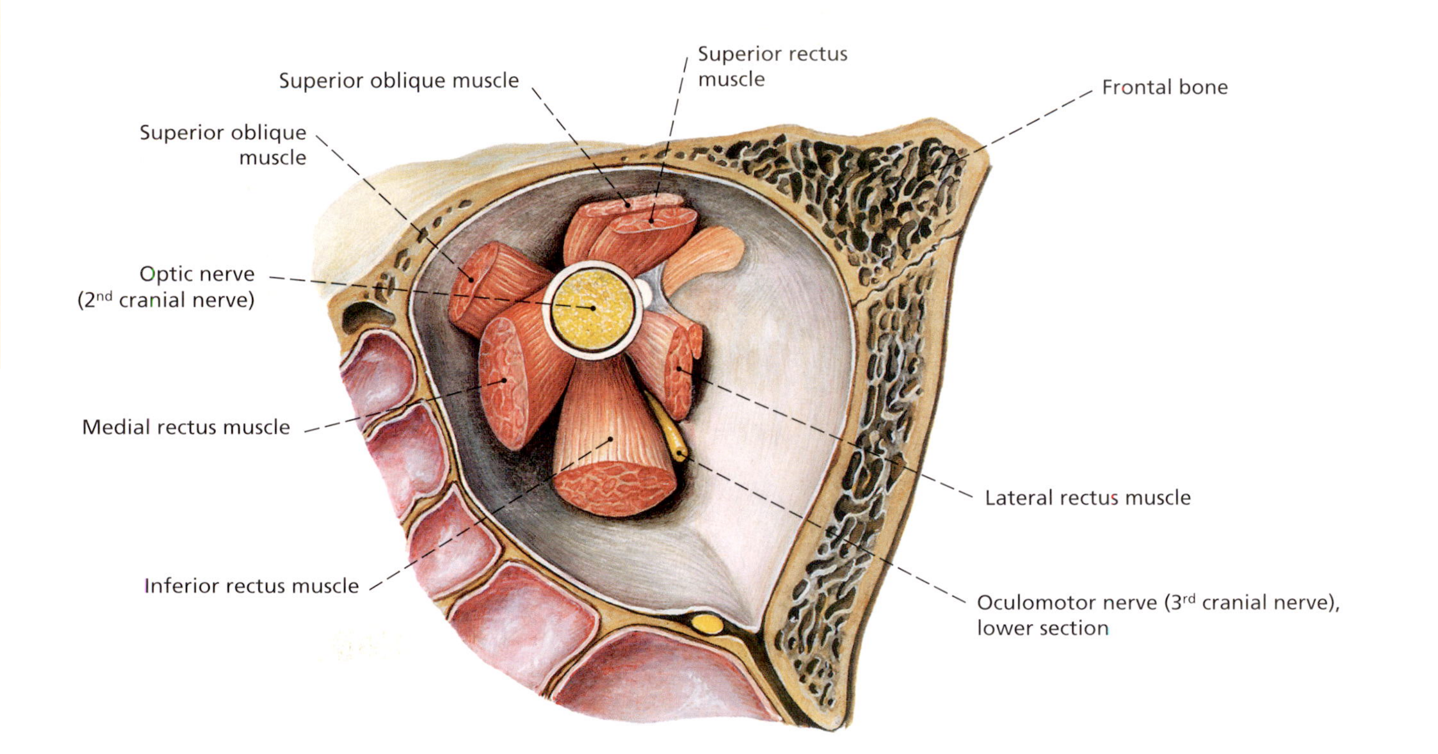

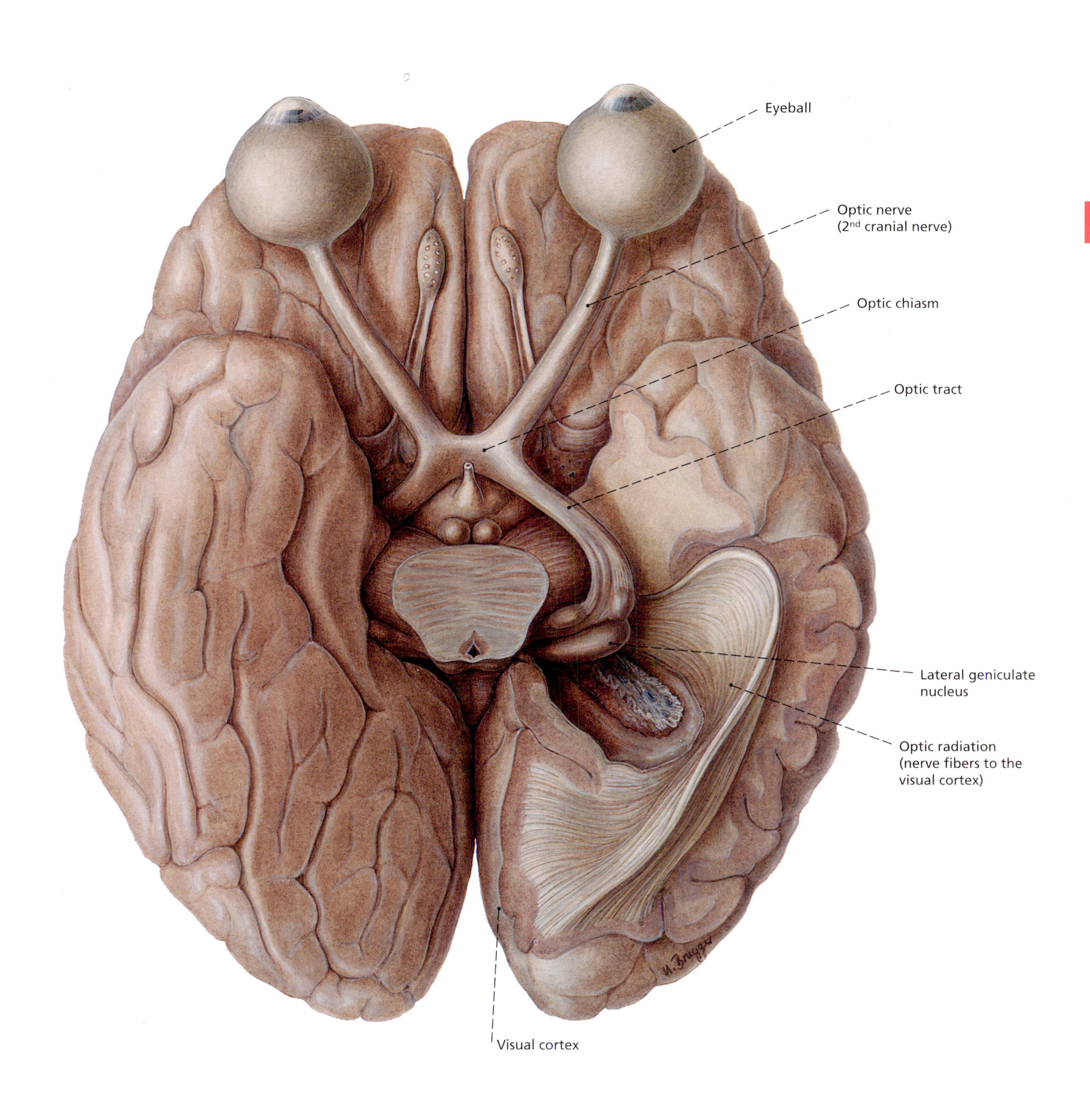

◆
Fig. 196
View of the optic chiasm from below, looking at the lower surface of the brain. Parts of the left temporal and occipital lobes have been removed so that the structure of the optic tract (see page 174) is visible.

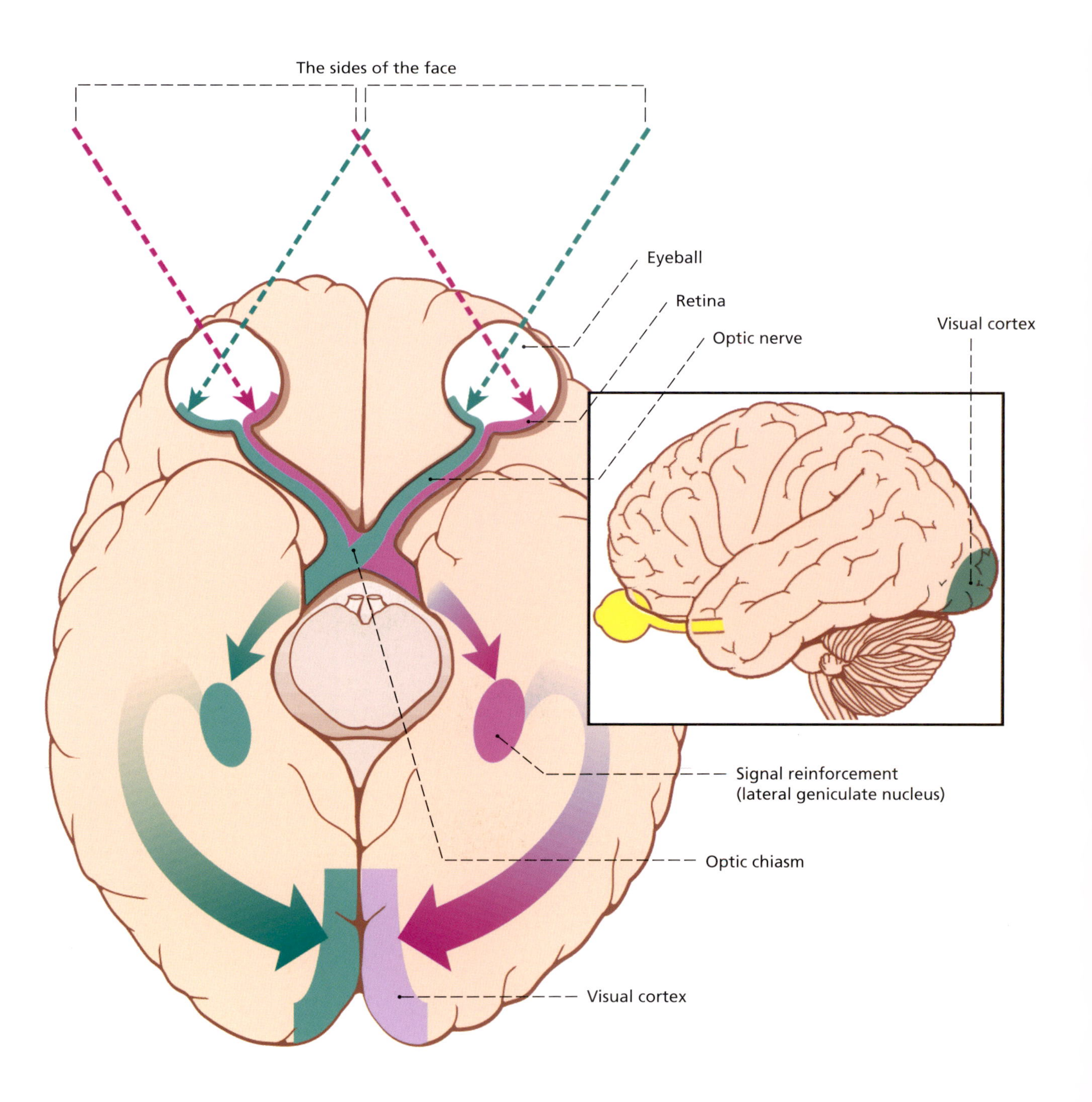

◆
Fig. 197
A graphic illustration of the path of the optic tract from the eyeball to the cerebral cortex. The visual impressions from both eyes are converted into electrical impulses by the retina at the back of the eye. These are conveyed via the optic chiasm to the nerve centers of the brain portrayed by the oval shapes (known as the lateral geniculate nuclei). Here they are reinforced and transmitted to the visual cortex where the impulses from both sides of the face are converted into a graphic image.

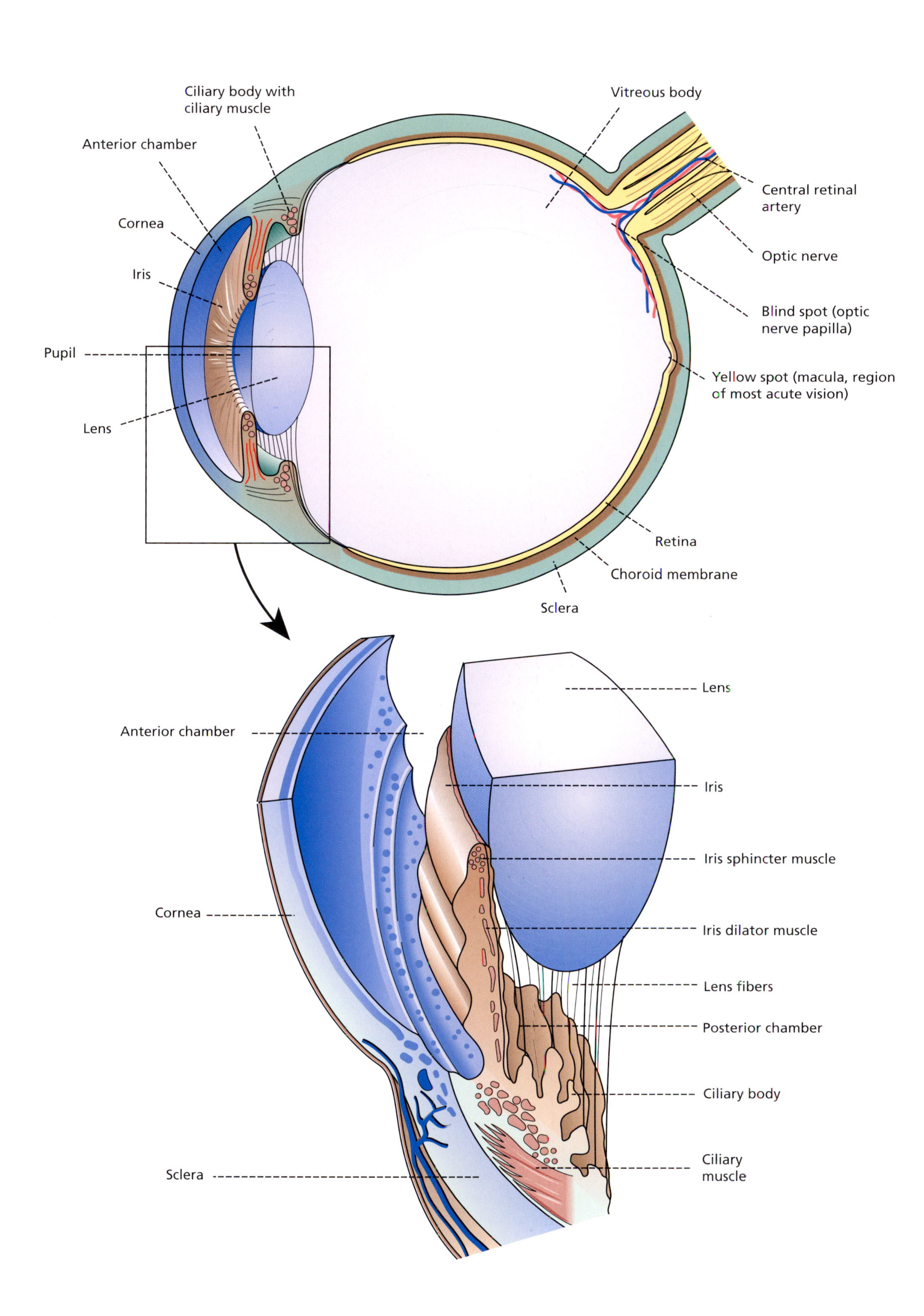

◆
Fig. 198
A cross-section of the eyeball (above), an enlarged section showing the cornea, iris and lens (below).

◆
Fig. 199
The retinal arteries and veins of the right eye. View of the back of the eye. The veins are colored blue and the arteries red.

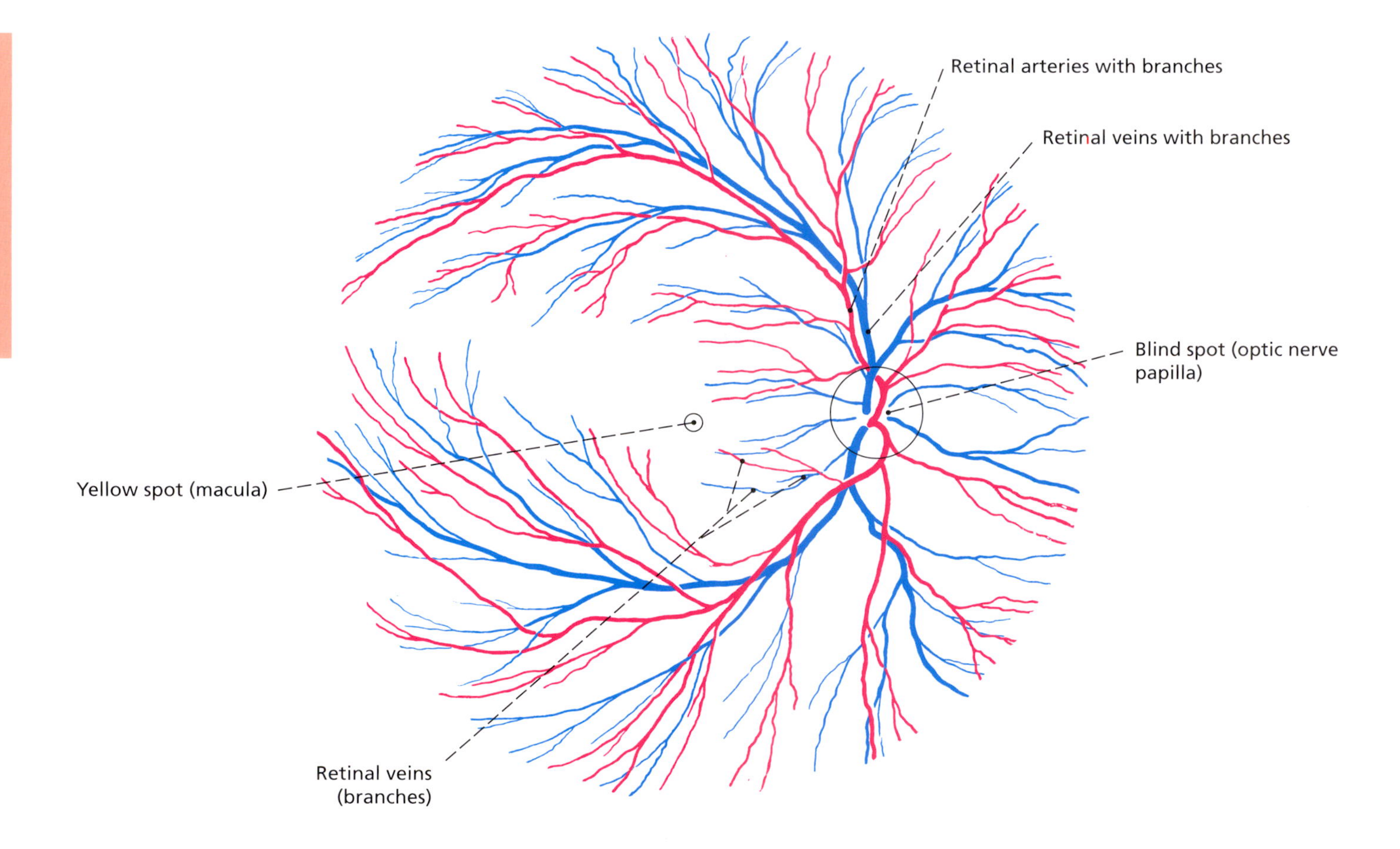

◆
Fig. 200
The fundus of the right eye as seen through an ophthalmoscope. The so-called yellow spot (macula) on the left of the image indicates the region of most acute vision on the retina, which has only cone cells in this region. The strong luminosity on the right is caused by the entry point of the optic nerve. This is also known as the "blind spot" because it contains no photoreceptors.

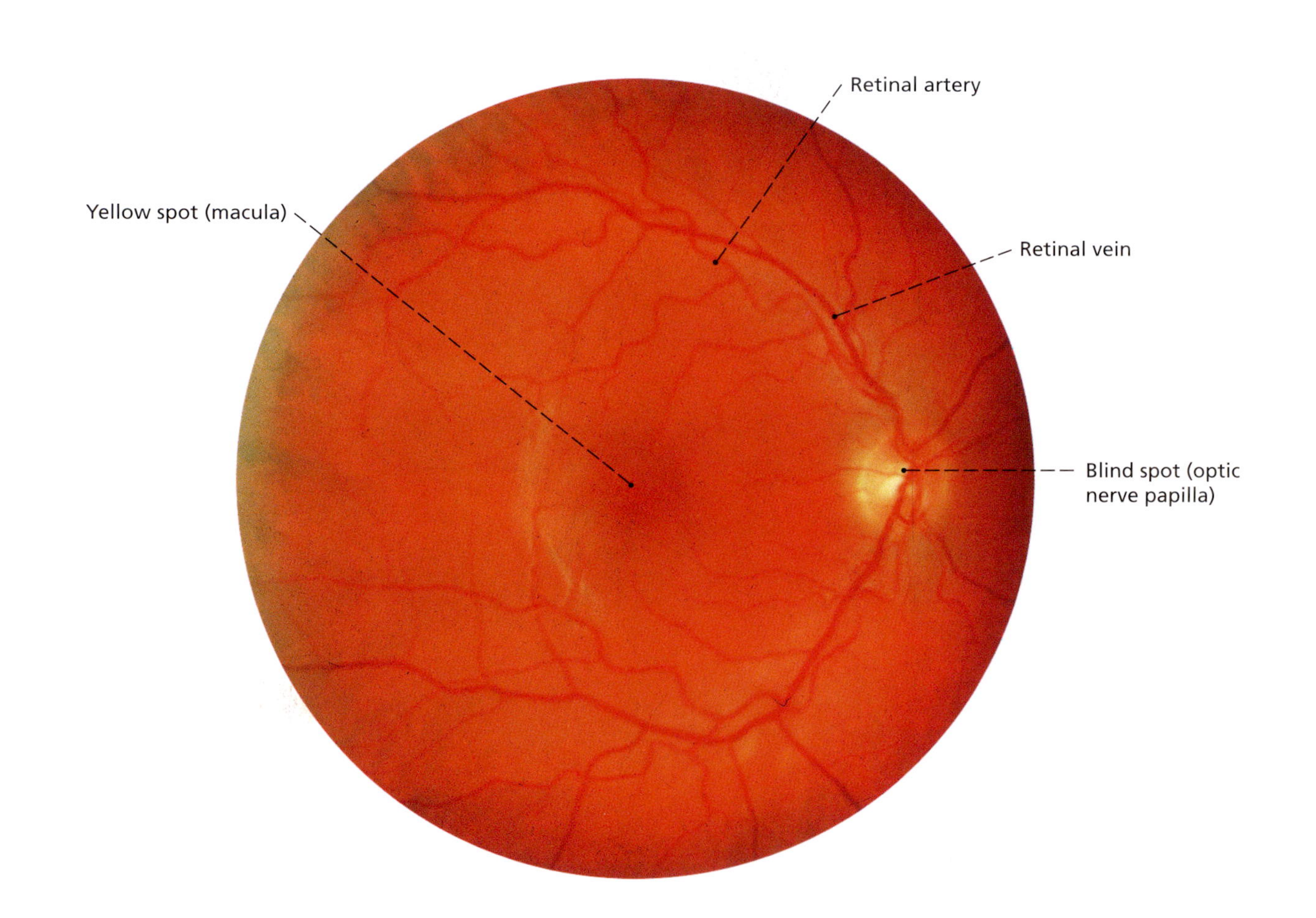

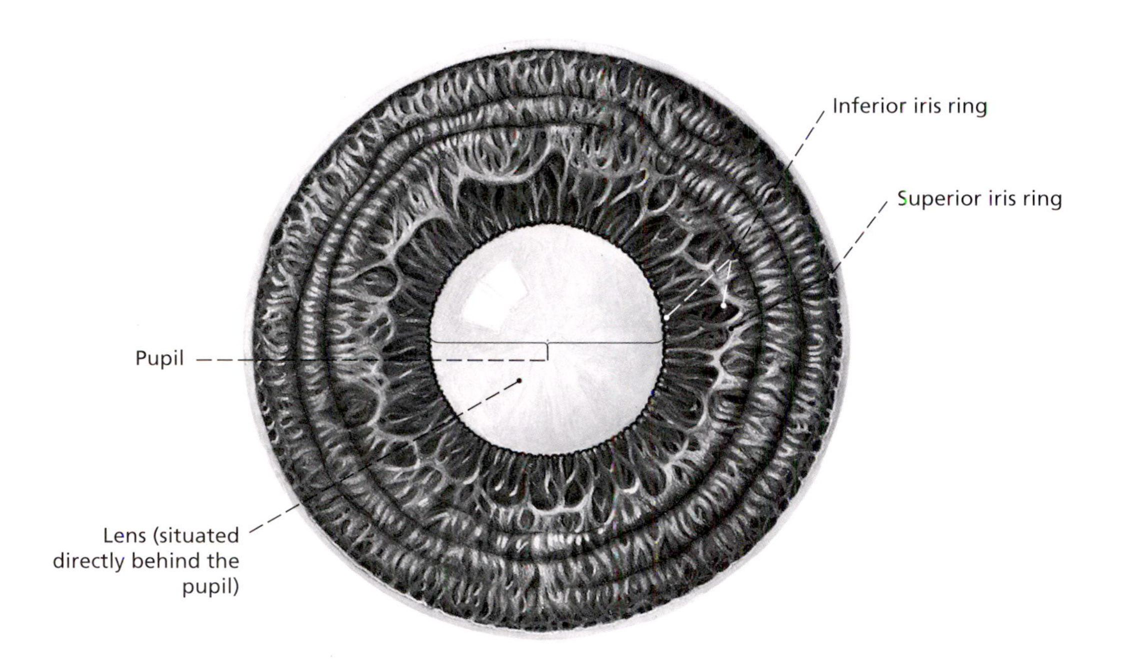

◆
Fig. 201
The iris following removal of the cornea, seen from the front.

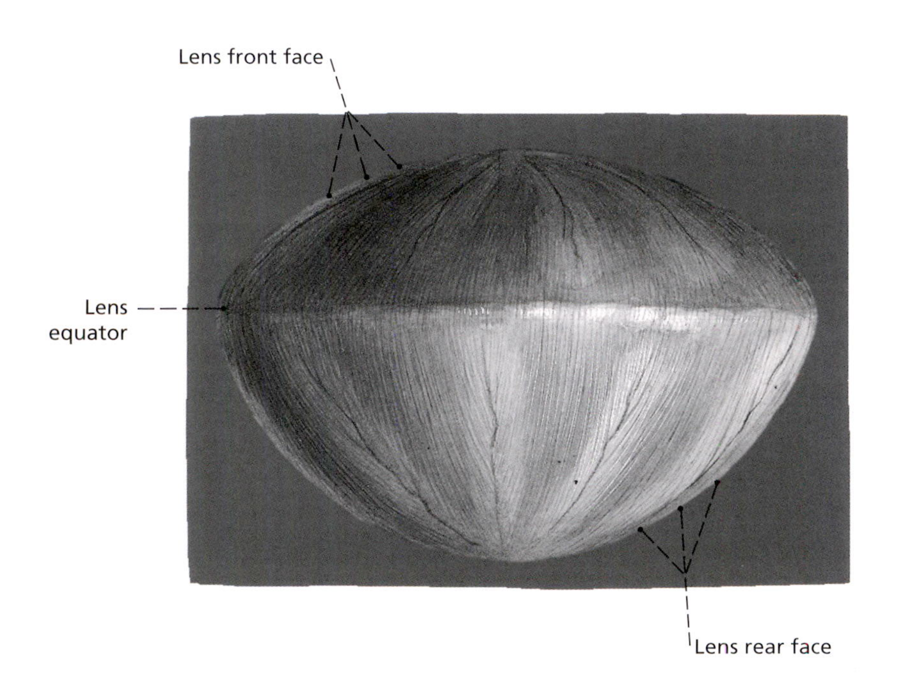

◆
Fig. 202
The lens of the eye viewed from the lens equator.

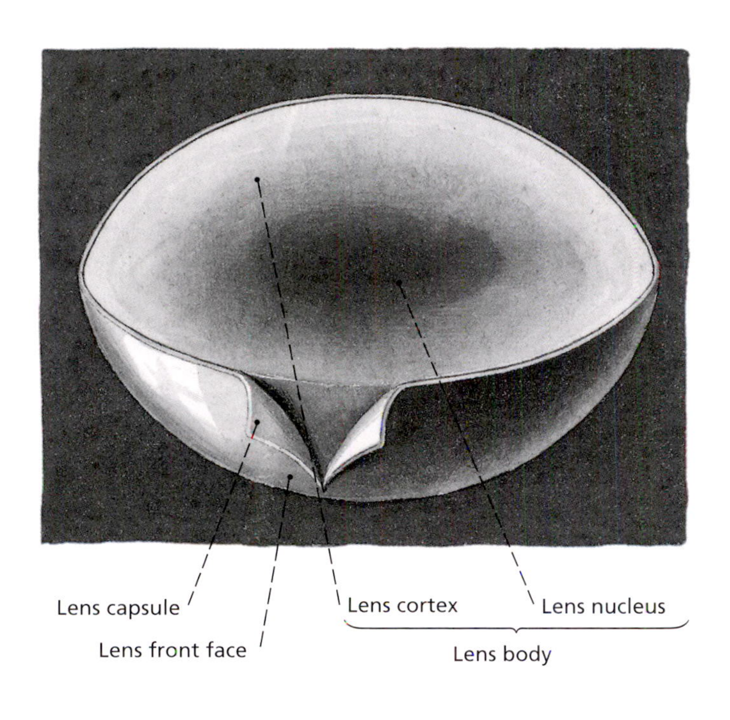

◆
Fig. 203
The lens of the eye in cross-section. The lower section of the lens capsule has been opened and folded back.

Symptoms and diseases

Visual disorders

Disorders relating to visual acuity, color vision, and spatial vision are the most common. Blurred vision is usually caused by changes to the eyeball. In the case of short-sightedness the eyeball is too long, while in long-sightedness it is too short. In astigmatism, the cornea of the eye is distorted.

• *Astigmatism*

In astigmatism, is a corneal irregularity leading to visual distortion, e.g., a dot is not seen as a dot, but rather as a dash or bar. The corneal irregularity leads to reduced visual acuity, sensitivity to glare, and headaches. It is usually inherited but can also appear following injury and glaucoma surgery. Astigmatism is corrected by special glasses or contact lenses. It is also possible to achieve permanent correction using laser treatment.

• *Short-sightedness (myopia)*

Blurred vision in the distance caused by excess eyeball length meaning that the concentration of the light rays entering the eye forms an image in front of, instead of on, the retina. Without correction by means of glasses or contact lenses, a short-sighted person is able to see close up objects clearly, but objects further away are blurred.

• *Long-sightedness (hyperopia)*

In long-sightedness, distant objects are seen more clearly than those close up. Hyperopia is caused by a reduction in the eye's refractive powers or by a congenital shortening of the eyeball. Presbyopia is also a form of long-sightedness and is caused by a loss of lens elasticity in old age.

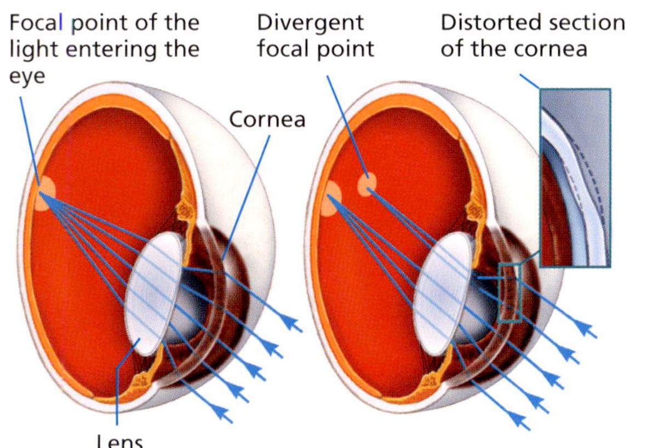

◆ **Astigmatism** The light entering a healthy eye is concentrated at the focal point (left). If the shape of the cornea is irregular (right), then the light rays are further deflected; a second focal point results and therefore a blurred image.

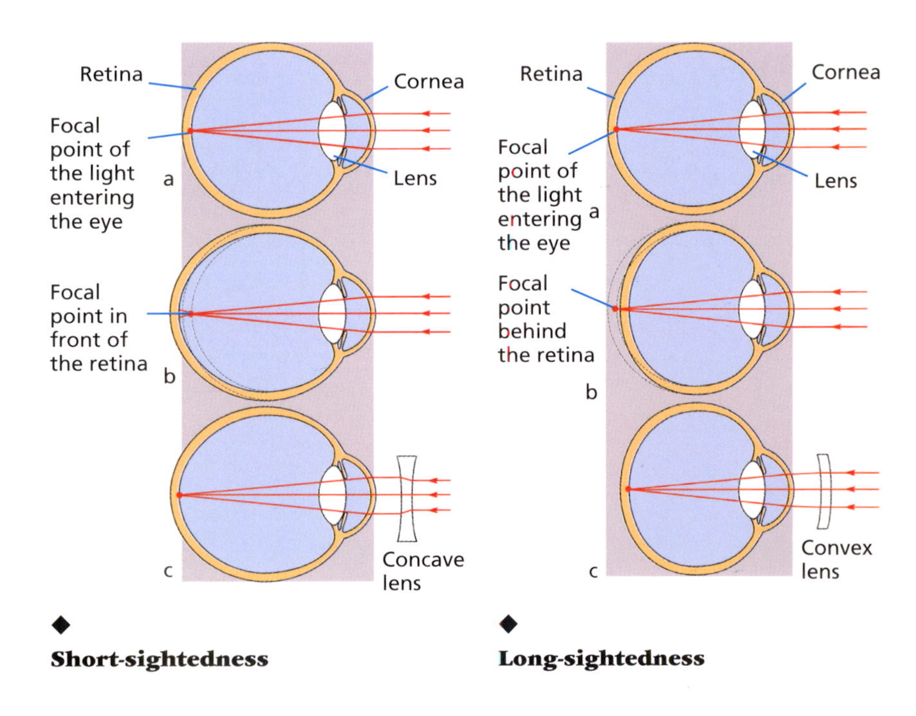

◆ **Short-sightedness**

◆ **Long-sightedness**

• *Strabismus*

Strabismus is caused by a disruption between the eyes or the eye muscles. In the case of uncorrected strabismus in one eye the blurred image is not picked up by the brain. The affected eye goes unused and drifts as a consequence. Strabismus can usually be corrected with glasses. In children, care needs to be taken that visual acuity develops to the same extent in both eyes. The squint eye also needs to be able to focus on objects in order to be able to learn to see and so the healthy eye is often covered with a patch or a covering on the glasses for certain period of time. This form of treatment does not heal the strabismus itself but improves visual acuity in the affected eye. The squint then has to be corrected separately.

Inflammations

• *Conjunctivitis*

An inflammation of the thin, transparent layer of skin covering the eyeball and the inside of the eyelids. It can be caused by hypersensitivity to pollen, draughts, overexertion of the eye (screen work), and chlorine (swimming pools), and also by infections or exposure to foreign bodies. A reduction in lacrimal fluid means that the eyes dry out, become itchy, very red, sensitive to light, and exude secretions. Treatment is with disinfectant, anti-bacterial eye drops. Foreign bodies need to be removed by a physician.

• *Scleritis*

The eye becomes red-blue in color and swollen. The inflammation may be caused by bacteria but can also be an accessory symptom to rheumatic complaints. It is treated with special eye ointment or drops.

• ***Keratitis***

An inflammation of the cornea, often occurring in connection with conjunctivitis; it is usually the result of a viral infection.

• ***Iritis***

An inflammation of the iris due to external influence (injury), ulcers, or to diseases of the cornea, sclera, or retina. Iritis can also be caused by burns, insect bites, toxoplasmosis, or syphilis. It sometimes occurs in association with rheumatic conditions.

Cataracts

A condition in which the lens of the eye becomes cloudy or opaque, instead of transparent and crystal clear. The pupil appears gray-to-white in color. The initial symptoms are decreased vision, sensitivity to glare, and misting.
Cataracts often appear after the age of 60. The lens increases in volume during the course of a lifetime, becoming thicker and its core becoming harder. This leads to clouding of the lens, which causes short-sightedness because it increases the eye's refractive power. Congenital cataracts also occur but have a tendency to be rare.
Cataracts can also be caused by eye injuries (e.g., bruising of the eye), high-voltage injuries, radiation exposure, specific forms of chemical poisoning, or occur as a result of metabolic disorders (e.g., diabetes). Conservative treatment methods, in the form of eye drops, only bring about a temporary halt to the clouding. **Cataract surgery**, in which the clouded lens is replaced by an artificial lens, is able to substantially restore vision.

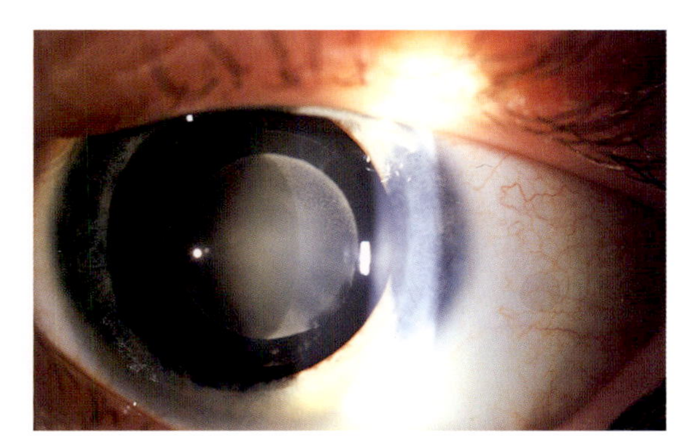

◆
Cataracts
The gray-white clouding of the lens can easily be seen through the fully dilated pupil.

Glaucoma

In glaucoma, the intraocular pressure is increased in one or both eyes. The increase in pressure leads to damage to the optic nerve. This causes a deterioration in visual acuity and a noticeable reduction in the visual field. Such damage is irreparable as no method of treatment exists that is able to heal the damaged nerve fibers.
A **glaucoma attack** is caused by a sudden change to the chamber angle and the resulting blockage of the aqueous humor drainage system. The symptoms include impaired vision, the seeing of colored rings, and ultimately sharp pains, nausea, and vomiting.
People with constitutionally narrow front eye sections and who also have a cataract often suffer from so-called **angle-closure glaucoma**. In such cases the swollen lens presses the iris located in front of it forwards against the cornea, thus closing the chamber angle.
In the case of **open-angle glaucoma** the chamber angle is open, but the drainage system is blocked, e.g., by inflammation-related agglutination or age-related changes. It develops slowly, usually over the course of years. The first signs are clouded vision, a gradual deterioration of vision, and restriction of the visual field.
Glaucoma is treated with medication, laser treatment, or with an operation.

Diagnostic methods

Eye pressure measurement

The painless measurement of intraocular pressure forms part of a routine examination by the ophthalmologist, and can detect the presence of glaucoma in its early stages. A puff of air is blown onto the cornea and the resistance measured: this indicates the intraocular pressure.

Examination of the ocular fundus

Examination of the fundus of the eye using a strong magnifying lens and a bright light. The pupil has to be dilated with the help of eye drops in order to gain an overview of the entire fundus. The rear sections of the eye and the optic nerve are examined. The examination is painless but afterwards the patient may be extra sensitive to glare, due to their dilated pupils, and vision is blurred: driving a car is therefore prohibited. This is an essential examination for patients with high blood pressure or diabetes, as these conditions may cause vascular changes.

Visual field measurement

The ophthalmologist assesses the size of the visual field as an important preventative measure to look for the early signs of glaucoma, using a device in which a variety of lights are shown briefly in different areas of the visual field.

Eyesight test

Ophthalmologists and opticians recommend regular eye tests as eyesight can often deteriorate without notice over the course of time. Visual acuity is checked using screens or panels displaying letters, numbers, or figures in a variety of standard sizes at a distance of about 6 meters. A refraction test is required for a precise correction of visual acuity. This determines any deviations from normal light refraction, which are then corrected with the appropriate glasses or contact lenses. An accommodation test checks how well the eye is able to adjust from distance vision to nearby objects, while a visual field measurement determines the area that the inactive eye is able to perceive.

Treatment methods

Orthoptics

Training to increase spatial vision in both eyes using special exercises and equipment. It can help severely visually handicapped people to regain their reading ability with exercises using visual panels and they can also learn spatial orientation.

Eye exercises

Visual training may be given to treat visual disorders previously diagnosed by specially trained orthoptists. For example, exercises for patients who have suffered strokes, severe brain injury, or whiplash injuries, which have caused disruption to the processing of visual impressions in the brain. The need to wear glasses cannot be eliminated by eye exercises.

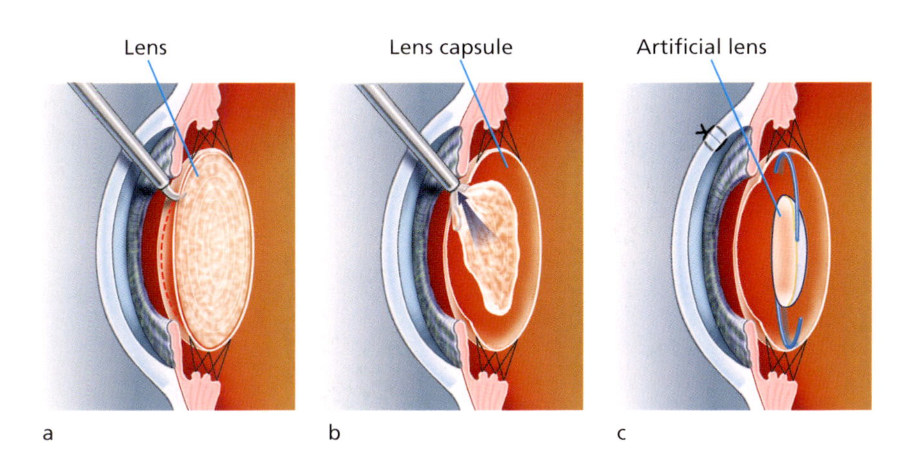

◆
Cataract surgery
In a cataract operation the lens capsule is opened through a cut in the cornea (a), the clouded lens body is extracted (b) and the artificial lens anchored in the lens capsule (c).

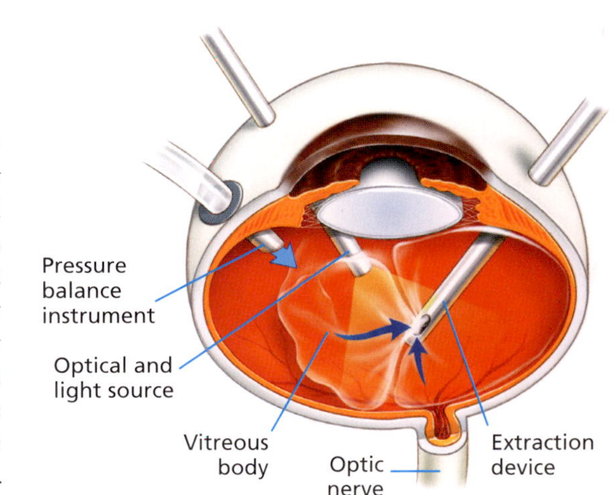

◆
Vitreous body removal
If the eyesight is restricted by a damaged or clouded vitreous body the latter is usually extracted under local anesthetic and the eyeball filled with saline solution.

Surgical methods

• ***Cataract surgery***

Surgery for the treatment of cataracts in which the lens of the eye is removed and replaced with an artificial lens. Using a special instrument, the lens body is extracted through a small cut in the upper edge of the cornea. An artificial implant is usually inserted immediately and eyesight is thus restored.

• ***Strabismus operation***

The straightening of an eye by moving or shortening the corresponding eye muscles. A strabismus operation can usually be performed on an outpatient basis and, in the case of adults, can be carried out under local anesthetic.

• ***Iridectomy***

The partial removal of the iris carried out to improve the eyesight, treat inflammation, or to treat a cataract.

• ***Cornea transplant (keratoplasty)***

The replacement of a defective cornea with a healthy cornea from an organ donor. Tissue rejection is rare following cornea transplants because the cornea is not supplied with blood and therefore has little contact with the organism's immune cells.

• ***Vitreous body removal***

The removal of the front section of the eye can be necessary following injury to the eye. In the case of vitreous body clouding, the rear section is removed and the cavity is filled with saline solution.

• ***Laser therapy***

In ophthalmology, lasers are used to treat retinal detachment, for the restoration of eyesight following a cataract operation, and to correct some forms of strabismus.   1 line over

The ear and organ of equilibrium

The organs of hearing and equilibrium are extremely sensitive organs and are housed in a well-protected antrum in the cranial bone (petrous bone). The organs are situated close together but have very different tasks. While the ear assimilates the external sound stimuli and relays them to the brain, the organ of equilibrium, or balance, is responsible for spatial orientation and upright posture.

The ear

The ear is divided into three sections: the outer ear (auricle) and outer auditory canal, the middle ear with the eardrum, tympanic cavity, auditory ossicles (small bones), and Eustachian tube, and the inner ear with the cochlea. This is also where the organ of equilibrium (known as the vestibular organ) is situated.

Outer ear

The auricle differs in shape and size in each individual. It collects sounds from the surroundings and relays them via the outer auditory canal to the eardrum. The outer auditory canal contains hairs and sebaceous glands that prevent foreign bodies from entering the ear.

Middle ear

The middle ear largely comprises an air-filled cavity, known as the tympanic cavity. It borders the outer ear with a connective tissue membrane, the eardrum, and is connected to the upper pharynx via the Eustachian tube. This connecting space is opened during swallowing and yawning, thus ensuring pressure equalization with the external air. Swallowing thus enables the elimination of an uncomfortable feeling of pressure in the ear, such as occurs when encountering large changes in altitude in airplanes or cable cars.
Sound is conveyed from the eardrum to the inner ear via the three auditory ossicles, which articulate with one another. They are known as the hammer, anvil, and stirrup, due to their shapes. Fixed to the eardrum, the hammer absorbs the sound waves and relays them via the anvil to the stirrup, which then relays them via the oval window to the cochlea and the inner ear. The acoustic pressure is controlled by two small muscles on the eardrum and stirrup.

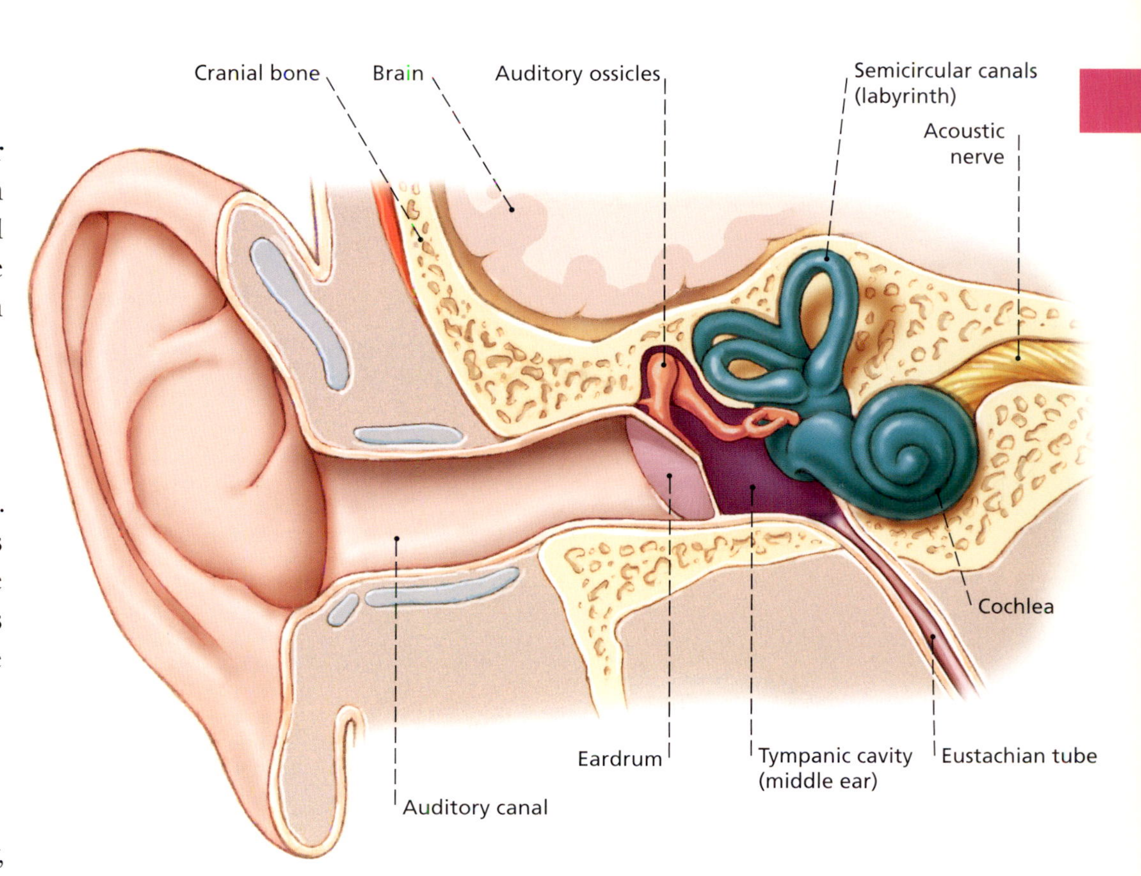

◆
Fig. 204
The auricle, auditory canal, and middle ear with the auditory ossicles, relay sound to the sensory cells of the inner ear located in the cochlea. The cochlea is linked to the semicircular canals, which contain sensory cells that register changes in the body's position. The inner sections of the hearing and balance organ are embedded in the petrous bone within the skull.

Inner ear

The inner ear with the sensitive auditory and equilibrium organ is located in a chamber within the cranial bone known as the bony labyrinth. The spiral-shaped, convoluted cochlea contains three fluid-filled chambers that contain a clear, lymph-like liquid. The upper chamber (scala vestibuli or vestibular canal) runs from the oval window to the tip of the cochlea where it merges with a second chamber (scala tympani or tympanic canal) which runs parallel back to the middle ear, ending in a membrane known as the round window. The auditory sensory cells are located in the middle, triangular-shaped, chamber, known as the cochlear duct. When the eardrum starts to vibrate as a result of a sound wave, the vibrations are relayed via the auditory ossicles and conveyed to the endolymph in the cochlea at the oval window. This stimulates the fine hairs of the sensory cells in the cochlear duct, which relay impulses to the brain via the axons and dendrites. Thousands of these nerve fibers make up the acoustic nerve and lead to the hearing center of the cerebrum. The sound waves only reach both ears simultaneously if their source is directly in front or behind. Otherwise they reach our ears at different times and this is what enables us to tell the direction from which a sound comes.

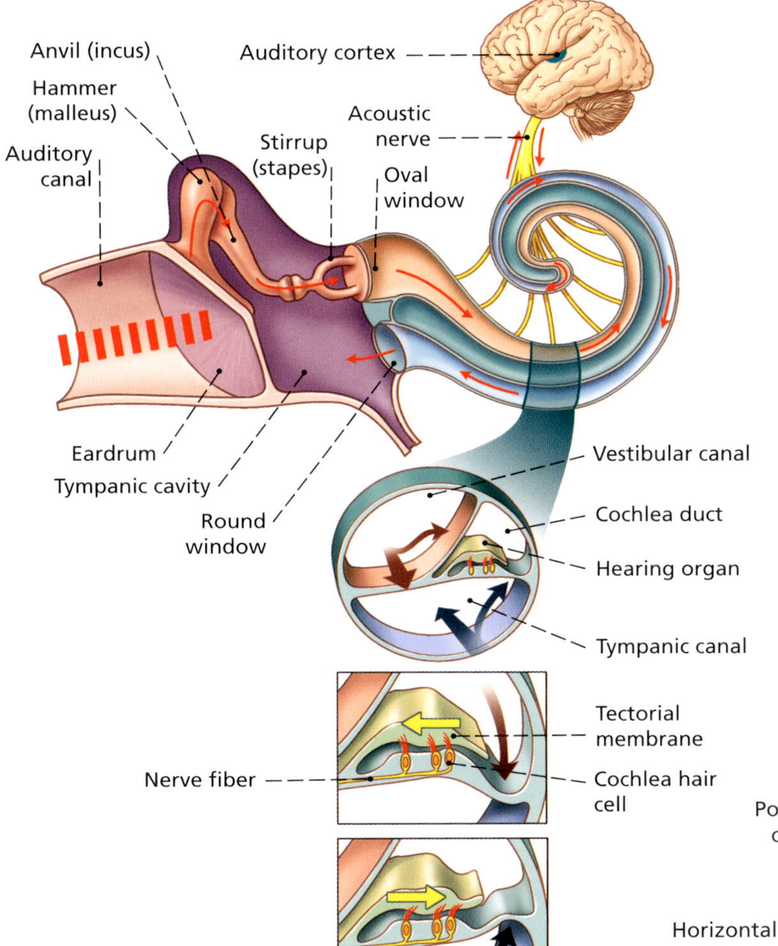

Fig. 205
Vibrations are relayed to the membrane of the oval window via the auditory ossicles. In the vestibular canal, the vibrations are converted into a liquid "wave," which continues along the tympanic canal *(central detail)*. The cochlea duct contains the sensory cells, the hairs of which extend up into a thin membrane (the tectorial membrane). The liquid wave moves this membrane, thus stimulating the sensory cells.

Organ of equilibrium

This organ comprises three bony semicircular canals arranged at right angles to one another on three spatial levels. These canals, which are filled with endolymph and equipped with sensory cells, are interconnected. The sensory cells react to waves in the liquid caused by movements of the head or of the body as a whole, and relay their impulses via the nervous system to the various centers of the brain. The fact that the semicircular canals cross different spatial planes means that the little hairs on their sensory cells are subjected to movement and/or stimuli with any changes in position. Their signals convey a conscious perception of circular motion and a sense of top and bottom. The impulses from the sensory cells are relayed not only to the cerebral cortex but also to the spinal cord (which controls the muscle reflexes), the cerebellum (which controls motor activity), and to other areas of the brain. This is also the reason that strong stimulation of the equilibrium organ, such as in an aircraft or on a ship, can cause nausea, dizziness, vomiting, and headaches.

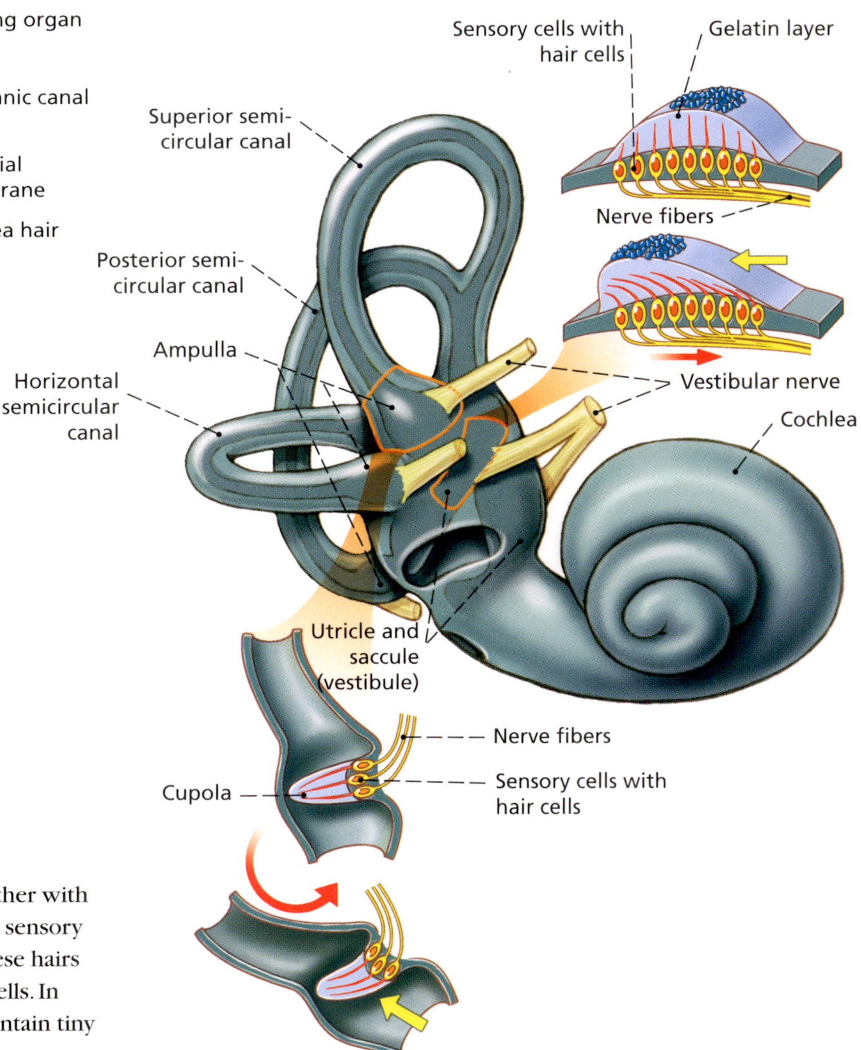

Fig. 206
The semicircular canals and their ampullae *(detail right)* together with the utricle and saccule of the vestibule *(detail above)*, contain sensory receptors with hair cells extending into a gelatin-like layer. These hairs move with every change in position, stimulating the sensory cells. In addition to the gelatin-like layer, the utricle and saccule also contain tiny calcium carbonate particles that amplify linear stimuli.

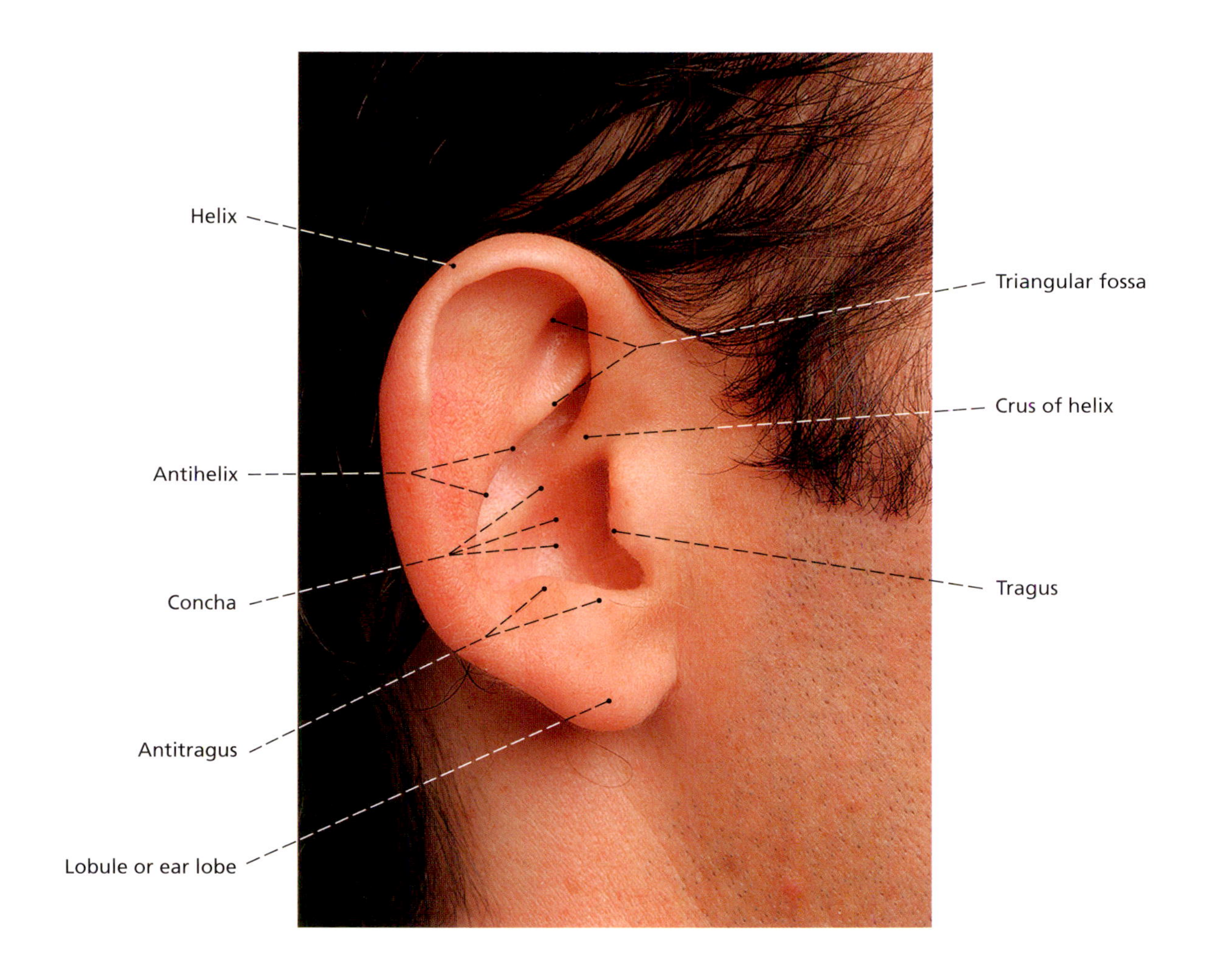

◆ **Fig. 207**
Visible anatomy of the outer ear.

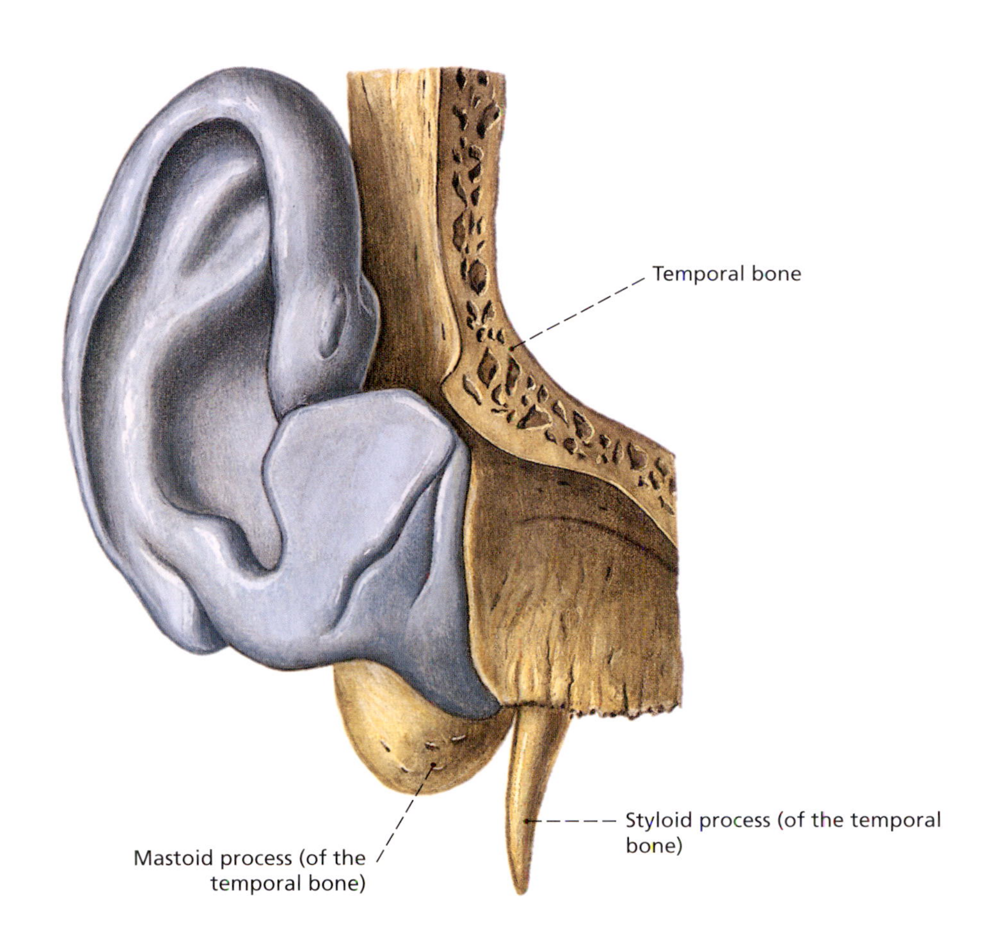

◆ **Fig. 208**
Transverse frontal view of the cartilage skeleton of the right ear.

◆
Fig. 209
Frontal view of the right ear muscles.

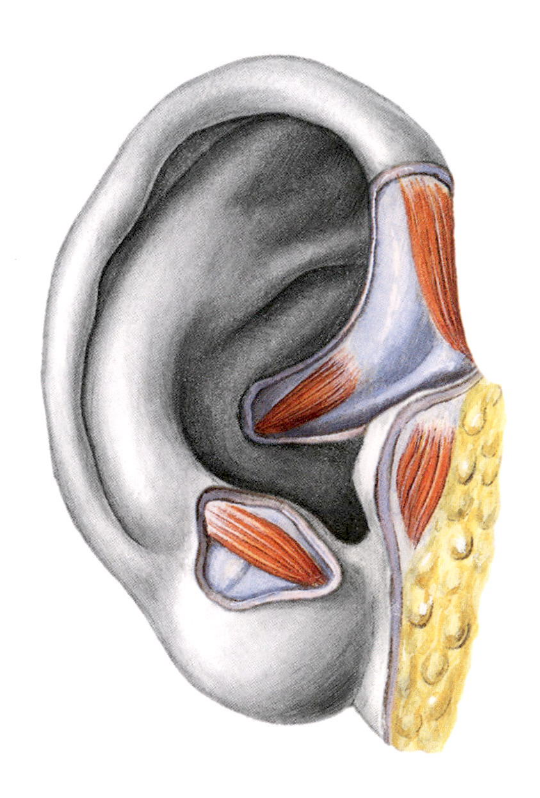

◆
Fig. 210
Rear view of the right ear muscles.The outer auditory canal aperture in the auricle has been removed (the ear is folded forwarded slightly).

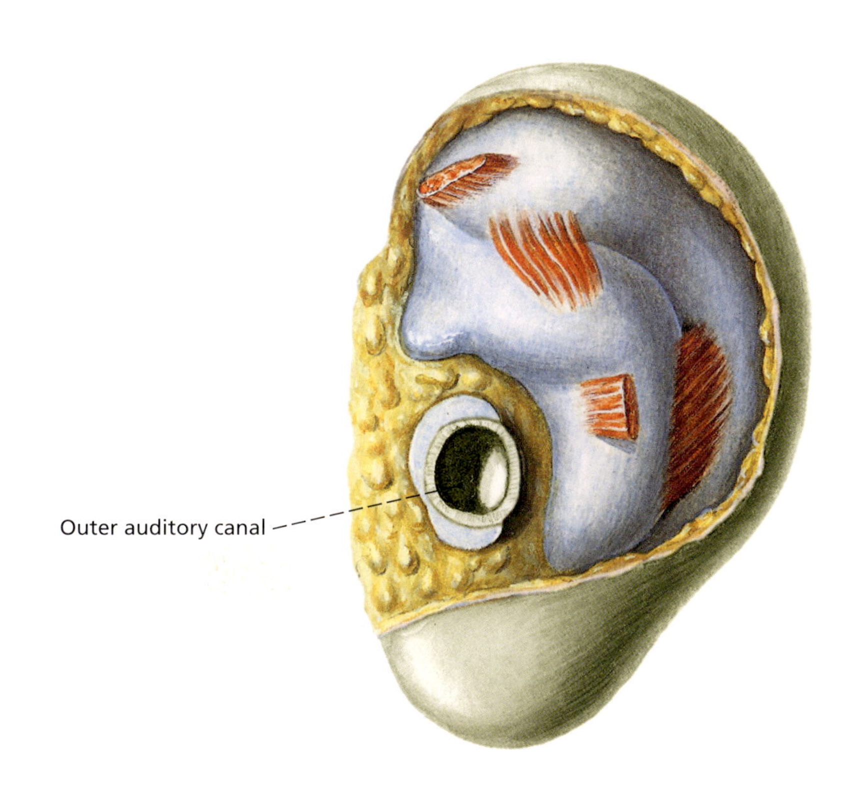

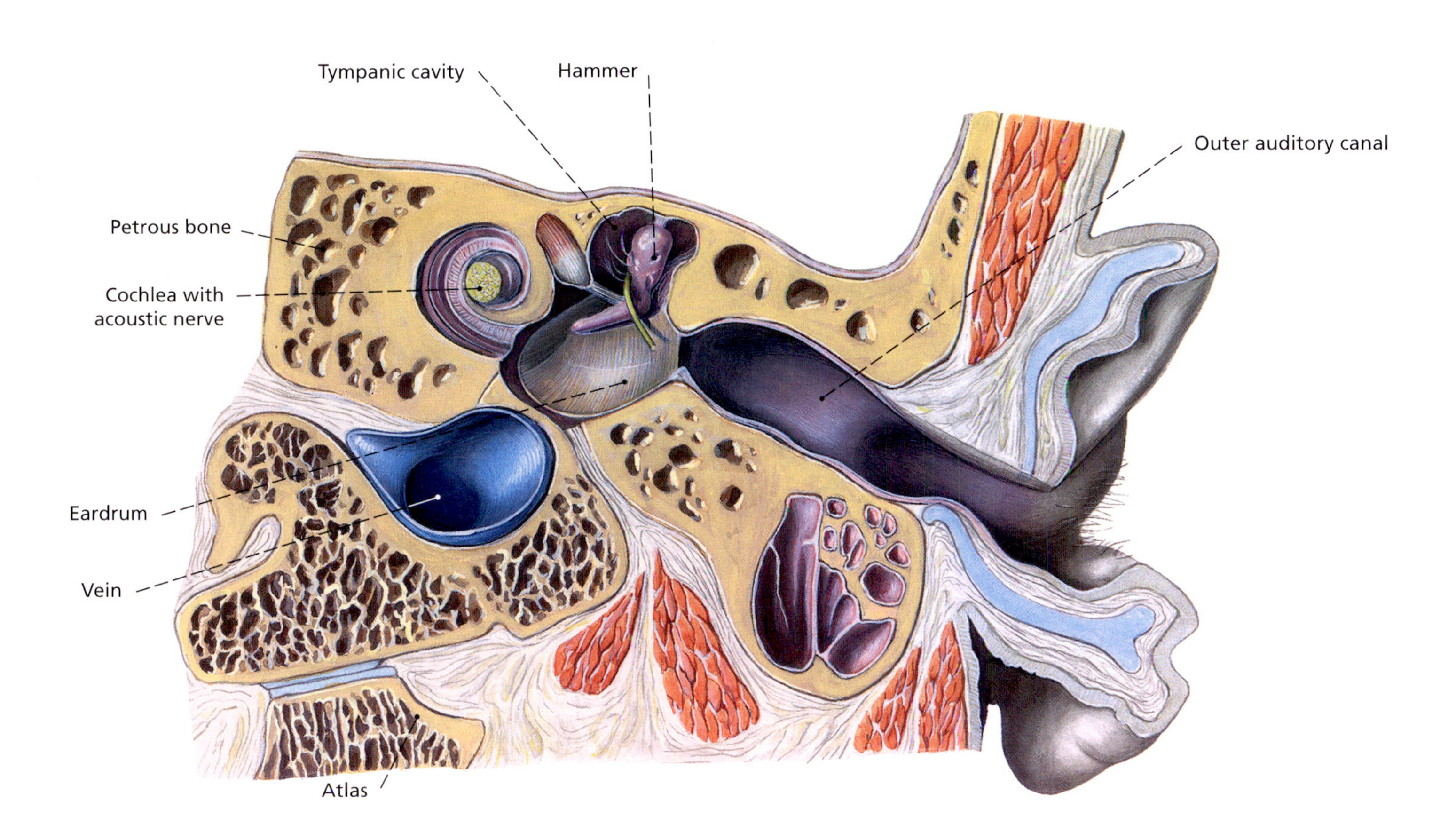

◆
Fig. 211
Longitudinal section through the outer auditory canal with the middle ear (tympanic cavity) and inner ear, viewed from the rear. The hearing organ is embedded in the petrous section of the cranial bone.

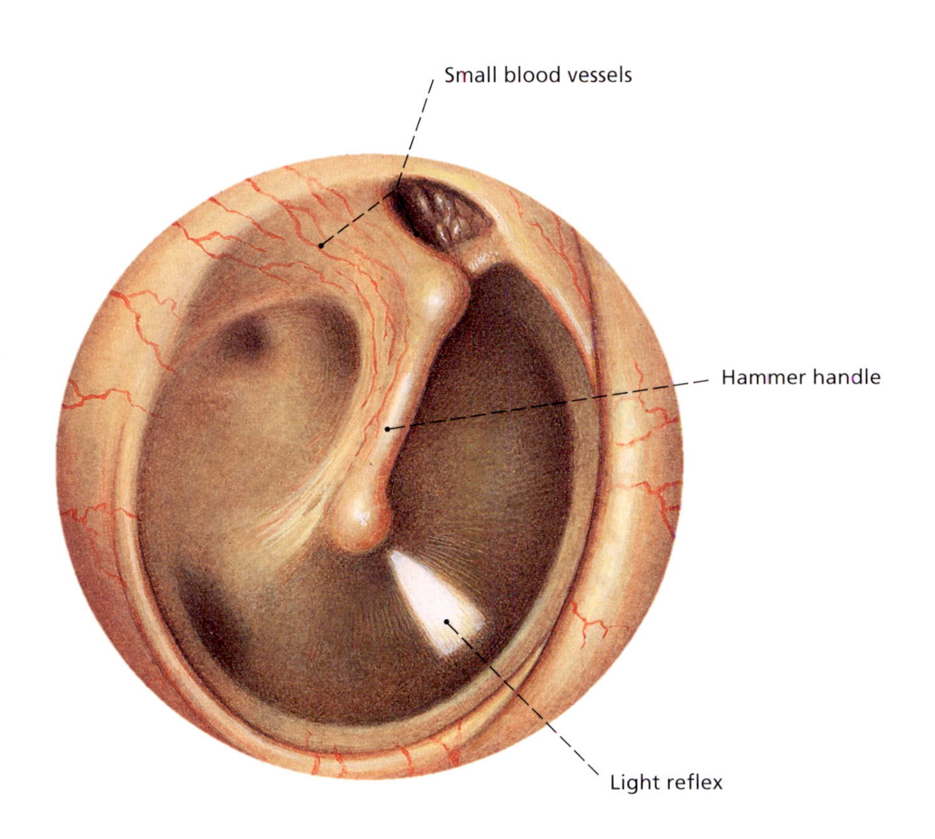

◆
Fig. 212
The eardrum. View through an otoscope. The hammer handle (middle of the image) is visible through the eardrum. The light reflex is less easily recognizable in illnesses of the middle ear.

◆
Fig. 213
The auditory ossicles in their natural positions.

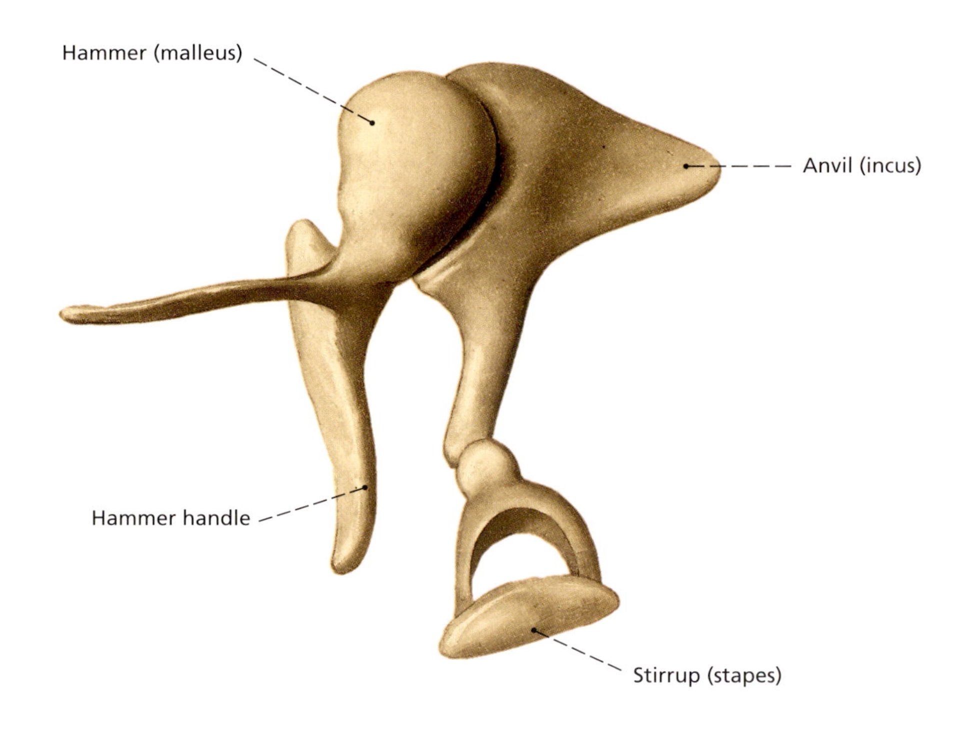

◆
Fig. 214
The auditory ossicles with the joints and ligaments.

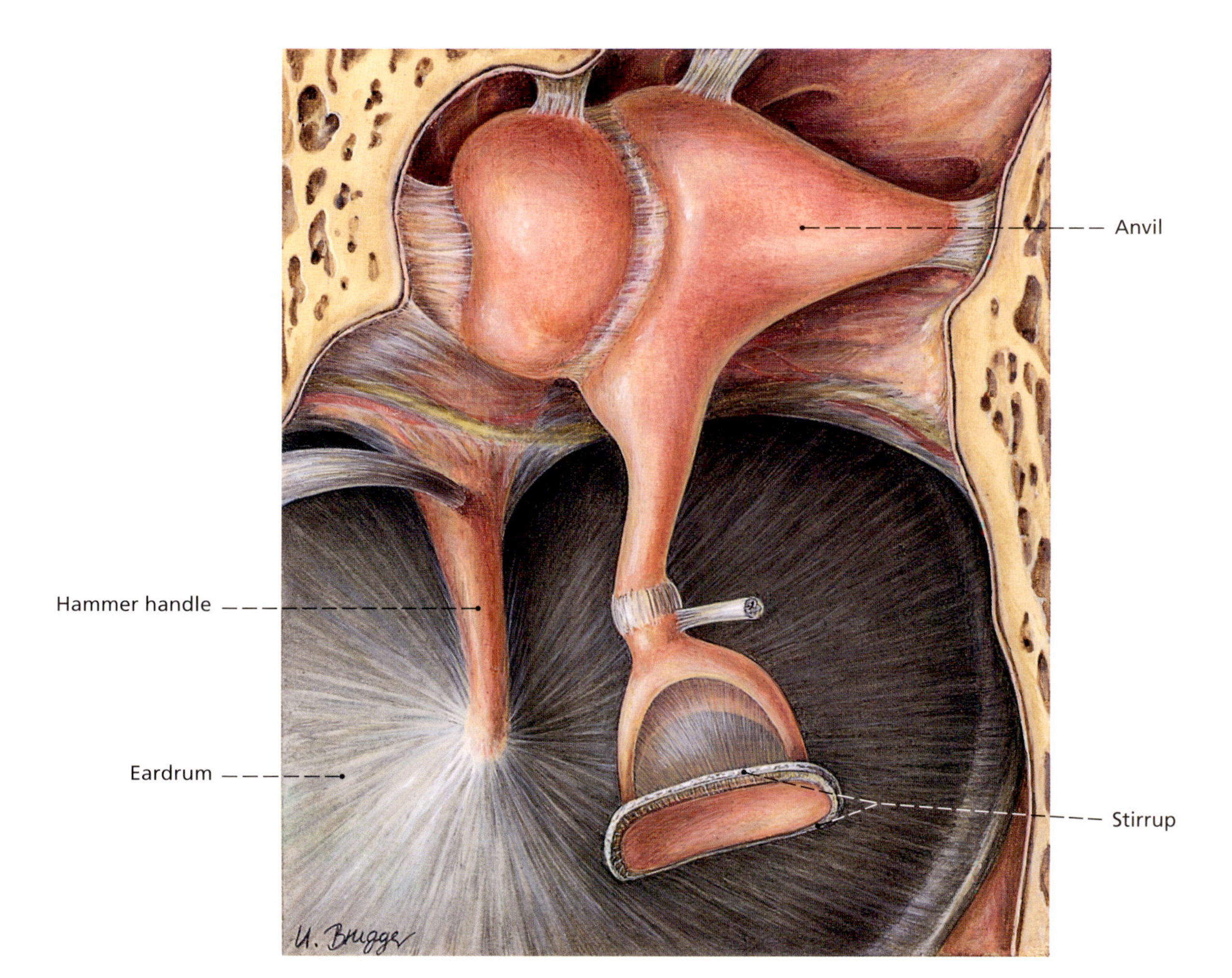

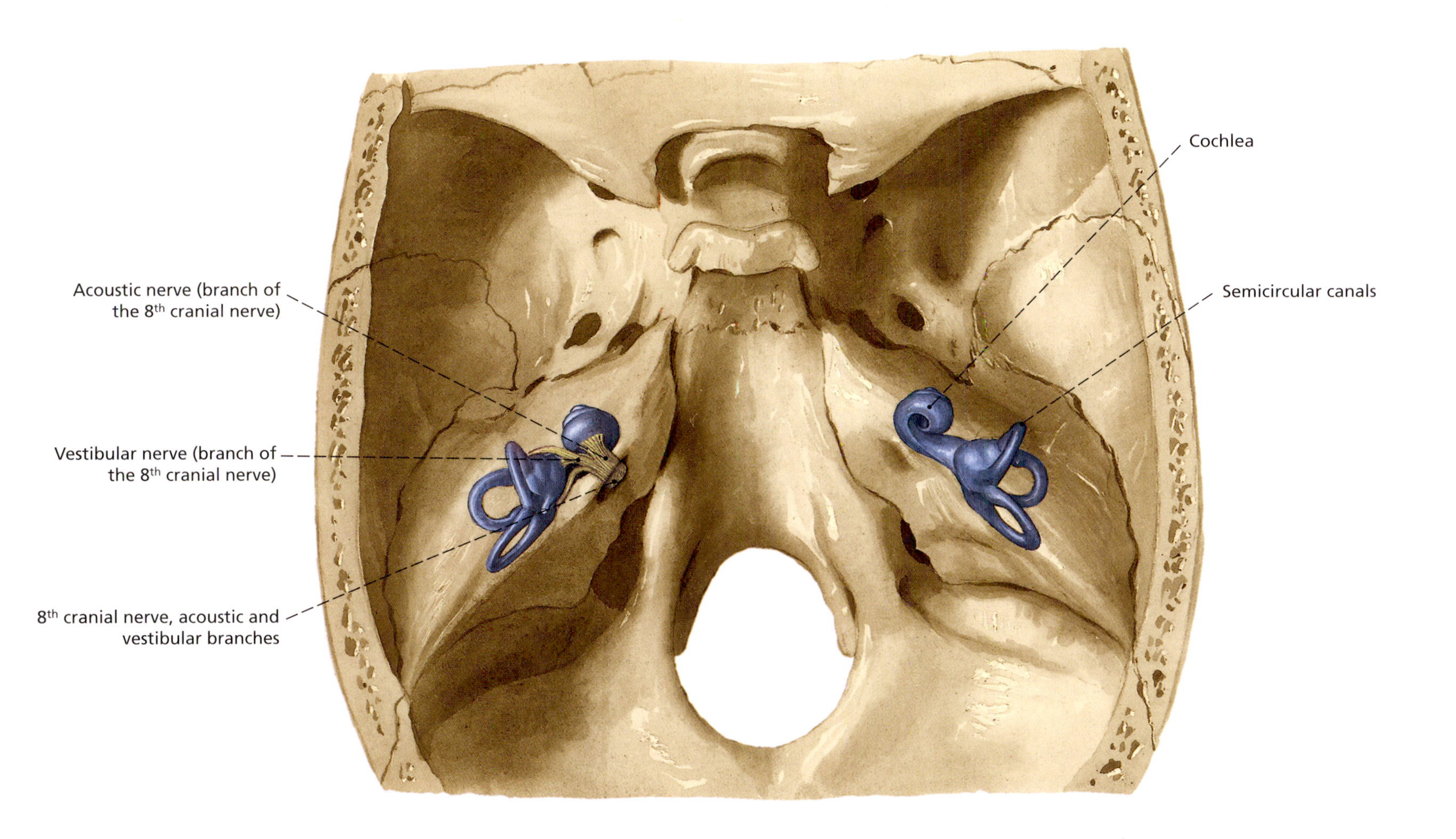

◆
Fig. 215
The location of the cochlea and the semicircular canals in the open skull, viewed from above. Both structures have been colored blue for better illustration and are depicted on the underlying petrous section of the cranial bone. In reality, they are embedded in the cranial bone.

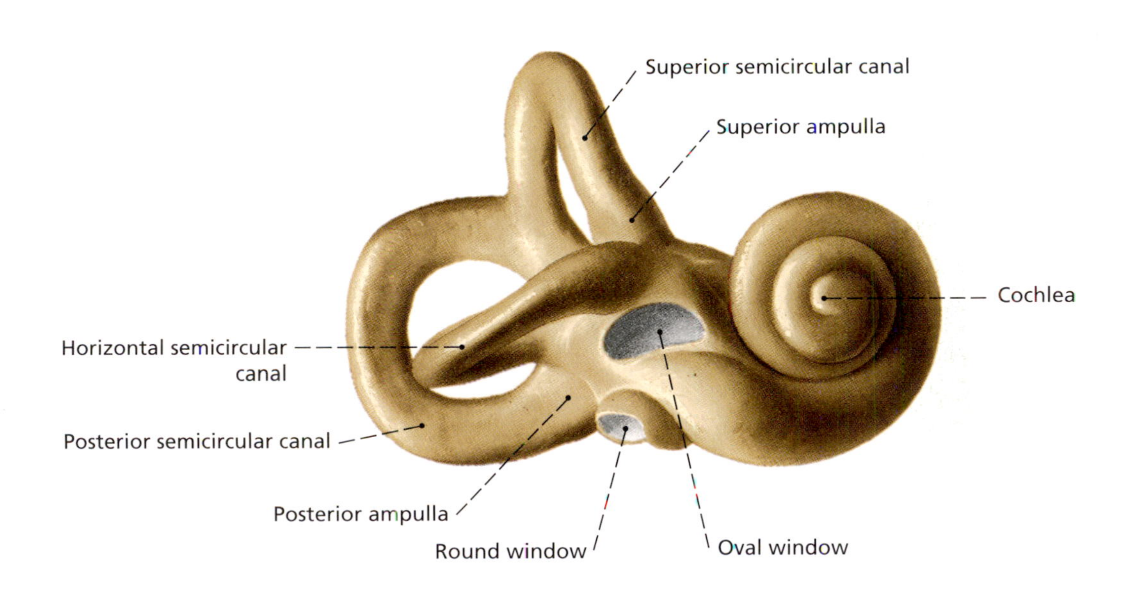

◆
Fig. 216
The semicircular canals and cochlea reamed out of the petrous bone, viewed from the front right.

Symptoms and diseases

Tinnitus

Disturbing noises, usually in the form of buzzing, swishing, ringing, or hissing, which occur in the ear itself.
The noises can be caused by disorders of the middle or inner ear, including circulatory disorders, blood pressure which is too high or too low, an ear wax plug, poisoning (mercury, carbon monoxide, etc.), or an ossification of the auditory ossicles. Tinnitus is often a psychosomatic symptom of mental or physical overexertion.
Where possible the underlying cause is treated. If this is unsuccessful, a medication to increase the blood supply to the inner ear can be administered. A so-called tinnitus masker can help in some cases: a special device produces noises that mask the disturbing tinnitus sounds.

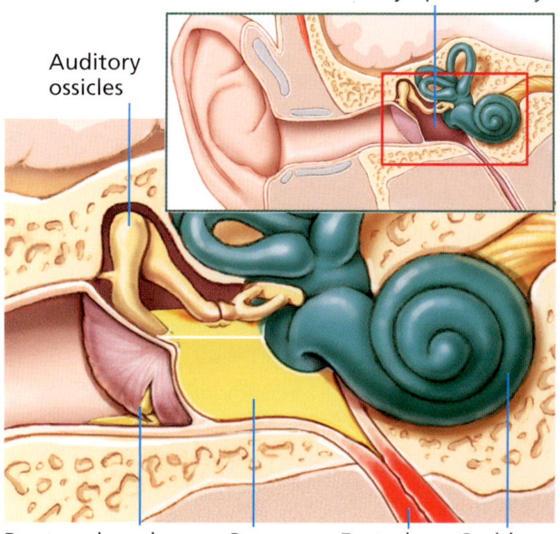

Otitis

A bacterial or viral infection of the hearing organ that can affect the outer auditory canal, the middle, and the inner ear.
An inflammation of the outer auditory canal can be caused by inappropriate cleaning (e.g., using cotton buds), as well as by bathing in contaminated water. Middle ear inflammation with strong earache, fever, and impaired hearing, often occurs in children, usually as an accessory symptom of a cold. Inner ear inflammation, on the other hand, is often a consequence of a middle ear inflammation or of another infectious disease such as measles. The typical symptoms include deafness and, if the equilibrium organ is also affected, vertigo, and vomiting.
While bacterial ear inflammations are treated with antibiotics, in the case of viral otitis, pain relief and fever reduction take priority. If the ear is affected by other infections then it is the underlying illness that is treated.

• *Otitis externa*

Bacterial or viral infection of the external auditory canal. The auditory canal and auricle become red and swollen and this is accompanied by pain and fever. The lymph nodes behind the ear are often enlarged. Often caused by swimming pool infections, fungi, foreign bodies, or a sebaceous gland infection. Treatment is with antibiotics and cortisone ointments.

• *Otitis media*

Infection within the tympanic cavity behind the eardrum. The middle ear is connected to the nasal and pharynx chambers via the Eustachian tube that normally drains off fluid from the middle ear. If it becomes blocked, due to a cold, for example, mucus and germs may build up behind the eardrum. This results in a feeling of pressure in the ear, earache, impaired hearing, nausea, vomiting, and fever. Sometimes the eardrum ruptures as a result of the increased pressure, releasing pus, which reduces the pain.
Infants are especially susceptible to blockage of the Eustachian tube, as it is still very narrow and becomes blocked more easily than in adults. The illness usually heals within a few days if treated quickly with antibiotics. A ruptured eardrum also grows over again quickly.
Delayed or incorrect treatment can lead to complications, e.g., to inflammation of the cranial bone behind the ear (mastoid process), or even to loss of hearing.

◆ **Otitis media**
In otitis media, pus gathers behind the eardrum and is unable to drain away.

• *Myringitis*

The eardrum separates the outer auditory canal from the middle ear. If it becomes inflamed this is almost always a sign of middle ear inflammation. This is usually caused by a respiratory tract infection spread via the Eustachian tube, which connects the nasal and pharynx chambers with the middle ear. Hearing is impaired and the patient suffers severe pain, particularly when swallowing. In addition to the treatment of the middle ear inflammation itself, eardrops are also administered to the outer auditory canal and an absorbent cotton dressing may be applied.

Perforation of the eardrum

The eardrum can be ruptured by external influences, such as being hit on the ear, pressure from explosions or when diving, by the insertion of pointed objects, or by a basal skull fracture. However, it is often also caused by a suppurative middle ear infection, in which pus forces its way out through the eardrum. The rupture causes a sharp pain, and blood or pus may be seen to flow out of the ear. Antibiotics, sometimes in the form of ear drops, are administered as treatment. If the perforation does not heal fully, hearing may remain impaired and pathogens are able to enter the ear more easily. Plastic surgery on the eardrum may be necessary in such cases.

Deafness

Deafness can have a variety of causes but they are all are related to the complicated hearing process, the path taken by the sound waves through the ear through to the relaying of impulses to the brain. A minor hearing impairment can be caused by a plug of earwax blocking the auditory canal. In this case, hearing returns to normal on removal of the wax.
The causes of deafness may also lie in the middle or inner ear, however. If a disturbance of the auditory canal is eliminated as a cause but deafness remains, then the cause may be found in the middle ear where the auditory ossicles are located. Here, an acute or chronic inflammation, an eardrum injury, or an excrescence of the small auditory ossicles, can hinder the relaying of sound. If there are no problems in the middle ear, i.e., sound is relayed but not properly perceived, then the cause of deafness may be in the inner ear. In this case, it is often the fine hair cells responsible for receiving sounds and relaying them to the acoustic nerve, which have been damaged. Constant noise exposure, for example, leads to the destruction of the hair cells and thus to deafness. A number of other causes such as infectious diseases (e.g., the common cold), acute hearing loss, or injury to the inner ear (e.g., through a basal skull fracture), or to the acoustic nerve itself, can also lead to deafness.
The treatment depends on the cause of the deafness. Medication can bring about the subsidence of an ear infection and/or lead to improved circulation in the blood vessels. Surgery may be necessary in some cases, for example, if a perforation of the eardrum does not heal or if the auditory ossicles are unable to relay sound. A cochlea implant may also be considered in a few cases. A hearing aid is usually required in the case of deafness of the inner ear.

Acute hearing loss

Acute hearing loss occurs suddenly and unexpectedly, usually within a matter of hours. There is either complete loss or severe impairment of hearing in the affected ear and this is often accompanied by noises in the ear and/or a feeling of pressure. It is rare that both ears are affected. This is an emergency situation and the quicker treatment begins, the better the chances of avoiding permanent damage such as deafness.
The causes are still not clear but it is suspected that, among other things, circulatory disorders and the related reduction in the supply of oxygen to the fine blood vessels in the ear, result in such hearing failures. Minor injuries, virus infections, or toxicity (e.g., the side effects of specific medications) are also possible causes. It has also been established that many patients with acute hearing loss are under severe psychological pressure at the time of diagnosis.
Acute hearing loss may be treated with infusions containing active agents that enter the ear via the circulatory system, in order to reduce inflammation, improve the blood flow, or thin the blood if necessary, so that any constricted blood vessels in the ear can be supplied with oxygen-rich blood again. Oxygen can also be administered. Prognosis is best when the patient is seen early.

Ménière's disease

The symptoms of Ménière's disease include bouts of dizziness or vertigo, often accompanied by tinnitus, hearing loss, oscillation of the eyes, nausea, and vomiting. Hearing can become impaired over longer periods of time. The disease is triggered by an increase of pressure in the inner ear, caused by an increase in the fluid contained in the equilibrium organ. Ultrasonic radiation, medications, and cold therapy may be helpful, as well as surgery to reduce the pressure.

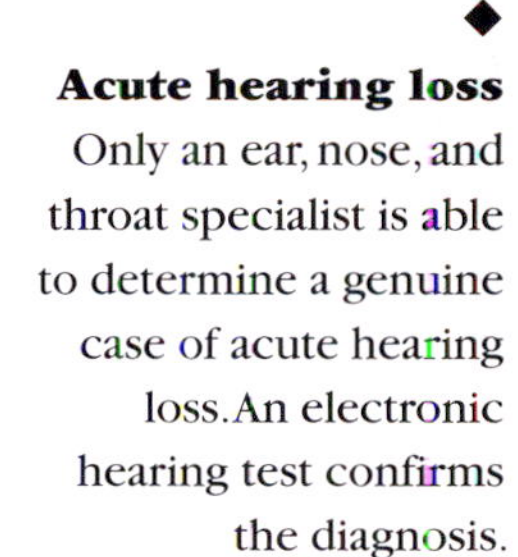

Acute hearing loss
Only an ear, nose, and throat specialist is able to determine a genuine case of acute hearing loss. An electronic hearing test confirms the diagnosis.

Diagnostic methods

Hearing test

Measurement of the sense of hearing. The patient is exposed to sounds of different intensities and levels, using earphones or via a loudspeaker in a sound-insulated room. The patient reacts by pressing a button as soon as he perceives a sound. The test results enable a precise assessment of individual hearing ability.

Otoscopy

Examination of the outer auditory canal and the eardrum using an otoscope. A small metal ear speculum is first inserted into the ear. Light is channeled into the auditory canal using an otoscope (a concave mirror with a central viewing hole), so that the auditory canal and eardrum are illuminated and can be examined by a specialist.

Treatment methods

Ear drops

Liquid medication applied as drops in the outer auditory canal. They usually contain a local anesthetic for pain relief and an anti-inflammatory agent or antibiotic. Their use must always be preceded by examination by a doctor because they should only be administered if the eardrum is intact.

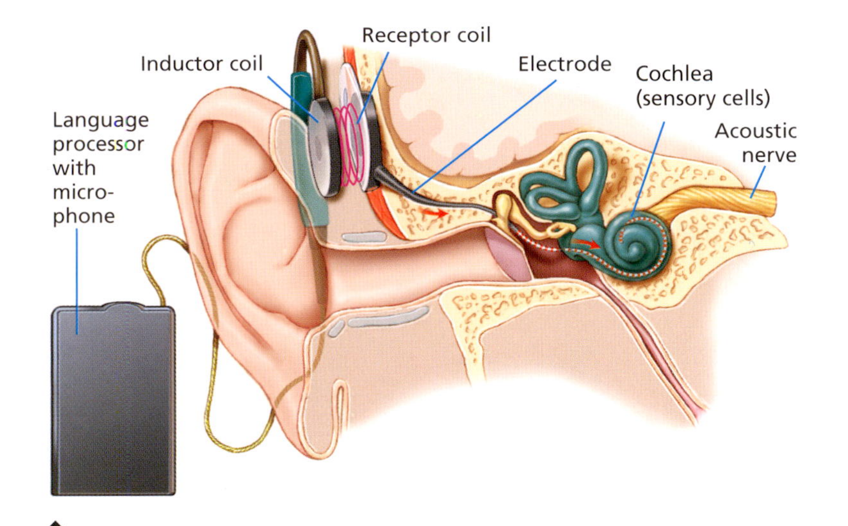

Cochlea implant
The language processor converts language into electric impulses and relays them to an inductor coil next to the auricle. These impulses are received by a receptor coil inserted under the skin. An electrode in the inner ear then relays the impulses directly to the sensory cells in the cochlea.

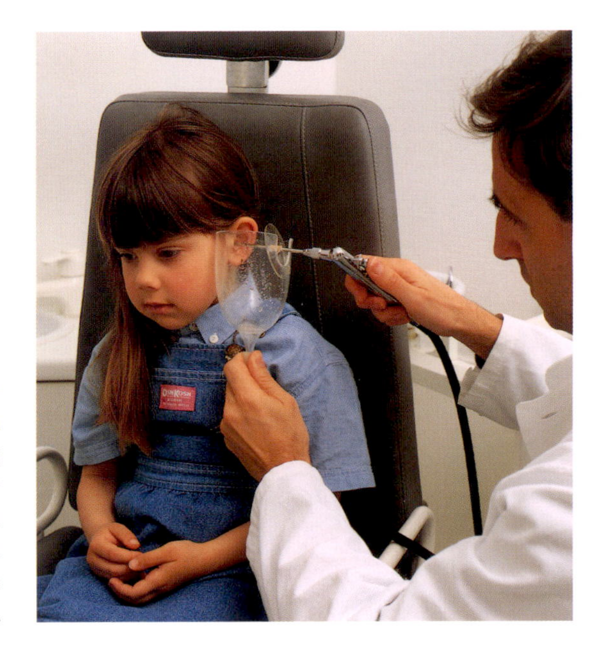

Ear irrigation
The thorough cleaning of the outer auditory canal using irrigation can only be carried out by an ear, nose, and throat specialist.

Ear irrigation

The flushing out of the outer auditory canal with lukewarm water in order to remove earwax, pus, or foreign bodies from the ear. The prerequisite for an ear irrigation, which should only be carried out by an experienced doctor, is that the eardrum is intact.

Cochlea implant

An electronic hearing prosthesis (inner ear prosthesis) implanted in cases of deafness of both ears due to cochlea damage. This method is only successful if the nerve supplying the cochlea is still intact. Electrodes are surgically inserted into the cochlea, via which the acoustic signals are relayed directly to the acoustic nerve. This enables the achievement of limited language comprehension with intensive hearing training.

Grommet

A synthetic tube inserted into the eardrum to drain off mucus and to ventilate the tympanic cavity. A grommet is commonly used in cases of chronic middle ear inflammation in children, when the viscous mucus is no longer able to drain away. The grommet is almost always inserted facing outwards into the auditory canal and the eardrum closes of its own accord.

Eardrum plastic surgery

The closure of a perforated eardrum by means of an operation, in which the eardrum is replaced by a tissue transplant. To do this the ear, nose, and throat surgeon removes a small piece of the muscle fascia from either the temporal muscle or from a muscle in the patient's hip. This is then implanted to replace the damaged eardrum.

Olfactory and gustatory organs

As well as giving us information about our environment, the senses of smell and taste act as important warning systems in human beings. An unpleasant smell, such as the smell of burning, is a signal to take measures to eliminate it or to get away as quickly as possible. A nauseating smell or taste invokes a defensive reaction and can cause vomiting in extreme cases. The pleasant smell or taste of a tasty meal, on the other hand, is inviting.

Sense of smell

Human beings perceive smells via special olfactory cells in the nasal mucosa. These are located in a small area measuring only about 2.5 square centimeters in the superior nasal concha, directly below the nasal roof section of the cranial bone. The 10–25 million cells have fine olfactory hairs extending into the nasal cavity where they come into contact with the aromas in inhaled air. This contact (and therefore the perception of smell) is most intense when air eddies are formed, e.g., by sniffing. The smell perceived is relayed to the brain via fine nerve fibers at the base of the olfactory cells. As there are nerve connections to the centers of the brain that stimulate the flow of saliva, and which are responsible for emotional perception, smell perception can have many effects, e.g., increased salivation in the case of smelling something good, or revulsion, if the smell is unpleasant.

The majority of people are able to smell several thousand smells but the majority of these are combinations of specific odor categories or "primary odors" (e.g., rotten, flowery, pungent). Frequent contact with a specific scent invokes a habituation effect such that the scent hardly registers any more.

People suffering from a loss of smell perception, e.g., following a head injury, find this extremely disturbing. This is because smell is an important component of our life experiences. Some smells enter the nose once only and are remembered for a lifetime. They can have either a positive or a negative effect on moods and thoughts. We have an exceptionally high sensitivity to some chemical substances, e.g., hydrogen sulfide compounds that smell like rotten eggs, meaning that they register in even the lowest concentrations.

Sense of taste

The taste buds, located on the tongue, tell the brain how something tastes. Taste buds, or gustatory papillae, are walled, foliate, and/or mushroom-shaped protrusions equipped with sensory cells (10,000 in babies, only 2,000 in old age). They come into contact with food via an orifice on their surface known as a gustatory pore. Tastes are perceived as stimuli by chemoreceptors in the sensory cells located in the gustatory pore, converted into an electrical impulse, and relayed to the brain via the fine nerve fibers of the 7th (facial nerve), 9th (glossopharyngeal nerve), and the 10th (vagus nerve) cranial nerves. The taste buds are only able to determine the taste quality of a substance when they come into contact with it in a dissolved state. Therefore, substances of any kind can only be tasted once the saliva has dissolved at least a small part of them.

There are four different taste perceptions: sweet, salty, sour, and bitter. "Sweet" is perceived on the tip of the tongue, "salty" at the front and sides of the tongue, "sour" on the back of the tongue, and "bitter" at the base of the tongue.

As taste perception involves, not only the balance between the sweet, sour, bitter, and salty receptors, but also the specific smell of a substance, disorders affecting the sense of smell also compromise taste perception. This is why someone suffering from a cold, for example, where the sense of smell is also affected, is barely able to taste his food.

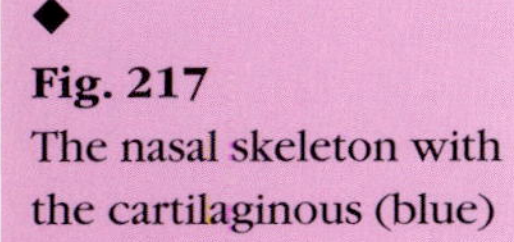

◆
Fig. 217
The nasal skeleton with the cartilaginous (blue) sections.

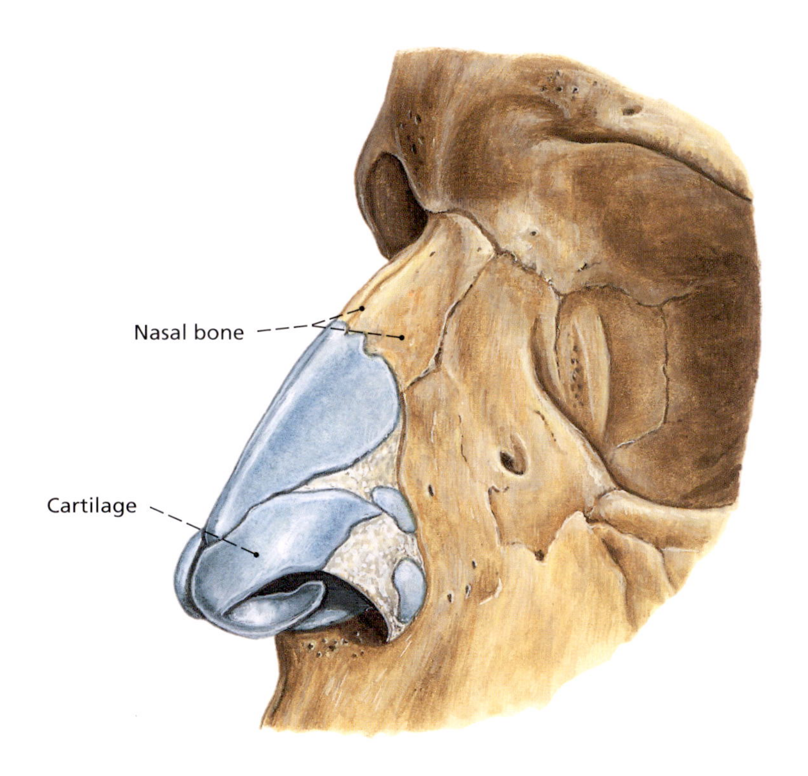

◆
Fig. 218
The left side of the nasal cavity. The mucous membrane has been partially removed from the upper section (olfactory zone).

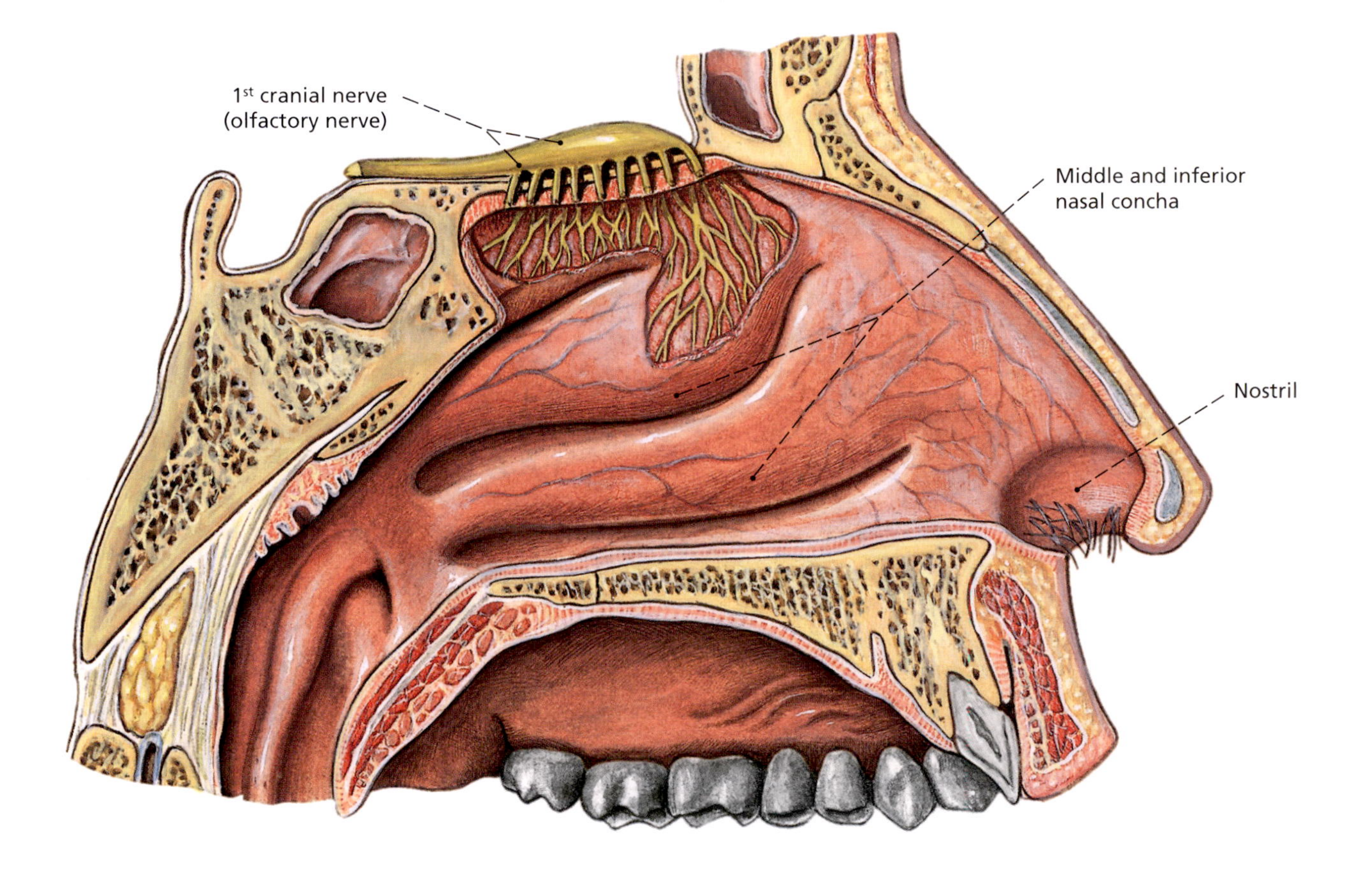

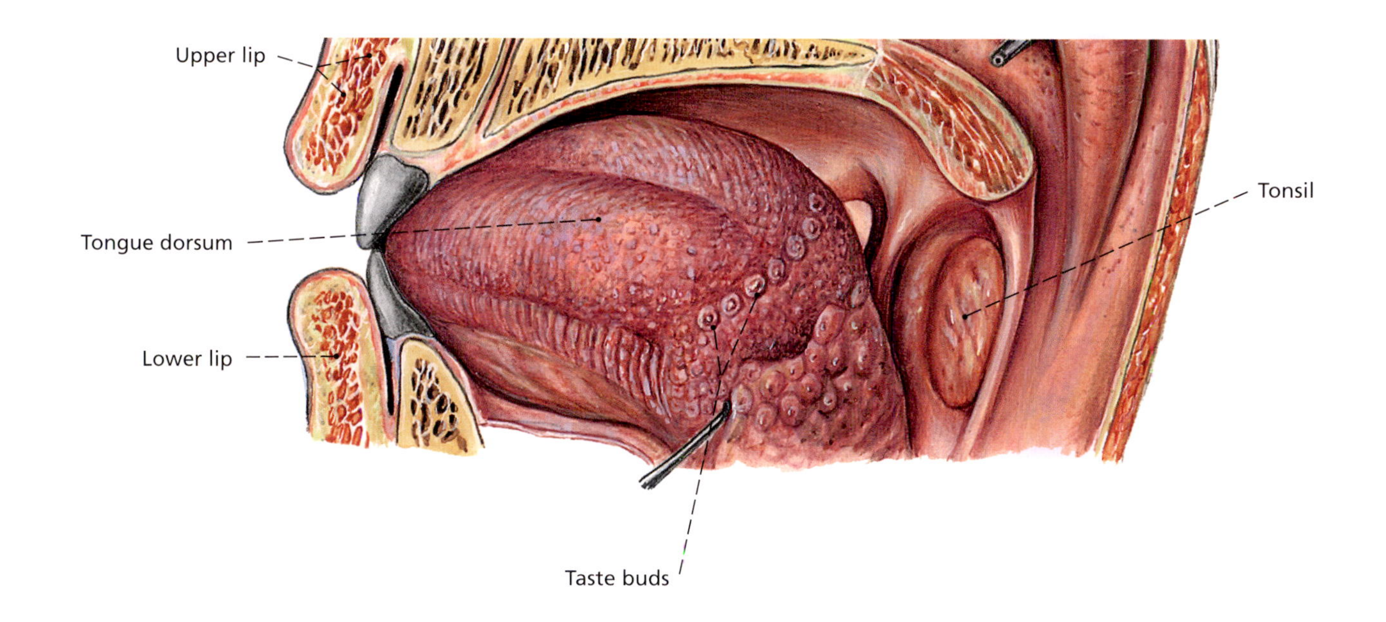

◆ **Fig. 219**
Cross-section of the oral cavity. The tongue has been pushed towards the right cheek slightly in order to show the taste buds on the edge of the tongue.

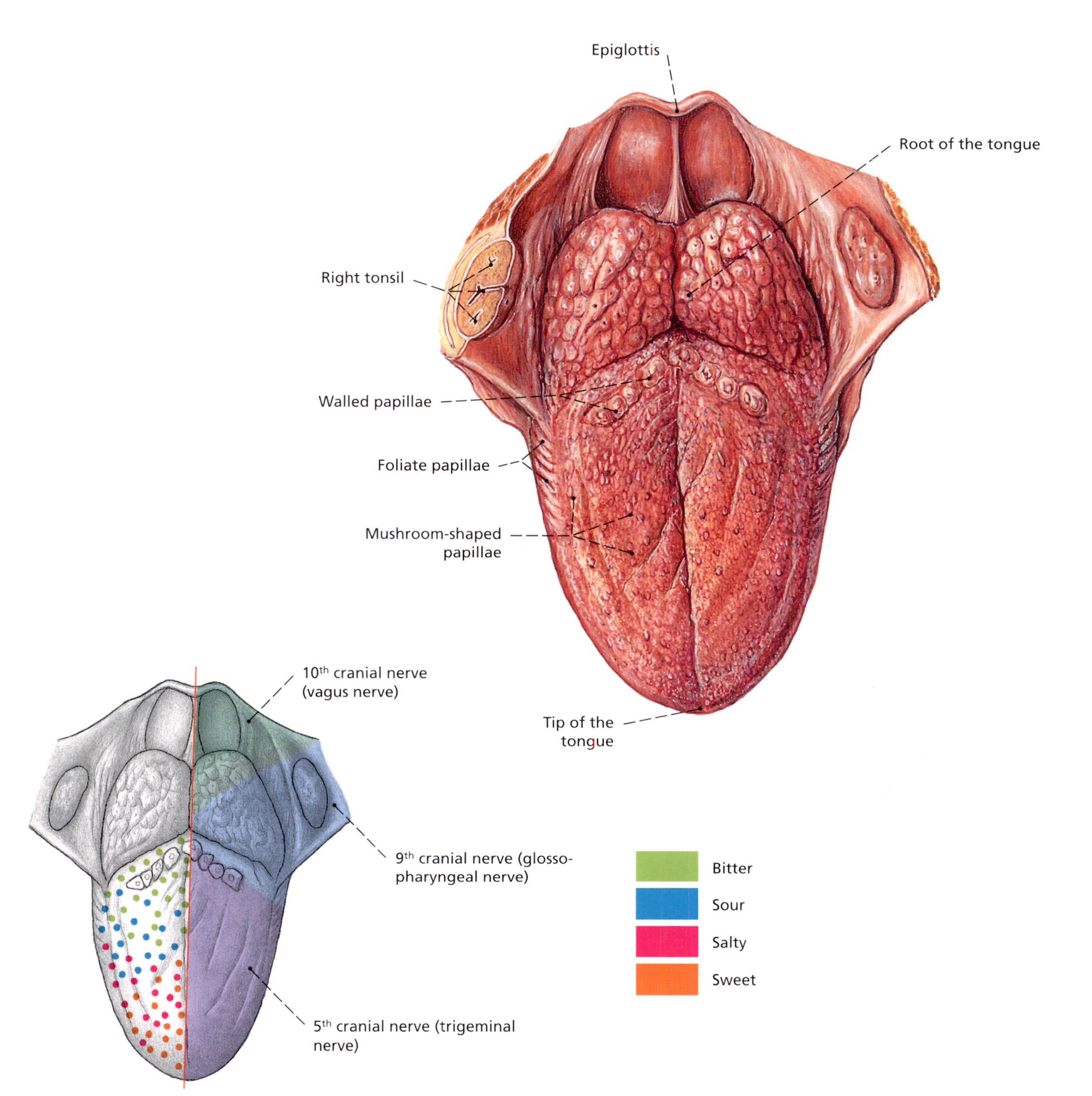

◆ **Fig. 220**
The tongue dorsum with the root of the tongue and tonsils. The right tonsil has been cut open. The walled, foliate, and mushroom-shaped papillae are taste-sensitive growths in the mucous membrane of the tongue.

◆ **Fig. 221**
The tongue indicating the taste zones (left half) and the associated nerves (right half).

Symptoms and diseases

Hay fever (allergic rhinitis)

One of the most common allergic illnesses is triggered by the pollen of specific trees, bushes, grasses, herbs, corn, and, less frequently, by flowers. In the case of pollen allergy, the immune system no longer treats the pollen as a harmless foreign substance, and the body's defense systems attack the foreign substances (allergens). The symptoms typical of this overreaction on the part of the immune system are: a runny nose, sneezing, a blocked nose, red, itching and streaming eyes, itching in the throat and ears, coughing and breathlessness, and even asthma. Hives and gastrointestinal problems can also occur.
If hay fever is suspected, the substances responsible (in this case, specific pollens), can be established using allergy tests and blood tests.
A treatment of the cause does not exist but a number of things can limit the discomfort. Medication in the form of eye drops, nasal and bronchial sprays, and tablets, either prevents the allergic discomfort from appearing in the first place or alleviates the symptoms by reducing the swelling of the nasal and bronchial mucous membranes and stopping the itching. Hyposensitization is also a successful means of treating allergic rhinitis. The most important thing is to avoid the allergens and this includes:

- Washing the hair frequently (pollen gets trapped in the hair) as well as changing clothes outside of the bedroom.
- Frequent vacuum cleaning (preferably with a pollen filter).
- Avoiding sleeping with the windows open. The release of pollens starts at about three o'clock in the morning.
- Undertaking walks and sporting activities shortly after it has rained, if possible. It is also worth heeding the pollen forecasts on the radio or in the newspaper.
- Taking holidays at the coast or in high mountain areas where the air contains less pollen and different, often shorter, flowering times mean that it is often possible to escape the main pollen season at home.

Nosebleeds

Nosebleeds can be the result of mucous membrane irritation due to sneezing, strong nose blowing, or injury from a foreign body, but can also occur in the context of specific illnesses such as high blood pressure, vascular disease, blood clotting disorders, and acute inflammation.

Common colds

Acute or chronic inflammation of the nasal mucosa usually caused by viruses, bacteria, or allergic stimuli. The nose secretes mucus, which may be infective, or may become blocked by swollen mucosa, resulting in difficulties with breathing. The acute common cold, which is the most common of all illnesses, is transmitted via droplet infection and is often accompanied by an infection of the nose and throat region. As a rule, medication only alleviates the symptoms and, in most cases, the cold goes away anyway after about a week.

Nasal polyps

Sac-shaped growths in the mucous membrane of the nose and in the nostrils. These benign mucous membrane growths can cause significant complaints, obstruct breathing through the nose and stimulate the mucous membrane to produce large quantities of a watery or infective secretion. In this case, the removal of the polyps by surgical means is strongly recommended.

Impaired sense of smell

A reduction or loss in the ability to perceive or sense odors. An impaired sense of smell can occur in one or both nostrils. The function of the olfactory cells is often temporarily impaired as a result of a common cold, influenza, or sinusitis. Brain concussion and head injuries with damage to the olfactory nerve can lead to longer-term impairment of the sense of smell. An initial loss of the sense of smell on one side, and subsequently on both sides, can be the first indication of a brain tumor. Disorders affecting the sense of smell also occur in the case of psychiatric illnesses. Asthma sufferers, on the other hand, are known to have a particularly acute sensitivity to smells.

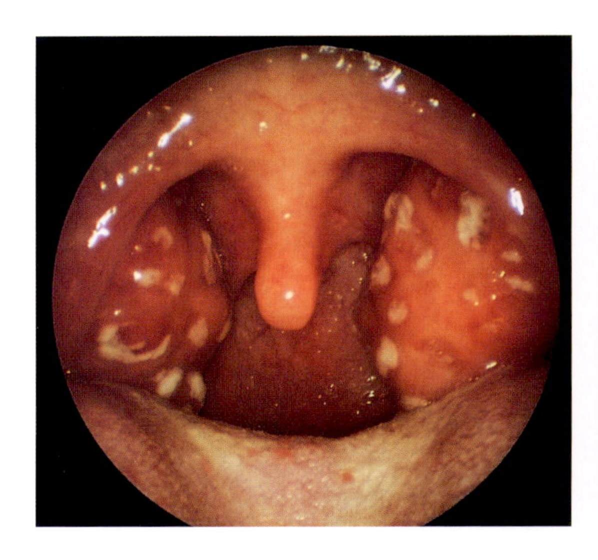

Tonsillitis
The swollen tonsils covered with white spots are usually easy to see when looking into the throat.

Tonsillitis

Inflammatory, painful swelling of the tonsils that is especially frequent in childhood. The cause is usually a bacterial infection and the symptoms include a severe sore throat and swallowing difficulties, as well as a slight fever. The swollen, red tonsils can be seen at the back of the throat, often covered with small, white spots. Tonsillitis can sometimes become chronic if the tonsils do not heal properly. The tonsils then remain permanently enlarged and become infected on a repeated basis. They are no longer able to fight off germs and, in some circumstances, they do the body more harm than good and surgery to remove them altogether is considered.

Sinusitis

Acute or chronic inflammation of one or all of the sinuses. Acute inflammation often occurs together with a common cold. The chronic form, on the other hand, is usually based on anatomical peculiarities (curvature of the nasal cartilage, nasal polyps, or nasal concha growths) that hinder or completely block the draining of secretions from the sinuses.
Typical complaints include a dull feeling of pressure above the affected sinuses, throbbing headaches that increase when the head is bent downwards, and fever. In order to prevent it from spreading to the eye orbits or to the brain and brain membranes, the infection needs to receive prompt treatment with nose drops, heat therapy (steam or red light), or irrigation of the sinuses to remove the build-up of secretions. In the event of chronic sinusitis and in especially severe cases, surgery may become necessary, e.g., to straighten the nasal cartilage or enlarge the drainage aperture in the sinuses.

Maxillary sinusitis

Temporary or ongoing inflammation of the mucous membranes lining the maxillary and frontal sinuses, often associated with the formation of mucus and pus. Pain occurs beneath the eyes, in the cheeks, and in the cheekbone region, with forehead headaches especially when bending down, as well as fever. Virus or bacteria are able to migrate from the nose into the maxillary sinuses causing inflammation and swelling, which hinder the draining of secretions. It can also be caused by dental root disorders and jaw injuries.

Harelip and cleft palate

Congenital cleft in the lip, jaw, and palate. In newborn babies the cleft palate is replaced by an artificial palate to close the cleft, when they are just a few days old. This encourages nasal breathing and facilitates food intake. The actual correction of the malformation requires a number of operations. The clefts in the bony tissue can only be closed from the age of about two and half, in order to prevent growth disturbances.

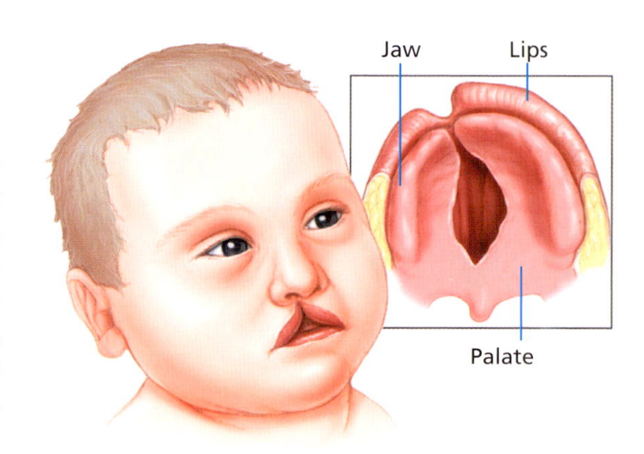

Harelip and cleft palate
The cleft means that the mouth and nasal cavity are directly connected, making breathing and food intake difficult.

Cancer of the tongue

One of the most common and most threatening forms of cancer in the mouth and throat region, tongue cancer is usually located on the edge of the tongue. Risk factors include long-term tobacco and alcohol consumption. The preliminary stage is often a thick white patch, known as a leukoplakia, which sometimes disappears when tobacco and alcohol are given up, without developing into cancer. The early stages of cancer of the tongue are not painful and can be treated surgically if discovered early enough. Any changes to the tongue such as hardening, growths, or white patches, which do not disappear within a few days, should therefore be examined straight away.

Treatment methods

The majority of infections in this region are treated with **antibiotics**. Other diseases are treated according to their causes. **Antihistamines** in the form of eye drops, nasal, and bronchial sprays, as well as tablets, can relieve the worst complaints associated with allergies such as hay fever. **Hyposensitization** or desensitization, in which the patient is exposed to the irritant in increasing concentrations using an allergen solution, can also be considered. Around 70% of patients treated in this way respond to the therapy. This is particularly important because nearly one in two individuals suffering from a pollen allergy runs the risk of developing allergic asthma over the course of time.

Respiratory tract and lungs

Human life would not be possible without oxygen. The lungs and respiratory tract are responsible for ensuring that the cells of the body are adequately supplied with oxygen. Their main function is to transport oxygen from the atmosphere to the alveoli in the lungs, and at the same time to eliminate carbon dioxide, the main end product of metabolism, by transporting it out of the lungs on breathing out.

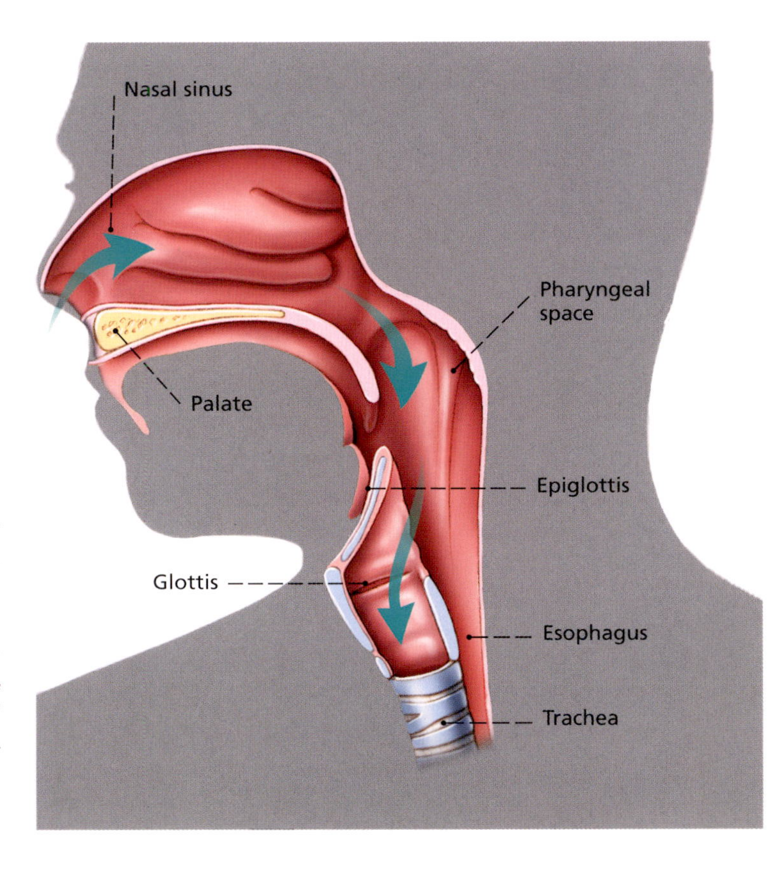

Fig. 222
Upper respiratory tract. The nasopharyngeal space extends from the nasal sinuses to the epiglottis. Its main function is to clean, moisten, and warm the air we breathe in.

When we breathe in (inhale), air enters the trachea via the nose, throat, and larynx. The trachea divides into two main branches (bronchi), which lead to the left and right lungs, respectively; each of these then divides and subdivides, like the branches of a tree, becoming ever smaller. The smallest branches of this bronchial tree end in small air sacs, called alveoli. The respiratory tract is divided into two sections:

- the upper respiratory tract, consisting of the nose and pharyngeal space, and
- the lower respiratory tract, including the larynx, trachea, and bronchial tree.

The upper and lower respiratory tracts provide the connection between the external atmosphere and the lungs. However, their function is not limited to the transport of oxygen to the lungs. They also clean the air we breathe by removing dust particles, warming it to body temperature (37°C), even when air temperatures fall below zero, and moistening it. The mucous membranes lining the nose, larynx, and bronchi are specially equipped for this purpose. Dust particles are trapped in the fine hairs inside the nose, while tiny hairs called cilia, that are found all along the airways, trap even smaller particles as well as bacteria. Mucus-producing cells ensure that the air is sufficiently moist.

Structure and function of the lungs

The two lungs fill the thorax almost completely. The heart lies between them, on the left side. Consequently, the left lung is smaller and consists of only two lobes, while the right lung has three lobes.
The outer wall of the lungs and the inner wall of the thorax are both covered with a tough membrane, called the pulmonary and the parietal pleura, respectively, which is coated with a film of moisture. Negative pressure between these two membranes ensures that the lungs, which consist of elastic tissue, remain firmly attached to the wall of the thorax, even on breathing out, and do not collapse. The film of moisture allows the two membranes to glide smoothly over one another during breathing movements.
Gaseous exchange, i.e., the absorption of oxygen and the elimination of carbon dioxide, takes place in the air sacs (alveoli), the surface area of which reaches a total of nearly 100 square meters. On breathing in, the alveoli are filled with oxygen. The alveoli are surrounded by a dense network of extremely fine blood vessels, known as the pulmonary capillaries. The blood vessels feeding these capillaries contain blood rich in carbon dioxide from the right side of the body. The pulmonary capillaries, the thin walls of which are in direct contact with the walls of the alveoli, release this carbon dioxide into the

alveoli, while at the same time absorbing oxygen from them. This exchange takes place in a fraction of a second. The blood leaving the lungs is now oxygen-rich, and is carried from the lungs to the left side of the heart, from where it is pumped, via the aorta, back around the body. On breathing out, the carbon dioxide that has been released into the alveoli, is exhaled.

With each breath we take, approximately half a liter of air is breathed in and out. The capacity of the lungs is far greater than this, however, at around five liters, with the result that there is plenty of capacity held in reserve for gaseous exchange if there should be an increase in demand for oxygen.

Breathing is controlled by the respiratory center in the brain. It receives its information via sensors (known as chemoreceptors) in the blood vessels and the brain, which constantly measure oxygen and carbon dioxide levels in the blood. If the oxygen concentration in the blood falls, the respiratory rate is increased; conversely, if the level rises, the respiratory rate slows down.

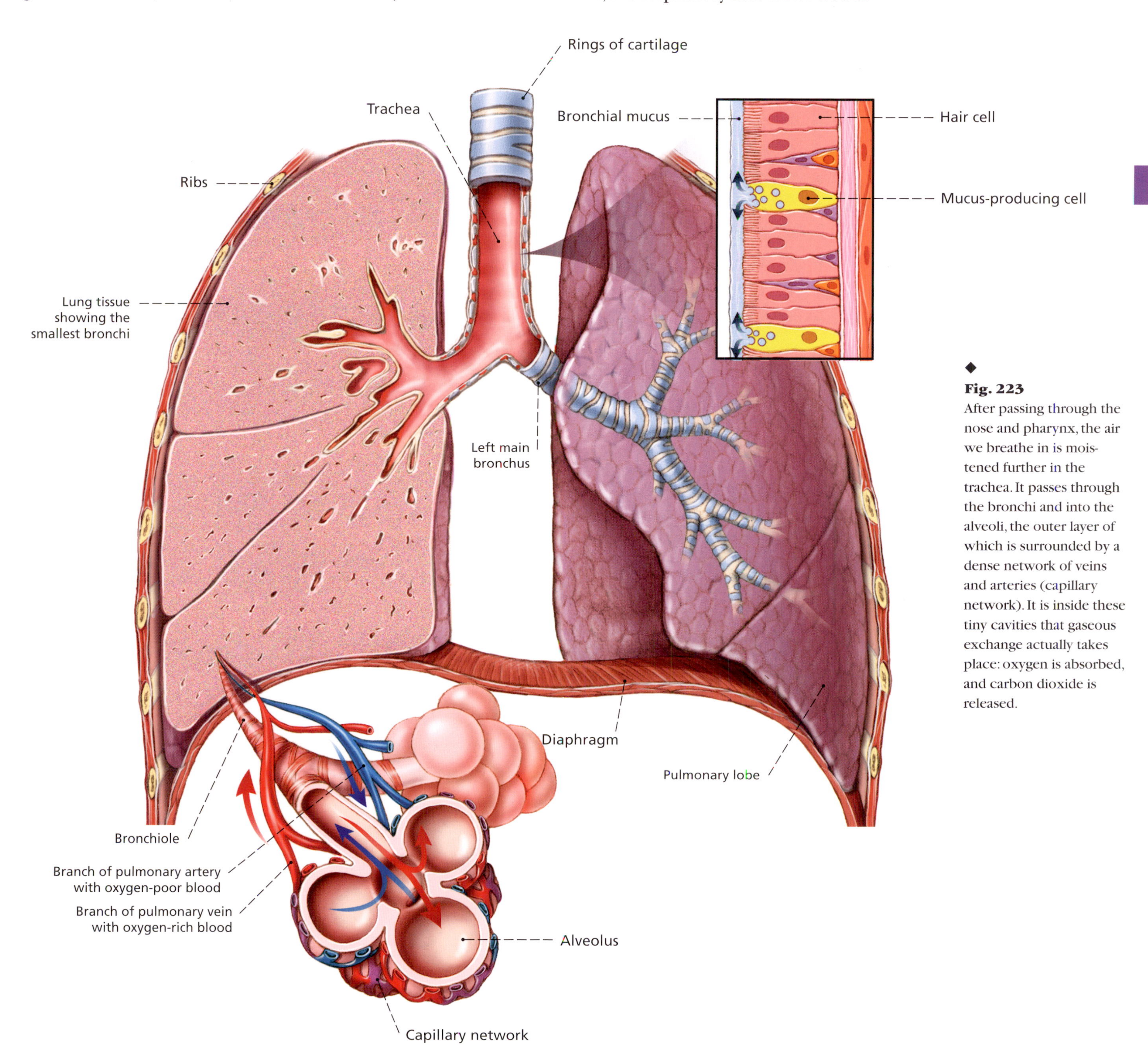

◆
Fig. 223
After passing through the nose and pharynx, the air we breathe in is moistened further in the trachea. It passes through the bronchi and into the alveoli, the outer layer of which is surrounded by a dense network of veins and arteries (capillary network). It is inside these tiny cavities that gaseous exchange actually takes place: oxygen is absorbed, and carbon dioxide is released.

◆
Fig. 224
Nasal and pharyngeal space, longitudinal section.

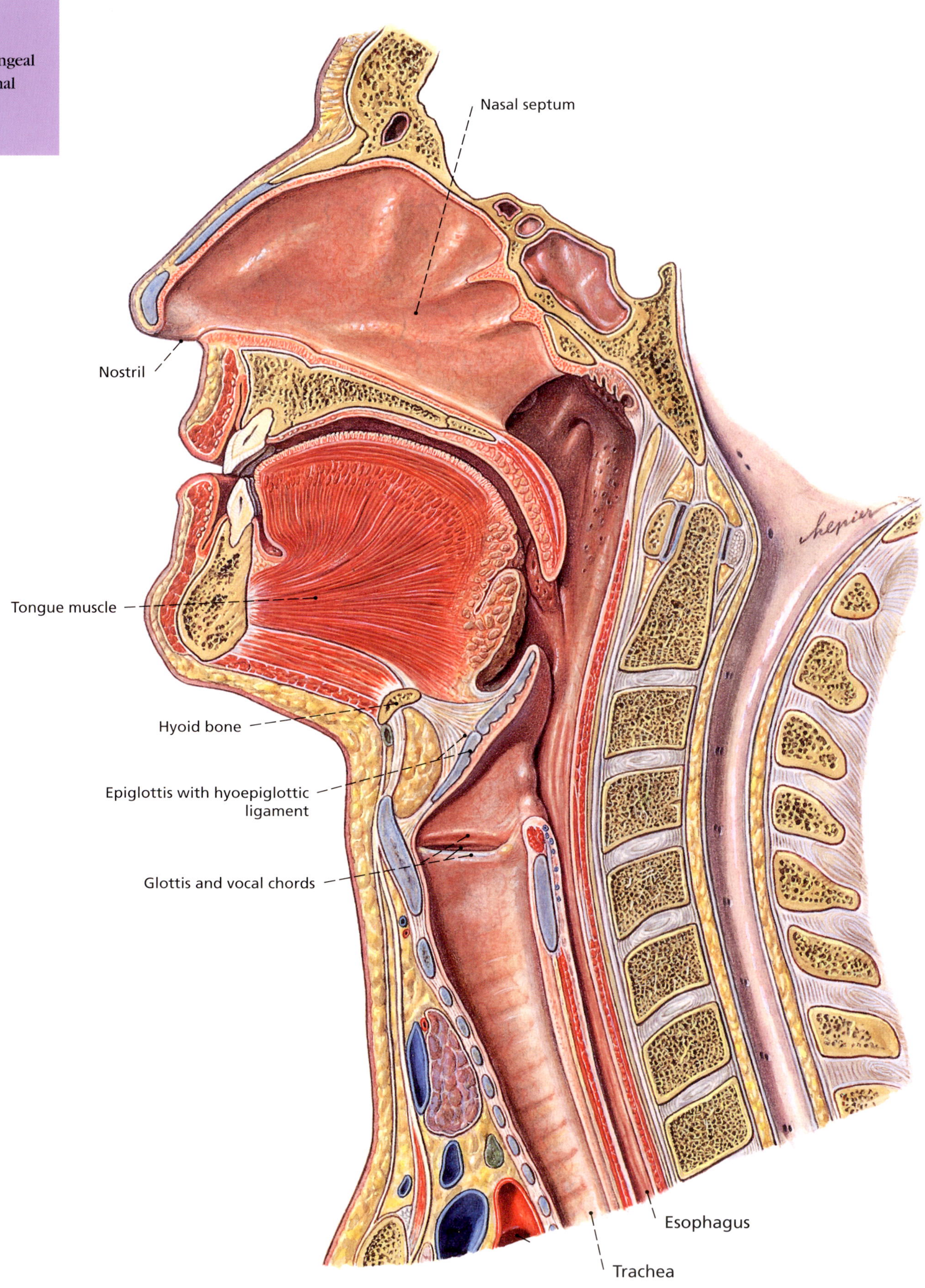

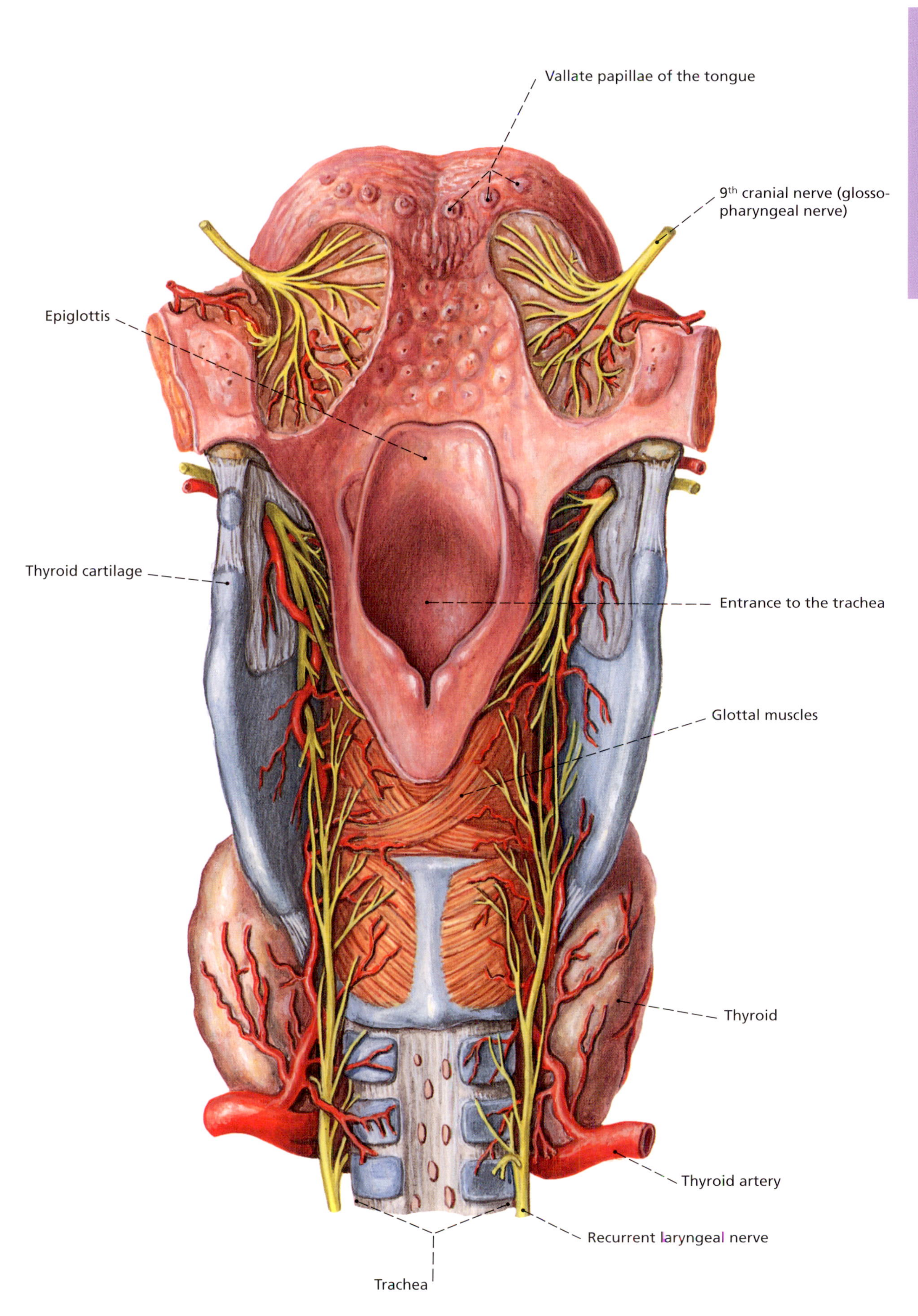

◆

Fig. 225
Parts of the larynx with the root of the tongue, viewed from the rear. In order to reveal the nerve supply to the tongue, the illustration shows the tongue with part of the mucous membrane cut away.

◆
Fig. 226
Examination of the larynx using a mirror. In indirect examination (a), the tongue is pushed forwards, making room for the laryngeal mirror. Direct examination (laryngoscopy) (b) is carried out using a thin metal tube fitted with an optical device.

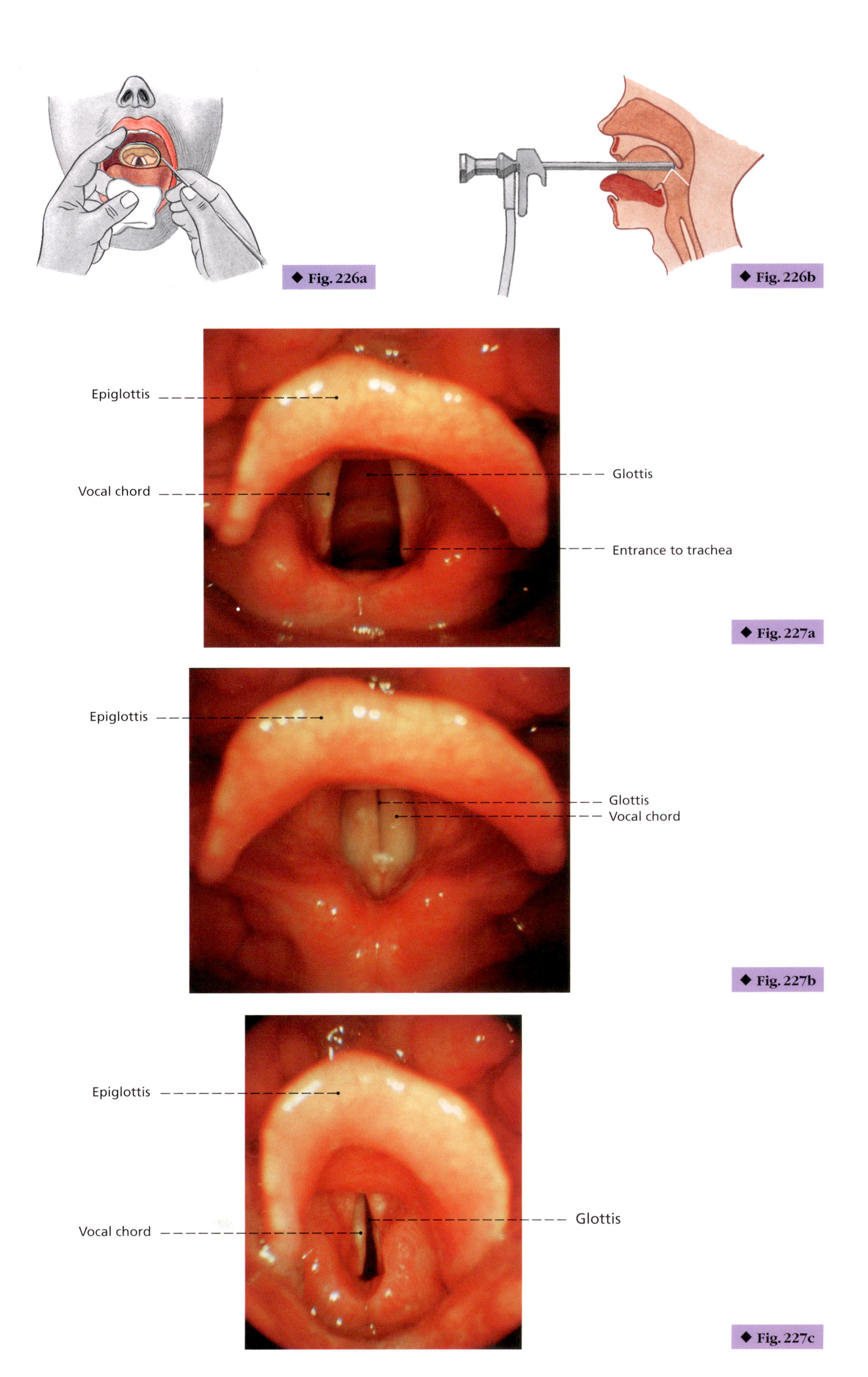

◆ Fig. 226a

◆ Fig. 226b

◆ Fig. 227a

◆ Fig. 227b

◆ Fig. 227c

◆
Fig. 227
Position of the vocal chords during direct examination of the larynx. (a) In the breathing position, the glottis is wide open; (b) in the speaking position, it is closed; and (c) in the whispering position, it is slightly open.

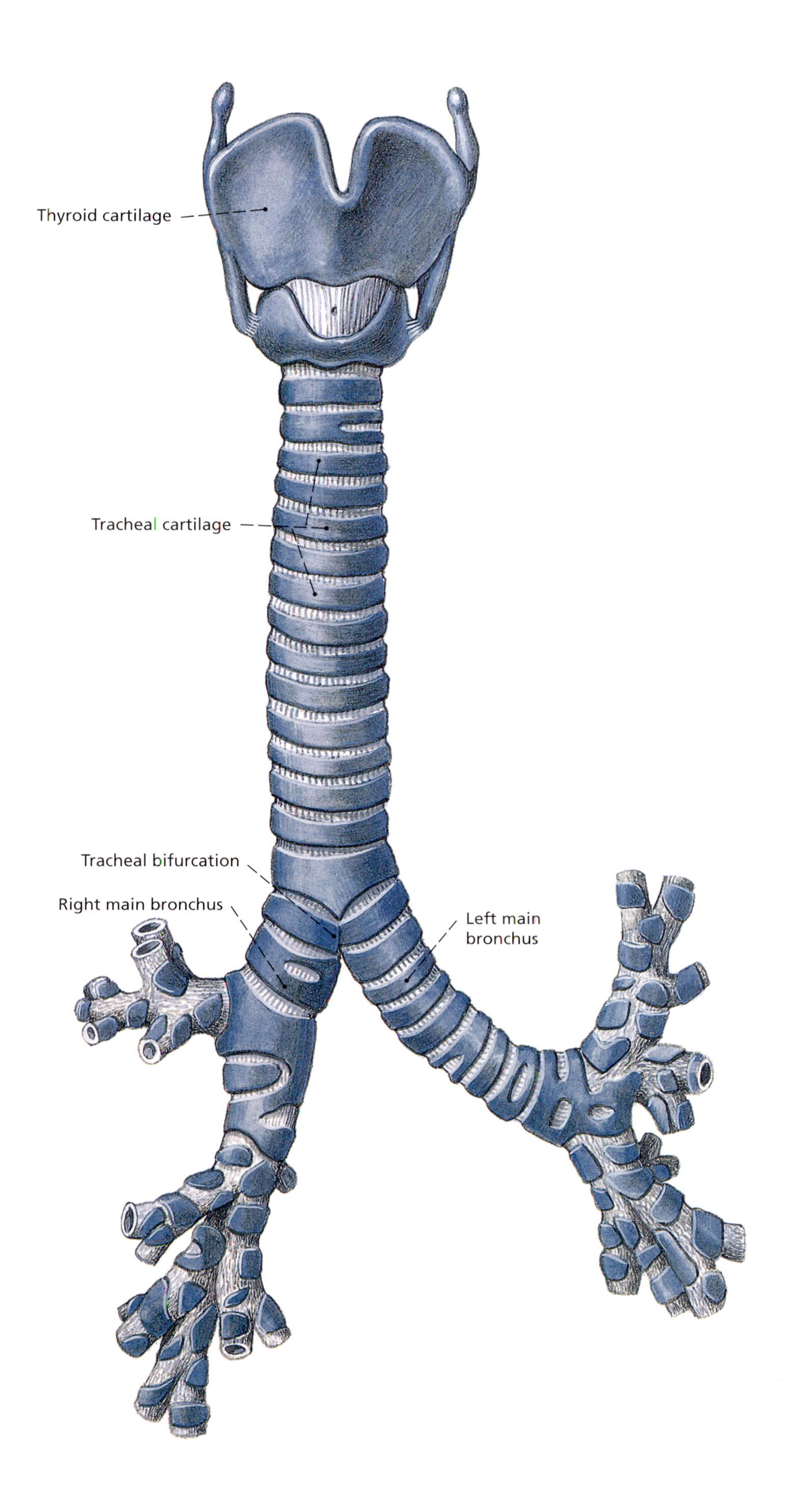

◆
Fig. 228
Chondroskeleton of the larynx, trachea, and bronchi, viewed from the front. Note: the path of the right main bronchus is at a considerably steeper angle than that of the left main bronchus.

◆
Fig. 229
Chondroskeleton of the larynx, trachea, and bronchi, viewed from the rear. The trachea and bronchi consist of rings of cartilage. Each of these is shaped like the letter "C," open at the back. Between the ends of these rings is a membrane, extending along the entire length of the trachea and consisting of connective tissue in which a ring-shaped muscle is embedded.
The membrane is also covered by a thin layer of mucous membrane (in the illustration this has been cut away in the upper third of the trachea). On the visible section of this mucous membrane, the small oval glands of the tracheal mucosa can be seen.

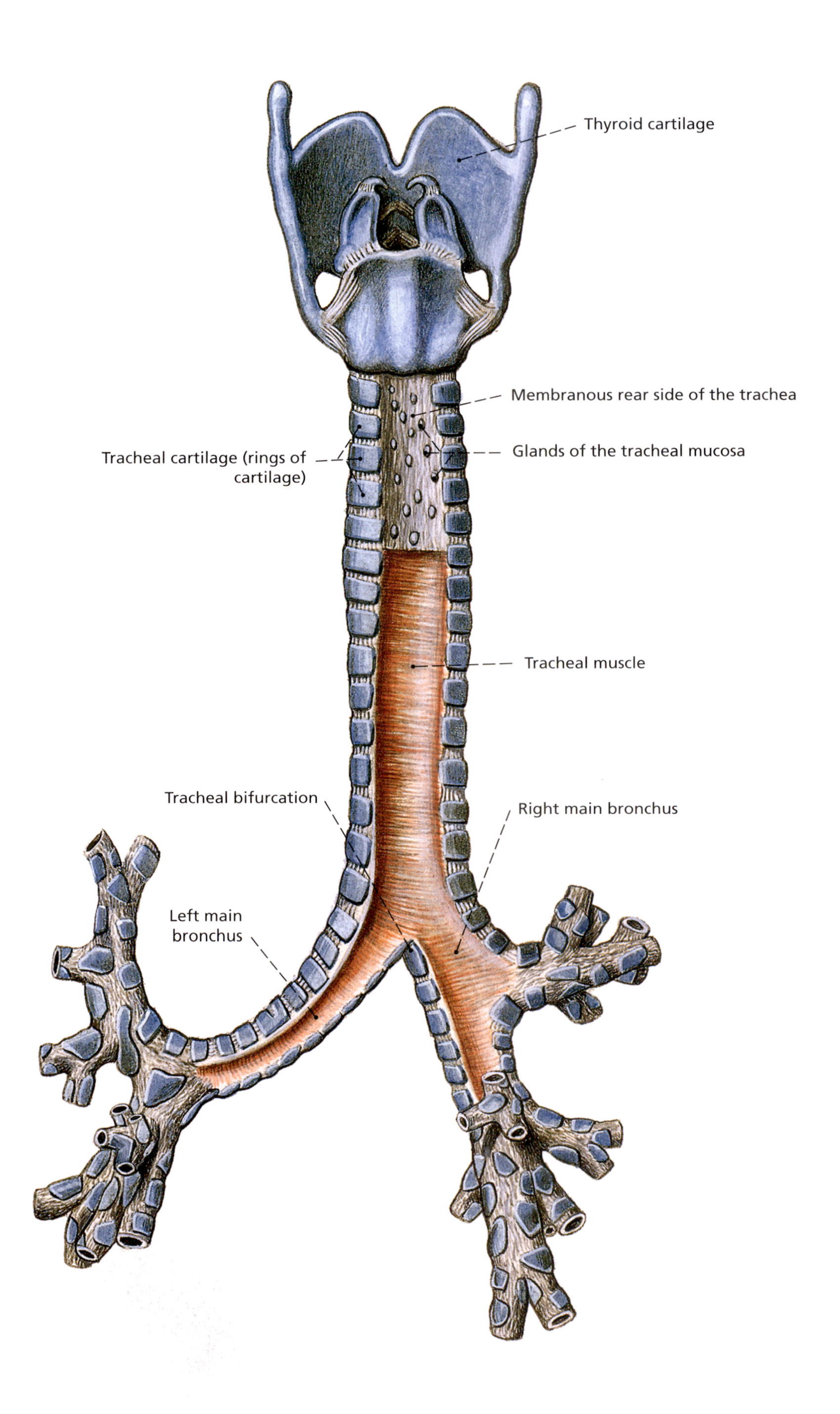

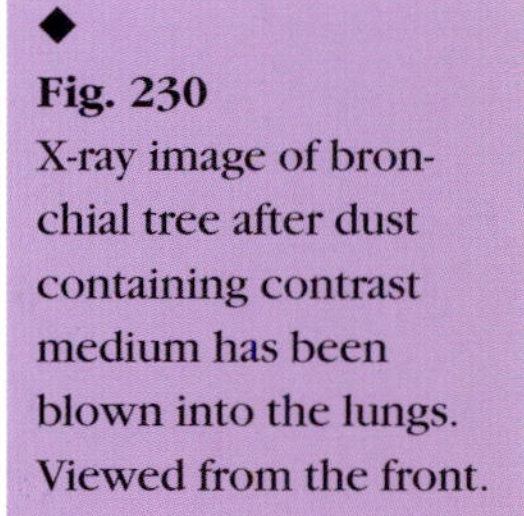

◆
Fig. 230
X-ray image of bronchial tree after dust containing contrast medium has been blown into the lungs. Viewed from the front.

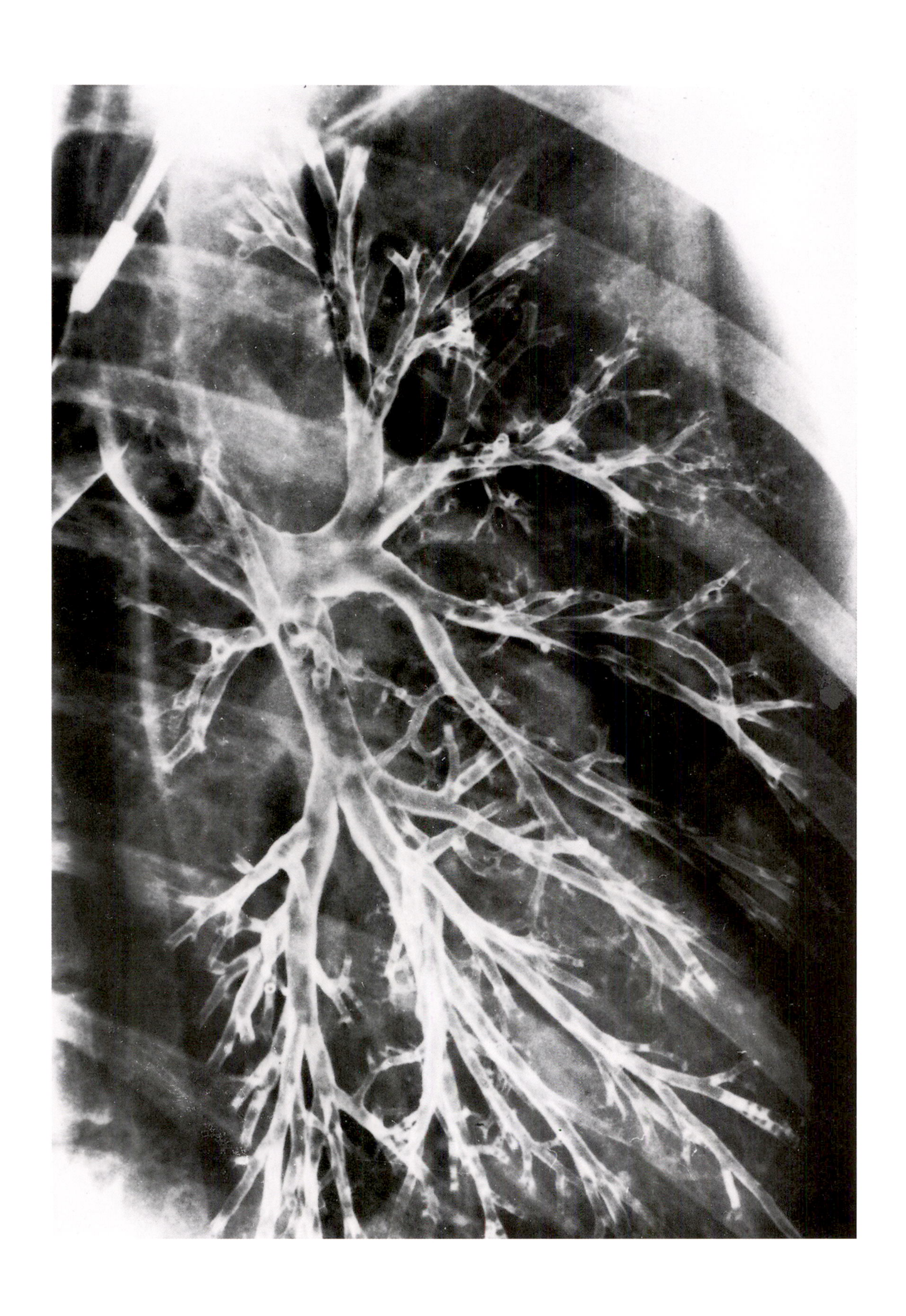

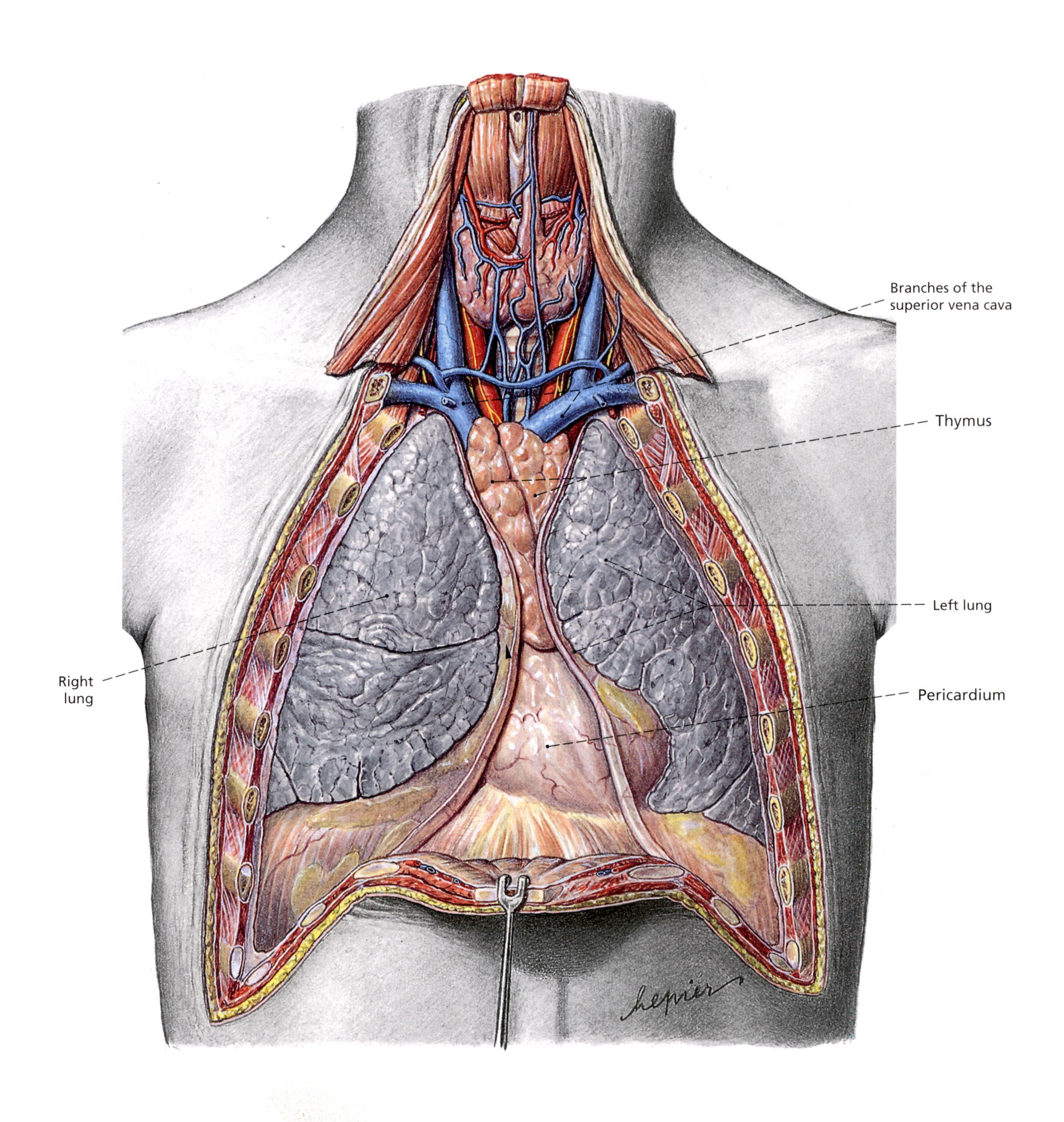

◆
Fig. 231
Position of the lungs in the thorax. The anterior thoracic wall has been opened.

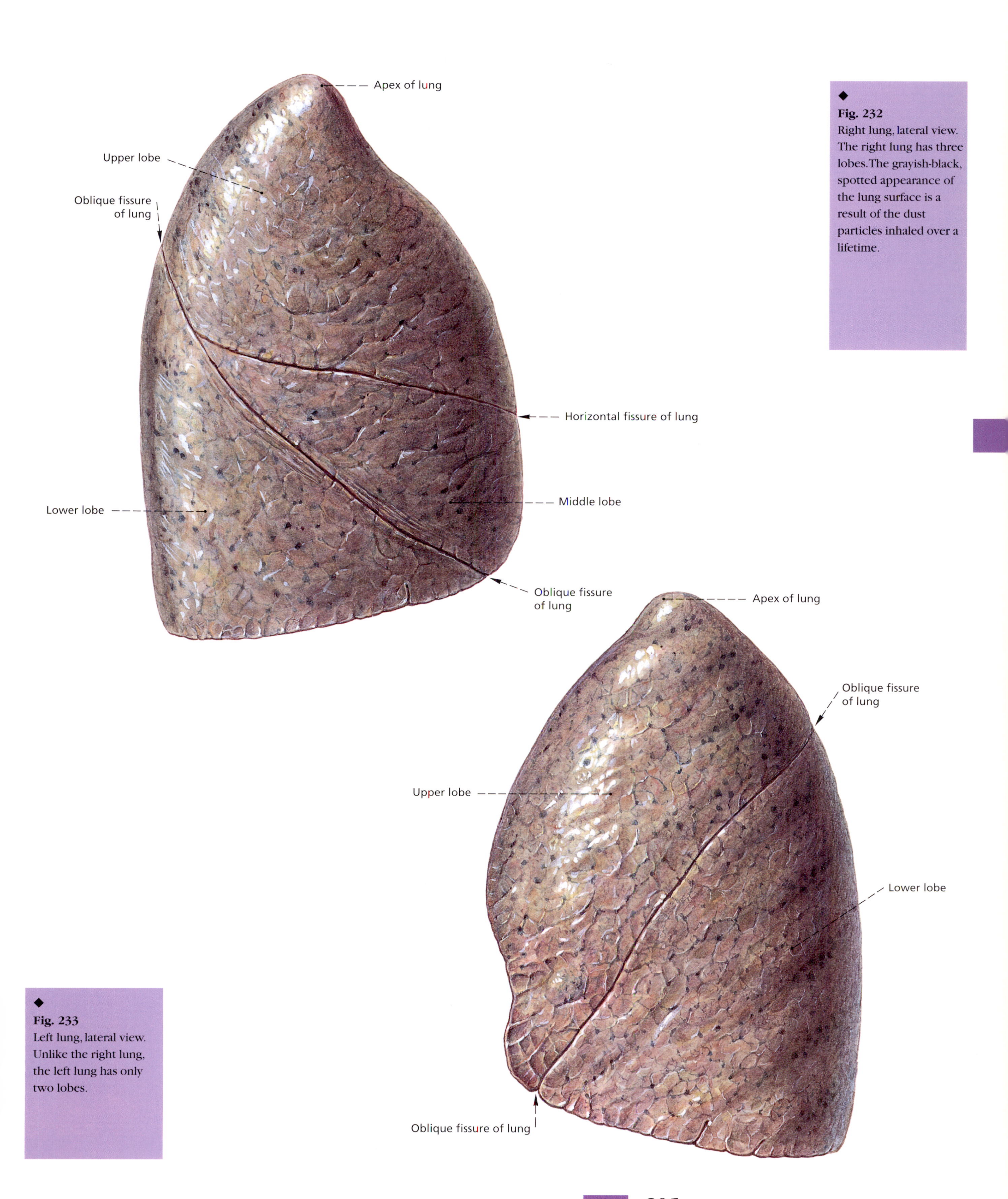

◆
Fig. 232
Right lung, lateral view. The right lung has three lobes. The grayish-black, spotted appearance of the lung surface is a result of the dust particles inhaled over a lifetime.

◆
Fig. 233
Left lung, lateral view. Unlike the right lung, the left lung has only two lobes.

◆
Fig. 234
Inner surface of the right lung, viewed from the left. The blood vessels (in section) are shown in red (pulmonary arteries) and blue (pulmonary veins).

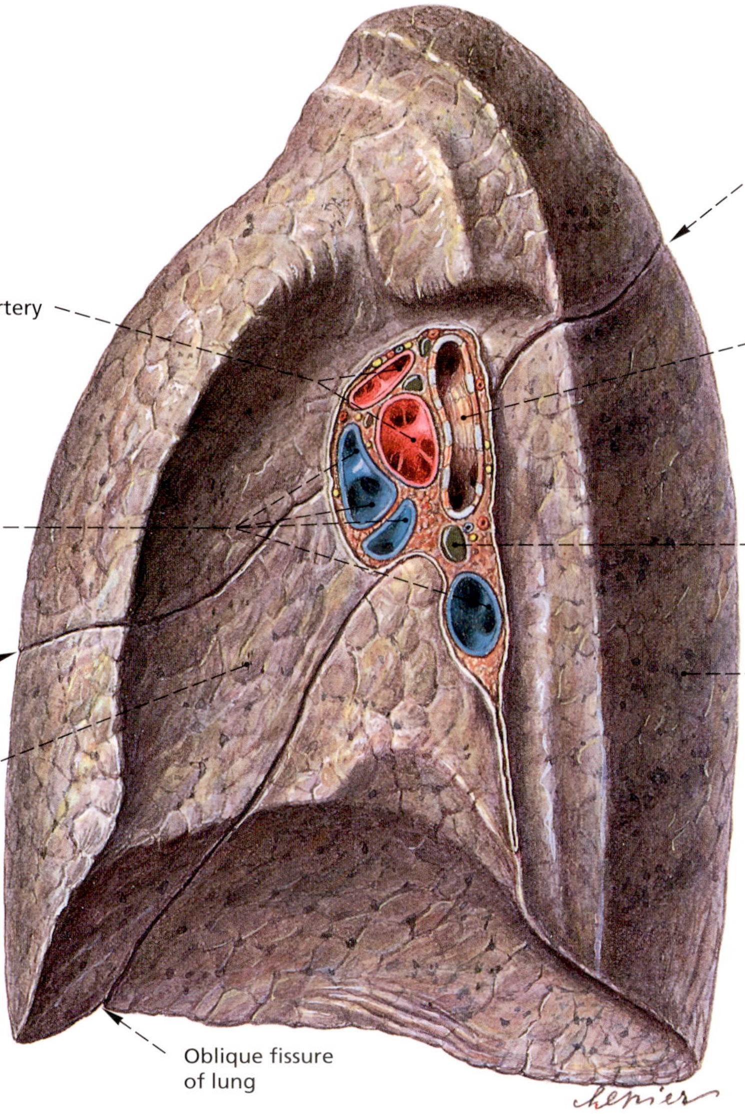

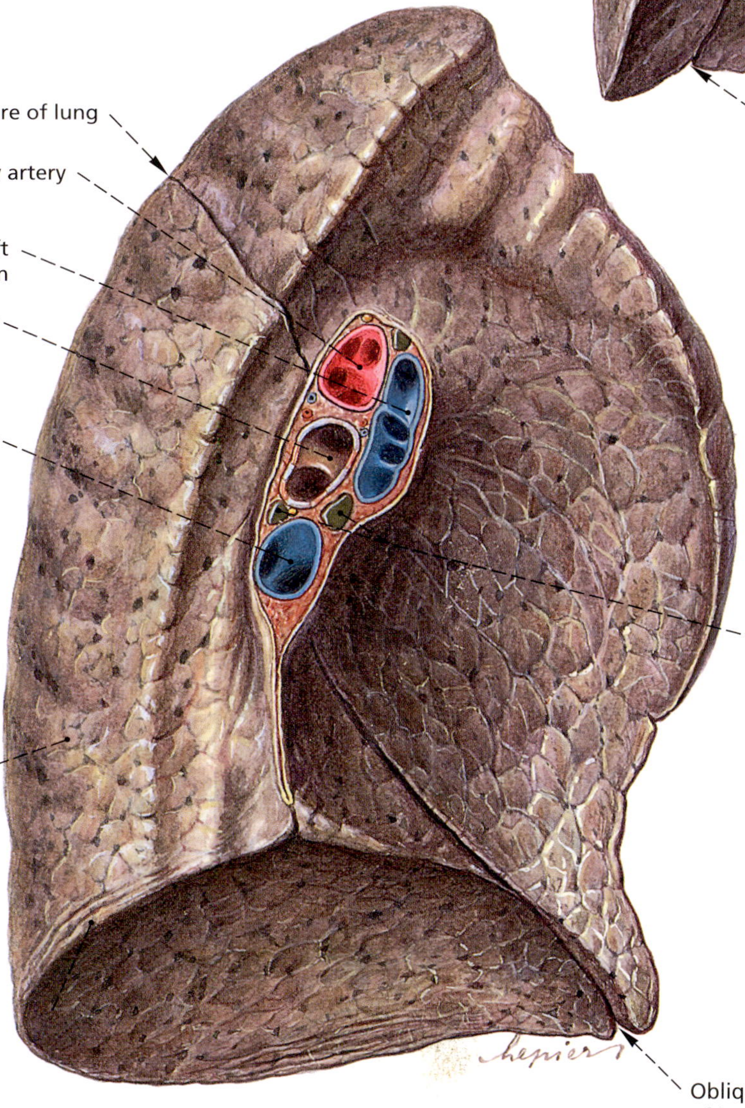

Oblique fissure of lung
Right pulmonary artery
Right main bronchus
Right fissure of lung
Lymph nodes
Horizontal fissure of lung
Lower lobe
Middle lobe
Oblique fissure of lung

Oblique fissure of lung
Left pulmonary artery
Superior left pulmonary vein
Left main bronchus
Left inferior pulmonary vein
Lymph nodes
Lower lobe
Oblique fissure of lung

◆
Fig. 235
Inner surface of the left lung, viewed from the right.

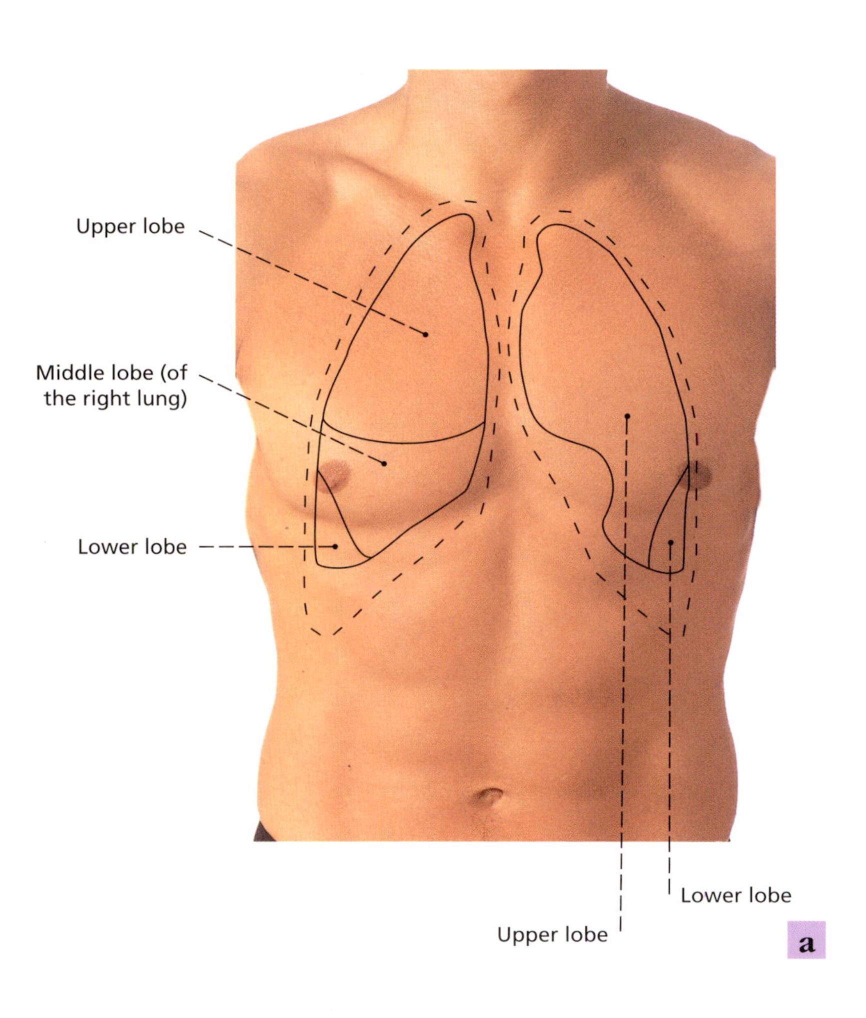

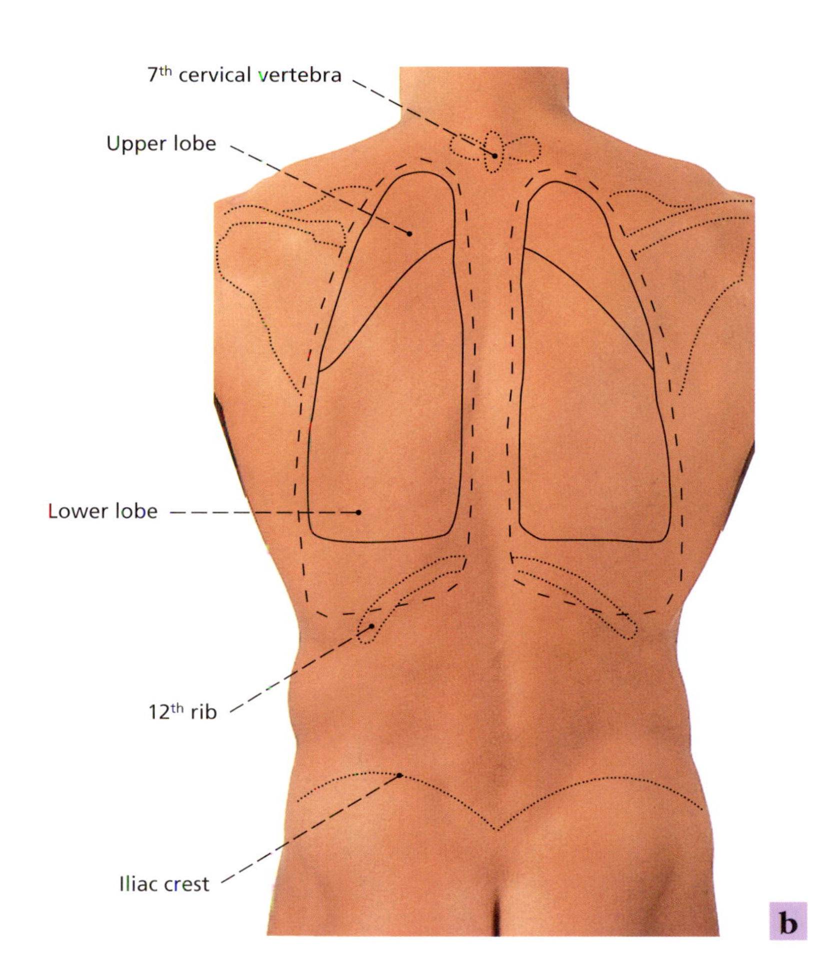

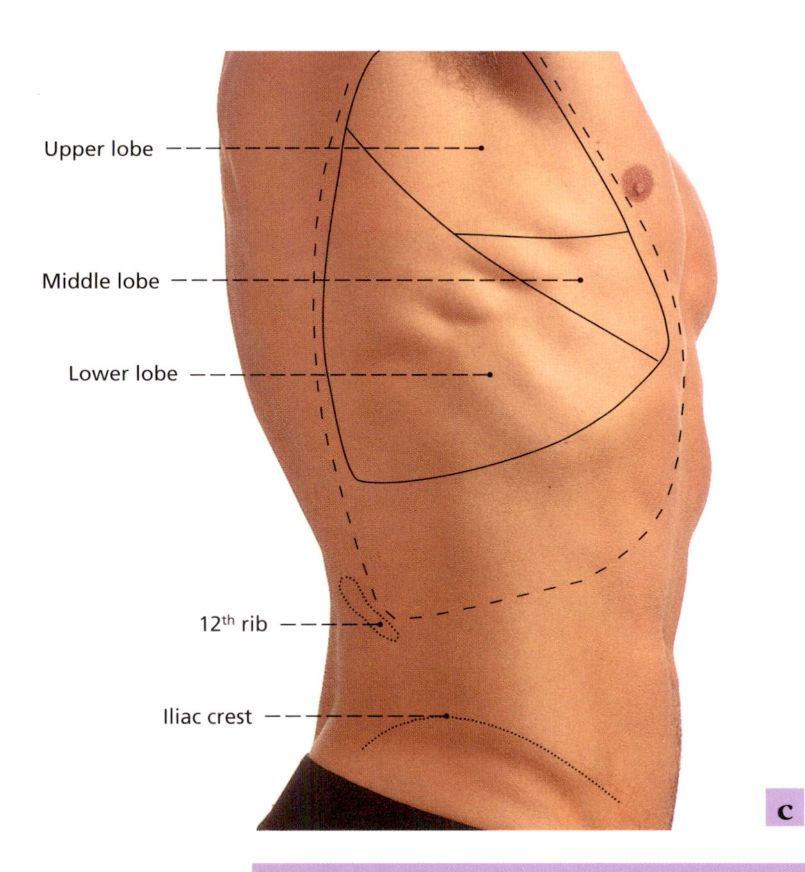

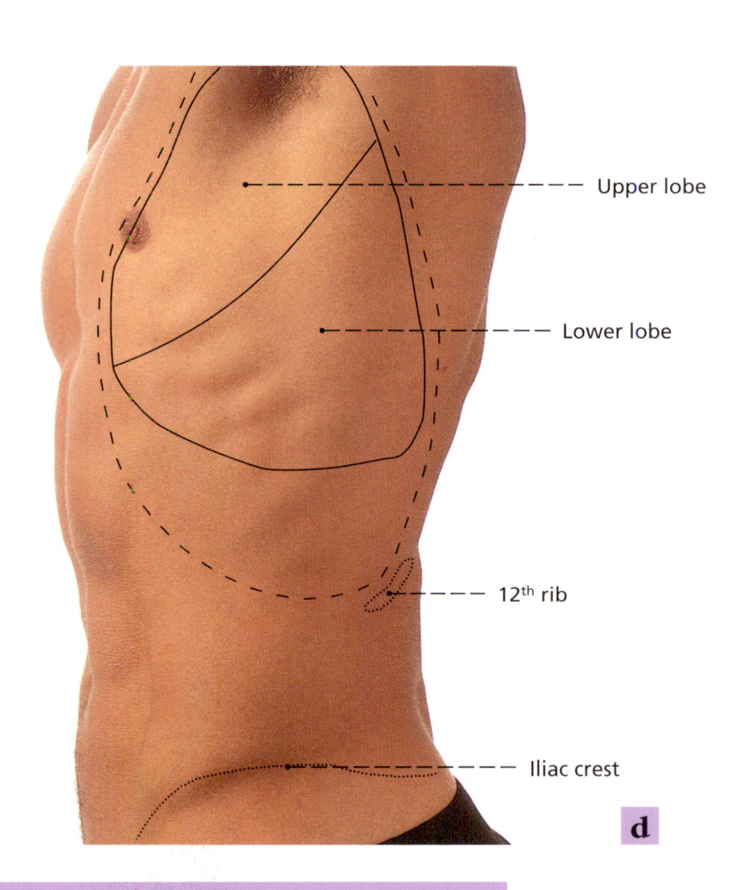

◆

Fig. 236
Expansion of the lungs and costal pleura with lines drawn on the skin (broken line = costal pleura, solid lines = pulmonary lobes): (a) on the front of the upper body, (b) on the back of the body, (c) on the right lateral thoracic wall, and (d) on the left lateral thoracic wall.

Symptoms and diseases

Respiratory paralysis (apnea)

Cessation of breathing. This can be caused by brain damage, poisons such as opiates, oxygen starvation, elevated carbon monoxide levels, stroke, or other injuries causing paralysis of the respiratory center. Alternatively, the respiratory muscles themselves can be impaired as a result of disease (e.g., infantile paralysis). General anesthesia for surgical interventions in the abdominal region involves respiratory paralysis being deliberately induced by drugs, in order to achieve total relaxation of the abdominal wall.

Dyspnea

Dyspnea or breathlessness is characterized by a decreased oxygen saturation in the blood. The patient tries to remedy the situation by breathing more rapidly, leading to increased respiratory effort. In children it is often accompanied by intercostal recession (hollowing between the ribs). It can be triggered when the passage of air to the lungs is obstructed, e.g., if a foreign body is blocking the trachea, or if the lungs have been so severely damaged by poisoning or inflammation that they can no longer absorb enough oxygen. It is always associated with anxiety. Dyspnea can also occur as a result of cardiac failure.

Respiratory arrest

Paralysis of the respiratory center in the brain results in total respiratory arrest. This is usually a result of the brain receiving too little oxygen, and can be caused by obstruction of the airways by foreign bodies, or by an inadequate supply of blood to the brain as a result of severe hemorrhage. A myocardial infarction, or certain types of intoxication, in which the heart muscles are damaged, can also result in too little oxygen reaching the respiratory center, because the heart is no longer capable of pumping enough blood to the brain. Other causes include direct damage due to pressure from a brain tumor or a head injury.

Inflammation of the lungs (pneumonia)

Inflammation of part or all of the lung tissues, usually of viral or bacterial origin. Pneumonia is the most common cause of death in children worldwide. It often occurs as a complication of other diseases (such as influenza). Insufficient ventilation of the lungs, which may occur when patients are confined to bed for long periods, also carries a risk of pneumonia. For this reason bed-ridden patients, and those who have undergone surgery, are encouraged to do breathing exercises every day.
The characteristic early signs of pneumonia are chills, followed by a rapid rise in temperature, a feeling of weakness, and in most cases a severe cough. These symptoms are often accompanied by chest pains and breathing difficulties. Blood tests, tests on sputum (or phlegm) the patient has coughed up, and a chest X-ray are required for a diagnosis. Antibiotics are usually administered, but the choice of treatment is primarily dependent on the nature of the pathogen.

Bronchitis

Bronchitis is an acute or chronic inflammation of the bronchial mucosa, caused by viruses, bacteria, or chemicals (e.g., in cigarette smoke); it can also be triggered by air pollution and allergies.
Acute bronchitis is associated with fever, cough, chest and back pains (due to coughing), and frequently with expectoration of mucus and wheezing. It usually improves after a few days without any specific treatment (humid air alleviates the symptoms).
In chronic bronchitis, symptoms last for months. Without treatment, the symptoms keep returning and the breathing problems become increasingly severe. The walls of the bronchi become thickened and the alveoli are enlarged. Chronic bronchitis is treated with antibiotics or cortisone, as well as inhaled oxygen.

COPD (chronic obstructive pulmonary disease)

An irreversible, progressive disease that develops over many years, starting initially as chronic bronchitis. During the course of disease, there is progressive and irreversible narrowing of the bronchi. Early symptoms include breathlessness on moderate exertion, and subsequently also when at rest, in addition to a feeling of tightness in the chest, coughing at night, expectoration of viscous mucus, and increased susceptibility to respiratory tract infections.
Although COPD is incurable, total abstinence from smoking and long-term treatment with drugs (including corticosteroids) can alleviate the symptoms and delay progression.

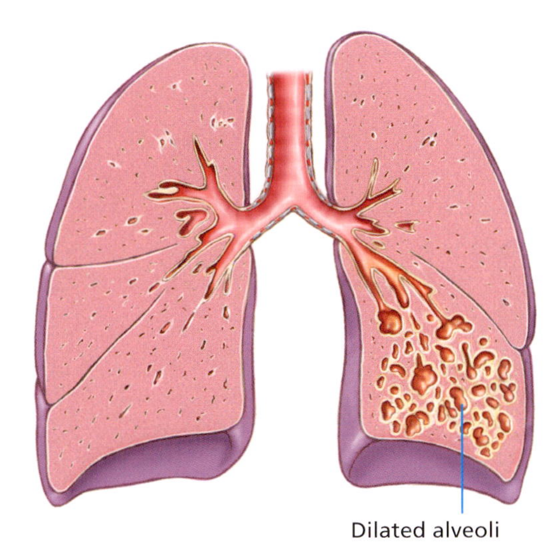

◆
Bronchiectasis
Dilated peripheral bronchi, or bronchiectasis (right), results in a cough with production of increased quantities of sputum.

Bronchiectasis

Bronchiectasis is a lung disease in which the branches of the bronchial tree dilate and form sacs, with consequent damage to their inner walls. The disease may be congenital or caused by acute or chronic bronchitis. Treatment consists of the administration of antibiotics.

Pulmonary emphysema (over-inflation of the lung)

Chronic overinflation of the lungs with destruction of the respiratory units, especially the alveolar walls. The bronchial mucosa become swollen due to chronic inflammation. When the patient breathes out, the thickened bronchial mucosa offer increased resistance, the elasticity of the lung tissue is reduced, and the area within the lung that can be used for oxygen uptake becomes smaller. The classic symptom of emphysema is marked breathlessness, which occurs even on the slightest exertion.

Pneumothorax (collapsed lung)

Collapse of part of the lung as a result of a change in pressure conditions in the thorax. Normally, there is negative pressure between the lungs and the thoracic wall (in the pleural sac). The lungs may collapse as a result of injury to the thorax, or diseases of the lungs (such as asthma), but also when no identifiable pathogen is present. Treatment consists of extracting the air that has entered the thorax by means of suction through a drainage tube inserted through a small incision in the thorax.

Pulmonary edema

An accumulation of fluid in the lungs, usually caused by impaired heart function. Congestion due to the presence of fluid prevents oxygen from the air breathed in from being released into the blood in sufficient quantities. The consequences are breathlessness, rapid, labored breathing, rattling noises on breathing in, exacerbation of breathlessness when lying down, and foam-like sputum, usually with a cough.
Prompt treatment is essential in order to counter the increasing oxygen deficit. The patient is given drugs to strengthen the heart as well as encouraging the kidneys to excrete water, with the result that the amount of fluid accumulated in the body rapidly diminishes.

Pulmonary embolism

Obstruction of a pulmonary artery by a displaced foreign body (embolus) or air (air embolus). The clinical significance of the embolism is referred to as major or minor depending on the size of the blood vessel affected. Predisposing factors include advanced age, obesity, inflammatory processes, operations, childbirth, and extreme climatic conditions. The symptoms are a severe radiating pain in the chest, which worsens on breathing in, a feeling of constriction, palpitations and anxiety, cold sweats, and sometimes a dry cough.
The diagnosis is confirmed by an X-ray of the lungs and X-ray imaging of the pulmonary vessels following injection of a contrast medium (angiography). As an embolism is usually caused by a blood clot (thrombus), the patient must be given drugs to dissolve the blood clot quickly, as well as medicines to inhibit blood coagulation and prevent the formation of additional clots. In some cases, surgical removal of the foreign body may be necessary.

Pulmonary tuberculosis

A disease of the lungs caused by the tuberculosis bacterium. Tuberculosis is a notifiable infectious disease. Although it has become less common in the western world, it has not been totally eradicated and prevalence in different countries changes with migration. Infection is spread by droplets, typically coughing by an infected person.
Tuberculosis is no longer life-threatening in industrialized countries, providing it is diagnosed promptly and treated with the appropriate drugs. In uncomplicated cases, this debilitating disease can be cured in six to nine months.

Coughs

Coughing is not always symptomatic of disease: it is often a useful mechanism to protect the body. We all cough now and then to clear mucus, dust, and other foreign bodies from the airways. As a symptom of illness, coughs frequently accompany colds, which are generally harmless. However, coughs can also be a sign of a serious disease.

The cough reflex is an important defense mechanism of the respiratory tract, triggered by various stimuli such as dust particles and microbes, and also by smoke, mucus, and cold air. Extremely violent coughing is caused by foreign bodies entering the airways, e.g. when swallowing liquids that "go the wrong way." These stimuli affect the nerves in the mucous membranes lining the respiratory tract, triggering an autonomic (unconscious) cough reflex.

Coughs accompanying colds

Respiratory tract infection is the most frequent cause of coughs. The cough reflex accompanying a cold is indicative of inflammation caused by viruses or bacteria. The cough starts as a dry cough (when no mucus is produced, it is also referred to as an "unproductive" cough). After a few days, this dry cough can turn into a "productive" chesty cough with sputum or phlegm.

Smoking—a frequent cause

Most smokers suffer from coughs, as cigarette smoke causes persistent irritation and damage to the respiratory organs. Typically, smokers have a productive cough in the mornings. As this cough develops slowly over a long period, and initially causes few problems, it is often dismissed as being harmless. Frequently, however, it develops into chronic inflammation of the respiratory tract (bronchitis).

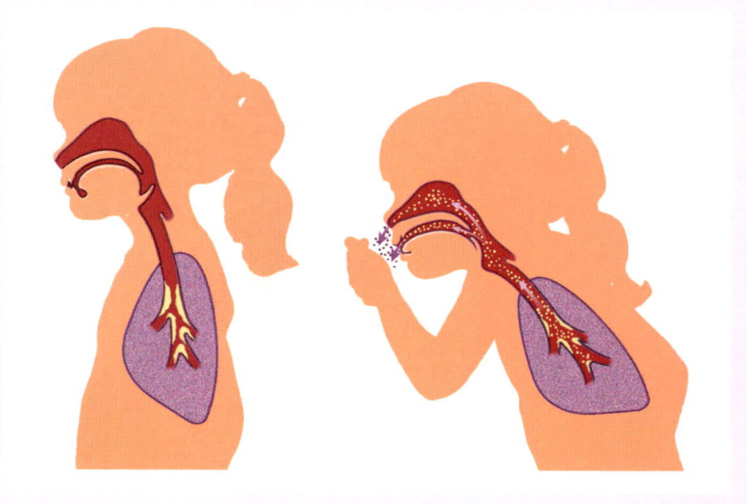

On coughing, the layer of excess mucus (yellow) is propelled out of the bronchi, clearing the airways and making breathing easier.

A common symptom

Increasingly, polluted air also plays a major role in the development of coughs. If the airways have an allergic reaction to certain substances, such as exhaust fumes, dust, pollen, or certain foodstuffs, this can also result in coughing, with or without sputum.

Patients with weak hearts tend to suffer from coughs, due to circulatory congestion in the pulmonary arteries. Coughing also occurs during asthma attacks, with production of viscous mucus. Paroxysmal coughing fits in infants, accompanied by difficulty breathing in, may be symptomatic of whooping cough.

Coughing as a warning signal

Sometimes a cough may be a symptom of a serious underlying disease, such as pneumonia, pulmonary tuberculosis, or cancer. Coughs require special attention when accompanied by other symptoms such as:

- painful breathing,
- fever,
- breathlessness,
- green, brown or bloody sputum.

In such cases medical advice should be sought without delay. It is also advisable to seek medical advice if a cough lasts for more than a week.

Long-term consequences

Chronic coughs always require treatment, because each coughing fit causes further irritation of the respiratory tract. In addition, the constantly irritated mucosa provides an ideal breeding ground for infections. In chronic bronchitis, the symptoms gradually worsen, as the extra mucus produced can no longer be completely transported away and eventually blocks the bronchi. Bacteria then colonize the mucus, causing constant inflammation, which in turn destroys the lung tissue. Overinflation of the lungs (pulmonary emphysema) can occur as a long-term consequence, accompanied by a chronic cough and breathlessness. Prompt treatment and removal of the trigger factors can prevent lasting damage to the lungs and respiratory tract.

Status asthmaticus (severe, prolonged asthma)

Severe acute asthma attacks, which follow one another in quick succession and can last for hours. When the attacks are so severe that the usual medication no longer helps, medical assistance or transfer to hospital is essential, as a life-threatening situation can rapidly develop. In addition to oxygen, an infusion of fast-acting drugs to stop the spasms and disperse the mucus is administered to arrest the attack. Such life-threatening situations can often be avoided by giving the asthma patient and his or her family adequate information, as well as administering prophylactic medication.

Sleep apnea

Temporary episodes of respiratory arrest, lasting about ten seconds each, while a person is asleep. These occur mainly in snorers, overweight men, cases of severe obesity with heart and respiratory problems, in various central nervous system disorders (e.g., disorders affecting the respiratory center in the brain), or in disorders of the pharynx (e.g., enlarged tonsils or polyps). Many sufferers feel tired and exhausted during the day and can only concentrate with difficulty. The oxygen deficit that occurs when the individual stops breathing can be life-threatening, such that, in many cases, mechanical ventilation is required. Additional risks include high blood pressure, heart failure, heart attacks, and strokes. Many sufferers can be helped simply by weight loss and drugs. The greatest danger lies in the fact that those affected are totally unaware when they stop breathing. Sleep apnea has been linked to sudden infant death syndrome.

Whooping cough

An infectious disease in children caused by bacteria, and characterized by violent bouts of coughing. Typically, whistling inspiration is followed by short, harsh coughing fits, often with the tongue hanging out, which are interrupted when the patient breathes in sharply, making a 'whoop' noise. After the next breath, there is another, weaker, coughing fit. These agonizing attacks are frequently associated with expectoration of mucus or vomiting. As the coughing is triggered by a toxin in the brain that is produced by the pathogen, ordinary cough remedies have no effect. Antibiotics only help in the initial stages of the disease. Life-threatening pauses in breathing can occur in affected babies, who should therefore always be treated in hospital.
The illness may last for several months and often weakens children considerably. Infants should therefore be vaccinated once they are three months old as part of a program of routine preventive measures.

Cystic fibrosis

Cystic fibrosis is the most common inherited disease in Caucasians, causing significant impairment of breathing and digestion. Due to a genetic defect, the secretions produced in the lungs are more viscous than in healthy children, so that mucus commonly blocks the fine branches of the bronchi, while secretions from the pancreas are lacking in certain digestive enzymes, making it difficult to provide adequate nutrition. Affected children suffer from frequent coughing attacks, breathing difficulties, digestive disorders, and emaciation. They perspire excessively and their sweat is extremely salty. The viscous mucus in the lungs provides an ideal medium for pathogenic microbes, so pneumonia and other infections are common. Today, thanks to intensive specialized treatment regimes, including special diets and dietary additives, in conjunction with removal of the mucus, the prognosis has improved enormously. This has resulted in the identification of additional problems, however, now that children frequently survive into adulthood. The earlier cystic fibrosis is diagnosed, the longer the life expectancy of the child. Treatment based on gene therapy may also be possible in the future.

Croup

Croup is a self-limiting viral infection of young children, also known as severe laryngotracheobronchitis. Symptoms include hoarseness, a deep, barking, raw cough, a characteristic high-pitch breath sound when breathing in (stridor), and whistling when breathing out. It is usually accompanied by anxiety and fever. The symptoms most commonly occur at night. In the first instance parents should ensure that the surrounding air is cool and moist, and reassure the child. Medical care is important, since breathlessness can occur in some circumstances. Croup is treated with corticosteroids to reduce the swelling of the mucosa.
A spasm of the glottis (e.g., due to an allergic reaction) may give rise to a similar respiratory sound and is known as pseudocroup; there may be a link between pseudocroup and air pollution, especially high concentrations of sulfur dioxide. It is more common in autumn and winter, accompanying an existing or approaching cold front, and in weather conditions in which there is an increase in the amount of dust and moisture in the air.

Asthma

Asthma is caused by hypersensitivity of the bronchial mucosa, which swell up in an exaggerated defensive response by the body to irritant stimuli. The airways go into a convulsive spasm, causing dramatic symptoms of acute breathlessness and fear of asphyxiation. However, asthma can be kept under good control with an appropriate lifestyle and medication.

If you were to spread out the mucous membrane lining all the branches of the respiratory tract (bronchi and bronchioles) in the lungs and alveoli of an adult, the surface area would come to 70–80 square meters. This large surface area is necessary to allow a sufficiently large volume of oxygen from the air we inhale to pass into the bloodstream via the cells of the alveoli, and for as much spent air as possible to be exhaled from the lungs. However, throughout our lives, this surface area comes into contact with all the substances we inhale. Since the quantity of air pollutants is on the increase, the lungs are exposed to ever greater stress, and the number of asthma sufferers is constantly rising.
About one in 15 people in the USA suffers from asthma. Usually the disease manifests itself in childhood. The cause is an allergy to a wide variety of substances, known as allergens, found in the air (pollen, house dust, animal hairs, and pollutants), which come into contact with the bronchial mucosa. Overactive defense mechanisms then trigger the changes typical of asthma. Less commonly, allergens that do not enter the lungs also cause asthma, e.g., some foodstuffs and medicines.

◆ Possible triggers of asthma.

Pathogenic changes

Each individual produces substances in their own body that trigger defensive responses. If the allergic individual comes into contact with an allergen, or irritant, these substances are produced in far greater quantities. The result is a hypersensitive (allergic) reaction:

- The mucous membranes in the bronchi swell up.
- The delicate musculature of the walls of the bronchioles go into spasm, narrowing the airways.
- The mucus-producing cells in the walls of the bronchioles produce increasing quantities of viscous secretions that block the airways.

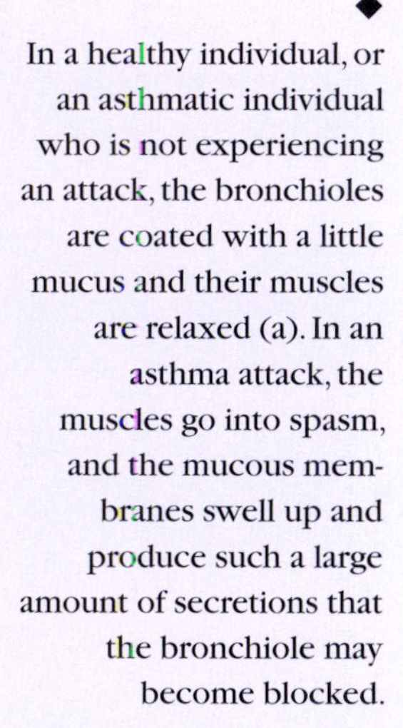

◆ In a healthy individual, or an asthmatic individual who is not experiencing an attack, the bronchioles are coated with a little mucus and their muscles are relaxed (a). In an asthma attack, the muscles go into spasm, and the mucous membranes swell up and produce such a large amount of secretions that the bronchiole may become blocked.

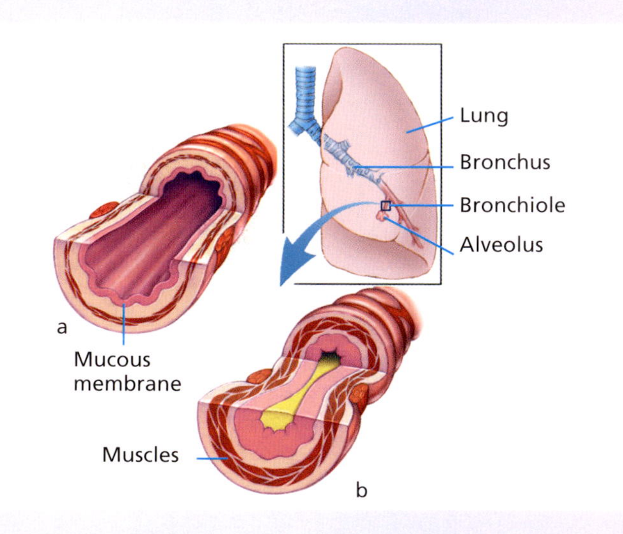

Symptoms

Typical symptoms are attacks of breathlessness marked, in particular, by difficulty in breathing out. This is due to the fact that the airways are in spasm and swollen, and can close like a valve, especially on expiration. Increased mucus production results in attacks of coughing with viscous sputum. The breathing difficulties lead to a reduction in the amount of oxygen absorbed, and the patient's lips may turn a bluish color. Anxiety and restlessness are caused by the shortness of breath.

Treatment

If the substances to which the patient is allergic are known as a result of tests, hyposensitization may be successful. This involves administering to the patient minute doses of the allergen over an extended period of time. The body becomes used to it and the allergic reaction consequently gradually diminishes in severity. In addition, drugs (antihistamines) can weaken the allergic reaction. In an acute asthma attack, drugs are usually inhaled, immediately dilating the airways.
An accurate diagnosis of the extent of the disease must always be made. A special instrument is used to measure peak flow. The patient breathes out into this instrument, which measures the force with which the air is exhaled. By measuring these values several times a day, both patient and doctor can identify the respiratory pattern during a 24-hour period. This provides the basis for deciding when, how often, and at what dosage, the patient must take his or her medication. It is essential that drugs to prevent the bronchioles going into spasm are taken regularly. Other drugs can reduce the viscosity of the increased lung secretions produced and make it easier to cough them up.
Corticosteroids "damp down" the allergic reactions of the mucosa and encourage healing of any inflammatory changes. Depending on the severity of the asthma, additional treatment with corticosteroids may be required. Since all drugs can cause undesirable side effects at too high a dose, the physician must determine the amount which is sufficient to free the patient from symptoms while avoiding undesirable side effects.

◆ The peak flow meter measures the force with which the patient breathes out to determine the severity of the asthma.

Prevention

A specially adapted lifestyle can minimize the problems caused by asthma by:

- avoiding contact with allergens,
- recognizing the signs of an imminent asthma attack at an early stage and taking steps to prevent it,
- always carrying an asthma inhaler,
- taking medication regularly,
- avoiding pollutants (in particular, not smoking),
- living in places with favorable climatic conditions (e.g., by the sea or in the mountains), or at least taking vacations and breaks there,
- physical exercise,
- good medical supervision.

Risks

If an asthma attack becomes semi-continuous (status asthmaticus), the life of the person involved is threatened, with the possibility of asphyxiation. In this situation, medication in the form of inhalers no longer helps, because the drug is unable to enter the airways, as they are in complete spasm. In this situation, the patient should be placed in a half-sitting position and reassured while emergency help is called.
Permanent damage may result if chronic asthma is not properly treated. If the air breathed in can only be exhaled with difficulty, increased resistance to breathing out leads to overinflation of the lungs (emphysema), which in turn increases the pressure on the pulmonary arterioles. The strain on the heart increases, as it has to pump the blood through these constricted arteries. The result is a chronically weakened heart.

◆ A metered aerosol spray is able to deliver a targeted dose of medication to the airways in the event of an attack.

Lung cancer

Lung cancer is one of the most common cancers in western industrialized countries, and its incidence is increasing. Formerly, it affected mainly men, but a significant increase has been observed in the numbers of women with lung cancer, due primarily to the steep rise in numbers of women smokers. Smoking cigarettes is the primary cause of lung cancer.

Most malignant lung tumors arise in the mucous membranes of the bronchi. Therefore the term "lung cancer" is usually synonymous with bronchial cancer or bronchial carcinoma. The disease usually strikes after the age of 40.

Causes

The close link between smoking and lung cancer is clear: 90% of all patients who develop lung cancer are smokers. The risk increases in proportion to the number of years the patient has been a smoker and the number of cigarettes smoked daily: the risk of developing lung cancer in a chain smoker is 30 times as great as in a non-smoker. The risk is also increased if the individual began smoking at an early age.
Chronic inflammatory conditions such as bronchitis, from which most smokers suffer, provide favorable conditions for the development of lung cancer. Passive smoking also constitutes a risk.
Not all smokers develop lung cancer, however, so there appear to be other factors involved in triggering the disease. An increased incidence of this type of cancer has been demonstrated in individuals whose jobs involve contact with asbestos, chromium, arsenic, nickel, or radioactive substances such as uranium. Increasing air pollution, due to traffic and industrial waste gases, also plays a significant role. When several of these factors are present in combination, the risk is even greater. For example, the carcinogenic effect of asbestos is considerably more potent when exposure to asbestos is accompanied by smoking. It has now also been proven that susceptibility is inherited, i.e., there is an increased risk if a parent has suffered from lung cancer.

Are there early symptoms?

The initial signs are not very typical, as they also occur in many other lung diseases. Early diagnosis is therefore difficult, as initial complaints are often attributed to chronic bronchitis. Valuable time can therefore be lost, and appropriate treatment delayed.

The following symptoms may be warning signs:
- stubborn cough,
- sputum containing blood,
- hoarseness,
- breathlessness,
- chest pain.

Recurring episodes of pneumonia, major weight loss, and reduced stamina are particularly suspicious symptoms. However, not all of these symptoms point unequivocally to a diagnosis of cancer.

Diagnosis

Only if lung cancer is diagnosed at an early stage is there a prospect of a cure. If there is the slightest suspicion, or even vague symptoms that may be dismissed by smokers as harmless, a physician should be consulted immediately.
One of the first investigations carried out is a chest X-ray. However, it is not always possible to make a definite diagnosis from an X-ray image, in which case a bronchoscopy will be necessary. In a bronchoscopy, an endoscope is introduced into the bronchi via the nose or mouth, enabling the bronchi to be viewed directly; tissue samples can be taken at the same time if necessary. Cancer cells can also sometimes be identified in the sputum.

◆
Cigarette smoke penetrates deep into the lungs and contributes to the development of lung tumors.

Treatment

Treatment depends on the nature, site, and stage of the tumor. The physician will determine the most appropriate treatment for each individual patient. The two most common types of lung cancer are small-cell and non-small-cell bronchial carcinomas.

Small-cell bronchial carcinoma accounts for 20–30% of cases of lung cancer. These cancers tend to grow rapidly, spread quickly to other organs, and have a less favorable prognosis. In this form of cancer, surgical removal is an option only in the early stages (small tumors without lymph node involvement); however the tumors do respond well to radiotherapy and chemotherapy.

Non-small-cell lung cancer is significantly more common (70–80% of cases) and has a better prognosis. The prospect of a cure is offered only by surgery in which the whole lung or part of the lung is removed. If the cancer has spread to the lymph nodes, or if some of the tumor remains in the lung after the operation, the patient will have to undergo radiotherapy. Chemotherapy can be administered in parallel with surgery and/or radiotherapy, or sometimes as the sole treatment.

The **rare pancoast tumor** is usually localized in the apex of the lung and spreads in the early stages to adjacent vertebrae, ribs or nerves (invasive cancer). Surgical removal of the tumor is performed only following radiotherapy. After the operation, further radiotherapy is usually administered.

Despite all the advances in medicine, lung cancer is still characterized by a poor prognosis, for which failure to make an early diagnosis and the aggressive growth of this type of tumor are responsible. These tumors tend to produce satellite tumors, known as metastases, which spread via the bloodstream and lymphatic system and may colonize the liver, brain, adrenal glands, and/or bones, in particular. Similarly, malignant tumors from other organs such as the stomach, bones, breast, and prostate may cause lung metastases. Only a small proportion of lung tumors is benign.

Complications, side effects and check-ups

The most important complication of lung cancer is malignant pleural effusion, in which fluid accumulates between the pleural membranes, resulting primarily in breathlessness. The pleural effusion can be removed by aspiration.

In vena cava syndrome, the tumor impinges on the superior vena cava and prevents blood flow, resulting in swelling of the arms or face, dizziness, and headache. This is usually treated by insertion of a small metal mesh tube to keep the vein open.

The possible side effects of treatment are mainly a consequence of chemotherapy and radiotherapy. Inflammatory conditions, increased susceptibility to infection, nausea, fatigue, and weakness are all possible, as well as skin changes in the irradiated areas, but these generally recede over time.

Regular check-ups, at short intervals in the early stages, are essential in order to be able to treat a recurrence of a tumor or metastases as early as possible.

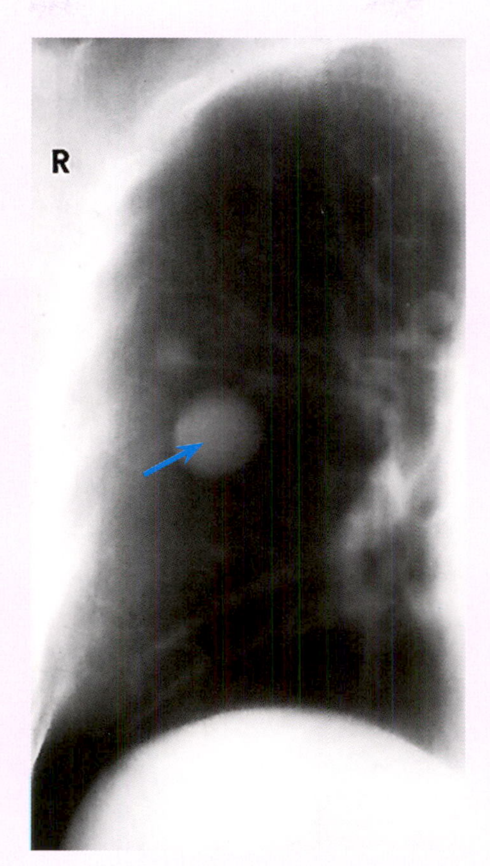

◆ It is not always possible to distinguish lung tumors clearly on X-rays.

Avoiding carcinogens

Giving up smoking is always worthwhile. After about ten years, the risk of a former smoker developing the disease is once again equivalent to that of a non-smoker. Individuals who are unwilling to give up smoking should at least undergo regular medical examinations.

It is also extremely important to comply strictly with workplace safety measures when handling carcinogenic substances.

Diagnostic methods

Auscultation (listening to internal sounds of the body)

A diagnostic method of identifying pathological processes due to internal organ dysfunction. Disturbances of heart activity, breathing (rattling sounds or "rales" in the lungs), or the peristaltic action of the intestine can be identified using a stethoscope. The pulsing of the blood in the major arteries is also audible. Auscultation can also pick up sounds from areas of turbulent blood flow caused by the narrowing of blood vessels.

Percussion

The part of the body under investigation is tapped using a percussion hammer, the fingers, or the hand. In the relevant organ beneath, vibrations are produced as a result of the impact, and the nature of the sound heard provides indications about the size of, and any pathological changes in, the organ in question. In physiotherapy, percussion is used on the back of the patient to encourage expectoration if the airways are clogged with mucus, in order to prevent pneumonia. Percussion can also relax muscles spasms in the back.

Bronchoscopy

Examination of the bronchi using an endoscope. This is carried out under general or local anesthetic; if a tumor is suspected, tissue and mucus can be removed for closer examination.

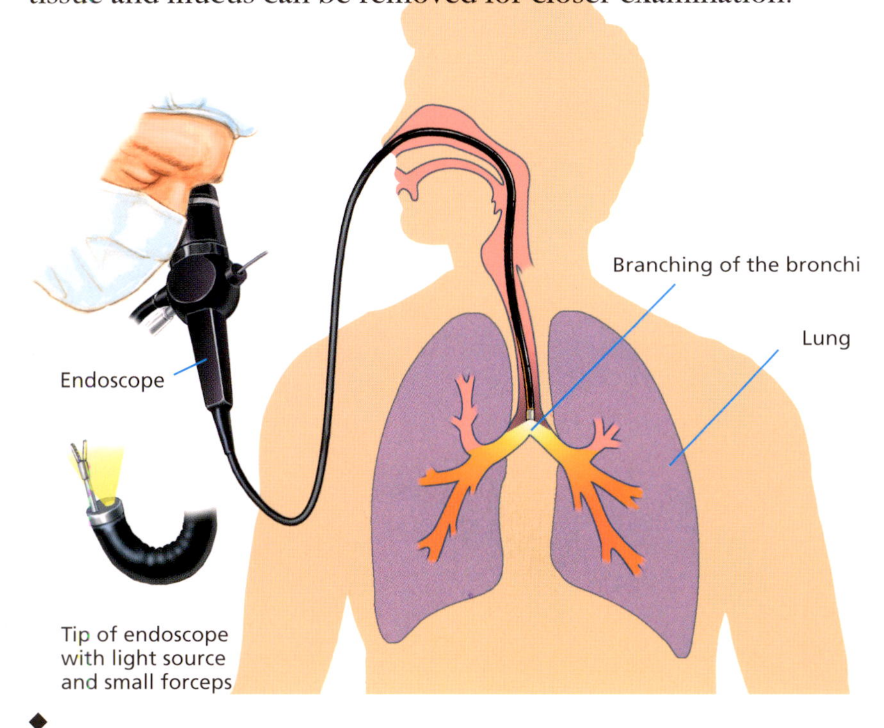

◆
Bronchoscopy
Changes in the mucosa can be identified at an early stage and tissue samples can be taken during bronchoscopy.

Bronchoscopy is also used to remove foreign bodies and aspirate secretions from the lungs.

Lung function testing

A blanket term for various test procedures to determine the capacity of the lungs at rest and under stress. The results provide information about lung diseases or localized disturbances in the lungs. The tests include measurement of the volume of the lungs, air flow rates, oxygen uptake, the carbon dioxide content of exhaled air, and the residual air remaining in the lungs after breathing out.

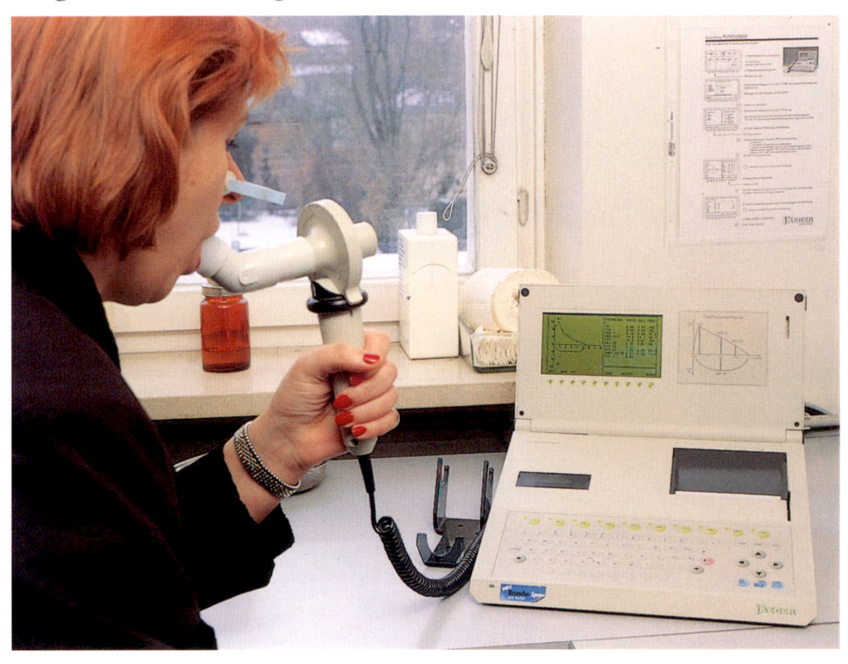

◆
Lung function testing
Lung function testing includes measuring the volume of the lungs by determining the amount of air breathed out and the force required to do so.

Imaging procedures

The chest **X-ray** is used to diagnose lung diseases occurring as a result of tissue proliferation (e.g., lung cancer) or fluid collections (e.g., pulmonary edema). Inflammatory changes appear lighter on X-ray image, while pulmonary emphysema appears darker. **Computed tomography** and **magnetic resonance imaging** are indispensable tools in the diagnosis of cancers.
In the event of narrowing of the bronchi, or if a pulmonary embolism is suspected, **pulmonary scintigraphy** (also known as ventilation or inhalation scintigraphy) may be used. This involves the patient breathing in an inert gas containing a radioactive marker so that the physician can see which parts of the lung are adequately ventilated, and which are not. **Perfusion scintigraphy** is also used in the diagnosis of pulmonary embolism. A substance containing a radioactive marker is injected into a vein in order to pinpoint the exact site of the obstruction.

Treatment methods

Anti-asthma drugs

Asthma medicines serve one of two purposes: quick relief of symptoms and prevention or control of symptoms. Most asthma medications for adults and older children are taken via a hand-held device called a metered dose inhaler or a dry powder inhaler. Corticosteroids are given to treat acute asthma attacks, while muscle spasm in the airways may be relieved by bronchodilators. Non-steroidal anti-inflammatories may be given to treat the airway swelling and extra mucus production that come with inflammation, while immunomodulators are used to suppress the antigen–antibody response in allergic asthma.

Cough medicines (anti-tussives)

Dry coughs are treated with suppressants, such as codeine, which acts directly on the brain, and should therefore only be used when prescribed by a physician. Productive coughs are treated with expectorants that loosen the mucus, e.g., N-acetylcysteine. Other proprietary medicines containing cough suppressants such as eucalyptol, menthol, or camphor, may be applied to the chest so that the active ingredients are inhaled.

Inhalation

Inhalation of popular remedies, such as Friar's Balsam, has been used for generations to ease respiratory disorders. Nowadays, modern medications are often delivered through hand-held inhalers. In both instances, the benefit is that the active ingredient reaches the lung tissue directly, rather than via the blood stream.

Respiratory physiotherapy

Special breathing techniques practiced by the patient with the help of a physiotherapist are particularly important for patients who are confined to bed, as they can prevent pneumonia. Regular breathing patterns are also important in existing lung diseases, or in malformations of the spinal column or thorax that make breathing difficult. Breathing exercises are also used in antenatal care and to promote relaxation. They usually form part of a more extensive treatment program.

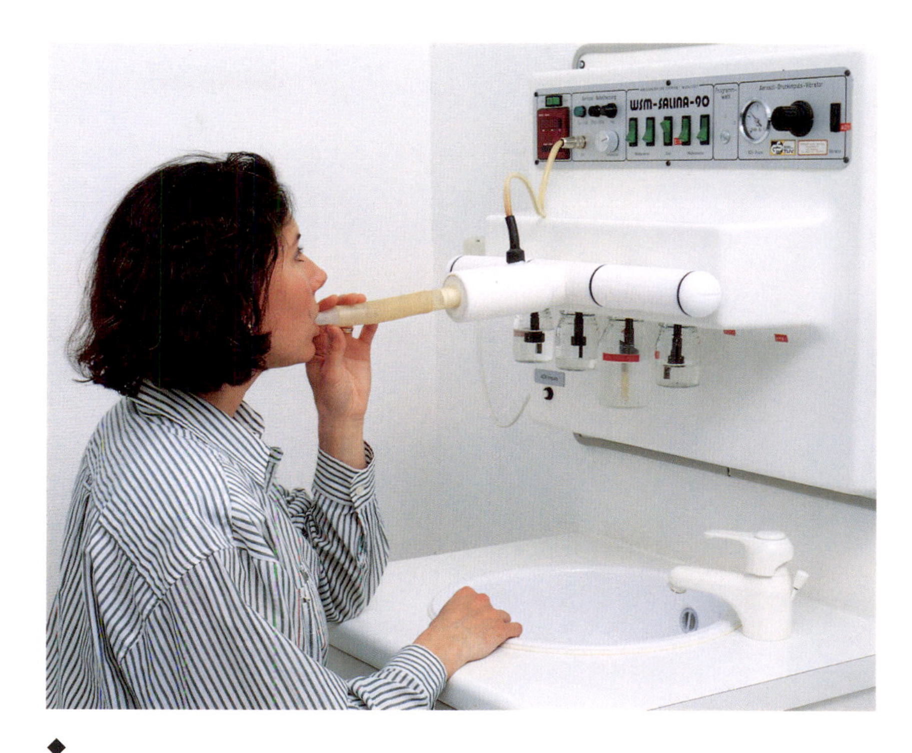

◆
Inhalation
Special inhalers mix the active substances to be inhaled with the air we breathe in.

Oxygen therapy

Patients may be given oxygen in any situation where the oxygen level in the circulating blood falls below normal levels (hypoxia). Causes of oxygen deficiency include respiratory diseases, trauma, conditions affecting the respiratory centers in the brain, and intoxication (e.g., by carbon monoxide or smoke). Hyperbaric (high-pressure) oxygen is used as an adjunctive therapy to treat a variety of medical conditions, e.g., decompression sickness.

Lung transplantation

A lung transplant may be necessary in some severe lung conditions. In severe congenital heart malformations, in which the pressure of blood in the heart is excessively raised, damaging the heart with potentially life-threatening consequences, a combined heart–lung transplant will be required. Today, around half of all patients undergoing a combined heart–lung transplant are still alive five years later. The prospects are especially good for younger people. A year after surgery they should be able to work again, living a practically normal life and even having children.

The heart

The heart is a powerful hollow muscle that constantly pumps blood through the body. Its performance is astonishing: at an average rate of 70 beats per minute, it contracts around 100,000 times a day and pumps 300–400 liters of blood around the body in just one hour.

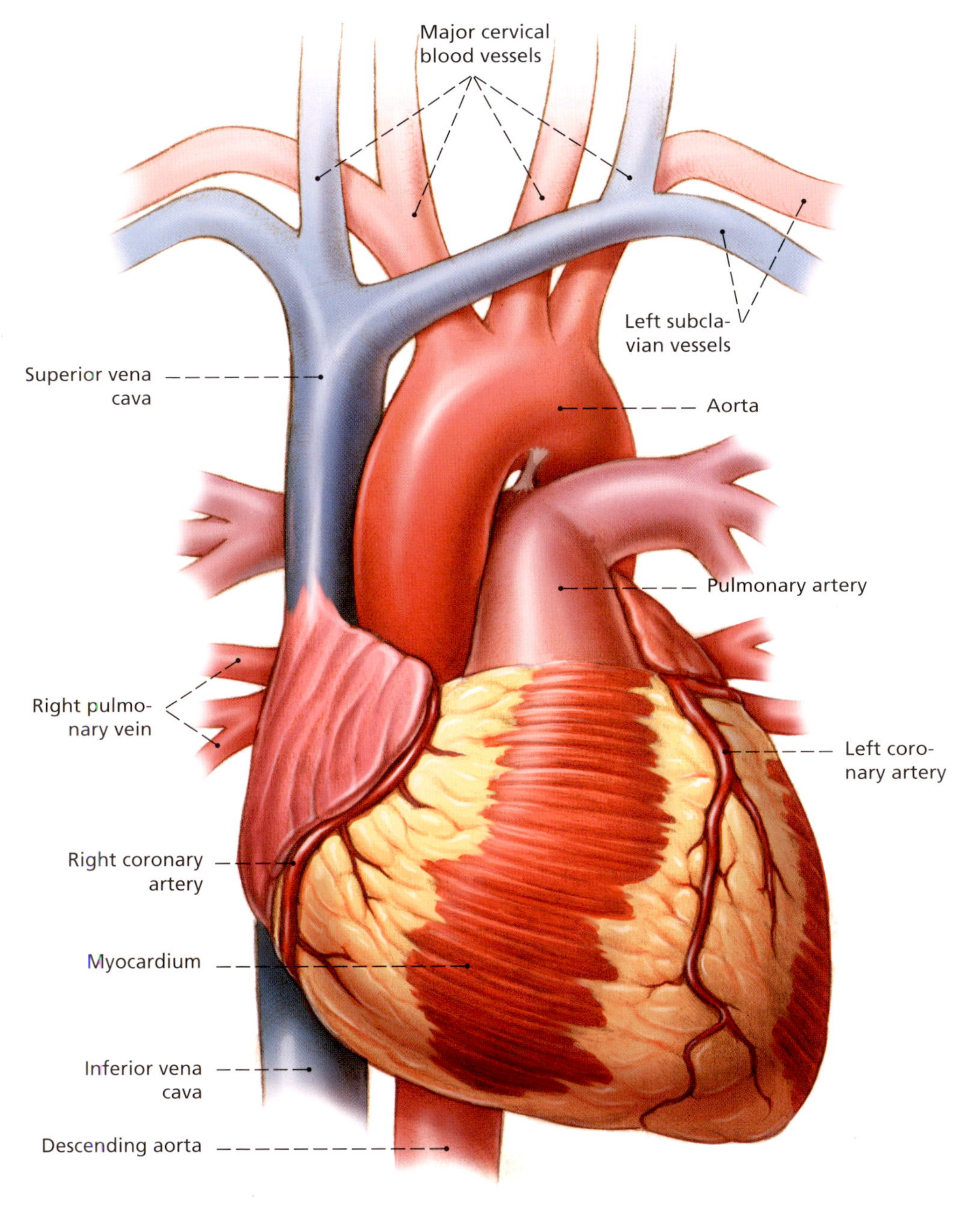

◆
Fig. 237
The following vessels originate from the hollow muscle of the heart:
1. the aorta
2. the superior vena cava and inferior vena cava
3. the pulmonary vessels, which are unique in that the pulmonary artery carries de-oxygenated blood away from the heart, while the pulmonary vein brings oxygenated blood to the heart. The tissues of the heart are themselves supplied with oxygenated blood by the three coronary arteries.

Position and function

Covered by a thin skin, called the pericardium, the heart is located in the left-hand side of the thorax, directly behind the sternum, and is largely covered by the two lungs. It is about the size of a clenched fist and, in adults, it weighs about 300 grams. Its posterior wall lies next to the esophagus, while the lowest point is next to the diaphragm. The longitudinal axis tilts forwards and to the left, so that the apex of the heart extends almost to the chest wall. In very thin people, the heartbeat can be felt, or even seen, at this point (the apical impulse).

The heart is divided into a right and left side by a wall known as the interventricular septum. The right side of the heart contains deoxygenated (oxygen-poor) blood, while the left side supplies the body with oxygenated (oxygen-rich) blood. Each side of the heart is additionally divided by means of sail-shaped valves into an atrium (the upper part) and a ventricle. The tripartite atrioventricular valve in the right side of the heart is called the tricuspid valve (from the Latin tricuspid = with three points), while the bipartite valve in the left side is called the mitral valve (from the Latin mitralis = with two points).

Both atria serve as collection points, the right atrium collects deoxygenated venous blood from the body, and the left atrium collects oxygenated blood from the lungs. From the atria, the blood is transferred into the ventricles, passing from the right ventricle into the pulmonary circulation and from the left ventricle into the systemic (general) circulation. The heart completes its pumping cycle approximately 70 times a minute, with a relaxation and filling phase (diastole) and a contraction phase (systole). Both the atrioventricular valves and the "pocket" valves at the ventricular outlets ensure that the blood can flow only in the right direction: from the right atrium (to which venous blood is transported from the body via the superior vena cava and the inferior vena cava) to the right ventricle, and from there to the lungs, where it is enriched with oxygen. The oxygenated blood from the lungs returns to the left atrium and is subsequently pumped from the left ventricle into the systemic circulation via the aorta.

Heart rhythm

The electrical impulses controlling the activity of the heart come from its own autonomous conduction system. The impulses are generated by two small nodes located in the wall of the right atrium. The main pacemaker is the sinoatrial node, which triggers 70–80 beats per minute, while the atrioventricular node (AV node) maintains the heart rate of 40–60 beats per minute that is essential to sustain life. The AV node only springs into action in the event that the sinoatrial node, the main pacemaker, has failed. The sinoatrial and atrioventricular nodes are joined by the bundle of His, also known as the atrioventricular bundle. This bundle of heart muscle tissue is a conductor of the electrical impulses, which are relayed over the entire heart via the two left bundle branches and Purkinje's fibers (fibrous tissue in the interventricular septum and the ventricular walls). If the heart is required to work less hard or harder than usual (for example, in the event of unusual exertion or stress), two nerves from the autonomic nervous system come into play: the parasympathetic nerve, which lowers the heart rate, and the sympathetic nerve, which increases the heart rate and hence increases the performance of the heart (see also Brain and nervous system, page 126).

Blood supply to the myocardium

The heart muscle (myocardium) is supplied with blood by the right and left coronary arteries, which both branch off from the aorta. If these vessels are narrowed, e.g., by calcium deposits, the myocardium is less well perfused with blood. This manifests itself initially as angina pectoris (literally, tightness of the chest), especially on exertion and/or stress. Sudden occlusion of a coronary artery by a blood clot (thrombus) results in a heart attack (myocardial infarction).

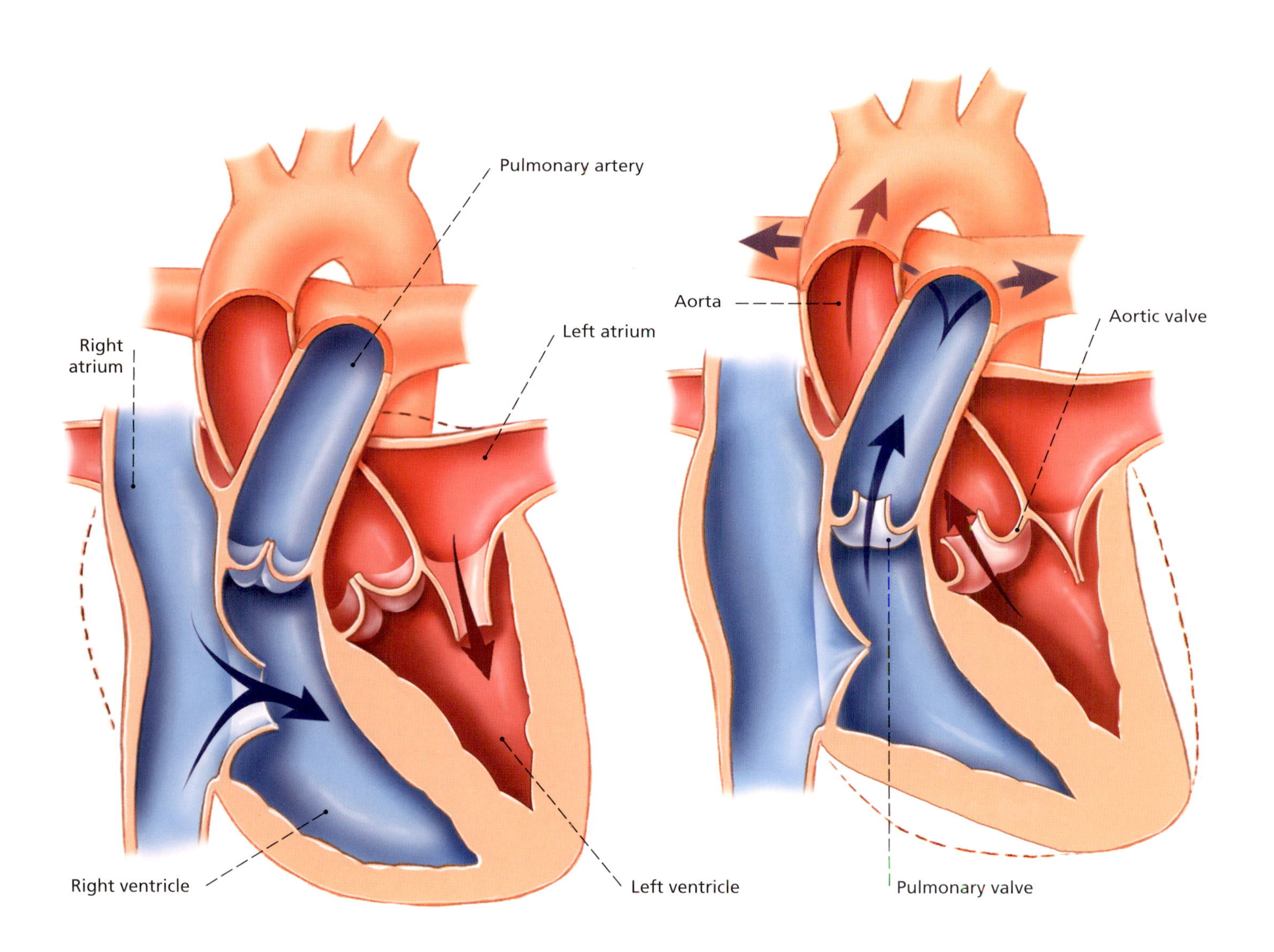

◆
Fig. 238
In the contraction phase of the heart's cycle (systole), the atria contract, forcing blood from the atria into the ventricles. Following closure of the atrioventricular valves, the ventricles contract, pumping the blood through the aortic valve into the aorta and through the pulmonary valve into the main pulmonary artery and on to the lungs, respectively. During the relaxation phase (diastole), the heart is refilled with blood.

◆
Fig. 239
The heart, with peri-cardium, in the thoracic cavity. The anterior wall of the thorax has been cut away. Also visible are the two vagus nerves, which pass through the cervical region before reaching the heart.

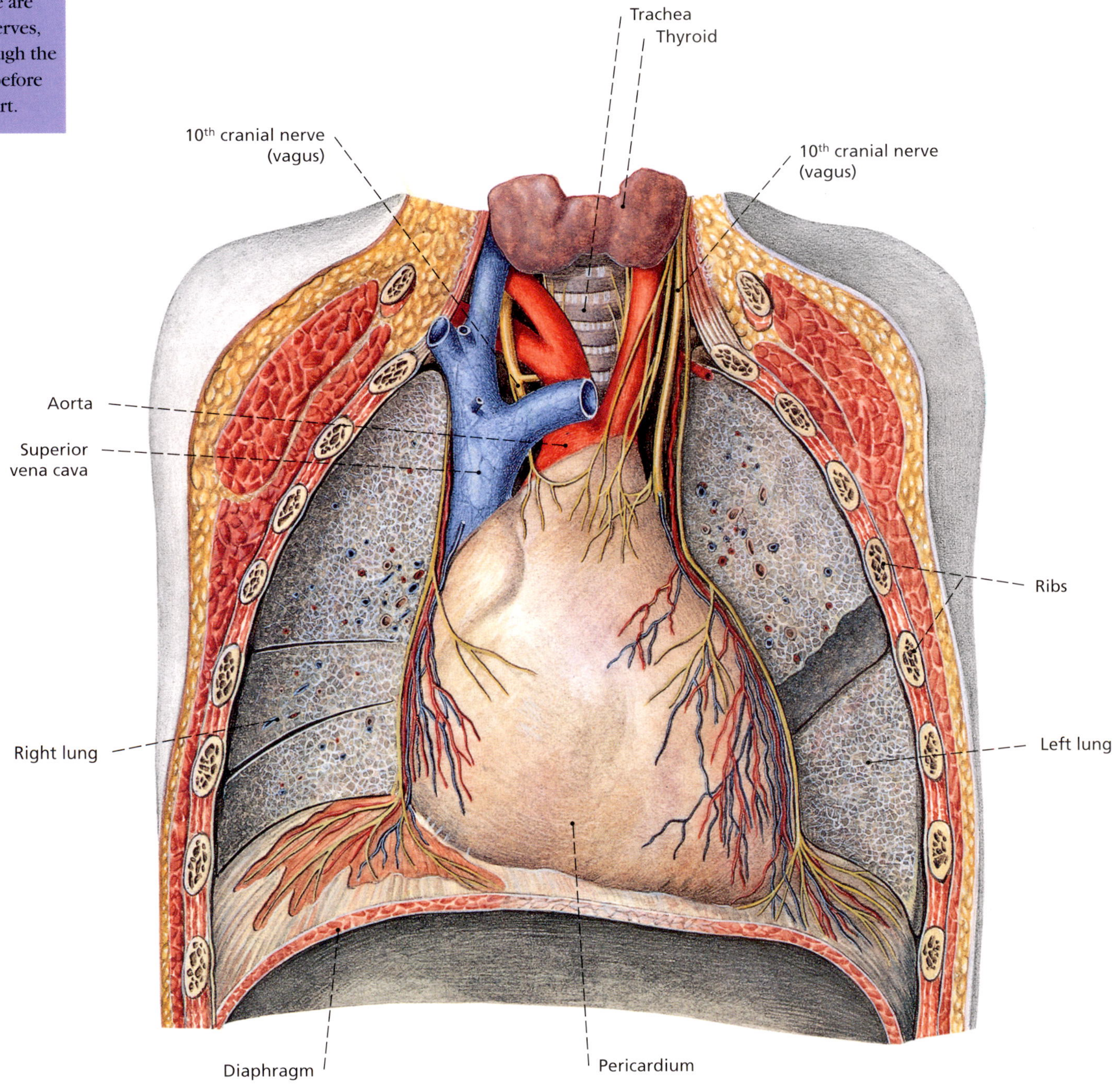

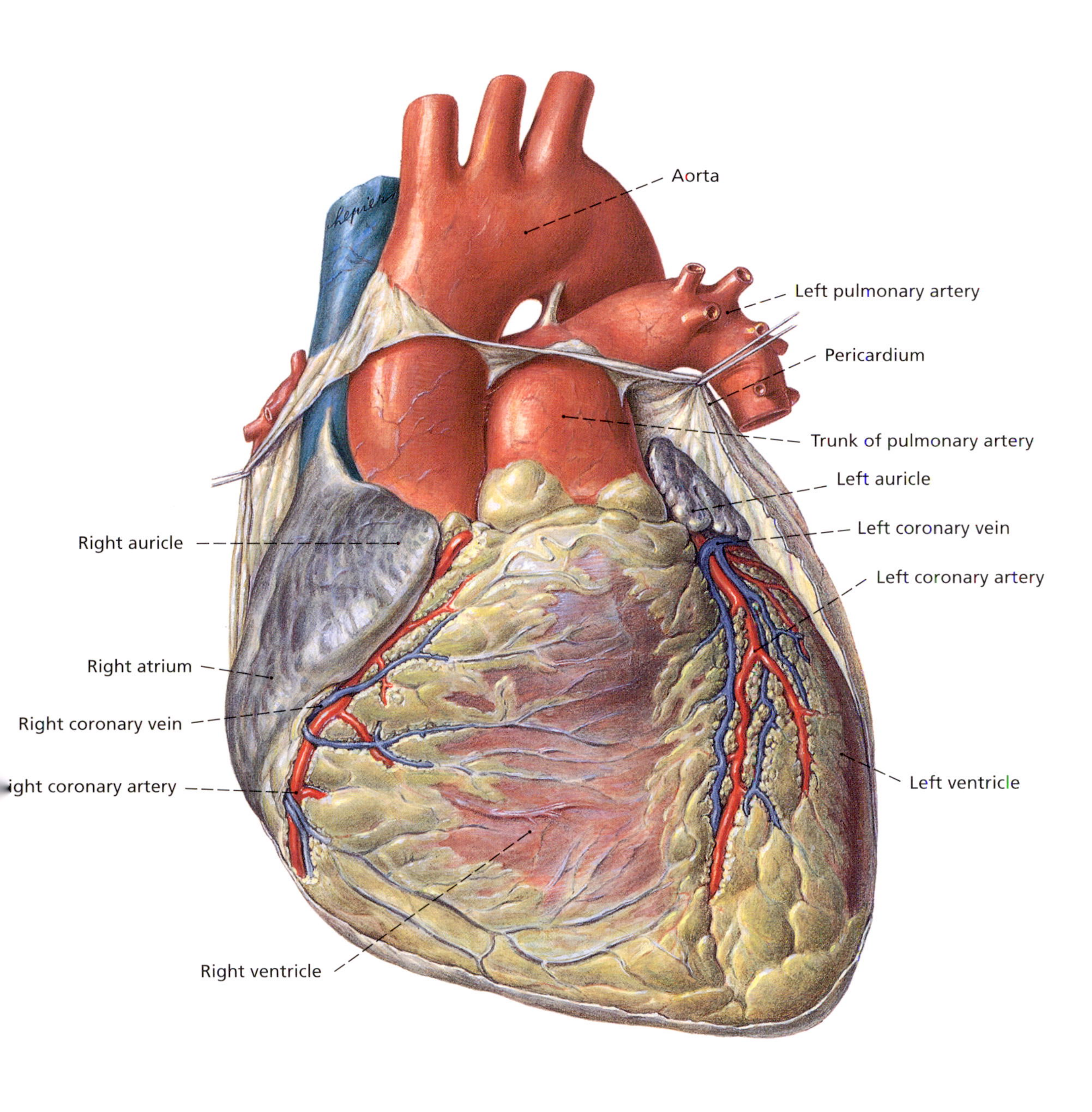

◆
Fig. 240
The heart after opening the pericardium. The right and left auricles are bulges in the right and left atria, respectively.

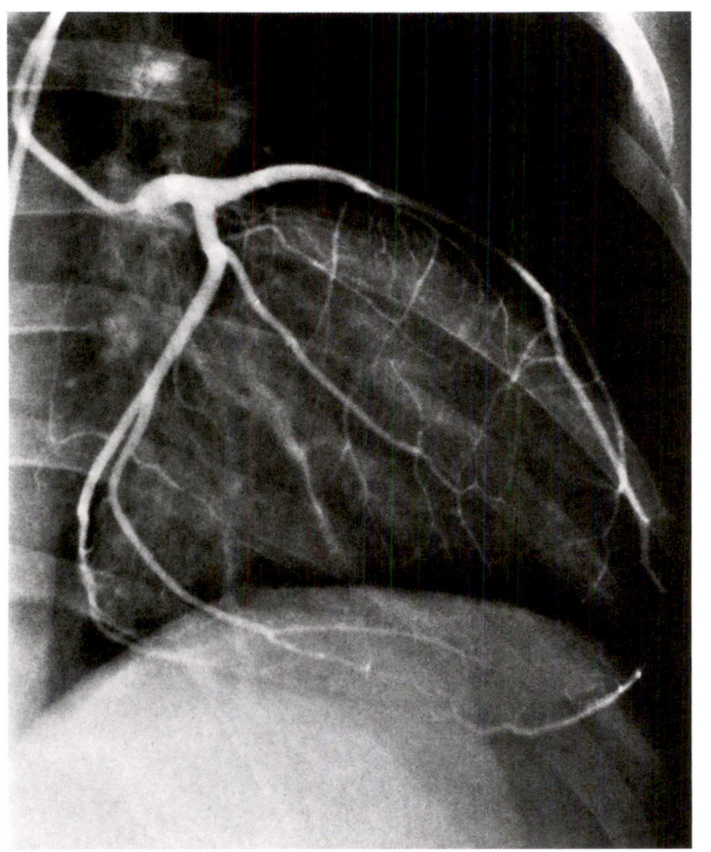

◆
Fig. 241
X-ray image of the coronary blood vessels, visualized by means of contrast medium.

◆
Fig. 242
Heart with coronary arteries (semi-diagrammatic representation), viewed from the front. In order to show the path of the left coronary artery, part of the trunk of the pulmonary artery (truncus pulmonalis) has been cut away. (An arrow indicates the transverse path of the reflection of the pericardium, between the posterior surface of the aorta and pulmonary artery trunk and the pulmonary veins.)

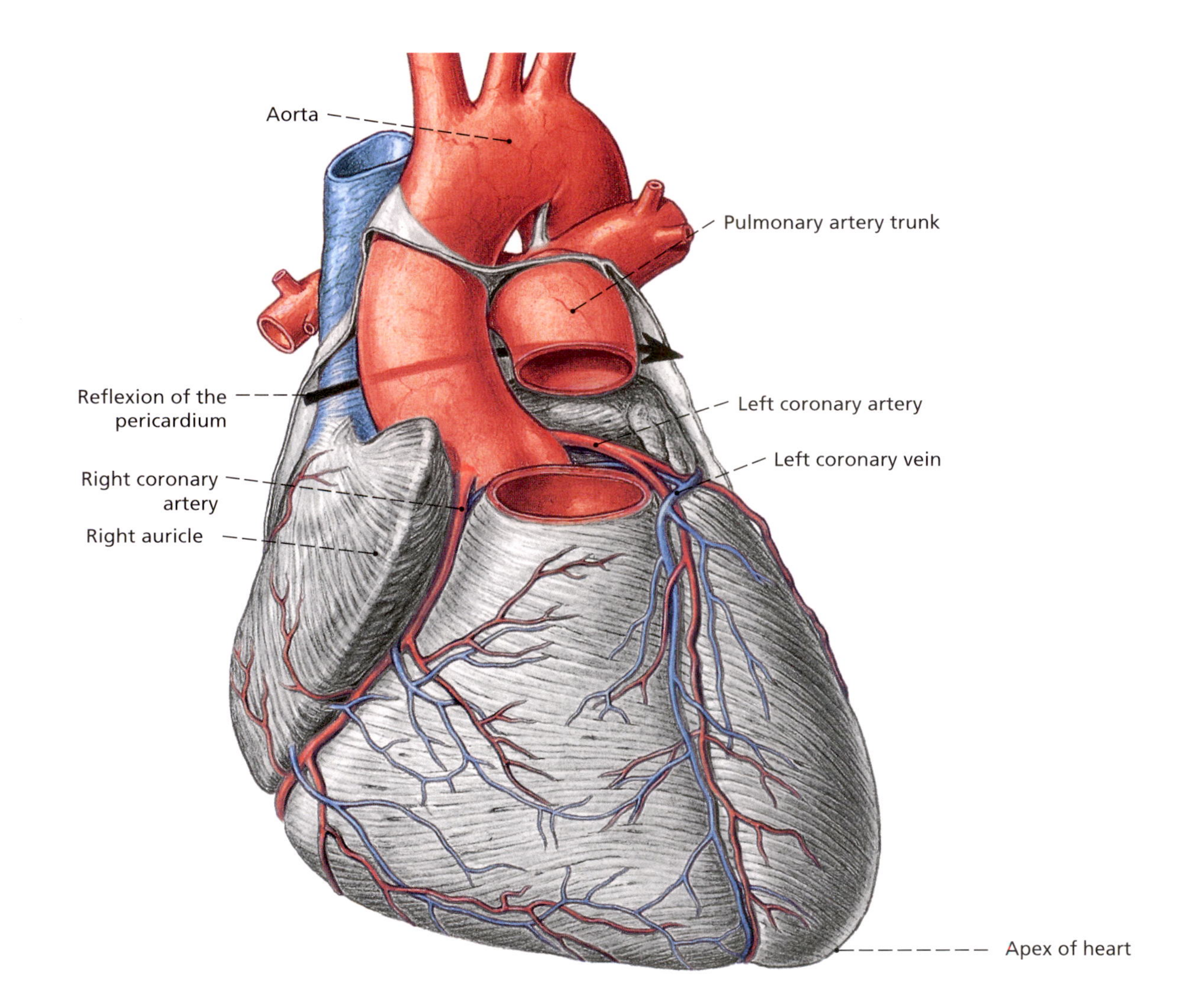

◆
Fig. 243
Coronary arteries, viewed from above. In order to show their path more clearly, the upper portion of the heart as far as the heart valves has been removed (cross-section through what is termed the valvular plane of the heart).

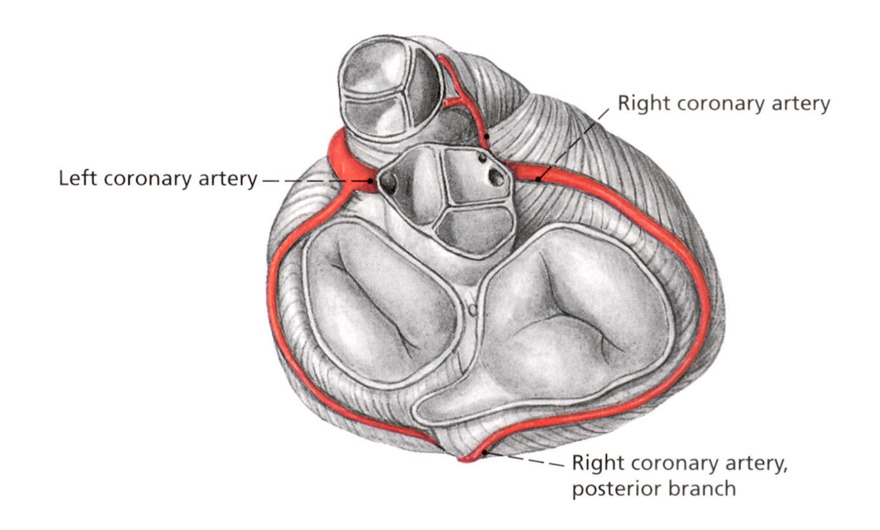

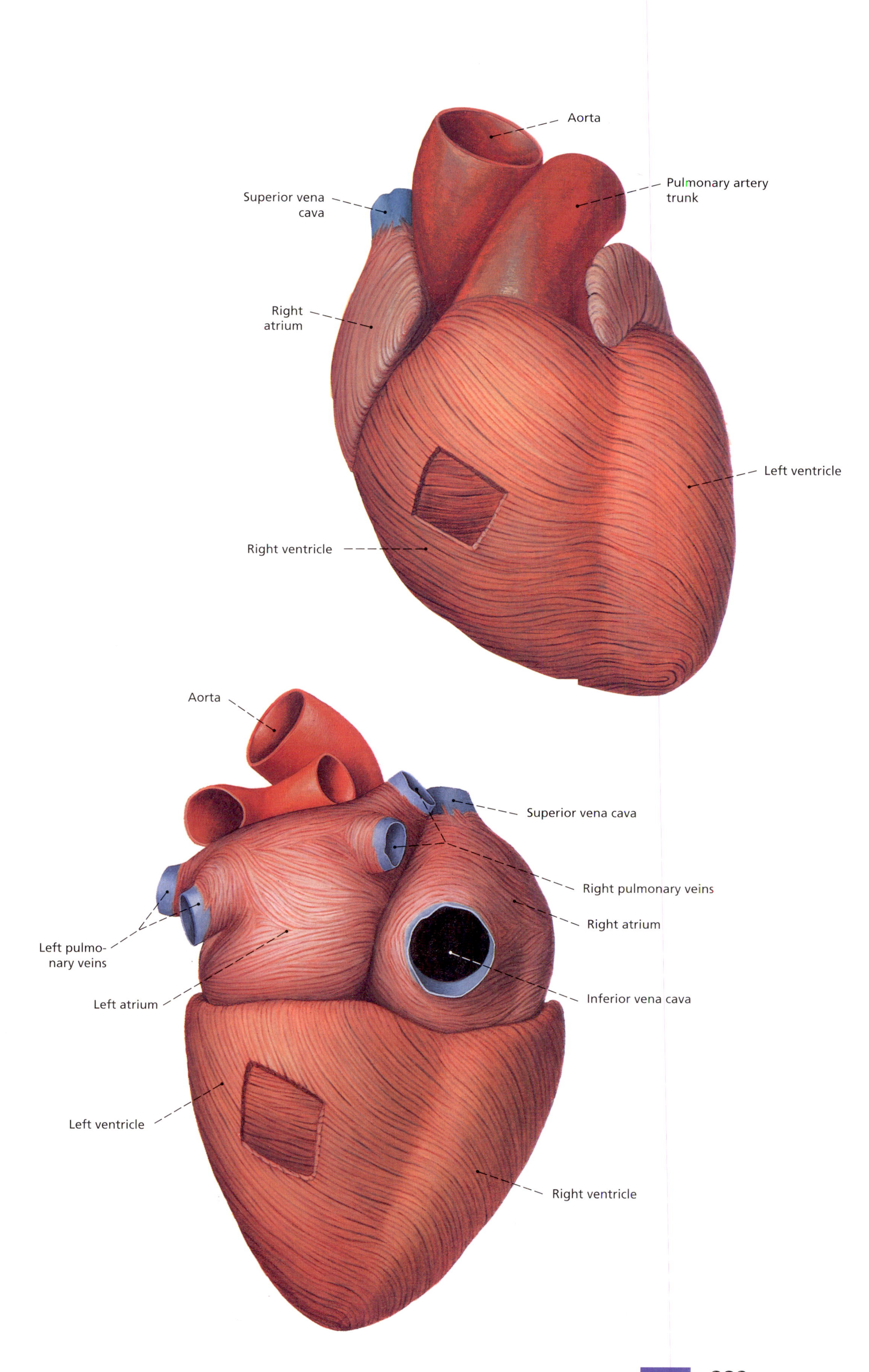

◆
Fig. 244
Orientation of the heart muscle fibers, viewed from the front. In order to show the different directions in which they run, a section of the superficial layer of muscle has been cut away above the right ventricle.

◆
Fig. 245
Orientation of the heart muscle fibers, viewed from the rear. In order to show the different directions in which they run, a section of the superficial layer of muscle has been cut away above the left ventricle.

◆
Fig. 246
The heart with its valves, in projection on the anterior wall of the thorax.

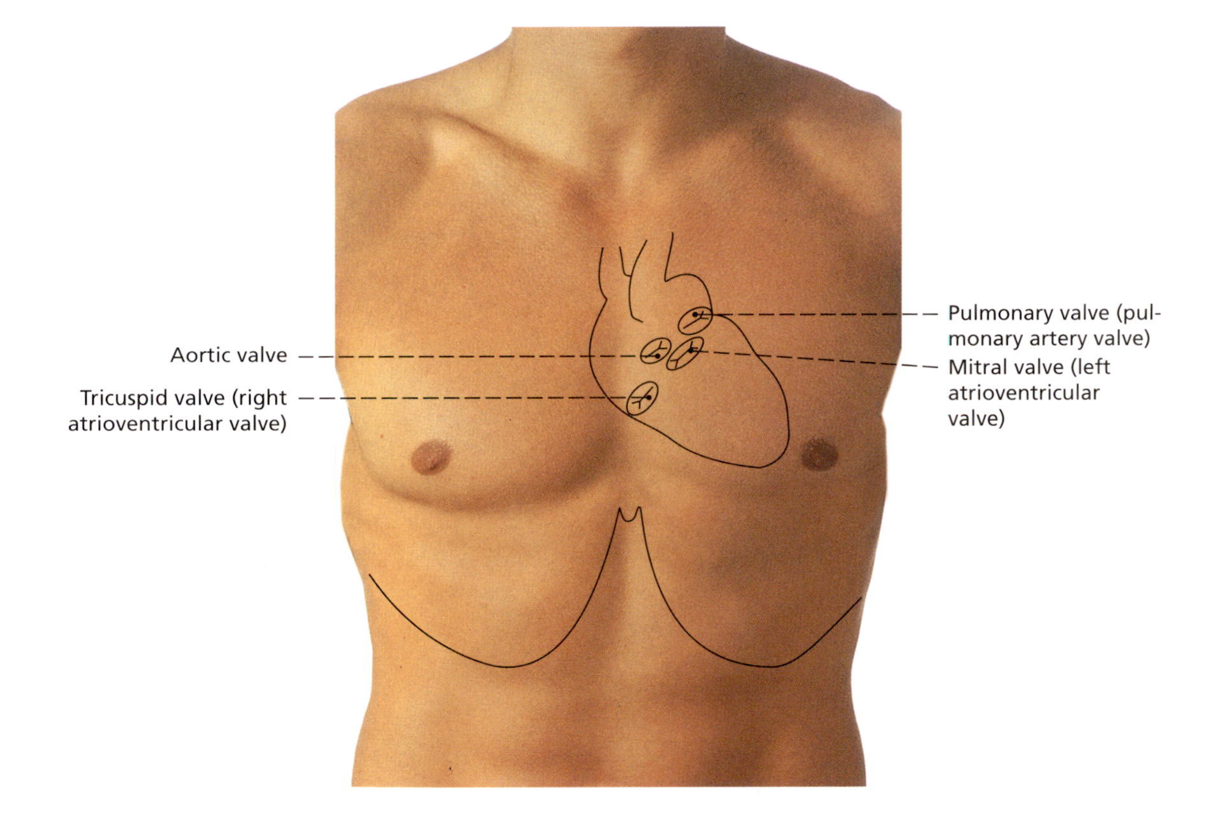

◆
Fig. 247
Ultrasound image of the heart (top), with the individual structures shown in diagrammatic form beneath: RV = right ventricle, LV = left ventricle, IVS = interventricular septum, RA = right atrium, LA = left atrium.

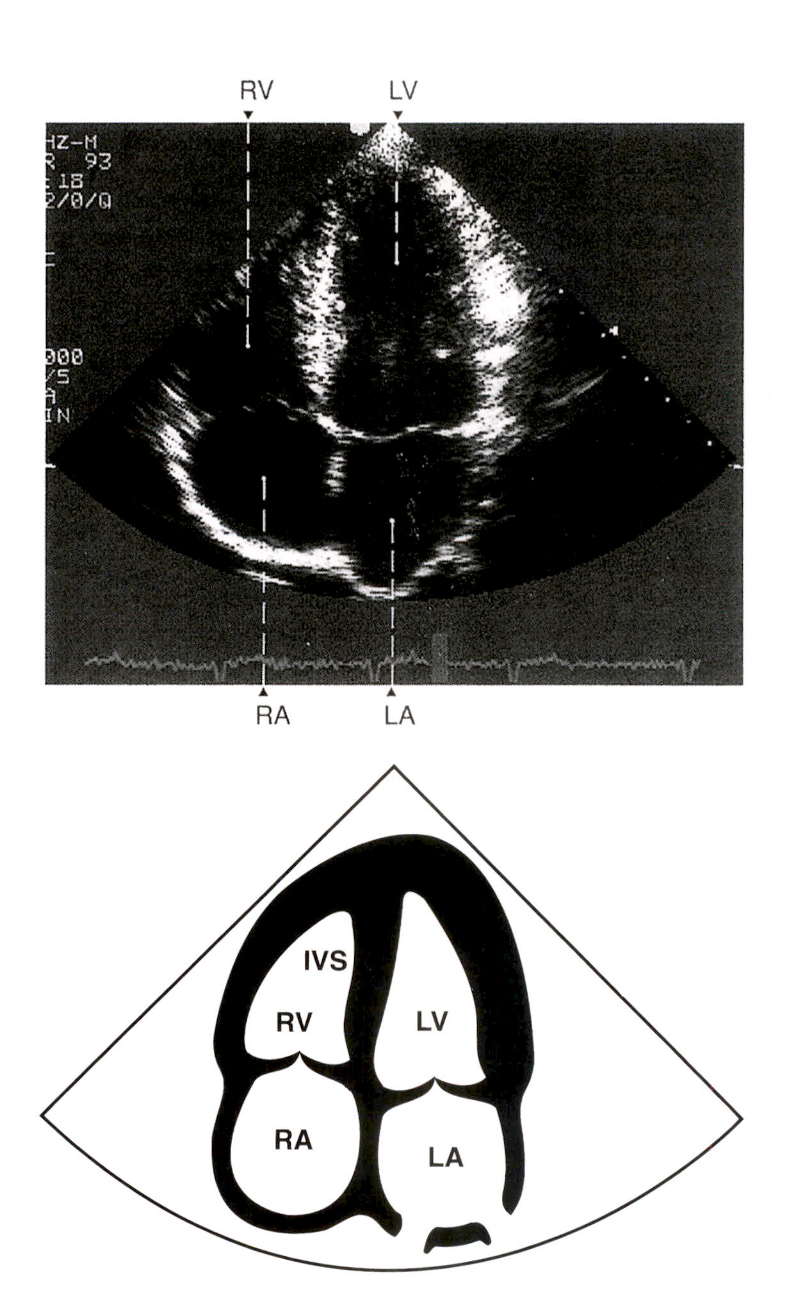

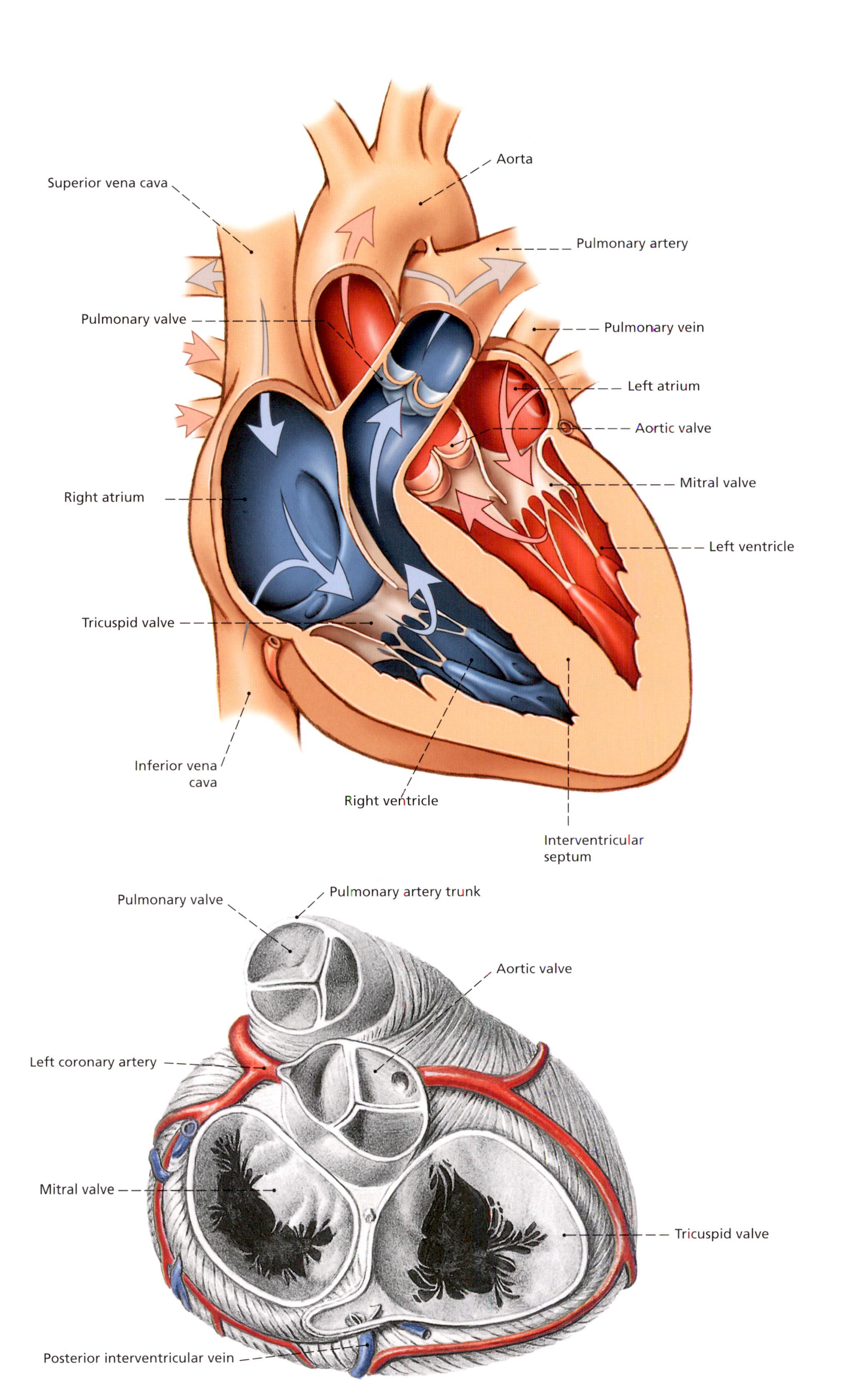

◆
Fig. 248
Heart showing heart valves. Semi-diagrammatic representation of blood flow, which is controlled by the heart valves (oxygenated blood = red, deoxygenated blood = blue).

◆
Fig. 249
Valvular plane of the heart, viewed from above. The atria have been cut away and the trunk of the pulmonary artery (truncus pulmonalis) divided. The heart muscle is relaxed and both atrioventricular valves are open.

◆
Fig. 250
Right ventricle and right atrium, viewed from the front (longitudinal section). The right tricuspid valve (atrio-ventricular valve) is open.

◆
Fig. 251
Papillary muscles and tendinous fibers of the right ventricle, viewed from the rear (window view, see sketch on right for orientation). Contraction of the papillary muscles causes the right ventricle to open; relaxation causes it to close.

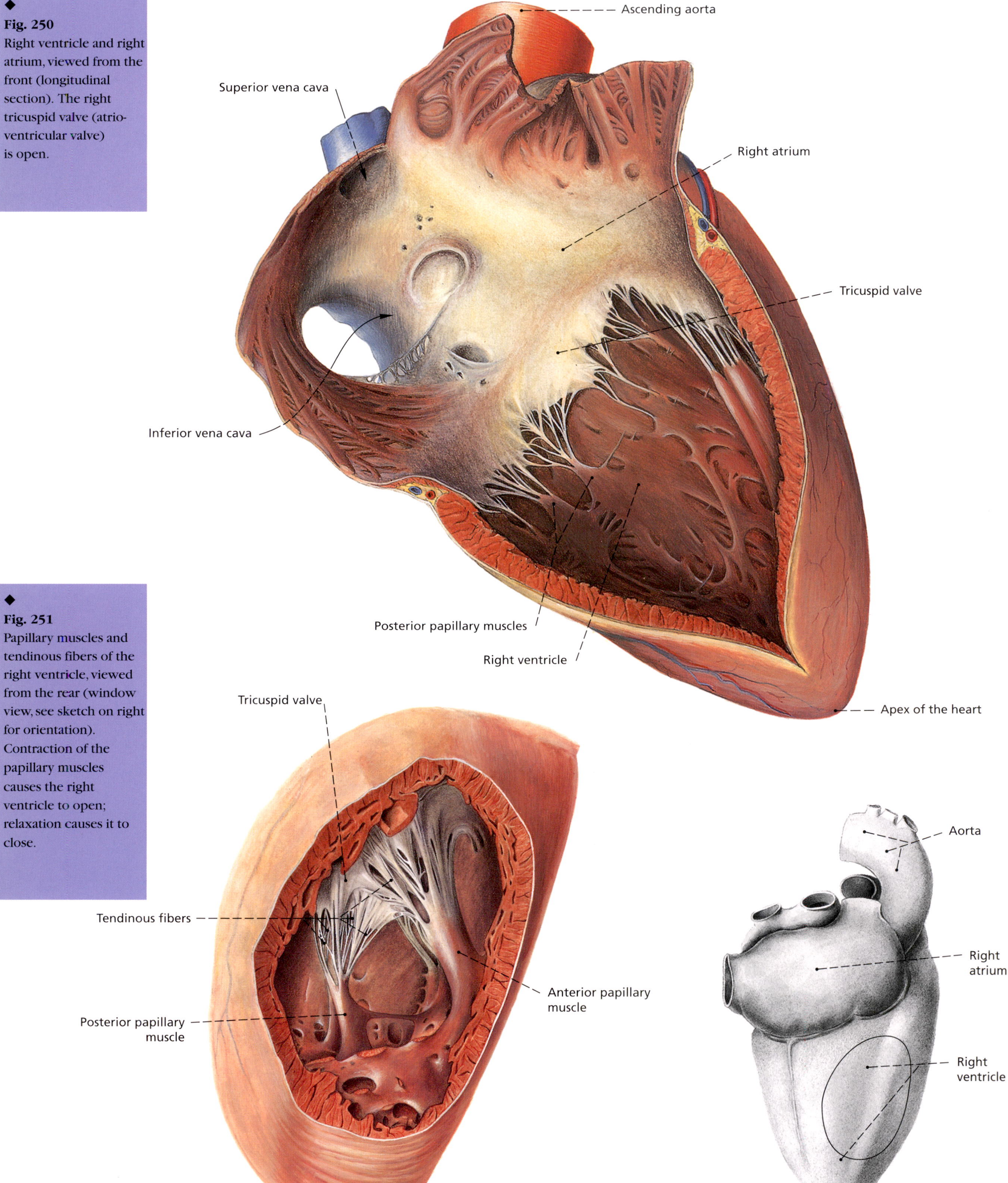

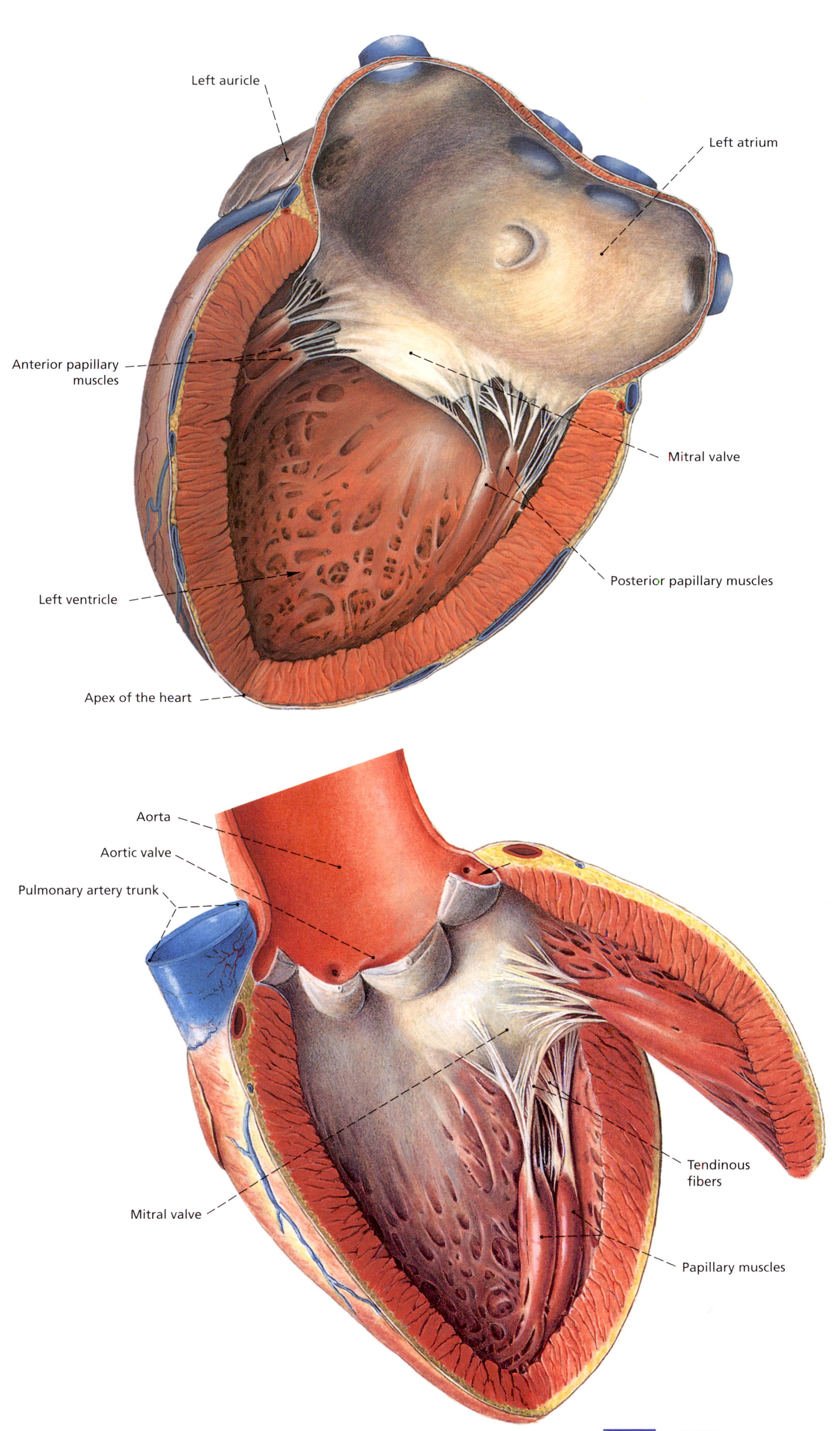

◆
Fig. 252
Left ventricle and left atrium, viewed from the front left (longitudinal section).The left mitral valve (atrioventricular valve) is open.

◆
Fig. 253
Left ventricle with aortic valve and left mitral valve (atrio-ventricular valve). Longitudinal section through the centre of the left ventricle, viewed from the front left and laterally.

◆
Fig. 254
Right and left ventricles (longitudinal section through the axis of the heart, viewed from front left, and laterally). The ventricles are divided by the interventricular septum. The muscular wall of the left ventricle is significantly thicker than that of the right ventricle (the left side of the heart faces greater resistance on contraction than the right side).

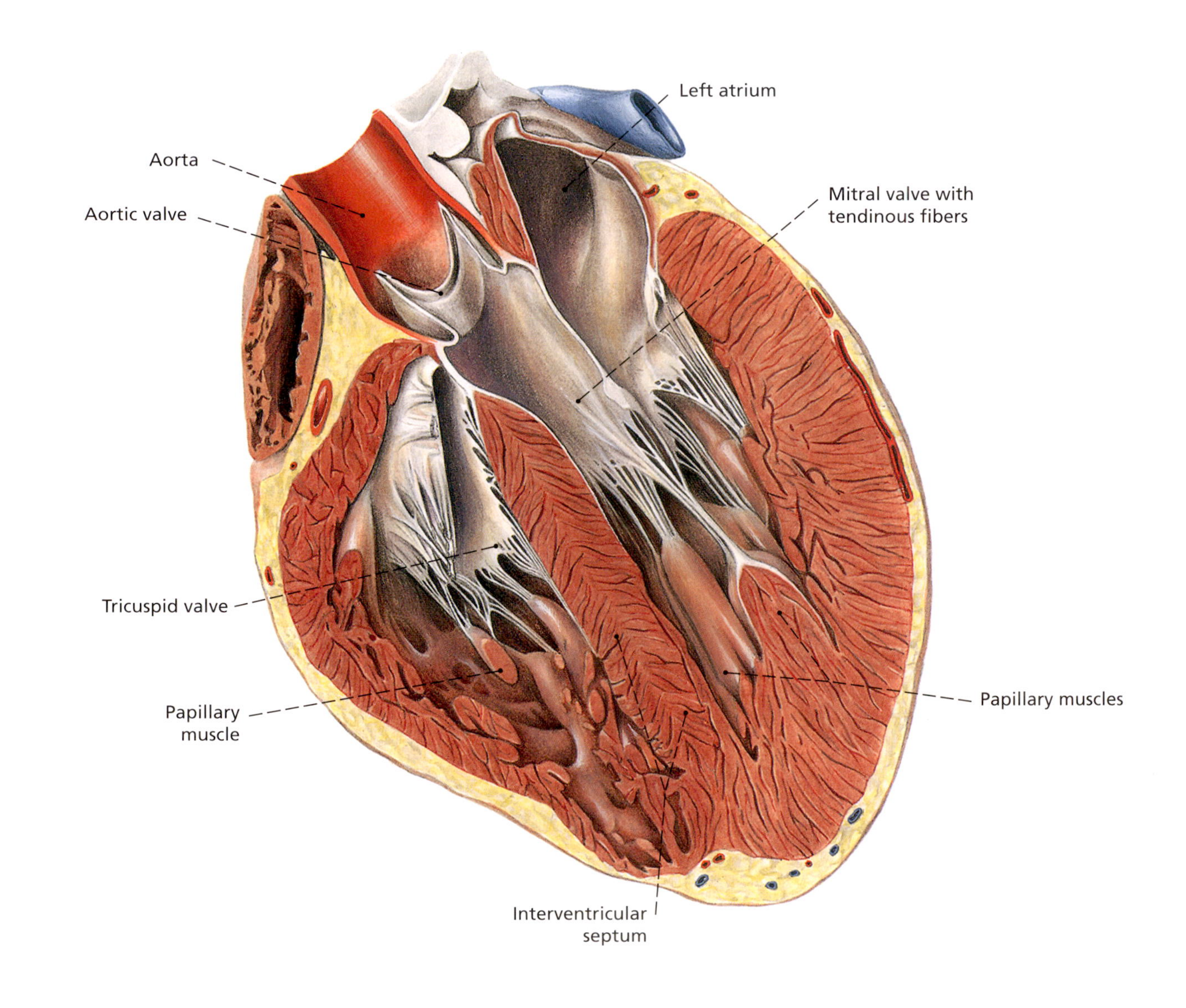

◆
Fig. 255
Right and left ventricles (cross-section at right angles to the axis of the heart, viewed from above).

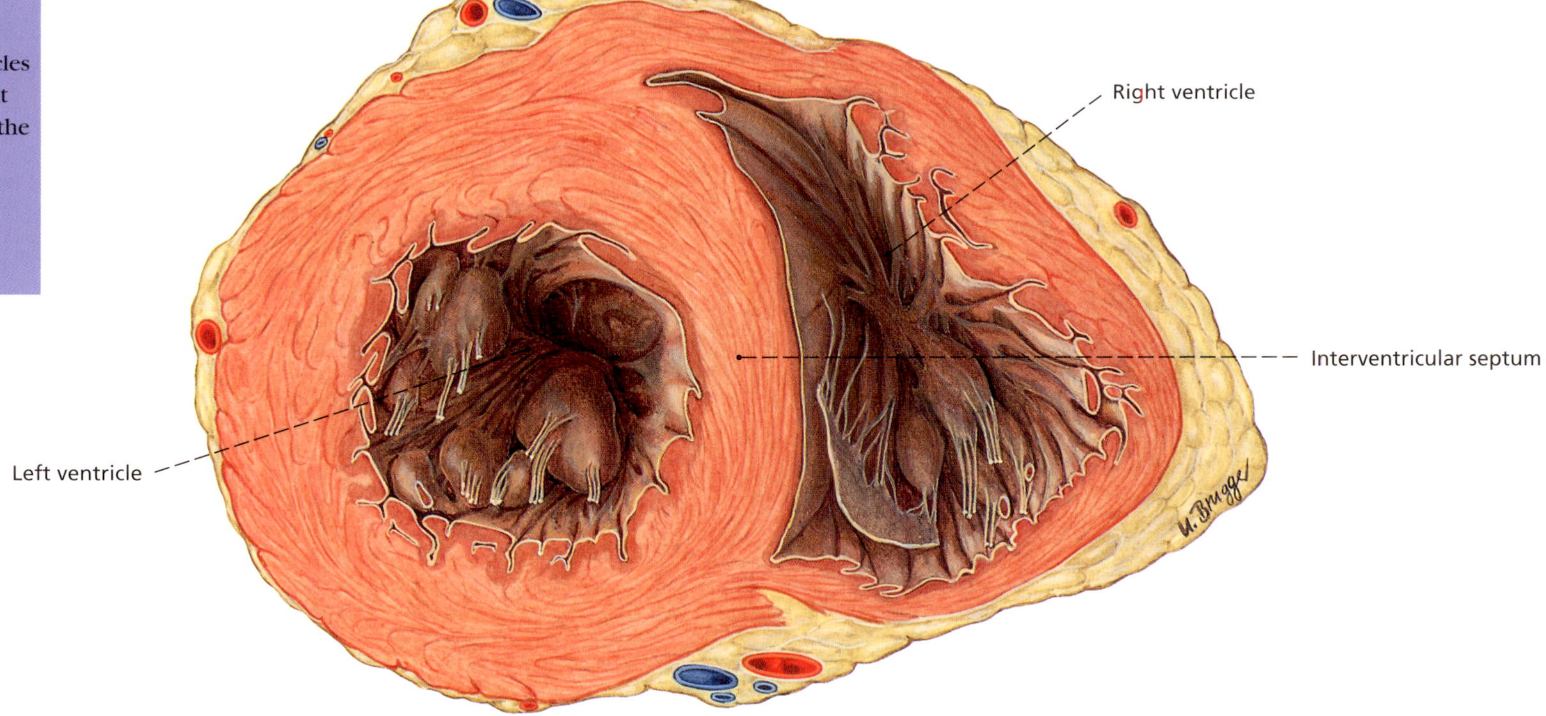

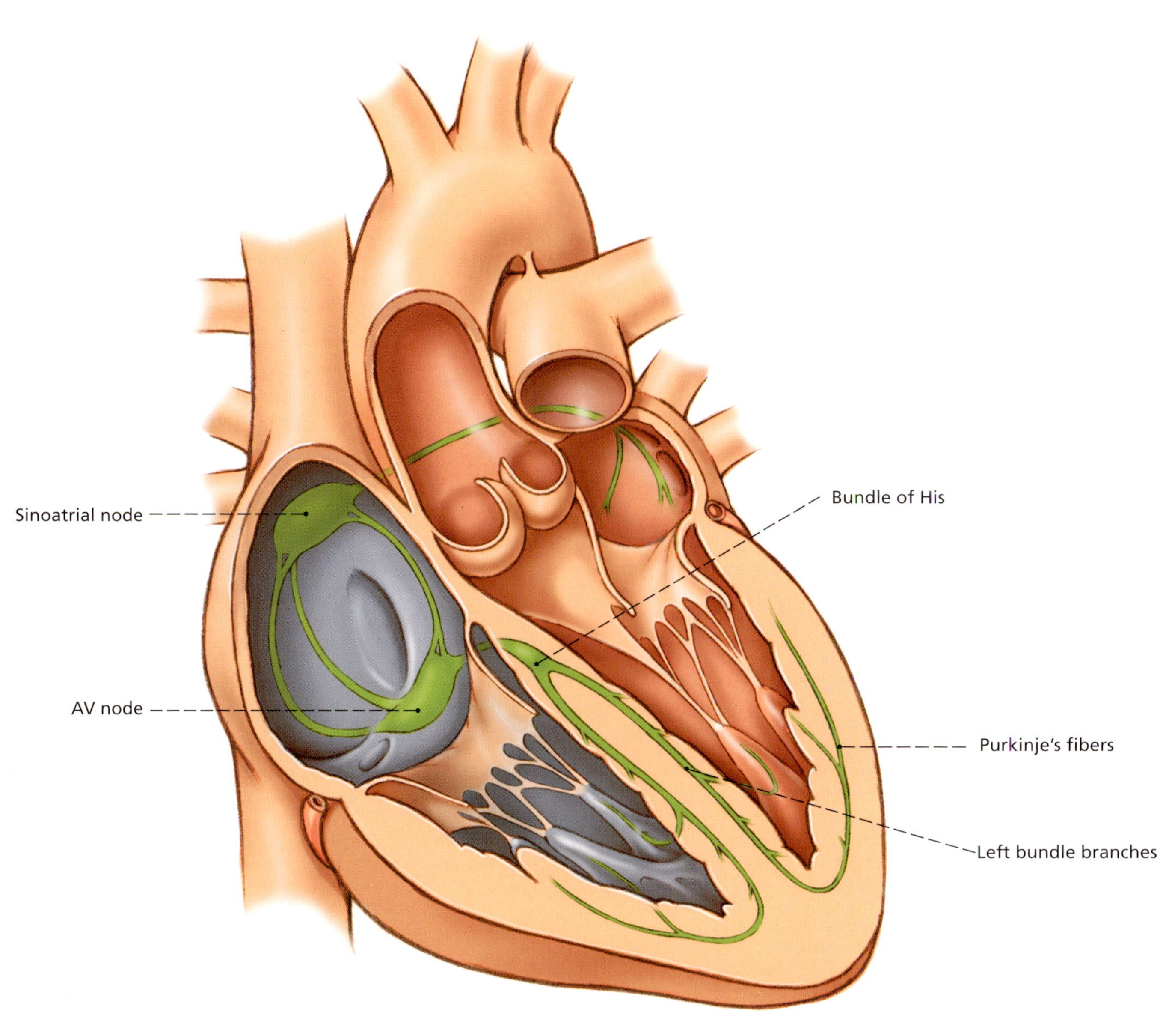

Fig. 256
Conduction system of the heart (semi-diagrammatic: the individual centers are shown in green).

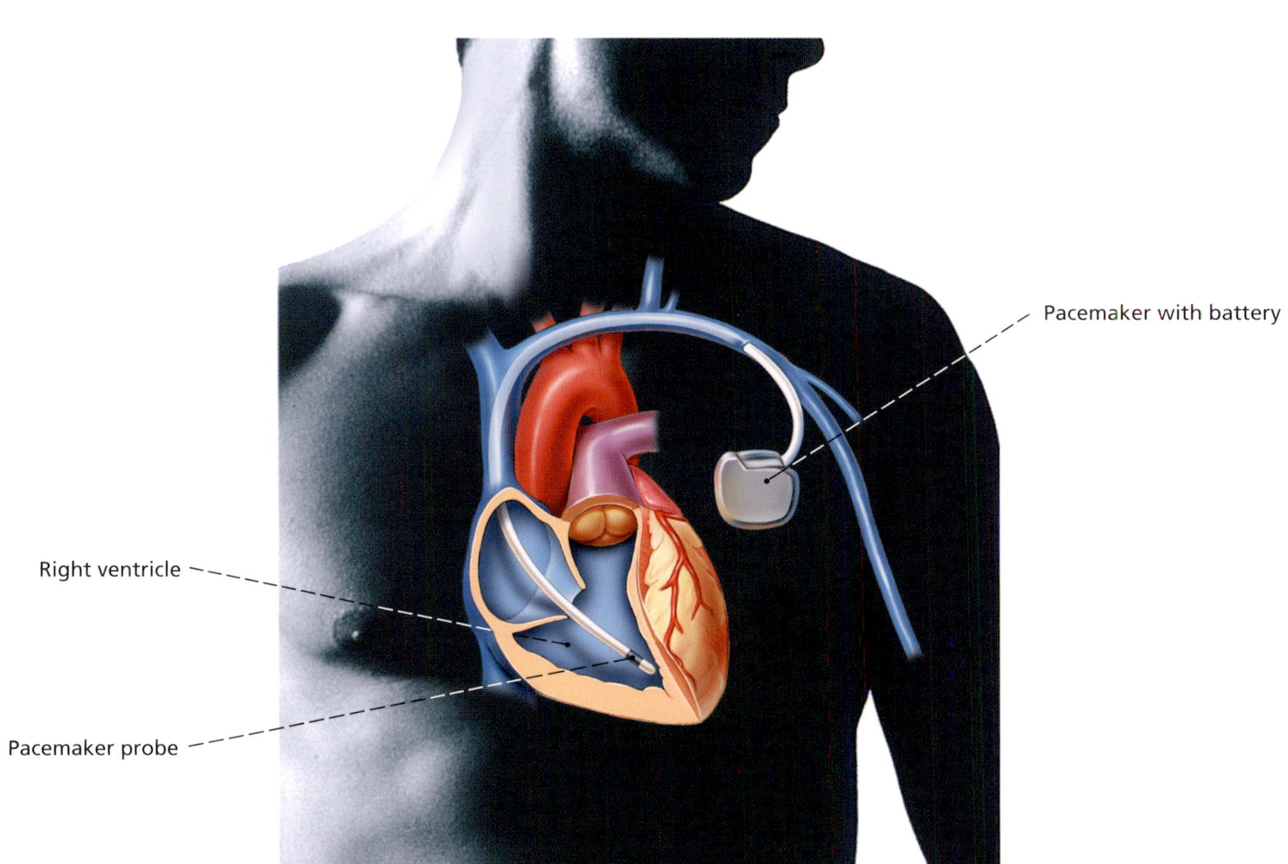

Fig. 257
A cardiac pacemaker in position. The pacemaker probe is advanced via the left subclavian vein into the right ventricle. Incorporated in the tip of the probe is an electrode, which sends electrical impulses to the myocardium, encouraging it to contract rhythmically. The pacemaker battery is sewn into a pocket of skin beneath the left clavicle.

Symptoms and diseases

Cardiac rhythm disturbances

An irregular, excessively fast (over 100 beats per minute), or excessively slow (below 60 beats per minute) heart rate. Cardiac rhythm disturbances are often the result of inadequate blood supply to the myocardium (heart muscle) due to narrowing of the coronary blood vessels. Any disturbance in the electrical conduction system of the heart can also cause cardiac rhythm disturbance. Other causes include intoxication or drugs (e.g., side effects). Excessive caffeine intake or nicotine use can also trigger heart rhythm disturbances, and accelerate the heart rate.
For extremely fit participants in endurance sports, a heart rate of 60 beats per minute is normal. Apart from a slower or more rapid pulse, cardiac rhythm disturbances frequently have no symptoms, although they can lead to pain, dizziness, episodes of fainting, and even life-threatening heart attacks.
Not all heart rhythm disturbances are serious or need treatment, but as a general rule they should be investigated by a physician. Depending on the cause and how pronounced they are, treatment may be administered in the form of anti-arrhythmic drugs or drugs to increase the blood supply to the heart.

• ***Palpitations***

Subjective perception of a stronger or more rapid heartbeat than normal. This is usually a product of circumstances, and harmless. Frequent palpitations with no apparent cause can, however, be a sign of organic disease (e.g., hormonal disturbance or heart disease) or psychiatric disturbance (cardiac neurosis).

• ***Bradycardia***

Slow heart rate, with a pulse below 60 beats per minute. This is a normal, non-pathological phenomenon in extremely fit athletes and during sleep. A slow heart rate can, however, also be a sign of heart disease or an under-active thyroid gland.

• ***Tachycardia (raised heart beat)***

A heart rate exceeding 100 beats per minute. A brief episode of tachycardia following physical exertion or accompanying anxiety or excitement is normal. The heart rate can also be increased in feverish states, after taking certain medicines or consuming large amounts of caffeine, or with heavy nicotine use. A temporarily or permanently increased heart rate without any obvious cause should always be investigated by a physician. The most frequent pathological causes are an overactive thyroid, coronary artery stenosis, or diseases of the myocardium.

• ***Extrasystole***

Irregular heartbeat. An otherwise regular pulse may be briefly interrupted by a premature or delayed heartbeat. This occurs frequently in completely healthy individuals (e.g., after drinking alcohol or coffee), but can also be indicative of heart disease.

• ***Irregular heart beat***

Heartbeats occurring outside the basic cardiac rhythm, which can be felt as an irregular pulse or demonstrated on an electrocardiogram (ECG). An irregular heartbeat is a sign of premature contraction of the heart, often involving only certain parts of the heart (e.g., atrium or ventricle). The cause is a disturbance in the heart's electrical conduction system. An irregular heartbeat can occur at regular intervals or sporadically. Though often harmless, it may be indicative of heart disease and needs to be taken seriously; it should always be investigated.

• ***Heart flutter***

Cardiac rhythm disturbance with distinctly increased (but still regular) contractions of an atrium (atrial flutter) or ventricle (ventricular flutter) at a rate of over 200–250 beats per minute. Flutter can turn into life-threatening ventricular fibrillation.

• ***Ventricular fibrillation (cardiac fibrillation)***

Irregular, overexcited activity of the ventricles of the heart, in which the fibers of the myocardium can no longer contract and relax in an orderly fashion. Because of the continuing spontaneous excitation of small sections of the heart, it becomes impossible for the heart to perform its pumping function. Excessive activity (over 300 beats per minute) may be followed by circulatory arrest with respiratory paralysis. This condition is fatal unless immediate resuscitation is performed.

• ***Cardiac arrest***

Arrest of the function of the heart. Symptoms of cardiac arrest include blue lips, skin pallor, absence of a pulse, and dilated, rigid pupils. Death will be the outcome unless resuscitation is performed immediately.

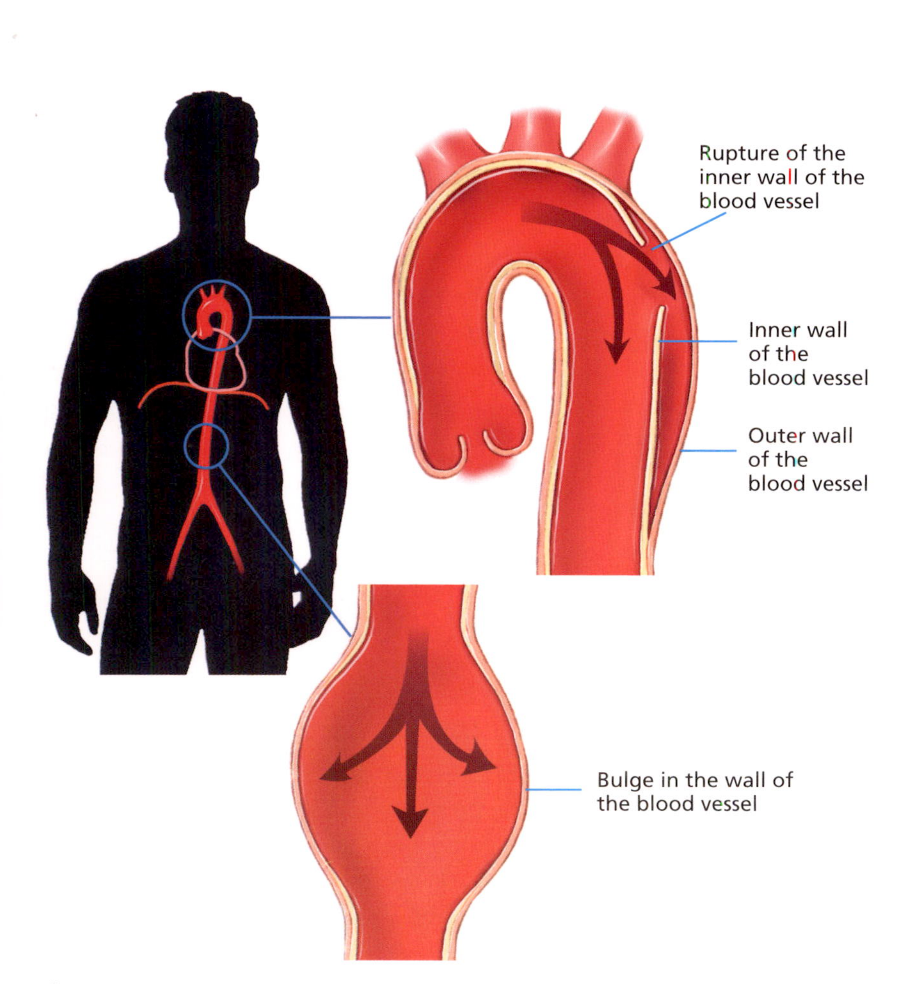

◆
Aneurysm
Aneurysms of the aortic arch and the abdominal aorta are frequent occurrences.

Aneurysm

A balloon-like bulge in the wall of an artery that does not disappear of its own accord. Aneurysms may be congenital, but usually occur where there is previous damage to the artery, e.g., after phlebitis or vascular surgery, arteriosclerosis, or (on the wall of the heart) after a myocardial infarction. They are diagnosed by means of a special ultrasound investigation and/or computed tomography. Aneurysms may burst, and for this reason they should be bypassed or removed surgically.

• ***Aortic aneurysm***

A sac or bulge in the aorta. This can be a congenital malformation, but can also arise as a result of inflammation of the vascular wall or calcification of the artery. If the aortic aneurysm occurs in the upper section of the aorta, it can impinge upon neighboring organs or adjacent nerve pathways and cause problems with swallowing, coughing, hoarseness, or unequally sized pupils. Aortic aneurysms must always be surgically repaired, as there is always a danger that they will burst and cause the death of the patient from internal bleeding.

• ***Cardiac aneurysm***

Bulging of the heart wall, usually in the region of the left ventricle. It is almost always the consequence of a severe myocardial infarction. As the affected part of the heart wall is weakened and no longer plays an active part in the work of the heart, the heart function is impaired. In addition, small blood clots (thrombi) may form in the bulge, entailing an increased risk of embolism.

Heart attack

An episode (attack) of cardiac rhythm disturbance with pain, breathlessness and an increase or decrease in blood pressure. The causes can include myocardial infarctions (see page 234), severe coronary artery stenosis, or even intoxication (e.g. excessive caffeine or nicotine consumption). The possibility of a myocardial infarction must always be considered in any heart attack, and an emergency doctor must always be called.

Angina pectoris

Pains and transient feelings of constriction in the chest, caused by restricted blood flow to the heart tissue. Angina should normally be regarded as a warning sign of a myocardial infarction or heart attack (see page 234). It is caused by narrowing of the coronary blood vessels, resulting in oxygen deficiency in the heart muscle, and may be triggered by excitement, physical exertion, or cold temperatures. Typical signs of angina, in addition to feelings of constriction, include breathlessness and chest pains. The chest pains may radiate out to the left shoulder and down the left arm; they may also radiate into the right shoulder. There may be upper abdominal discomfort and nausea. If these symptoms are present, medical advice must always be sought immediately in order to identify an incipient myocardial infarction. Angina can be confirmed by means of a stress ECG. As acute treatment, glyceryl trinitrate spray or tablets can be used to dilate the blood vessels and thus provide rapid relief from pain. Beta-blockers can be given to slow down the heart rate and decrease the force of contraction so that the heart requires less blood. A wide range of treatments may be prescribed for chronic use. Angina may be mistaken for a heart attack because the symptoms are similar.

Stenosis of the coronary blood vessels

• ***Aortic stenosis***

Narrowing of the aorta. Many heart defects are attributable to aortic stenosis (narrowing), which may occur at the point where the aorta leaves the left ventricle, or form due to scarring from previous inflammation of the aortic valve. If the aorta itself is

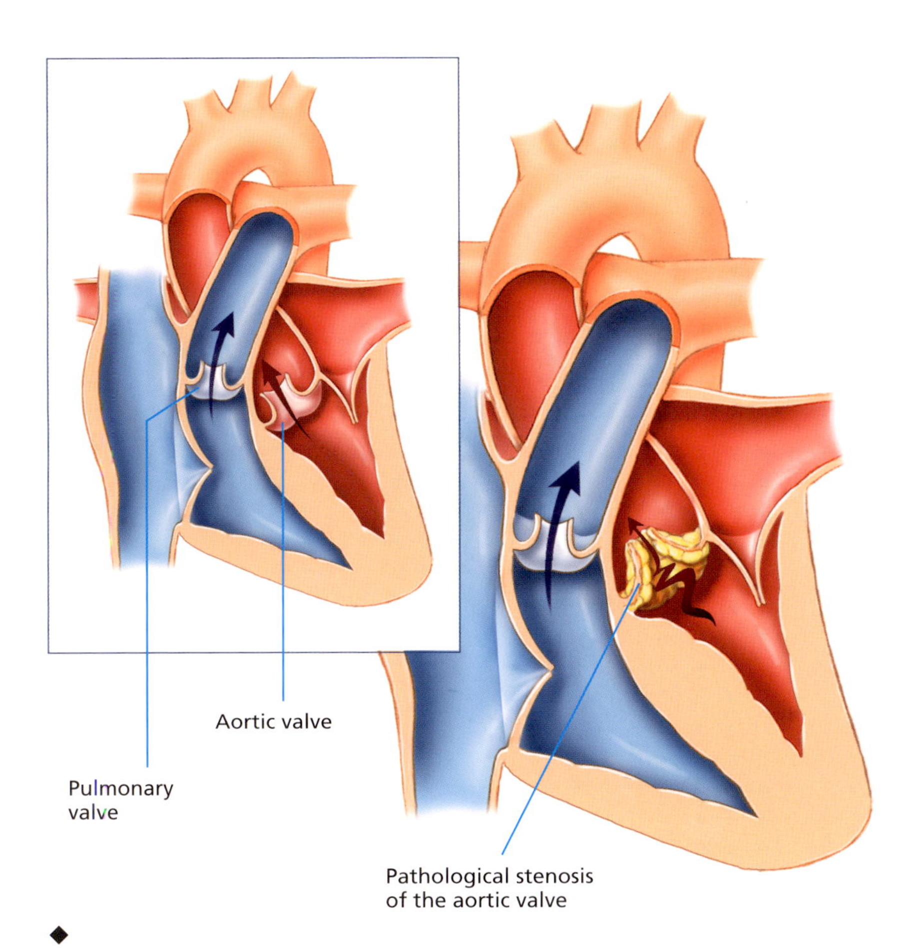

◆
Aortic stenosis
In a healthy heart (left), the blood flow from the left ventricle into the aorta is unobstructed. In the presence of stenosis, the heart valve can no longer open properly (right).

stenosed in the region of the aortic arch, this is termed stenosis of the aortic isthmus. This frequently occurs in conjunction with other heart defects and results in inadequate blood flow to the lower half of the body. Aortic stenoses are sometimes discovered only by chance when investigating high blood pressure. The ECG shows excessive strain on the left ventricle, which also appears enlarged on X-ray. Other signs are reduced blood supply to the brain with dizziness or fainting; the patient may tire rapidly, and suffer palpitations, breathlessness, and cardiac rhythm disturbances. Aortic stenoses are either treated surgically by dilating the blood vessel, or the aortic valve may be replaced with an artificial one.

• ***Triple vessel coronary heart disease***

Atherosclerosis or calcification of the arteries is often widespread. The more vessels are affected, the more serious the disease. Thus, if the three main branches of the coronary artery are affected, the patient is said to have triple vessel coronary artery disease. Without treatment, this disease is often fatal. Surgery is always required, in which the stenosed vessels are bypassed or replaced. If this is no longer possible, the only remaining option is a heart transplant.

• ***Coronary artery stenosis***

Narrowing of the coronary arteries. Symptoms, such as stress-related pain behind the sternum (angina pectoris), usually only appear when the diameter of a coronary vessel at any point is narrowed by at least 70%. If untreated, stenosis of the coronary blood vessels can result in a myocardial infarction. Surgical bypass of the stenosed vessel using a section of vein offers good prospects of recovery.

• ***Calcification of the coronary blood vessels***

Thickening, hardening, and loss of elasticity of the coronary vessel walls. This results in narrowing of the coronary blood vessels, obstructing the blood supply to the heart. High levels of fat (lipids) in the blood, high blood pressure, smoking, and lack of exercise all encourage calcification of the coronary blood vessels. If a blood clot forms at an affected site, completely occluding a vessel that is already narrowed, this results in a myocardial infarction.

Inflammatory conditions

• ***Inflammation of the inner wall of the heart (endocarditis)***

The chambers of the heart and the heart valves are lined on the inside with a smooth layer of connective tissue, the endocardium.

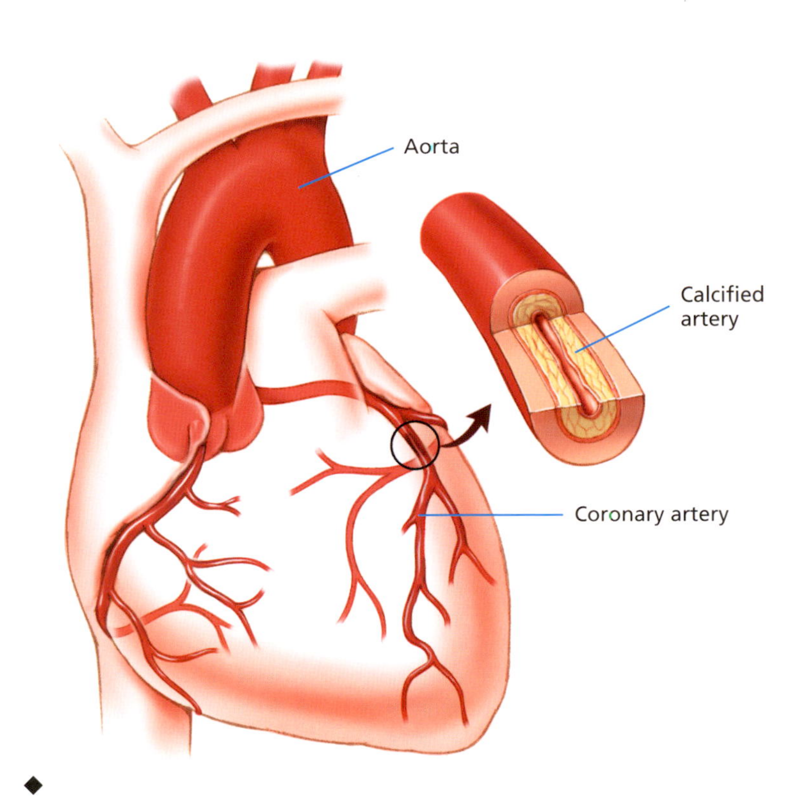

◆
Stenosis of the coronary blood vessels
If the blood flow to the coronary vessels is disturbed, the tissue downstream of the blockage ceases to receive an adequate supply of oxygen.

Endocarditis is usually the result of an infection, carried by microbes in the bloodstream, and most commonly affects the heart valves. Endocarditis frequently also occurs in combination with rheumatic diseases (inflammation of the joints), and a few weeks after an infection (e.g., rheumatic fever). Endocarditis itself usually only produces nonspecific symptoms, such as exhaustion, racing heart, feelings of constriction, and fever. Some forms progress slowly and creep up over a period of several weeks, while others are more aggressive. It is not unusual for scars to form on the endocardium when it heals, which can permanently affect the function of the heart valves. In the acute stage, treatment is with antibiotics and corticosteroids; surgery may be required if there is damage to the heart valves.

• ***Pericarditis***

Acute or chronic pericarditis (inflammation of the pericardium), often accompanied by a pericardial effusion containing blood or pus, which can be seen on X-ray. The principal symptoms are breathlessness, heart pain associated with breathing, and fever. These may be caused by the spread of viral, bacterial, or fungal infections, or arise as a complication of a myocardial infarction, inflammation of the myocardium, severe rheumatic disease, lung diseases, or tumors growing into the pericardium from neighboring organs. Often no plausible explanation is found, in which case it may be suspected that the cause is an unidentified viral infection. While inflammation caused by infections can be treated with appropriate drugs, in other cases the underlying disease must be treated.

• ***Inflammation of the heart muscle (myocarditis)***

Acute or chronic inflammation of the muscle of the heart (myocardium). This usually arises as a result of infection with bacteria or viruses, either affecting the heart muscle directly or via toxins carried in the blood. Occasionally, myocarditis may be due to a defective response of the immune system to the body's own heart muscle tissue. Possible symptoms are heart rhythm disturbances, signs indicating a weak heart, and fever. The inflammation is treatable with antibiotics.

Heart defects

An umbrella term for all congenital malformations and functional disturbances of the heart caused by disease. Heart defects are not usually genetic, but develop in the embryo as a result of unfavorable conditions during pregnancy (e.g., due to infection with rubella or excessive alcohol or nicotine consumption, or nicotine abuse). They are sometimes not identified until adulthood. A moderate to severe congenital heart defect is present in about six out of every 1,000 newborn babies.

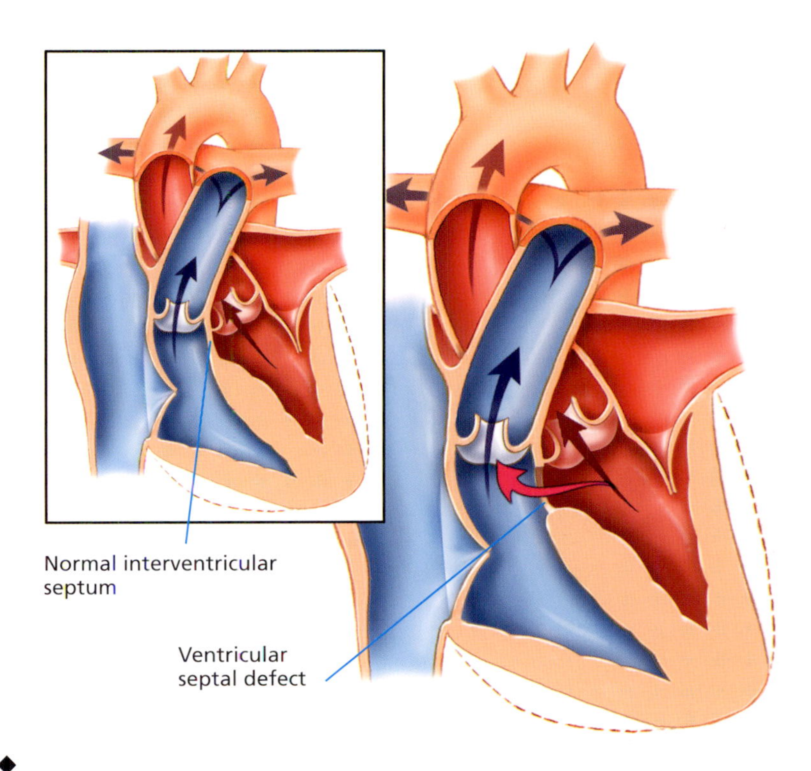

◆
Heart defects
A hole in the interventricular septum (ventricular septal defect) means that blood from the systemic circulation mixes with blood from the pulmonary circulation.

One of the most common congenital defects is a hole in the interventricular septum separating the left and right ventricles (ventricular septal defect). As a result, oxygenated blood flows back into the right ventricle, and the body is inadequately supplied with oxygen.
Following a bacterial infection or a rheumatic disease affecting the joints, the heart valves can seize up, making it difficult to pump the blood out of the ventricles.
Many heart defects are associated with functional disturbances of the heart, which may be compensated for by the heart working harder. If this continues for too long, however, there may be a risk of consequential damage, including overstretching of the heart muscle, leading ultimately to a failure to supply the body with enough oxygen. Common symptoms are breathlessness, tiring quickly on physical exertion, and a bluish tinge to the skin, especially around the lips (the sign of oxygen deficiency). A low oxygen supply may occasionally also cause growth disturbances, characteristic deformities of the fingers and toes, and an increased risk of infections such as pulmonary infections. Most heart defects can now be surgically repaired; prompt intervention eliminates the risk of subsequent damage.

Myocardial infarction

If sudden occlusion of an artery results in inadequate blood supply to a region of the heart, the muscle tissue in that region dies. Infarctions like these can be very dangerous, but the right treatment, administered swiftly and calmly, can ensure a good outcome, even in the case of life-threatening infarctions.

With every heartbeat, powerful contractions pump the blood into the arteries that transport it all over the body. Some of the blood from the heart is also pumped into the coronary vessels, consisting of two major arteries (one on the right, one on the left) which branch off into other vessels spreading all over the heart. These blood vessels keep the heart muscle, which is constantly in action, adequately supplied with oxygenated blood and important nutrients. In addition to supplying the whole body with blood, therefore, the heart also supplies itself.

Causes

A range of factors may be responsible for the occlusion of a coronary vessel, and in most cases a combination of several factors is involved. With advancing age or due to a genetic predisposition, for example, the connective tissue in the blood vessel walls can become thicker and hardened (arteriosclerosis). The internal lining of the vessel may tear at a particular point, resulting in the formation of lesions, which may also be bacterial in origin. Most commonly, however, fatty substances and calcium crystals from the blood are deposited on the artery walls, making the blood vessel rigid, brittle, and increasingly narrow (coronary artery stenosis). The blood no longer flows easily along a smooth vascular wall; instead, blood components adhere to the obstructions. A small foreign body or a thrombus is then enough to cause a complete blockage of the clogged blood vessel.

Risk factors

As well as aging and genetic predisposition, other factors encourage the arteriosclerotic changes in the vascular walls that may trigger an infarction. These include:

- high blood pressure,
- elevated levels of cholesterol and triglycerides in the blood,
- severe obesity,
- heavy smoking,
- lack of exercise.

A hectic lifestyle may also play a role, as the body produces larger amounts of substances when under stress, which narrow the blood vessels, make the heart beat faster, and hence increase its oxygen requirement, a fatal combination when the coronary vessels are already narrowed. This is known as coronary heart disease. Developing slowly, this disease is the most common precursor of an infarction.

What happens during an infarction?

Ultimately, many factors can lead to the total occlusion of a coronary vessel. An acute shortage of oxygen and nutrients arises in the area of the myocardium, which would normally be supplied with blood by the now occluded vessel. The heart muscle cells are starved of oxygen and die. The consequences of this in terms of cardiac function depend on the region of the heart that is affected and how big it is. If parts of the conduction system are affected, interference with the conduction of electrical impulses can lead to heart rhythm disturbances. If the damaged area is large, the function can be impaired to such an extent that the heart can no longer pump sufficient blood around the body. In the most severe cases, this can lead to sudden death due to heart

failure. Fortunately, though, most infarctions are not on this scale. If promptly recognized and swiftly treated, it is possible to survive even a severe infarction.
Over time, the muscle tissue that has died forms scar tissue, which can no longer perform the function of healthy heart muscle cells. In rare cases the area that has died is so small that the patient does not notice it, in which case it is known as a 'silent infarction'. This important warning signal is often only discovered later on an ECG carried out during the course of routine investigations.

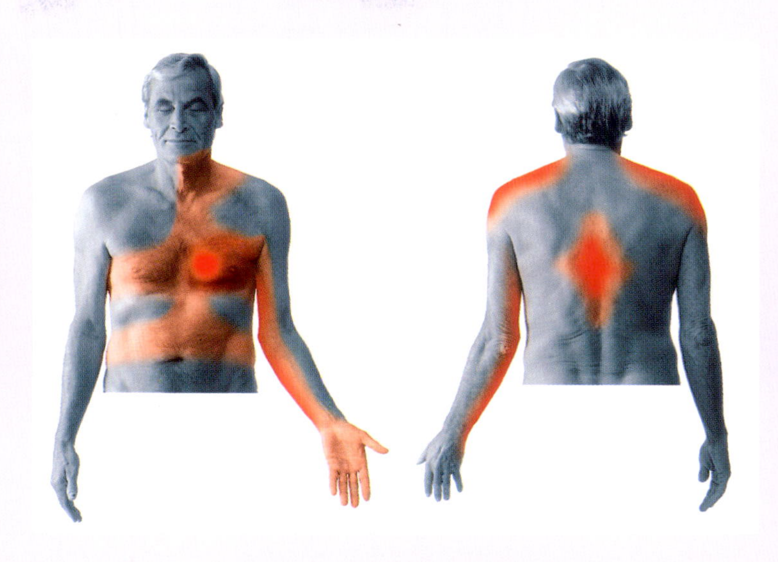

◆
Zones of pain in a myocardial infarction
The pains radiate throughout the thorax and abdomen and even into the lower jaw and the whole of the left arm.

Symptoms

The classical symptoms of a heart attack, or myocardial infarction, are crushing central chest pain (very severe angina), which may radiate into the left arm, right arm, the shoulders, and the lower jaw. Patients experience shortness of breath, dizziness, nausea, and restlessness and are clammy and sweaty, with a gray complexion. Patients often fear they may be dying, and avoid moving or taking deep breaths. An emergency physician must be called immediately.
Initial symptoms may not be at all typical, however, and there is a danger that valuable time is lost before treatment is started. At first, many patients have only back or upper abdominal pain, while nausea and vomiting are often initially misdiagnosed as a gastrointestinal complaint. Pallor and cold sweats as a result of the circulatory disturbance usually remove all doubt.

Treatment

While waiting for a physician or paramedic to arrive, the patient should lie down with the upper body raised in a half-sitting position. All physical activity must be avoided. Talking to the patient quietly can reduce anxiety. The physician will first inject strong pain-killing and sedative drugs to reduce the amount of oxygen used by the heart, which is beating too rapidly due to anxiety and pain. Oxygen is administered during transfer to hospital and drugs are given to strengthen and relieve the burden on the heart.
At the hospital, the severity of the infarction can be determined accurately, and targeted treatment is continued under supervision. Blood-thinning drugs prevent clots forming in the blood vessels and so limit further occlusion of the coronary vessels. Bed rest is essential at first. As soon as circumstances permit, however, a program of light physical exercise is started (usually after only a few days).

If the acute infarction has been successfully treated, the stenosis in the coronary vessel can in some cases be widened without surgery, using balloon dilatation. In other cases, bypass surgery may be necessary. The obstructed vessel is bypassed by a diversion created by grafting a short section of blood vessel (usually taken from the lower leg). This restores a good blood supply to the area of tissue in the heart that was previously inadequately perfused.

Prevention, and life after an infarction

It is clear from a review of the causes of myocardial infarctions how these can be prevented. Effective, long-term treatment must be given for underlying diseases causing arteriosclerosis, such as high blood pressure, disturbances in the metabolism of fats, and diabetes. Action can be taken by the individual to remedy obesity, smoking, stress, and lack of exercise, by means of lifestyle changes.
In the recovery phase after a myocardial infarction, the patient should be taught how to improve cardiac and cardiovascular fitness, and given information about healthy living, the necessity of a diet low in fat and cholesterol, and the importance of taking preventive medicines such as anticoagulants (e.g., aspirin) and statins.

• ***Mitral valve defects***

Functional disturbance of the heart valves in the left heart, e.g., defective opening and closing of the valve between the ventricle and the atrium.
In **mitral valve stenosis**, the valve is narrowed and the blood flow from the left atrium to the left ventricle is obstructed. The left atrium and the blood vessels from the lung become congested, with consequent enlargement of the left atrium, and cardiac rhythm disturbances. There is also an increased risk of emboli arising from the damaged valve.
In **mitral valve insufficiency**, the mitral valve does not close fully, so that some of the blood flows back into the left atrium. This results in too little blood reaching the cardiac circulation, so that the performance of the heart and its oxygen supply are reduced. The left ventricle becomes enlarged and its performance begins to deteriorate, resulting in breathlessness and atrial fibrillation.

• ***Aortic valve defects***

Aortic valve stenosis is only manifest at a late stage. Because of the narrowing of the valve, extra force is needed to push the blood through the aortic valve. The musculature of the left ventricle becomes thickened and eventually the supply of blood through the coronary blood vessels becomes inadequate. The consequences may include angina pectoris, breathlessness, dizziness, and fainting. The aortic valve is replaced surgically.
In **aortic valve insufficiency**, blood from the aorta flows back into the left ventricle. This results in thickening of the musculature, with a reduction in the performance of the heart and enlargement of the ventricle. Aortic valve insufficiency is also treated surgically.

• ***Tetralogy of Fallot***

A congenital heart defect. The heart valve to the pulmonary artery is narrowed, with consequent enlargement of the right ventricle due to the extra force needed to pump the blood to the lungs. In addition, there is a hole in the interventricular septum between the left and right ventricles; the aorta overrides the muscular part of the interventricular septum and therefore carries both oxygenated and deoxygenated blood. This results in the tissues of the body receiving an inadequate supply of oxygen, which is manifest in a blue coloration of the lips (cyanosis). In newborn babies, it is this symptom, together with breathlessness, that is the first indicator of a heart defect. In many cases it can be treated surgically.

• ***Ductus Botalli***

In unborn babies there is a connection between the pulmonary artery and the aorta, through which blood flows to the lungs before these are functional. This duct normally closes with the first breath after birth, but sometimes it fails to close (known as an open ductus Botalli). A relatively common heart defect in children, the condition requires surgery while the patient is still a baby.

Myocardial disease (non-ischemic)

An umbrella term for a range of conditions involving damage to the heart muscle. Depending on severity, the pump function of the heart may be increasingly impaired, and signs of heart failure begin to appear. X-rays often reveal a severely enlarged heart. Myocardial diseases are most commonly attributable to malformation of the individual myocardial fibers, to disturbances in the metabolism of the myocardial cells, or a stenosed outlet to the aorta. The causes of these defects, which are often congenital, are largely unknown. Infections, auto-immune responses affecting the myocardial tissue, intoxication (including chronic alcohol consumption), and long-term vitamin or mineral deficiencies, can all result in myocardial disease.

Heart pains

Burning, oppressive, transient (in the case of coronary stenosis), or persistent (in the case of myocardial infarction) pain behind the sternum, often radiating leftwards into the shoulder, arm, and lower jaw. This is usually caused by an inadequate blood supply to the heart or, in rare cases, by another heart disease. Sporadic stabbing pains and pains lasting several seconds in the same region that are not exercise-related, usually have nothing to do with the heart and are harmless.

Heart failure (cardiac insufficiency)

Severe restrictions of the pump function of the heart. The cause is usually a permanently increased resistance in the circulatory system (e.g., due to high blood pressure or severe chronic bronchitis), which makes the heart work harder. The heart adapts to the greater demands made on it by increasing in size: the heart walls thicken and the heart rate increases. Over time, however, the capacity of the heart to adapt is exhausted, and heart failure becomes apparent, frequently manifesting itself as breathlessness. Initially this occurs only on exertion, but the threshold level of exertion gradually drops and breathlessness is eventually present when the patient is at rest. Additional causes of heart failure include congenital heart defects, death of myocardial tissue following a severe infarction, and other diseases of the myocardium. Wherever possible, the causes should be treated first. In addition, avoidance of physical over-exertion and drugs such as digitalis to strengthen the heart, may be recommended. In the final stage, a heart transplant is needed.

Coronary insufficiency

Inadequate blood supply to the coronary blood vessels, due to damage, blockage, or stenosis of these vessels. The myocardium receives fewer nutrients than it needs, ultimately resulting in damage. The first signs are usually heart pains on physical exertion, later also when at rest.

Diagnostic methods

Electrocardiogram (ECG)

Active heart muscle cells generate electrical potential. This can be measured at various points on the body by means of small metal plates (electrodes). The electrical potential picked up by these plates can be represented graphically as a series of traces. A healthy heart has a characteristic pattern on these traces, and any abnormalities may be typical of underlying disease or previous damage, e.g., due to an infarction. Cardiac rhythm disturbances can also be identified. A stress ECG can provide information on the performance of the heart.

• ***Stress ECG***

An ECG performed immediately after or during physical exertion, complementing an ECG performed at rest. It shows problems with the blood supply to the heart which cannot be seen on the resting ECG.
The most frequently used form of stress ECG is **cycle ergometry**, in which the patient, with electrodes attached to his or her body, sits on a stationary cycle and pedals against a gradually increasing resistance. Meanwhile, pulse and blood pressure are recorded, and the body's performance capability and cardiovascular status are observed. Afterwards, the ECG trace is used to evaluate the ability of the heart to recover. Instead of cycling, the stress can be induced by running on a treadmill or even climbing up stairs.

• ***Continuous ambulatory cardiac monitoring***

A trace of the electrical impulses of the myocardium over 24 hours. Adhesive electrodes are affixed to the sternum of the patient and connected to a small recording device attached to a belt worn by the patient. During the 24-hour recording, the patient keeps a record of activities, symptoms, and drugs taken. This information is used when evaluating the ECG. A 24-hour ECG may be performed in the case of heart symptoms where the cause is unclear, or to identify asymptomatic cardiac rhythm disturbances and inadequacies in the cardiac blood supply. A different type of long-term ECG is an event recorder, which is started by the patient himself when he becomes aware of the onset of symptoms. The event recorder is usually worn for several days.

Ultrasound investigations

• ***Cardiac ultrasound (echocardiography)***

Ultrasound investigation of the heart. Like light waves, ultrasound waves are reflected or interrupted when they hit a surface, with different tissue types causing reflections of varying degrees of strength. These physical properties are harnessed in echocardiography, in which an ultrasound probe is moved over the patient's thorax, which has been covered in contact gel, in the region of the heart. The ultrasound waves are transmitted and reflected, and then converted electronically into an image of the heart. It is then possible, from the screen or from a printout, to tell whether the heart valves are working properly.
In rare cases, the ultrasound probe is introduced through the esophagus (transesophageal cardiac ultrasound).

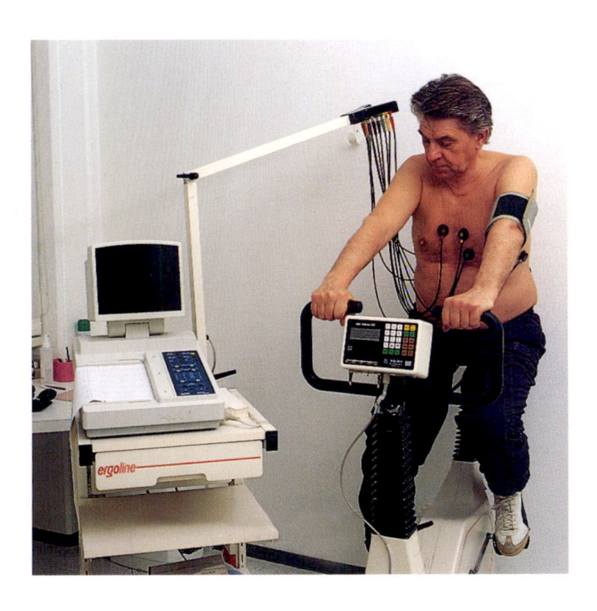

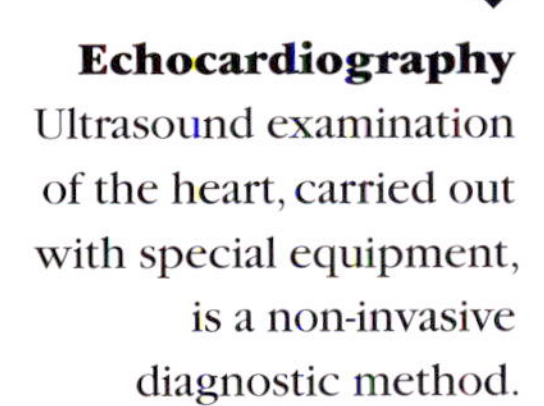

◆ **Stress ECG**
With the aid of a stress ECG, it is possible to tell whether the heart is capable of adapting well under physical stress.

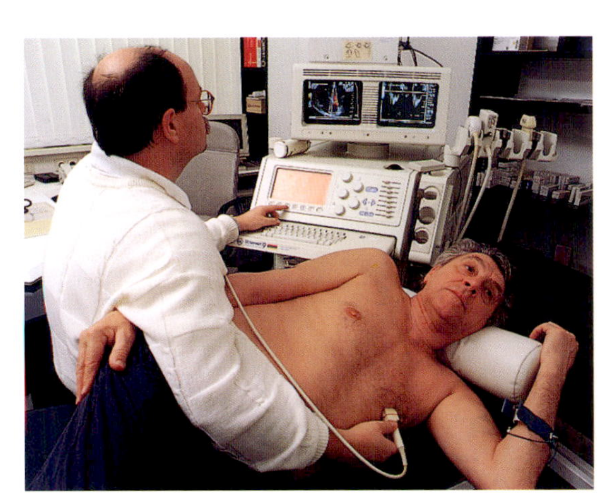

◆ **Echocardiography**
Ultrasound examination of the heart, carried out with special equipment, is a non-invasive diagnostic method.

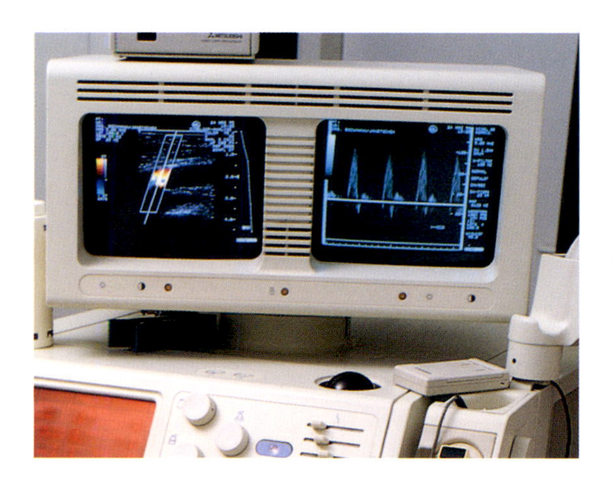

Color doppler
This special type of ultrasound investigation provides information about the flow properties of blood in the veins and arteries.

• ***Doppler ultrasound***

Doppler ultrasound is used to determine the direction and rate of flow of the blood.

• ***Duplex ultrasound***

Duplex ultrasound makes it possible to visualize tissues and blood flow at the same time.

• ***Color duplex ultrasound***

The colored image on the screen gives information about the direction of blood flow, the approximate speed of the flow, and the location and extent of any turbulence. The examination is most frequently performed if narrowing of the vessels (stenosis) or dilation of the vessels (aneurysm) is suspected.

X-rays

• ***Chest X-ray***

Chest X-rays, pictures of the thorax from the rear and from the side, are used to assess the size and shape of the heart and the status of the lungs, pulmonary arteries, and aorta.

• ***Cardiac catheterization***

A thin, flexible plastic tube, known as a catheter, is introduced into an artery in the groin from where it is advanced as far as the heart. Contrast medium is then injected via the catheter, so that X-rays of the functioning heart may be taken. It is also possible to take a sample of tissue from the heart using this technique, and to measure blood pressure and oxygen levels.
The most common form of cardiac catheterization is left heart catheterization; right heart catheterization has been rendered obsolete with the advent of ultrasound. Left heart catheterization serves to evaluate the pump function of the heart and the function of the heart valves, as well as to provide images of the coronary vessels.

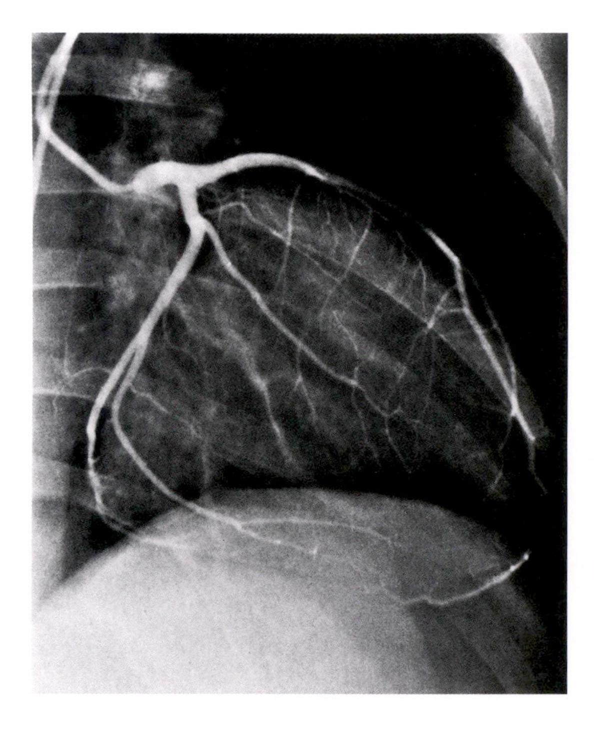

Imaging of the coronary vessels
It is possible using contrast medium to make the coronary vessels visible on an X-ray.

• ***Imaging of the coronary vessels (coronary angiography)***

An X-ray procedure to investigate the blood vessels supplying the heart (coronary vessels). Under X-ray guidance, a heart catheter is used to inject contrast medium into the main trunk of the coronary vessels. On X-ray, the contrast medium is visible as it spreads, so that the patency of the coronary vessels can be evaluated and any stenoses or occlusions identified.
A coronary angiography is performed if symptoms (such as stress-related chest pain) or ECG findings suggest there might be narrowing of these vessels, or to investigate a myocardial infarction. It is also performed before all heart surgery.

Acoustic cardiograph

A recording of the heart sounds and heart murmurs using special microphones and amplifiers.

Treatment methods

Anti-arrhythmics

Drugs to treat cardiac rhythm disturbances, such as beta-blockers, certain potassium salts or digitalis preparations (foxglove).

• ***Beta-blockers***

An abbreviated name for beta-receptor blockers, these drugs are used to treat high blood pressure and prevent diseases caused by high blood pressure. Beta receptors are cells in the blood vessels or myocardium which cause the blood pressure to rise in response to a stimulus; if they are blocked, the blood pressure returns to normal or is lowered.

- ***Potassium***

An essential mineral, potassium is an indispensable component of all cells. It regulates the water content of the cell and, in conjunction with sodium, controls the transmission of electrical stimuli to nerves and muscles to enable these to function. The activity of the heart is also influenced by potassium. If there is a surplus, e.g., as a result of inadequate excretion via the kidneys, dangerous cardiac rhythm disturbances may result.
A deficiency of potassium is equally dangerous. This may be caused in particular by excessive use of laxatives, but also as a result of taking diuretics, dehydrating drugs, or of frequent vomiting and diarrhea. It causes general muscular weakness, affecting both the heart muscle and the intestinal muscles; the consequences are irregular heartbeat and increasingly sluggish intestinal function.

- ***Digitalis therapy***

Treatment of heart patients with drugs obtained from the foxglove plant (digitalis). The active substances contained in digitalis that affect the heart are known as cardiac glycosides.
Treatment with cardiac glycosides is necessary when the heart can no longer contract adequately and is therefore unable to pump blood into all regions of the body. Cardiac glycosides increase the force with which the heart contracts and restores the capability of the heart to function economically; it fills with blood more rapidly and pumps the blood out of the ventricles into the main arteries faster and with greater force. As a result, the number of heartbeats per minute also drops. Because of the improved performance of the heart, physical symptoms such as breathlessness, cyanosis, and hemostasis (which can cause edema) recede, and the patient feels physically fitter.

◆
Digitalis therapy
Substances obtained from the leaves of the foxglove strengthen the heart.

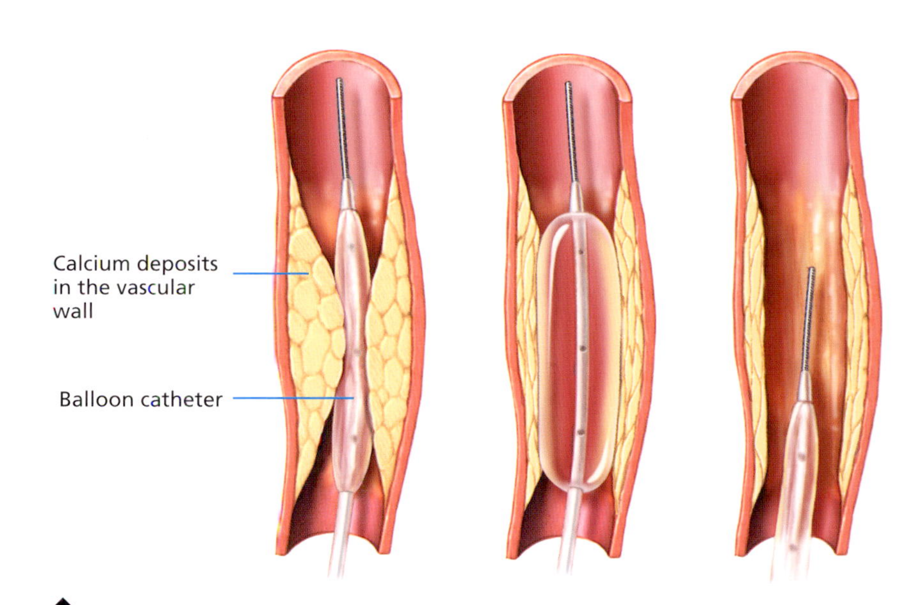

◆
Balloon dilatation
Balloon dilatation is used to stretch narrowed blood vessels. It is often used following a myocardial infarction to restore the patency of occluded coronary vessels.

Balloon dilatation (angioplasty)

Angioplasty is a vascular surgical procedure in which a stenosed or occluded blood vessel is expanded using a balloon catheter (see above). At the end of the catheter is a balloon advanced to a point just before the section of the blood vessel that has been stenosed due to calcium deposits or obstructed by a blood clot. The balloon is then inflated to exert pressure on the internal wall of the stenosed or obstructed blood vessel (e.g., a coronary vessel), thereby widening it and restoring its patency. In order to prevent the stenosis or occlusion recurring, a stent (a small metal tube) may be inserted to keep the artery open after removal of the balloon. In time, the artery wall grows over the stent thereby providing an internal scaffolding. Nowadays, most of these stents contain drugs which are eluted from the stent over a period of time, to prevent re-stenosis.
In another variation of the technique, a rotating circular knife is advanced through the catheter to the affected spot to remove the blockage.

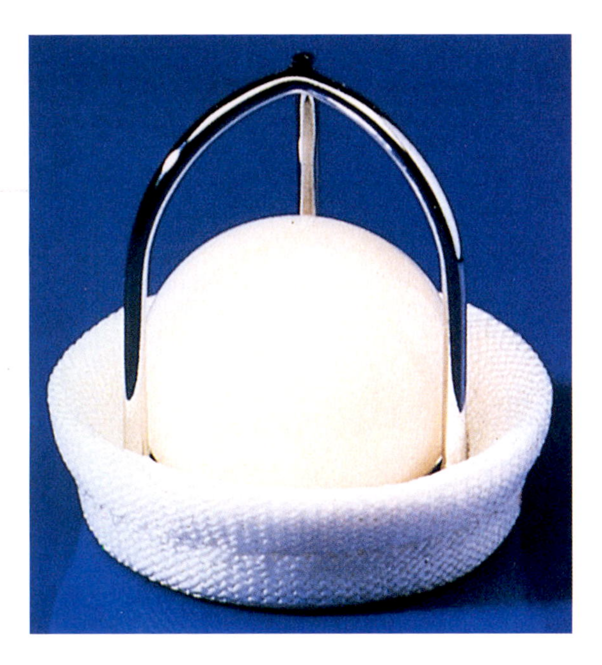

◆ **Heart valve prosthesis**
The insertion of artificial heart valves is now a routine procedure in cardiac surgery.

◆ **Bypass surgery**
In a cardiac bypass operation, the blocked blood vessel is usually bypassed using a section of a leg vein.

Heart valve operation

A surgical intervention to repair heart valve defects. Depending on severity, defective heart valves are either repaired or removed and replaced by a heart valve prosthesis. Some heart valve repairs can be performed without the need for open surgery, by means of a balloon catheter.

• *Heart valve prosthesis*

A replacement heart valve. Depending on which is most suitable, defective heart valves can now be replaced by plastic valves, valves formed from animal heart valve material, by complete animal heart valves, or by heart valves from organ donors.

• *Cardiac valvotomy*

Correction of heart valve stenosis by dilatation of scar tissue around the edges of the valve. In the surgical method, the chest wall and heart are opened, and the stenosed heart valve is dilated with a finger or special instrument. A heart valve can also be dilated using a balloon catheter advanced via a major blood vessel as far as the defective heart valve and then inflated, so that the opening of the valve is enlarged. The same procedure is used to dilate stenosed blood vessels.

Cardiac bypass surgery

Now the most common type of surgery in the USA, bypass surgery is a method of cardiovascular diversion. A cardiac bypass involves the surgical grafting of blood vessels to replace occluded or stenosed blood arteries in the heart. A section of the patient's leg vein is most commonly used to create the bypass. Cardiac bypass surgery is the most common form of bypass surgery used nowadays; parts of between one and four diseased coronary arteries are bypassed in order to prevent a threatened myocardial infarction.

Cardiac pacemaker

A device to maintain the proper rhythm of the heart. In healthy people, the heartbeat is regulated by a mechanism in the right atrium of the heart (the sinoatrial node). Regular electrical impulses are generated in the sinoatrial node, conducted via the heart musculature and make the heart contract. Cardiac rhythm disturbances due to sinoatrial node malfunction or conduction problems may be compensated for by an artificial pacemaker. An electrode is inserted in the heart and connected to a battery-operated unit implanted beneath the skin, usually on the chest, that sends regular small electrical impulses to simulate the function of the sinoatrial node.

Heart transplantation

Transfer of the healthy heart from an organ donor to someone with a severe, incurable heart disease. Although heart transplants are now regarded as routine in medicine, and survivors may have

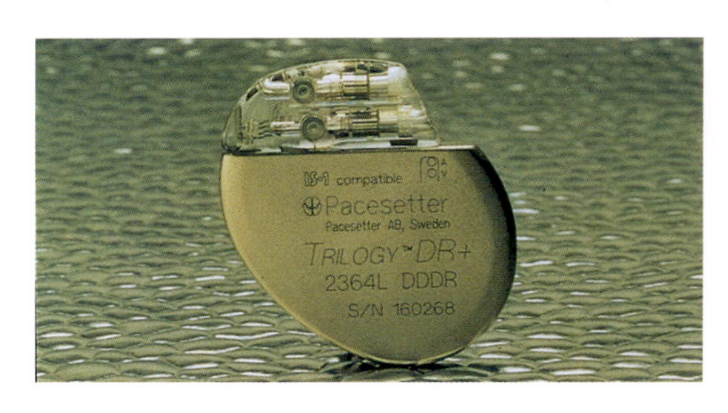

◆ **Cardiac pacemaker**

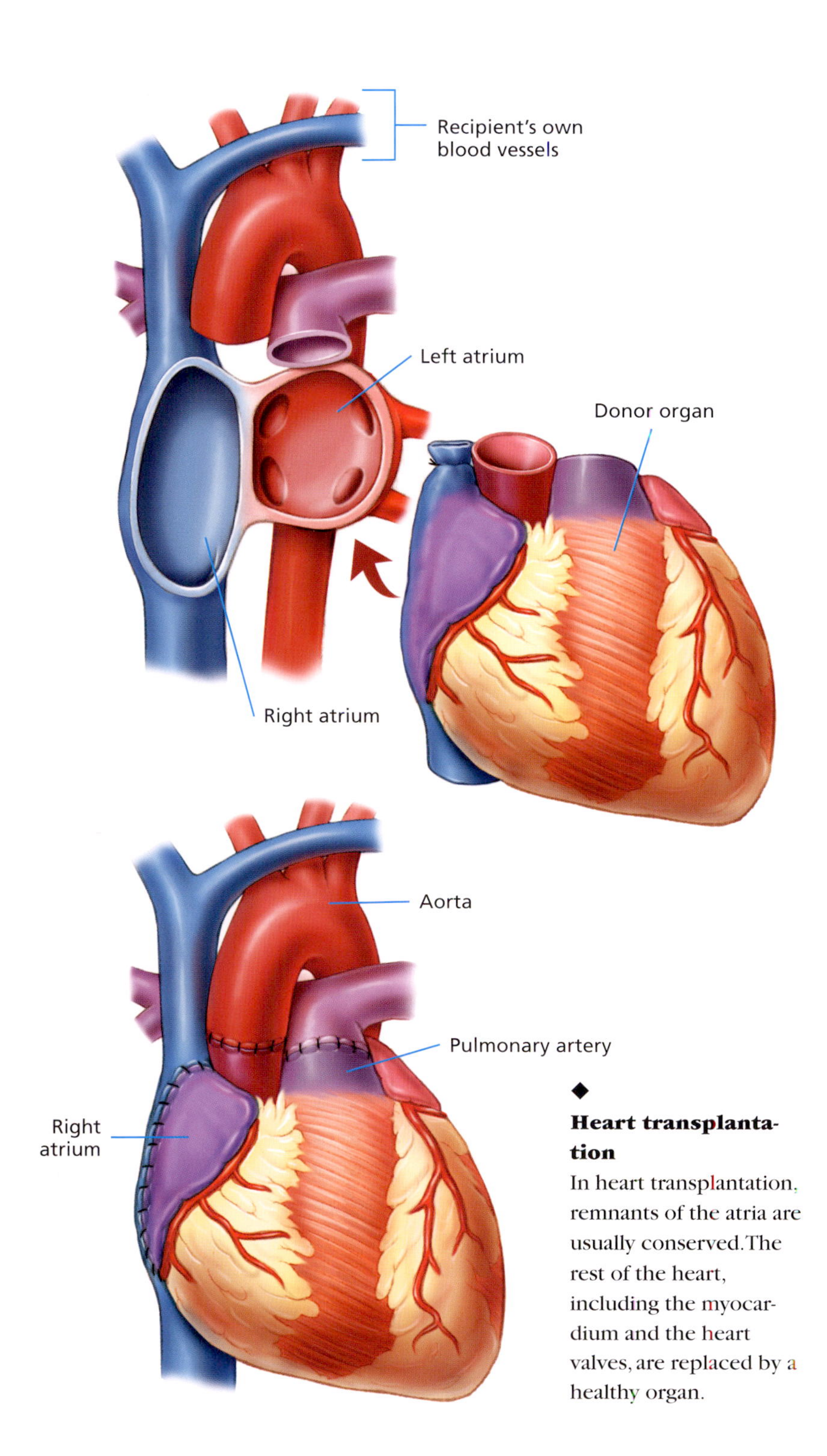

◆ **Heart transplantation**
In heart transplantation, remnants of the atria are usually conserved. The rest of the heart, including the myocardium and the heart valves, are replaced by a healthy organ.

a significantly longer life expectancy with improved quality of life, the problems are greater than with other transplants. The mythological view of the heart as the source of life and the seat of the emotions means that recipients of hearts suffer psychological problems more frequently than recipients of other organs.
Because of the shortage of organ donors, fewer donor hearts are available than are needed. While shortages of donor kidneys can be circumvented by use of an artificial kidney, even in the longer term, heart patients have strictly limited options. Artificial hearts are only capable of keeping patients alive for a short time while they are waiting for a transplant.

Cardiac massage

A first aid procedure (rhythmic, strong pressure applied to the sternum) to maintain the circulation temporarily in the event of cardiac arrest.

Defibrillation

A technique to halt a severe, life-threatening, cardiac rhythm disturbance such as ventricular fibrillation, in which the heart muscle stops contracting rhythmically and pumping blood into the general circulation and instead 'fibrillates' or flutters rapidly in a wave pattern with very high frequency. In such cases, where it is impossible to restore normal rhythm by administering drugs (anti-arrhythmics), the normal cardiac rhythm can be restored by means of electric shocks from a device called a defibrillator.

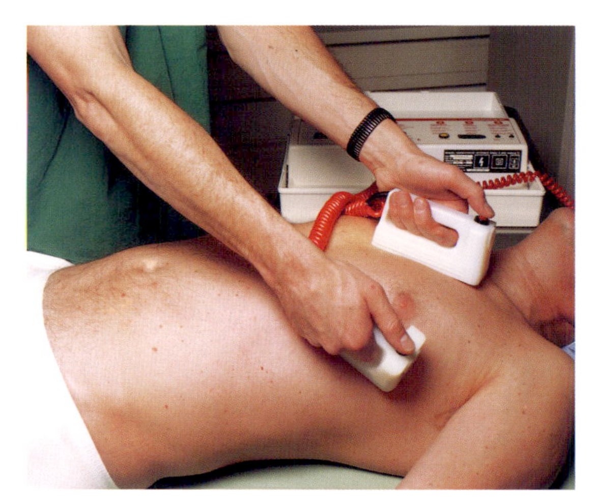

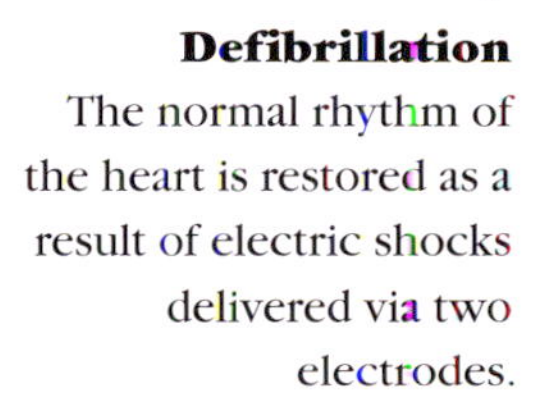
◆ **Defibrillation**
The normal rhythm of the heart is restored as a result of electric shocks delivered via two electrodes.

Circulatory system and blood vessels

In conjunction with the heart, the blood vessels form a closed system, the cardiovascular system, whose function is to supply oxygen and nutrients to all the cells in the body. It is also responsible for transporting waste products, such as carbon dioxide, away from the cells.

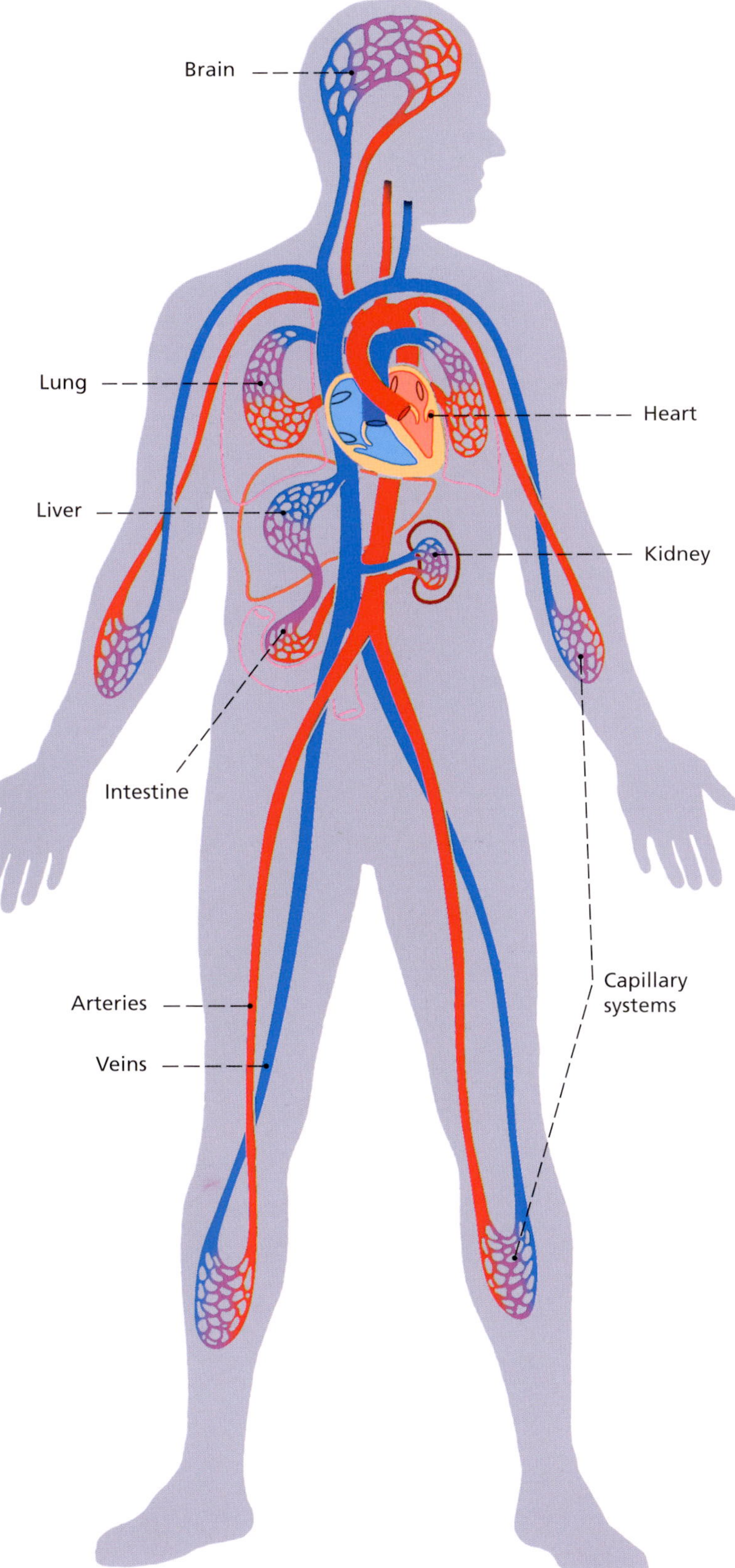

Fig. 258
The blood vessels and heart form a functional unit. A distinction is drawn between the systemic circulation and the pulmonary circulation. In the systemic circulation, oxygenated blood (red) is pumped away from the heart via the arteries to the peripheral areas of the body, while deoxygenated blood (blue) flows from the body back to the heart via the veins. In the pulmonary circulation, the situation is reversed: deoxygenated blood flows to the lungs in the pulmonary arteries, and returns to the heart, laden with oxygen, via the pulmonary veins.

Circulatory system

In the pulmonary circulation, deoxygenated blood flows to the lungs from the right ventricle of the heart in the pulmonary arteries (the only arteries in the body to contain de-oxygenated blood) and returns to the left atrium of the heart, laden with oxygen, via the pulmonary veins (the only veins to contain oxygenated blood). This oxygen-rich blood then enters the systemic circulation. It is pumped from the left ventricle of the heart into the largest artery in the body, the aorta. It then travels through increasingly small, constantly branching arteries (arterioles) until it reaches the capillary network of the tissues. The exchange of materials between blood and tissue takes place via these minute blood vessels: oxygen and nutrients are absorbed by the cells, while carbon dioxide and the by-products of cellular metabolism, are collected into the venous part of the capillary system. Capillary blood collects in the smallest veins (venules), which flow into increasingly large vessels eventually merging to form the superior and inferior vena cava, through which the blood flows back into the right atrium of the heart. It is then pumped back to the lungs to collect more oxygen, thus completing the cycle.
The circulatory system can be divided into two parts:
- the (lesser) pulmonary circulation: blood flows through the lungs on its way from the right to the left side of the heart, and
- the systemic (greater) circulation: blood vessels carry blood from the left side of the heart to all parts of the body and back again to the right side of the heart.

Arteries and veins

All the blood vessels that transport blood away from the heart are called arteries, while those carrying blood back to the heart are called veins. Because of their different functions, the vascular walls of arteries and veins are structured differently. While the walls of arteries consist of a thick layer of muscle cells and elastic fibers, because of the high pressures inside them, the layer of muscle in the walls of the veins is considerably thinner. The inner

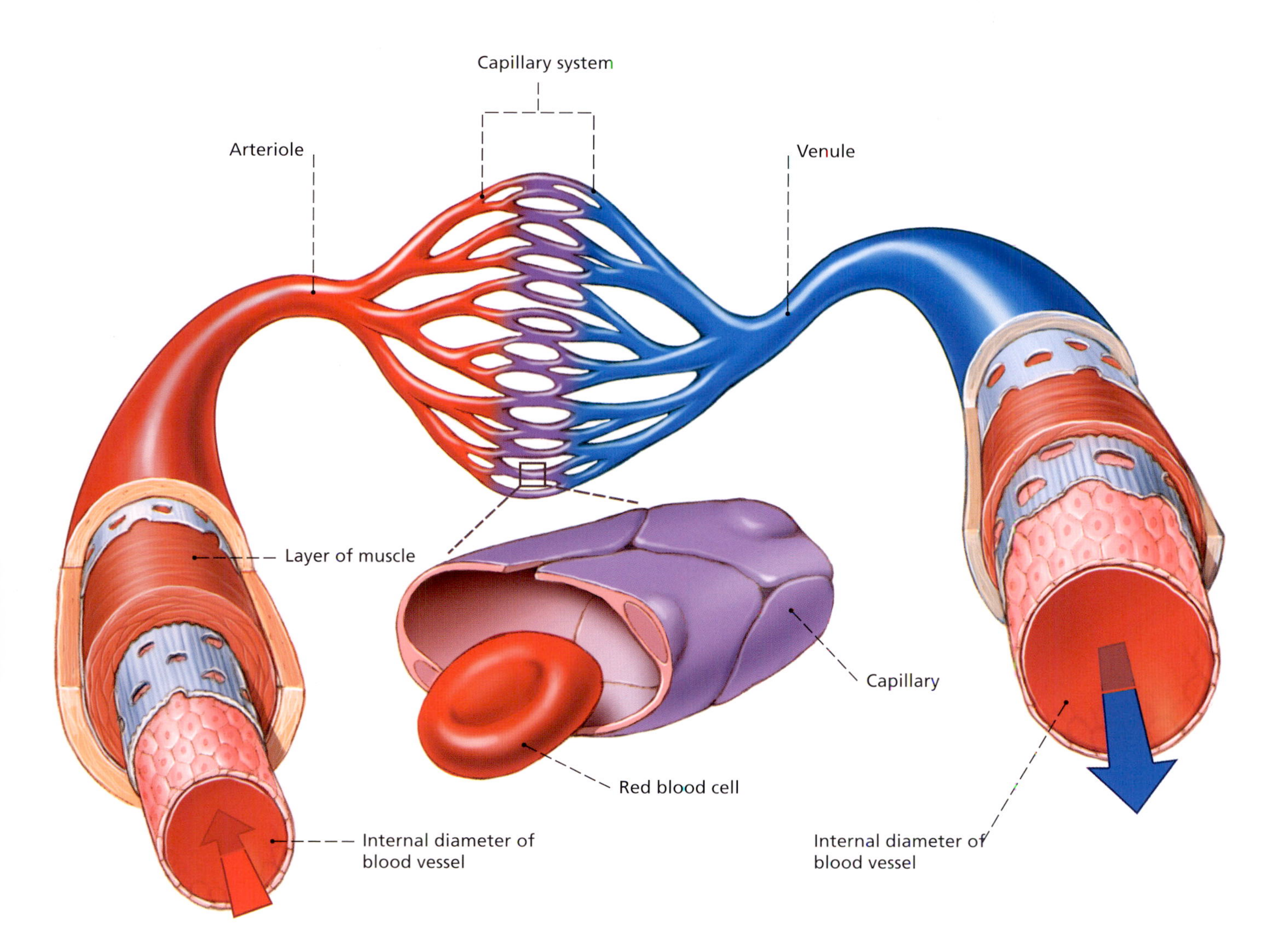

Fig. 259
The smallest unit in the circulatory system is the capillary system in the tissues, in which it is possible to distinguish between an arterial (left) and a venous (right) arm. In the arterial arm, oxygen and nutrients are released into the tissues, while the venous arm collects deoxygenated blood and waste products. In cross-section, a capillary is hardly bigger than a red blood cell.

walls of most veins contain valves, which direct the flow of blood towards the heart and stop it flowing back in the direction of the tissues. Blood flow is supported by the contraction of the (skeletal) muscles, in which the deepest veins are embedded (muscle pump function).

Regulation of blood flow

In order that the flow of blood in the various organs of the body can be adapted to its needs at any one time, the body has various regulatory mechanisms at its disposal:

- The concentration of waste products from cellular metabolism and the concentration of carbon dioxide in the tissues control the width of blood vessels inside the organs, increasing or decreasing blood supply.
- Special nerve cells in the blood vessels (pressure sensors) relay information about the current blood pressure, via their fibrous connections, to the circulatory center in the brain. If the blood pressure drops, the control center sends a nerve impulse to narrow the vessels; if the blood pressure rises, the control center sends a signal that dilates the blood vessels.
- The hormones epinephrin (adrenaline) and norepinephrin (noradrenaline) raise the blood pressure, thus ensuring elimination of by-products of metabolism.

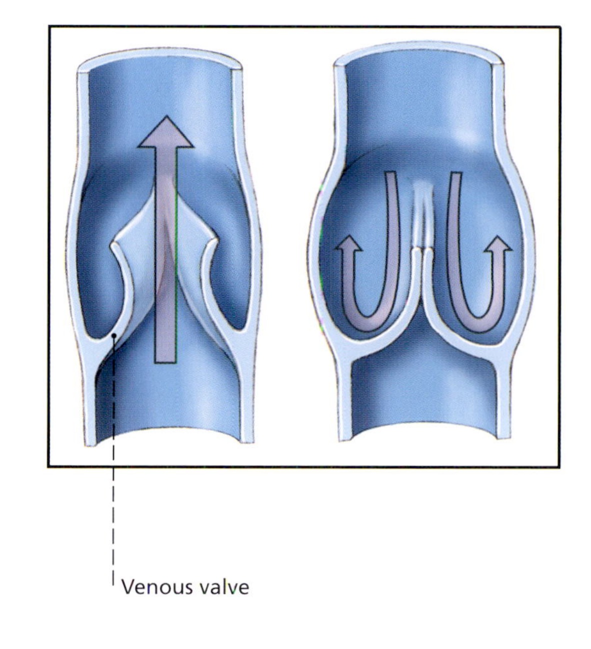

Fig. 260
Valves inside the veins prevent the blood flowing back in the wrong direction.

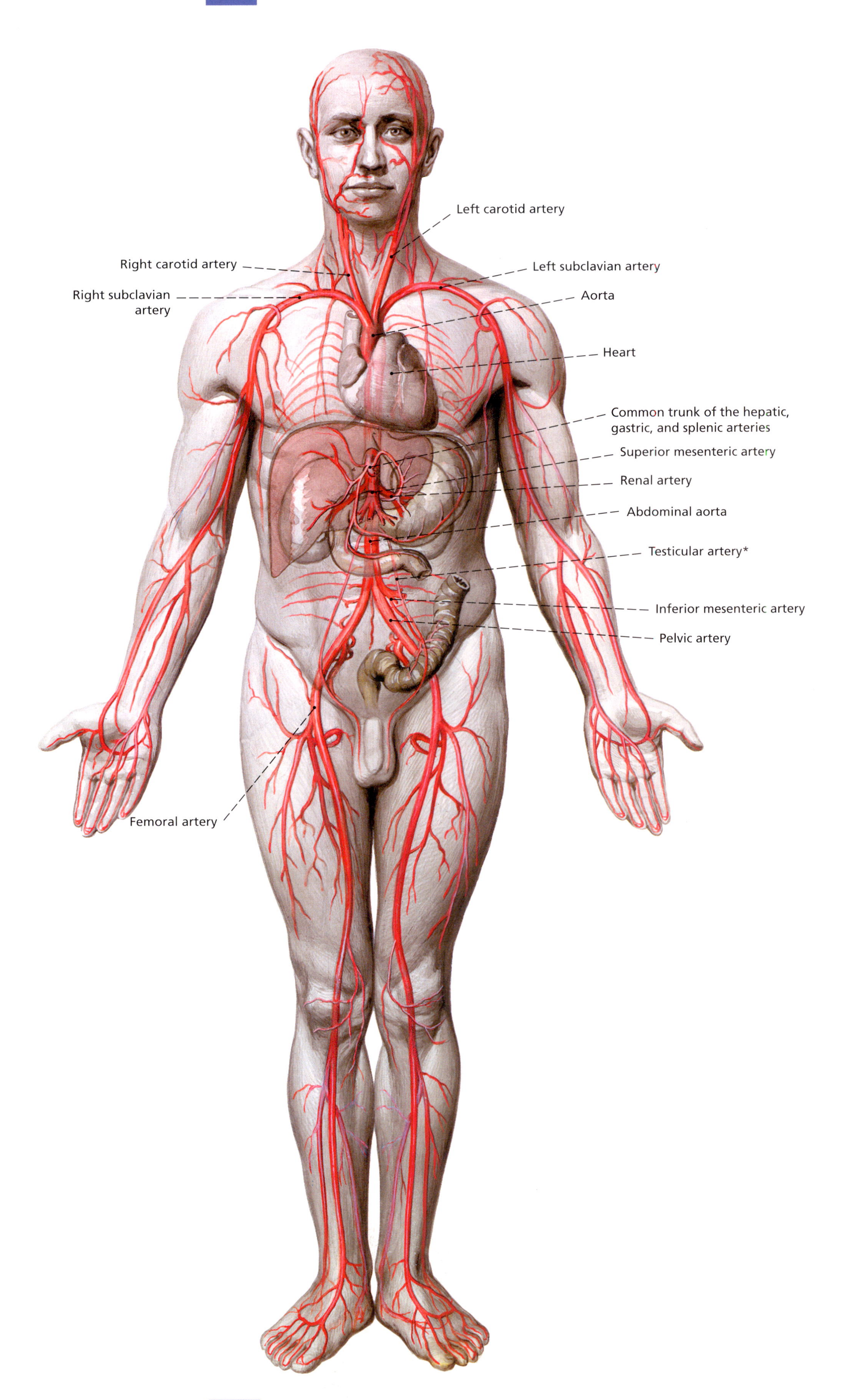

◆
Fig. 261
Arterial system of the body.
* Ovarian artery in females

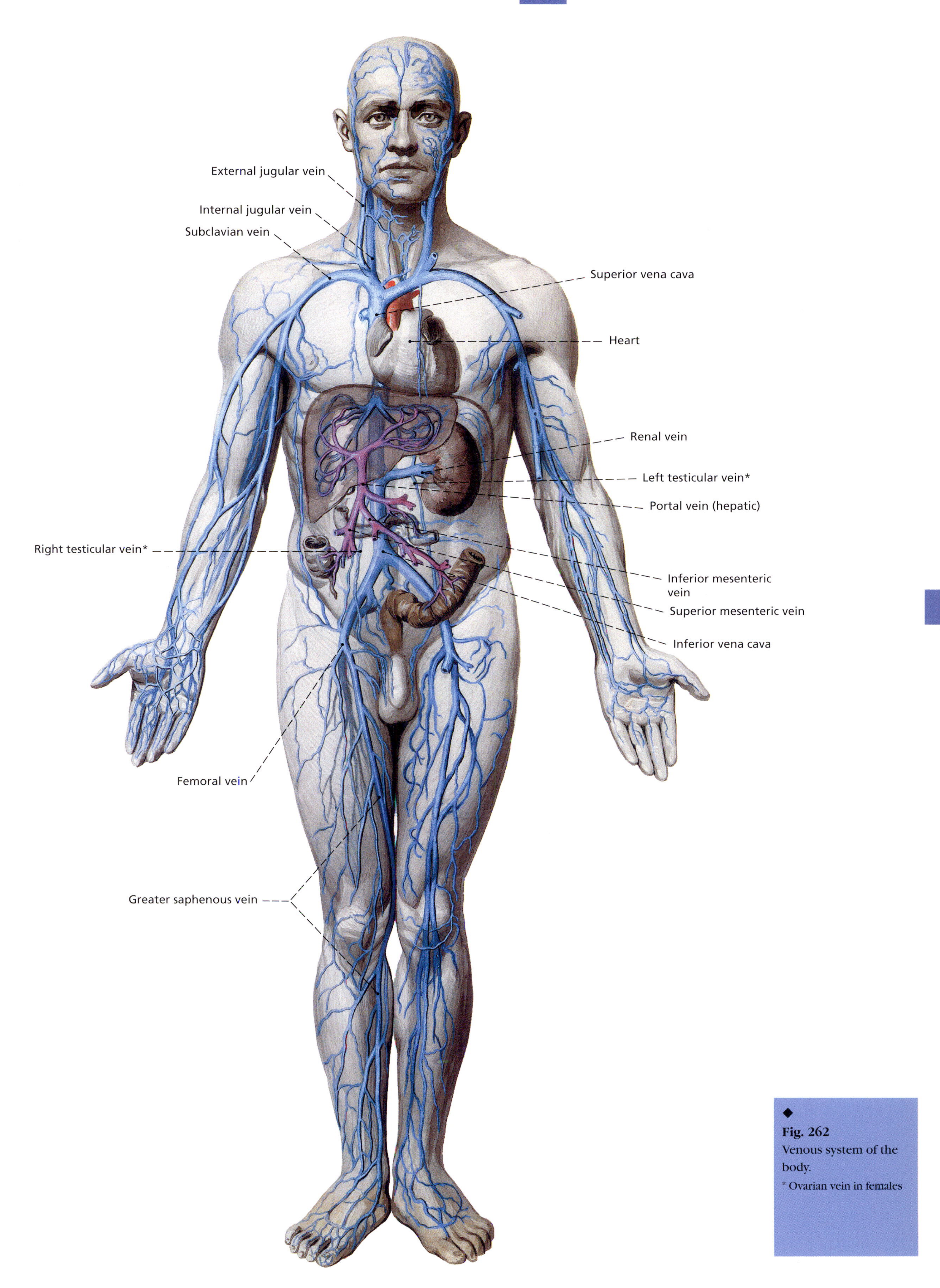

Fig. 262
Venous system of the body.
* Ovarian vein in females

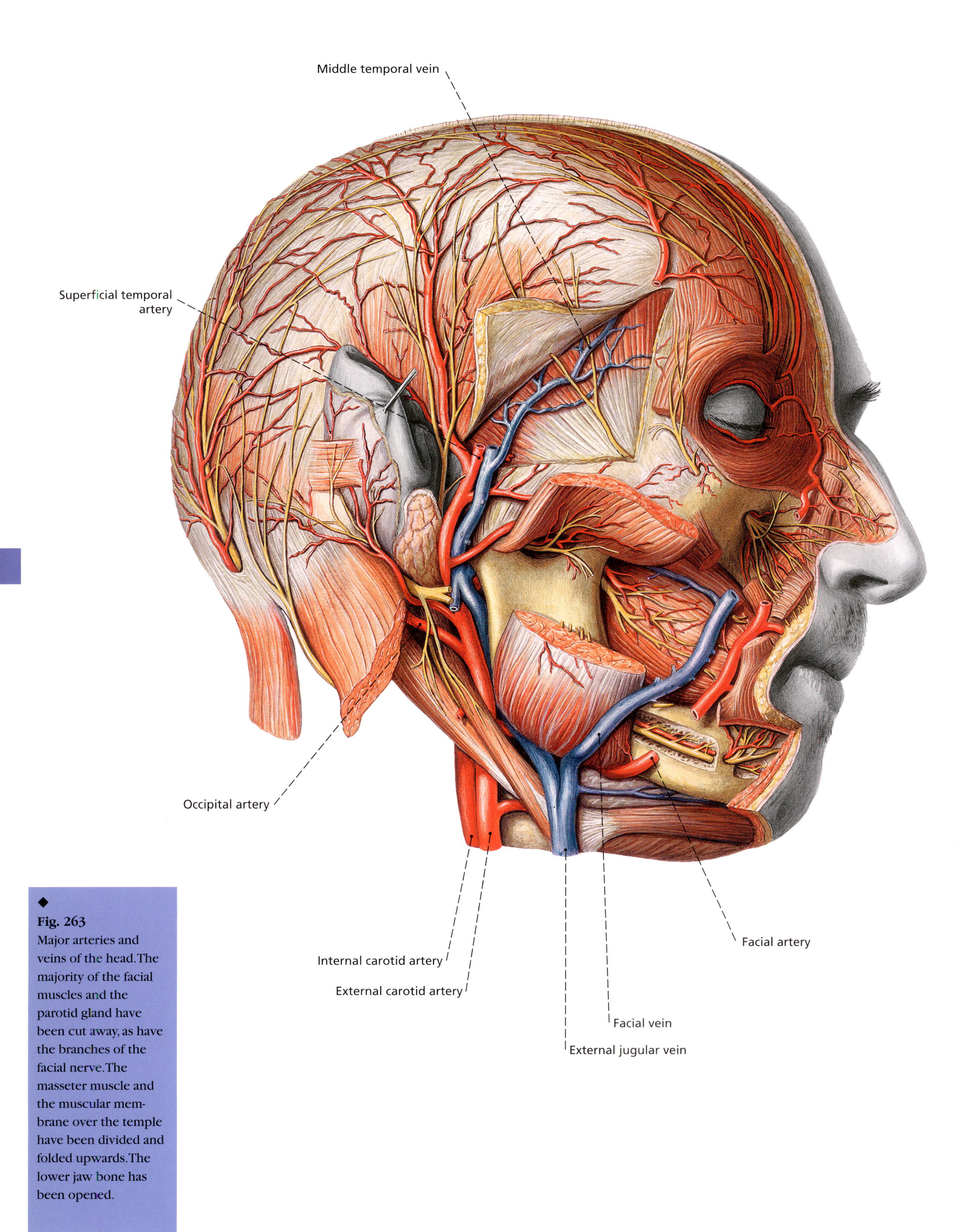

◆
Fig. 263
Major arteries and veins of the head. The majority of the facial muscles and the parotid gland have been cut away, as have the branches of the facial nerve. The masseter muscle and the muscular membrane over the temple have been divided and folded upwards. The lower jaw bone has been opened.

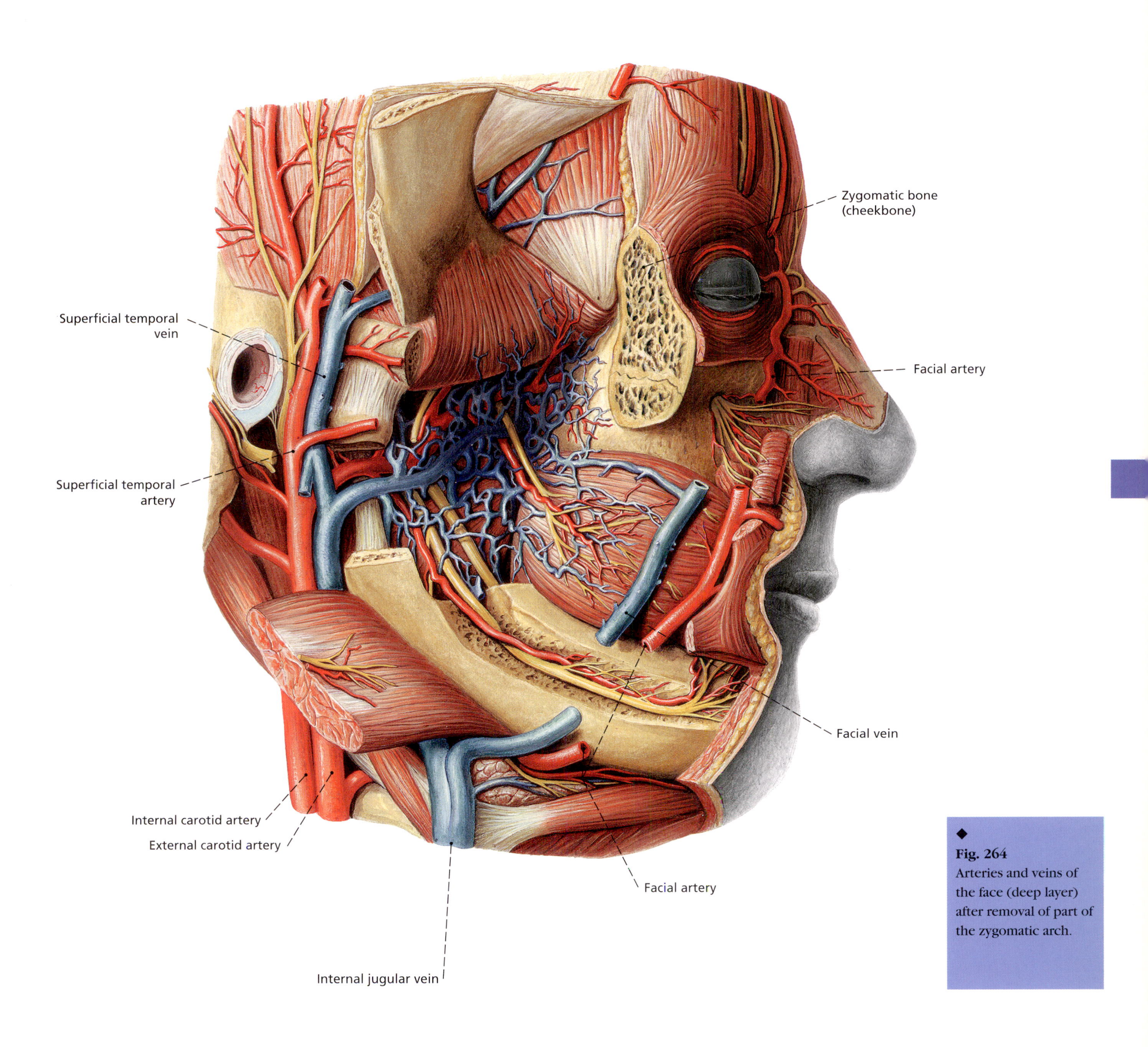

◆
Fig. 264
Arteries and veins of the face (deep layer) after removal of part of the zygomatic arch.

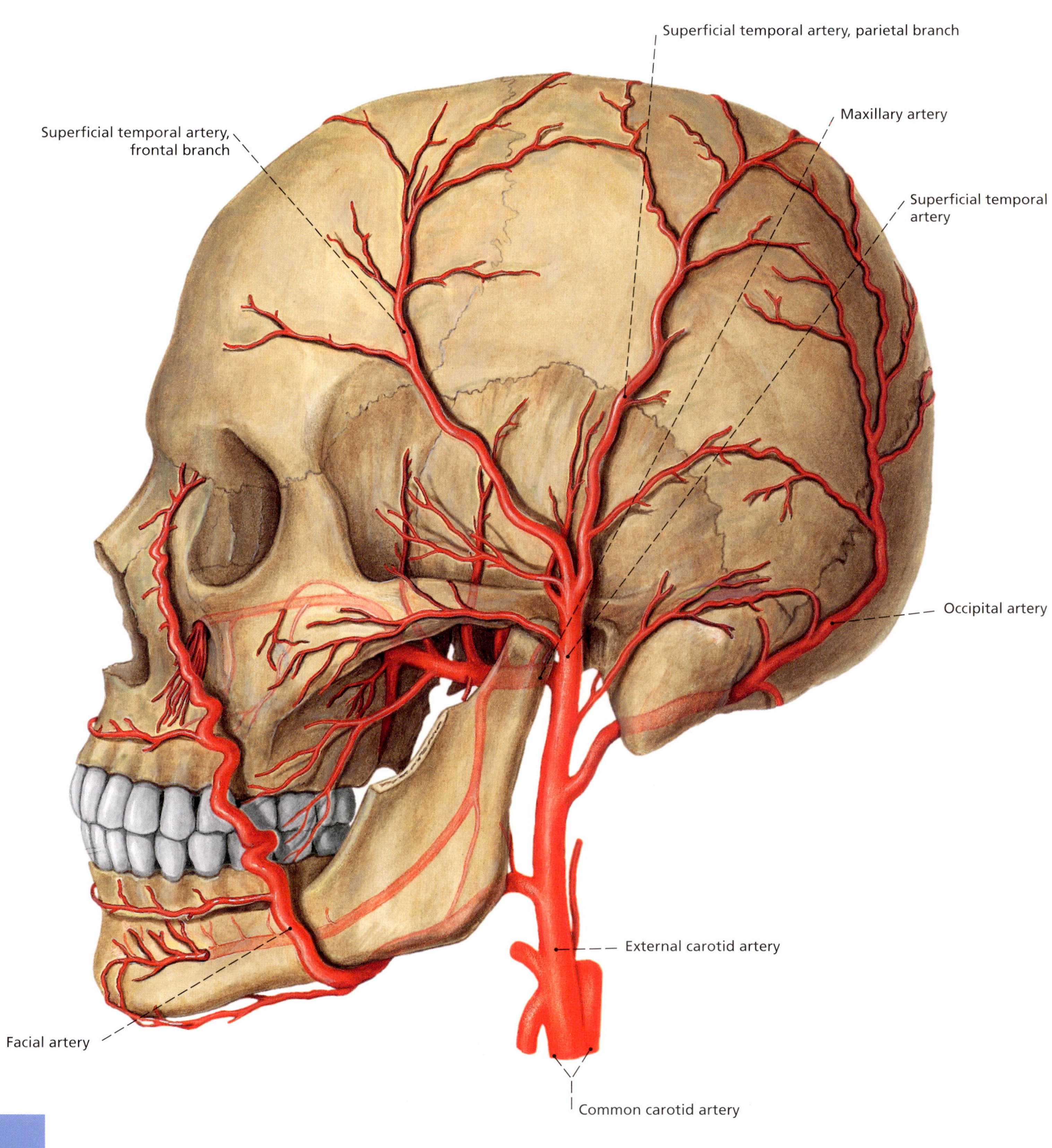

◆
Fig. 265
The external carotid artery (arterial carotis externa) with its branches, viewed from the left side.

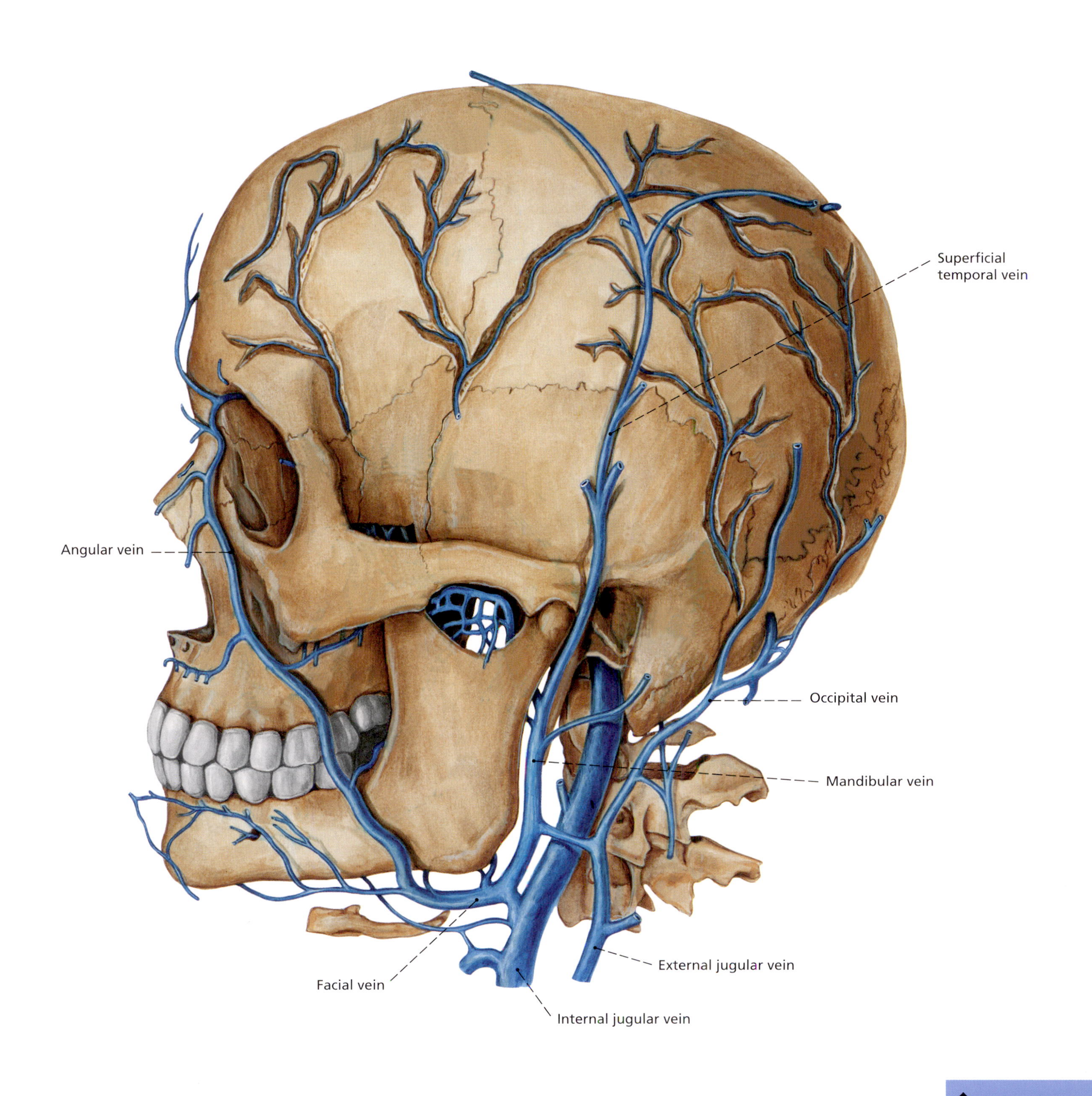

◆
Fig. 266
The internal jugular vein (vena jugularis interna) with its branches, viewed from the left side.

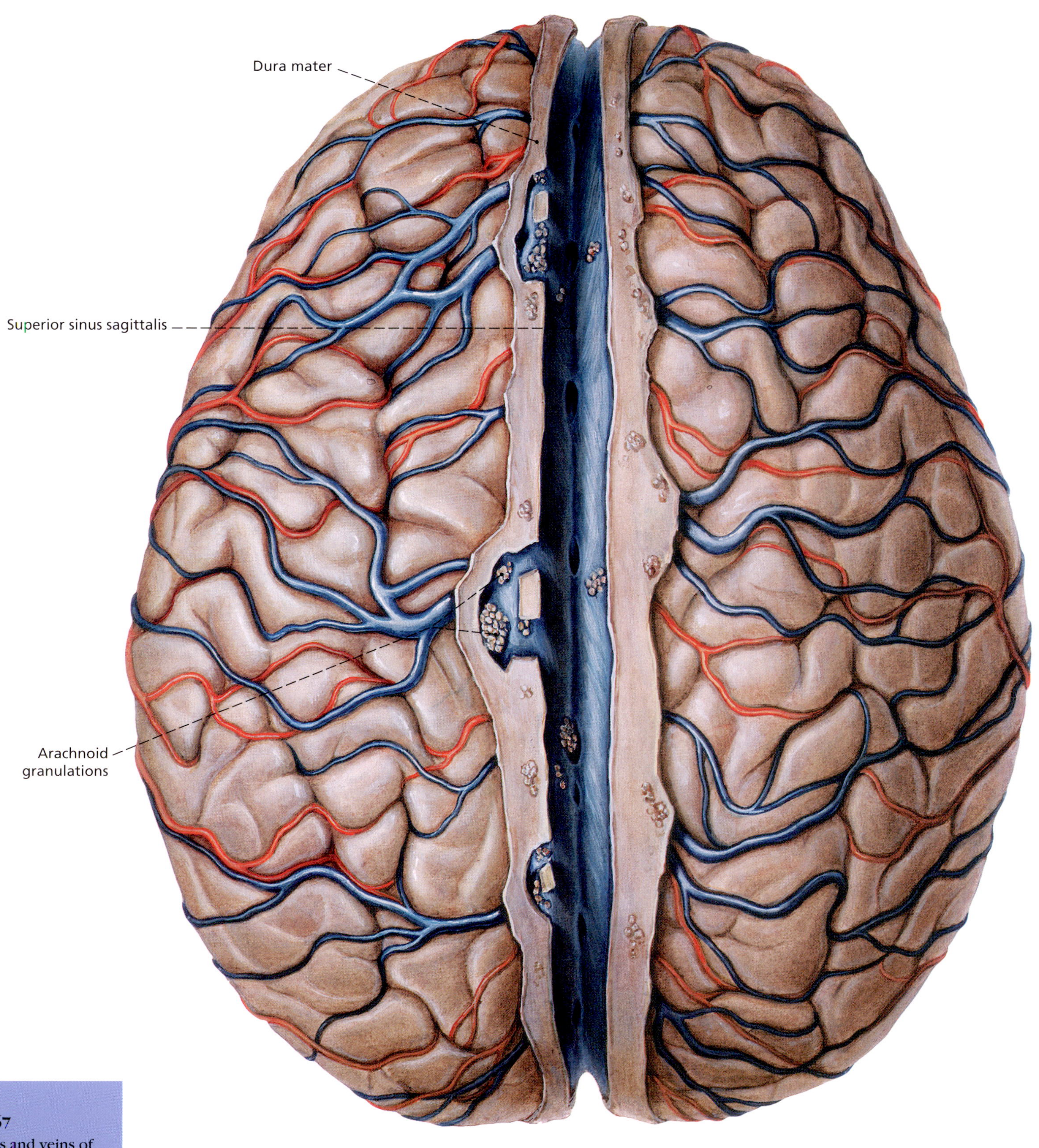

◆
Fig. 267
Arteries and veins of the surface of the brain (cerebral cortex), viewed from above. The dura mater and sinus sagittalis have been opened. Excess cerebrospinal fluid drains into the arachnoid granulations.

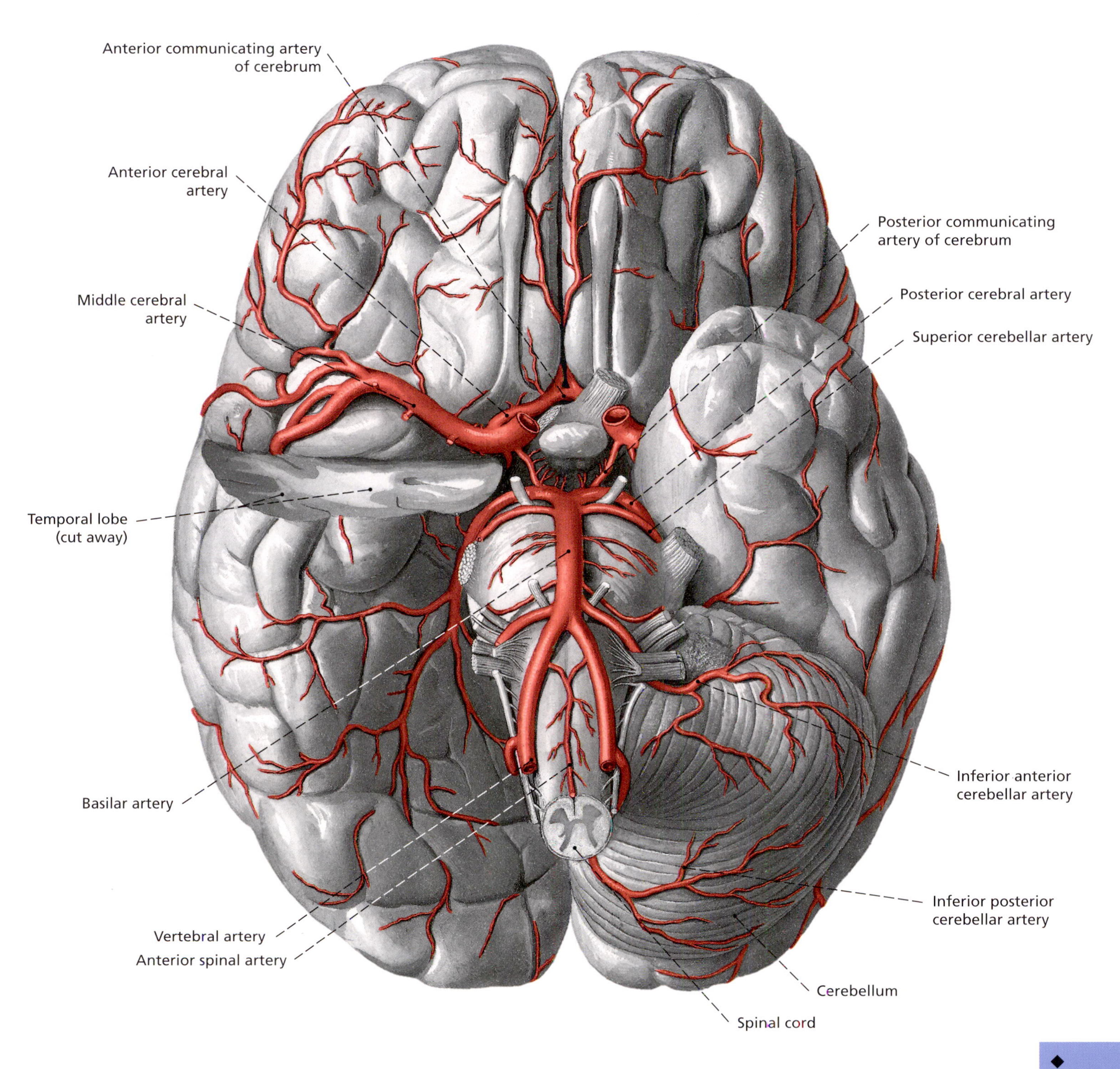

◆
Fig. 268
Cerebral arteries, viewed from beneath. Part of the right temporal lobe and the right half of the cerebellum have been cut away.

◆
Fig. 269
Arterial blood supply to the spinal cord (diagram).

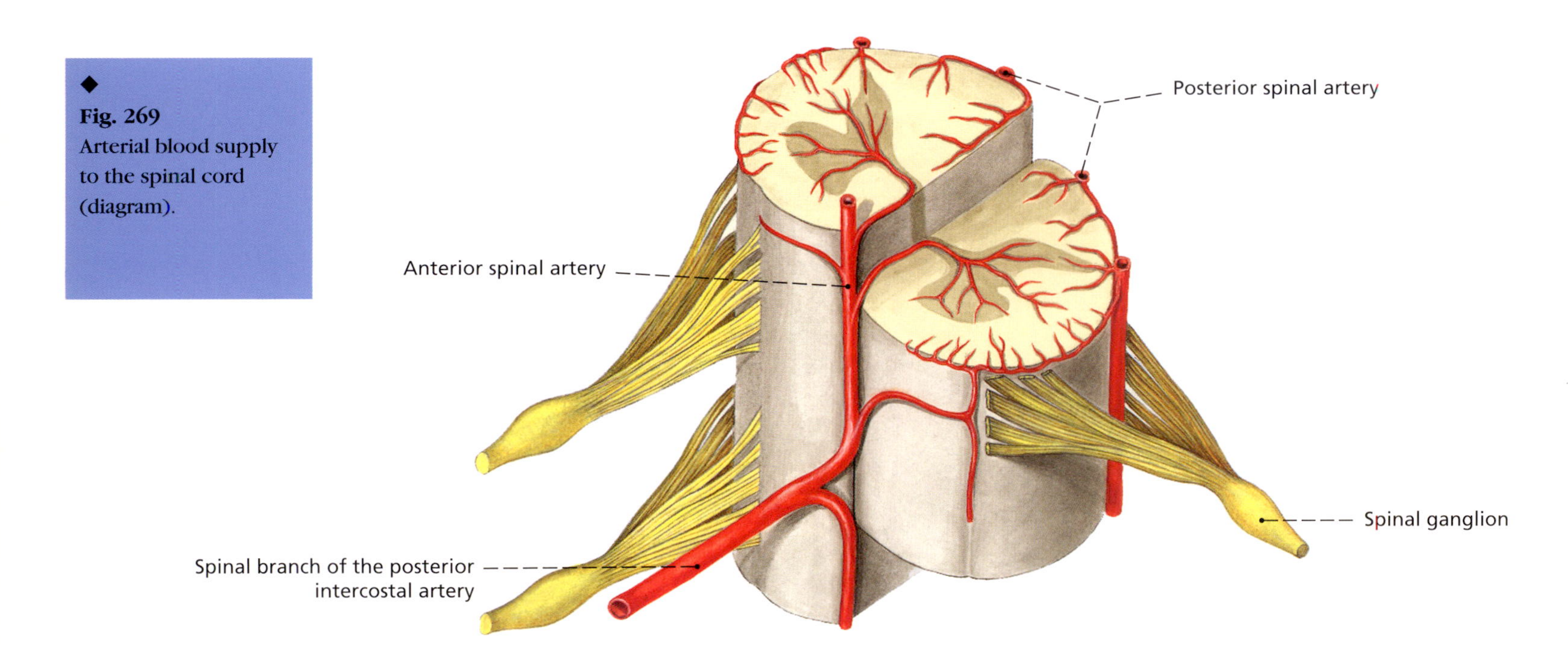

◆
Fig. 270
Venous plexus of the spinal canal in the lumbar region.

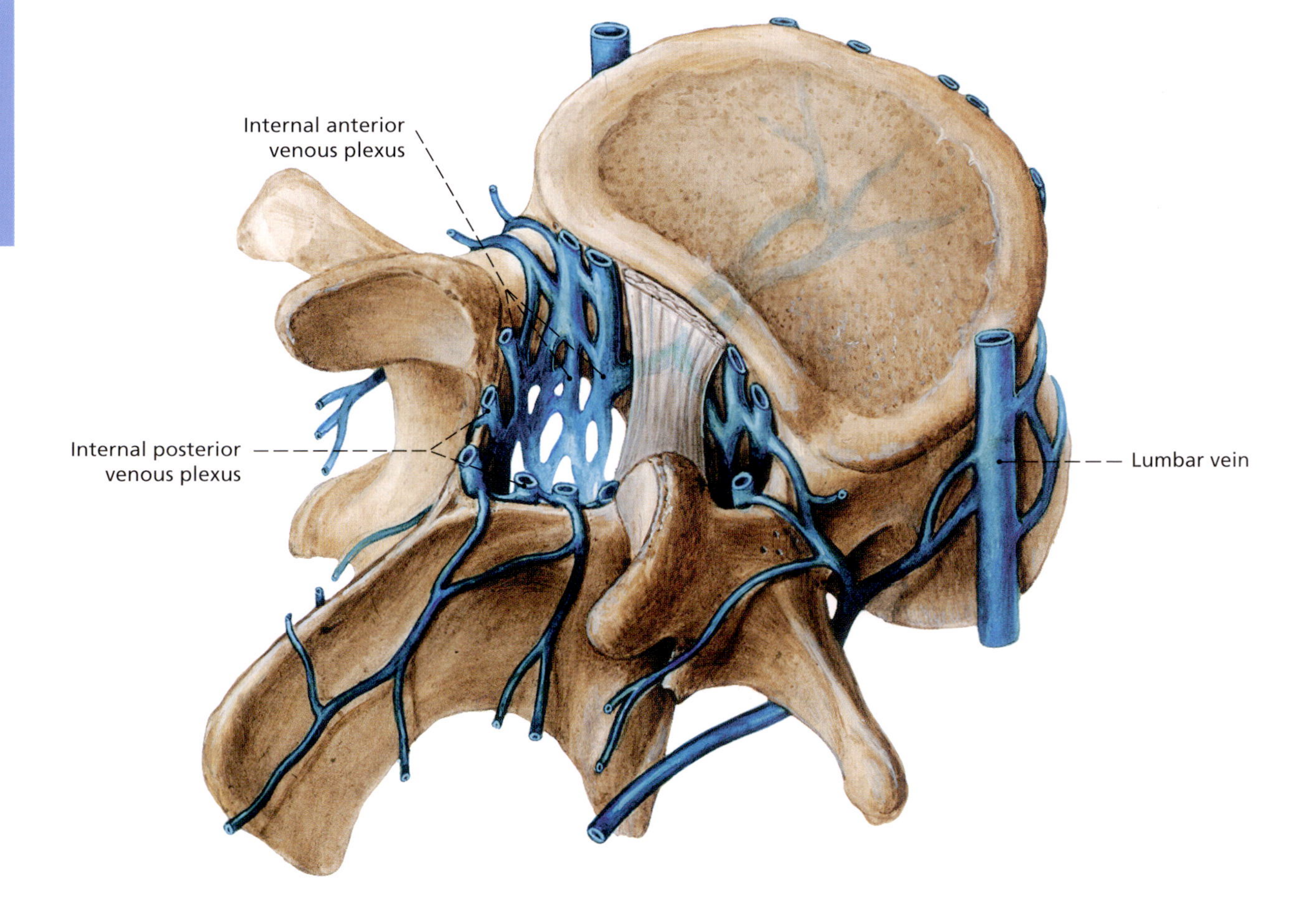

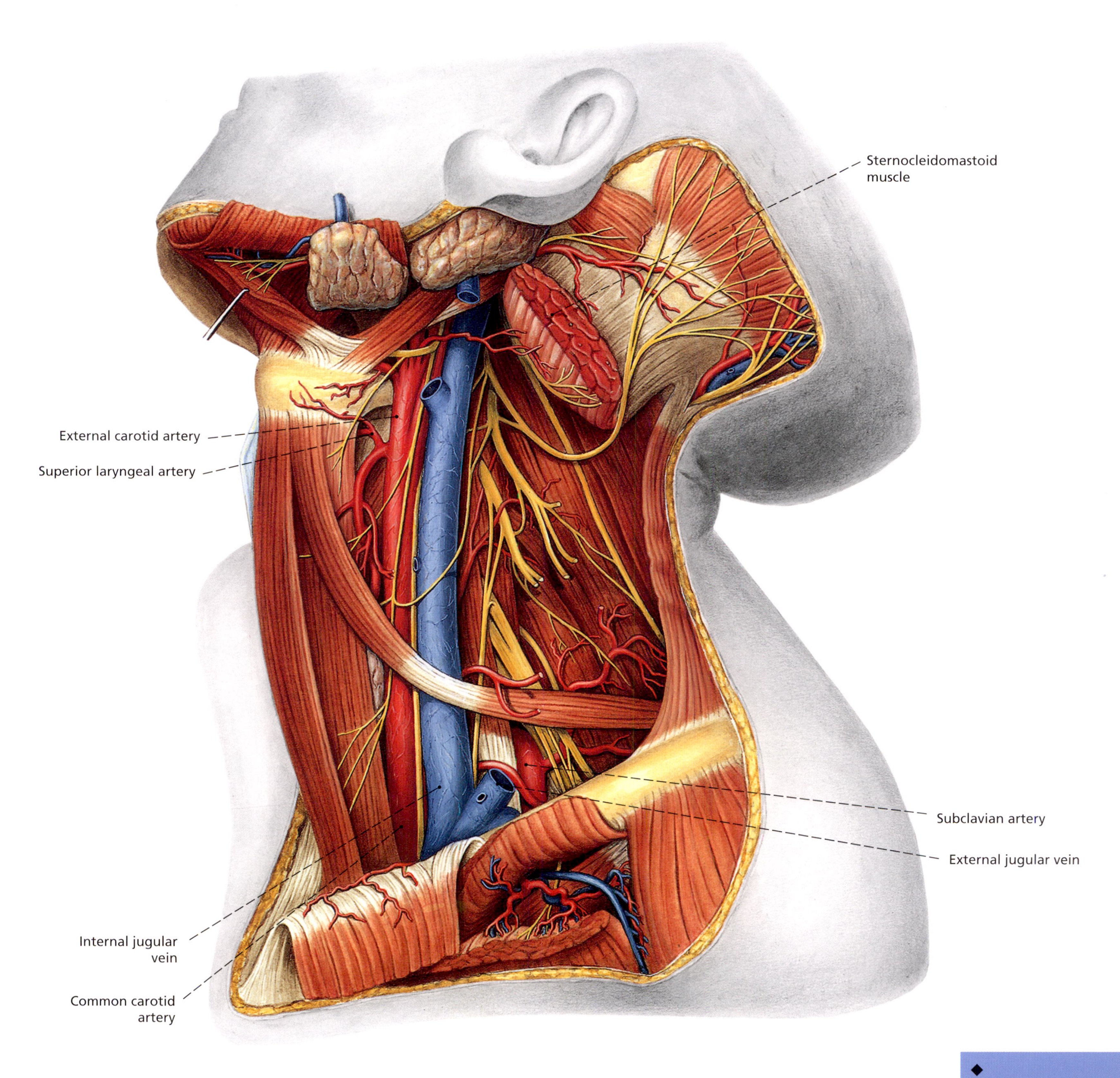

◆
Fig. 271
Blood vessels in the cervical area, viewed from the left side. Most of the sternocleido-mastoid muscle has been cut away.

◆
Fig. 272
Blood vessels of the neck and upper thoracic region, viewed from the front. The superficial layer of the neck musculature and the upper part of the thoracic wall have been cut away.

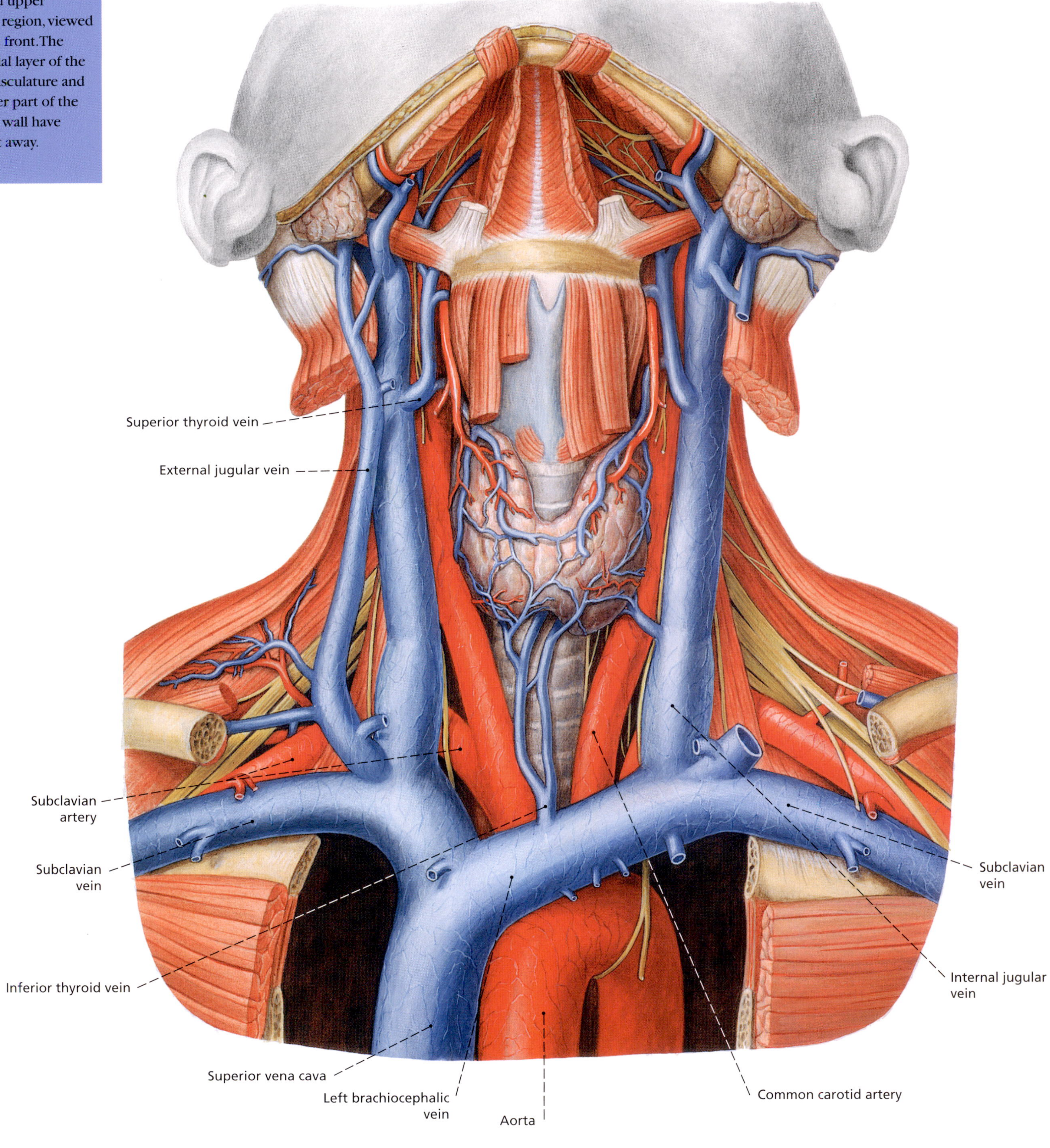

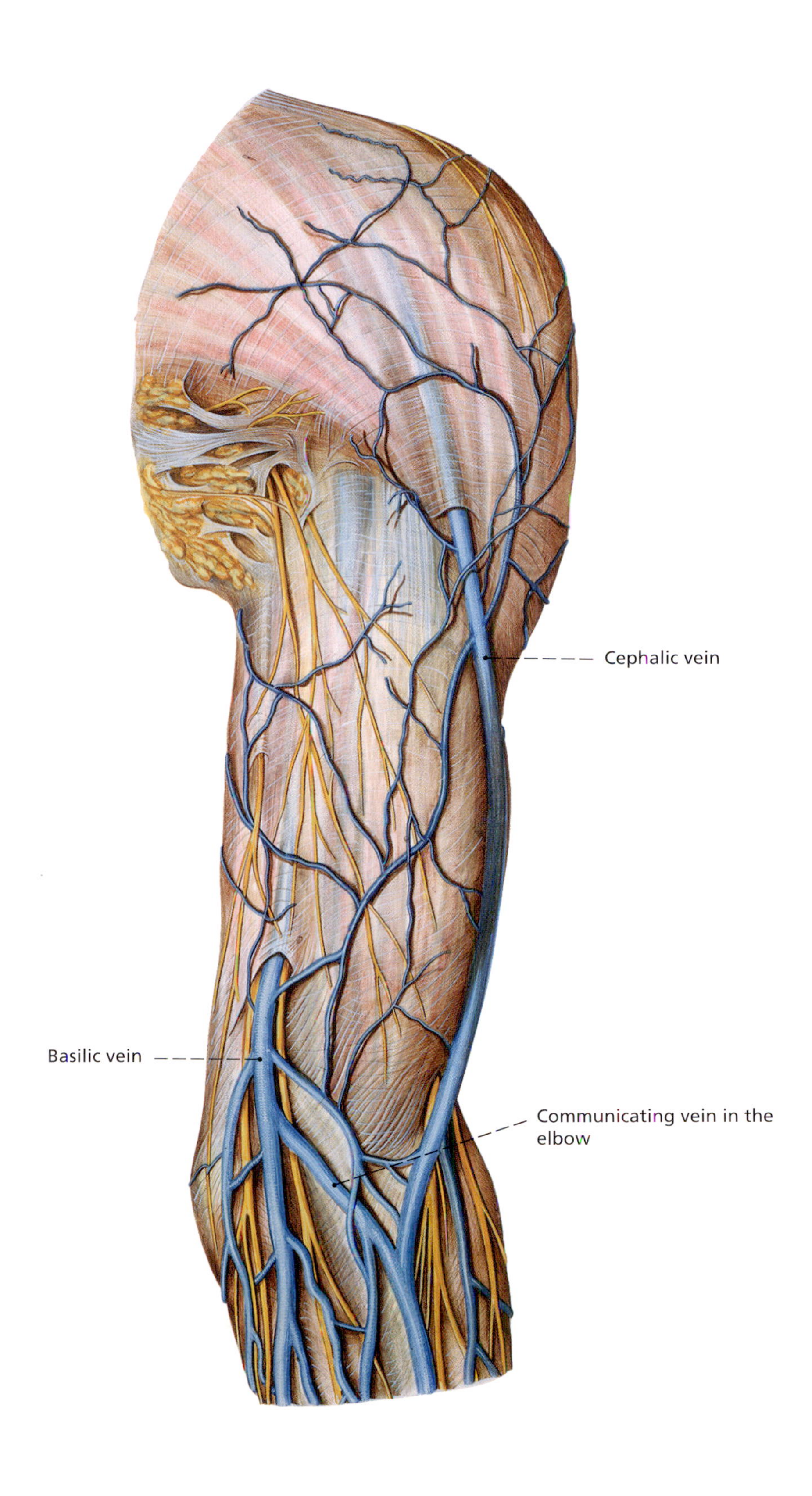

◆
Fig. 273
Superficial veins of the left upper arm and the elbow, viewed from the front. (The cutaneous nerves are shown in yellow.)

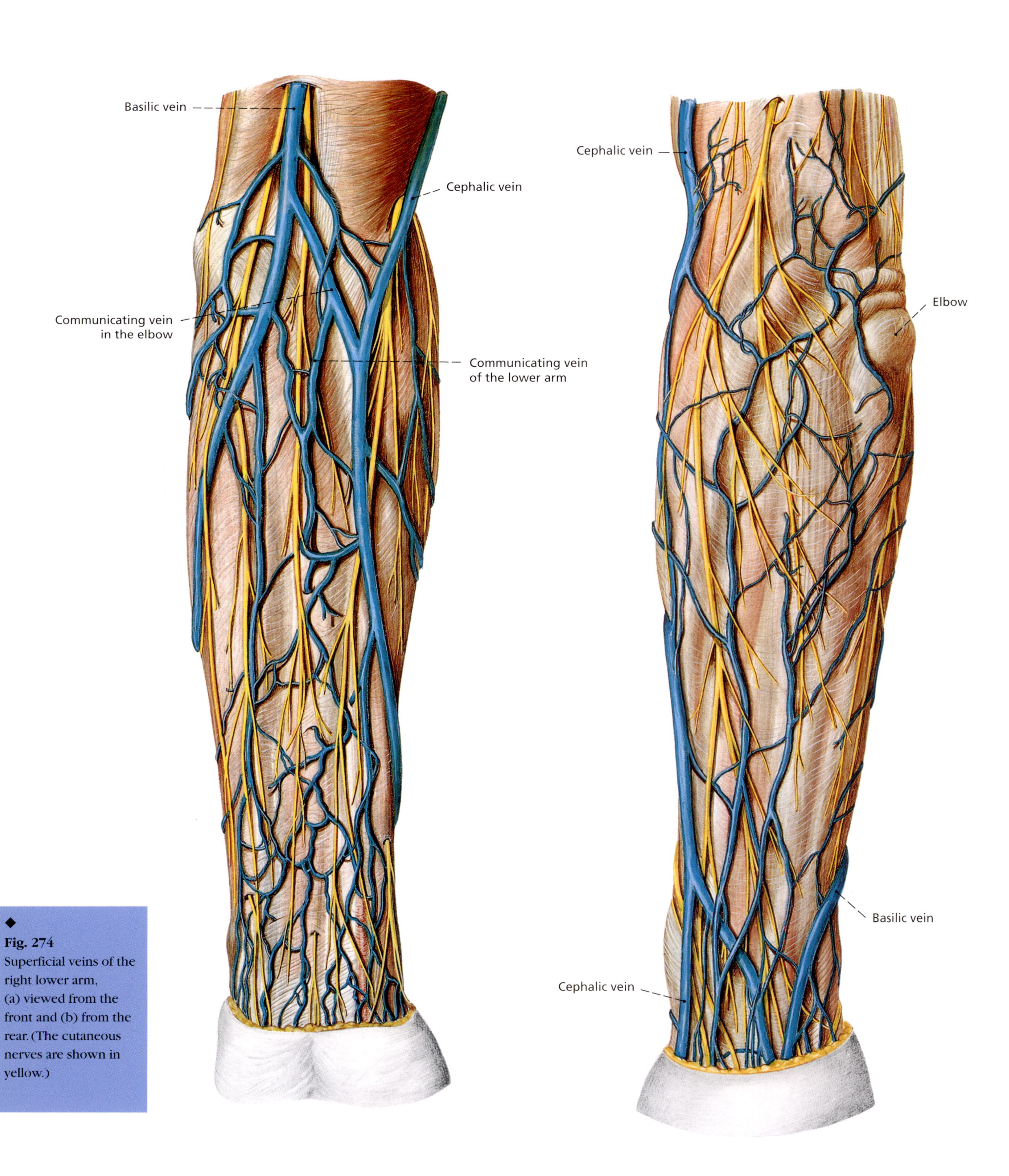

◆
Fig. 274
Superficial veins of the right lower arm, (a) viewed from the front and (b) from the rear. (The cutaneous nerves are shown in yellow.)

◆ **Fig. 274a**

◆ **Fig. 274b**

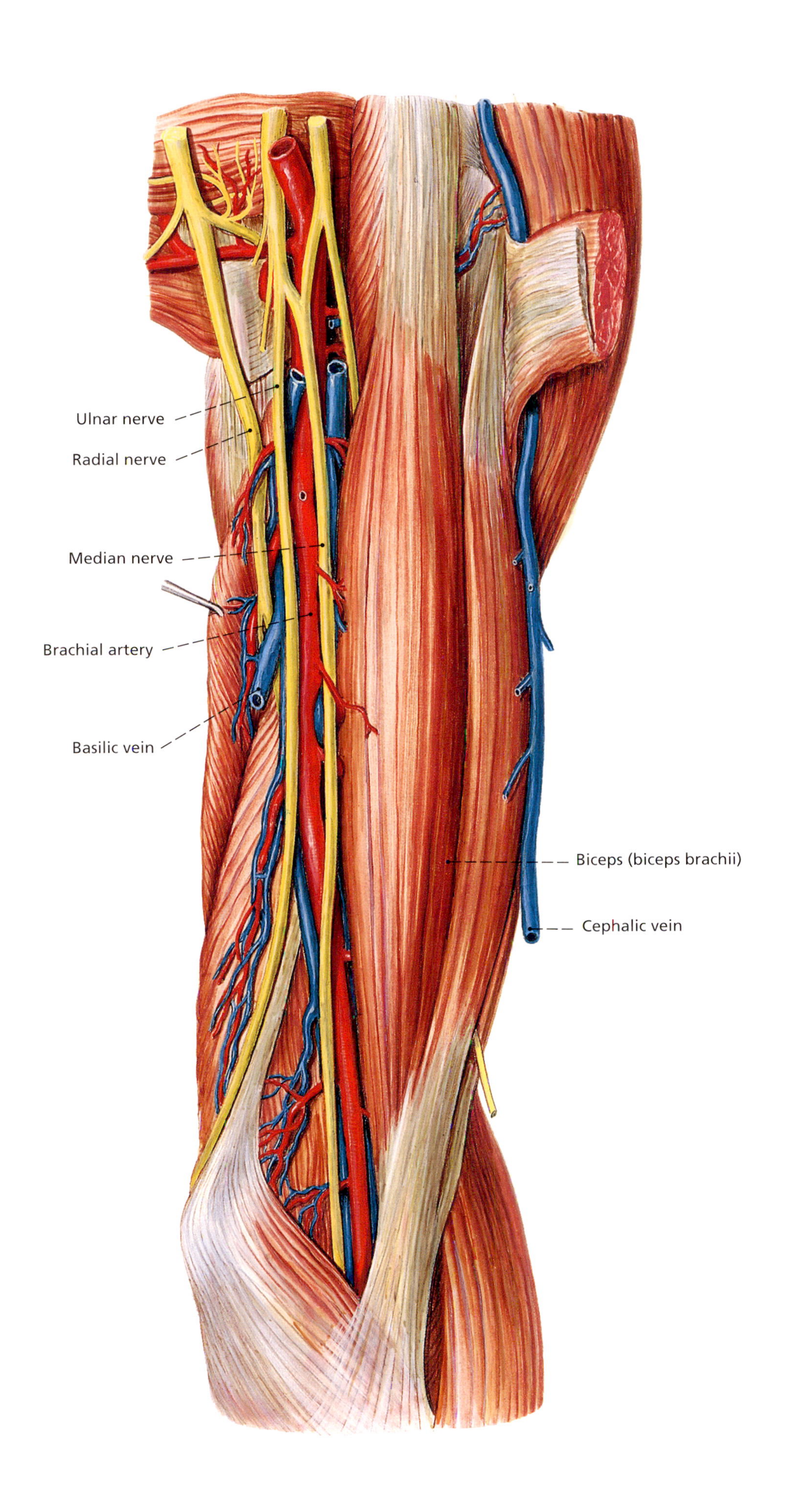

◆
Fig. 275
Blood vessels of the left upper arm, viewed from the front. (The nerves are shown in yellow.)

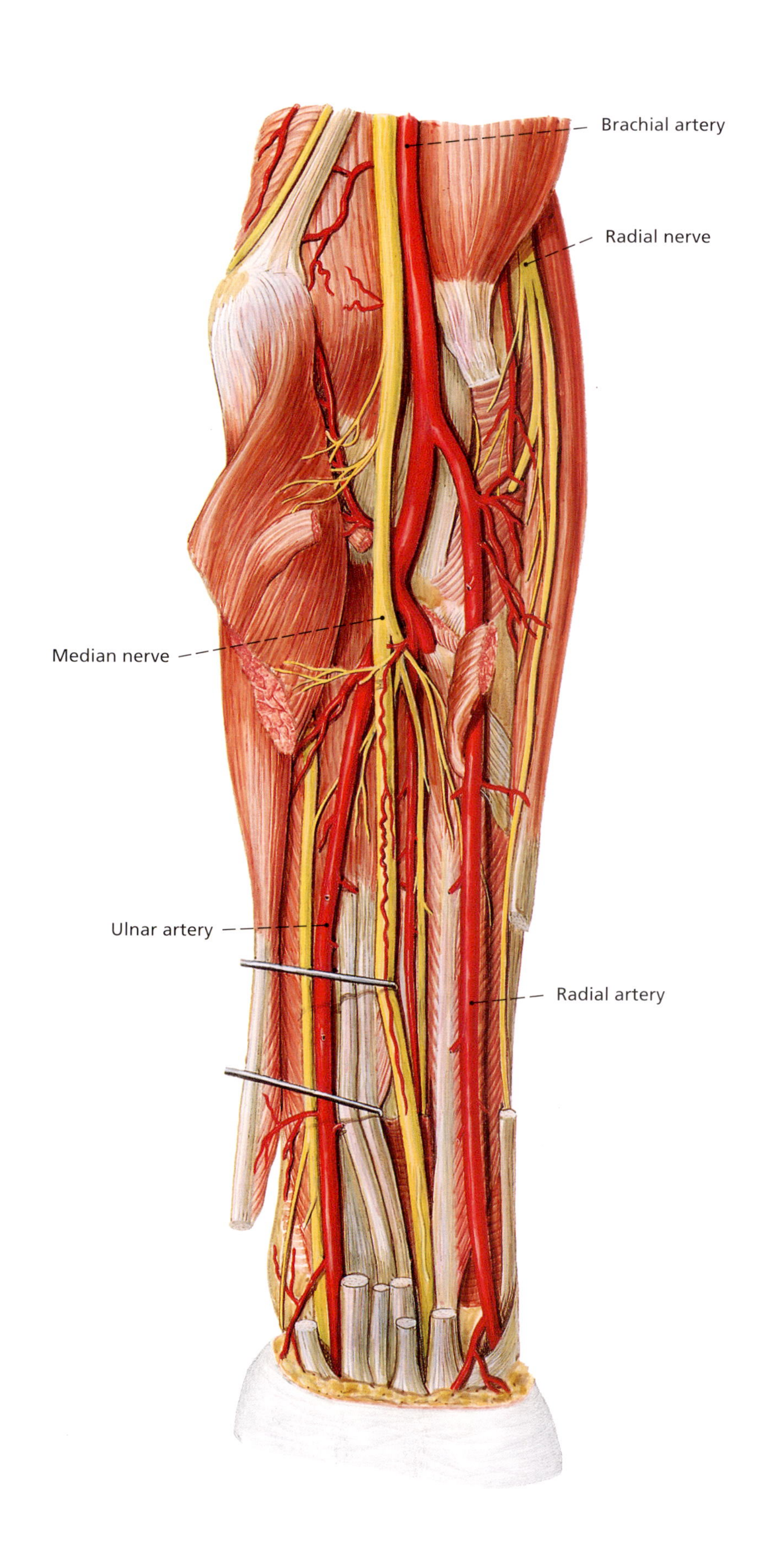

◆
Fig. 276
Arteries of the left lower arm, viewed from the front. The superficial muscles of the lower arm have been partially cut away. (Nerves are shown in yellow.)

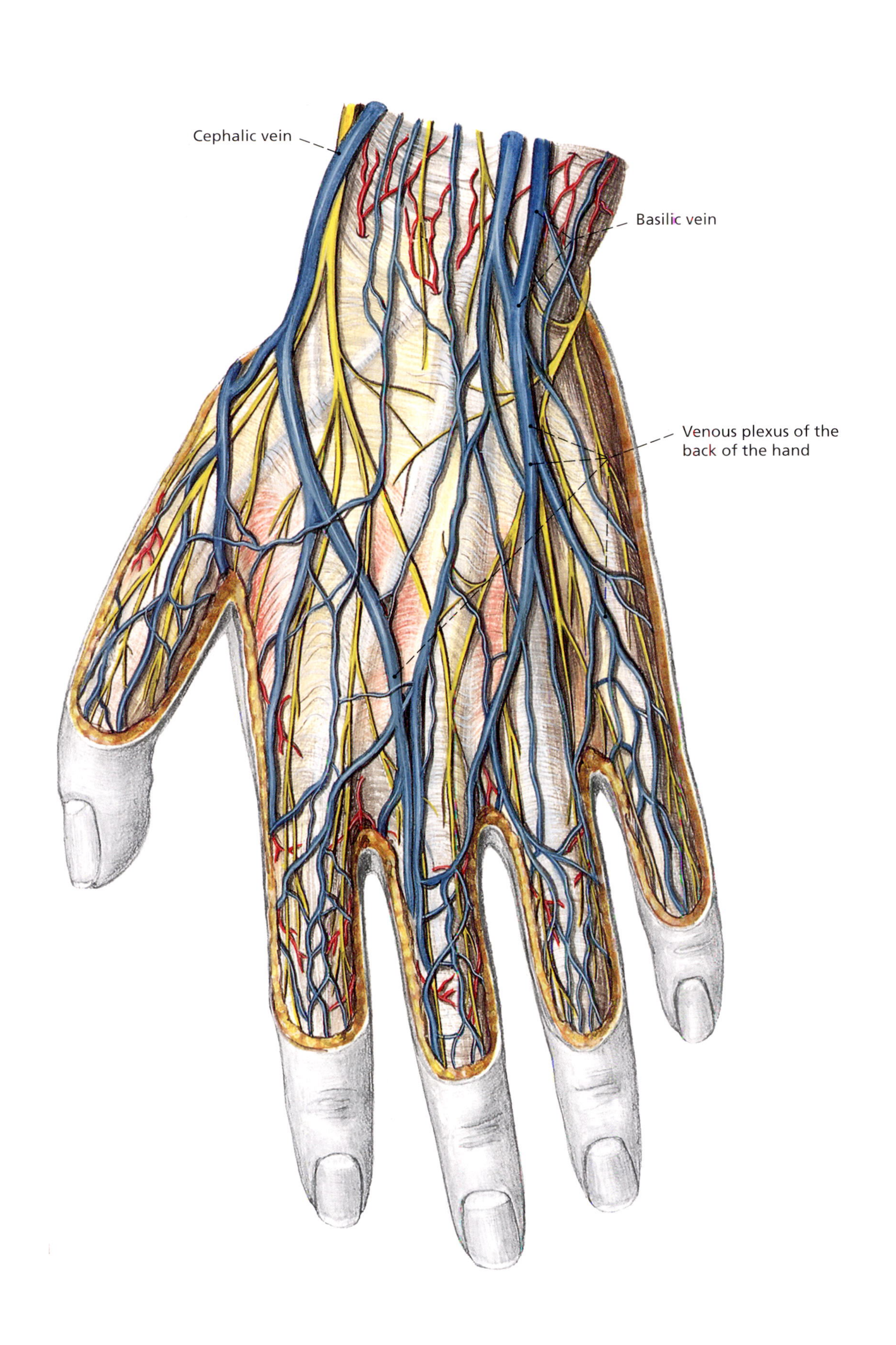

◆
Fig. 277
Venous plexus of the back of the left hand. (Cutaneous nerves are shown in yellow, small arteries in red.)

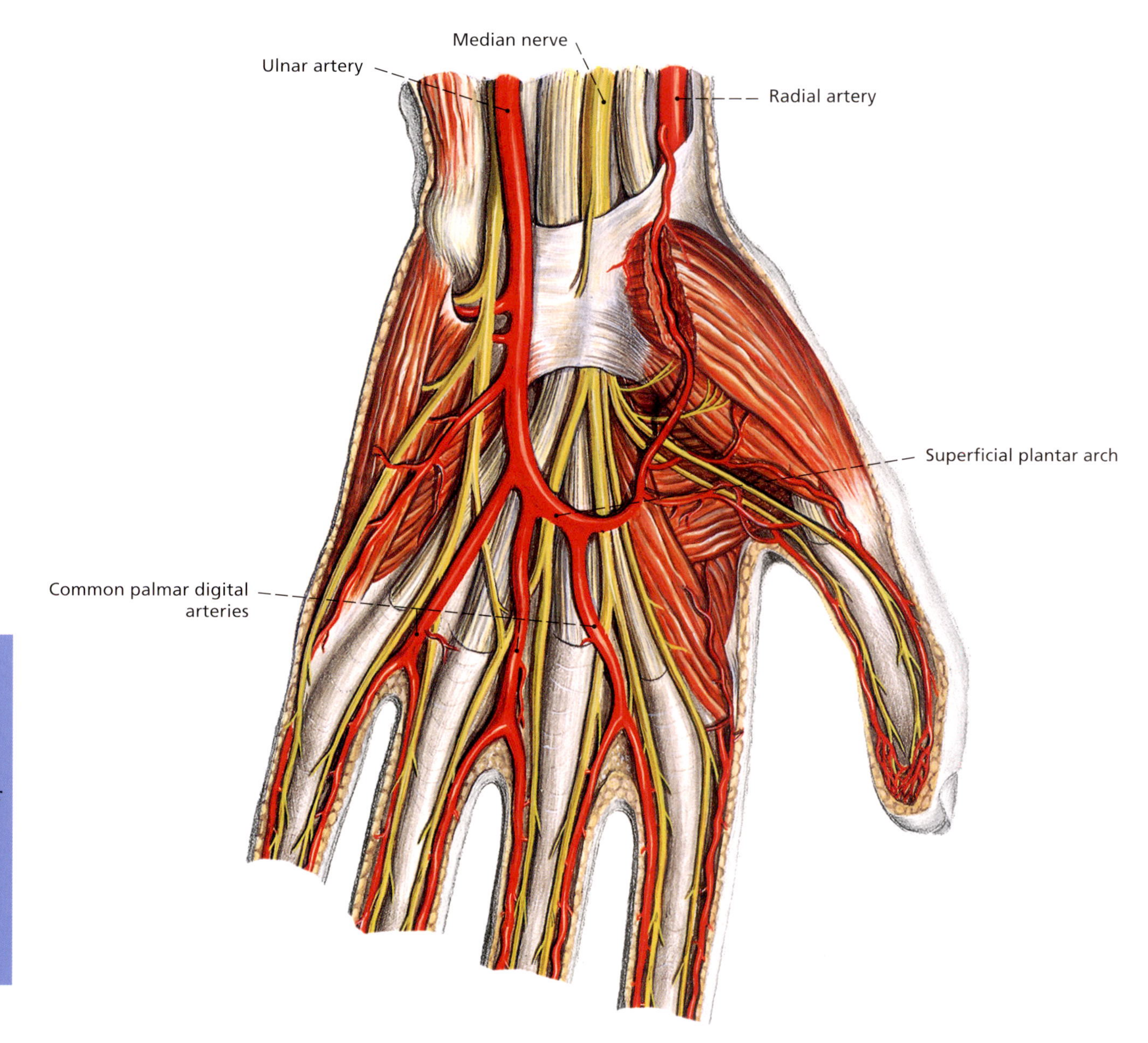

◆
Fig. 278
Arteries of the palm of the hand (deep layer). The tendinous layer of the palm and the flexor muscles of the fingers have been cut away. (Nerves are shown in yellow.)

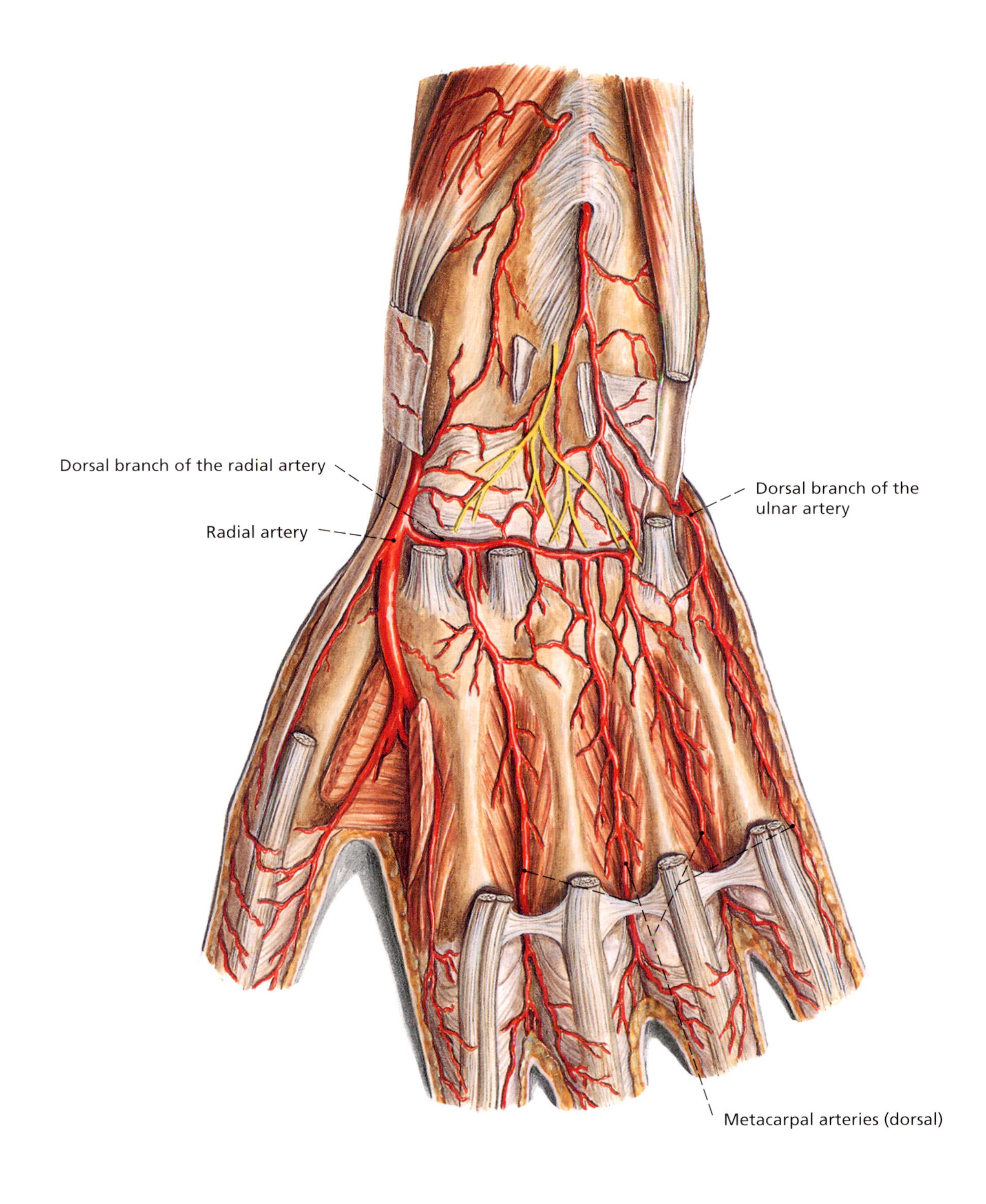

◆
Fig. 279
Arteries of the back of the left hand. Tendons and muscles have been cut away.

◆
Fig. 280
Configuration of the thoracic and abdominal aorta. The thoracic organs are not shown. The trachea and esophagus have been divided.

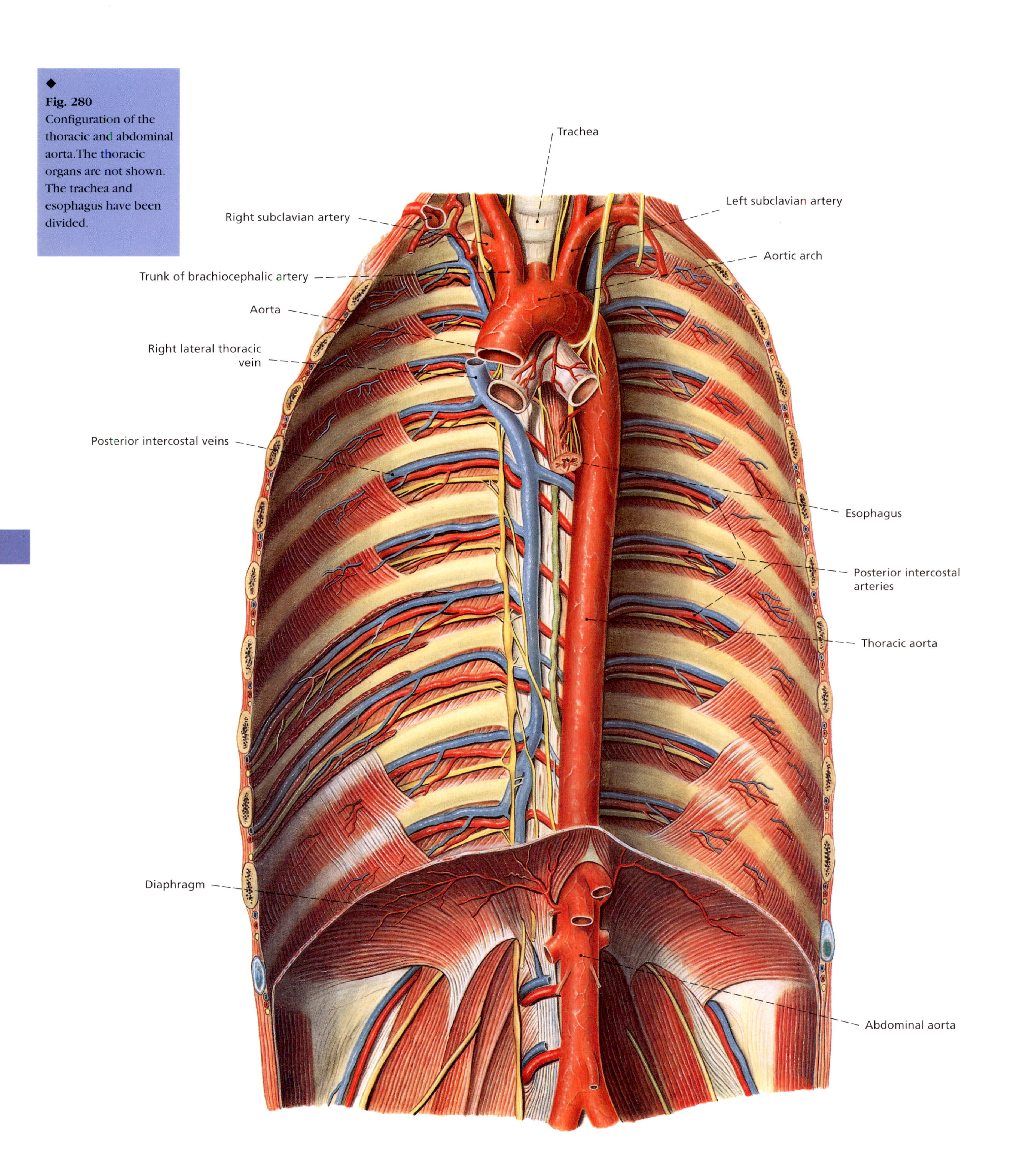

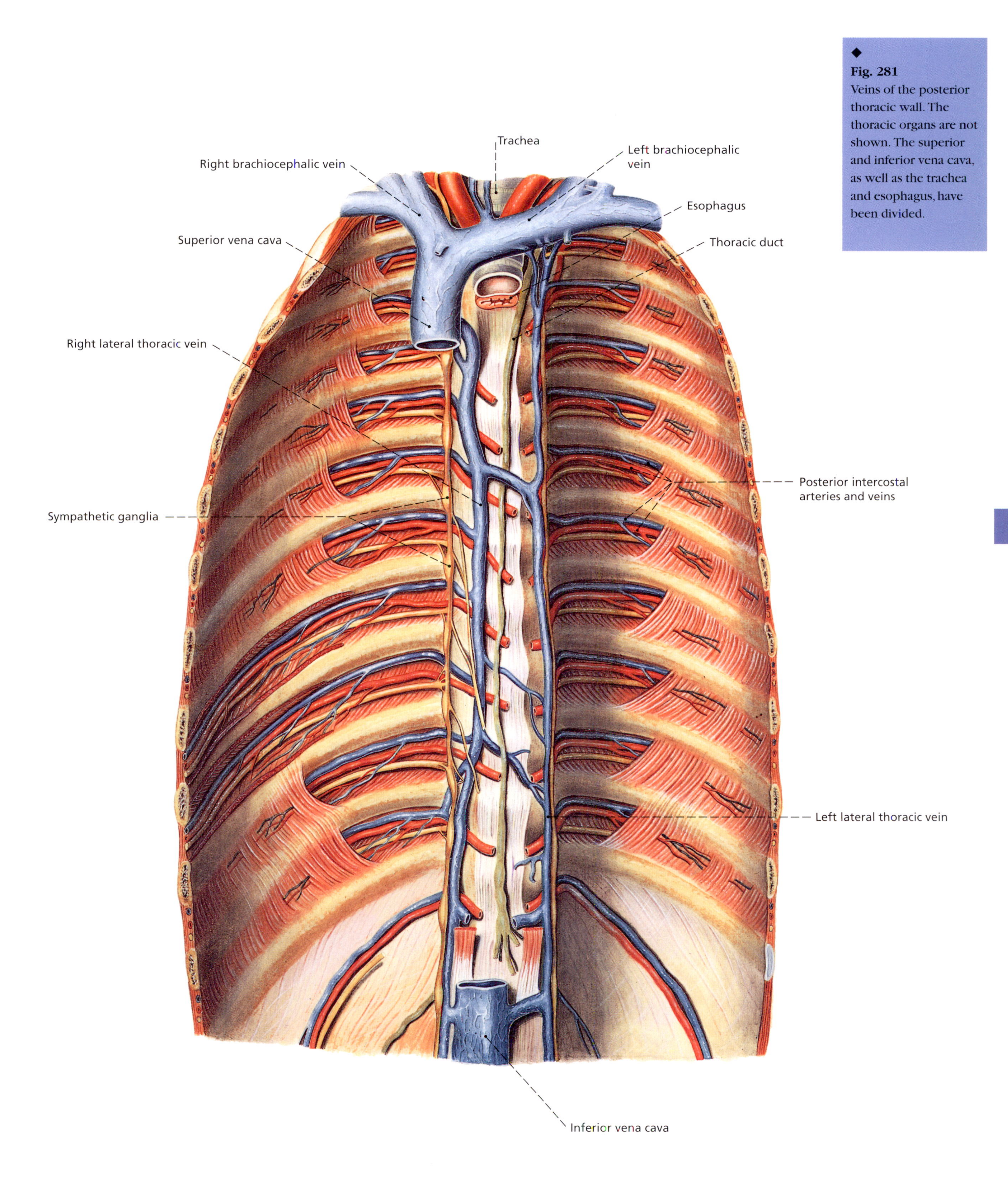

◆
Fig. 281
Veins of the posterior thoracic wall. The thoracic organs are not shown. The superior and inferior vena cava, as well as the trachea and esophagus, have been divided.

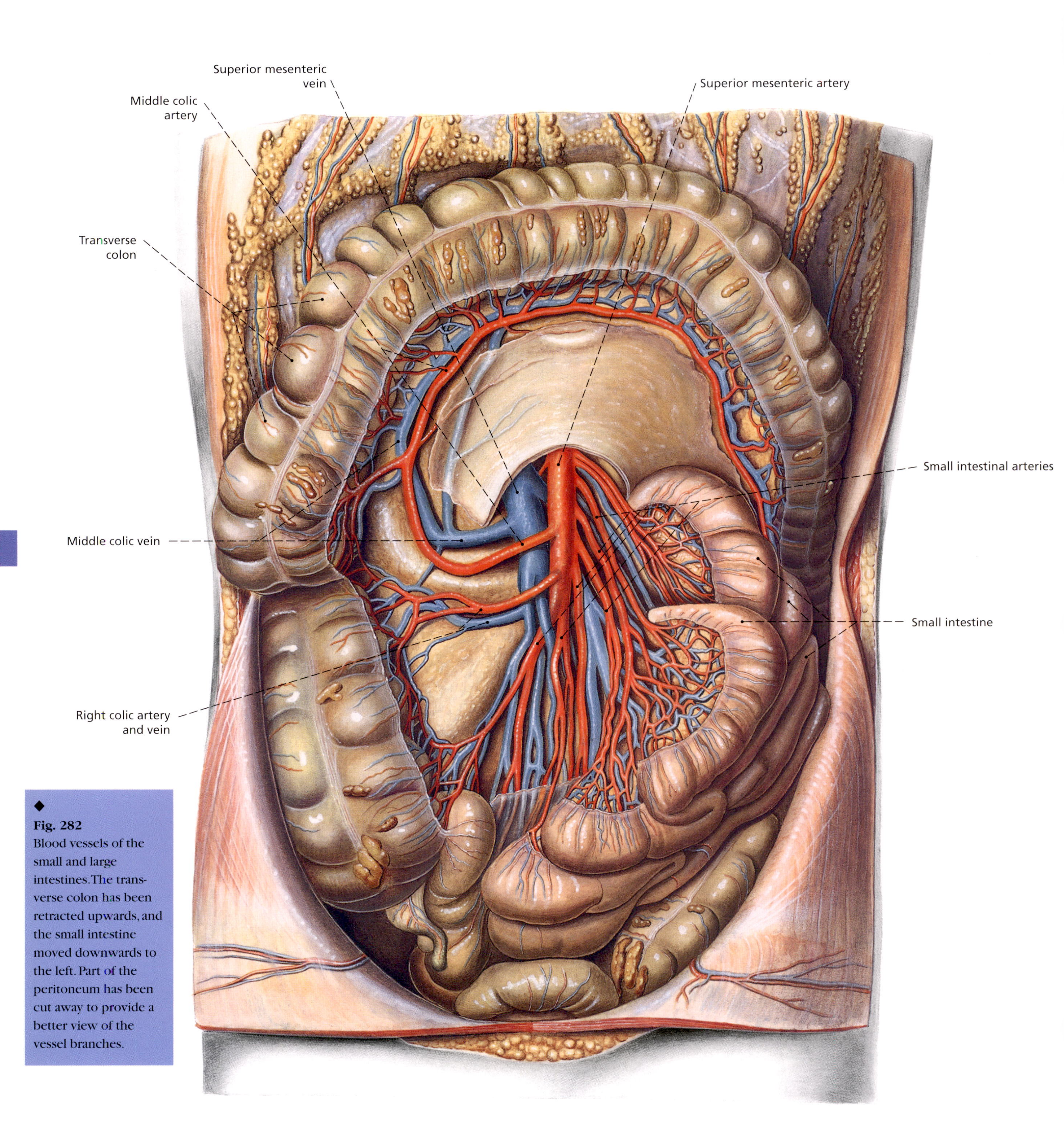

◆
Fig. 282
Blood vessels of the small and large intestines. The transverse colon has been retracted upwards, and the small intestine moved downwards to the left. Part of the peritoneum has been cut away to provide a better view of the vessel branches.

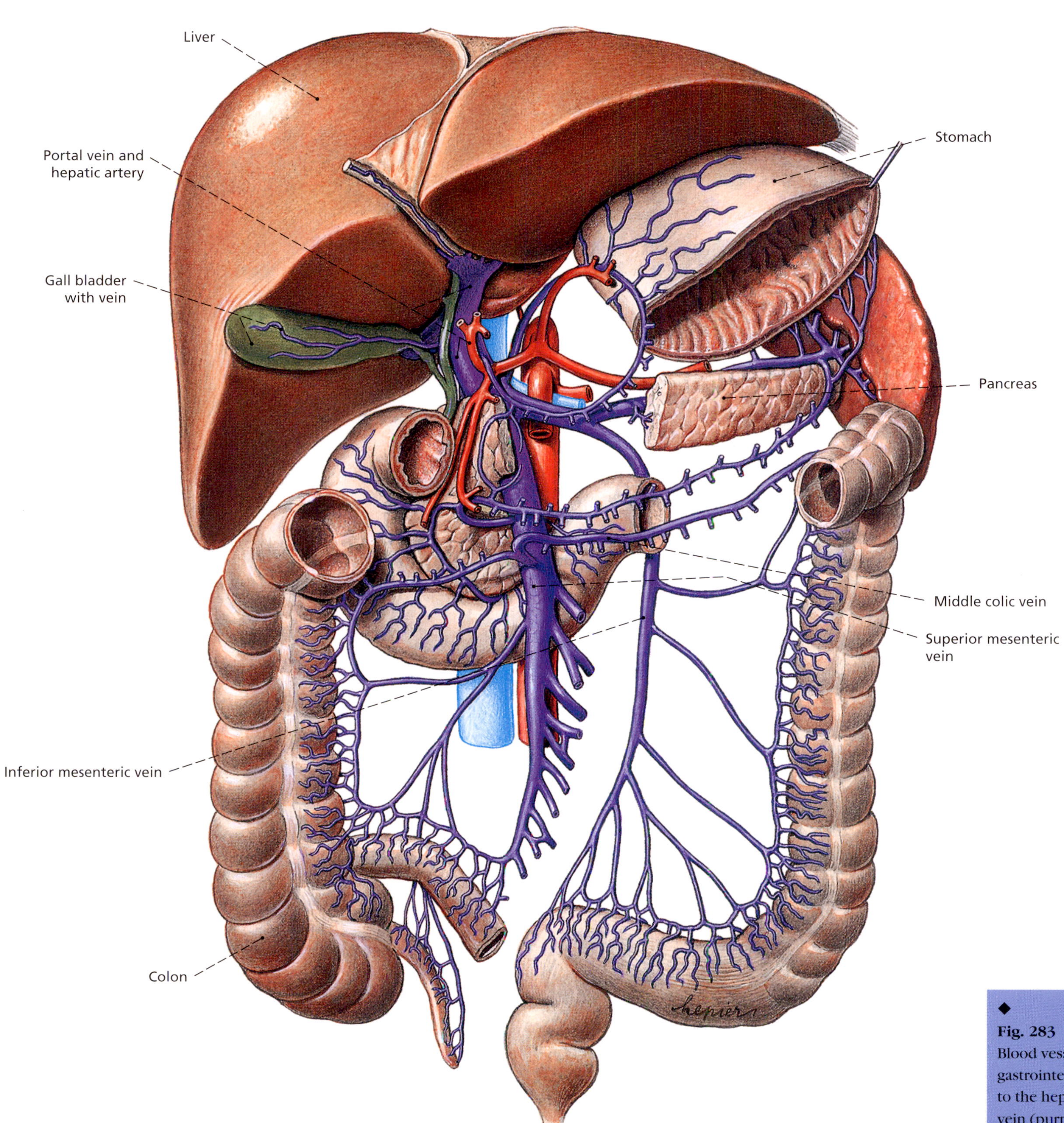

◆
Fig. 283
Blood vessels from the gastrointestinal region to the hepatic portal vein (purple; *see also* page 231). Part of the stomach and the pancreas, the transverse colon, and the entire small intestine have been removed. View from the front.

◆
Fig. 284
Arteries and veins of the pelvic organs in the male. Longitudinal section through the axis of the body, viewed from the left side. Most of the peritoneum has been cut away.

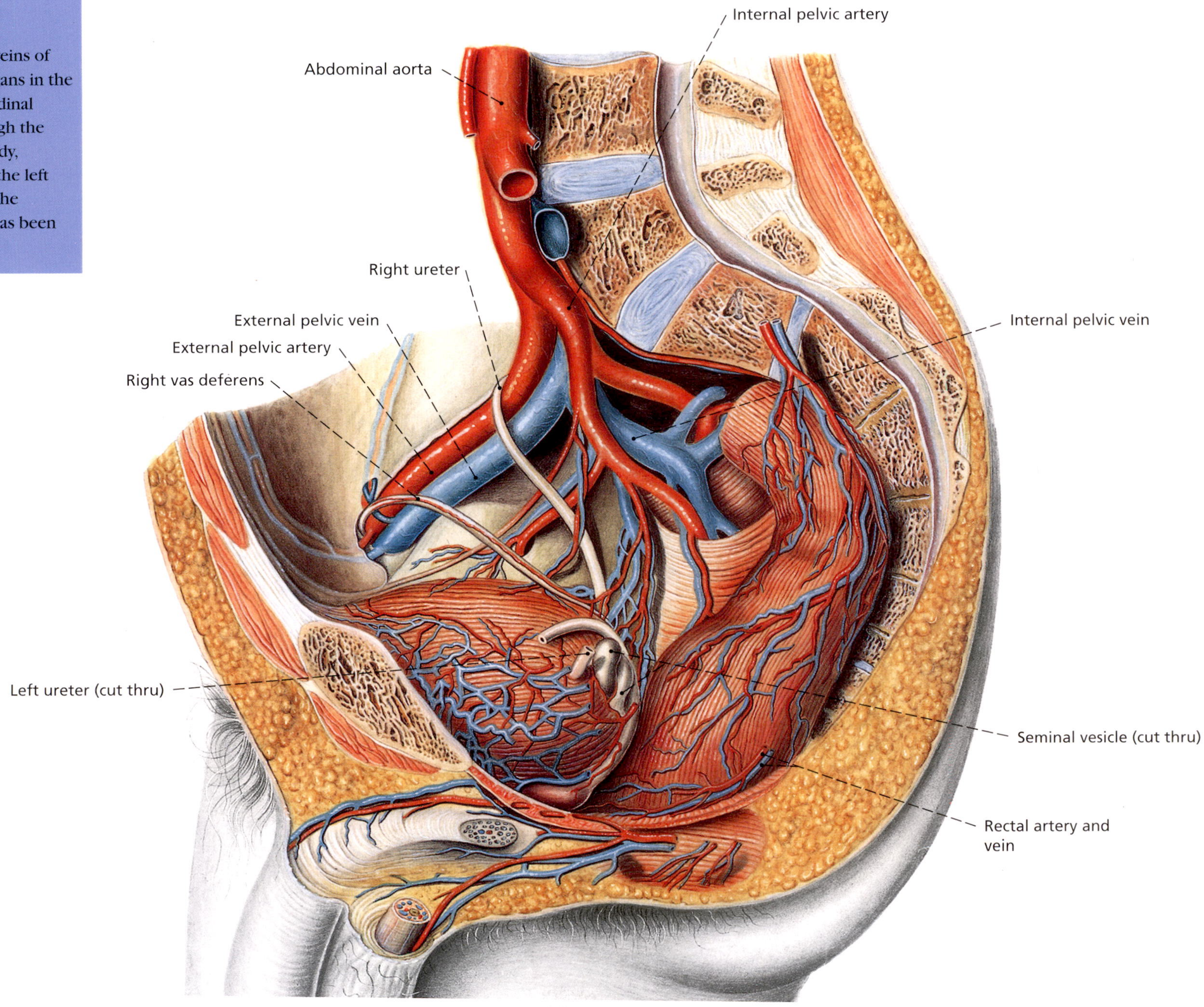

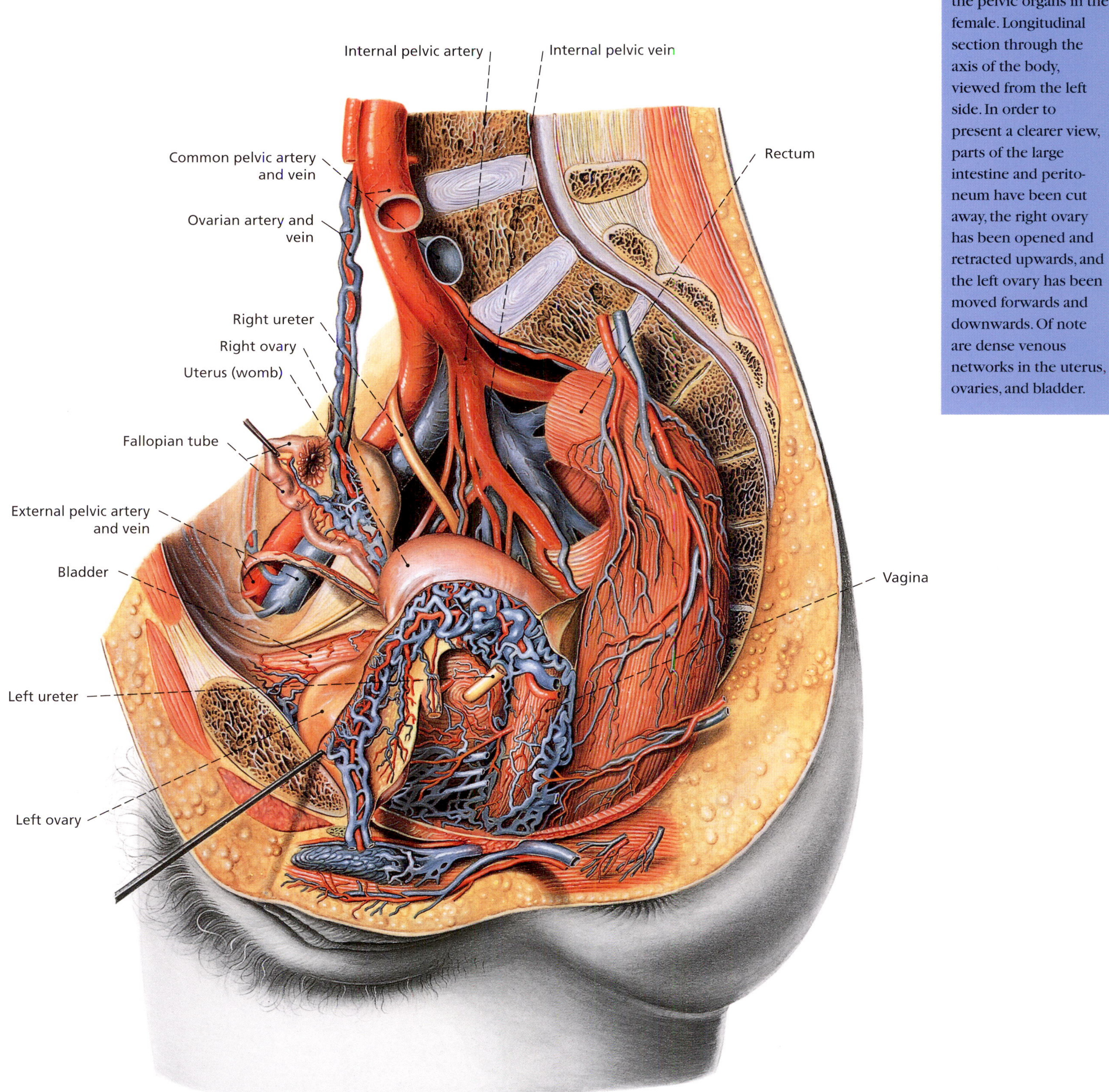

◆
Fig. 285
Arteries and veins of the pelvic organs in the female. Longitudinal section through the axis of the body, viewed from the left side. In order to present a clearer view, parts of the large intestine and peritoneum have been cut away, the right ovary has been opened and retracted upwards, and the left ovary has been moved forwards and downwards. Of note are dense venous networks in the uterus, ovaries, and bladder.

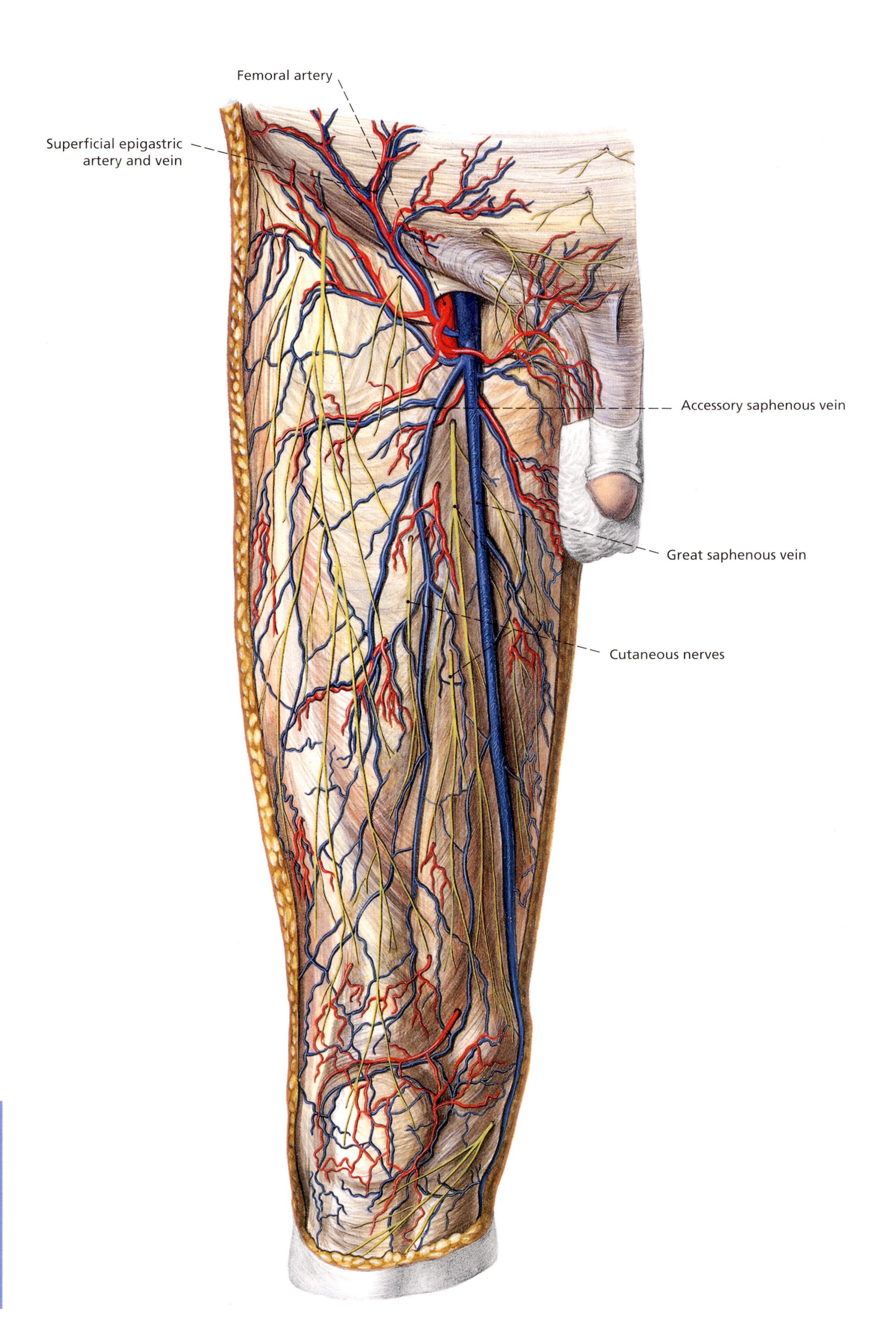

◆
Fig. 286
Superficial veins of the right thigh. (Cutaneous nerves are shown in yellow, arteries in red.)

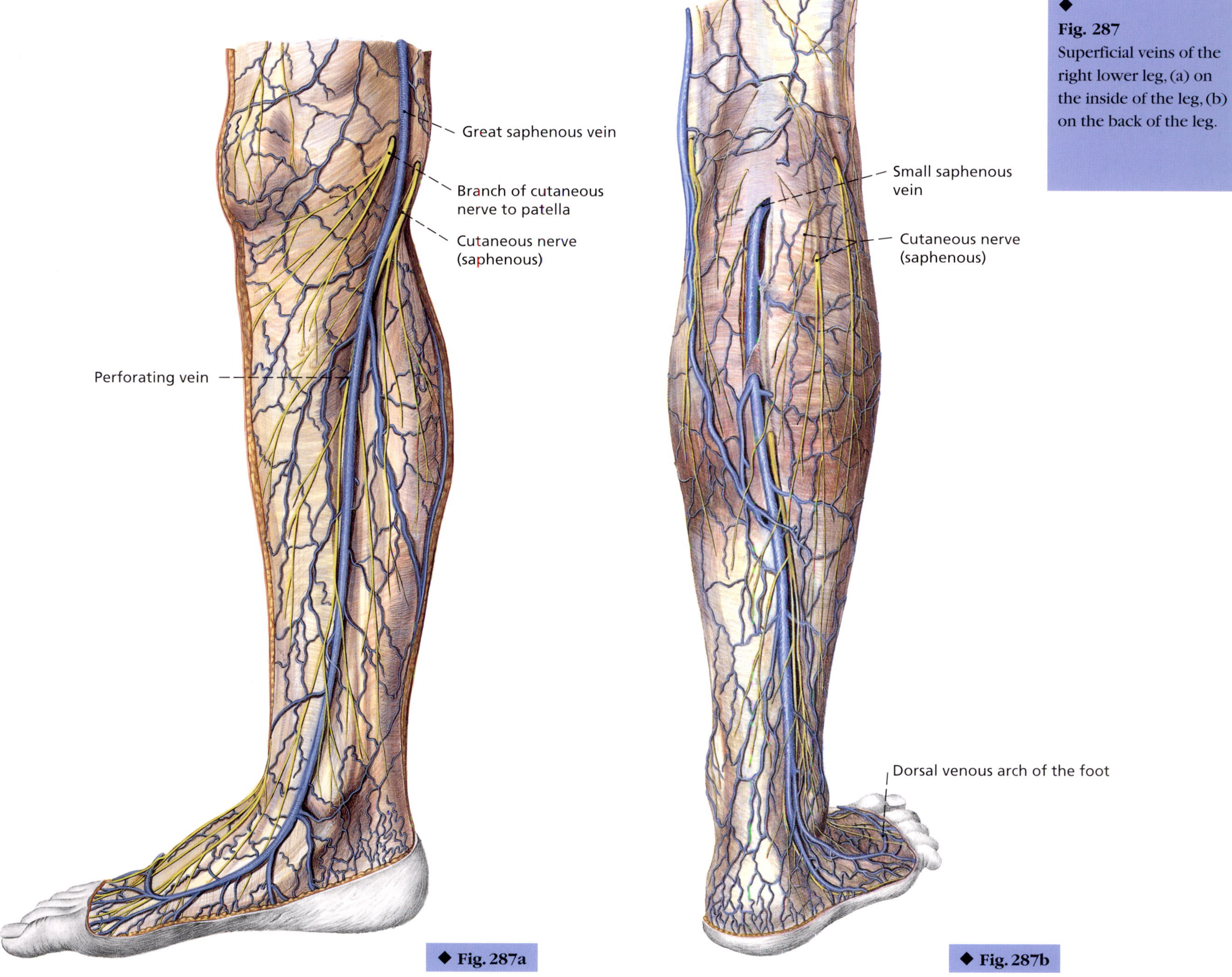

◆ Fig. 287a

◆ Fig. 287b

◆
Fig. 287
Superficial veins of the right lower leg, (a) on the inside of the leg, (b) on the back of the leg.

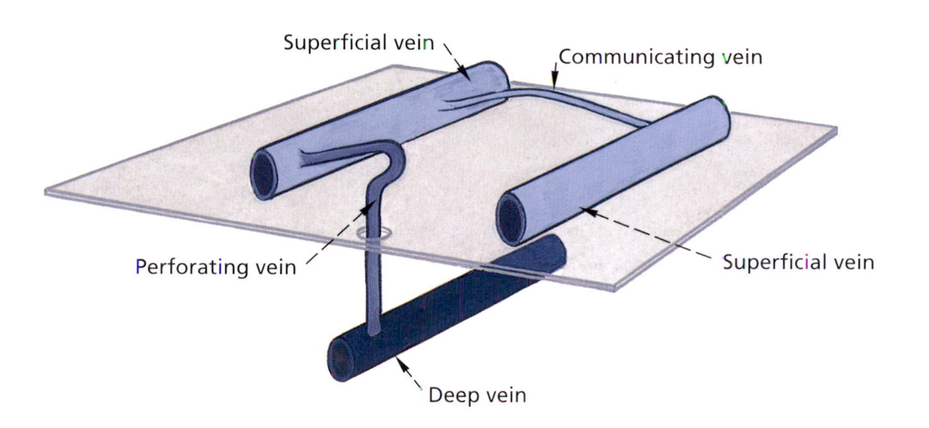

◆
Fig. 288
Diagrammatic representation of the venous flow system in the legs. In the event of occlusion of the superficial veins or one of their communicating branches (e.g., by a blood clot), the venous flow is maintained via the perforating veins (from the Latin perforare, meaning to penetrate). The perforating veins flow into the deep leg veins.

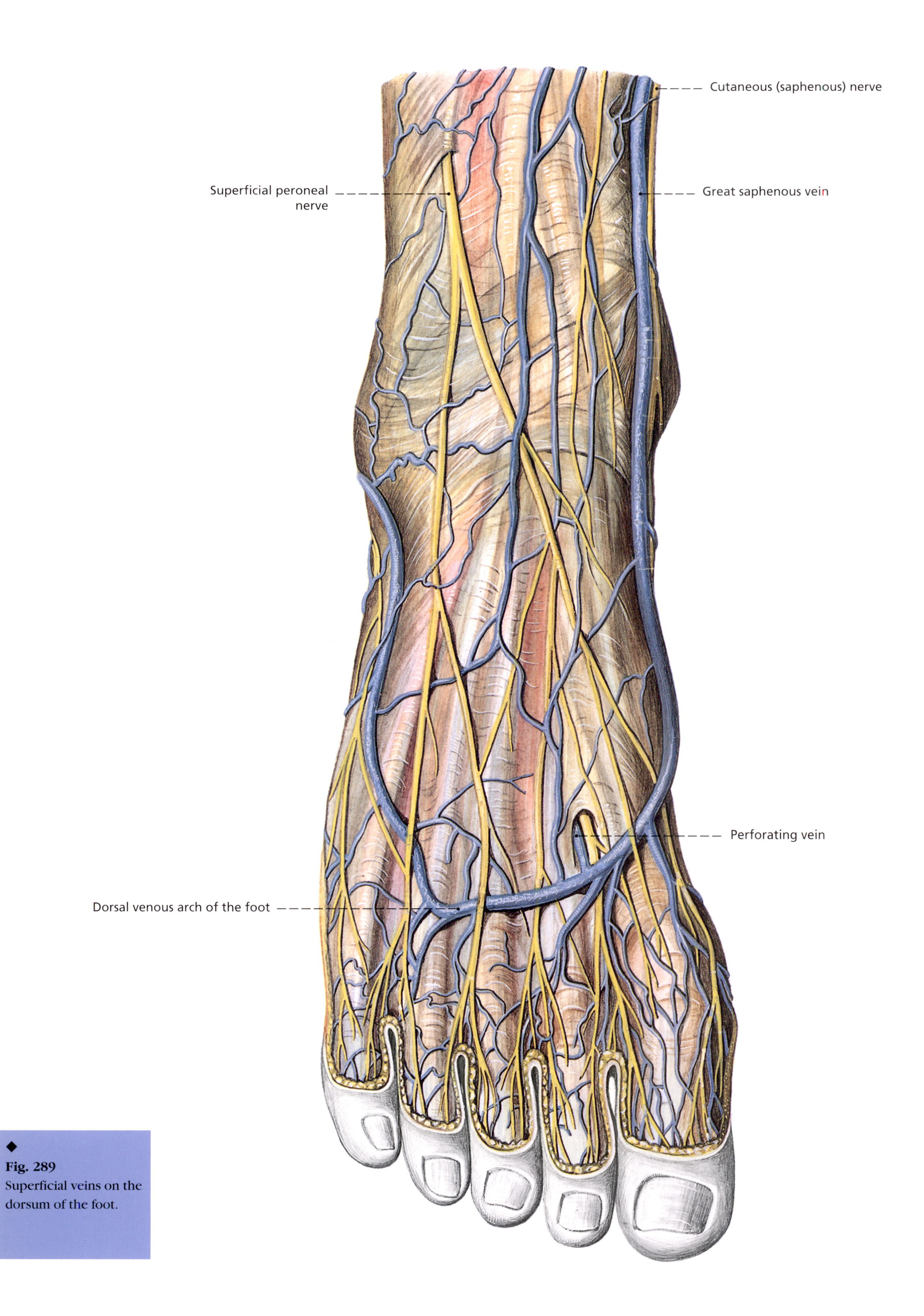

◆
Fig. 289
Superficial veins on the dorsum of the foot.

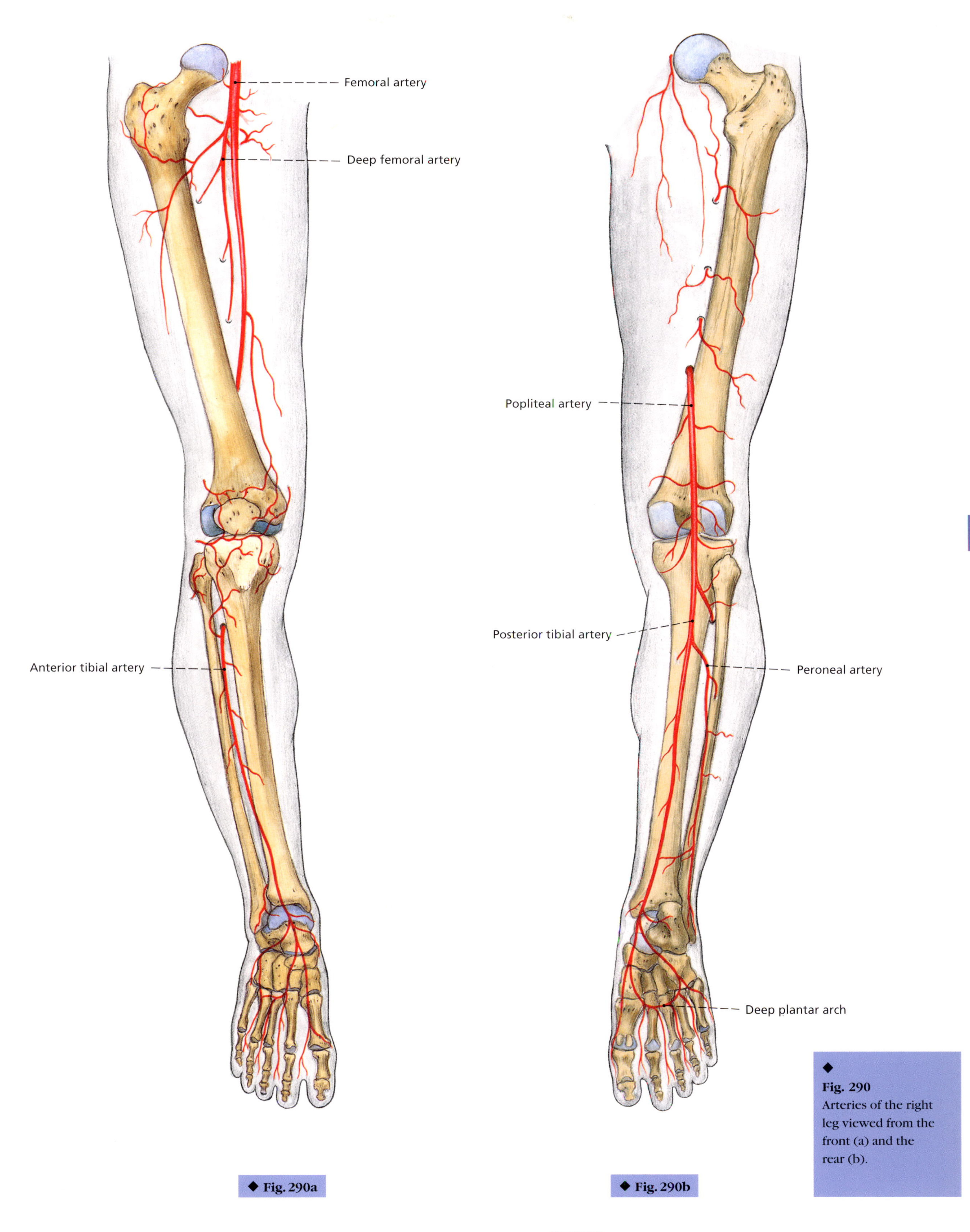

◆
Fig. 290
Arteries of the right leg viewed from the front (a) and the rear (b).

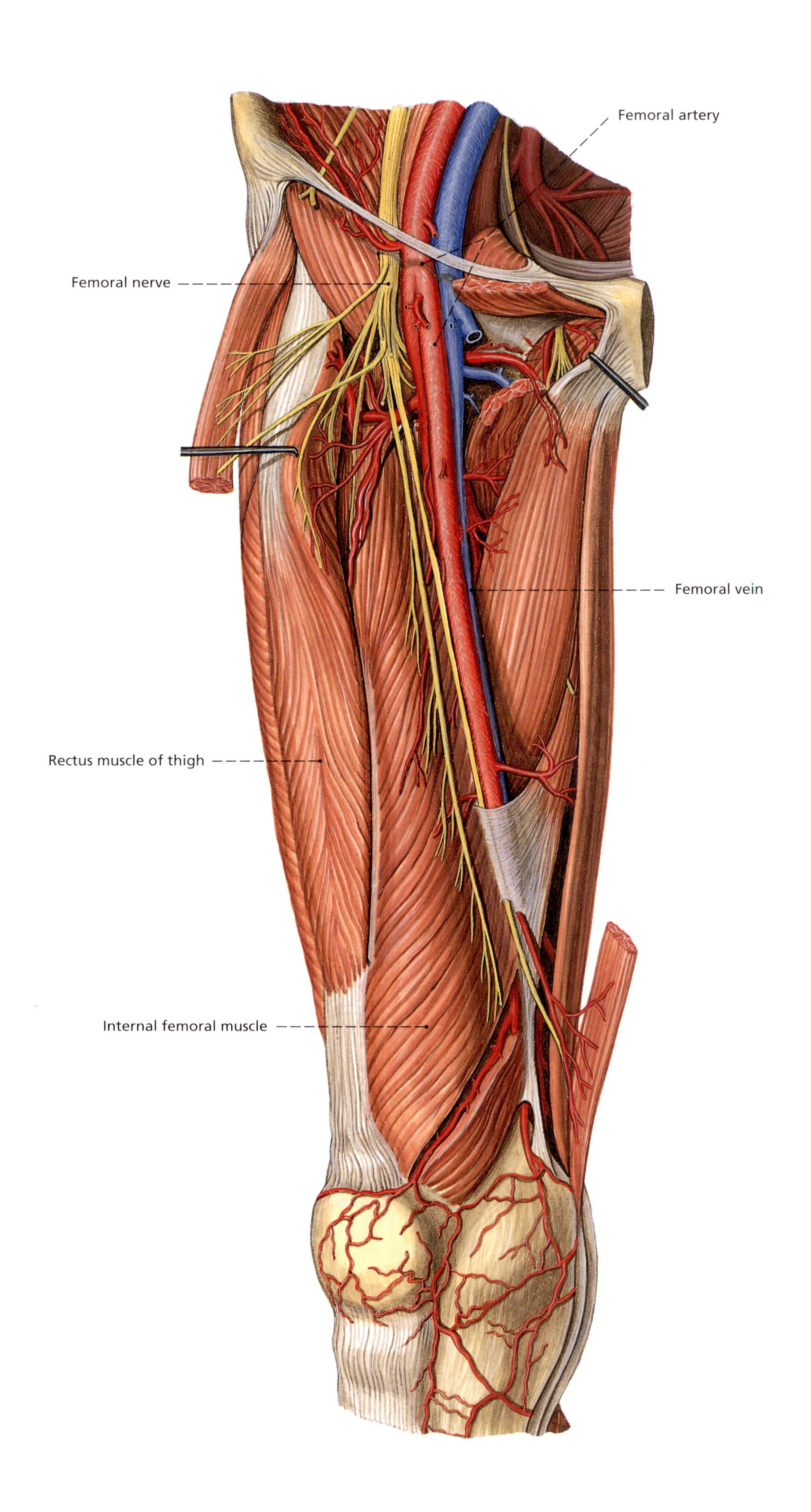

◆
Fig. 291
Arteries and veins of the right thigh (deep layer), viewed from the front. Part of the musculature of the thigh has been cut away.

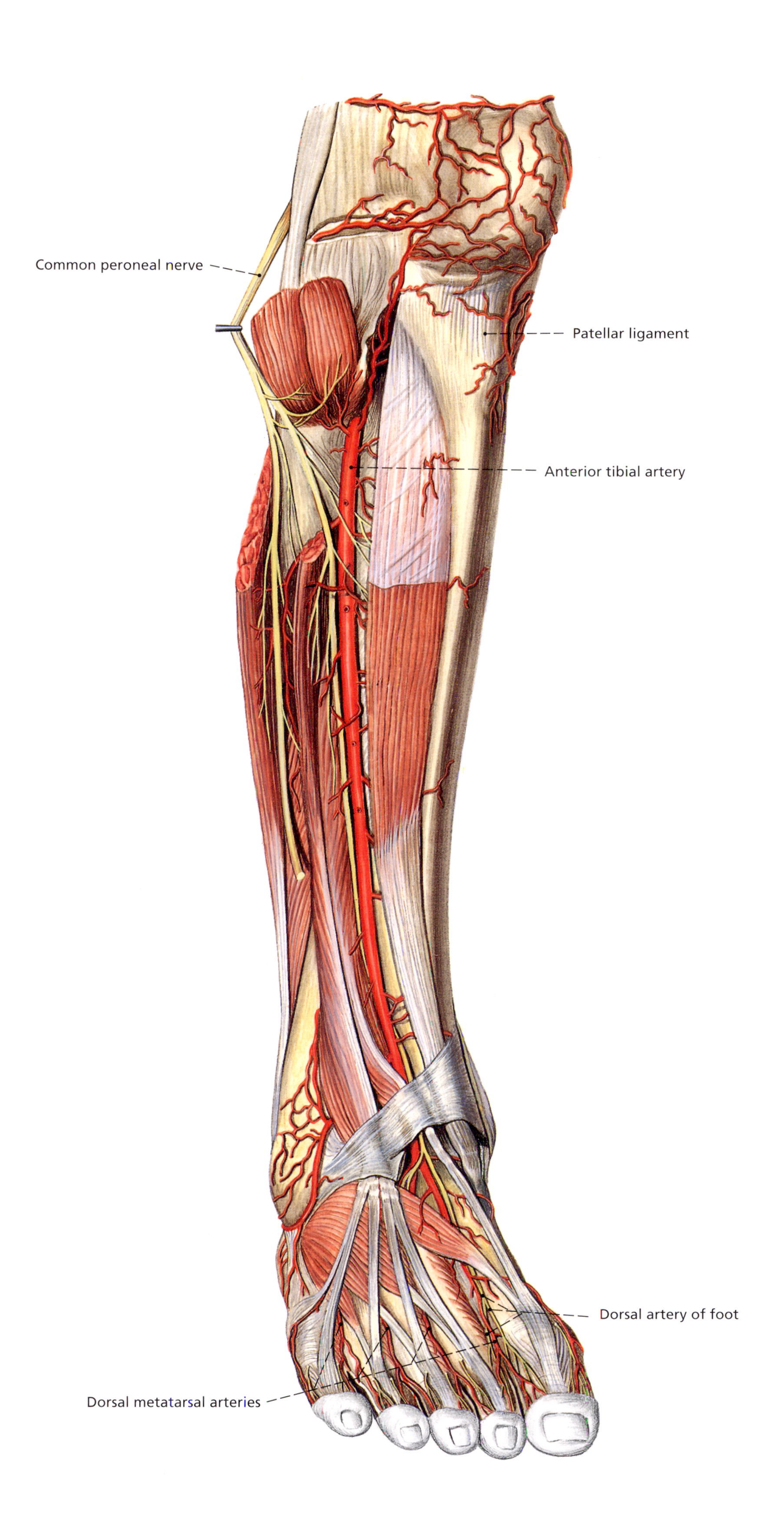

◆
Fig. 292
Arteries of the right lower leg (deep layer), viewed from the front. Part of the anterior lower leg musculature has been cut away.

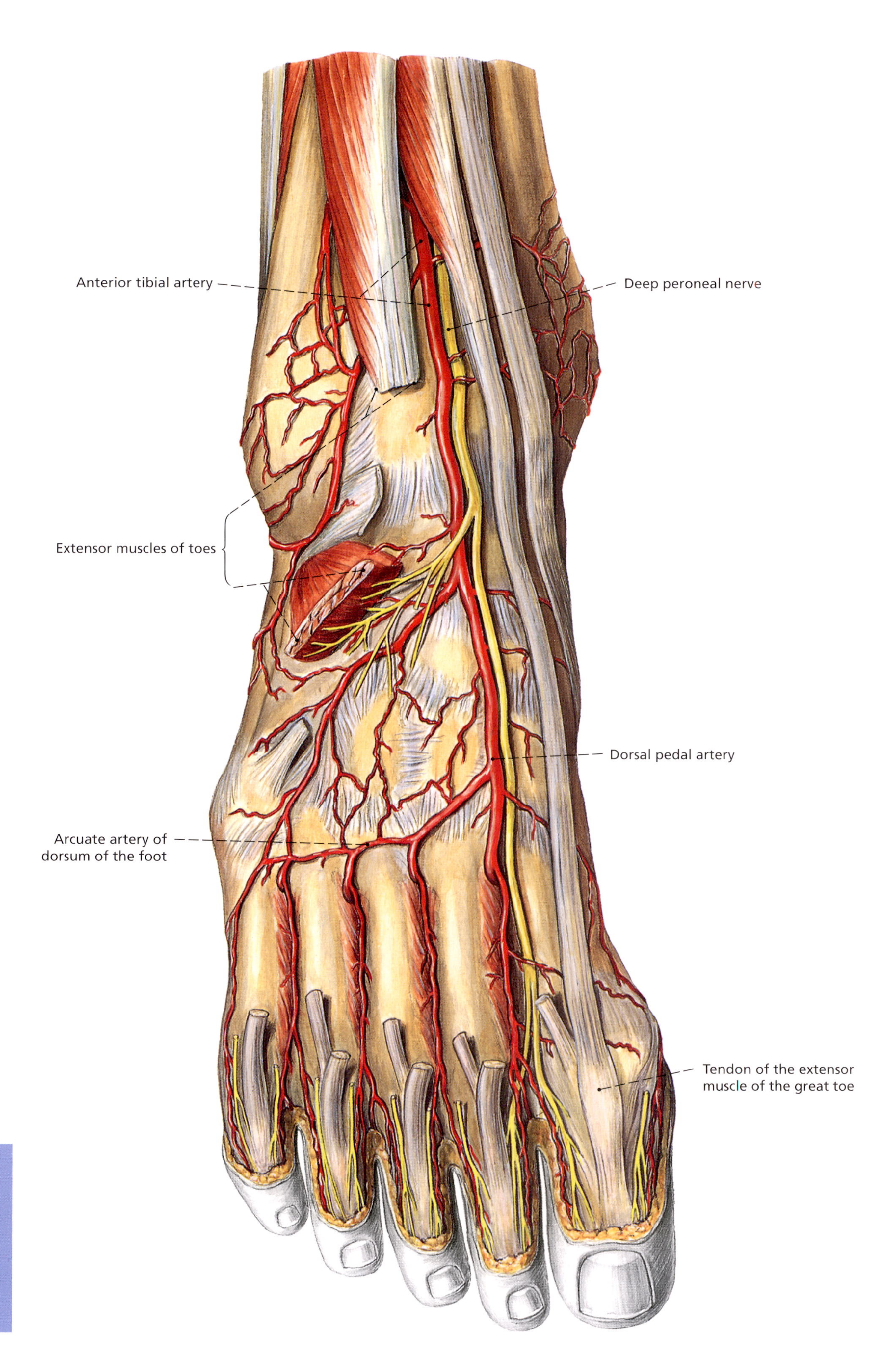

◆
Fig. 293
Arteries of the dorsum of the right foot. The metatarsal flexor muscles have been cut away.

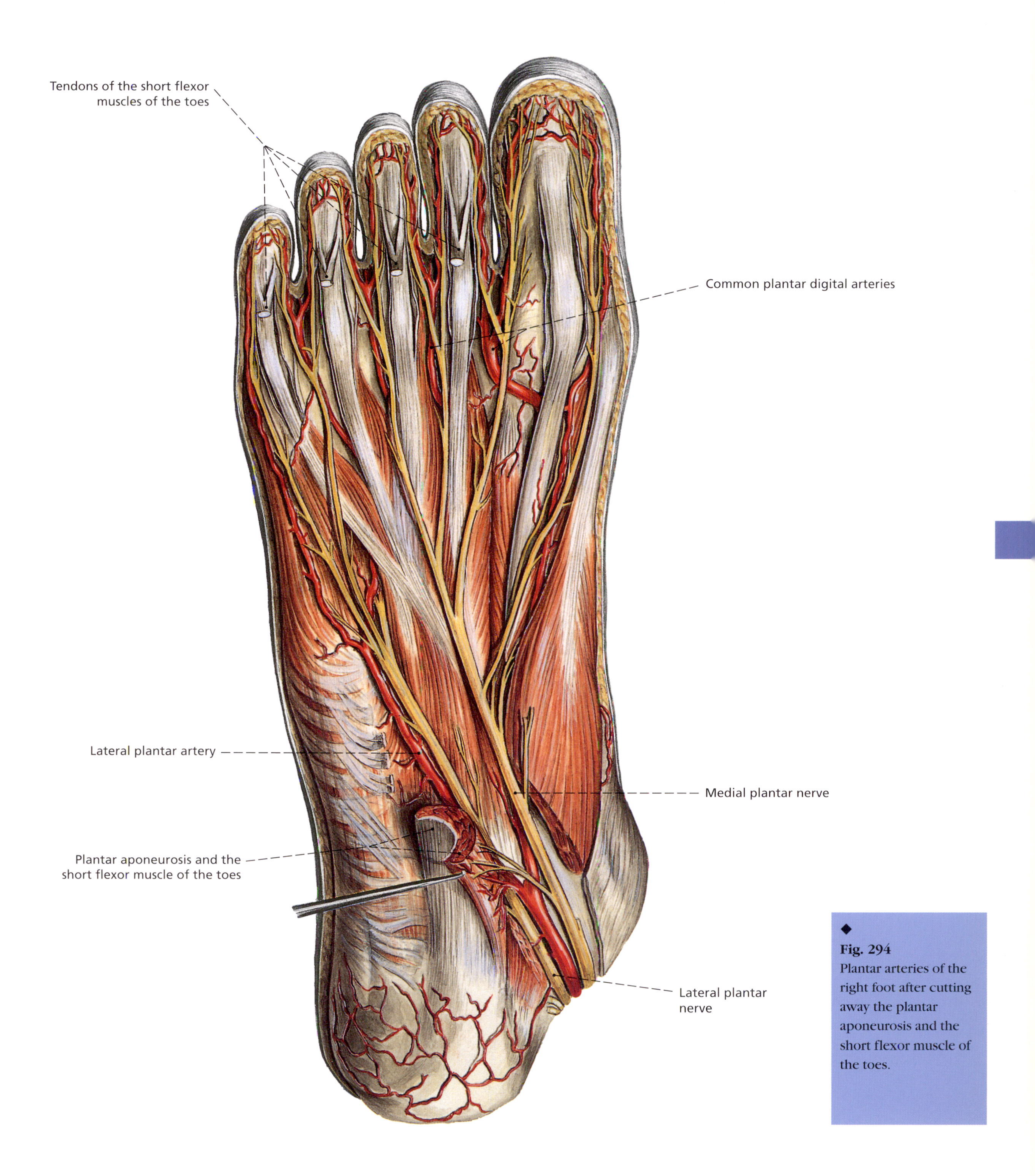

◆
Fig. 294
Plantar arteries of the right foot after cutting away the plantar aponeurosis and the short flexor muscle of the toes.

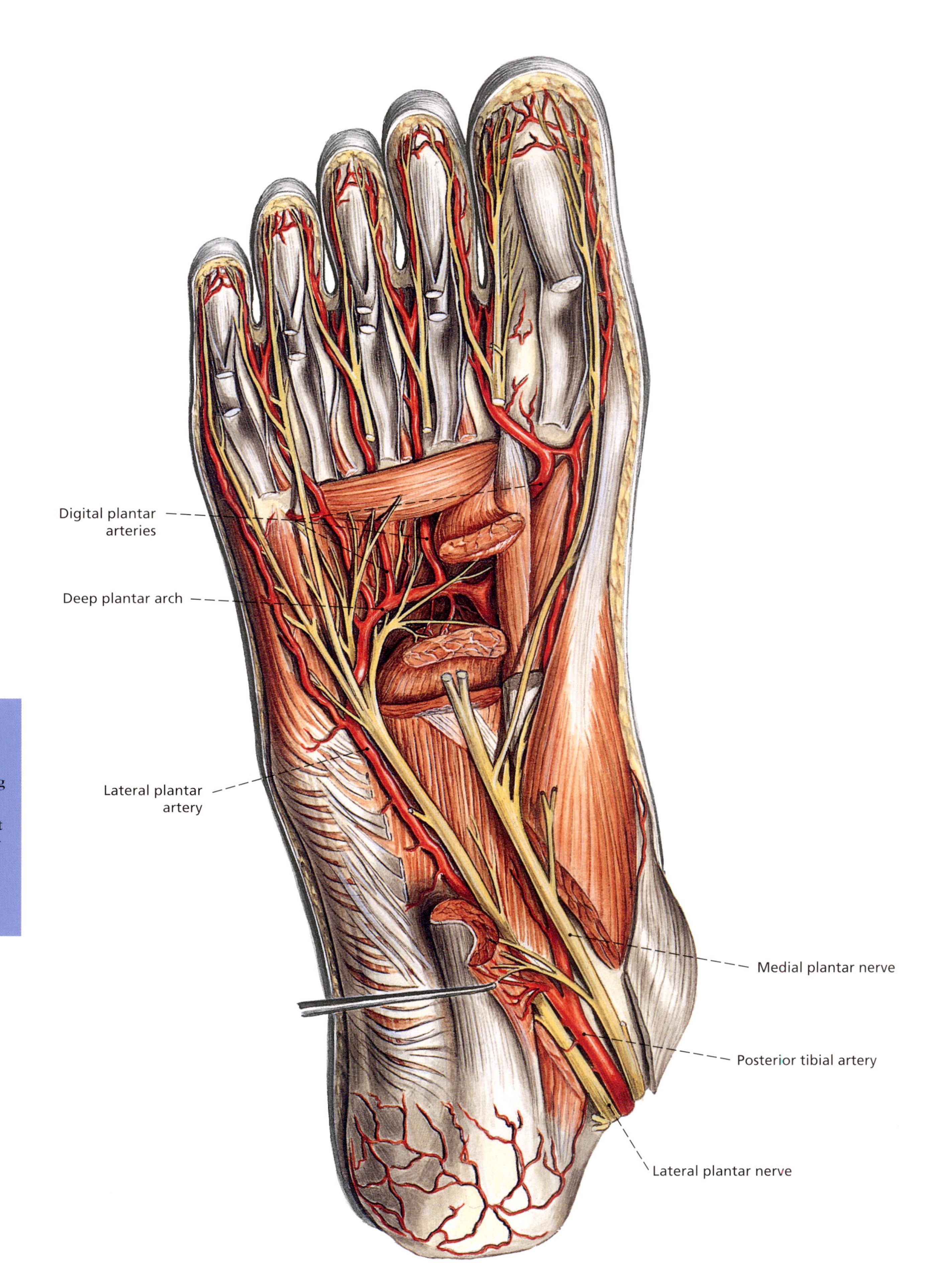

◆
Fig. 295
Plantar arteries of the right foot after cutting away the plantar aponeurosis and most of the musculature of the toes.

Symptoms and diseases

Arterial occlusion

Complete obstruction of blood flow in the arteries is caused by pre-existing stenoses, such as are frequently found in smokers, for example, or as a result of embolisms, high blood pressure, vasospasm, or vascular changes accompanying diabetes mellitus. In 80% of all cases, however, it is the result of arteriosclerosis.
The effects depend on the site of the occlusion, and are manifested in the legs, for example, as severe pain, skin pallor, a reduction in temperature, and sensory disturbances. In the eye they may even result in loss of vision. No pulse is palpable or can be auscultated in the affected part. Vascular occlusions are also diagnosed by means of ultrasound scans and X-ray investigations. In order to avoid death of the tissue downstream from the obstruction, drugs that promote blood flow, dilate blood vessels, and have an anticoagulant effect are administered immediately. If it proves impossible to clear the blockage, bypass surgery may be necessary.

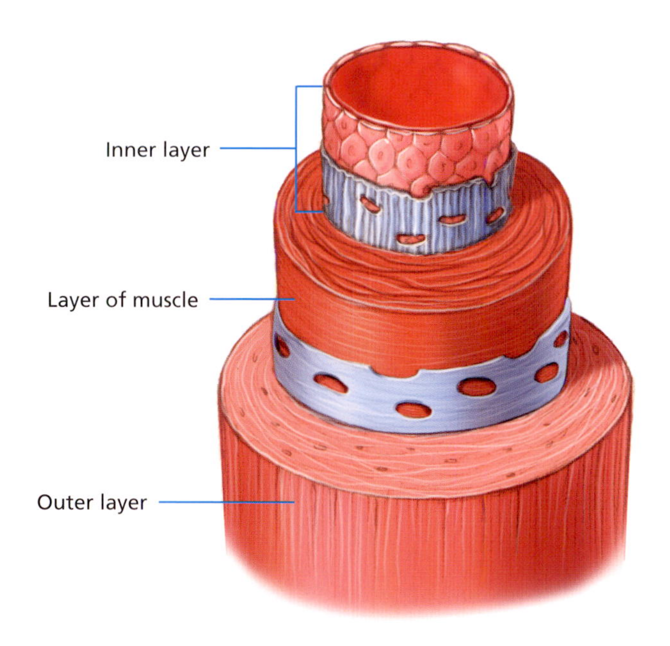

◆
Artery
The layer of muscle inside the artery wall reinforces the pumping action of the heart.

Stroke (apoplexy)

A stroke is triggered by a disturbance of the blood flow in the brain. In the vast majority of cases (approximately 80%) it is due to obstructed blood flow caused by occlusion of one of the arteries in the brain (cerebral infarction); it can, however, also occur as a result of a hemorrhage in the brain tissue (cerebral hemorrhage).

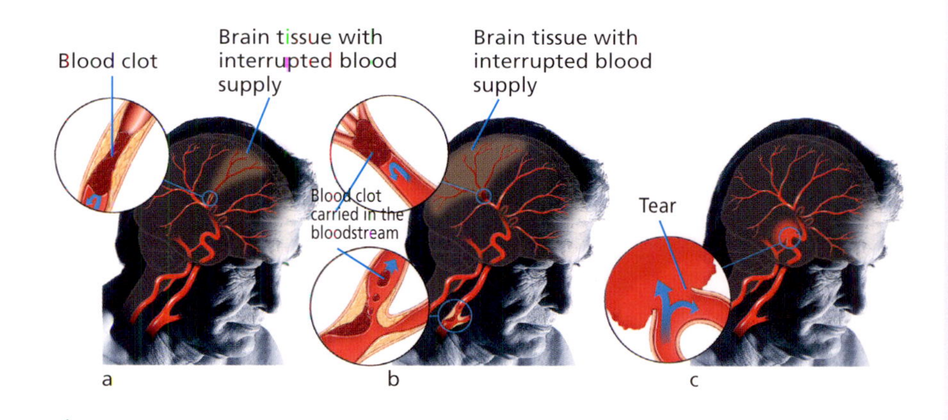

◆
Stroke
The three major causes of strokes are: (a) a sclerosed artery is occluded by a blood clot that forms at the exact point where the vessel is stenosed, preventing blood flow to the area of the brain beyond; (b) blood clots formed in other locations in the body are carried in the bloodstream to the brain, where they obstruct a blood vessel, and (c) if there is a tear in the wall of a cerebral artery, the blood supply to the areas beyond it is reduced while the area adjacent to the tear is damaged by the blood leaking into it.

A **cerebral infarction** results when an artery in the brain is occluded by a blood clot. The blood clot may have been formed at the site of the occlusion (thrombosis), or it may have been carried to the brain from elsewhere in the body (embolism). Sclerosed arteries are often responsible for blood clots being carried to the brain.
In younger people in particular, strokes are often caused by a **cerebral hemorrhage**. Cerebral hemorrhages occur more commonly in people with high blood pressure. In chronic high blood pressure, the walls of the arteries, especially where they branch, are damaged so badly that they become brittle and may burst. The brain is then damaged in two ways: by the interruption of the blood supply to the region downstream of the tear, and by the pressure of the leaking blood. Cerebral hemorrhages are often triggered by an additional surge in blood pressure caused, for example, by stress or physical exertion.
The decision to treat a stroke with drugs or surgery depends on the cause. In cerebral hemorrhages, particularly in younger patients, every effort is made to remove the accumulation of blood before the surrounding brain tissue is permanently damaged. In patients with severe stenosis of the carotid arteries due to arteriosclerosis, the calcified deposits can be surgically removed (thrombectomy). Anticoagulant drugs and drugs to improve blood flow are available for medicinal treatment of strokes.

Arteriosclerosis

You are as old as your blood vessels—a statement that sounds banal, but is extremely pertinent. Arteriosclerosis (calcification of the arteries) begins at an early age, and the arteries of a 45-year-old may be as severely affected as those of a 90-year-old. The positive aspect, however, is that the process is reversible, and the rate at which our blood vessels become sclerosed is partly up to us as individuals.

Calcification of the arteries, also known as arteriosclerosis, begins when fatty substances from the blood are deposited in the vascular walls. High blood levels of lipids, in particular cholesterol and its 'bad' carrier protein, low-density lipoprotein (LDL), result in an increase in these fatty deposits.

Fatty deposits in the arterial wall do not necessarily result in arteriosclerosis, however. More significant are the deposits of calcium in these islands of fat, which are thus transformed into islands of calcium. Since this is an ongoing process, the vascular walls become thicker and harder over the course of time, and the diameter of the vessels becomes increasingly narrow. The arteries close up and lose their elasticity, with the result that increasingly high blood pressure is needed to pump the required volume of blood through these rigid, stenosed vessels. In a vicious circle, the high blood pressure in turn increases the sclerosis.

In the long term, it becomes impossible, even with high blood pressure, to pump enough blood through the severely stenosed arteries, and the relevant parts of the body receive an inadequate supply of oxygen. The resulting circulatory disturbances initially occur only on exertion (when more oxygen is required), but later also at rest. The situation becomes critical when one of these sclerosed vessels tears or bursts. In the blood vessels in the brain, this causes a stroke; in the coronary blood vessels in the heart, it results in a myocardial infarction. Generally speaking, any artery, and even the smallest blood vessels, can become sclerosed. There are, however, sites at which sclerosis is more common due to the flow characteristics of the blood in those locations, for example, where blood vessels branch. The blood vessels most often affected are the cerebral and coronary vessels, the aorta, the renal arteries, and the arteries in the legs.

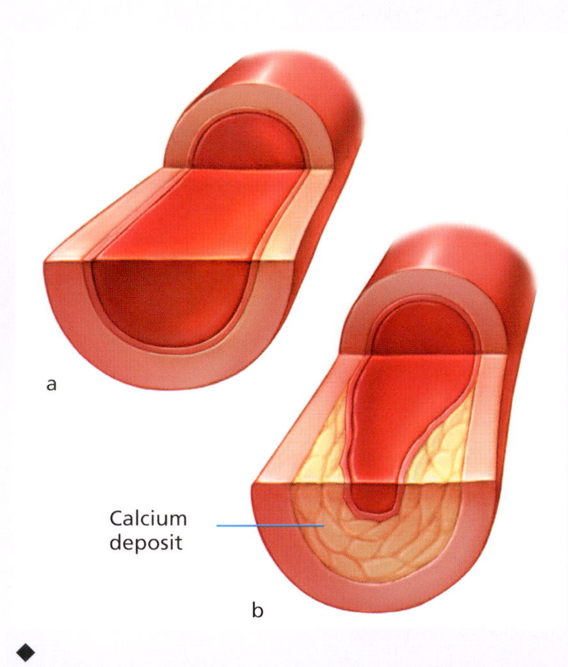

◆ Blood can flow unobstructed through a healthy artery (a). Because of its calcium deposits, a sclerosed artery is narrower and the blood flow is obstructed (b).

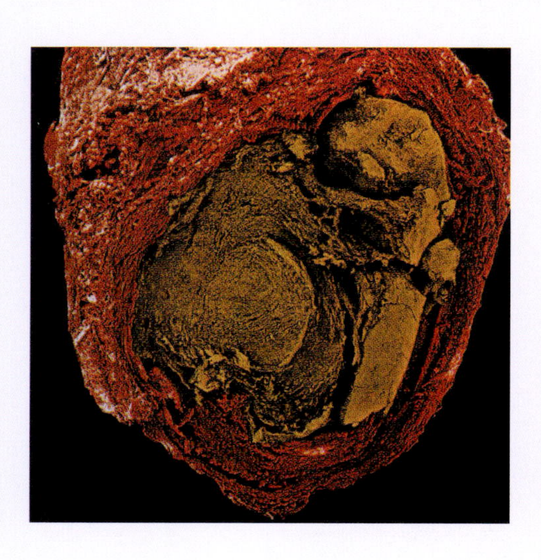

◆ This cross-section shows sclerosis at an advanced stage. The layer of calcium (yellow) has occluded the artery (red) almost completely.

Causes

The onset of arteriosclerosis is not related to any single cause, but to the interplay of a variety of risk factors. Fortunately, most of these are avoidable or can at least be minimized. The main risk factors are increased blood lipid levels, smoking, high blood pressure, stress, obesity, lack of exercise, and diabetes mellitus.

Symptoms

It is usually years or even decades before sclerosed arteries start to cause problems. The first indications may be pins and needles in the toes, cramp in the calves, or becoming easily tired. Symptoms such as heart pains or episodes of dizziness on exertion are unequivocal signs of inadequate blood circulation, while a stroke or myocardial infarction results when a blood vessel is completely occluded.

Diagnostic procedures

To date, it has only been possible to diagnose arteriosclerosis in its advanced stages. More sensitive diagnostic methods, however, are currently being tested. Of these, ultrasound scans, visualization of the blood vessels by means of an X-ray (angiography), and examination of the back of the eye (where the arteries are close to the surface and highly visible) have proved invaluable.

Preventive measures

Arteriosclerosis is not only preventable if the risk factors outlined above are excluded, it is also reversible. In many cases this requires changes to be made in the patient's lifestyle. Smokers must give up, or at least cut down considerably, and those who are obese must lose weight and take more exercise.
The right diet is extremely important. Animal fats in the diet (e.g., in butter, pork) should be restricted, and replaced by vegetable fats and oils (e.g., olive oil, sunflower oil, and wheat germ or safflower oil), which contain mostly unsaturated fatty acids, and are healthier. This applies also to fish (in particular, sea salmon, trout, mackerel, herring, and salmon), which should be on the menu at least once a week. Salt too should be used sparingly. When choosing dairy products, it is advisable to opt for low-fat products such as quark, cottage cheese, or buttermilk.

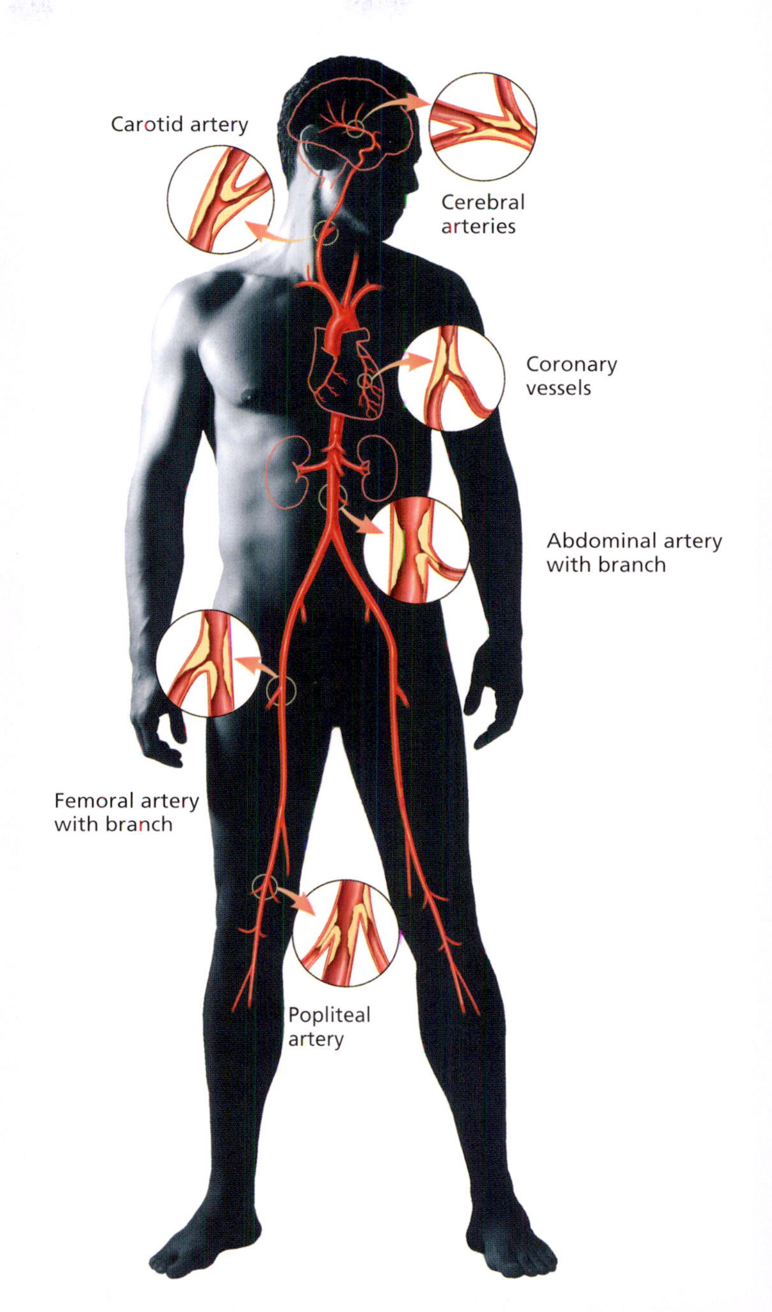

◆ The points at which the major blood vessels branch are frequently affected.

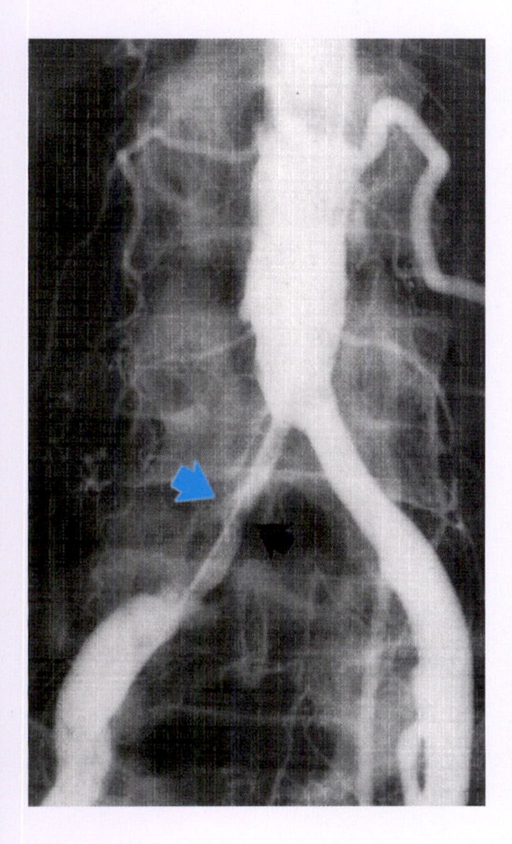

◆ The X-ray image shows severe stenosis of the right pelvic artery due to calcium deposits (see arrow).

It is crucial also to bring high blood pressure under control by means of diet and drugs. Similarly, it is essential to obtain treatment for diabetes mellitus.

Treatment

Drugs are generally only given to treat arteriosclerosis once all the preventive measures have proved fruitless. Drugs are available which lower the levels of lipids in the blood (e.g., statins) and improve blood flow properties (e.g., aspirin, anticoagulants). Garlic and gingko extracts are also reputedly effective. If all else fails, localized arteriosclerosis may be treated by dilating the affected artery, thus restoring its patency, or alternatively the affected section of the vessel may be excised and replaced by a vascular prosthesis.

Angiostenosis

Narrowing of a blood vessel caused by advanced calcification of the arteries (arteriosclerosis) or by a blood clot, which builds up inside the blood vessel and reduces its lumen until eventually it is completely occluded. It is also possible for stenosis to be caused by external pressure on the vessel wall, for example by tumors in neighboring organs and tissues. A blood vessel that is no longer patent can lead to a dangerously inadequate supply of blood to the area of the body affected.

Vascular occlusion

Occlusion of the inside of the blood vessel by a blood clot, arteriosclerosis, or inflammation. Acute occlusion (embolism) by a blood clot results in severe pain in the region supplied by the blood vessel in question, due to inadequate supply of oxygen. Arteriosclerosis is a condition that progresses slowly, as deposits on the vascular walls may gradually block the vessel completely.

Blood clots

Solidified, coagulated blood consisting principally of platelets (thrombocytes) and fibrin. A clot forms when the blood vessels are damaged. Platelets aggregate and fibrin is produced from components in the blood, under the influence of clotting factor. Since a blood clot can completely occlude a blood vessel (thrombosis), patients whose blood tends to form clots may be treated with anticoagulant drugs.
A clot can also be formed as a consequence of injuries. It seals the wound and protects against blood loss and infections.

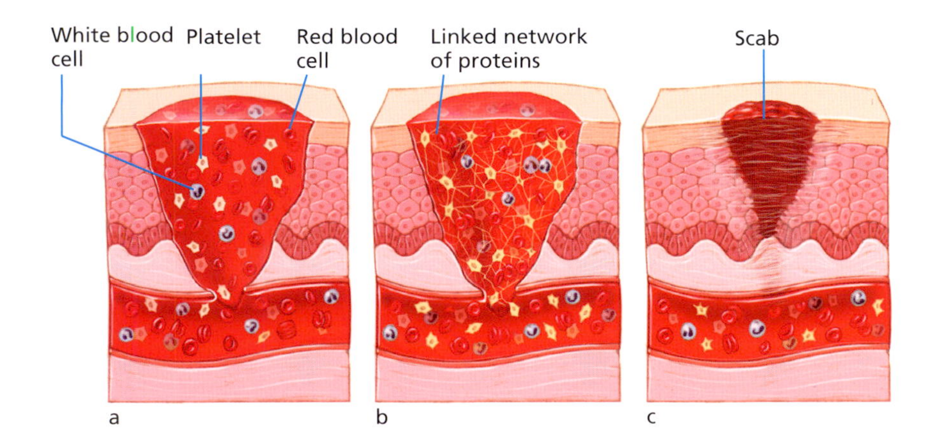

◆
Blood clots
In the event of injury, a blood clot forms on the surface, closing the damaged blood vessels. In a recent wound there is an accumulation of blood (a); gradually, as a result of the action of the platelets, a network of proteins in the blood are linked together to seal the damaged vessel (b). By the time the superficial injury has healed, the blood vessel will have completely regenerated (c).

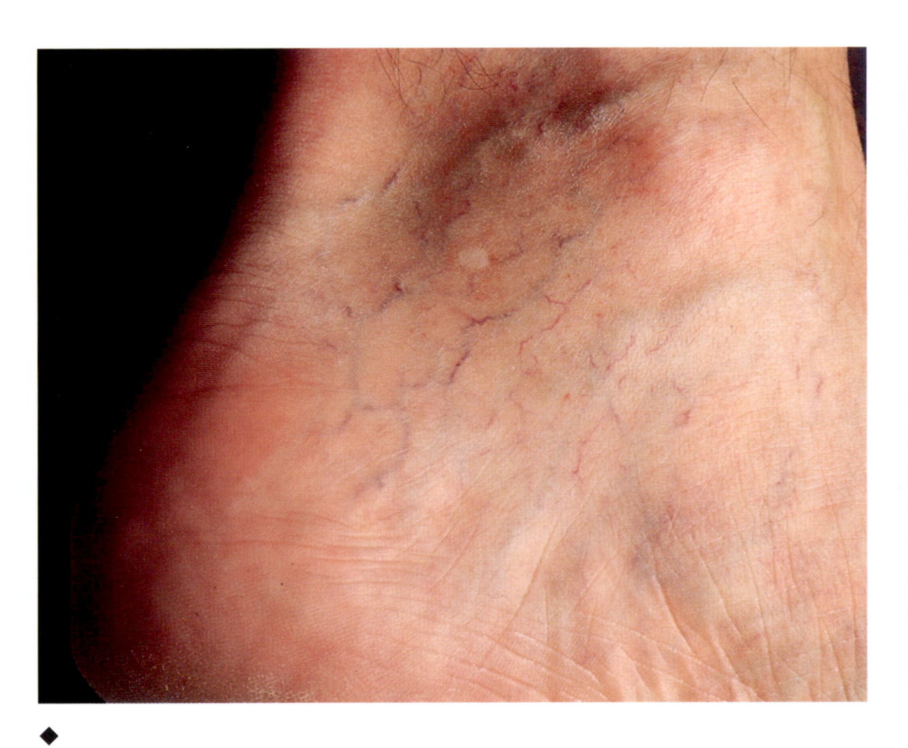

◆
Spider veins
Dilated veins, shown here on the ankle, are more common in women than in men.

Thrombosis

Formation of a blood clot in a blood vessel. Many liver diseases, and also the contraceptive pill, can increase the tendency of the blood to clot. Prolonged immobility may slow the blood flow in the vessels, while changes in the vascular walls, for example, as a result of varicosity or arteriosclerosis, can also cause turbulence in the blood flow. Clots can form and establish themselves in such conditions. If they increase in size, there is a danger that they will block the vessel. One possible complication is that a piece of the clot may break off and be carried in the bloodstream (thromboembolism). Drugs can dissolve clots (thrombolysis), but sometimes a clot will need to be removed surgically (thrombectomy).

Intermittent claudication

In this condition, cramp-like pains in the calves caused by chronic circulatory disturbance and lack of oxygen in the leg muscles on walking may mean that the patient has to keep stopping. After a few minutes, the symptoms disappear, but reappear when the patient starts walking again. This condition is triggered by arteriosclerosis in the legs. If untreated, it can result in critical circulatory disturbance in the affected leg, with the result that it has to be amputated. Drugs and physiotherapy can prevent progression to a critical stage.

Diseases of the veins

• *Spider veins*

Tiny dilated veins just beneath the skin, found particularly on the thighs and feet. Spider veins can indicate the start of venous disease. They do not require treatment.

• *Inflammation of the veins (phlebitis)*

The vascular wall of a vein can become inflamed as a result of injury or pathogens circulating in the blood. Redness and swelling in the affected area are signs of phlebitis. The pathological changes can result in the formation of blood clots (thrombi). Anti-inflammatory drugs combat the inflammation, and a pressure dressing may prevent thrombosis. If blood clots have already formed, additional drugs may be prescribed to thin the blood.

• *Venous thrombosis*

Occlusion of a vein by a blood clot (thrombus). **Pelvic venous thrombosis** arises in the deep great veins in the legs, abdominal wall, pelvic viscera, or in the lumbar region (e.g., following surgery in the lower body or after giving birth).
It manifests itself as pain in the affected leg, a slight bluish discoloration of the skin, and a raised temperature. One of the complications is that the rapidly inflammatory process may spread to the femoral vein and, more rarely, into the vena cava in the abdomen. In this case in particular, there is a danger that the blood clot may be transported to the lungs. Treatment is administered in the form of heparin, which inhibits coagulation and can dissolve blood clots.

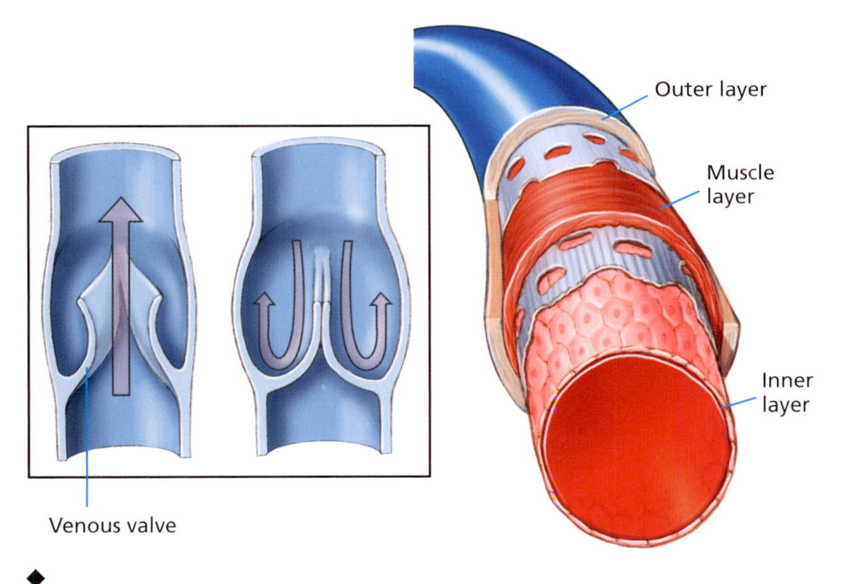

◆
Veins
In contrast to the arteries, muscles in the walls of the veins are inadequate to support the flow of blood. Valves inside the veins ensure that the blood flows only in one direction.

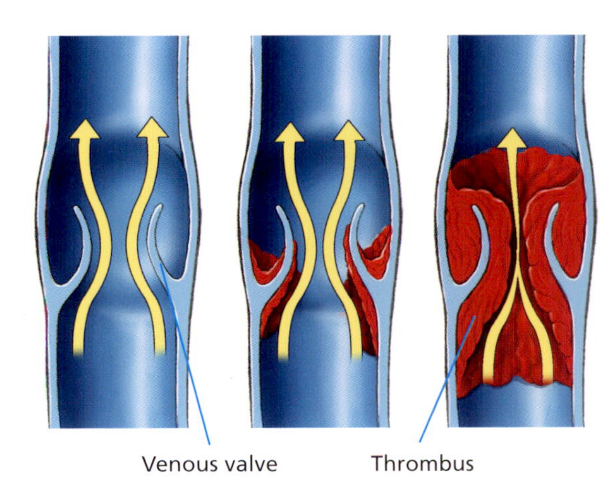

◆
Venous thrombosis
Solid deposits from the blood often accumulate around the venous valves, ultimately resulting in narrowing or even occlusion of the vein.

Venous thrombosis of the legs occurs most commonly in the deep veins, particularly in the lower leg, often spreading rapidly and causing venous thrombosis in the thigh.
The risk of thrombosis in the legs is greatest in those who are confined to bed, after surgery and accidents, or after giving birth, or in those who are susceptible to or already have a venous disease. It usually begins with sudden calf muscle cramps, then the leg becomes heavy and painful when standing and walking; later, the limb swells and turns a bluish color. The blood clot is dissolved with the aid of drugs. In addition, compression stockings need to be worn.

• *Varicocele*

Formation of varicose veins in the scrotum due to pathological dilation of the veins of the spermatic cord. In places, the testes are covered with many small reddish blue blood vessels filled with blood. A mass of tangled veins may also be formed. If the condition does not improve as a result of drugs, it is treated surgically.

Circulatory collapse

Breakdown of the circulation due to a severe reduction in the performance of the heart, blood loss, or sudden dilation of many small blood vessels as a result of an allergic reaction or hyperthermia. The weakness or unconscious state that results is caused by an inadequate supply of oxygen to the brain.
In this situation, the body has recourse to a number of emergency measures: the heart beats faster in order to compensate for the drop in blood pressure, the blood vessels in the skin and muscles contract, the blood flows back deep inside the body to the crucial organs (the heart, kidneys, and brain); and weakness and the threat of fainting oblige the patient to lie down, thus relieving the strain on the heart. In addition, it may be useful to elevate the legs, so the blood can flow back to the heart more easily. Minor circulatory collapse often soon passes without the need for further measures.

Hypertension

High blood pressure does not hurt and is completely normal, in fact necessary, on physical exertion. Only when it remains too high for too long a period does it become dangerous. Unrecognized and untreated, it can cause severe disease. However, identifying high blood pressure is extremely easy, and there are a number of treatment options available to prevent organ damage.

When the heart contracts (systole), it pumps blood through the arteries to supply the whole body, creating a pressure in the arteries known as systolic blood pressure. When the heart relaxes and expands to fill with blood again (diastole), the pressure in the arteries drops to what is termed the diastolic blood pressure. When the physician measuring your blood pressure tells you that your blood pressure is "120 over 80," this means that the systolic pressure is 120 mmHg and the diastolic pressure is 80 mmHg. Even in healthy people, various factors determine how high these blood pressure values are. According to guidelines of the World Health Organizations (WHO), high blood pressure (hypertension) is defined as a systolic pressure of over 160 mmHg and a diastolic pressure of over 95 mmHg.

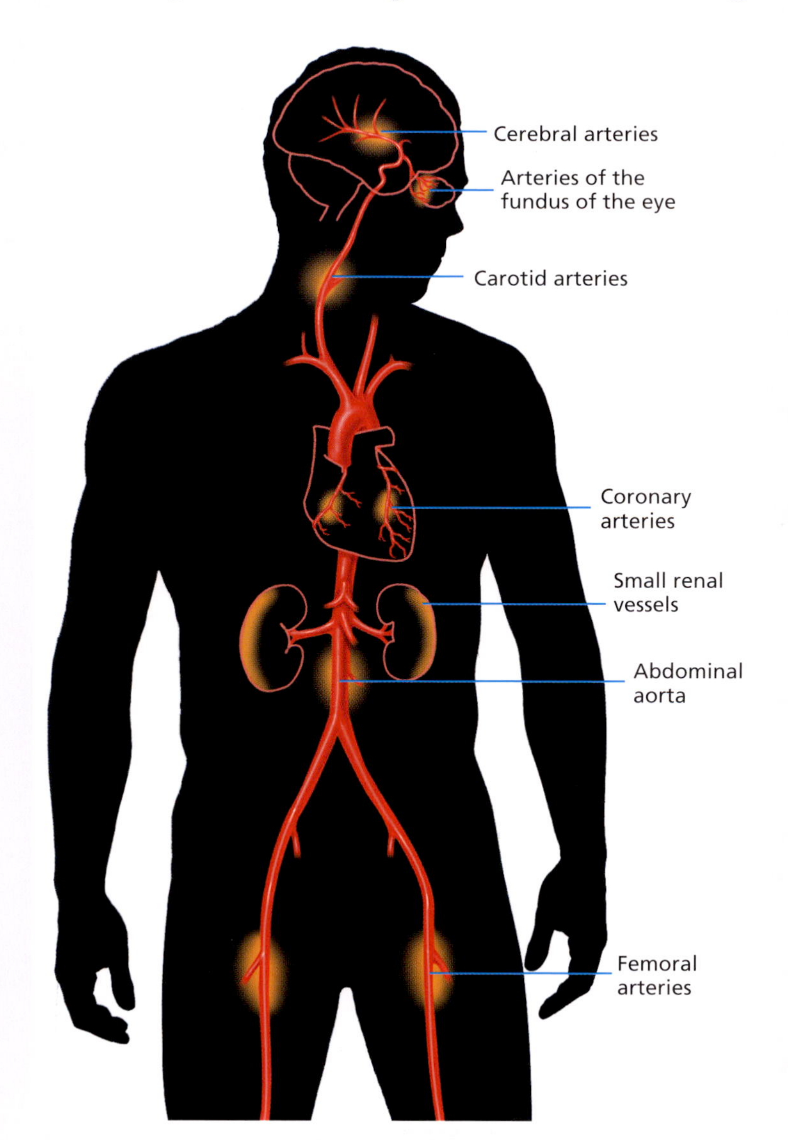

High blood pressure over a long period damages the arteries, particularly in the locations shown here.

Regulation

Blood pressure must be regulated in such a way that it is appropriate for the prevailing circumstances at any particular time. The circulatory center in the brain is responsible for regulating blood pressure. It sends nerve impulses to the muscles in the walls of the blood vessels. Depending on what is required, the blood vessels are then dilated (producing a drop in blood pressure) or narrowed (raising the blood pressure). In addition, this center controls the pump function of the heart, on which the blood pressure level also depends. Pressure sensors in the blood vessels send messages back to the brain about the current blood pressure status. The kidneys and the adrenal glands are involved in this signal, response, and feedback system. They produce messenger substances (hormones), which also dilate or narrow the blood vessels. Minerals circulating in the blood, such as potassium and calcium, also affect blood pressure. The finely-tuned interaction between these various regulatory mechanisms causes the blood pressure to fluctuate during the course of a day. During a heated argument, for example, the blood pressure is higher than when sleeping.
If there is a disturbance anywhere in this system, or if the blood vessels are no longer capable of being dilated or narrowed (e.g., as a result of disease), the normal adaptation of the blood pressure to the situation no longer functions.

Causes

It is usually possible to establish the cause unequivocally. It is known with certainty that the following factors can trigger persistent high blood pressure:

- Calcium deposits in the walls of the vascular system (arteriosclerosis) cause the blood vessels to become rigid and lose their elasticity, so that they are no longer able to respond to fluctuations in blood pressure by dilating or becoming narrower.
- Kidney disease (e.g., inflammation).
- Disturbances in the production of hormones responsible for blood pressure regulation.

Symptoms and diagnosis

High blood pressure is one of the most common diseases and is the cause of more deaths in industrialized countries than cancer. Often, high blood pressure is only discovered by accident during a medical examination. Although headaches, dizziness, nervousness, depression, frequent nosebleeds, and palpitations may be indicators, they do not necessarily confirm the presence of high blood pressure (hypertension).

Elevated blood pressure readings on a single occasion, though suspicious, can often be explained by the patient's anxiety. If this is accompanied by other suspicious test results, such as an electrocardiogram (ECG) with typical changes, or laboratory results indicating a disturbance in renal function, 24-hour blood pressure monitoring can be used to confirm the diagnosis. The patient is sent home wearing a small device attached to the body that measures the blood pressure at short intervals throughout the day and night. This evaluation shows whether blood pressure is permanently elevated or whether the high reading was an isolated instance.

Consequences

High blood pressure that persists for any length of time can damage several of the organs.

Vascular changes such as arteriosclerosis can be both a cause and a consequence of hypertension. The constant high pressure damages the inner lining of the vessels, forming small lesions in which fatty substances and calcium crystals are deposited. The blood vessels thus become increasingly narrowed, to the extent that in some places they may be completely blocked. Since the coronary blood vessels supplying the myocardium with nutrients are likewise affected, high blood pressure is a particular risk factor for a myocardial infarction.

Long-term strain on the heart can also gradually cause the blood vessels in the lungs supplying the heart to become congested with blood. The blood arriving in the heart can no longer be pumped away in sufficient quantities, and congestion (hemostasis) in the pulmonary vessels makes breathing and the exchange of carbon dioxide and oxygen difficult.

As well as the heart, the kidneys may be damaged, to such an extent that their function is impaired. If particular areas of the brain are no longer adequately perfused with blood, this may lead to a stroke. Possibly the most dangerous consequence of severe long-term hypertension is cerebral hemorrhage, when a brittle blood vessel in the brain, previously damaged by high blood pressure, bursts; the consequent bleeding into the brain is often fatal. The retina in the eye is also at risk. Inadequate blood supply or bleeding from a burst blood vessel can have serious repercussions in terms of vision.

◆ A diet high in salt and fats is one of the most common causes of high blood pressure.

Treatment

There are many causes of hypertension, and equally there are many different treatment options. Many of the causes, such as disturbances in fat metabolism, diabetes mellitus, and kidney diseases, can be directly and successfully treated.

There are also many drugs available that lower blood pressure; some of these dilate the blood vessels, others reduce the pumping function of the heart, or affect the hormonal regulatory mechanism in such a way as to reduce blood pressure with no other effects.

As table salt increases blood pressure, it may be advisable, in addition to following an appropriate diet, to take drugs to promote the excretion of salt via the kidneys. Successful treatment often consists in a combination of several drugs. The most important rule is to take these medicines exactly as directed by the physician.

What preventive measures can be taken?

It is possible to influence many causes of high blood pressure yourself by adopting an appropriate lifestyle:

- Avoid foods with a high salt and fat content. Fruit, vegetables, poultry (low in fat), and high-fiber foods are more suitable.
- Avoid stress.
- Strengthen the cardiovascular system with physical activity.
- Keep nicotine and alcohol consumption low.

Varicose veins

In industrialized countries, up to half of all women and a quarter of all men suffer from varicose veins. As thick as a pencil, these veins wind their way tortuously down the lower legs. The soles of the feet, the calves, backs of the knees, and the thighs are all painful. Occasionally severe complications can occur, making varicose veins more than a cosmetic problem.

The veins transport deoxygenated blood back to the heart. The heartbeat, which can only be felt in the arteries, does not have sufficient power to pump the blood on its return journey; furthermore, the blood in the legs has to be transported upwards against gravity. A complex system is required to achieve this:

- When the calf muscles and thigh muscles contract, the veins are compressed, and the blood is pushed upwards. This muscle pump function is also assisted by the pulsing of the arteries frequently located nearby, and by the slight suction pressure exerted by the heart.
- Valves inside the veins allow the blood to travel only in the direction of the heart.
- Elastic connective tissue fibers in the vascular wall allow the veins to expand and contract.

In varicose veins, the connective tissue fibers in the venous walls are weak and overstretched, with the result that the valves inside the veins are no longer able to close. The blood flows backwards, pools and stretches the already slackened vascular walls even more.

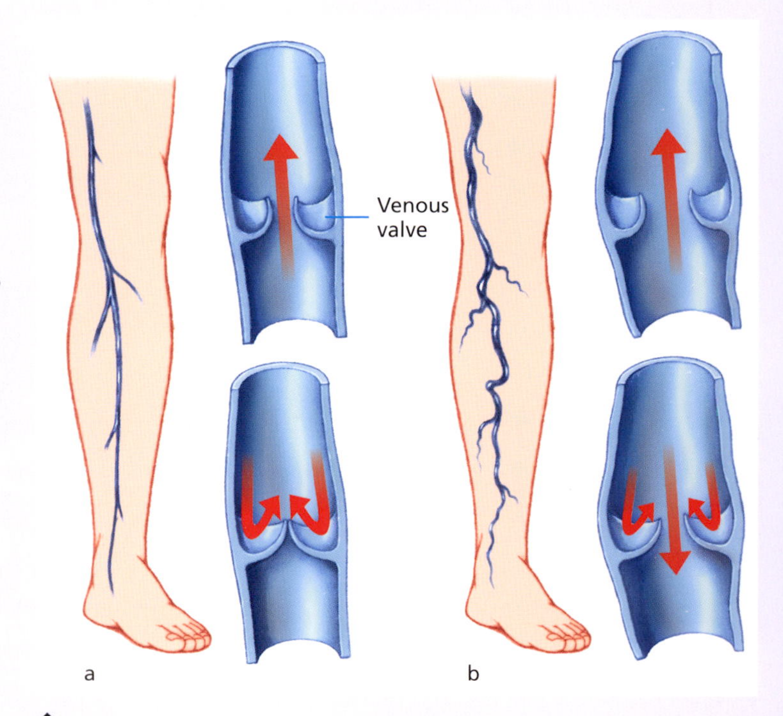

The valves inside the veins prevent the blood flowing backwards and stagnating on its return journey to the heart. The flow of blood is encouraged during walking by the pumping action of the leg muscles (a). Varicose veins are formed when the vessels are overstretched and the valves are no longer able to close the vein properly (b).

Hardened, tangled veins on the lower leg are typical of varicose veins.

Causes

An inherited weakness of the connective tissue is often responsible for varicose veins. As in many other diseases of the civilized world, however, obesity, smoking, and lack of exercise, are among the risk factors. Those whose jobs require them to spend a long time standing or sitting are particularly at risk. During pregnancy, hormonal changes can weaken the vascular wall and exacerbate the condition. The contraceptive pill causes similar hormonal changes. Therefore, for women with vascular disease who are on the pill there is an increased risk of formation in the veins of blood clots (thrombosis), which may block the blood vessels.

Accompanying symptoms and risks

Initially, the individual complains only of heavy, tired legs, swollen feet and joints, and sometimes cramp-like, dragging pain.
The damaged inner wall of the vein becomes slightly inflamed, and here too blood clots may form; these may be transported in the bloodstream to other regions of the body or to organs such as the lungs, where they may block a blood vessel (embolism).
In addition, a considerable amount of fluid may accumulate in the affected leg (edema), and there is even a possibility of open leg ulcers.

Primary or secondary varicose veins?

Depending on the cause, a distinction is drawn between primary and secondary varicose veins.

• ***Primary varicose veins***

These are by far the most common and more harmless forms of varicose veins, in which the blood flow is only impaired in the visible, superficial venous system.

• ***Secondary varicose veins***

If there is a disorder of the deeper veins preventing the unobstructed flow of blood back to the heart, the blood tries to find another route and flows through the superficial veins. If these are unable to cope with the additional strain posed by the increased blood volume and hemostasis develops as a consequence, the resulting varicose veins are termed secondary varicose veins. They occur mainly as a result of occlusion by a blood clot of a deep leg vein or of deposits on the vascular walls.

Investigational methods

A physician can tell whether varicose veins are of the primary or secondary type using tests involving percussion, Trendelenburg's test (also known as the tourniquet test), and one or more of the following:

- ◆ X-ray images of the veins using a contrast medium (phlebography),
- ◆ measurement of blood flow using an ultrasound device.

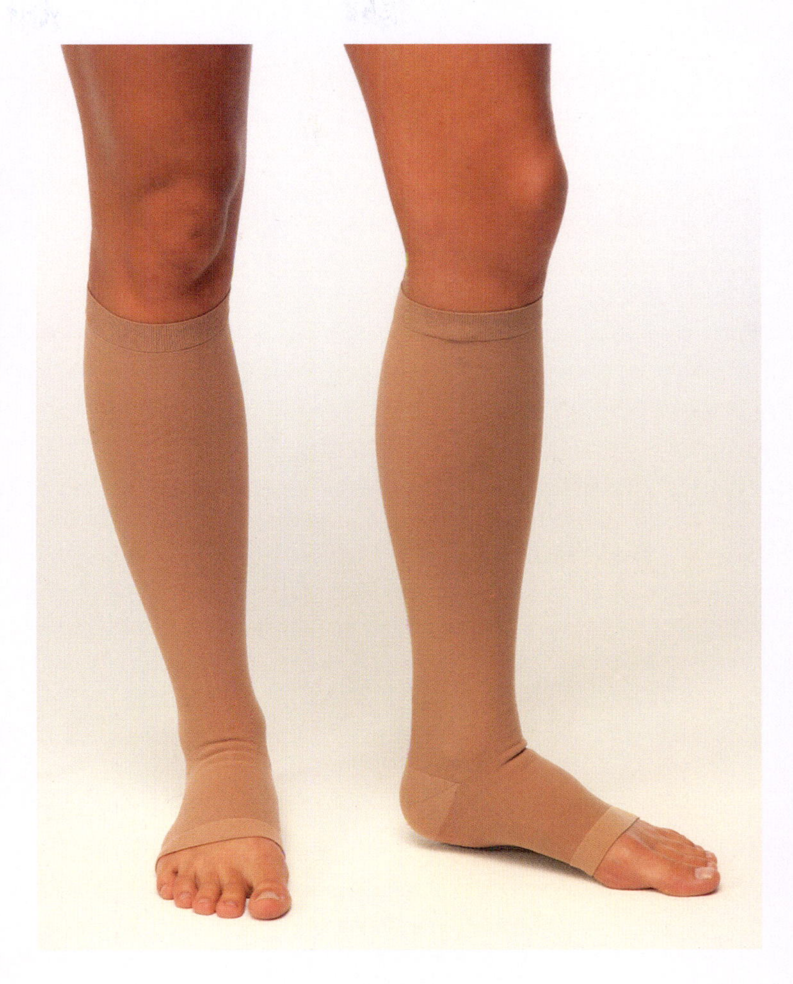

◆ Elastic support stockings assist the flow of blood from the legs back to the heart and are particularly recommended for those whose occupations are sedentary or require long periods of standing. They are often recommended for use on long flights.

Prevention and treatment

Massaging the legs with both hands in the direction of the heart encourages blood flow. Those obliged to spend a great deal of time on their feet should avoid wearing shoes with high heels. Moving the foot repeatedly up and down so that the toes and heel are lifted alternately in a seesaw motion can activate the muscle pump function. The legs should be elevated frequently and the end of the bed raised at night. The wearing of support stockings suited to the individual and careful wrapping of the legs in elasticated bandages also provide relief, as do regular exercises, hydropathic treatments, walking, cycling, climbing, and swimming. A normal body weight is a very effective medicine! In milder cases, active substances such as horse chestnut may help, but if the condition is more severe or if signs of inflammation develop, medical advice must be sought. Direct treatments are possible in the case of primary varicose veins:

- ◆ Sclerotherapy, in which an agent is injected into the vein that causes the vascular walls to stick together, thus closing the vein. The blood then tries to find a route through the healthy deep leg veins. Sclerotherapy is undertaken on an outpatient basis and without anesthesia. It is largely painless.
- ◆ Surgical treatment, known as stripping, in which the diseased veins are removed.

Diagnostic methods

Blood pressure measurement

Having your blood pressure measured whenever you visit your physician is immensely important. If a reading is high, it is a good idea to monitor your blood pressure at regular intervals. Specially designed devices for the layman are available for home use. Regular home blood pressure checks are also important for patients with high blood pressure in order to check that the treatment is working.

Angiography

An X-ray of the blood vessels, primarily of the arteries, but occasionally also of the veins and lymphatic vessels. After injecting a contrast medium, the route taken by the contrast medium through the vessels is shown in successive X-ray images. From these images it is possible to identify changes such as stenoses or damage to the blood vessels.

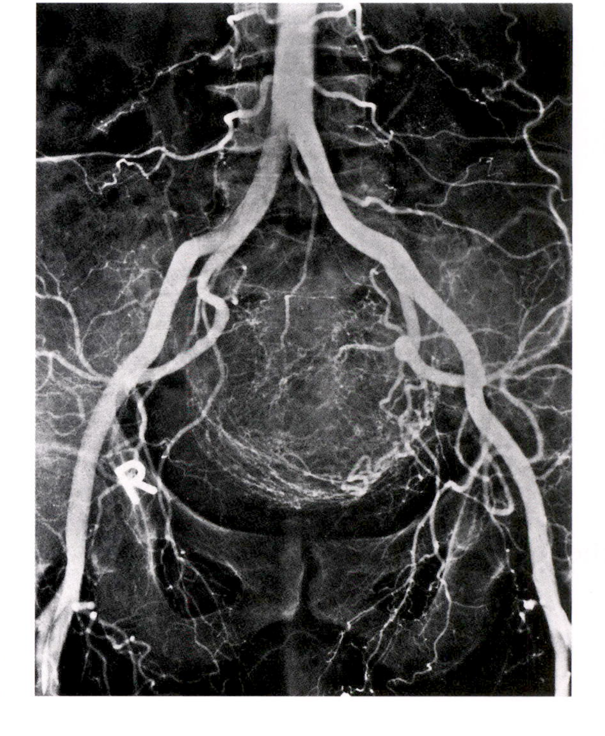

Angiography
It is possible to show blood vessels on an X-ray with the use of contrast medium. This image shows the branching of the aorta into the two pelvic arteries.

Quick test (prothombin time)

A blood test to check blood coagulation. In patients who have to take anticoagulant drugs to thin their blood, the Quick test is performed regularly. The results show whether the levels of coagulant in the blood are adequate and have not been excessively inhibited. As blood coagulation is influenced by the vitamin K content of the blood and by liver function, Quick test results can also be changed in the presence of liver disease.

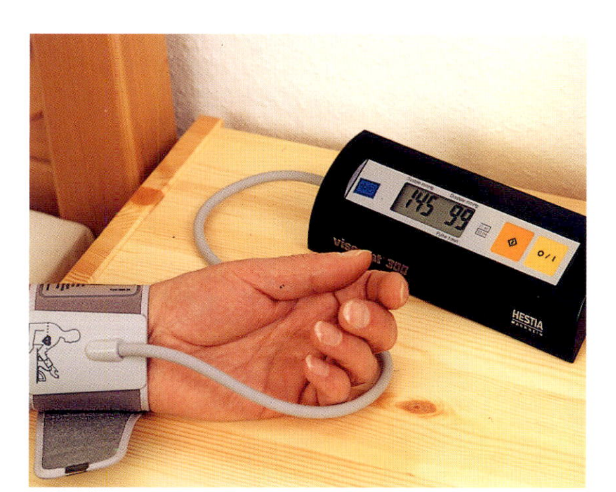

Measuring your own blood pressure
Devices that are easy to use are now available for regular measurement of your blood pressure at home.

Treatment methods

Drugs to improve the circulation

Drugs to improve the circulation include those which increase blood pressure and those which are mild stimulants and generally increase the activity of the heart.

Anti-hypertensives

Drugs to treat high blood pressure and to prevent diseases caused by excessively high blood pressure (myocardial infarction, stroke). These drugs include angiotensin-converting enzyme (ACE) inhibitors, beta-blockers, calcium antagonists, diuretics, and substances that dilate the blood vessels.

Beta-blockers

An abbreviation of beta-receptor blockers, drugs to treat high blood pressure and to prevent diseases caused by high blood pressure. The beta-receptor cells are cells in the blood vessels or myocardium which, when stimulated, increase the blood pressure; blocking them produces a normalization or lowering of the blood pressure.

Anticoagulants

Drugs that impair the normal coagulation of the blood. The most widely known substances are heparin and dicoumarol. Heparin is injected, in particular in the short term, in patients who have recently undergone surgery or in bed-ridden patients whose immobility means they have a greater risk of thrombosis. Dicoumarol is usually prescribed as long-term therapy, particularly in patients who have suffered a myocardial infarction.

Vascular bypass surgery

Bypass surgery is a method of cardiovascular diversion. Usually a bypass involves the surgical grafting of blood vessels to replace occluded or stenosed blood vessels (mainly arteries). The aim of the diversion is to restore the blood flow to an inadequately supplied area of tissue. A section of the patient's leg vein is most commonly used to create the bypass.

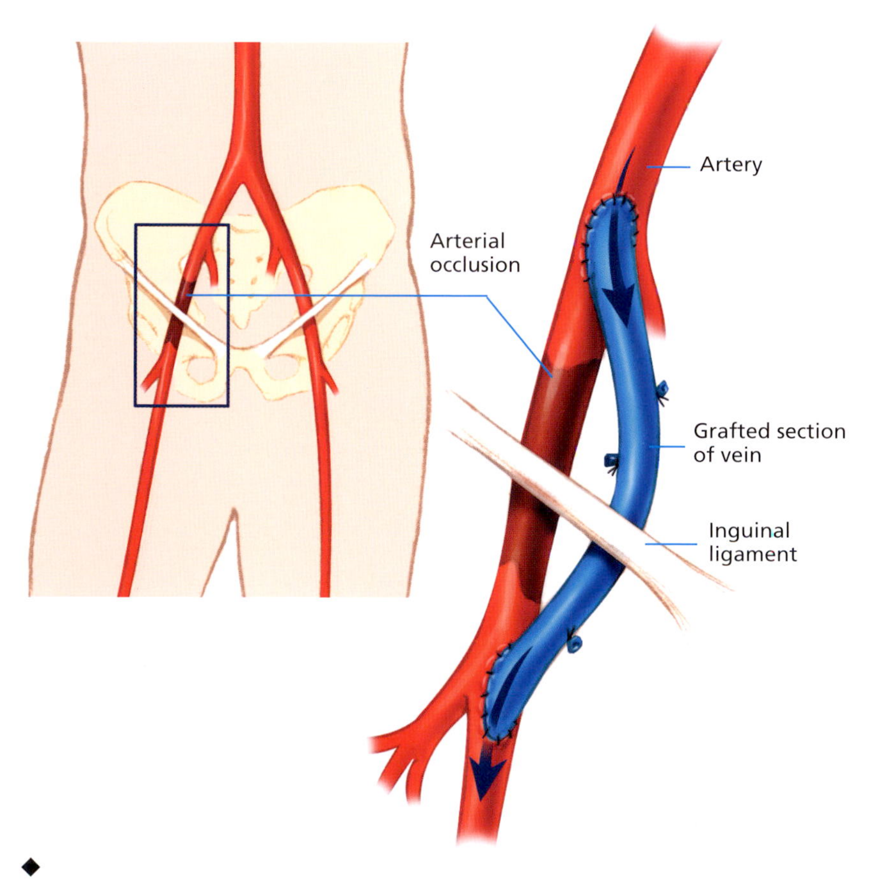

Bypass
If it is impossible to restore blood flow through a blocked artery, the blood is diverted around the blockage, flowing through a grafted section of vein.

Vascular graft

The surgical substitution of a damaged or occluded blood vessel by a replacement vessel, taken either from the patient or a donor, or made from synthetic materials. Autologous grafts (the patient's own vessels) are generally taken from the lower leg. This procedure is usually performed in the event of severe circulatory disturbance in the area of the coronary vessels, the aorta, or the cerebral arteries. If important arteries are occluded, a diversion is created using artificial prostheses. Similarly, part of a blood vessel can be replaced or a damaged vascular wall can be repaired using a small graft.

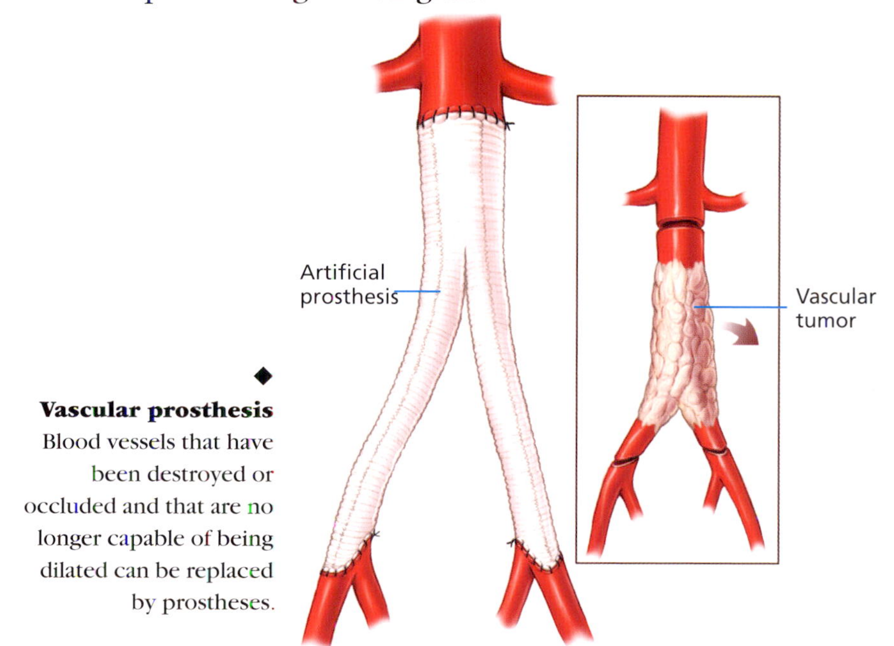

Vascular prosthesis
Blood vessels that have been destroyed or occluded and that are no longer capable of being dilated can be replaced by prostheses.

Vascular prostheses

Vascular prostheses are implanted surgically where there are severe circulatory disturbances, vascular stenoses, or damage to the blood vessels. The prosthesis consists of a plastic tube and is used in the region of the heart, legs, the major arteries, and the vessels supplying the brain, especially when an autologous graft of vein from the patient's own body is unavailable.

Sclerotherapy of varicose veins

Closure by artificial means of individual varicose veins. A sclerosing agent is injected into the varicose vein, causing inflammation of the vascular walls inside the vein. The leg is then bandaged so as to exert pressure on the vein. As a result of the inflammation, which later fades away, and the pressure from the outside, the inner walls of the varicose vein stick together so that no more blood can flow through it. If a supporting compression dressing is worn for several weeks, the varicose vein will largely disappear, and the existing healthy veins will take over the job of transporting the blood back to the heart.

Varicose vein surgery

The surgical removal of varicose veins is usually carried out only when the blood vessels affected are the two main external leg veins, or when sclerotherapy has proved ineffective. The intervention is performed in hospital under general anesthetic and is only possible where the deep veins inside the leg are sufficiently patent to be able to return the blood from the feet and lower legs to the heart by themselves. This must be confirmed prior to the operation by an X-ray investigation, in which the deep veins are visualized by means of contrast medium. If they are patent, and deemed capable of returning the blood to the heart unaided, the external varicose veins are removed. Small incisions are made at the ankle, behind the knee, and in the groin, and the diseased veins are pulled out. A firm bandage is then applied and must be worn until the leg has healed. In order to prevent blood clot formation, it is advisable to start walking again soon after the operation. The hospital stay usually lasts for about a week.

Thrombectomy

Surgical removal of a blood clot (thrombus). The blood vessel is opened and the blood clot is removed; the blood vessel is then closed again. If the thrombus is not easily accessible, a catheter is pushed past it. At the tip of this catheter is a small balloon, which can be inflated from outside. When the catheter is retracted, it pushes the blood clot in front of it and out of the blood vessel.

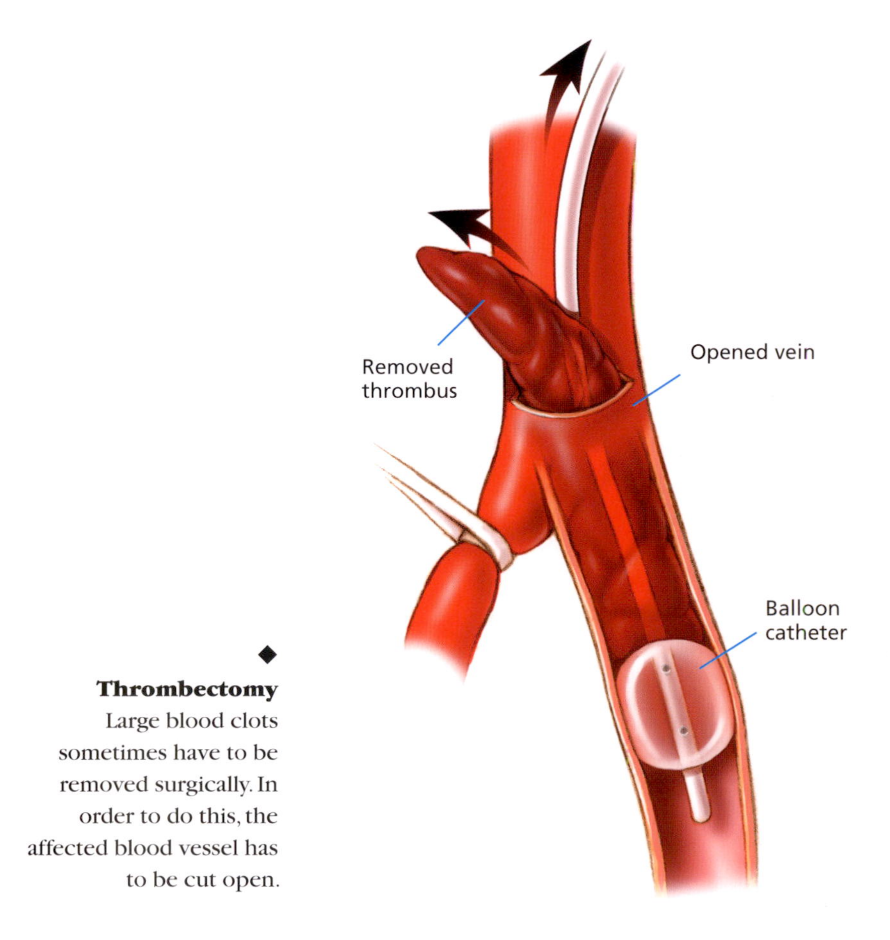

Thrombectomy
Large blood clots sometimes have to be removed surgically. In order to do this, the affected blood vessel has to be cut open.

Thrombolysis

In thrombolysis, a blood clot is dissolved using drugs injected into the bloodstream. In emergencies, a fine tube can be advanced through the blood vessel under X-ray guidance, and the drug can be targeted directly at the clot through this tube. This method is used, for example, in the event of a myocardial infarction.

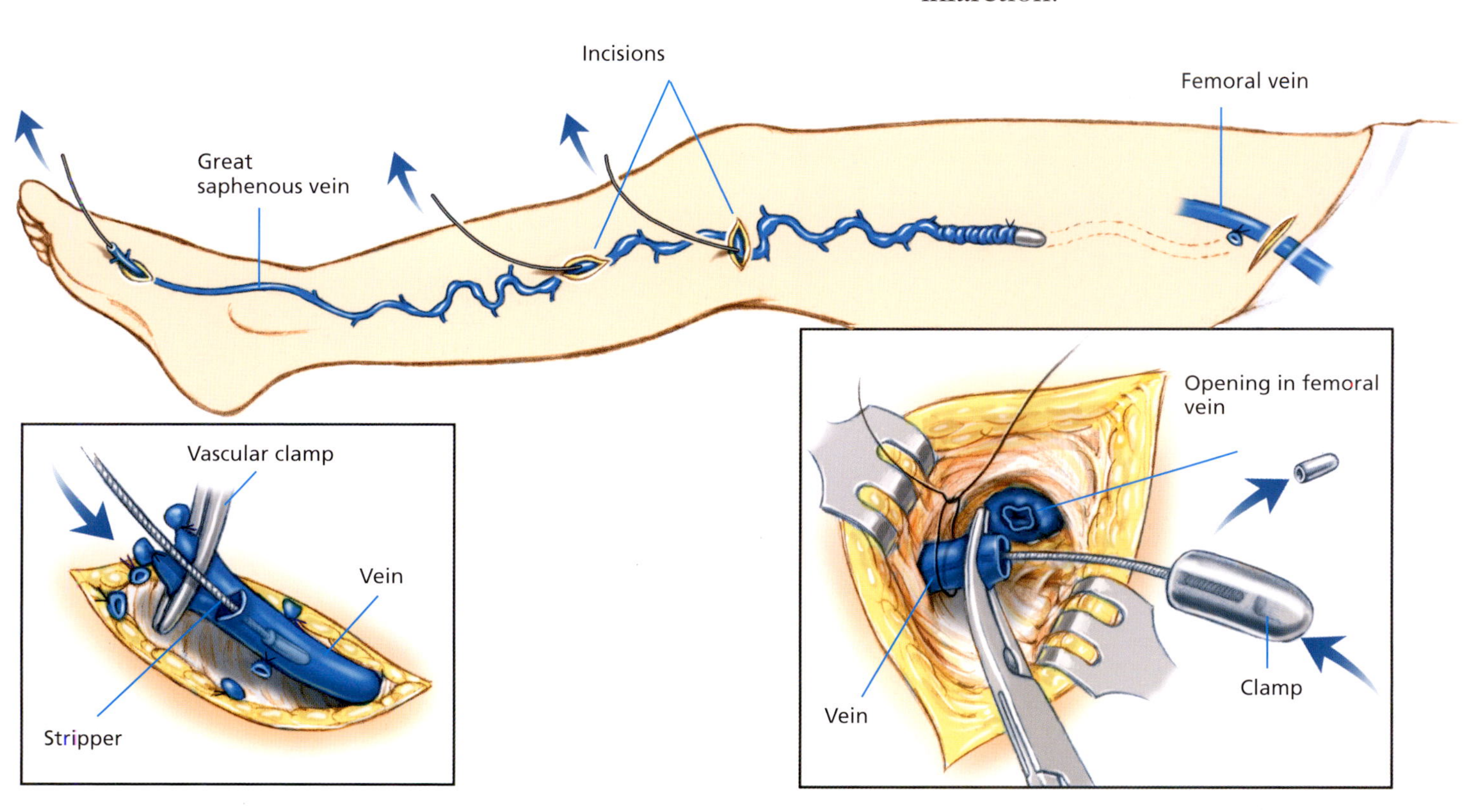

Varicose vein surgery
Varicose veins are removed from three locations on the affected leg using a metal probe, called a stripper, with a clamp on its end that grips the end of the diseased vein.

Lymphatic system

As well as possessing a system of blood vessels that supply the tissues with nutrients, our body also has a drainage system via which the tissue fluid, the lymph, is removed for further processing. This is known as the lymphatic system and comprises lymphatic vessels with interconnected filter stations, the lymph nodes. The task of the lymphatic system is not restricted to transporting the lymph. Together with the lymph nodes it also plays an important role in the body's defense against impurities and pathogens.

Lymph vessels

The smallest of the lymph vessel branches, the lymphatic capillaries, originate in the tissue spaces throughout the body. They reabsorb part of the liquid that has been passed from the blood vessels into the tissue in order to supply the body cells with nutrients. The removal of the tissue fluid is largely carried out by the veins but these are supported by the lymphatic vessels, to avoid fluid being stored in the tissue.

The small lymphatic capillaries combine to form larger vessels in a structure similar to that of the venous system. They, too, have valves to determine the direction in which the fluid flows. The lymph fluid ultimately gathers in a large vessel, the thoracic duct, which empties into the superior vena cava.

Nearly two liters of tissue fluid are drained off by the lymph vessels each day. The lymph consists of blood plasma, lipoproteins, cells, foreign bodies, and bacteria, and has a yellowish color due to its fat content.

Lymph nodes

The cleaning and defense work of the lymphatic system takes place in the lymph nodes, which act as filter stations. They are small, bean-shaped organs which differ greatly in size. While some are not visible to the naked eye, others are about the size of a hazelnut. Some lymph nodes lie just underneath the skin, such as in the throat and neck area, the armpits, and the groin. Here they can be felt when they are enlarged.

The tasks of the lymph nodes are:

- to relieve the lymph of impurities and dead cells
- to form defensive cells (lymphocytes)
- to attack exogenous agents (antigens), such as bacteria and pollen

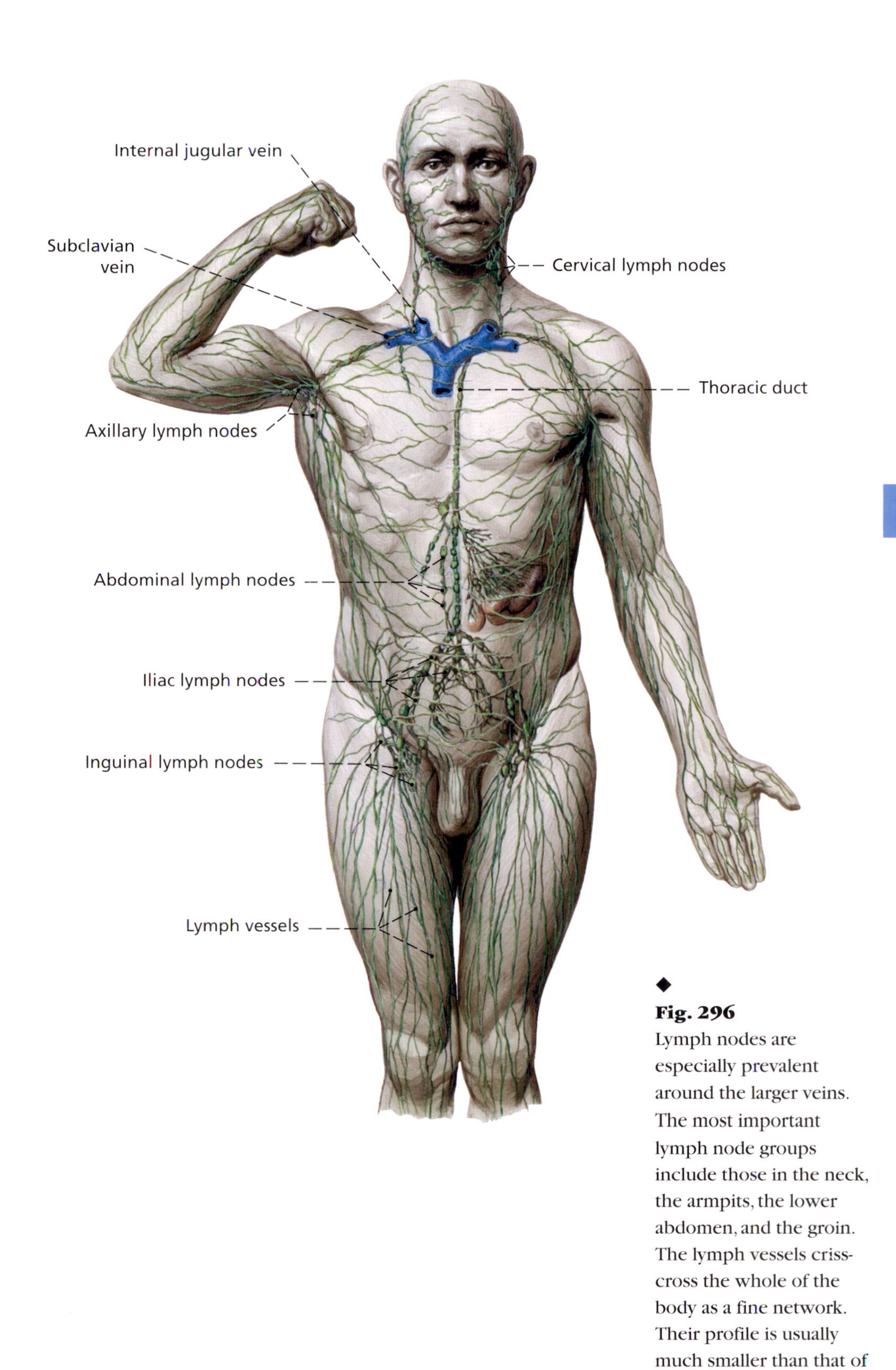

Fig. 296
Lymph nodes are especially prevalent around the larger veins. The most important lymph node groups include those in the neck, the armpits, the lower abdomen, and the groin. The lymph vessels crisscross the whole of the body as a fine network. Their profile is usually much smaller than that of the blood vessels.

The lymph nodes have a specialized construction which enables them to fulfill their tasks. Their loose, sponge-like tissue is interwoven like a mesh and so forms a filter for foreign bodies and pathogens. As the lymph passes slowly through this network, there is sufficient time for pathogens to be recognized and eliminated. An inflammation of the tonsils (tonsillitis), for example, presents a challenge for the lymph nodes in this region which need to "immobilize" the pathogens. This may cause the lymph nodes to swell, which can sometimes be painful. The lymph nodes also provide a barrier against tumor cells, initially preventing them from inundating the whole body. This function leads to a palpable enlargement of the lymph nodes in the affected region.

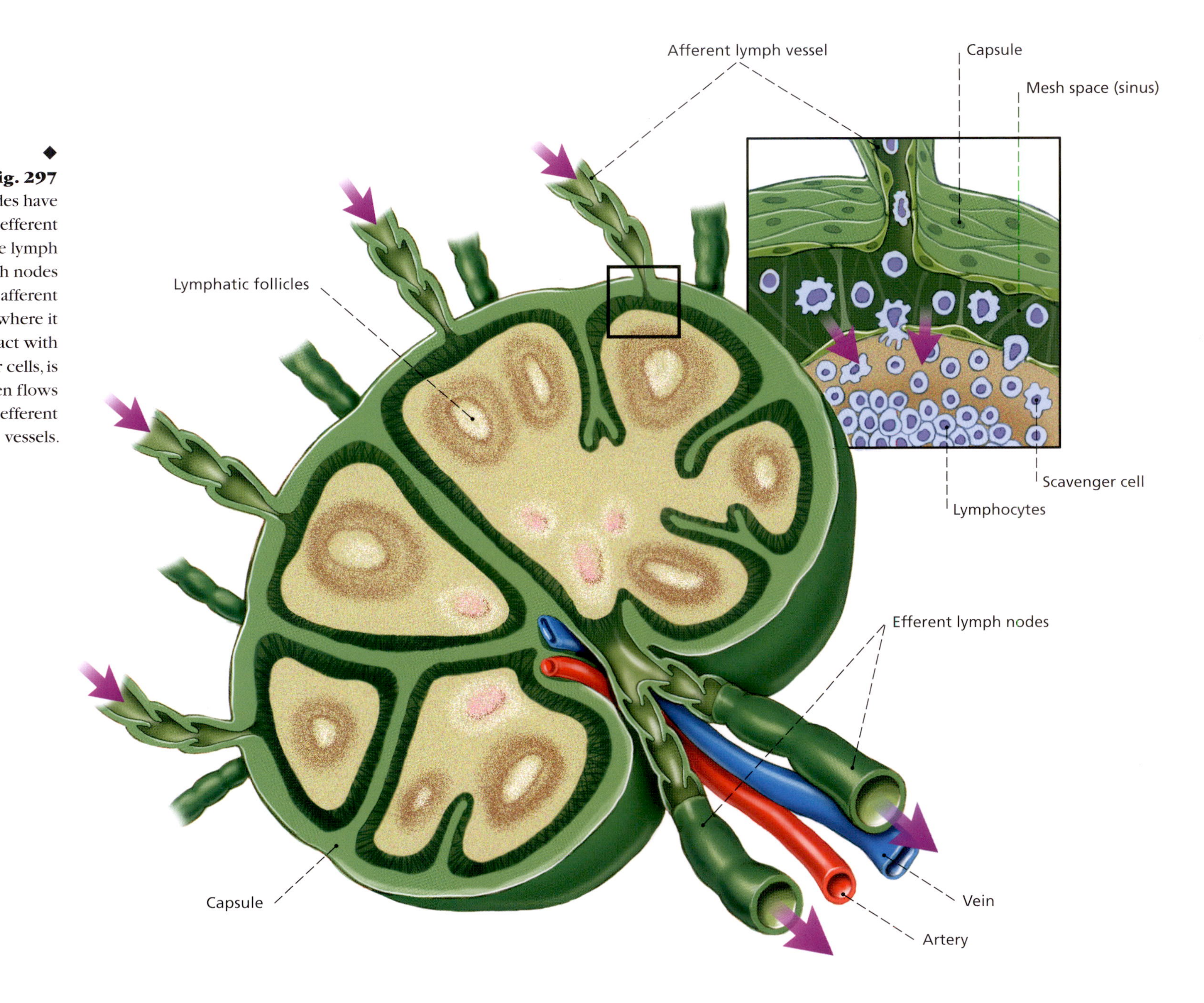

Fig. 297
The lymph nodes have both afferent and efferent lymph vessels. The lymph enters the lymph nodes through the afferent vessels (detail) where it comes into contact with the scavenger cells, is filtered, and then flows out again via the efferent vessels.

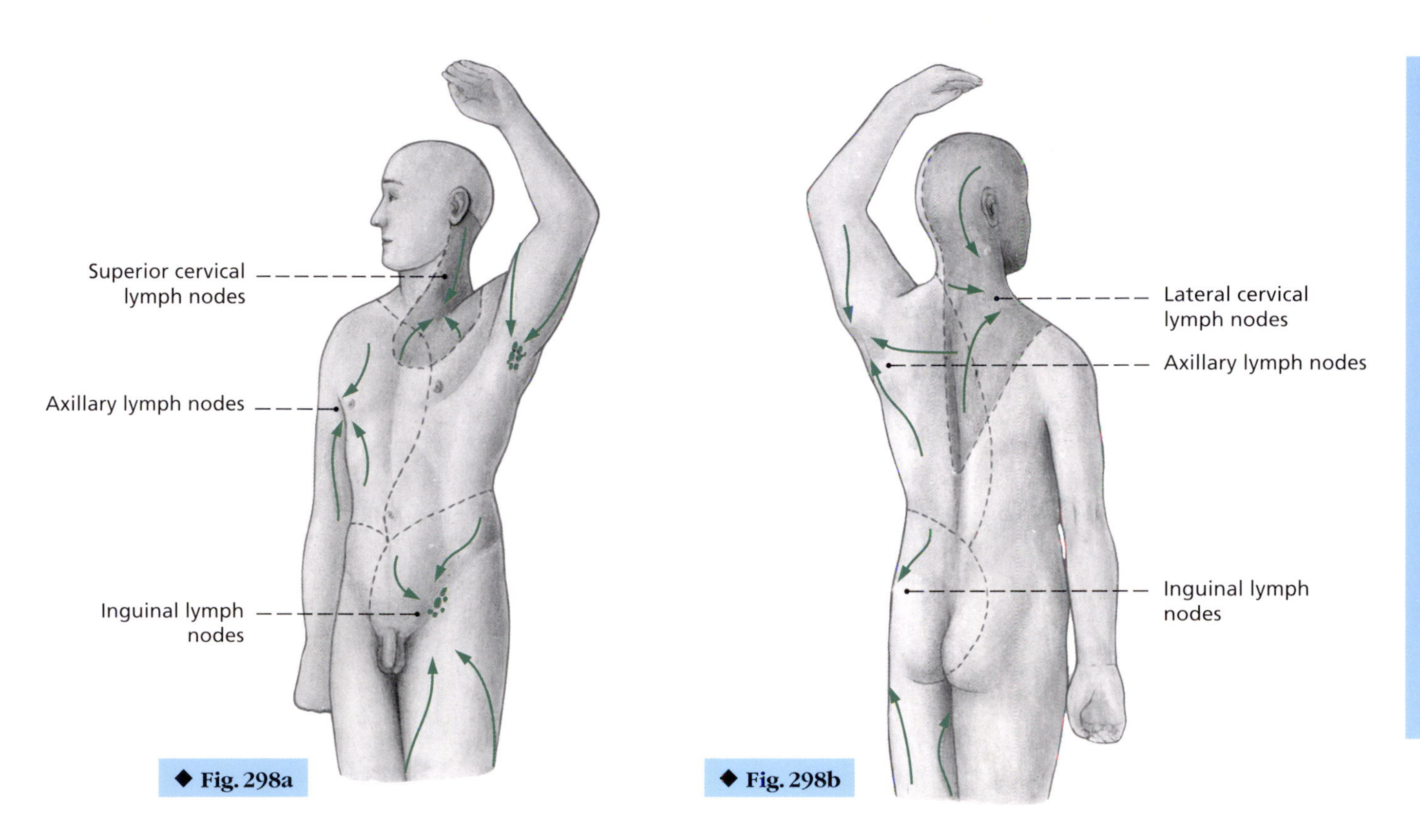

◆ Fig. 298a

◆ Fig. 298b

◆
Fig. 298
The drainage areas of the body's lymph node groups: (a) laterally and to the front, (b) laterally and to the back. Cervical lymph nodes: head, cervical, subclavian and upper dorsal region. Axillary lymph nodes: shoulder, arm and thoracic region up to the body axis, dorsal region. Inguinal lymph nodes: hypogastric region, buttocks, leg, and hip areas.

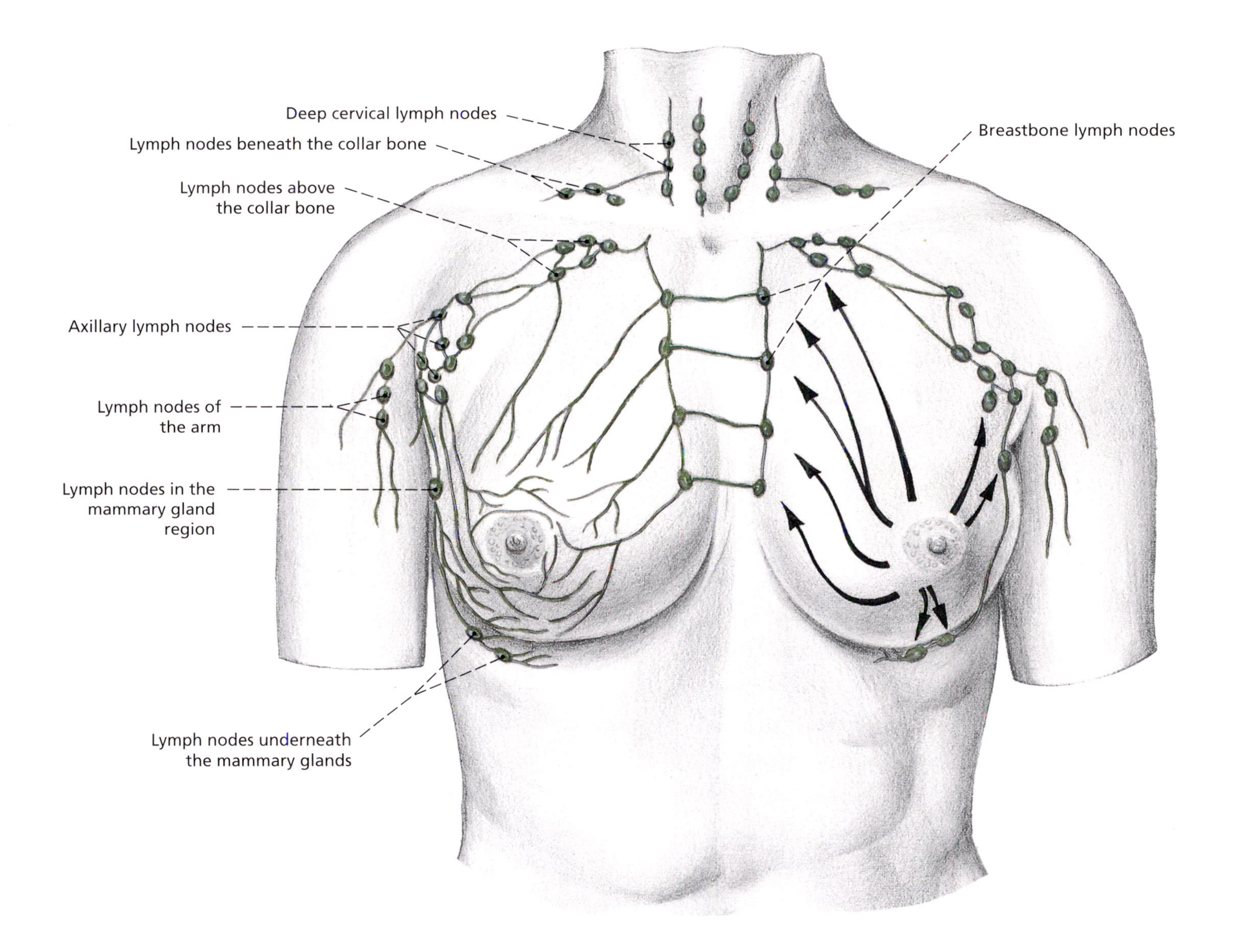

◆
Fig. 299
Lymph drainage paths in the female breast. There are lymph vessel connections linking both sides of the body.

◆
Fig. 300
The lymph nodes and lymph vessels of the lateral neck area (left).

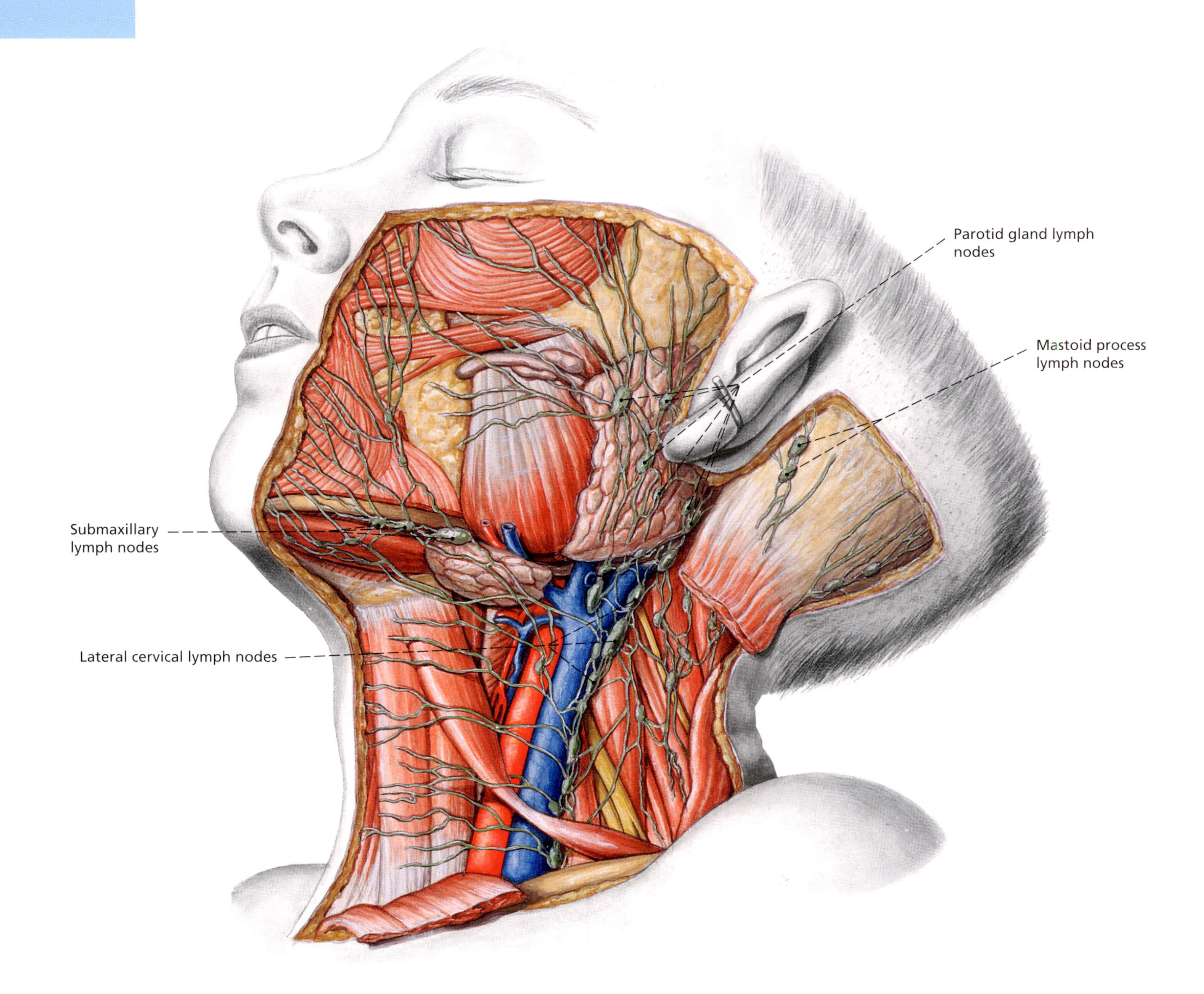

◆
Fig. 301
The armpit lymph nodes with inflow regions from the side of the thoracic wall and the arm (left).

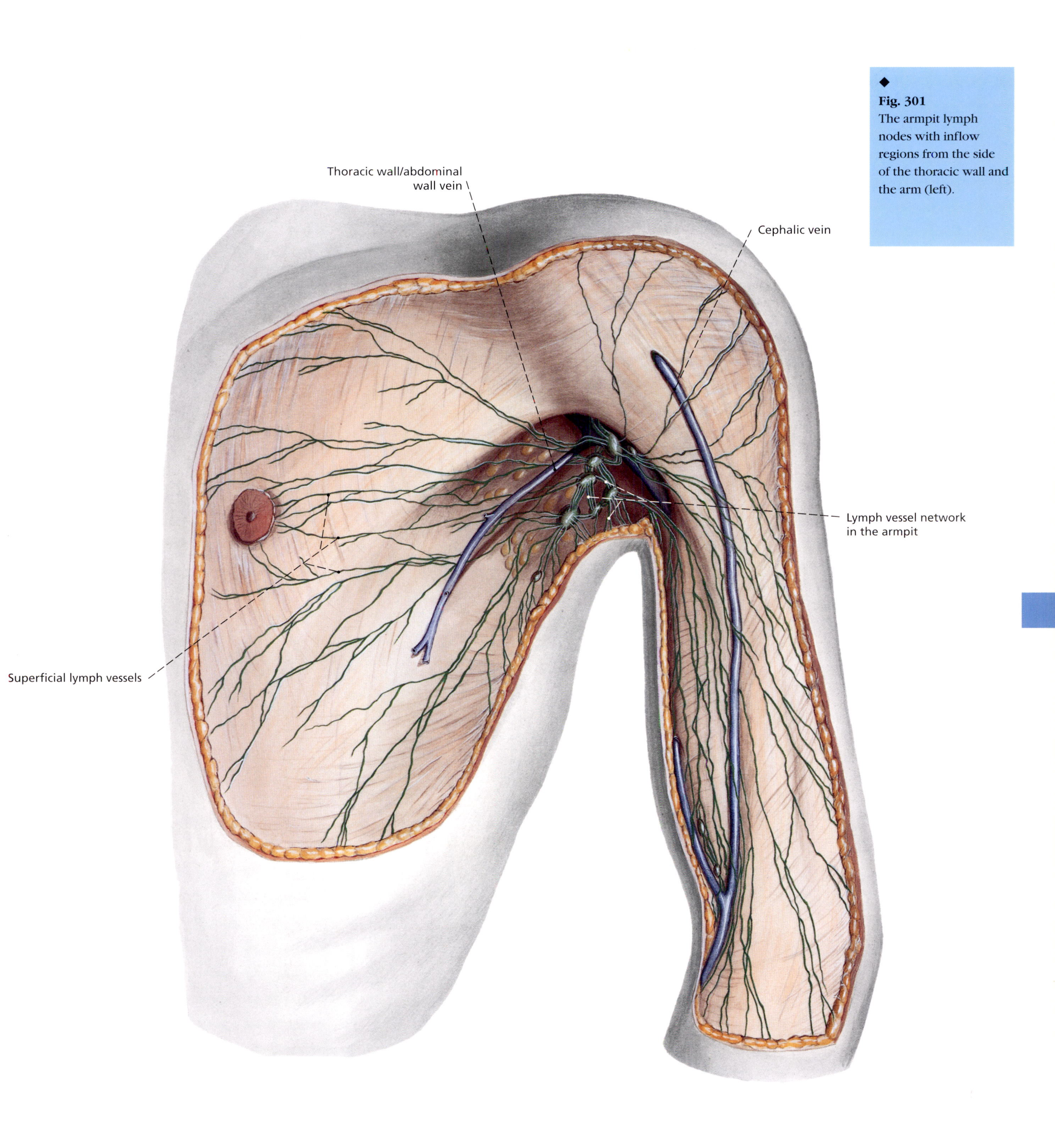

◆
Fig. 302
The lymph nodes of the armpit and the deep lateral neck region. The clavicle (collarbone) has been partly removed, and the chest muscles have been cut through and folded back.

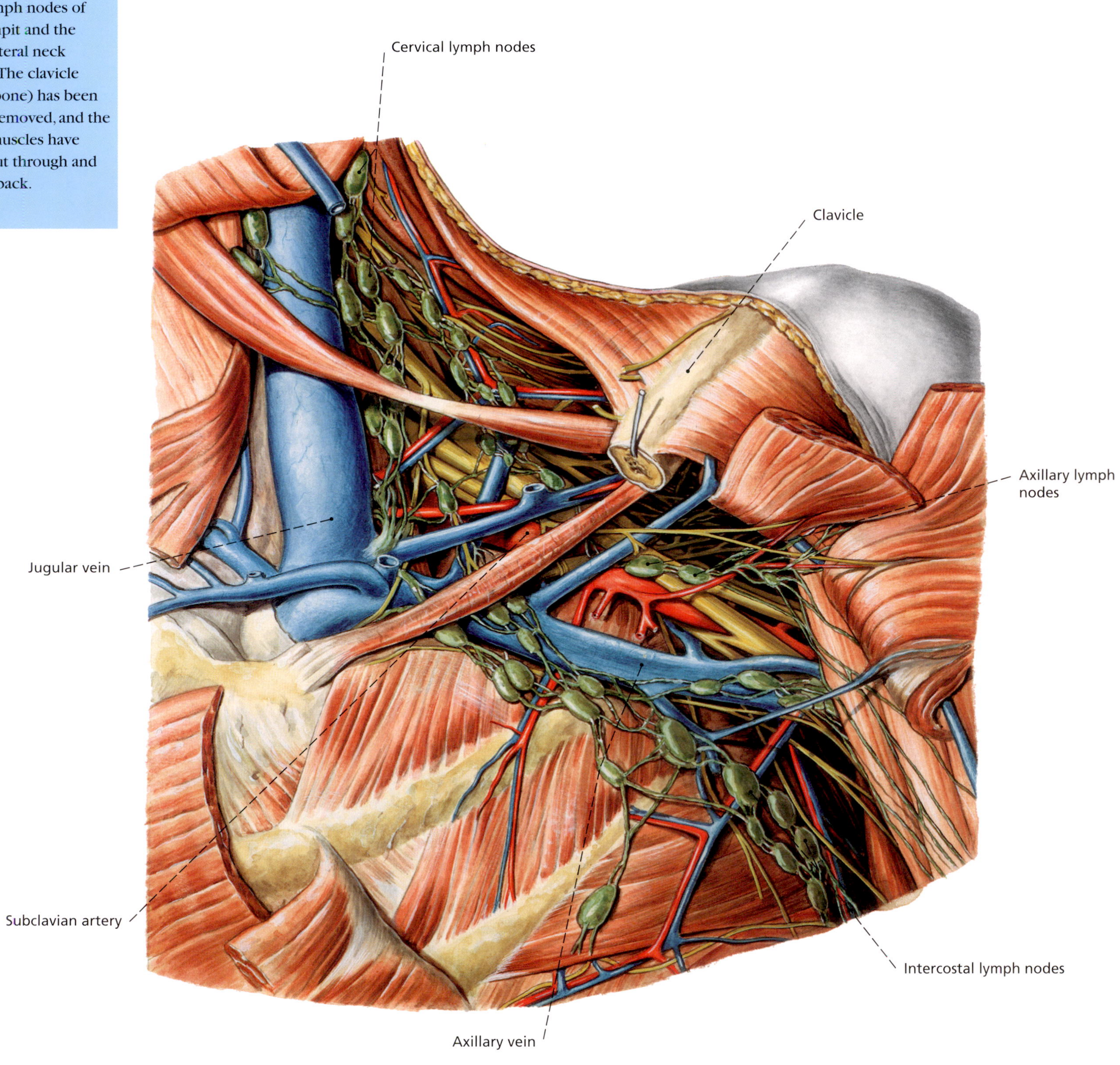

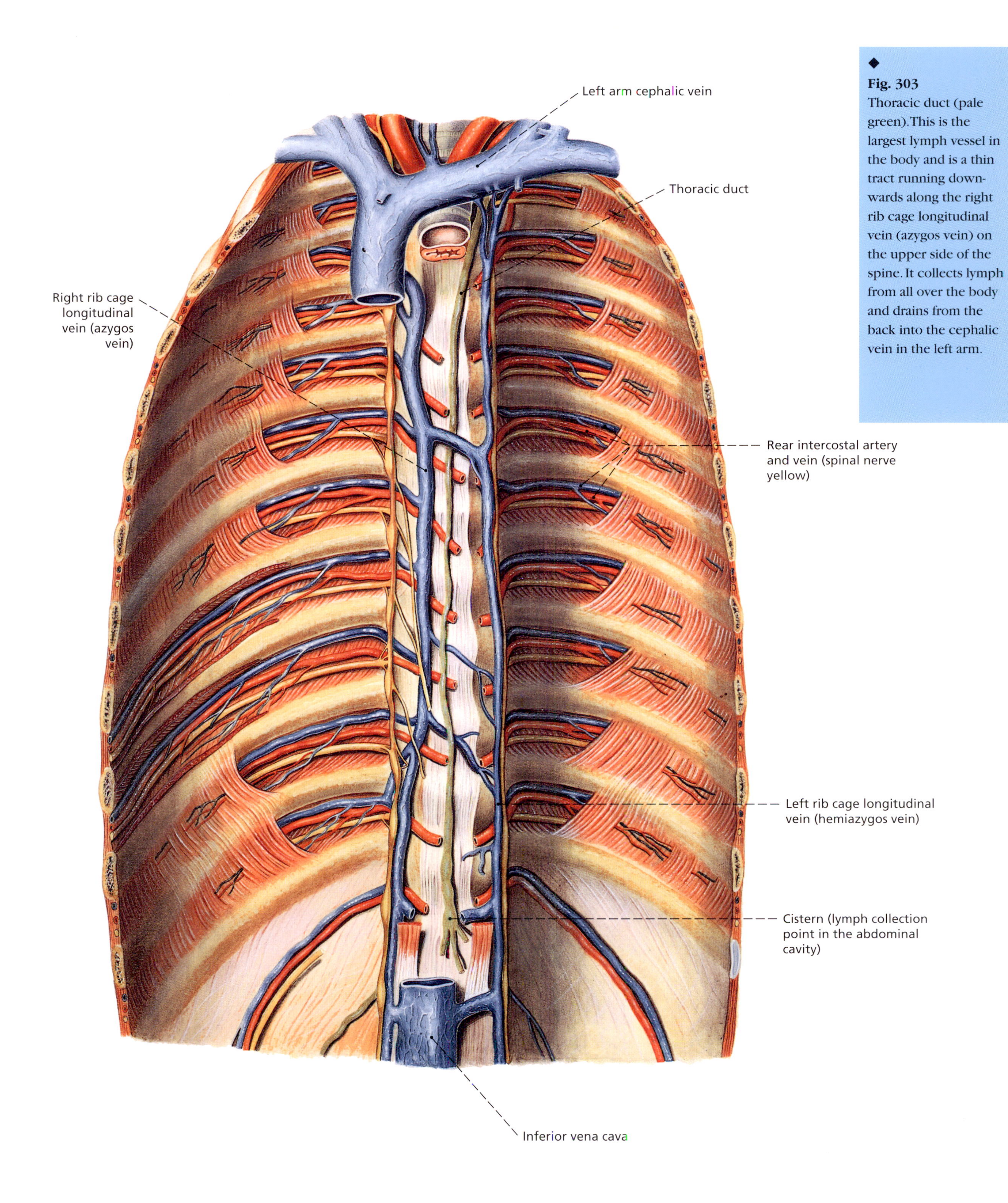

◆
Fig. 303
Thoracic duct (pale green). This is the largest lymph vessel in the body and is a thin tract running downwards along the right rib cage longitudinal vein (azygos vein) on the upper side of the spine. It collects lymph from all over the body and drains from the back into the cephalic vein in the left arm.

◆
Fig. 304
The lymph nodes in the heart and trachea area, here greatly enlarged. Rear view. The bronchi have been cut through and removed prior to their branching into the lungs.

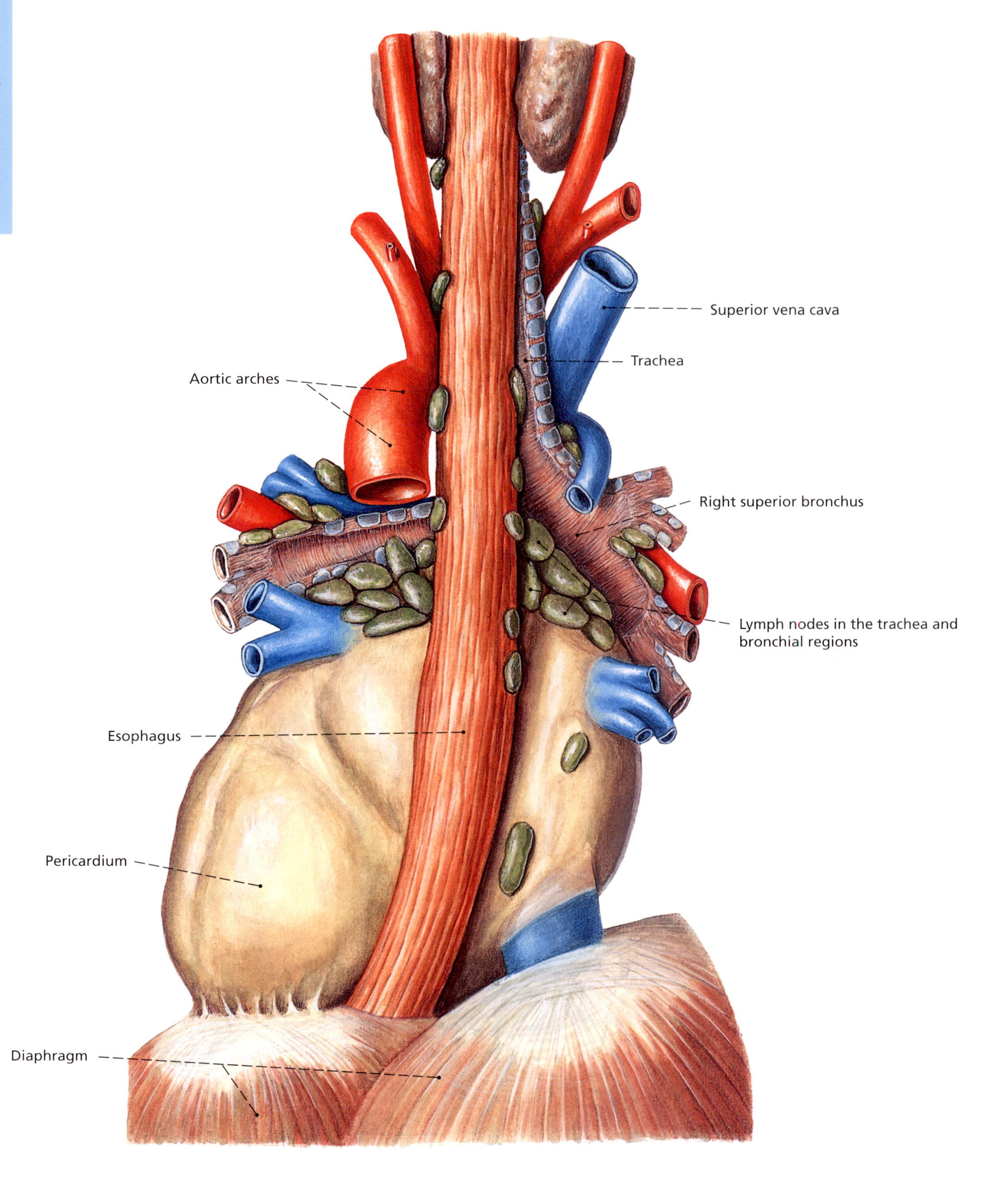

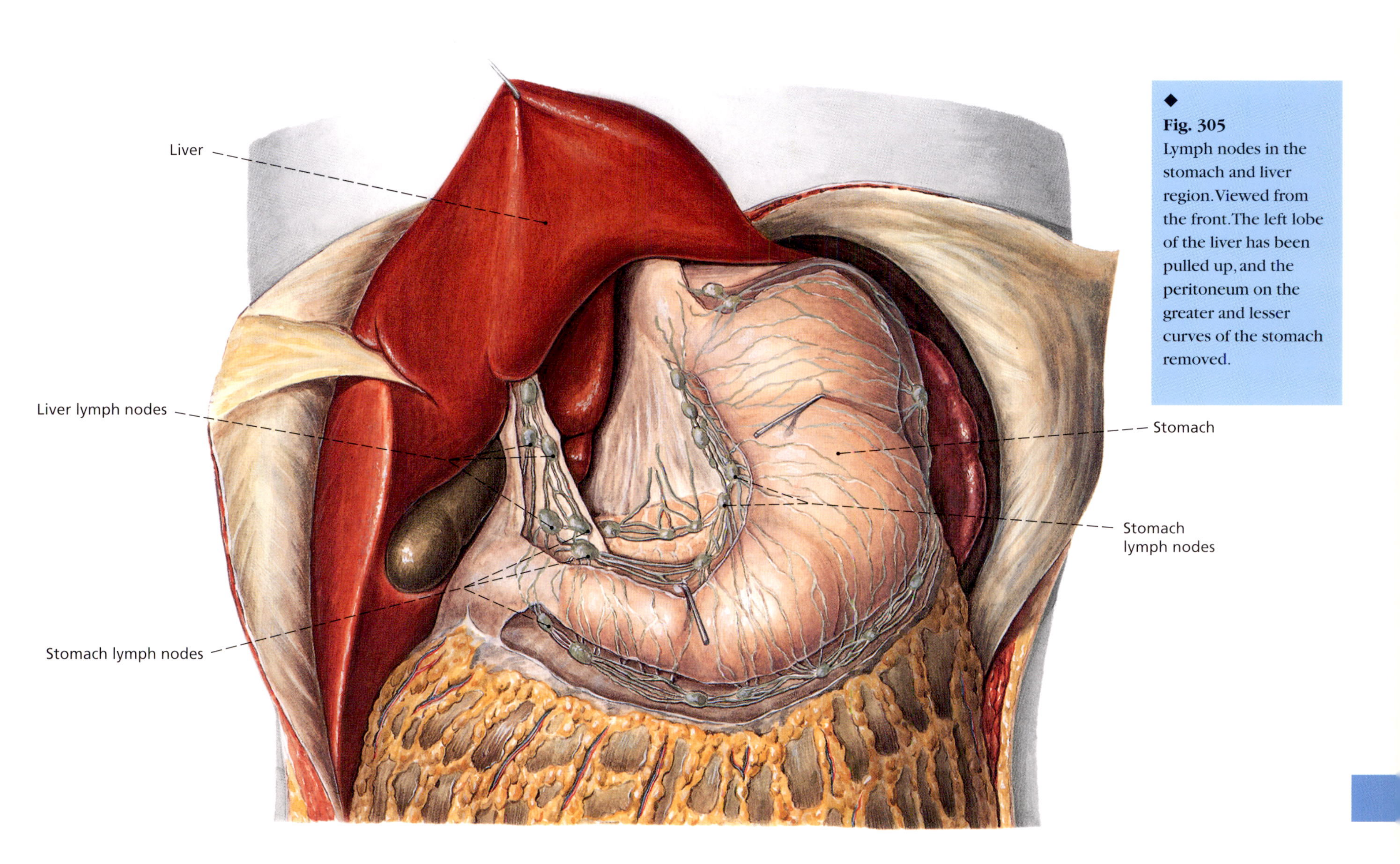

Fig. 305
Lymph nodes in the stomach and liver region. Viewed from the front. The left lobe of the liver has been pulled up, and the peritoneum on the greater and lesser curves of the stomach removed.

Fig. 306
The wall of the small intestine with lymph vessels in the villi (schematic representation).

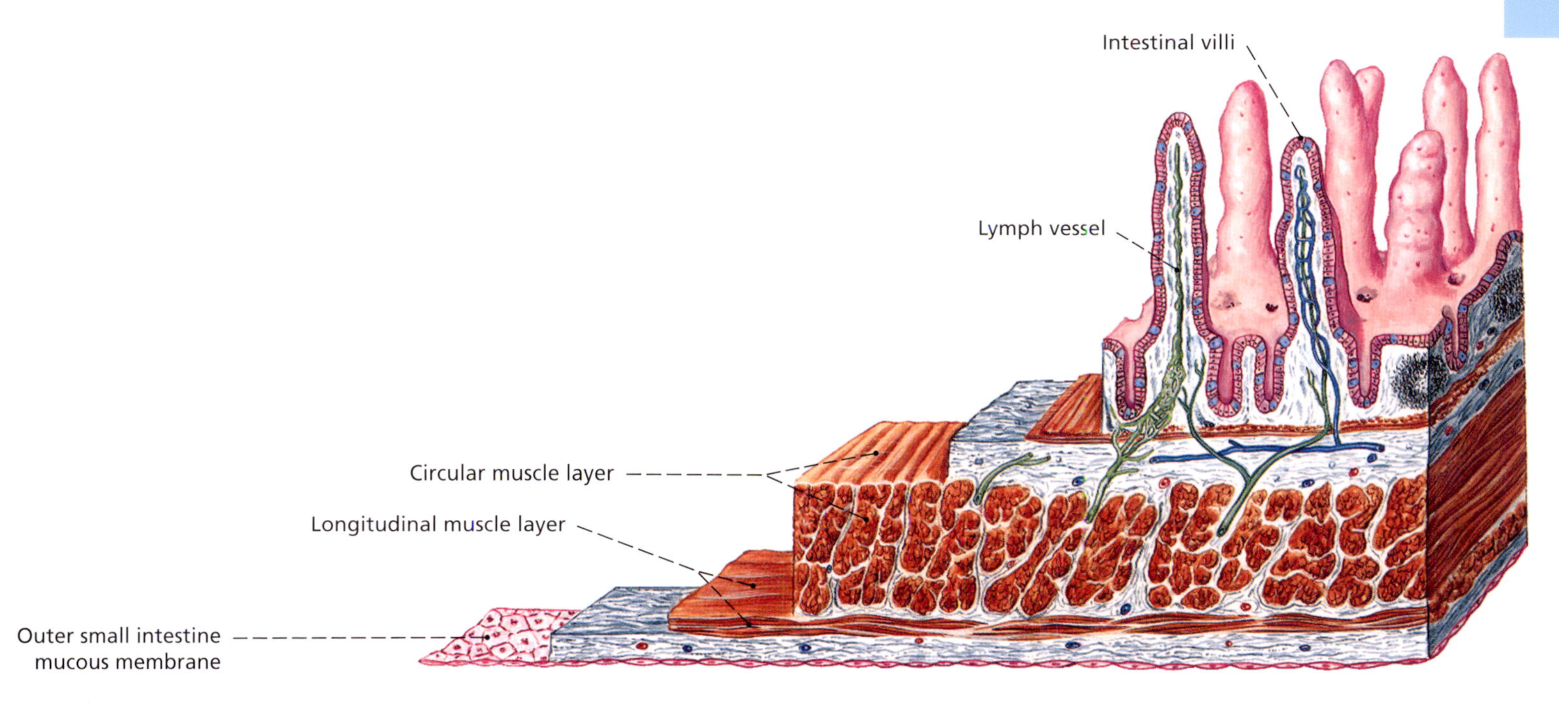

◆
Fig. 307
Lymph nodes and lymph vessels of the abdominal cavity and the lumbar region from the front. The abdominal organs have been removed.

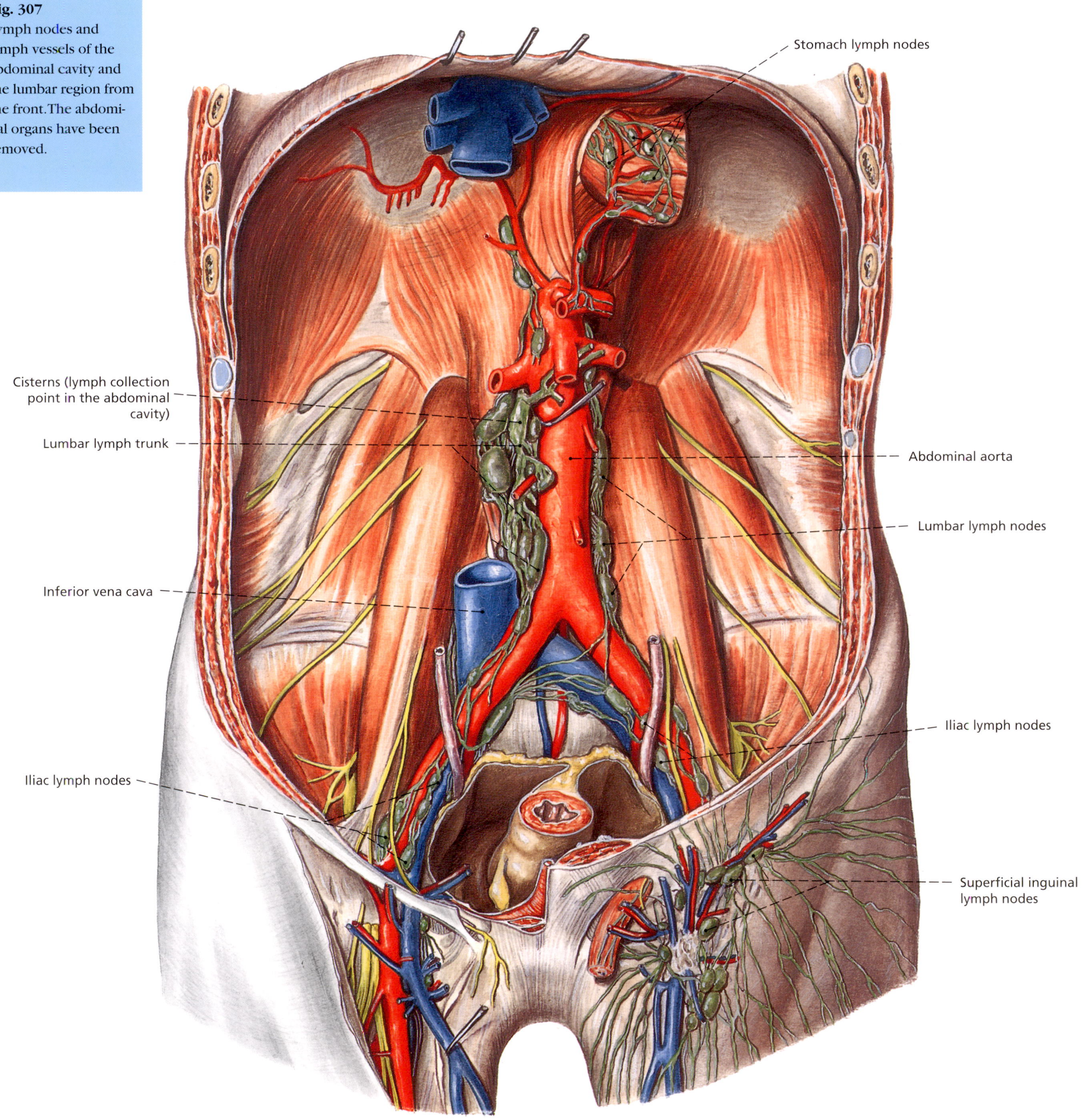

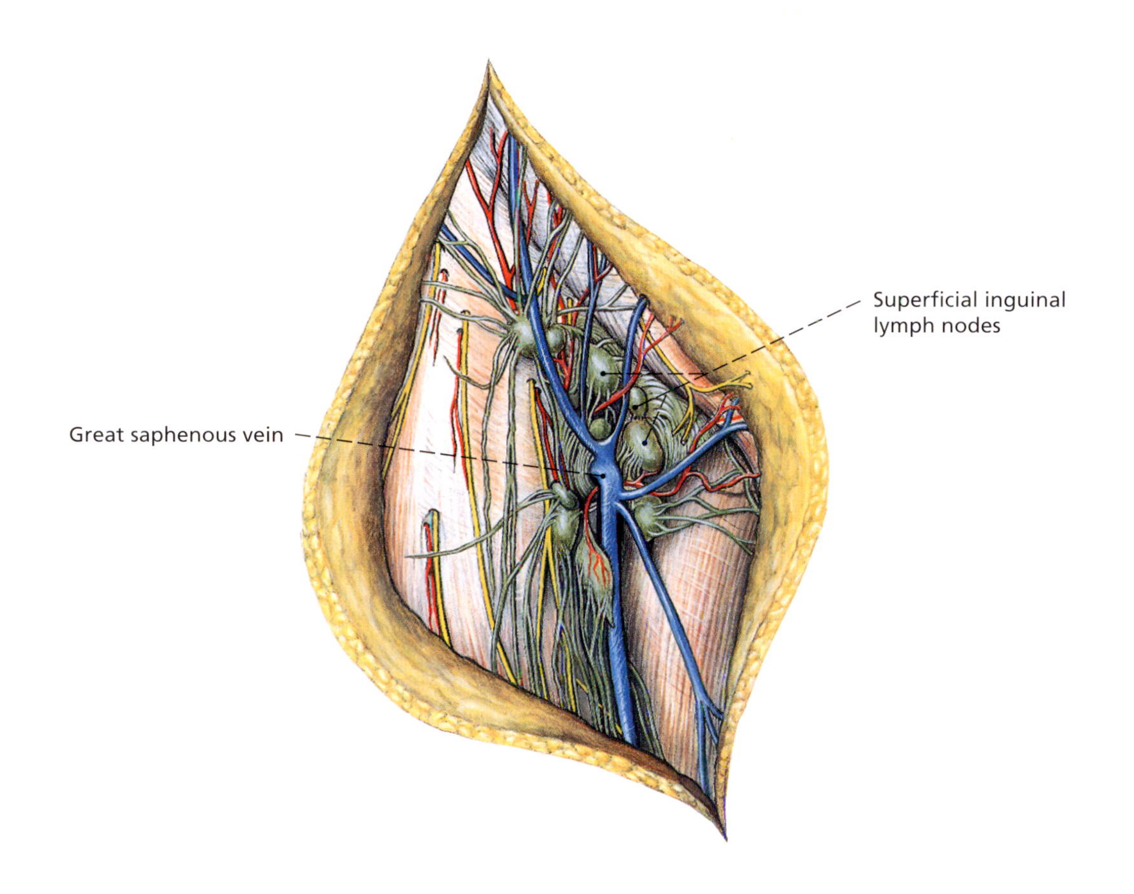

◆
Fig. 308
Lymph nodes and lymph vessels of the right lumbar region seen from the front.

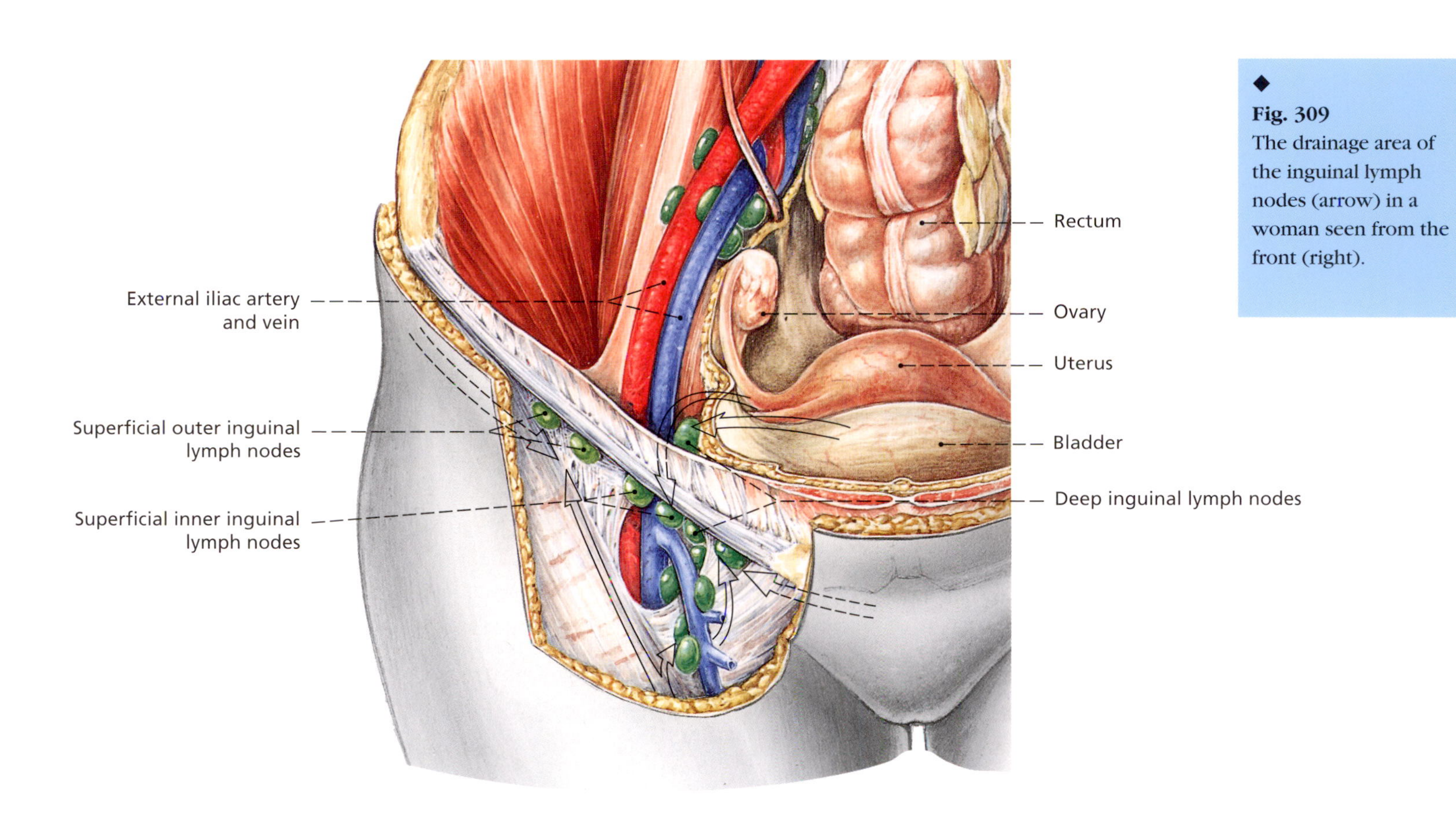

◆
Fig. 309
The drainage area of the inguinal lymph nodes (arrow) in a woman seen from the front (right).

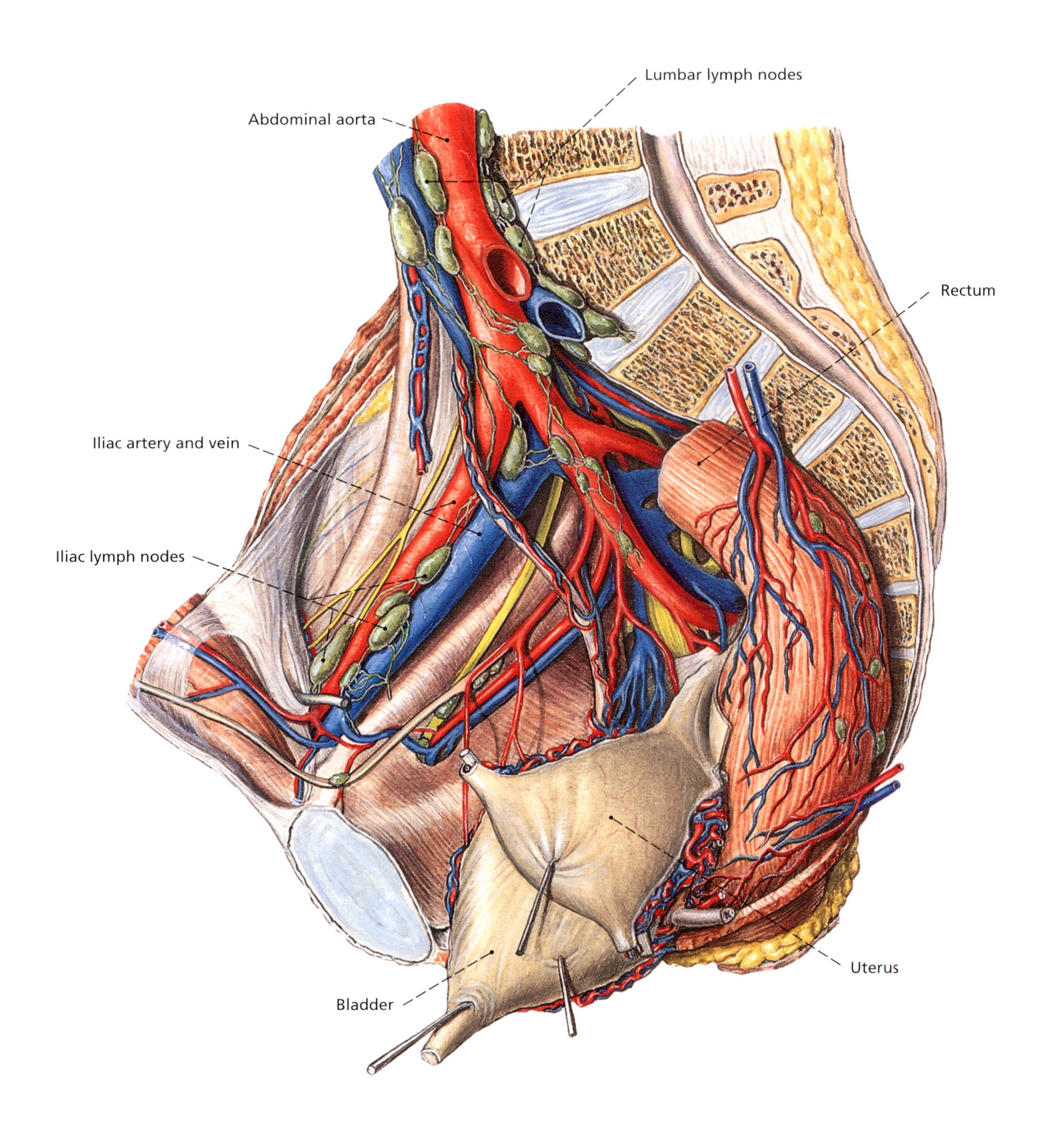

◆
Fig. 310
Lymph nodes and lymph vessels of the female pelvis. Cross-section of the pelvic bone and the lower spine, seen from the left. The lymph nodes are generally much smaller than indicated here.

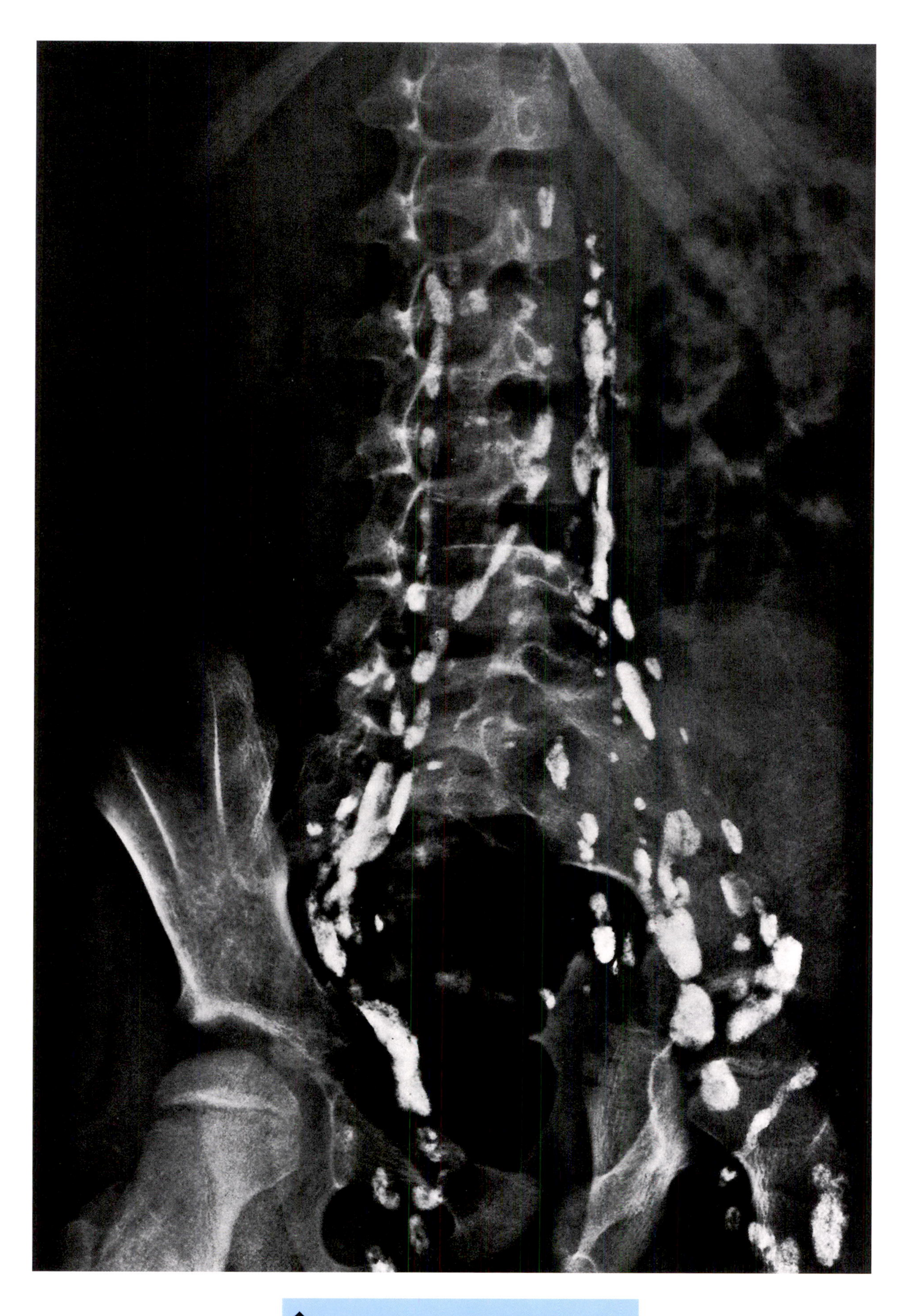

◆
Fig. 311
Lymphogram of the stomach and pelvic region (X-ray image). See the auxiliary image on page 224 for the individual structures.

◆
Fig. 312
Lymph node representation of the abdominal and pelvic region. Auxiliary image for the X-ray image (lymphogram) page 224.

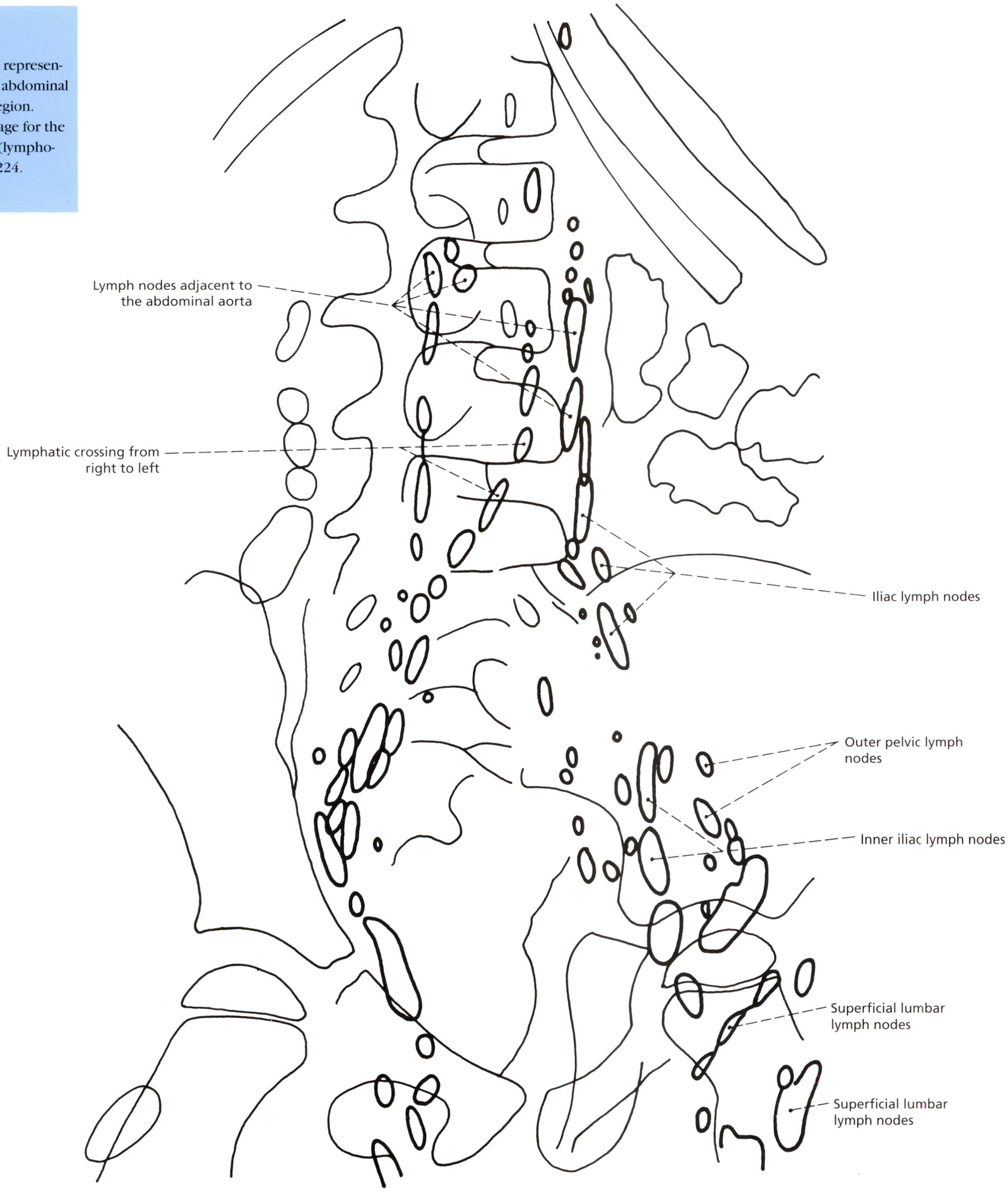

Symptoms and diseases

Swollen lymph nodes

Swelling of the lymph nodes (e.g., beneath the ear), can be an indication of disease. Since the lymph nodes function as barriers against pathogens and their toxins, their swelling is indicative of specific immune system activity. The location of the enlarged lymph nodes may indicate which organ or tissue is affected.

Lymphoma

Rare, malignant disease in which the cells in the lymph nodes or the spleen proliferate unchecked. They can be divided into two types, Hodgkin's and non-Hodgkin's lymphoma. The initial symptoms may be swollen lymph nodes in the neck, armpits, or lumbar region. The disease affects more and more lymph nodes over the course of time, leading to increasing impairment of the immune system. Affected individuals feel unwell; some have fever, loss of appetite, and suffer from night time sweating, as well as a loss of weight. Examination of a tissue sample from one of the enlarged lymph nodes serves to establish the type and stage of the disease. Many patients recover if treated in good time (usually by irradiation).

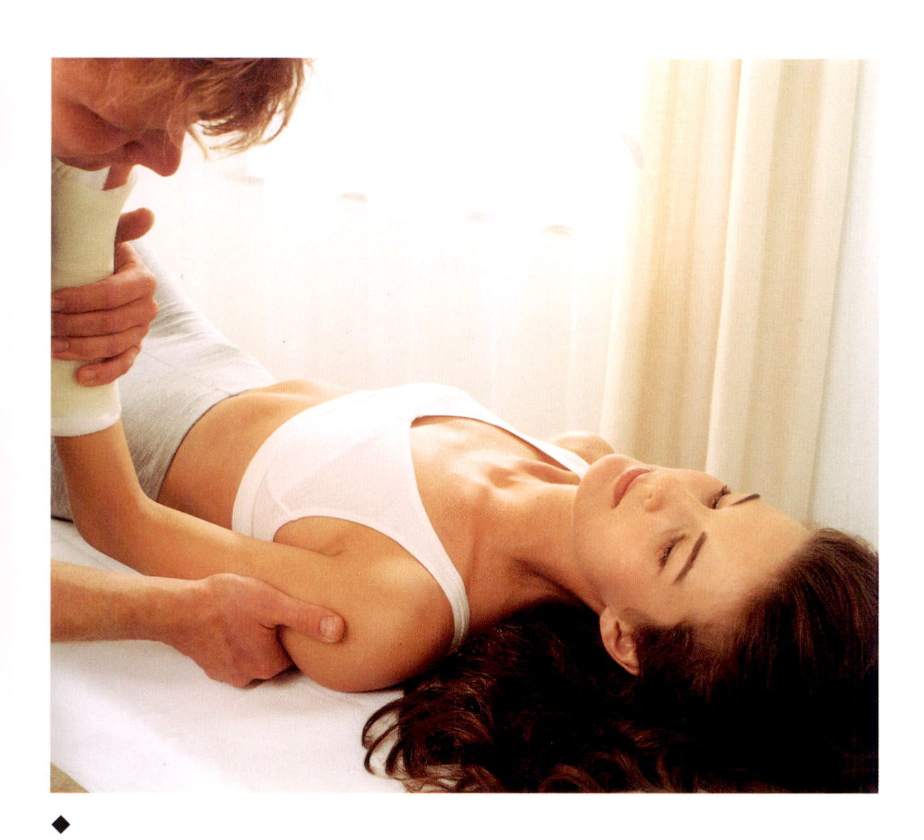

◆
Lymph drainage
Massage, carried out by physiotherapists or massage therapists, may stimulate the draining of the lymph fluid out of the congested lymphatic system.

Lymphedema

A swelling of the tissue caused by a build-up of lymph. Lymphedema occurs when the lymph fluid can no longer be transported from the lymph vessels quickly enough, and thus gathers in the subcutaneous tissue. Malignant tumors are a frequent cause, either squeezing the lymph drainage paths from the outside or growing inside the lymphatic system. Radiotherapy or major surgery can also result in damage to the lymph vessels, leading to a build-up of lymph. Less harmful causes include inflammation, injuries and operations, infections, and vein congestion. If the lymphedema remains untreated, the tissue in the surrounding area slowly changes to become fibrous and hard. Treatment usually consists of advising the patient to wear a pressure gradient dressing; manual lymph drainage may also be helpful.

Diagnostic methods

Lymphography

X-ray of the lymph nodes and the lymph vessels after injection of a contrast medium, impervious to X-rays, into the lymphatic system of the region, so that the system can be visualized.

Treatment methods

Lymph drainage

Gentle massage method that encourages the built-up tissue fluid to drain away. Rhythmic, circular movements of the fingers laid flat on the skin can push the fluid towards the lymphatic system.

Lymph vessel transplant

Transplantation of lymph vessel sections to improve drainage. This procedure is used after breast cancer surgery in particular, in which the axillary lymph nodes are removed and the region irradiated, as well as after other operations resulting in a blockage of lymph flow.

The spleen

The spleen is the largest organ in the immune system. It serves as a production facility for immune cells and antibodies against pathogens, and also breaks down aging red blood cells. The organ has a particularly high blood flow: the body's entire blood volume flows through the spleen approximately 500 times a day.

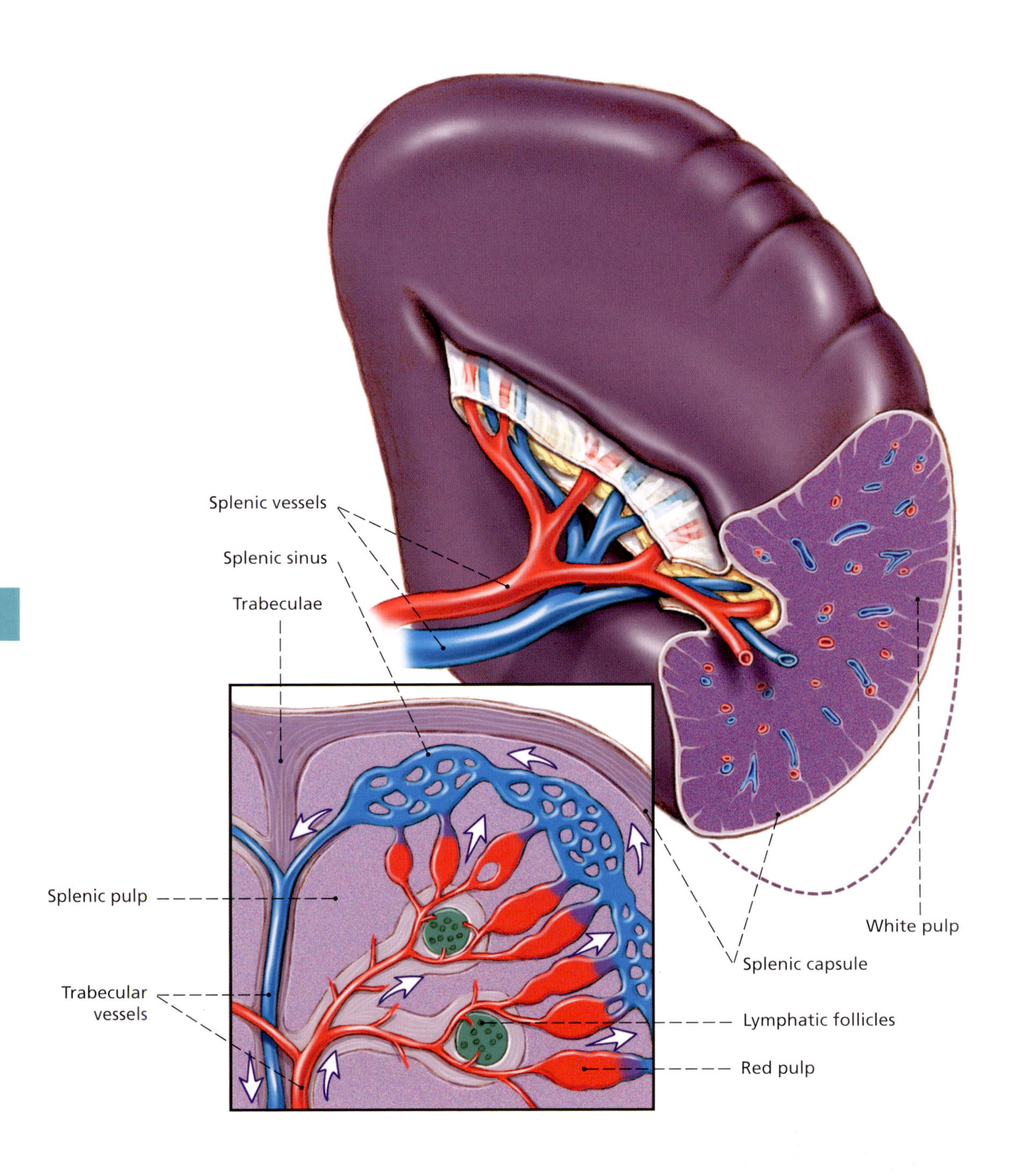

◆
Fig. 313
The splenic vessels branch out into the trabeculae of the spleen and then pervade the white and red pulp in a dense network *(detail)*. The white pulp contains small lymphatic follicles. The blood filtered in the spleen then flows out again via the trabecular vein.

Structure and function

A red-brown organ weighing 150–200 grams, the spleen is located on the left side of the upper abdomen, directly below the diaphragm. The outer casing consists of a compact connective tissue capsule, from which strands of connective tissue (trabeculae) extend into the inner organ.
These strands build a framework for the spleen tissue, or pulp. There are two different types of tissue in the spleen:

- the white pulp, consisting of lymph tissue. Immune cells (lymphocytes) and antibodies (immunoglobulins) are formed here.
- the red pulp, consisting of a dendritic network of blood vessels with enlarged areas in places (known as splenic sinuses). Aging red blood cells are broken down here, and cell remains, pathogens, and foreign agents are also taken out of circulation.

Despite these important tasks, the spleen is not a vital organ. Should it be injured in an accident, for example, and need to be removed, its tasks are taken over by other organs. The breakdown of the red blood cells then takes place in the liver, while the formation of immune cells and antibodies takes place in the bone marrow and the lymph nodes.

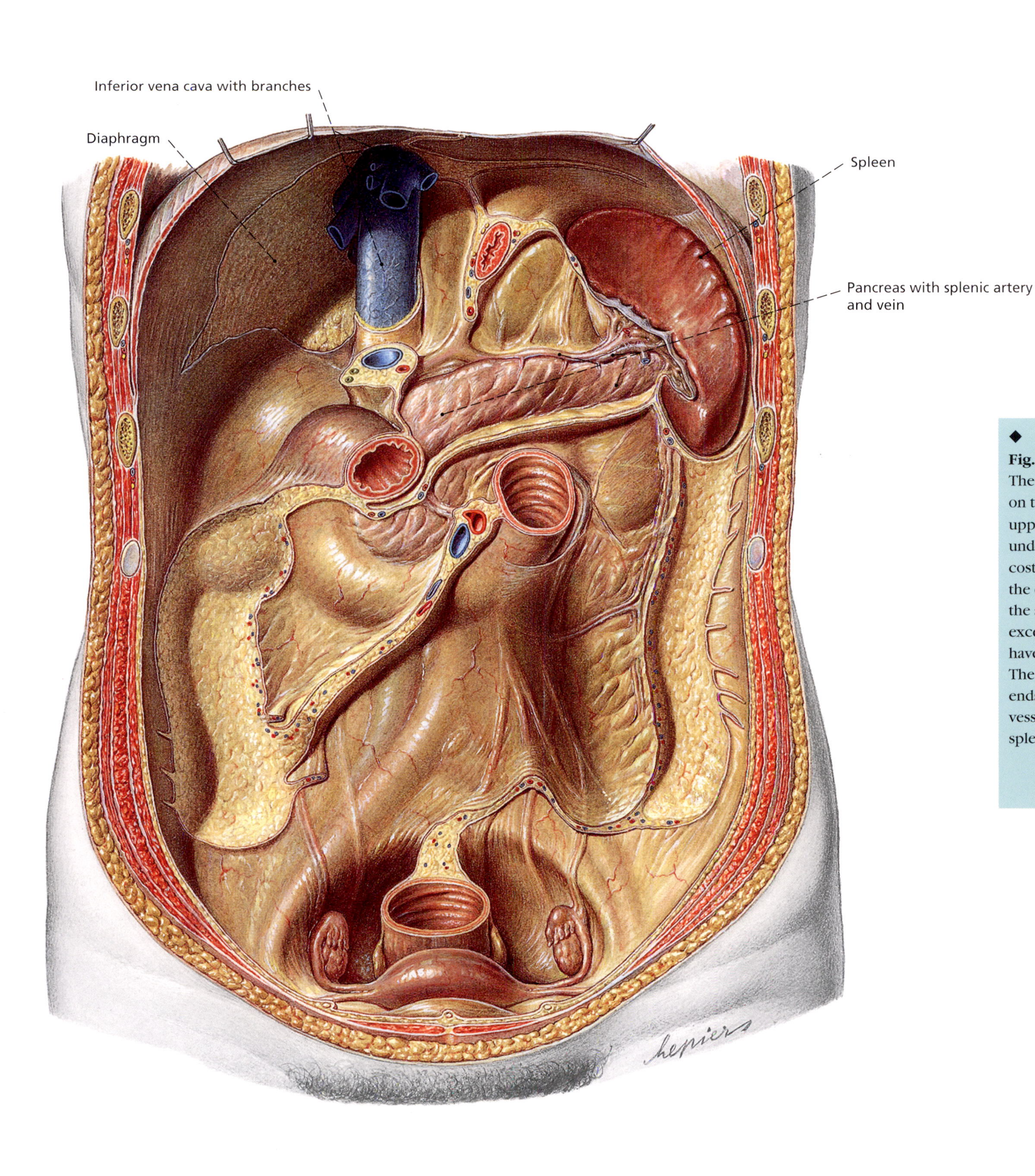

◆
Fig. 314
The spleen is situated on the left side of the upper abdominal cavity underneath the left costal arch and beneath the diaphragm. All of the abdominal organs except the pancreas have been removed. The tail of the pancreas ends where the splenic vessels enter the spleen.

◆
Fig. 315
The under surface of the spleen with the splenic artery and vein.

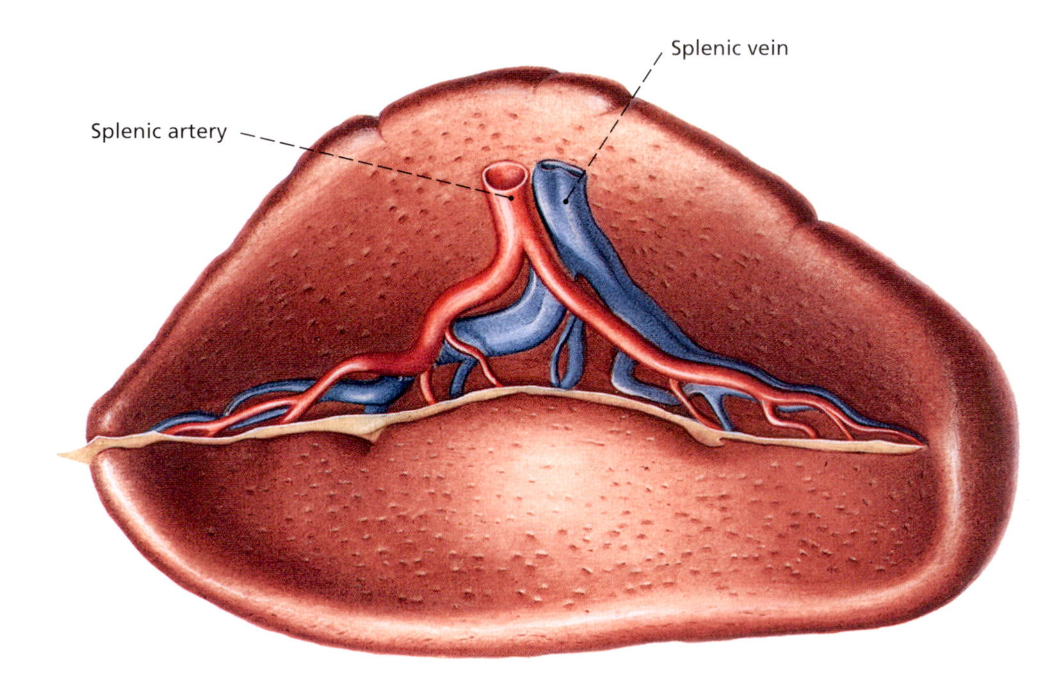

◆
Fig. 316
The outer surface of the spleen. It lies directly underneath the diaphragm.

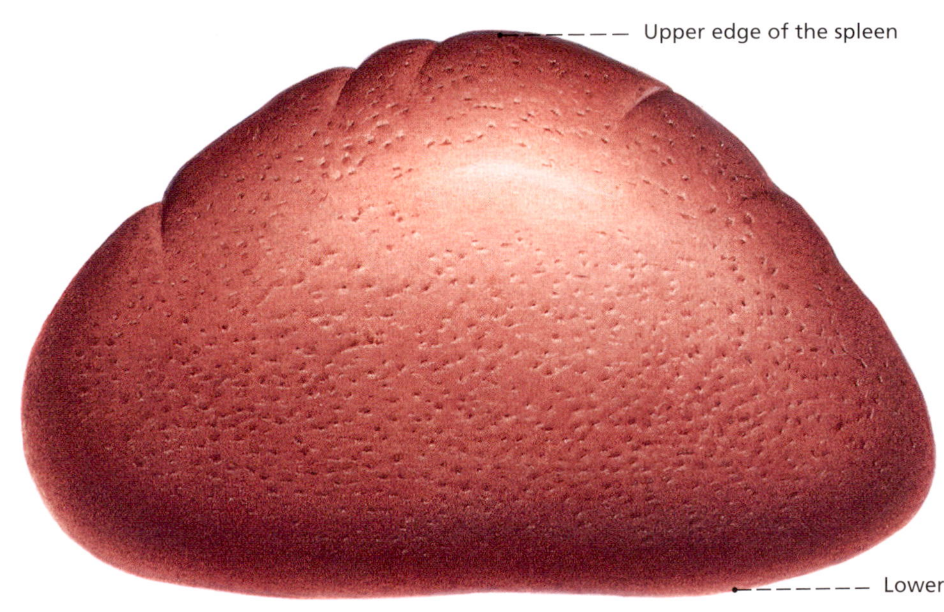

◆
Fig. 317
Cross-section of the spleen, viewed from below. The capsule, the splenic trabeculae, and the cross-sectional blood vessels (blue and red) are easily identifiable.

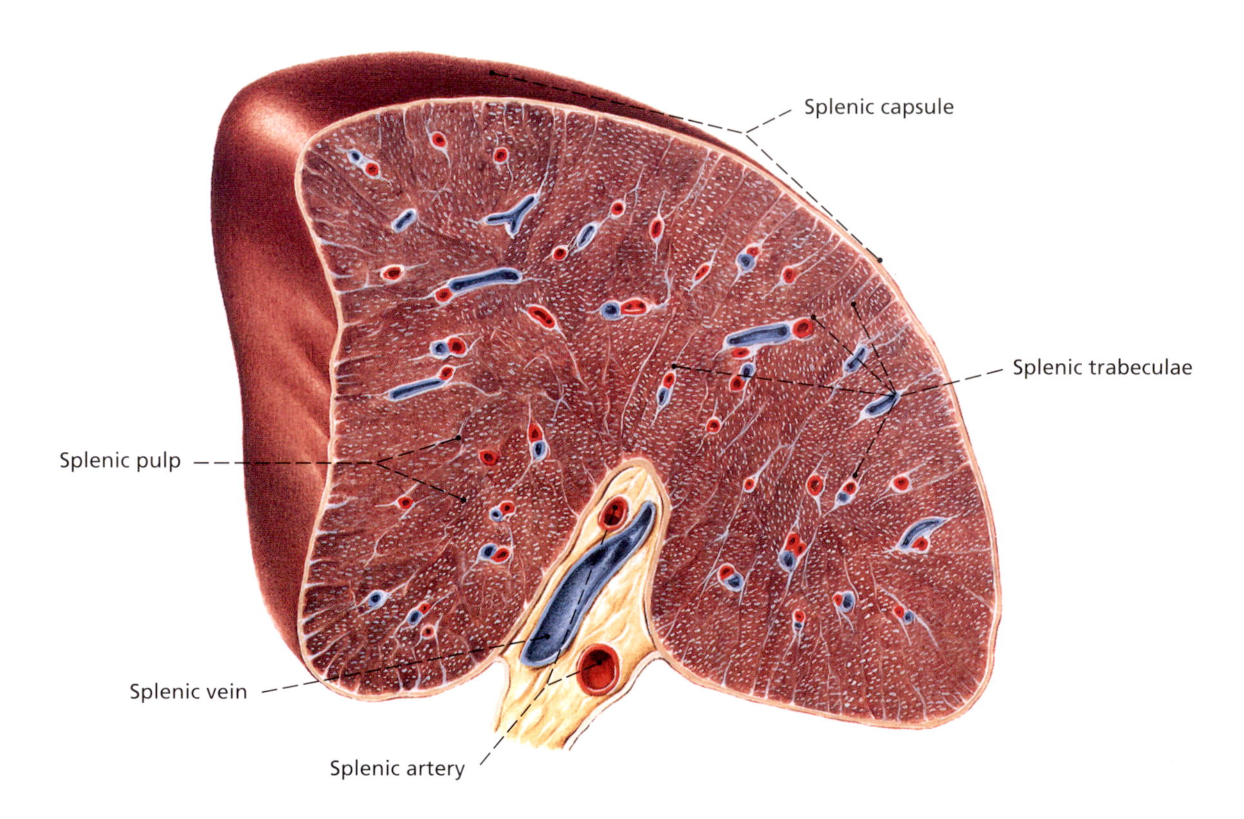

Symptoms and diseases

Ruptured spleen

Of all the abdominal organs, the spleen, with its thin connective tissue capsule, is the one most frequently injured by the application of blunt force to the abdomen. The blow does not have to hit the spleen directly; the spleen ruptures more frequently as a result of blows to other areas of the abdomen. If the spleen is enlarged it can rupture under very little impact. Broken ribs can also pierce the spleen. A ruptured spleen occasionally causes pain that can extend to the left shoulder. Attention is often only drawn to a ruptured spleen by the shock resulting from the loss of blood: the blood pressure drops and the pulse quickens; there is a loss of consciousness, and the affected individual passes out. Sometimes the splenic capsule does not rupture immediately, but does so only when sufficient blood has gathered in the spleen that the tension becomes too great (known as two-sided splenic rupture). This is especially dangerous because affected individuals only notice symptoms some time after the accident. Bleeding inside the splenic capsule does not require surgery itself, but if the capsule ruptures, surgery will be required Every effort is made to save the spleen, especially in children, as it is important for immune defense.

Swollen spleen, splenic tumor (splenomegaly)

Usually the consequence of other organ diseases or infections. As a general rule, without symptoms; at most, affected individuals report a feeling of pressure and mild pain in the left upper abdomen. The causes can include infections such as Pfeiffer's disease or glandular fever, malaria, benign and malignant tumors, or anemia. The enlargement of the spleen is established using ultrasound. The treatment depends on the cause; the removal of the spleen is occasionally necessary.

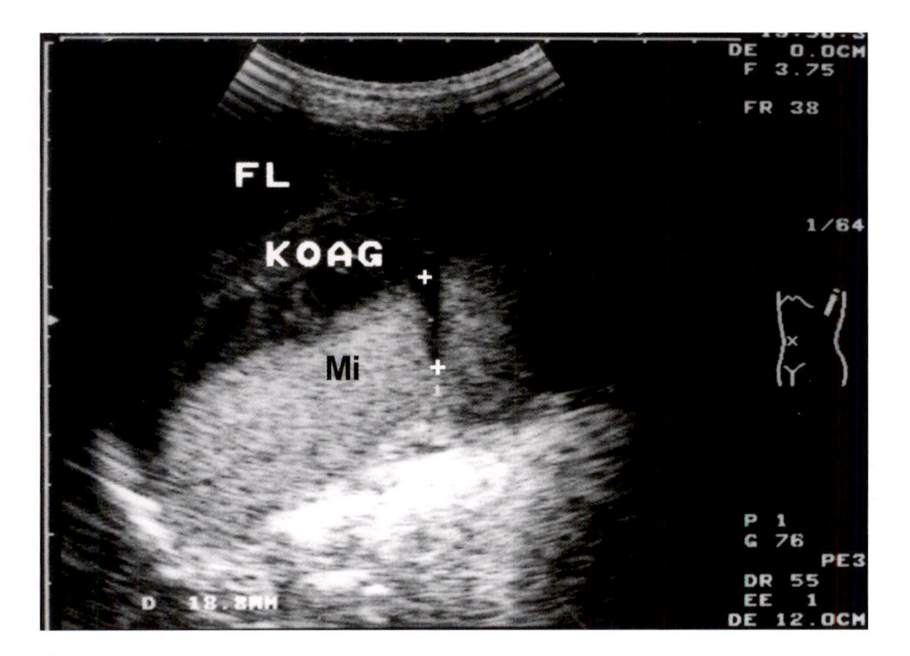

◆ **Ultrasound in the case of splenic rupture**
The spleen can easily be seen via ultrasound.

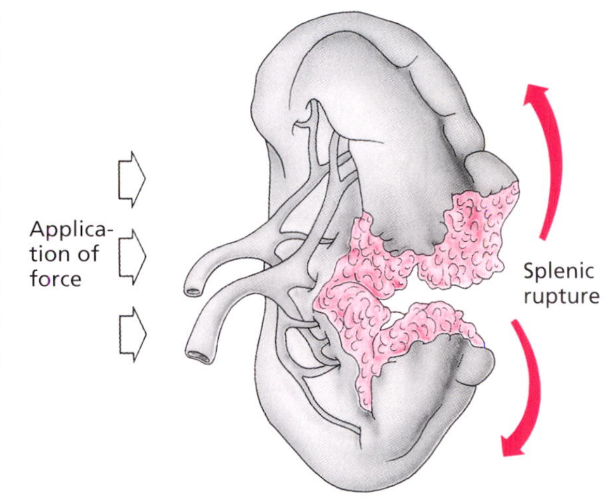

◆ **Splenic rupture**
The spleen is often affected by the application of force to the abdomen. It is not the force itself which leads to splenic rupture, but rather the resulting increase of pressure in the abdomen.

Diagnostic methods

Ultrasound (sonography) is the most important means of assessing the spleen. A diagnosis of ruptured spleen, as well as an enlarged spleen, can be established with complete certainty by this method. Ultrasound is painless and there is no exposure to radiation so it is also safe to use during pregnancy. A diagnosis of splenomegaly will necessitate laboratory tests after ultrasound, in order to determine the cause of the enlargement.

Treatment methods

Removal of the spleen (splenectomy)

The removal of the spleen is the only option in the case of an extensive rupture or splenic tumor. With children, an attempt is made to save at least part of the spleen in order to assist with immune defense. With adults it is usually removed entirely, as the adult body is better able to protect itself against infections without a spleen than a child's. Nevertheless, the immune system does need a while to readjust, and extra protection from infection is especially important following the removal of the spleen. All of the main vaccinations, e.g., against influenza, pneumococci, and meningococci, should be carried out, and antibiotics should be taken as a precaution in the event of imminent surgery, e.g., dental restoration work.

Digestive system

The path from the bite of bread to the excretion of its indigestible components in the stools is a long one. First the bread is broken up into small pieces in the mouth and mixed with saliva. The resulting food pulp then slides down the esophagus into the stomach where it is digested further. During their subsequent passage through the small intestine the food components undergo chemical changes that enable them to pass into the bloodstream through the intestinal villi. The contents of the intestine are then concentrated by elimination of water in the large intestine and the waste products are ultimately excreted via the anus.

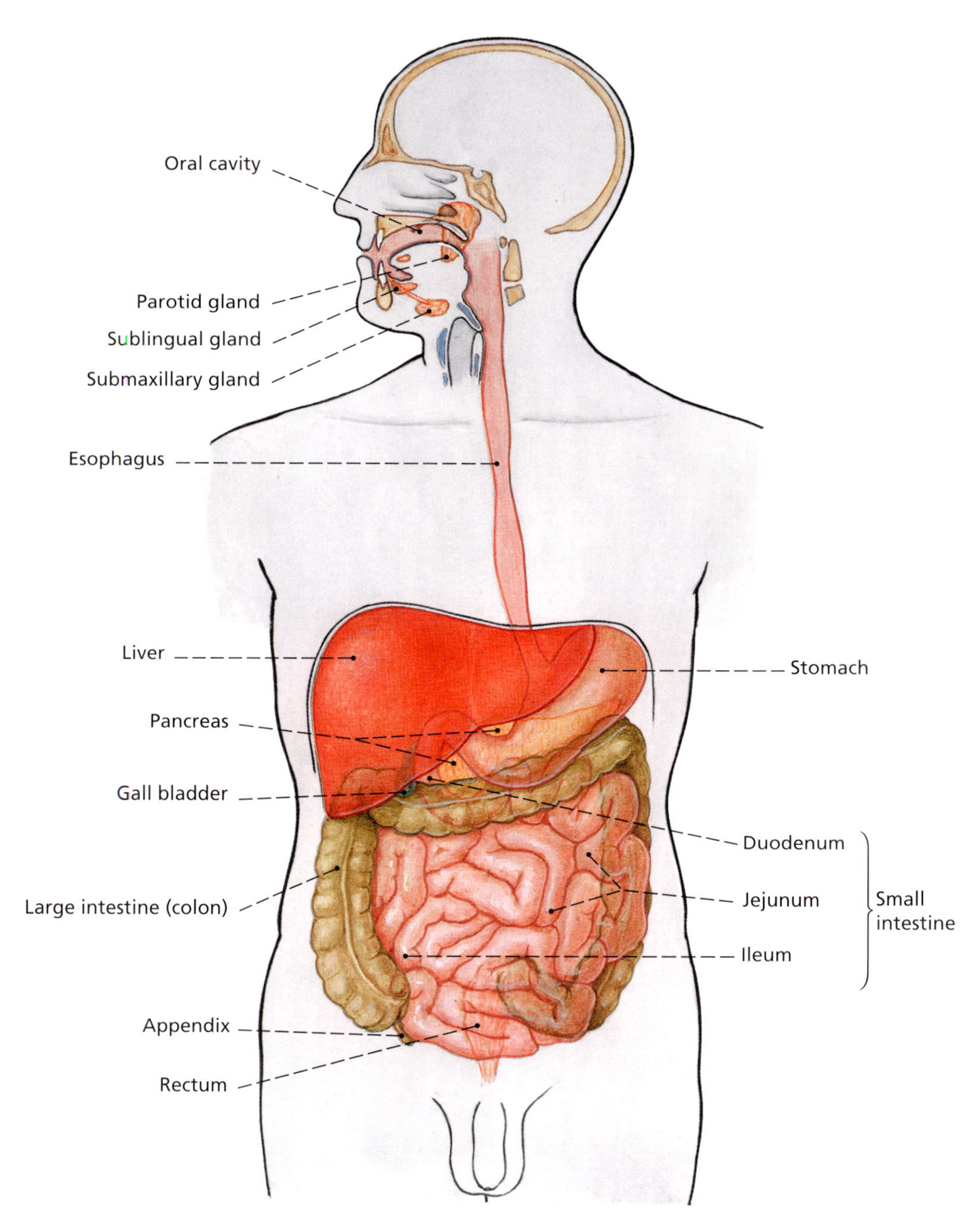

◆
Fig. 318
The digestive system consists of the oral cavity with the pharynx, esophagus, stomach, and intestines. The excretory ducts of a series of glands involved in the digestive process open into these hollow organs, as well as the main bile duct of the liver. In the oral cavity these are the salivary glands, while in the abdominal region they are the glands of the liver and pancreas.

Oral cavity

The excretory ducts from three pairs of salivary glands open into the oral cavity:

- the parotid gland, located on the masseter muscle below the ear
- the submaxillary gland, on the inside of the mandible
- the sublingual gland, in the floor of the mouth

Not only does the saliva lubricate the broken-down food, it also starts the digestive process, beginning with the breakdown of carbohydrates. Saliva also contains agents, which kill off bacteria, thus forming a barrier against infections. Last but not least, it protects against dental caries.

Esophagus

This tube of muscle, about 25 centimeters in length and lined with mucous membrane, connects the oral cavity with the stomach. No digestive processes take place here. Its elasticity allows the esophagus to double the size of its opening, e.g., when swallowing a large mouthful. However, the diameter of the esophagus is a limiting factor at three anatomical constrictions where it is unable to expand, such that swallowed foreign bodies or large pieces of food may get stuck.

Stomach

Food is stored in the stomach, mixed with the gastric juices, and digested further: proteins and carbohydrates are broken down into smaller molecules, and fats are fragmented into the smallest fat droplets. The wave-like movements of the muscles in the wall of the stomach ensure that the stomach contents are thoroughly kneaded together before being carried through to the small intestine in portions. The length of time the food pulp spends in the stomach depends on its composition. Carbohydrates are most rapidly transferred to the intestines, while fatty foods can remain in the stomach for several hours. The glands in the gastric mucous membrane, which are located deep in the gastric pits, produce about 2 liters of gastric juice daily for the digestion of nutrients. This juice consists of gastric acid, digestive juices (containing digestive enzymes), and gastric mucus (which covers the mucous membrane and protects it from the aggressive gastric acid). Gastric acid not only has a digestive function, it also destroys pathogens entering the stomach with the food. The production of gastric juices is controlled by the autonomic nervous system. The quantity produced depends on the type and extent of the food ingested, as well as on psychological factors. Production is stimulated by stress and irritation, as well as by the smell, sight, or even just the thought of tasty food.

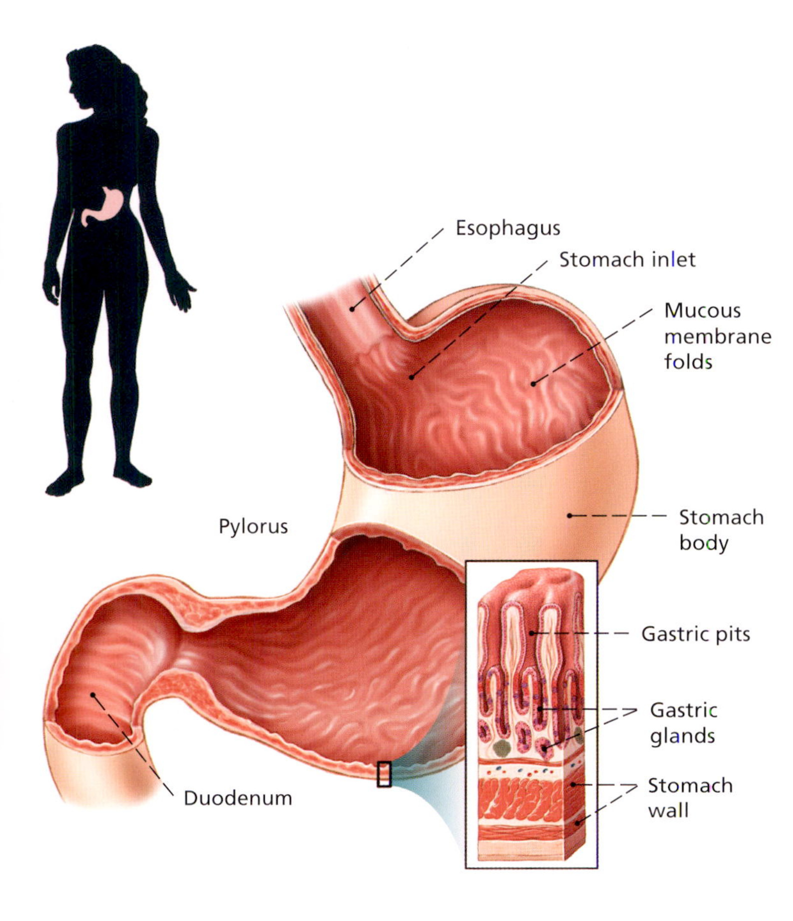

◆
Fig. 319
Tiny gastric pits are embedded in the folds of the gastric mucous membrane. The gastric glands excrete mucus, containing gastric acid and digestive juices, into these gastric pits *(detail)*.

Pancreas

The pancreas is a vital organ located directly behind the stomach. About 15 centimeters in length and weighing some 100 grams it is composed of two different types of glandular tissue:

- one type (exocrine) produces about 1 liter of digestive juice daily, which it delivers to the small intestine via an excretory duct.
- the hormones insulin and glucagon are formed in the other type of tissue (endocrine) and are delivered directly into the bloodstream.

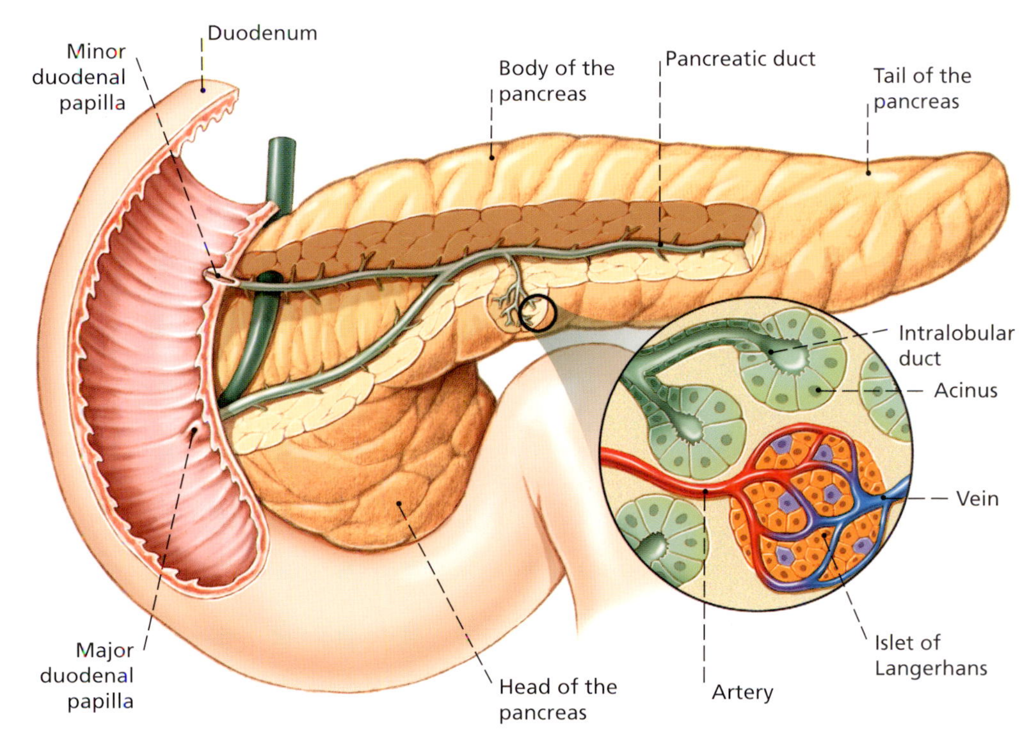

◆
Fig. 320
The pancreas contains two different types of glandular tissue: exocrine tissue containing pyramidal secretory cells, the ducts of which carry digestive juices into the duodenum, and endocrine glands, the Islets of Langerhans, which release the hormones insulin and glucagon directly into the bloodstream.

The digestive juice is produced in small, berry-shaped glandular lobules known as acini and gathers in the pancreatic duct which branches out in the middle of the organ, opening into the duodenum via two openings (the major and minor duodenal papilla), the lower branch usually combining with the bile duct. Fats, proteins, and carbohydrates are broken down in the duodenum by the enzymes in the digestive juice. The hormones insulin and glucagon are formed in the adenocytes, which are distributed throughout the organ like islands (known as the Islets of Langerhans). These two hormones regulate blood sugar levels: insulin facilitates the absorption of glucose, the most important energy source, from the blood into the cells of the body, thus reducing the level of sugar in the blood, while glucagon stimulates the regeneration of glucose, thus raising the level of sugar in the blood. Insulin production is dependent on the carbohydrate content of the food, which in turn determines the level of blood sugar.

Higher levels of insulin are produced when there is a high glucose concentration in the blood and vice versa. It is this regulatory mechanism that is defective in diabetics.

Liver and gall bladder

The liver, which weighs about 1.5 kilograms in adults, is located on the upper right side of the abdomen directly beneath the diaphragm. It is usually covered almost completely by the right costal arch. Two large blood vessels (the portal vein and the hepatic artery) enter the liver on the lower side via the porta hepatis, at the point at which the bile duct exits the organ. The hepatic artery carries oxygen-rich blood to supply the liver tissue. The portal vein carries nutrient-rich blood from the digestive organs to the liver. The hepatic vein collects the blood from the liver cells and carries it to the right side of the heart via the inferior vena cava.

The liver tissue is comprised of numerous hepatic lobules, in which the cells are arranged in a structure known as the corpus callosum. The lobules are surrounded by the fine dendrites of the hepatic artery, portal vein, and bile ducts, so as to ensure intensive metabolic exchange between the blood, the bile, and the liver cells. At the center of every hepatic lobule is a branch of the hepatic vein that carries blood from the liver cells to the venous circulatory system. The bile formed in the liver cells gathers in the bile ducts, through which it is channeled, via the porta hepatis, to the main bile duct and into the gall bladder or directly into the duodenum, the uppermost section of the small intestine.

The liver is the central metabolic organ of the body and has numerous tasks, including:

- the conversion, break-down, and storage of carbohydrates, proteins, and liposomes, all of which enable the extraction and/or storage of energy. The body uses some parts of the proteins from the food to form enzymes, antibodies, and blood clotting agents.
- detoxication of the body (removal of waste products and foreign substances). Waste products are

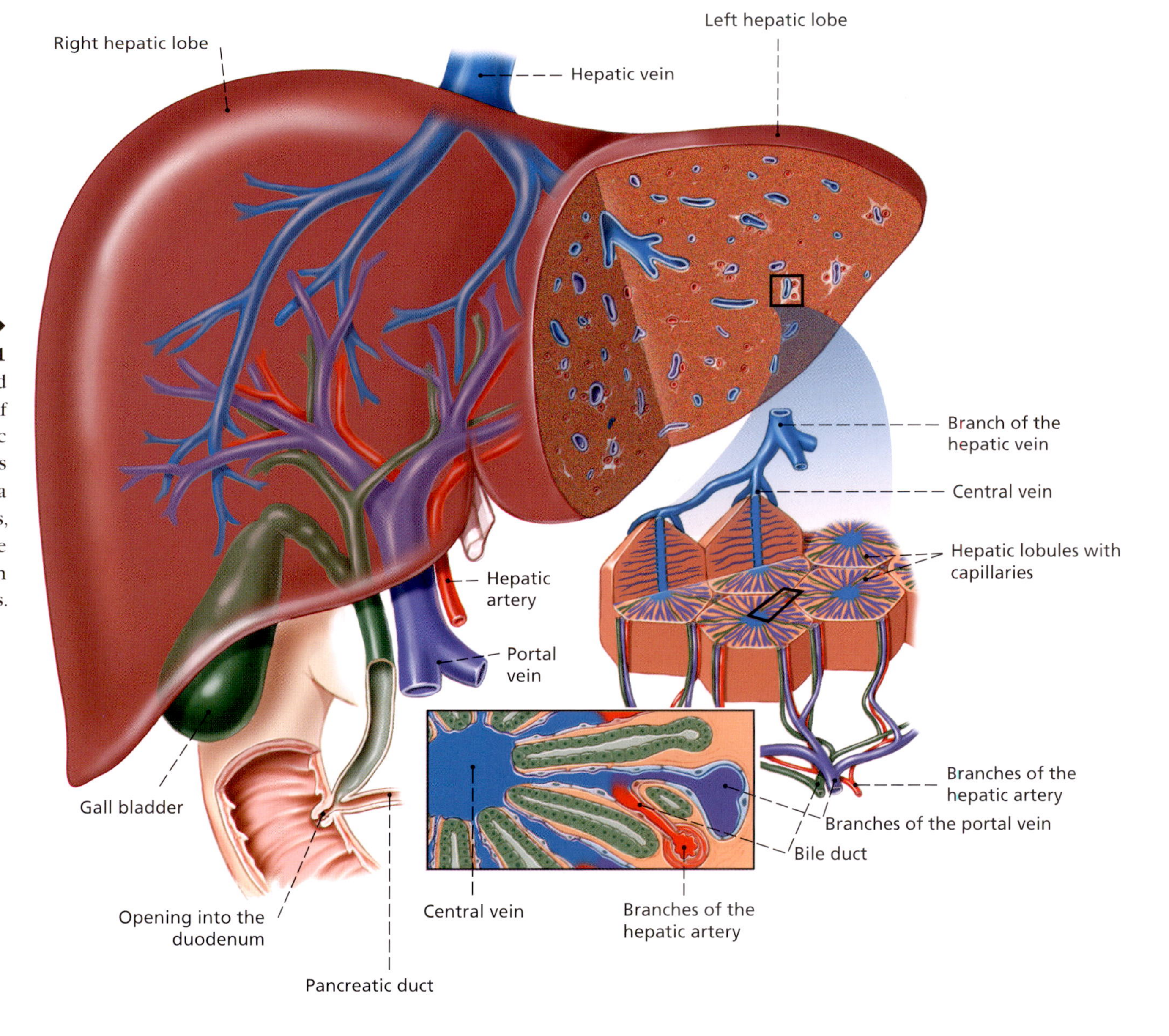

Fig. 321
Arterial and venous blood enters the lower side of the liver. In the hepatic lobules *(detail)* it washes into the liver cells via finely branched vessels, and then gathers in the central veins that open into the portal veins.

either excreted by the liver cells into the bile and then into the small intestine from where they are ultimately excreted in the stool, or else they are modified by the liver cells, absorbed into the blood, and excreted via the kidneys in the urine.

- formation of bile. Bile is required for the processing of fats in the small intestine so that these can be absorbed into the blood system. The gall bladder is a reservoir for bile; it is here that the heavily concentrated bile is "stored" temporarily until it is needed, for example, after a fatty meal. It is possible to manage without the gall bladder. In this event the bile is excreted into the small intestine directly via the main bile duct.

Intestine

From the stomach, the partially digested food (chyme) is initially transferred to the small intestine, which is about 3 meters in length and divided into three sections: the duodenum, jejunum, and ileum. While the majority of the duodenum—the name deriving from the latin duodeni, meaning 12 fingers, which represents its length—is attached to the back of the abdominal wall, the other parts of the small intestine are very flexible. The thin, moist membrane which encases the intestinal loops and which ensures their problem-free positioning next to one another, contributes to this flexibility. This outer membrane covers the muscle layer of the gut that comprises longitudinal and circular muscles. This muscle layer enables the wave-like movements of the intestinal wall (peristalsis), which are brought about by alternating relaxation of the longitudinal muscles and tensing of the smooth circular muscles, and ensures the further transportation of the intestinal contents. The interior of the intestine is lined with a mucous membrane, the surface area of which is significantly extended by many folds and cone-shaped protuberances (villi). The glandular cells of the mucous membrane produce digestive juices, which, together with the juices from the pancreas and the bile, complete the digestive processes that began in the oral cavity and the stomach. The mucus produced in the goblet cells of the small intestine villi lubricates the contents of the intestine. The intestinal wall also contains numerous blood and lymph vessels, as well as nerve fibers.

The main task of the small intestine is to break down the still undigested parts of the chyme and absorb the resulting nutritional components into the blood. Protein and carbohydrate components are transported directly into the intestinal blood vessels and then, via the portal vein, to the liver where they are processed further. The fat components, on the other hand, are absorbed and transferred by the lymph vessels.

The large intestine (colon), which measures about 1.5 meters in length, comprises the blind gut, the colon, and the rectum. Colloquially the blind gut is often incorrectly referred to as the appendix; in fact, the blind gut is the "blind" end of the large intestine, which includes the appendix.

A valve-like flap (Bauhin's valve) at the point where the small intestine enters the large intestine prevents the intestinal contents from flowing backwards. The processed chyme is concentrated in the large intestine by the extraction of water and salts. The goblet cells in the mucous membrane of the large intestine secrete a mucus that keeps the intestinal contents lubricated while becoming increasingly solid. The intestinal bacteria (intestinal flora) ensure the breakdown of indigestible food components through fermentation and decomposition, with the remains being discharged via the anus as stools.

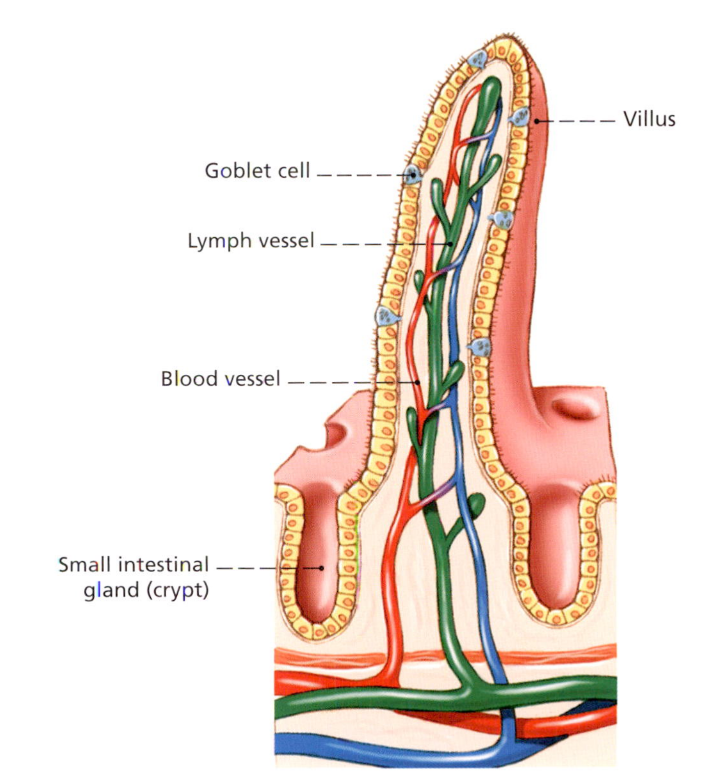

◆
Fig. 322
The villi of the small intestine extend into the intestinal cavity as tiny, finger-like protuberances. They have a thin layer of cells on their surface with the mucus-forming goblet cells. The villi tissue below contains blood and lymph vessels through which the pre-digested nutrients are conveyed to the liver and the larger lymph vessels. The small intestinal crypts with their mucus-producing glands are located in the deep spaces between the villi.

◆
Fig. 323
The structures of the oral cavity.

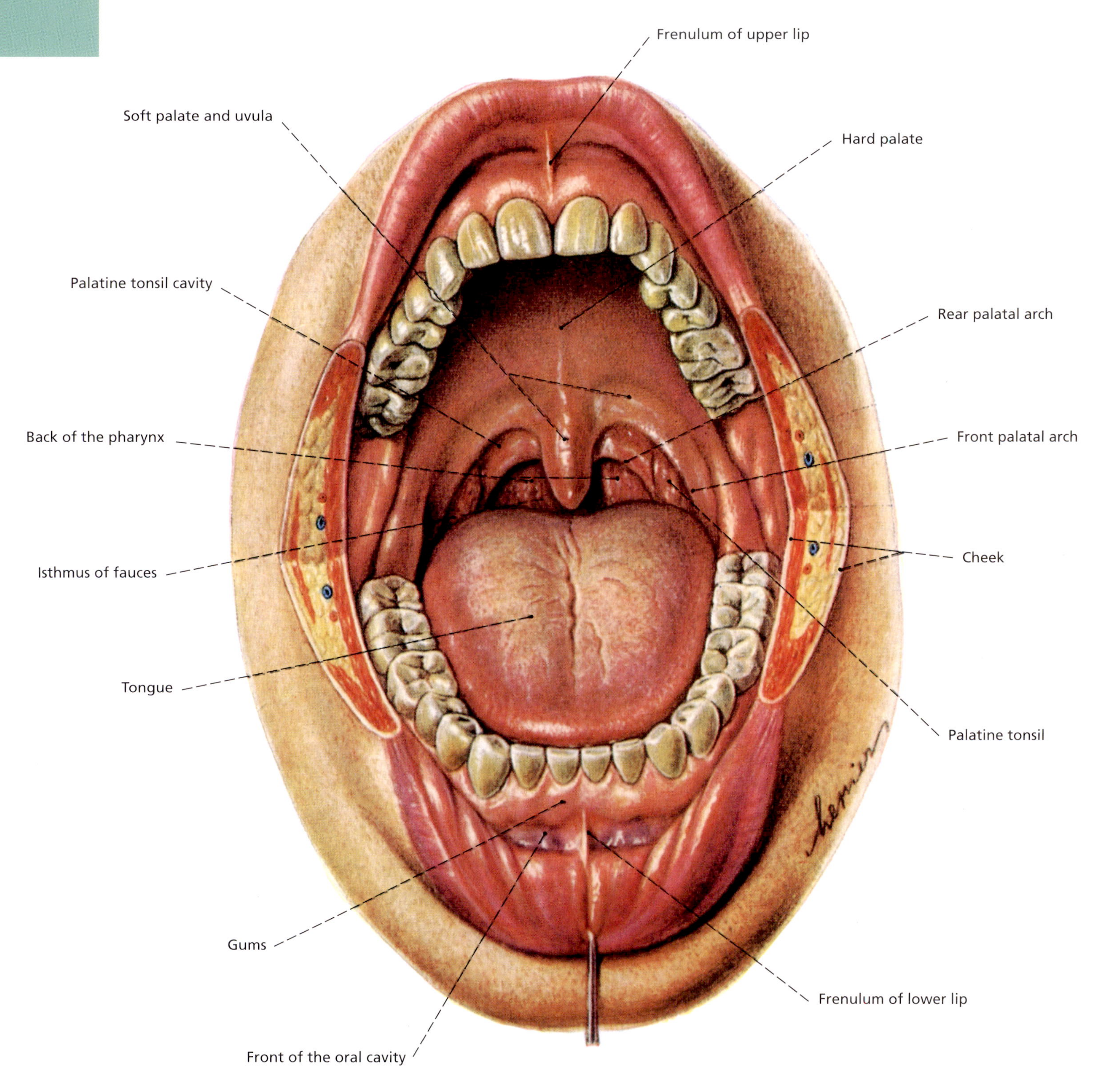

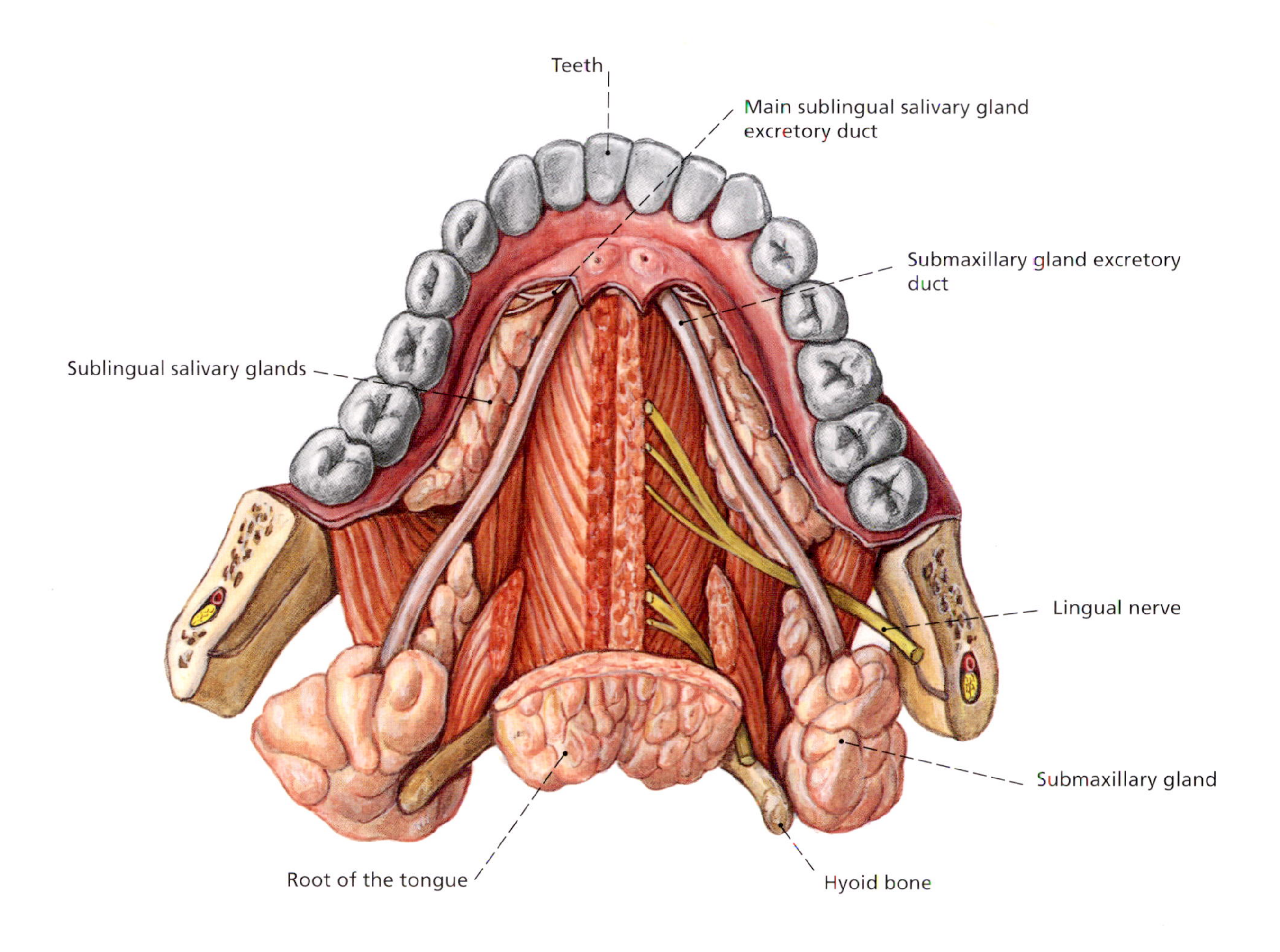

◆
Fig. 324
Submaxillary and sublingual salivary glands. View of the floor of the mouth from above. Part of the musculature on the floor of the mouth and the tongue has been removed. The excretory ducts of the glands open out into the oral mucosa beneath the inner surface of the incisors.

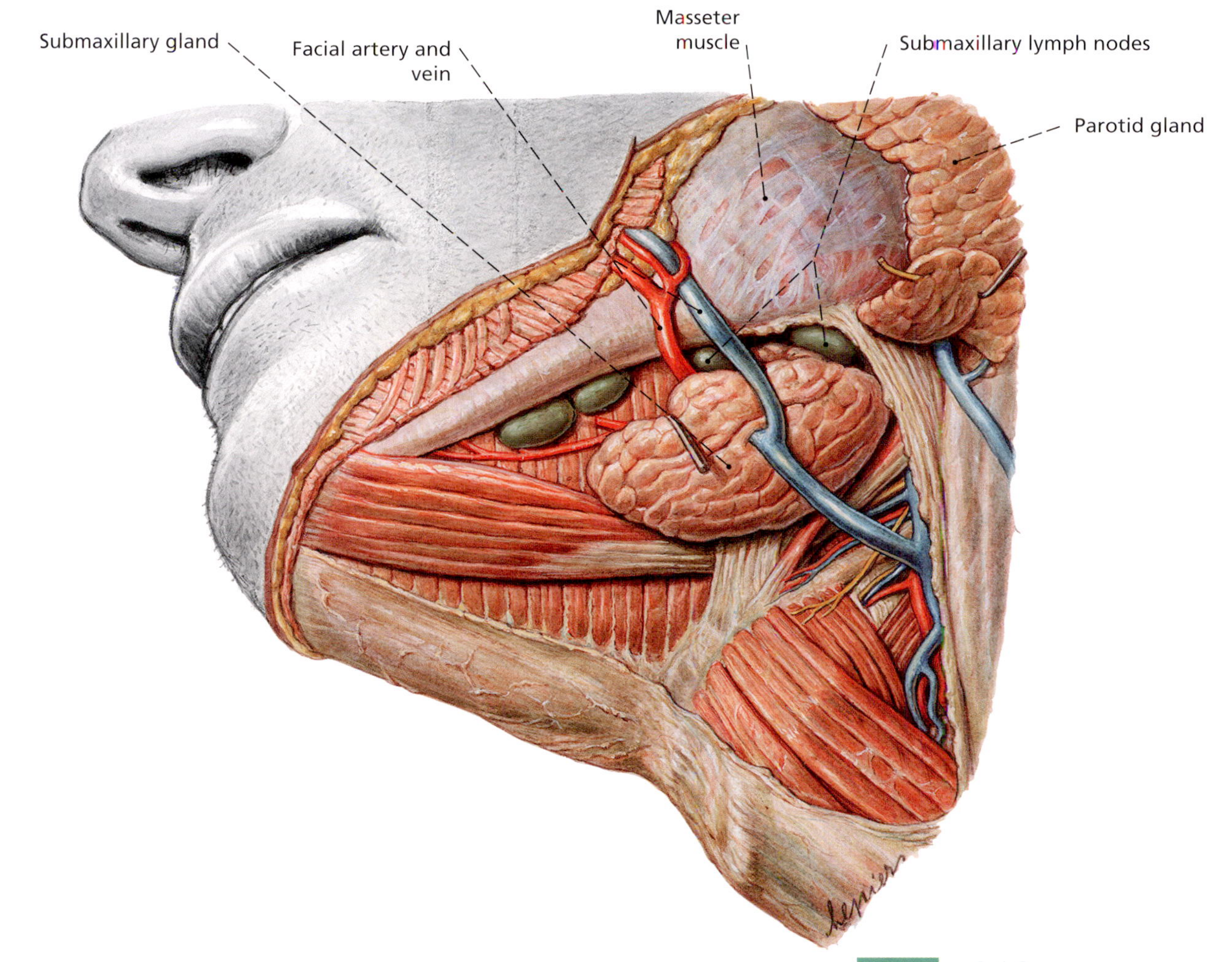

◆
Fig. 325
Submaxillary glands. Side view from below. The skin and skin musculature have been removed. The parotid gland is also visible at the submaxillary angle.

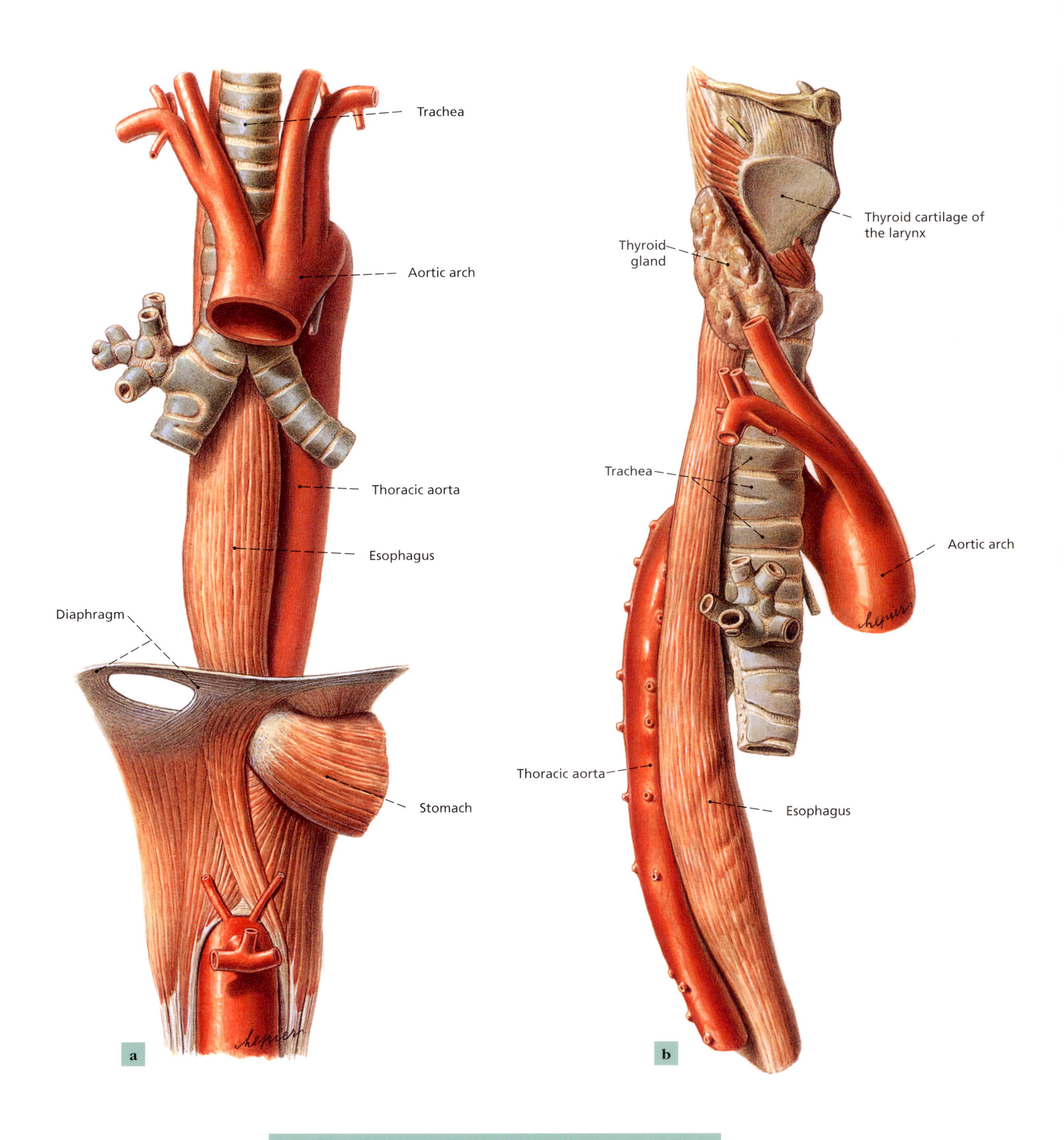

◆

Fig. 326
The esophagus without the abdominal organs seen from the front (a) and from the right (b) in relation to the trachea and the aorta (the largest artery in the body). The aorta and the trachea have been cut through in both illustrations. Image (a) shows the passage of the esophagus through the diaphragm, while the upper section of image (b) also shows the larynx with the thyroid gland.

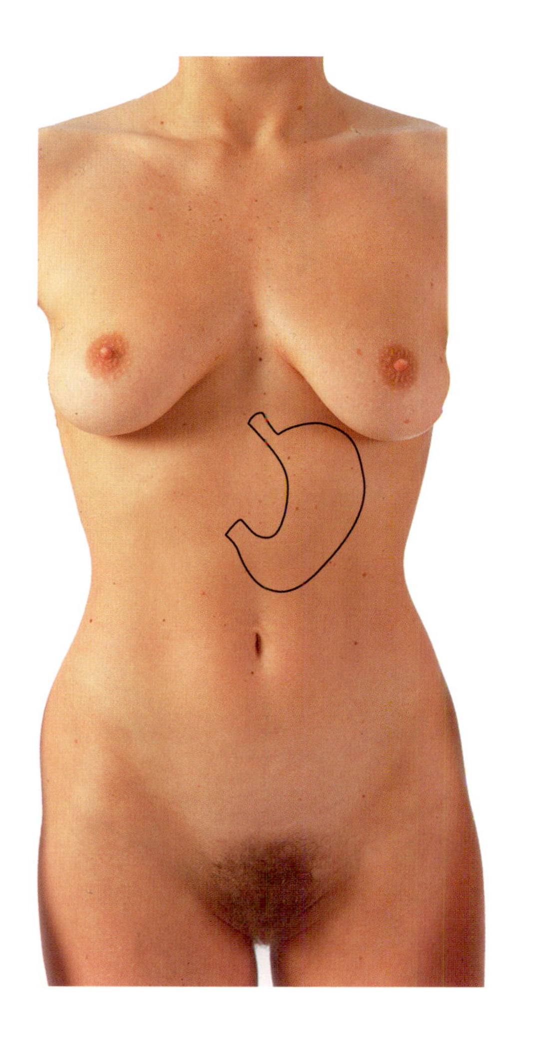

◆
Fig. 327
The position of the stomach, outlined on the abdominal skin.

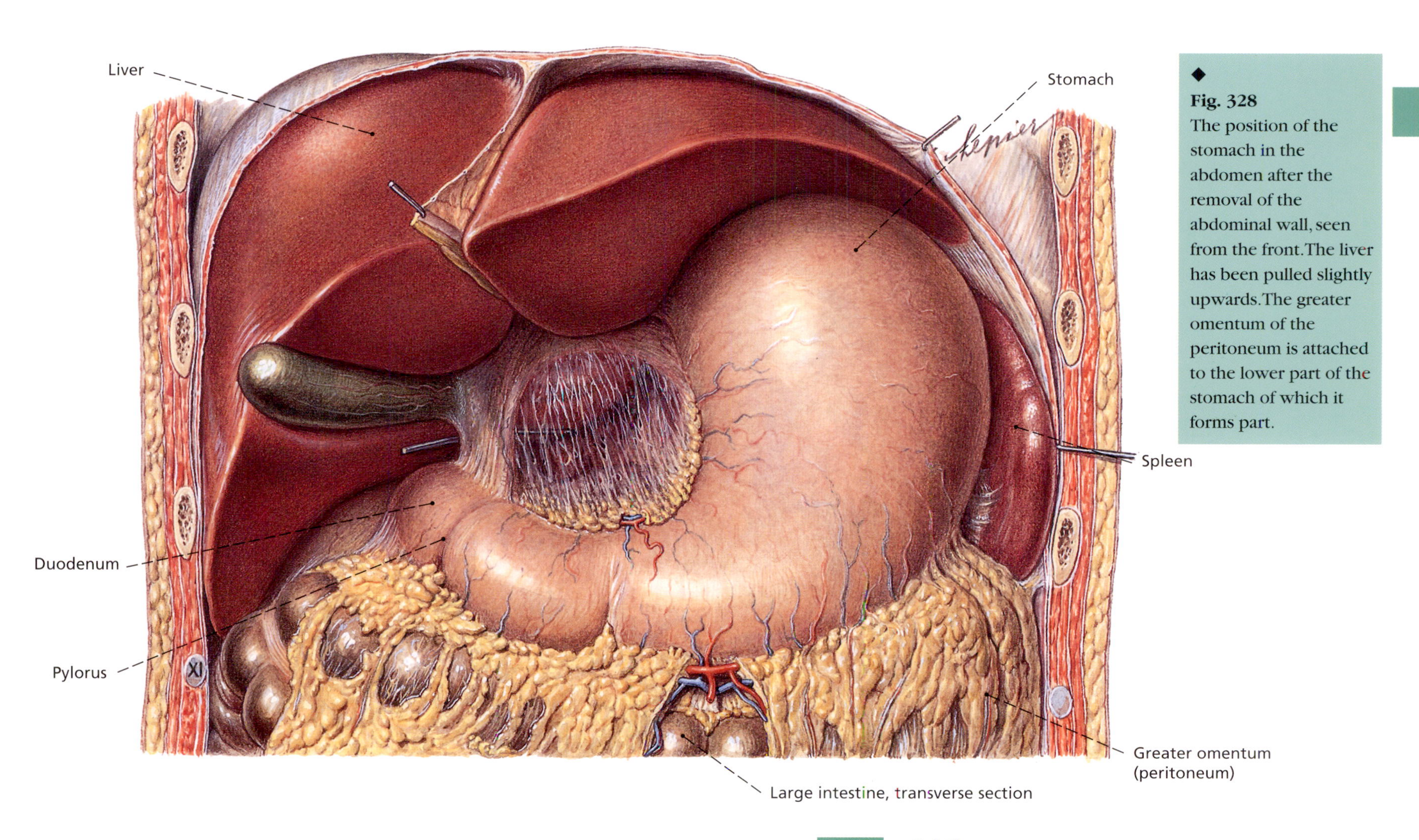

◆
Fig. 328
The position of the stomach in the abdomen after the removal of the abdominal wall, seen from the front. The liver has been pulled slightly upwards. The greater omentum of the peritoneum is attached to the lower part of the stomach of which it forms part.

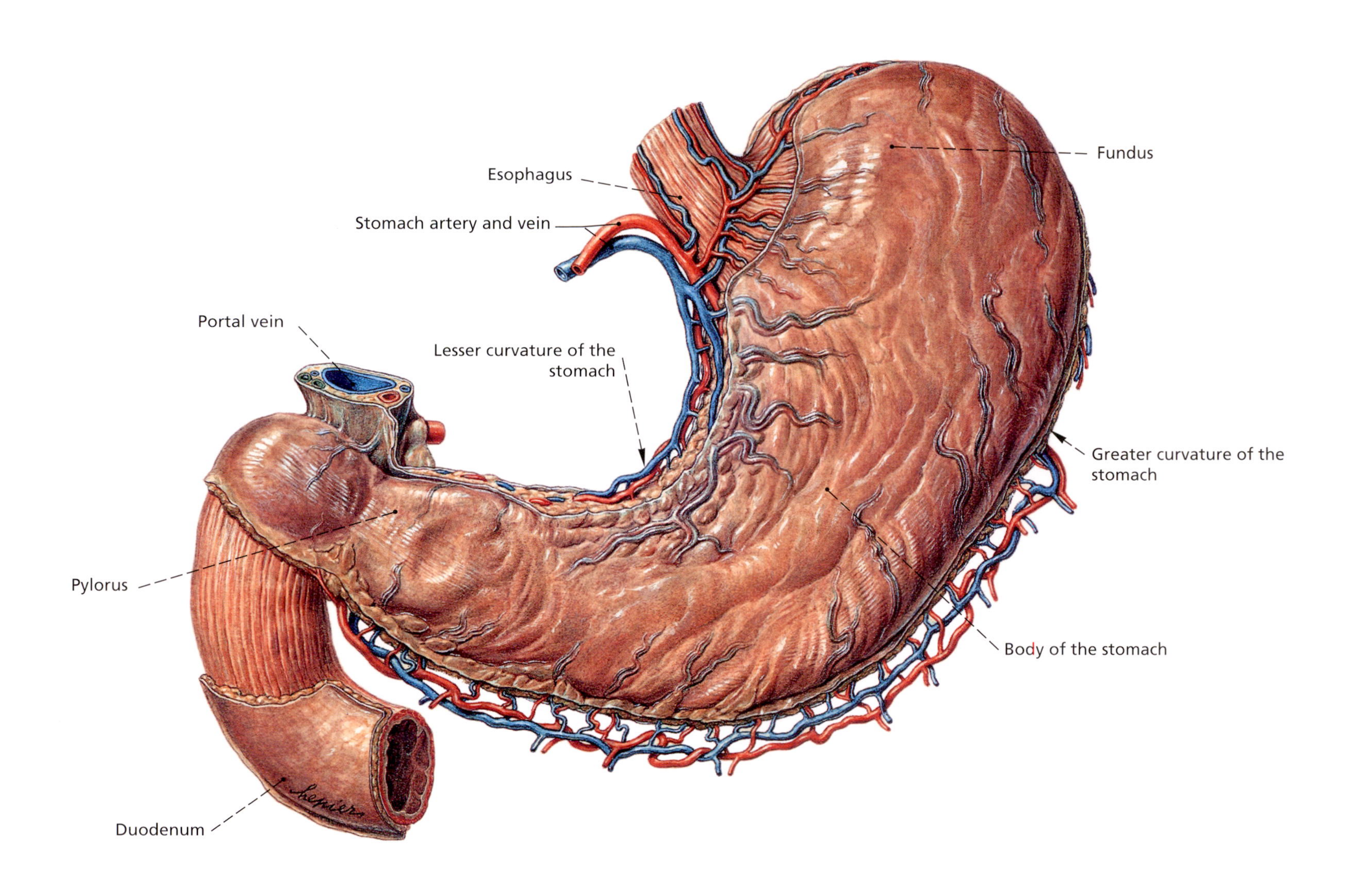

◆
Fig. 329
The stomach with the upper section of the duodenum. The blood vessels, which enter the gastric mucosa in the lesser and greater curvature regions of the stomach, are clearly visible. Part of the mucous membrane layer on the duodenum has been rolled back so as to better depict the intestinal musculature.

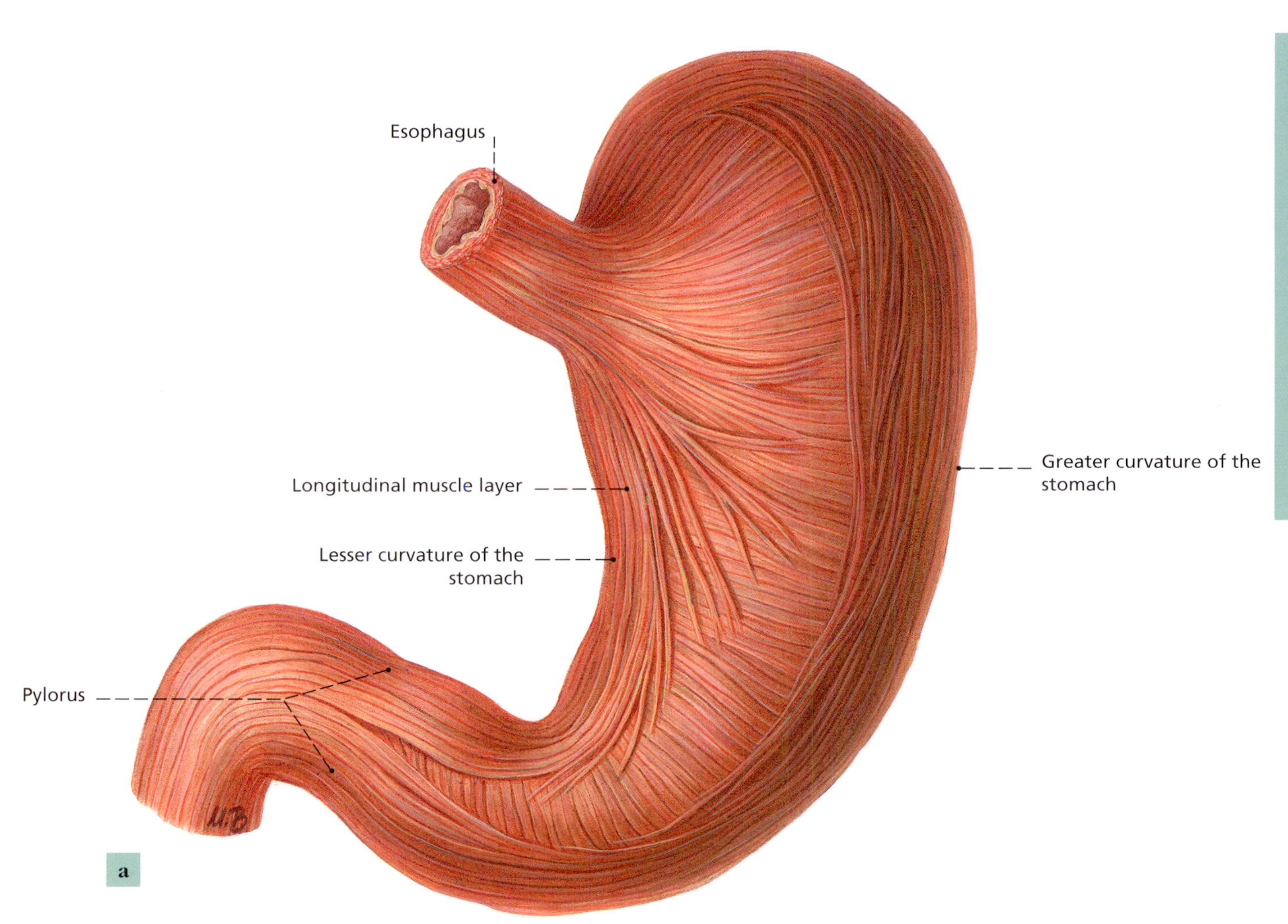

◆
Fig. 330
The stomach musculature. Illustration (a) shows the musculature after removal of the outer mucous membrane layer, while in illustration (b), the muscle layer sections have been removed to show the different directions of the muscle fibers.

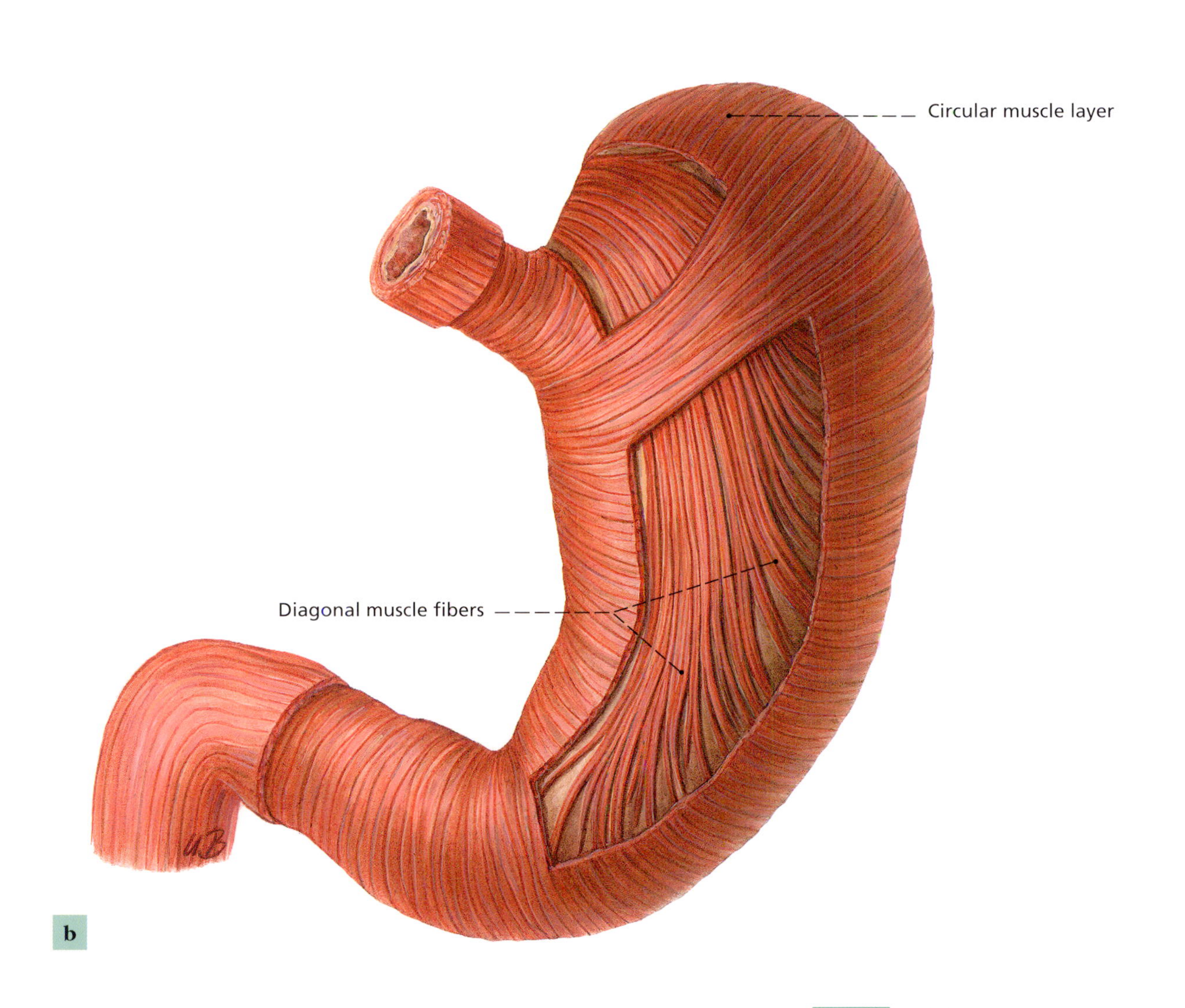

◆
Fig. 331
A longitudinal section of the stomach. The anterior wall of the stomach has been removed. The folds of the gastric mucosa are clearly visible on the inside of the stomach. The upper section of the duodenum is also depicted. The sphincter-like musculature is visible in the pyloric region.

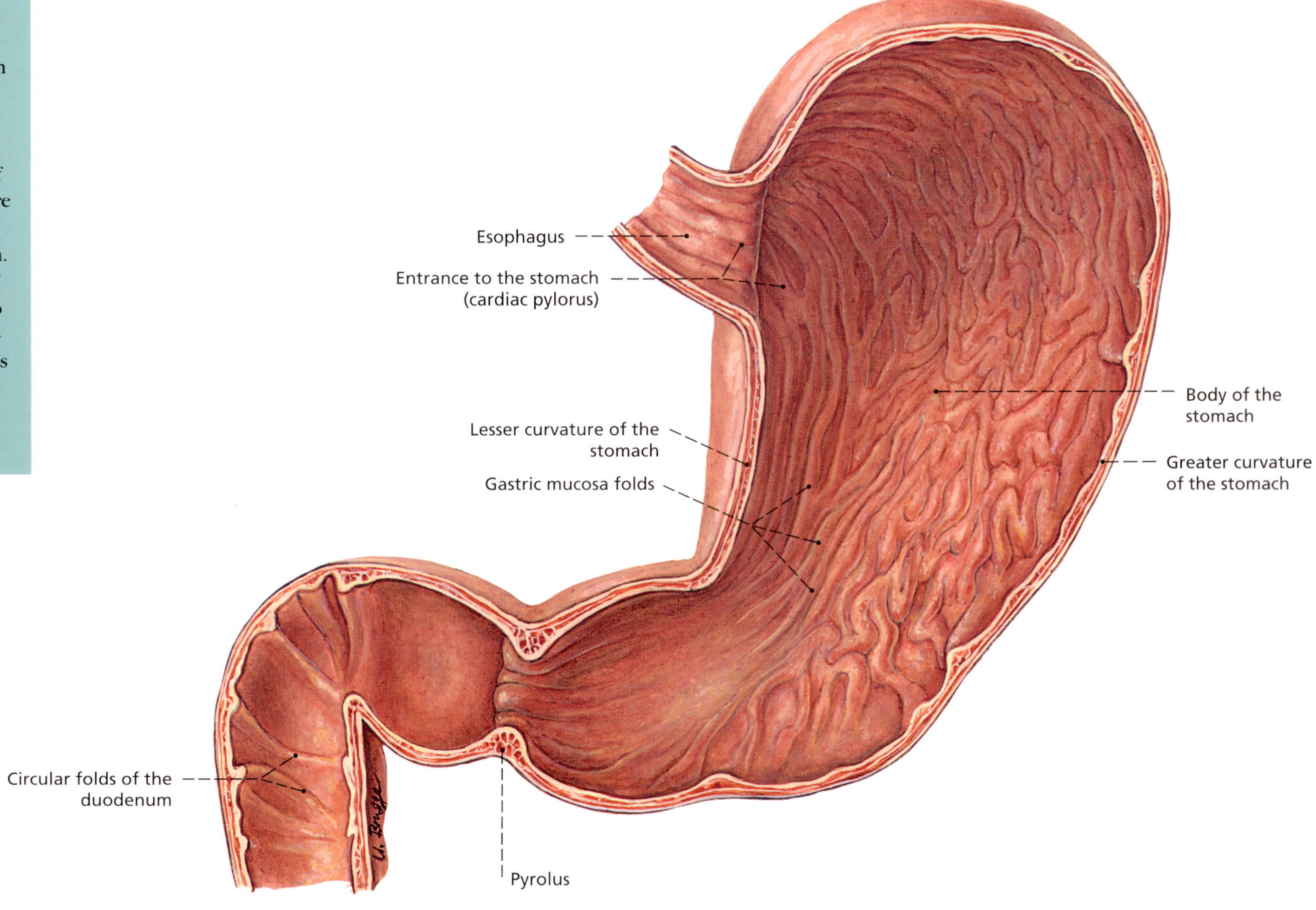

◆
Fig. 332
Structure of the stomach wall (schematic representation). The gastric surface and gastric pits are located on the inner mucous membrane layer of the stomach, while the outer layer forms part of the peritoneum.

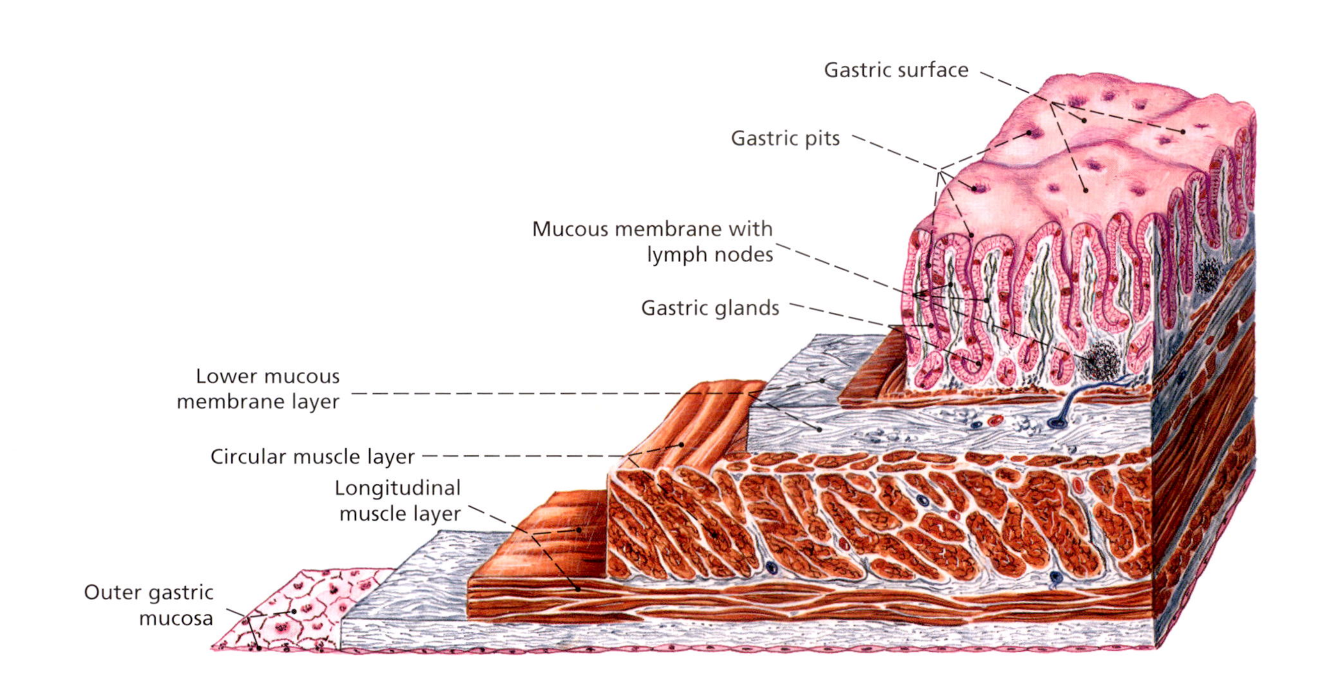

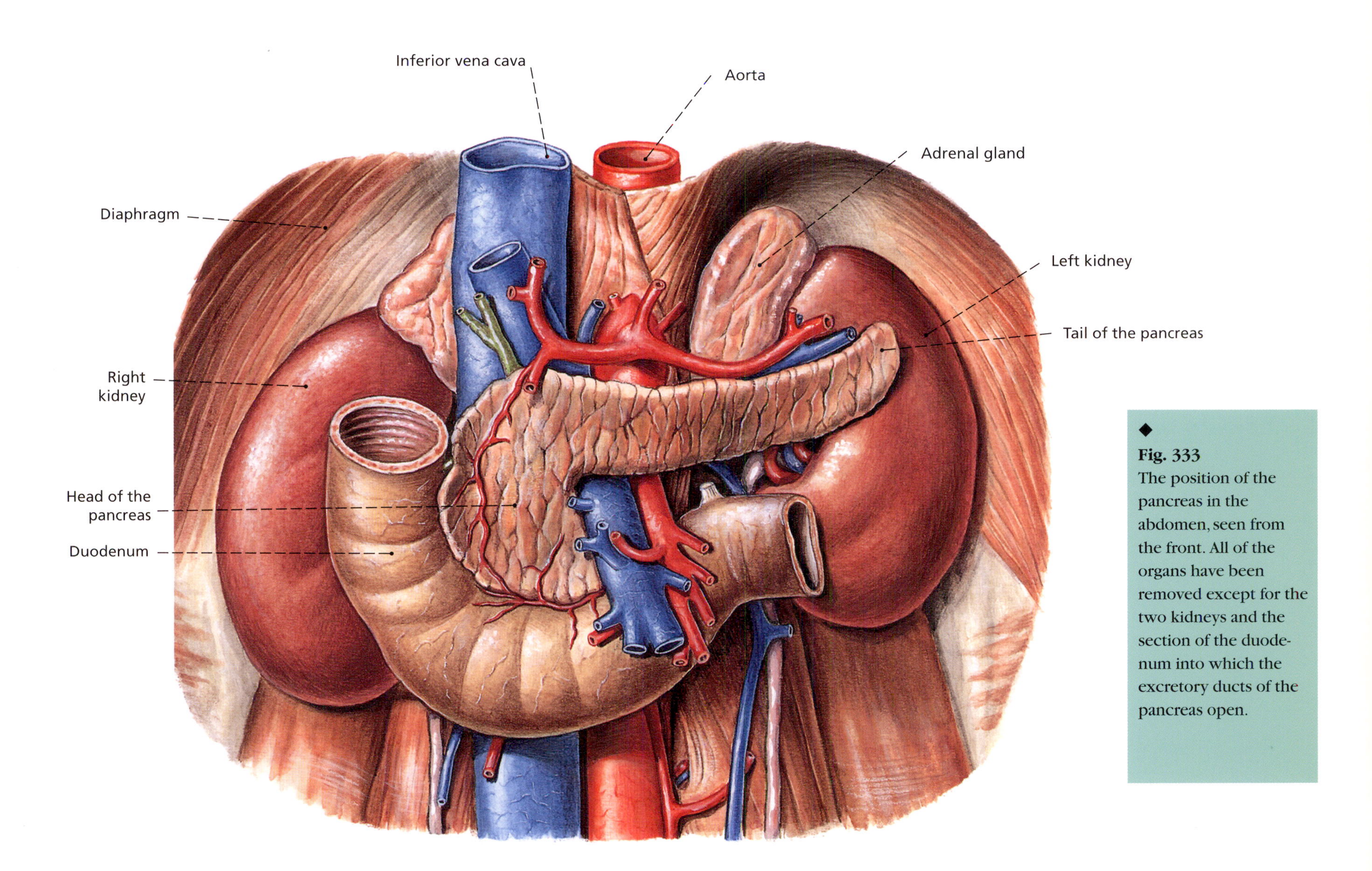

◆
Fig. 333
The position of the pancreas in the abdomen, seen from the front. All of the organs have been removed except for the two kidneys and the section of the duodenum into which the excretory ducts of the pancreas open.

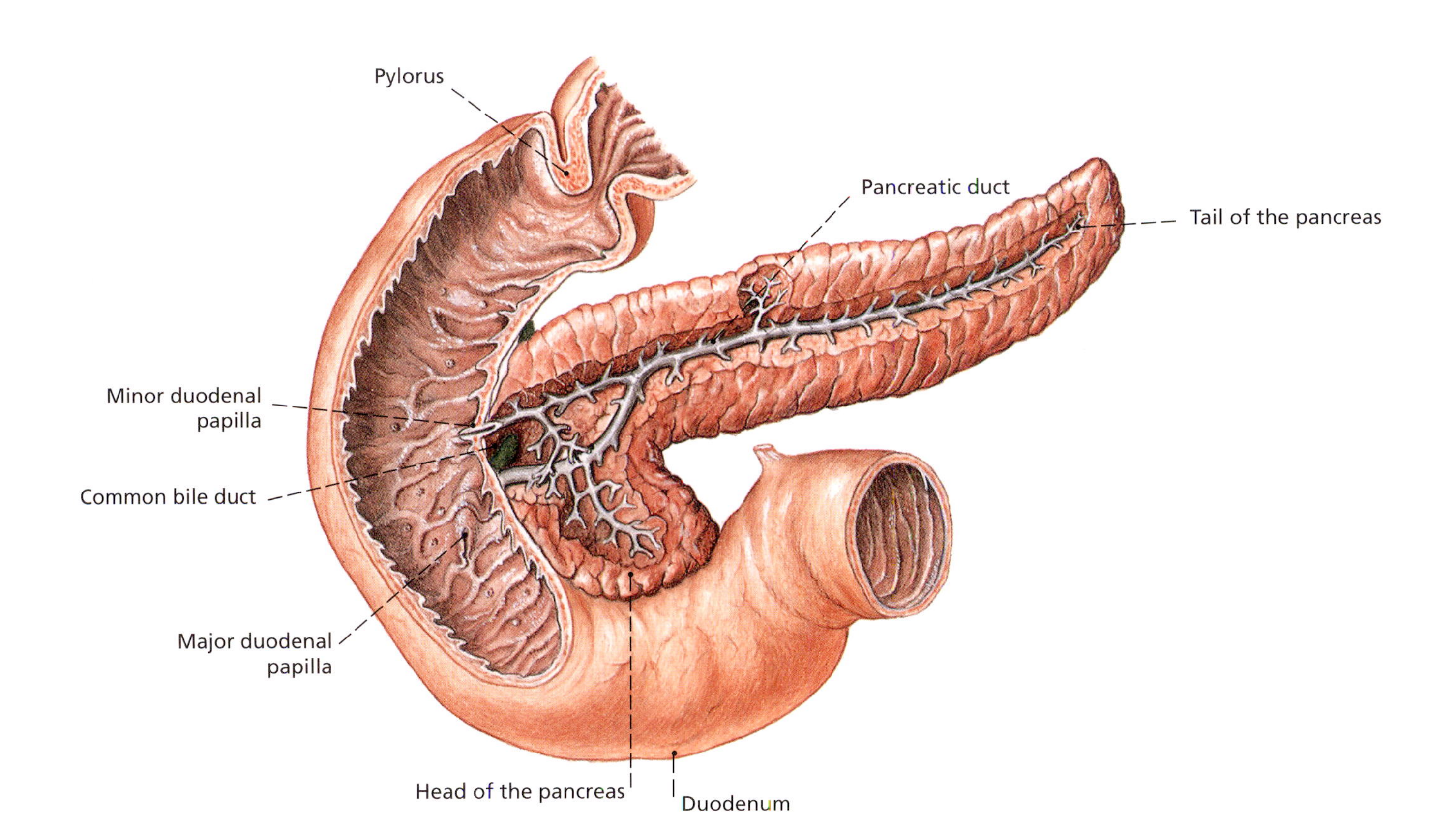

◆
Fig. 334
The pancreas together with the upper section of the duodenum, seen from the front. The body of the pancreas and the intestinal section have been opened in order to show the excretory ducts of the pancreas and the folded structure of the duodenum.

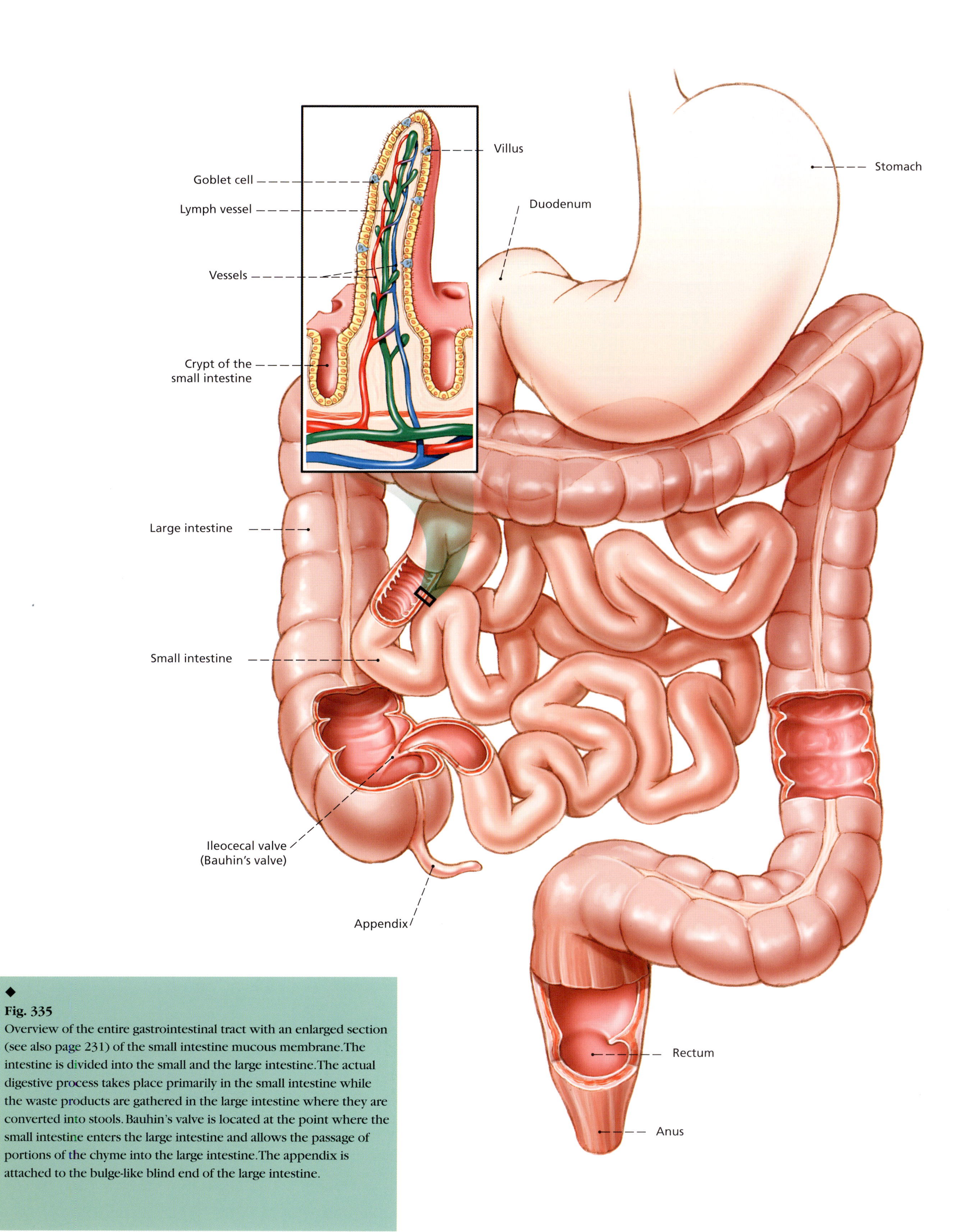

◆
Fig. 335
Overview of the entire gastrointestinal tract with an enlarged section (see also page 231) of the small intestine mucous membrane. The intestine is divided into the small and the large intestine. The actual digestive process takes place primarily in the small intestine while the waste products are gathered in the large intestine where they are converted into stools. Bauhin's valve is located at the point where the small intestine enters the large intestine and allows the passage of portions of the chyme into the large intestine. The appendix is attached to the bulge-like blind end of the large intestine.

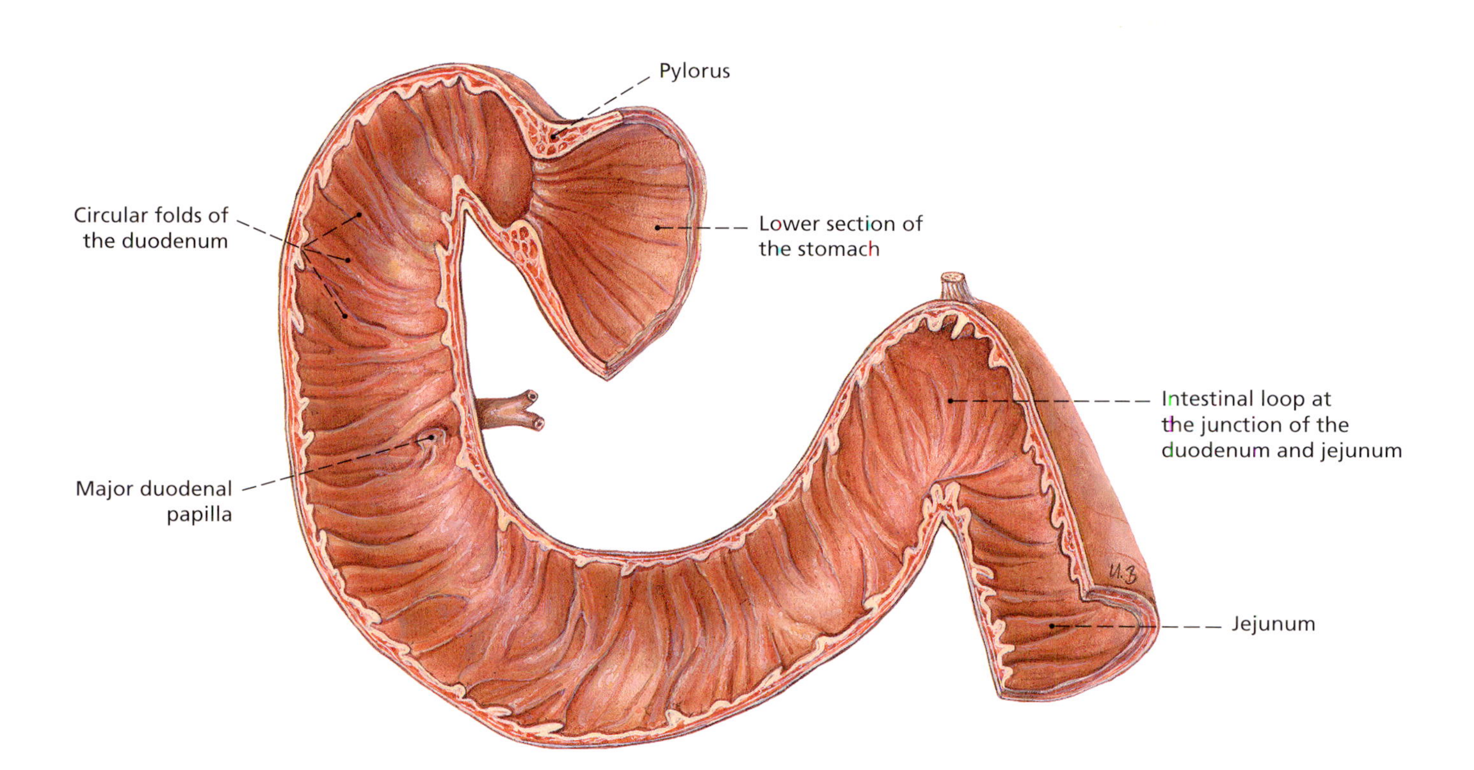

Fig. 336
The duodenum from the front. The anterior wall has been opened in order to show the folds of the mucous membrane and the opening of the pancreatic duct (major duodenal papilla). The characteristic C-shape of the duodenum is easily recognizable.

* The name duodenum (Latin: twelve fingers) refers to the length of the organ, i.e., the length of twelve fingers.

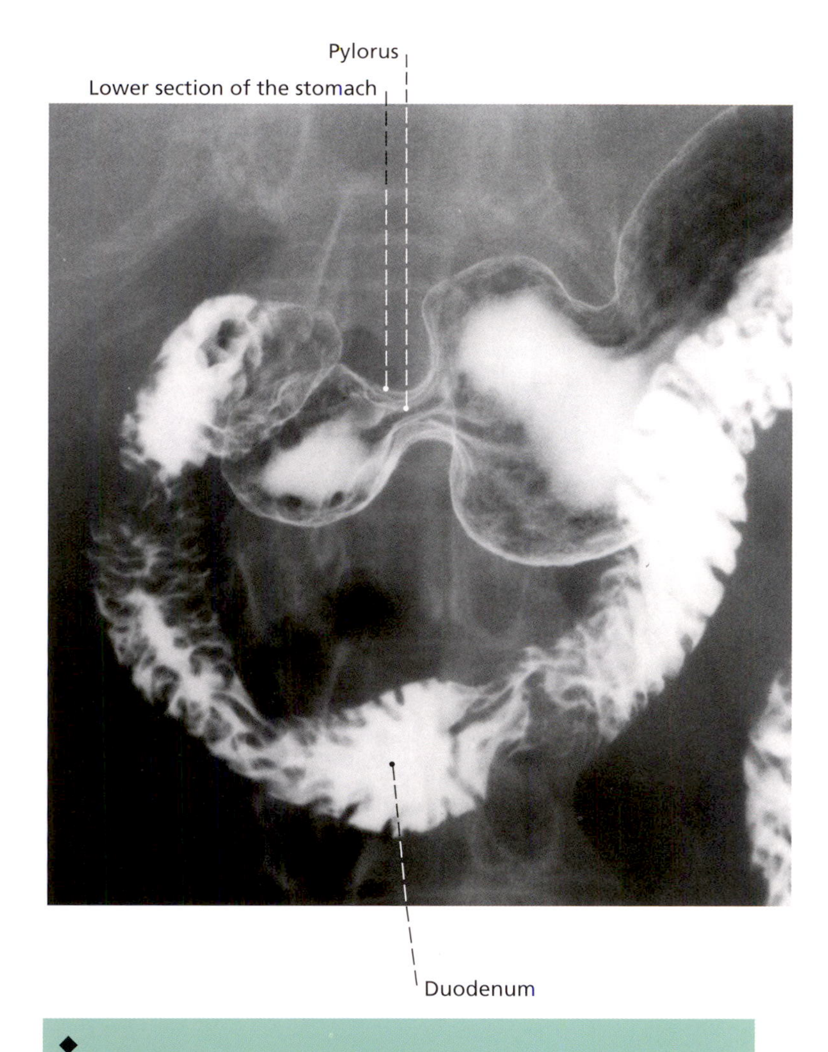

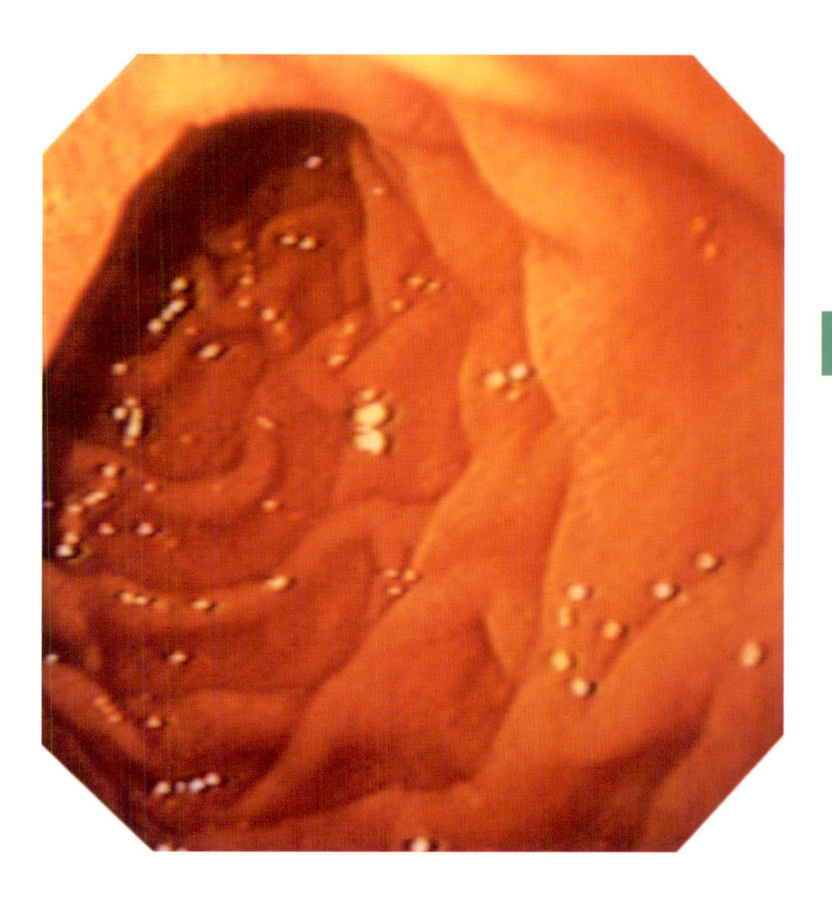

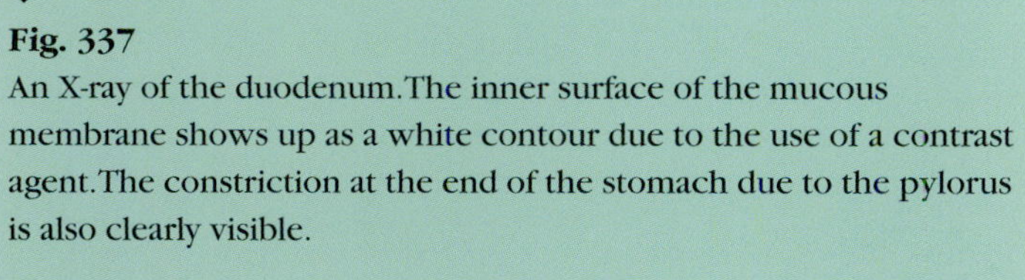

Fig. 337
An X-ray of the duodenum. The inner surface of the mucous membrane shows up as a white contour due to the use of a contrast agent. The constriction at the end of the stomach due to the pylorus is also clearly visible.

Fig. 338
The duodenum without any pathological changes viewed through an endoscope. It is the structure of the mucous membrane in particular which is examined during a colonoscopy.

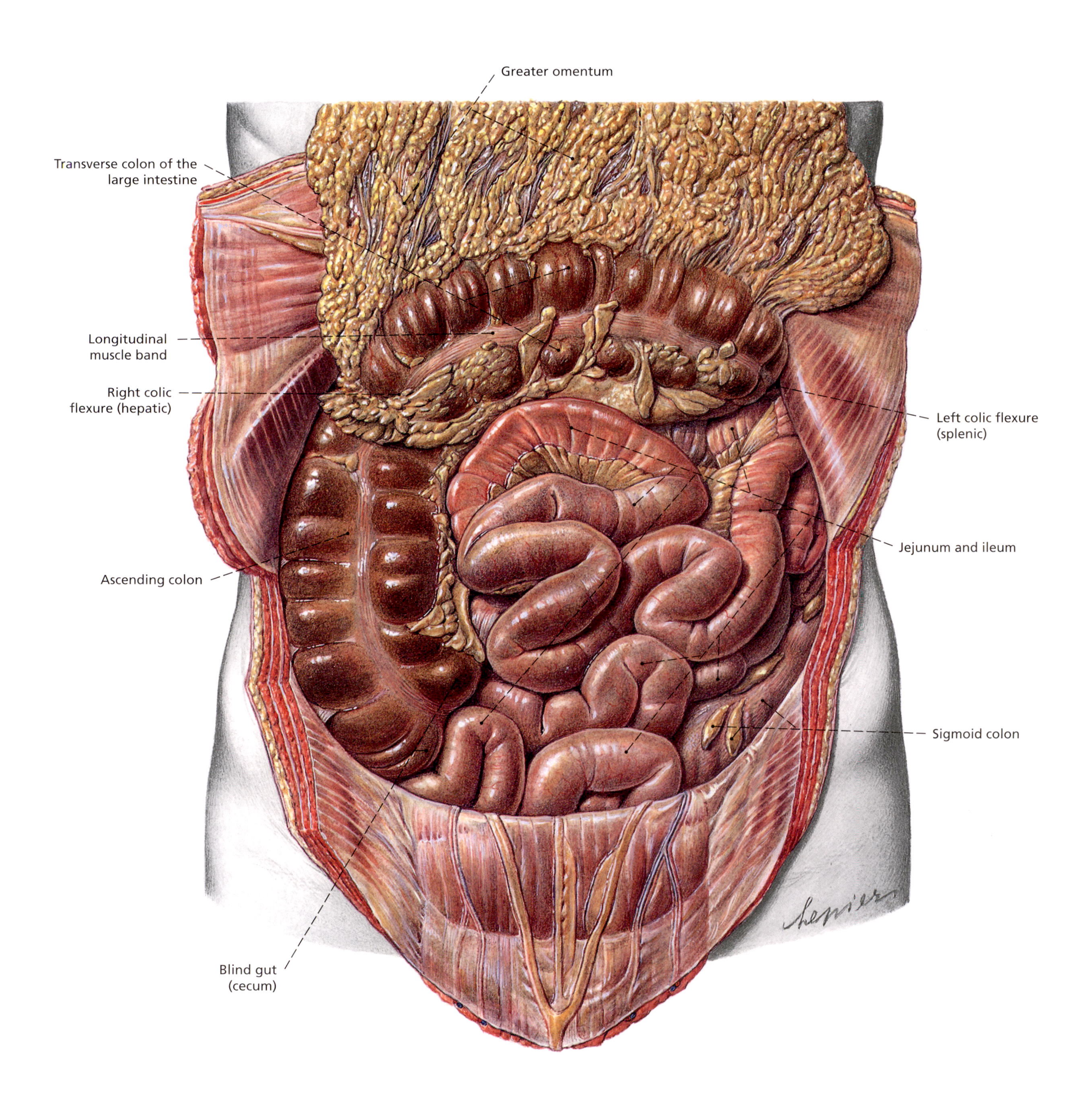

◆

Fig. 339

The intestines with the abdominal wall open. The greater omentum and the transverse colon of the large intestine have been pulled upwards. The loops of the small intestine fill almost the entire abdominal cavity. The ascending colon section of the large intestine can be seen on the right of the lower and central abdomen.

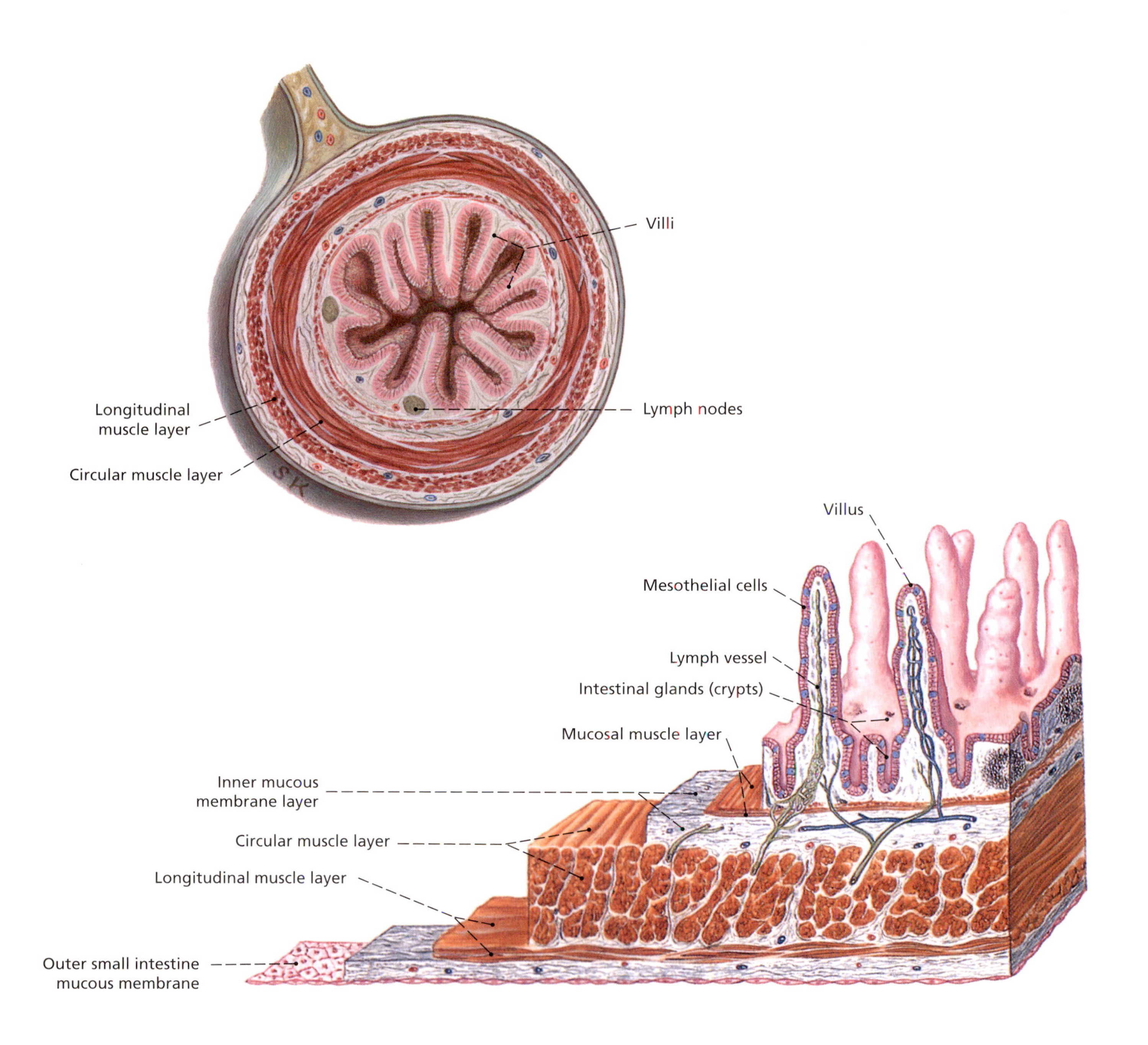

◆
Fig. 340
Cross-section through a loop of the small intestine. The inner circular and outer longitudinal muscle layers are visible, as is the structure of the mucous membrane extending into the intestinal cavity.

◆
Fig. 341
The structure of the small intestine wall (schematic representation).The mucous membrane layer facing the intestinal cavity contains the villi while the outer mucous membrane layer forms part of the peritoneum.

◆
Fig. 342
Villi in the small intestine. Scanning electron microscope image.

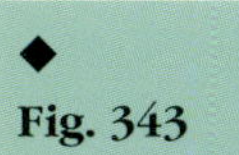

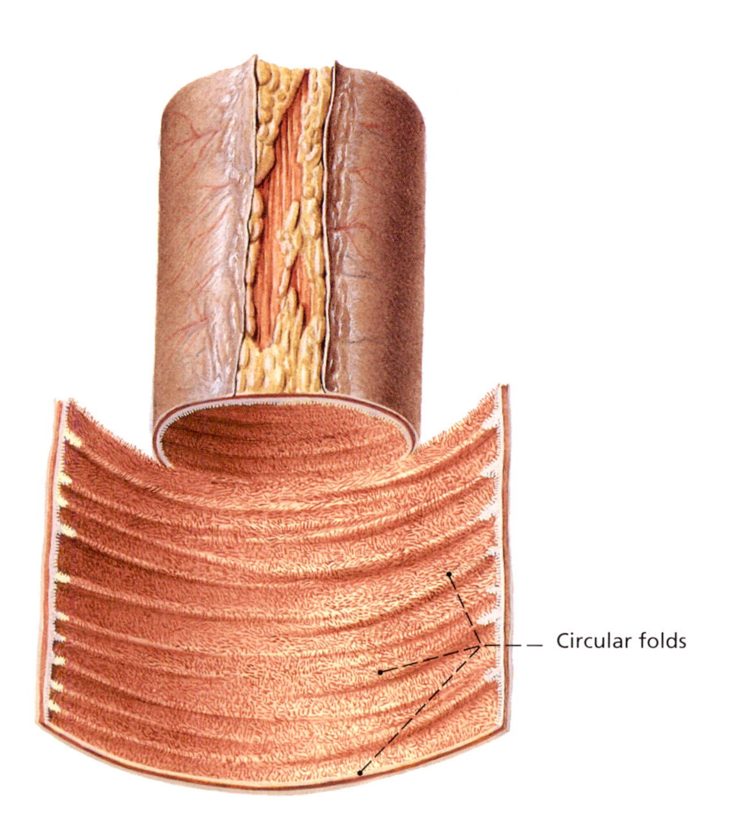

◆
Fig. 343
Section of the jejunum (upper section of the small intestine). In contrast to the mucous membrane structure of the ileum, the circular folds here are close together.

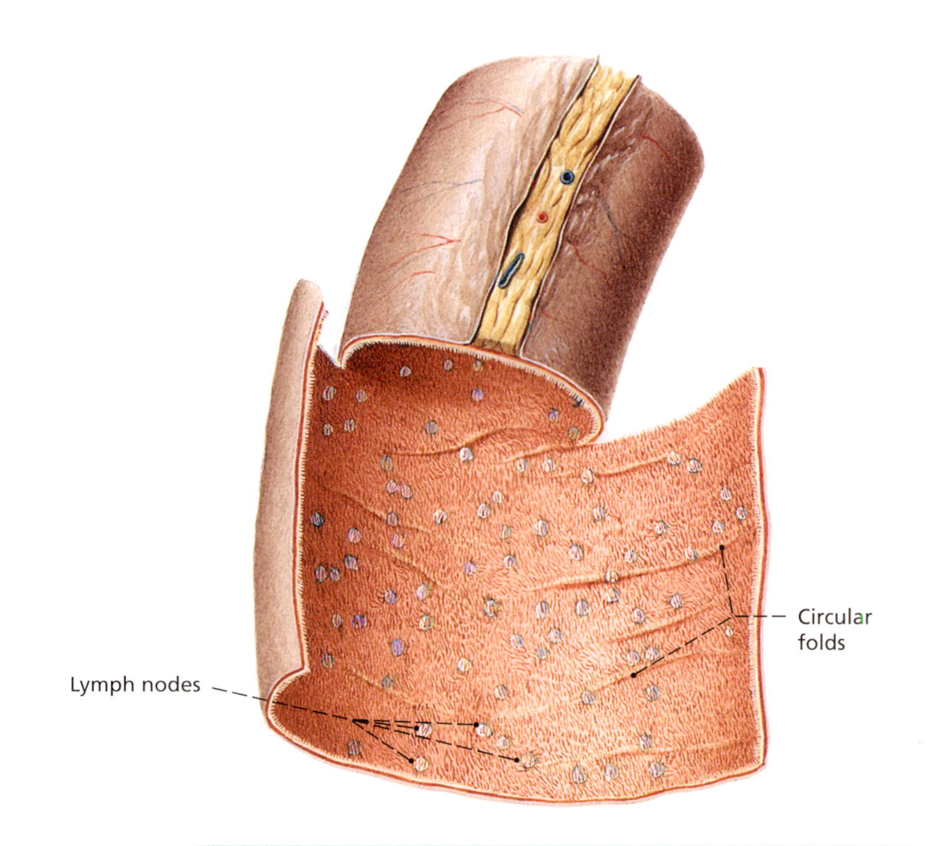

◆
Fig. 344
Section of the ileum (lower section of the small intestine). The circular folds are much further apart than those in the jejunum.

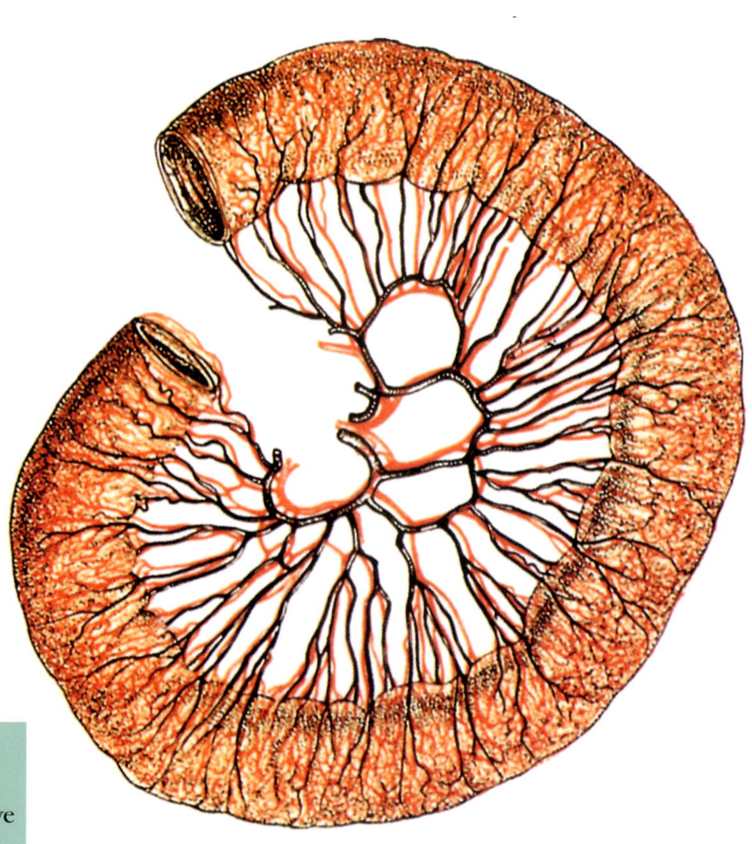

◆
Fig. 345
The blood supply in a section of the intestine. The blood vessels have been dissected from the enveloping folds of the peritoneum.

Fig. 346
The position of the large intestine (colon) in the abdomen, seen from the front with the abdominal wall completely removed. The whole of the small intestine has been detached at the peritoneal appendage. The transverse colon section of the large intestine and the greater omentum of the peritoneum have been pulled upwards. The ascending, descending, and sigmoid colon are visible.

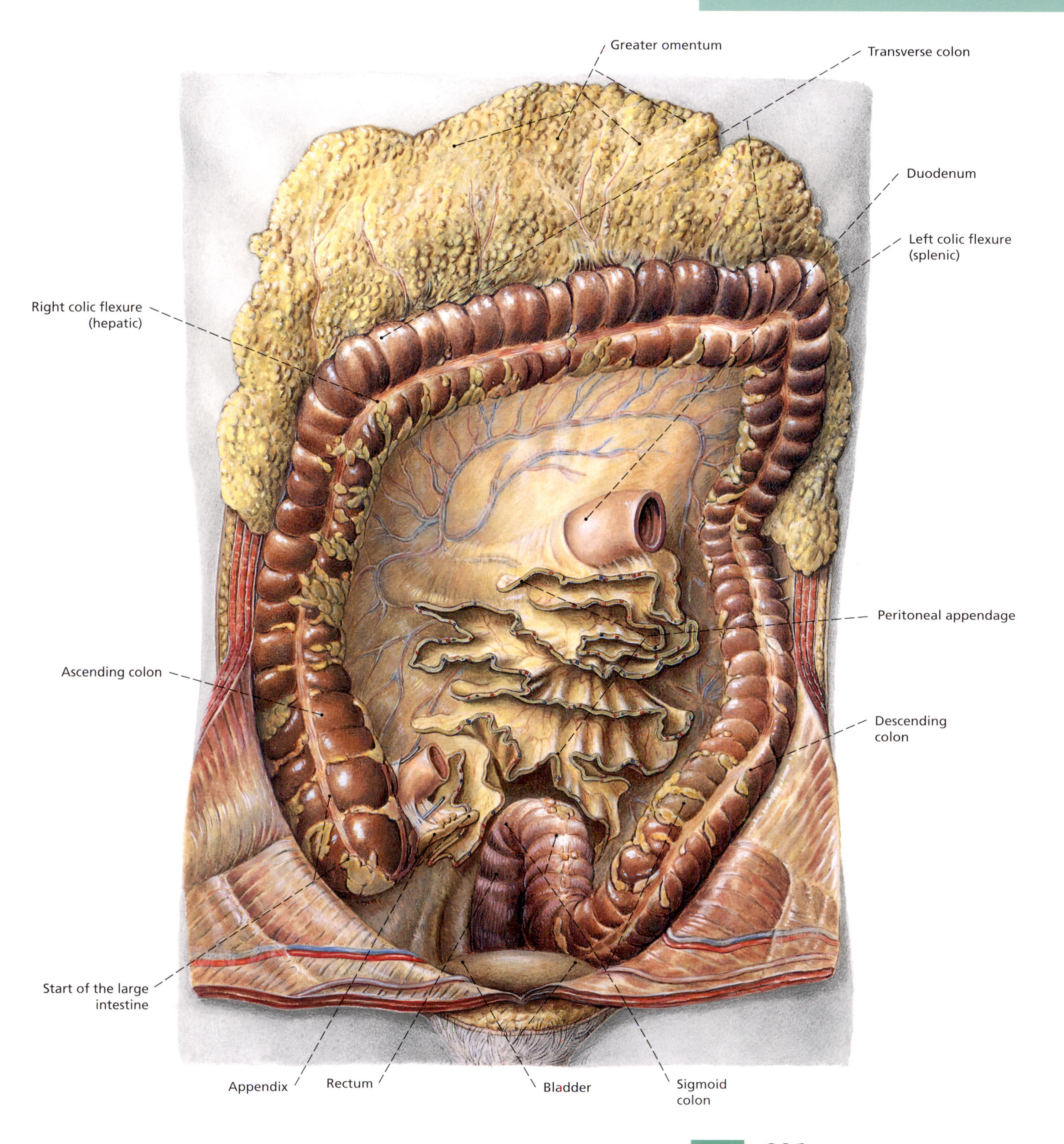

◆
Fig. 347
Section from the transverse colon of the large intestine. The intestine has been opened on the right side. The widely dispersed, crescent-shaped folds on the inside of the mucous membrane, the intestinal bulges, and the longitudinal muscle bands on the side are characteristic of the large intestine.

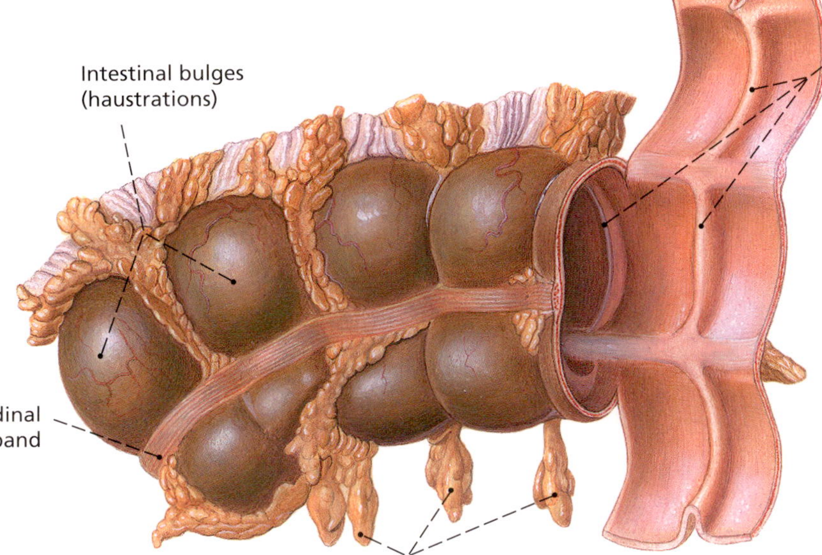

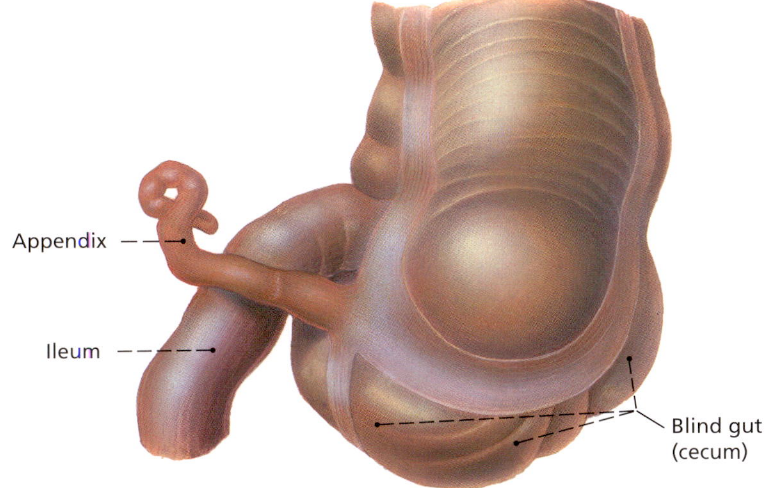

◆
Fig. 348
The start of the large intestine with the appendix*. View from the rear. The entry point of the much thinner ileum into the large intestine is clearly visible behind the appendix.
* The appendix is often incorrectly referred to as the "blind gut." To anatomists, the blind gut refers to the "blind" end of the large intestine, or cecum, which includes the appendix.

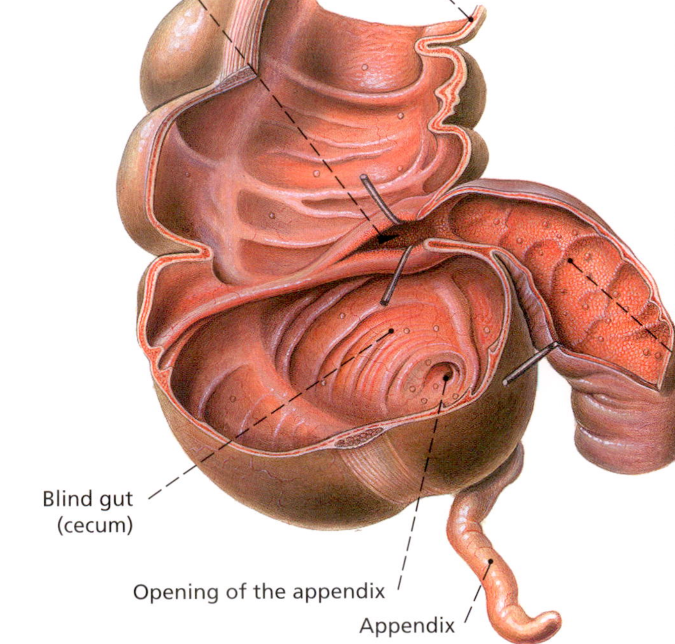

◆
Fig. 349
The start of the large intestine (as in Fig. 348), this time from the front with the anterior wall removed. Bauhin's valve at the junction of the ileum and large intestine is held open by two small pegs.

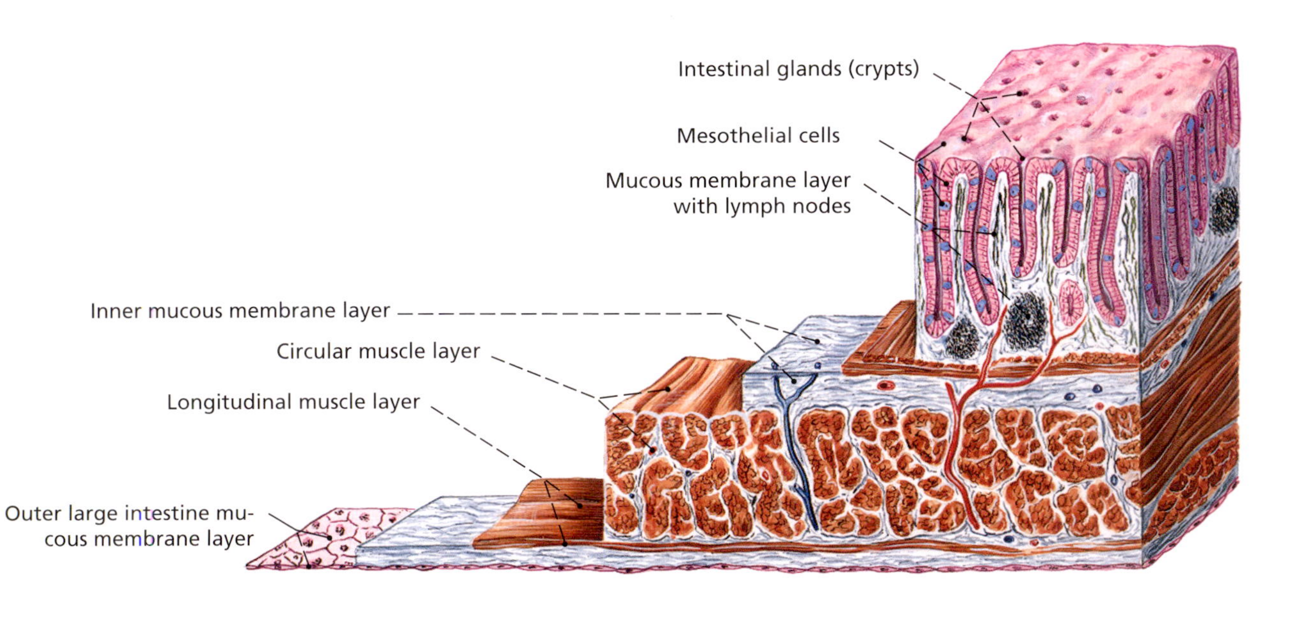

◆
Fig. 350
Structure of the large intestine wall (schematic representation). The mucous membrane facing the intestinal cavity contains the intestinal crypts (the intestinal gland openings). The circular and longitudinal muscle layers are also visible.

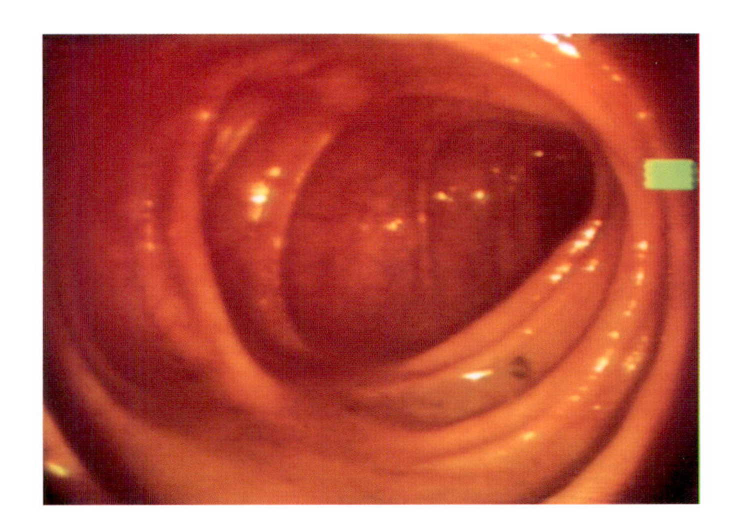

◆
Fig. 351
The ascending colon section of the large intestine seen through an endoscope during a colonoscopy.
The transverse folds and bulges characteristic of the large intestine are clearly visible.

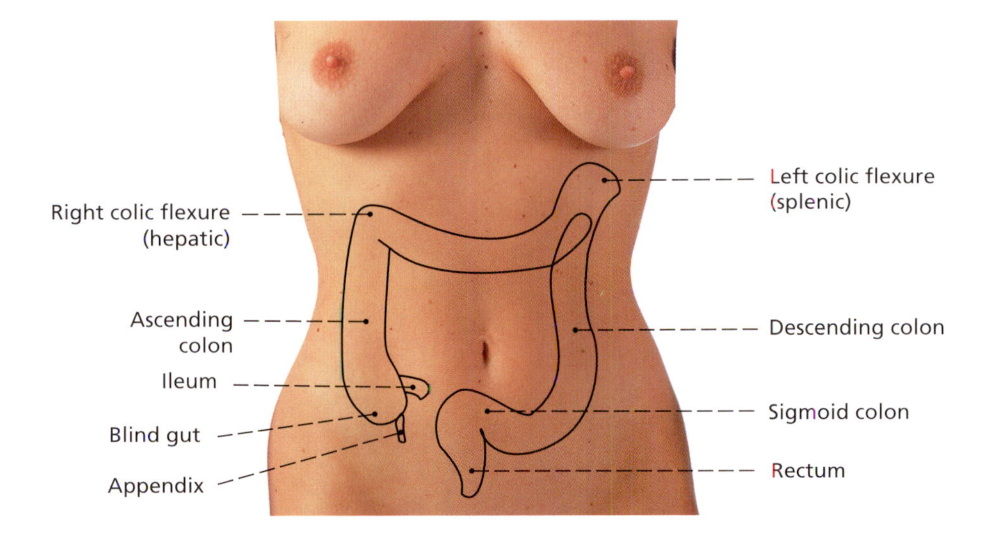

◆
Fig. 352
The position of the large intestine in the abdomen. The outline has been drawn on the skin.

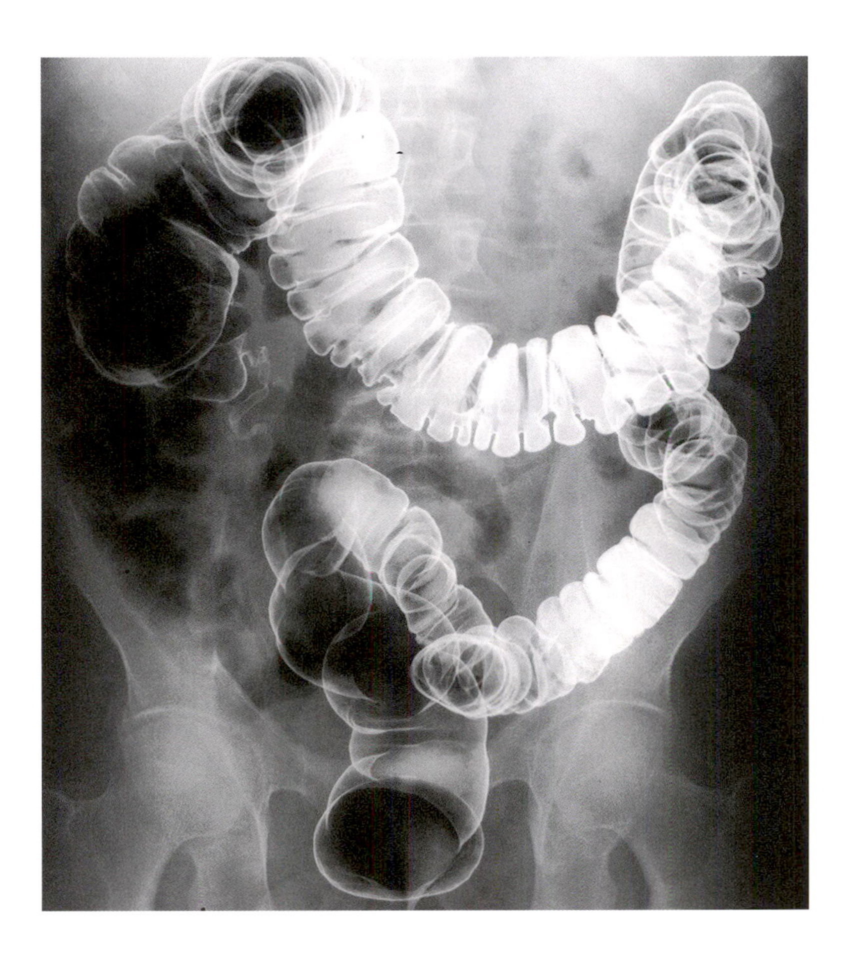

◆
Fig. 353
X-ray image of the large intestine. The inner surface of the mucous membrane appears as a white contour due to the use of a contrast agent.

◆
Fig. 354
The rectum, longitudinal section. The columnar puckering of the mucous membrane in this region is characteristic of the anus. The levator ani and the sphincter muscles are also clearly visible.

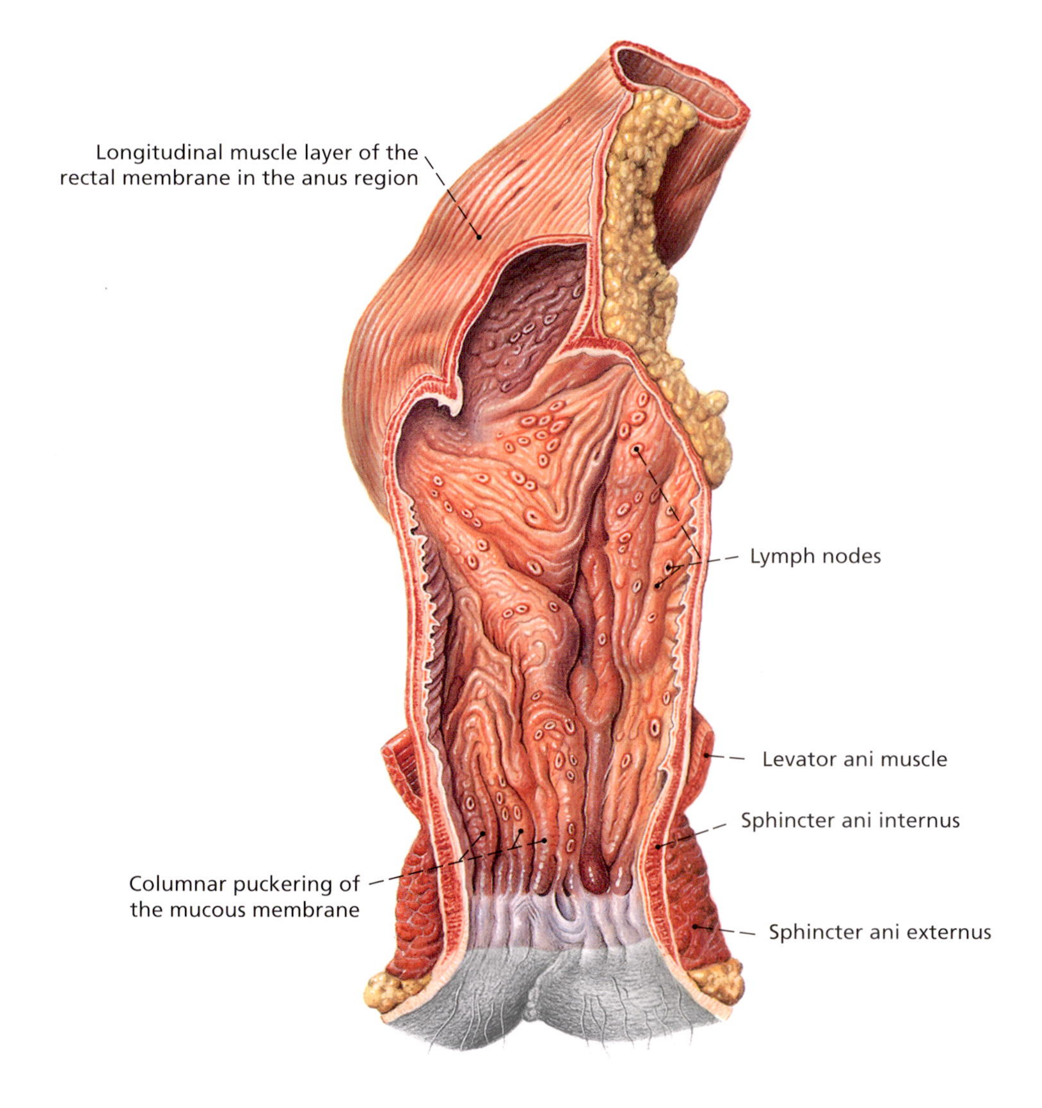

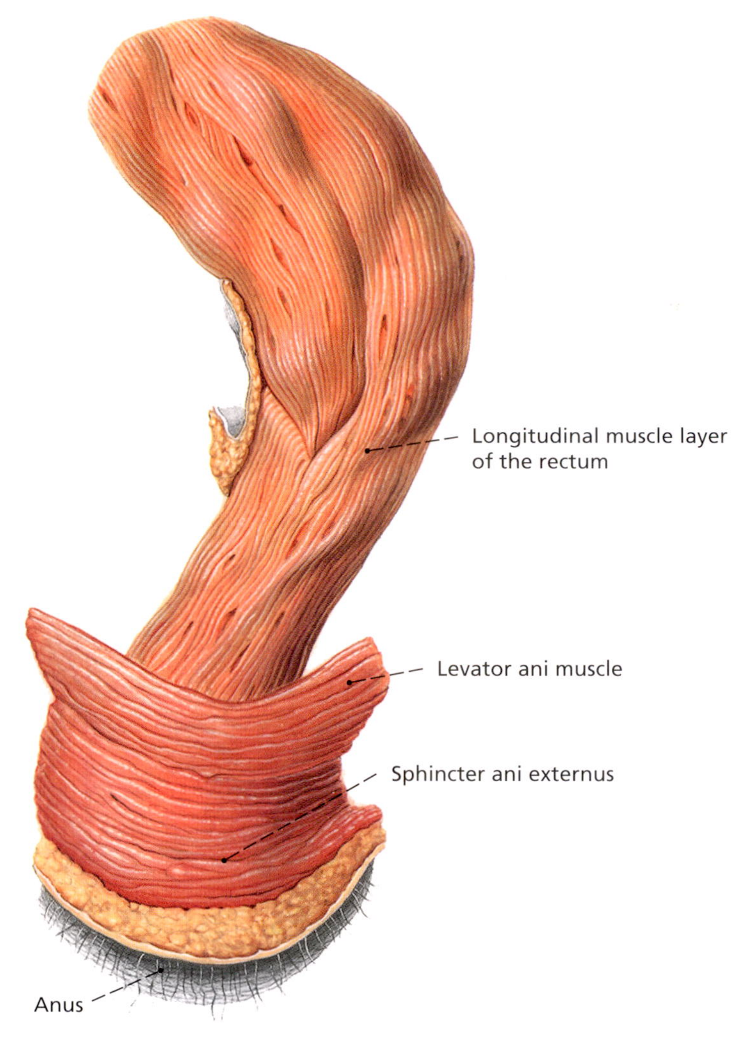

◆
Fig. 355
The musculature of the rectum.

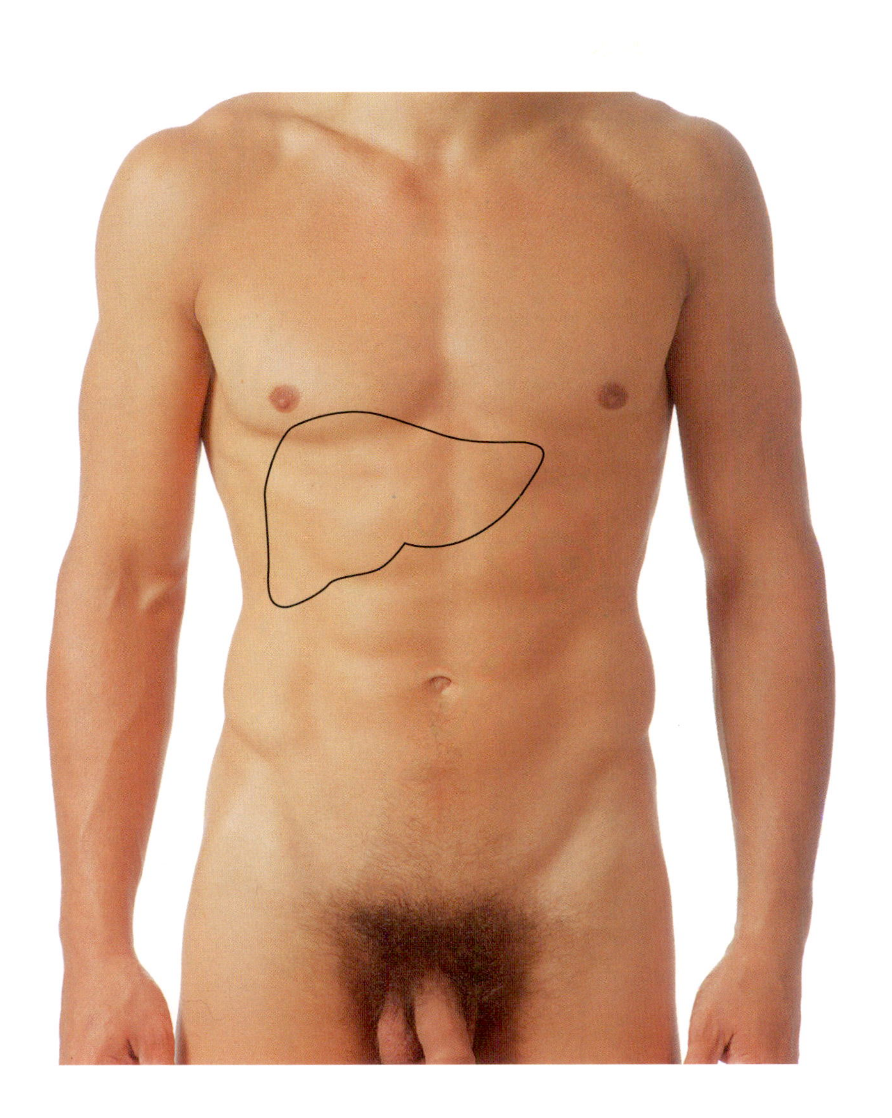

◆
Fig. 356
The position of the liver in the abdomen. The outline has been drawn on the skin.

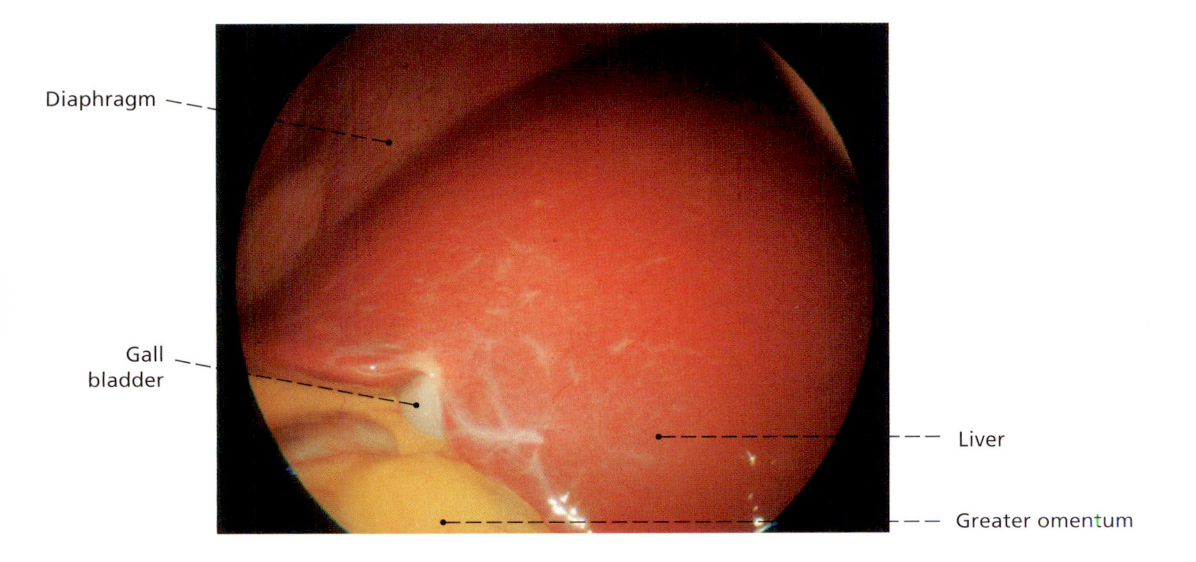

◆
Fig. 357
The anterior surface of the liver. Seen through the endoscope during a colonoscopy, which allows examination of the color, surface condition, and shape of the liver, as well as the condition of the gall bladder.

◆
Fig. 358
The liver, seen from the front. A distinction is made between the left and the right hepatic lobes. The upper surface of the liver is joined to the diaphragm from which a sickle-shaped ligament extends between the two hepatic lobes. The gall bladder can be seen at the lower edge of the right hepatic lobe.

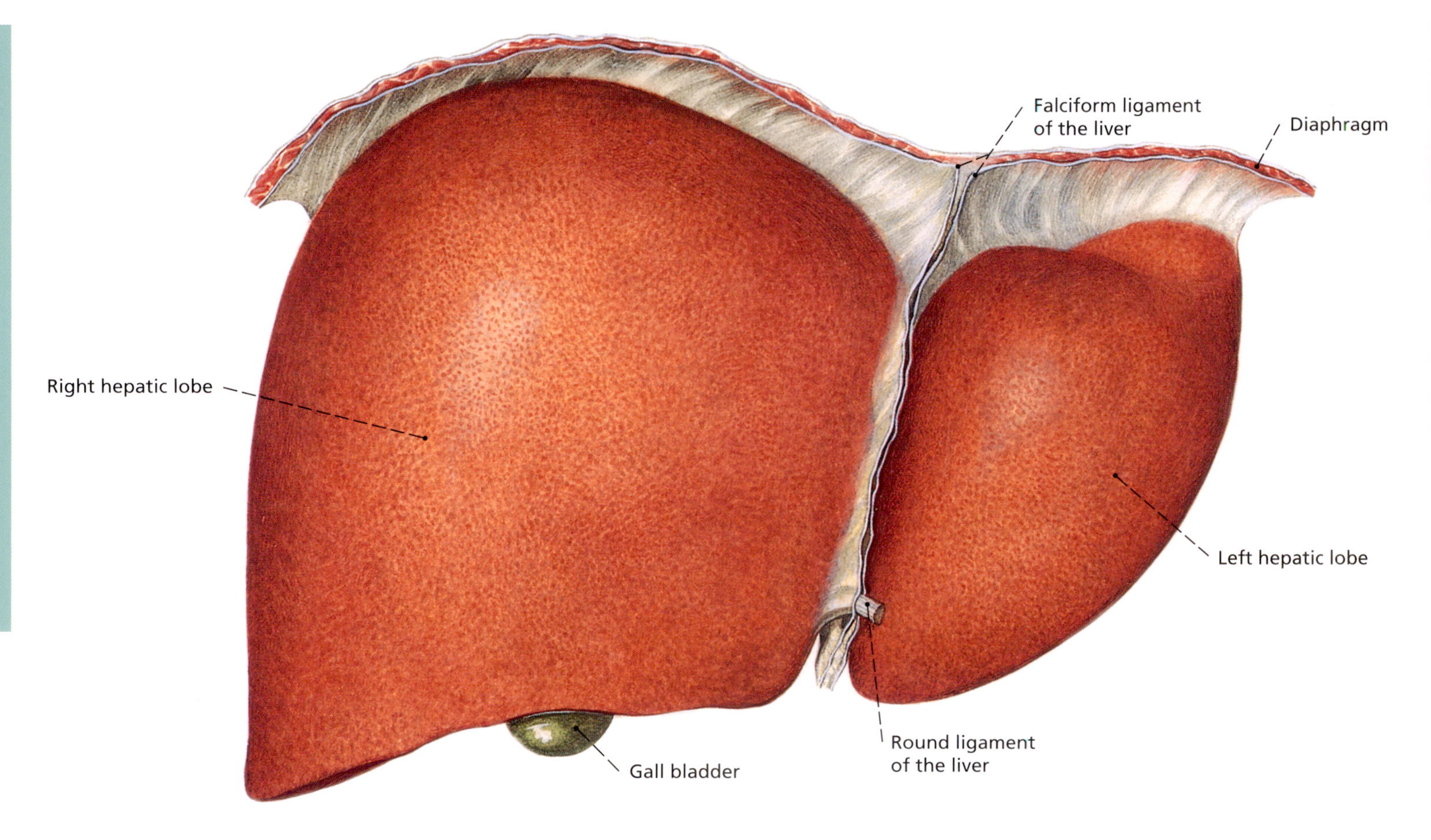

◆
Fig. 359
The liver, seen from the rear. The left and right hepatic lobes are visible, as is the gall bladder with the common bile duct, which opens into the liver. The vessel extensions are depicted in color.

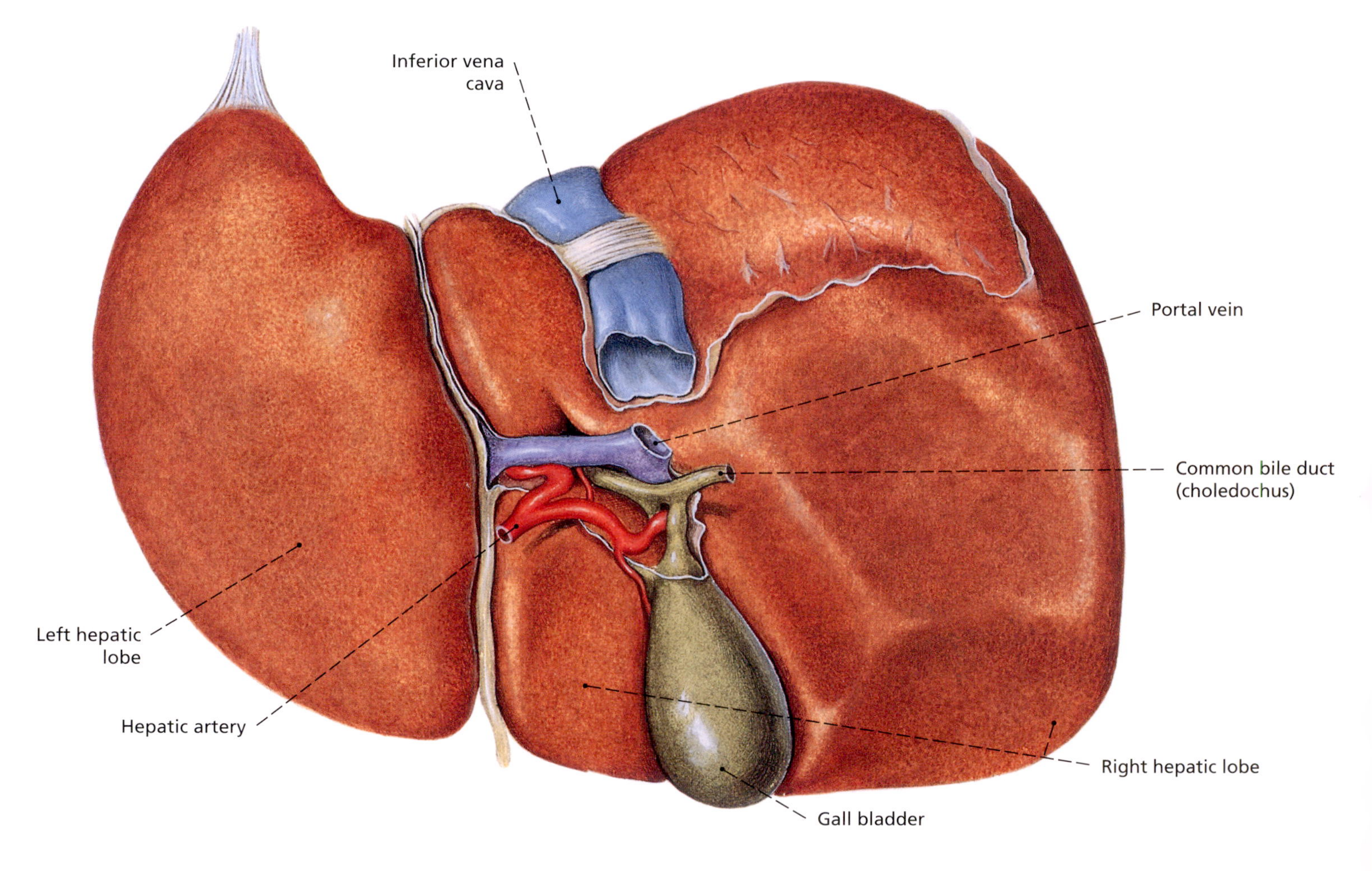

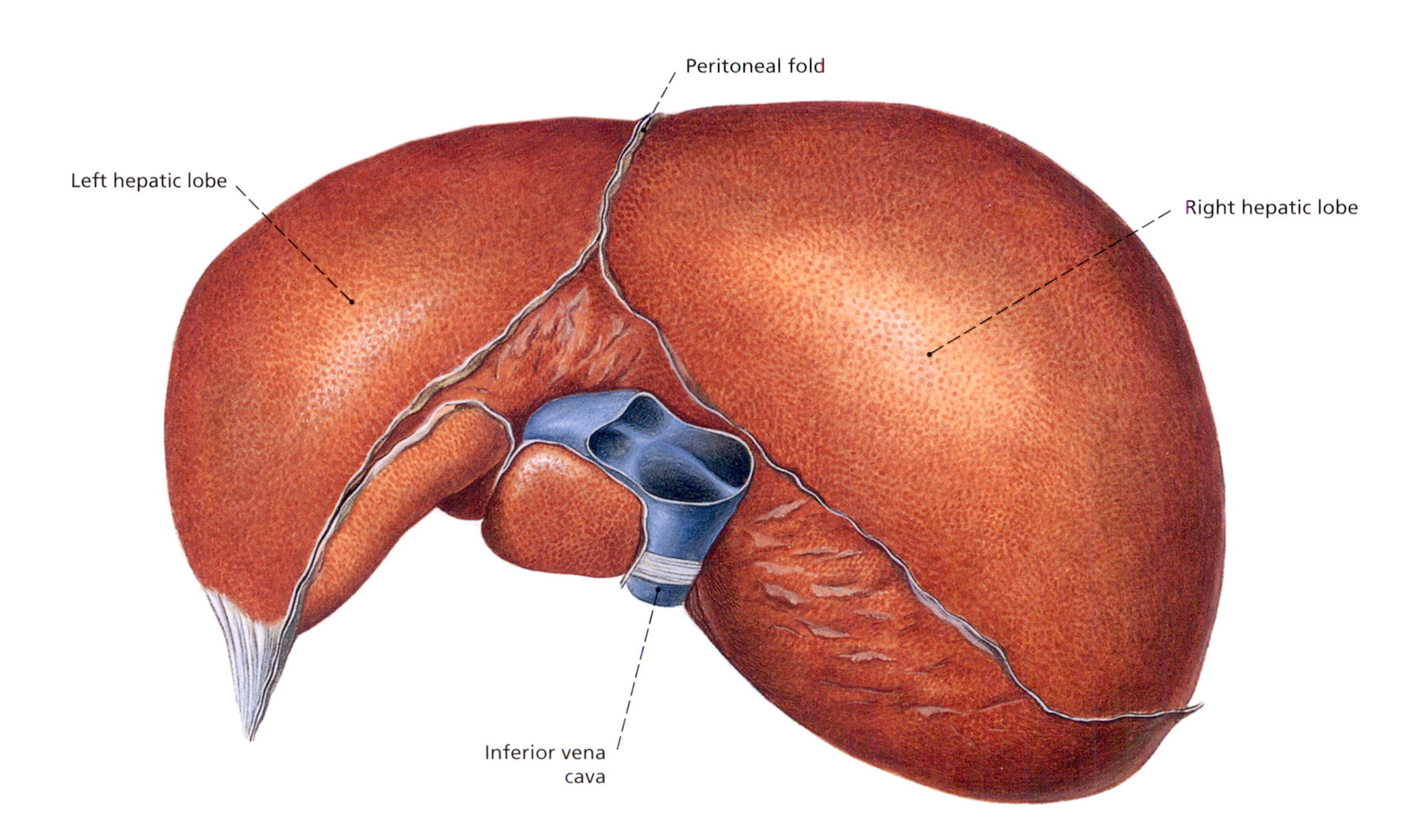

◆
Fig. 360
The liver, seen from above. The peritoneal folds which cover the liver as a thin, matt-finish membrane have been cut through. The opening of the inferior vena cava with the branches of the hepatic veins is visible between the two hepatic lobes.

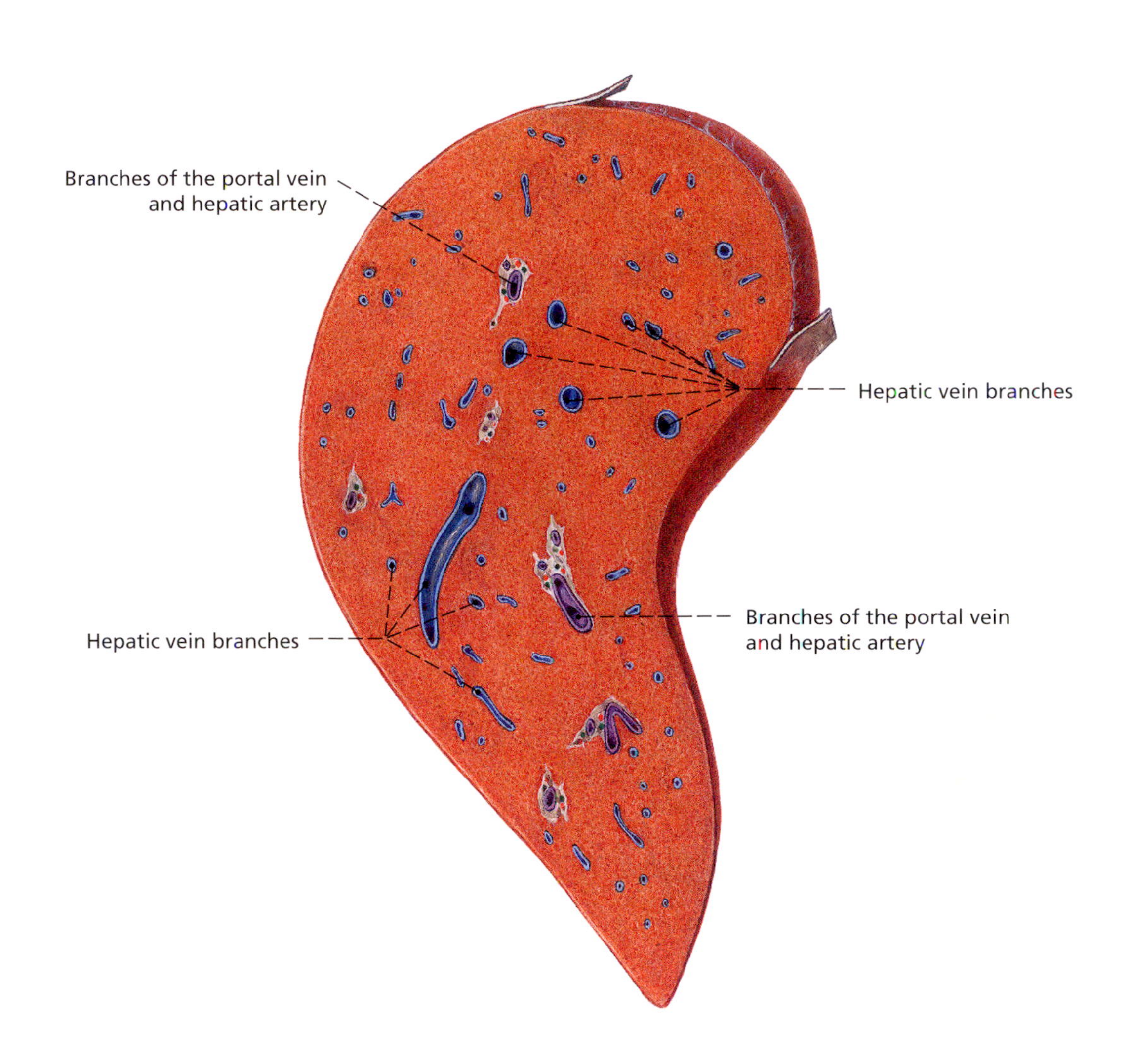

◆
Fig. 361
Cross-section of the right hepatic lobe. The numerous hepatic veins and portal vein branches have been colored blue or lilac, the small arterial branches red.

◆
Fig. 362
The gall bladder with the bile ducts. Parts of the anterior wall of the gall bladder, the bile ducts and the duodenum have been removed. The mucous membrane on the inner surface of the gall bladder contains numerous small indentations. The common duct, formed by the common bile duct and the pancreatic duct entering the duodenum (major duodenal papilla), is visible in the lower section of the illustration.

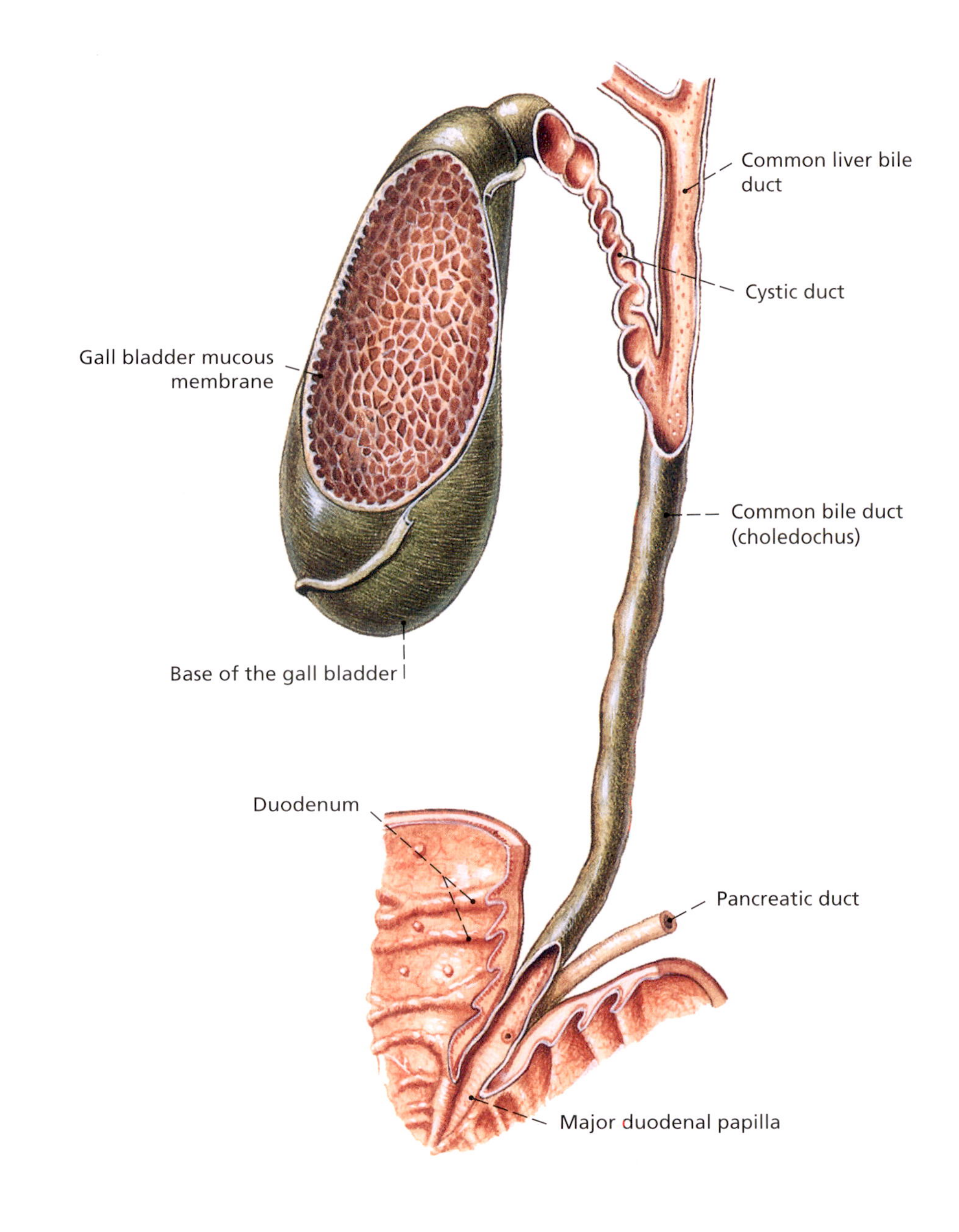

◆
Fig. 363
X-ray of the gall bladder with the bile ducts (contrast agent image), seen from the front. The vertebrae can be seen to the right of the image.

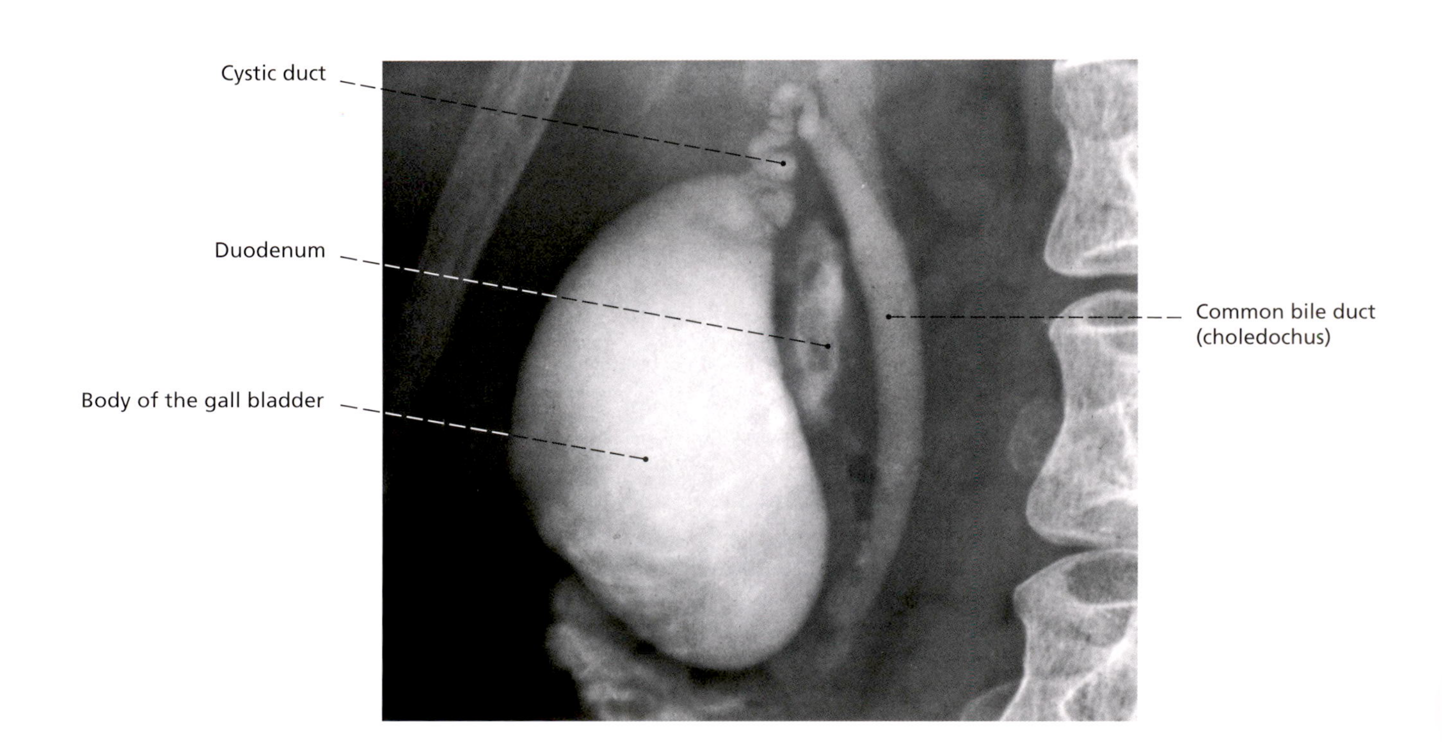

Symptoms and diseases

Esophagus

• *Inflammation of the esophagus*

Inflammation of the esophagus is most frequently caused by corrosive stomach contents flowing back up the esophagus, for example, in hiatus hernia, when part of the stomach is pushed upwards into the thorax and the gastric acid flows back into the esophagus. Medication to inhibit acid production in the stomach or disable the damaging influence of gastric acid is usually sufficient to treat this condition, although surgery may be necessary if this treatment fails. Other possible triggers include the intake of any substance damaging to the mucous membrane, chemical burns, scalding by food that is too hot or spicy, or excessive alcohol consumption.

• *Esophageal varices*

Caused by circulatory congestion and the subsequent widening of the veins in the lower section of the esophagus and sometimes also in the upper region of the stomach. The congestion is caused by increased blood pressure in the vein leading to the liver (portal vein) and is usually related to cirrhosis of the liver. This disease of the liver hinders or completely obstructs the blood flow, so that the blood gathers in the abdominal veins from where it flows into the veins of the esophagus. It builds up here and stretches the vessels to such an extent that they are likely to burst.
Heavy bleeding from the esophagus always constitutes a life-threatening emergency requiring prompt medical attention if death due to internal hemorrhaging is to be prevented. A balloon-like catheter is inserted into the esophagus and inflated in order to compress the varices and halt the hemorrhaging.
The veins are usually sclerosed, i.e., closed permanently, by an injection of medication, in order to prevent further hemorrhaging. Surgery to create a link between the portal vein and the inferior vena cava, thereby reducing the pressure in the portal vein system, can be performed as a preventative measure.

• *Esophageal cancer*

Cancer of the esophagus usually occurs in old age and more frequently in men than women, although the reason for this is not entirely understood. It affects alcoholics and heavy smokers in particular.
The first warning signs include persistent pain when swallowing, which can lead to an inability to swallow. There is also a feeling of pressure or pain behind the sternum (breastbone), bad breath, hoarseness, weight loss, nausea, and vomiting. The cancer is liable to spread quickly through metastases in the lungs and liver. Pain and general condition can be improved with chemotherapy and irradiation. If surgery is still possible, the tumor may be removed from the affected sections of the esophagus. Prostheses, or a tube made from stomach or small intestine tissue, are used to bridge or replace these.

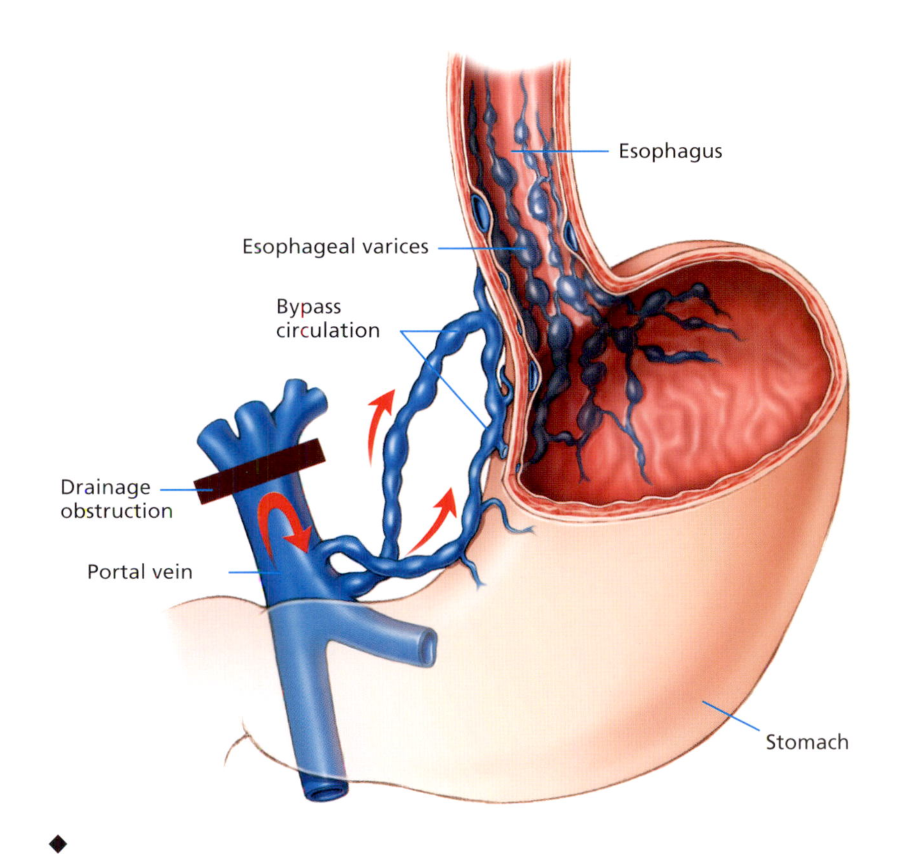

◆
Esophageal varices
If the blood is unable to flow unobstructed through the portal vein into the liver it seeks an alternative route through the veins of the stomach and the esophagus. These are unable to cope with the increased quantities of blood and become distended.

The stomach

• *Gastroenteritis*

Simultaneous inflammation of the mucous membrane in the stomach and that of the small intestine. Usually caused by a bacterial (e.g., salmonella) or viral infection, but sometimes by food or alcohol poisoning. The characteristic indications of gastroenteritis are nausea, vomiting, diarrhea, abdominal pain, and slight fever. With small children and older patients in particular, there is a risk of losing too much fluid through vomiting and diarrhea. It is therefore important to drink plenty of liquids (e.g., tea with sugar) and to visit the doctor quickly.

• ***Gastritis***

Acute gastritis is usually the consequence of irritation by harmful substances such as alcohol and nicotine, medication, acids, alkalis, or contaminated food. Stress can also cause gastritis. The symptoms include pain in the stomach region, nausea, vomiting, and loss of appetite. The damaged mucous membrane may also hemorrhage, in which case blood may be vomited and/or black-colored stools passed. In this event a medical opinion should be sought immediately. More minor complaints usually disappear quickly if the stomach is treated with care, i.e., by not eating or eating only small portions of bland food (e.g., toast or rusks) for a number of hours and drinking water or tea. The causes of **chronic** gastritis include specific bacterial infections (usually **Helicobacter pyloris**), alcohol abuse, and the reflux of bile and stomach contents. Those affected frequently do not experience any discomfort, although a feeling of pressure and bloating in the stomach region, burning pain, and loss of appetite can sometimes occur. A bland diet and the avoidance of alcohol and specific medication are usually recommended.

• ***Gastric ulcer (ulcus)***

Gastric ulcers may be caused by damage to the mucous membrane of the stomach, which extends into the deeper layers of the stomach wall, or excess production of gastric acid, which attacks the mucous membrane of the stomach. Mental and physical stress, bacterial infections, smoking, and the intake of certain medications can also be contributory causes. The usual symptoms include dragging pains and a feeling of pressure in the stomach region directly after meals. The diagnosis is usually made by means of a gastroscopy or an X-ray examination. If a gastric ulcer remains undetected it can lead to gastric bleeding or a perforation of the stomach. A gastric ulcer is treated with acid-inhibiting medication as well as with a bland diet that is gentle on the stomach. Surgery is seldom necessary.

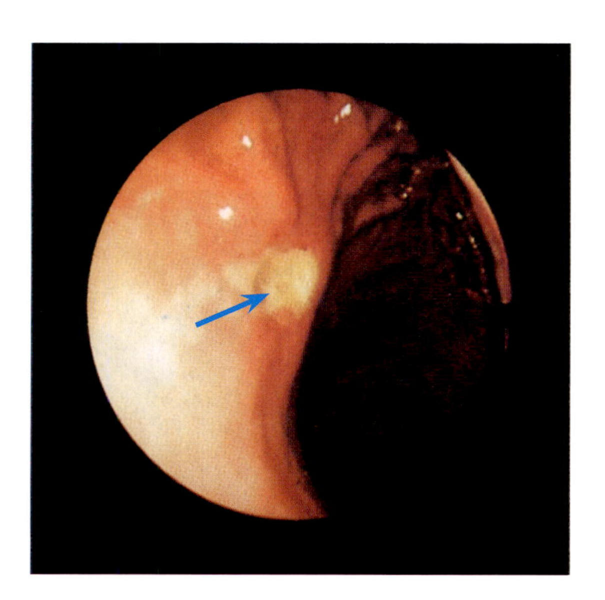

◆
Gastric ulcer
A mucous membrane defect or gastric ulcer is clearly visible during an endoscopic examination.

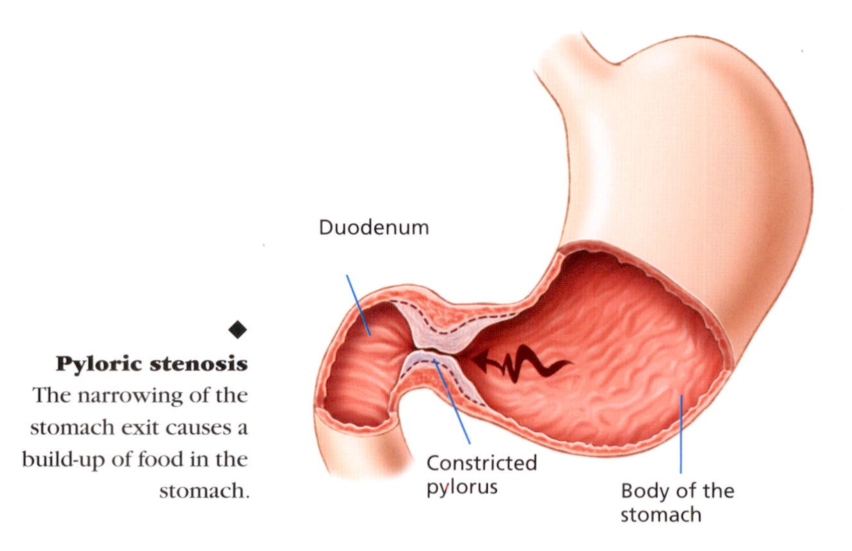

◆
Pyloric stenosis
The narrowing of the stomach exit causes a build-up of food in the stomach.

• ***Perforation of the stomach***

Perforation of the stomach wall and the emptying of the stomach contents into the abdominal cavity is usually caused by a gastric ulcer growing through the stomach wall. The patient suffers from sudden, severe pains in the upper abdomen. The abdominal wall is tightened and this is accompanied by shock with cold sweats, a quickened, weak pulse, and a pale skin color. An emergency physician should be called immediately and surgery will be required.

• ***Gastric carcinoma***

A malignant tumor in the mucous membrane of the stomach. The tumor may remain restricted to the mucous membrane for a long time and then suddenly begin to spread rapidly. In the initial stage the disease is often accompanied by no symptoms at all or else by non-specific complaints such as a feeling of pressure and bloating, burning, and pains in the upper abdominal region. The symptoms usually increase to include a loss of appetite, an aversion to meat, vomiting, and weight loss. The diagnosis is made on the basis of a tissue sample taken from the tumor by means of a gastroscopy. If confirmed, the affected section of the stomach will need to be removed surgically.

• ***Pyloric stenosis***

Narrowing of the stomach exit. The lower end of the stomach is enclosed by a circular sphincter muscle (pyloris) controlling the passage of the chyme into the small intestine. If this duct becomes blocked, the stomach contents build up and are ejected by projectile vomiting. In babies, the cause of such a constriction may be a congenital swelling of the sphincter muscle, which manifests itself at about 2–4 weeks of age with heavy, ongoing vomiting, weight loss, and developmental disorders. In adults, constriction is usually due to a spasm of the pylorus or an ulcer at the stomach exit. The stenosis is treated either with medication or surgery, depending on the cause.

Pancreas

• ***Pancreatitis***

The **acute** inflammation is accompanied by sudden, strong pains in the upper abdomen, with a hardening of the abdominal wall and an abnormal build-up of gas in the gastrointestinal region, radiating back pain, nausea, and vomiting. Movement causes an increase in pain. Signs can also include a build-up of fluid in the abdomen, shock (accompanied by a drop in blood pressure), and an increased heart rate. The most frequent cause is long-term alcohol abuse, with gallstones or viral infections being less common causes.
Chronic inflammation can be entirely pain-free or accompanied by repeated pains. The symptoms and causes are otherwise the same as those of acute infection, together with weight loss and fat intolerance. The consequence of chronic inflammation is permanent damage to the organ, with an increased risk of developing pancreatic carcinoma. In addition to antibiotics and painkillers, the treatment also includes medication to prevent autodigestion of the pancreas. Surgery is required if it is feared that the pancreas is already dead or if the inflammation has been caused by gallstones.

• ***Pancreatic carcinoma***

Tends to occur in men over the age of 50 and often only causes discomfort once subsidiary tumors (especially in the lymph nodes and peritoneum) have already formed. The initial signs are pains in the upper abdomen radiating from the back and increasing jaundice caused by the direct proximity of the pancreatic duct to the bile duct. The tumor can be removed surgically if it is detected in time.

Intestine

• ***Intestinal incarceration***

Incarceration of an intestinal loop following a hernia, in which part of the intestine has bulged out like a pouch through a weak point in the abdominal wall, frequently in the inguinal region. If the intestine cannot be quickly pushed back into its original position the incarceration can lead to the constriction of the affected intestinal loop and thereby to intestinal obstruction. The blood supply is often so severely restricted in this part of the intestine that it dies off, releasing its contents into the abdominal cavity. This can lead to life-threatening peritonitis. Surgery is required if there is even the slightest suspicion of intestinal incarceration.

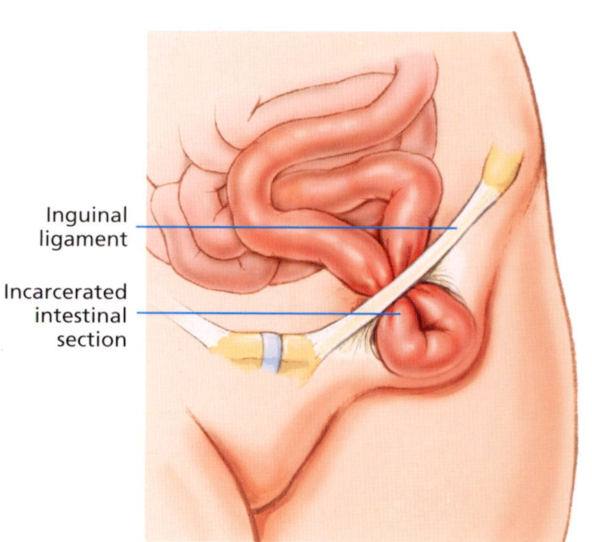

◆ **Intestinal incarceration**
Intestinal incarcerations often occur in the inguinal ligament region as a result of a hernia. Immediate treatment is essential as the intestine can be obstructed or even burst as a result of the incarceration.

• ***Enteritis***

An inflammation of the intestinal mucosa accompanied by diarrhea, and often by nausea, vomiting, and fever. It usually affects the small intestine but can extend to the mucous membrane of the stomach (gastroenteritis) and even spread to the large intestine. The causes range from nutritional failings and bacterial or viral infections through to allergens, toxins, and the side-effects of medications (e.g., antibiotics). Treatment is usually conservative, i.e., a bland diet and adequate fluids.

• ***Intestinal ulcer***

A far-reaching defect of the intestinal mucosa, usually in the upper section of the intestine. Intestinal ulcers can be caused by stress, infections (e.g., typhus, tuberculosis, dysentery), and poisoning, as well as injury. Intestinal ulcers also occur in the context of chronic intestinal illnesses such as Crohn's disease or ulcerative colitis. They are usually treated with antacids or acid-reducing medication (e.g., H2-blockers), and only require surgery in rare cases (i.e., of hemorrhage or perforation).

• ***Viral enteritis***

Intestinal illness caused by a virus infection. Although accompanied by influenza-like complaints (generally feeling unwell, headaches and rheumatic pains, chills, and fever) the main symptoms are intestinal tract complaints such as nausea, vomiting, and diarrhea. This illness does not generally require special treatment, but in cases of severe diarrhea and vomiting the often significant loss of fluids from the body should be compensated for.

Diabetes

At the beginning of the 20th century, diabetes was still a lethal condition, bringing an agonizing death to many patients. Since the discovery of insulin, however, the illness can now be well controlled and those affected are able to live almost normal lives.

The diabetes epidemic

Insulin, which is produced in the pancreas, regulates the transport of our main energy source, sugar, to the body cells. The level of sugar in the blood increases when there is a lack of insulin due to the fact that sugar can no longer be absorbed by the cells. All diabetics have a lack of insulin, to varying degrees.

Type I refers to the type of diabetes that occurs when the pancreas has been damaged by disease (e.g. by a viral infection) or when the body itself has damaged the exocrine functions of the pancreas for reasons which are as yet unknown.

Type II is largely considered to be a disease of modern civilization. While its occurrence is extensive, the determining factors for the onset of the illness are often bad eating habits: too much, too sweet, and too fat. With this type of diet the pancreas often reduces the production of the all-important insulin or else even stops it completely. Some 90% of Type II diabetics are overweight.

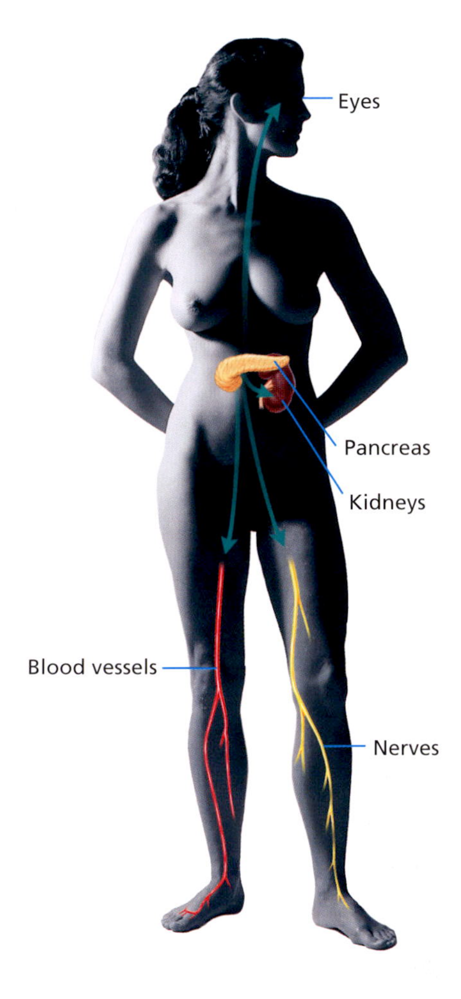

The organs and regions of the body directly affected by diabetes. Damage may also occur in most other regions of the body, as a lack of insulin means that the absorption of the vital energy source, sugar, by the cells is inadequate.

Insulin—the sugar regulator

Insulin is a hormone that allows the utilization of absorbed sugar in the organs and cells of the body. Insufficient quantities of insulin in the blood mean that many cells remain undernourished, while others may be damaged by excess sugar circulating in the blood.

Sugar—the energy source

Sugar is one of the body's most important energy sources. Not, of course, white household sugar but sugars contained in foods such as bread, potatoes, vegetables, and fruit, in the form of carbohydrates. The body converts these carbohydrates into a utilizable form of sugar, namely glucose, and this enters the blood as an energy source. Blood sugar levels normally rise after a carbohydrate-rich meal and then drop again once the glucose reaches the organs and cells. This is precisely the point at which insulin plays a key role.

Initial signs

Insulin is normally released by the pancreas following the intake of food, to ensure adequate uptake and utilization of the sugar produced after the food has been digested. If too little or no insulin is released, the blood sugar level increases so that large quantities of sugar are excreted via the kidneys in the urine. The body loses a great deal of fluid in the process, as well as the minerals dissolved therein. This leads to emaciation as well as to the simultaneous acidification of the body through substances containing the toxin acetone. Diabetes is often indicated by:

- excessive urination with extraordinary thirst,
- weight loss despite adequate nutrition,
- general fatigue, tiredness, and possibly impaired vision.

Diabetic coma and shock

In a **diabetic coma**, the patient becomes unconscious as a result of excess sugar in the blood stream (hyperglycemia). At its worst, this condition can result in fatal consequences. Diabetic coma does not occur in well-controlled diabetics. It is only when the sugar balance

has been seriously disrupted that a diabetic coma may result, in which event the patient must be taken to hospital immediately. The warning signs are: nausea, vomiting, abdominal pain, and the smell of acetone on the breath.
The opposite scenario is equally dangerous; if the diabetic has not eaten enough (e.g., has skipped a meal), injected too much insulin, or is suffering from a gastrointestinal disorder with diarrhea and vomiting, the blood sugar level sinks to such an extent that the nervous system and the brain in particular no longer receive an adequate supply of sugar. This results in **shock** caused by hypoglycemia. The signs include a sensation of pins and needles, pallidness, shaking, palpitations, and sweating (in the case of severe hypoglycemia: vertigo, double vision, a reduced state of consciousness, a lack of concentration, impaired speech, and unconsciousness). An experienced diabetic takes glucose immediately upon noticing the initial signs. It is essential to call a physician immediately in the event of unconsciousness.

Vascular complaints and other secondary disorders

Diabetes can cause potentially dangerous vascular complaints. The narrowing of the major arteries is no different to the arterial calcification in a non-diabetic but it occurs more frequently in diabetics. This can lead to cardiac infarction if the coronary vessels are affected, or to a stroke if the damage is in the brain.
Damage to the small blood vessels is characteristic of diabetes and the blood supply to the legs is often disrupted. This often goes unnoticed by the sufferer because diabetes is also associated with sensation deficits. This can result in the tissue dying off. In order to avoid this, any noticeable changes, e.g., redness or suppuration, must immediately be brought to the attention of a physician. Thorough, regular care of the feet is especially important because small wounds can become infected quickly as a result of the impaired blood supply. Changes to the blood vessels of the eye can lead to retinal detachment and to blindness, so regular ophthalmologic examinations are also important.
The kidneys should also be checked frequently because vascular damage can lead to restricted urination, with the consequence that excess sugar cannot be eliminated.

Treatment: From diet to injections

Initially, the physician checks the sugar levels in the blood and the urine, and then determines the appropriate treatment for the patient based on the results.
There are primarily three treatment methods, assuming the patient is not overweight (obese patients should lose weight), depending on the severity of the condition:

- control of dietary carbohydrate,
- blood-sugar-reducing tablets (oral anti-glycemics),
- regular administration of insulin.

A balanced ratio between food intake and/or sugar supply and the available insulin is important. A diabetic is referred to as "well-controlled" if the blood sugar level is maintained within the prescribed range. It is sensible, however, to always have a pack of glucose to hand in order to be able to raise the blood sugar level quickly in the event of sudden hypoglycemia. For safety's sake a diabetic should always carry a diabetic identity card with them.

Self-monitoring at home

Diet is the basis of any treatment, irrespective of whether the diabetic requires insulin or not. Nutritionally, carbohydrates are measured in bread units (BU) taken from the relevant tables. Diabetics learn very quickly how many BUs they may eat at what time of day and when they need to take or inject insulin.
Daily checking of the blood sugar level is nevertheless indispensable. Meters make it possible to determine the current blood sugar level from a drop of blood, while dipsticks are used to measure sugar or acetone levels in the urine. The more precisely the diabetic carries out this self-monitoring and records the results, the better the doctor is able to determine the treatment.

Living with diabetes

There is no social stigma attached to being diabetic and well-controlled diabetics are largely able to live normal lives.

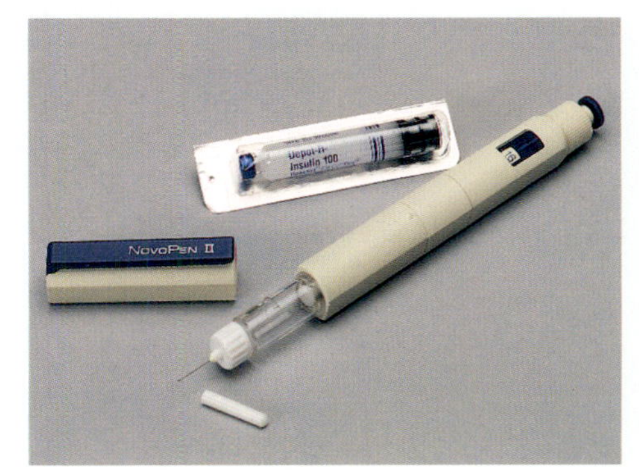

◆ If the body does not produce sufficient quantities of insulin, this may be administered by syringe, or by means of modern, easy-to-use injection devices which allow for a very precise dosage to be given.

• ***Intestinal polyps***

Swellings in the intestinal mucosa extending into the intestinal cavity. The polyps either spread out over the mucous membrane or else are attached to it via a shaft. Intestinal polyps are almost always non-malignant. There are specific types, however, which can undergo malignant change and become precancerous. This group includes the large intestine polyps (familial intestinal polyposis) that are prevalent in some families. Patients suffering from this illness should therefore attend regular examinations. An operation is required to remove the polyps as soon as there is the slightest suspicion that they might become malignant.

• ***Intestinal obstruction***

A condition in which the contents of the intestine cannot be transported further due to a constriction or complete obstruction of the intestinal cavity.
The intestinal passage can be interrupted for a variety of reasons: through paralysis of the intestinal musculature preventing the further transportation of the intestinal contents (known as a paralytic ileus), through the constriction of an intestinal loop due to a rupture in the abdominal wall (hernia), or by a tumor, gallstone, or hard, dried-out impacted feces (known as mechanical ileus). Scarred adhesions, which form after abdominal surgery, can also lead to intestinal obstruction.
The characteristic complaints include a lack of bowel movement and a swollen abdomen, nausea and vomiting of the stomach contents, followed later also by regurgitation of the intestinal contents. The patient's condition usually deteriorates so rapidly that surgery to remove the cause of the intestinal obstruction needs to be carried out immediately.

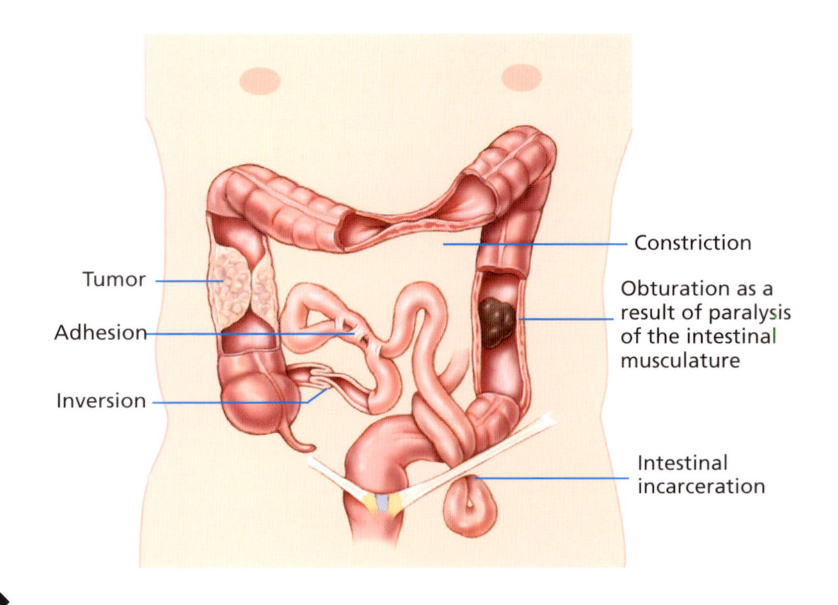

◆
Intestinal obstruction
There are different forms of intestinal obstruction depending on the location and cause.

• ***Crohn's disease***

Inflammation affecting any part of the gastrointestinal system. Crohn's disease most commonly occurs at the lower end of the small intestine, known as the ileum. Typical symptoms include pain, ulcers, and diarrhea. Sufferers lose weight, become pallid, and suffer from general weakness. The formation of fistulas or abscesses is a frequent complication. A conglomerate tumor is said to be present if the inflamed intestinal loops adhere and grow together. Scarring carries the risk of intestinal obstruction. Crohn's disease is diagnosed by means of stool tests, blood tests, biopsy, and endoscopy, either of the colon via the anus or of the upper part of the intestinal tract via the mouth. A barium X-ray of the large intestine (barium enema) or small intestine (barium meal) may also be used.
Medical treatment (corticosteroids, sulfasalazine) in combination with a low-residue diet can be successful if complications have not yet developed, but there is no specific, successful therapy as yet and surgery to remove part of the gut is often required. Patients may have to be fed by infusion for several weeks during flare-ups, in order to circumvent the digestive tract and allow recovery.

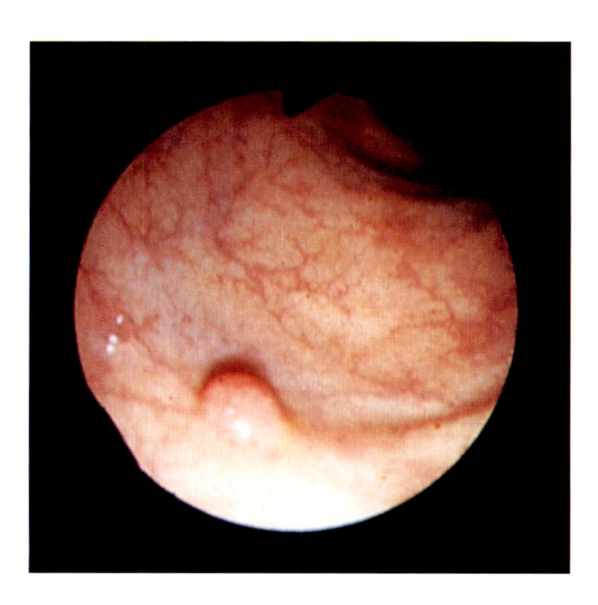

◆
Intestinal polyps
Intestinal polyps can be detected and sometimes removed through an endoscope during a colonoscopy.

• ***Ulcerative colitis***

Chronic inflammation of the large intestine that comes in phases with gradual disintegration of the intestinal wall. It can either begin inconspicuously with diffuse pains, or acutely with up to 20 bowel evacuations daily, accompanied by cramp-like pains. The stool is slimy and bloody and often contains pus. The intestinal mucosa bleeds slightly and is covered with suppurative tissue defects. These wounds do heal again but scars or so-called pseudopolyps are often formed, which constrict the intestinal canal and may lead to intestinal obstruction. A rupture of the intestine, followed by peritonitis, is a life-threatening complication. Ulcerative colitis can also develop into intestinal cancer.

Patients with ulcerative colitis are often malnourished. The causes of the disease are not clear.
Psychological factors and stress may play an important role in the development of ulcerative colitis. The diagnosis is made by means of a colonoscopy, when a tissue sample may be removed, as well as by means of a contrast X-ray image. Milder forms are treated with a special diet. The food should be low in indigestible dietary fiber and rich in calories. In severe cases nutrition has to be administered via an infusion in order to relieve the intestine. If these measures do not bring relief, part of the large intestine may have to be surgically removed. Psychological therapy can provide treatment support.

• *Appendicitis*

Appendicitis is the most common illness requiring urgent abdominal surgery (appendectomy). Appendicitis does not usually involve the cecum. The appendix has a small diameter and is barely able to dilate. A mass of hardened feces, or a foreign body such as a cherry pit, can block the entrance, enabling bacteria to proliferate there and causing inflammation. The inflammation may sometimes disappear by itself.
The symptoms are often characteristic: pain develops in the navel region, before spreading towards the right side of the lower abdomen. The pain is accompanied by nausea and a feeling of sickness, as well as fever characteristically measured as being a degree or two higher in the anus than in the armpit. The main risk posed by appendicitis is that of rupture. This occurs most frequently in young children and older people, usually on the third or fourth day following the start of the discomfort, and life-threatening peritonitis can be the consequence.

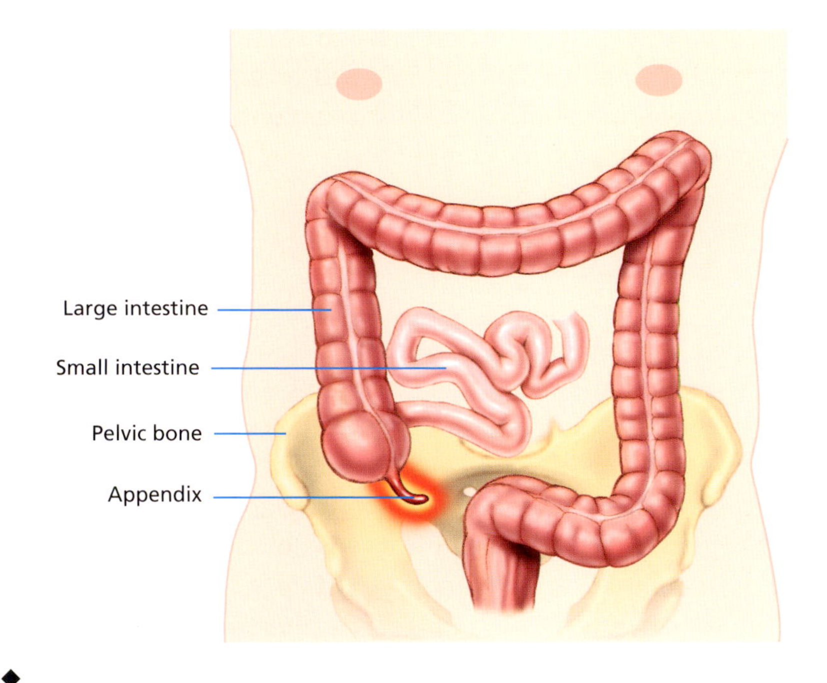

◆
Appendicitis
It is usually not the cecum but the appendix that is inflamed.

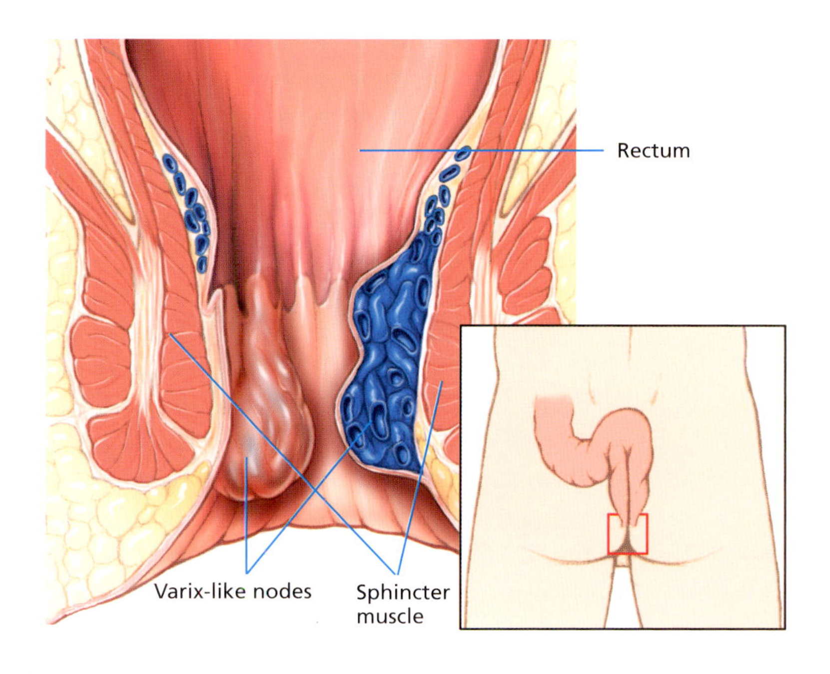

◆
Hemorrhoids
Hemorrhoids narrow the anus in the sphincter muscle region. This hampers stool evacuation and can be painful.

• *Hemorrhoids*

Nodular dilation of the veins in the mucous membrane of the anal canal. Internal hemorrhoids are located above the internal anal sphincter and are therefore not visible initially. They do tend to grow, however, and can then be pushed outwards when evacuating stools. Such bulging internal hemorrhoids are distinct from outer hemorrhoids that are located below the anal sphincter from the outset.
Hemorrhoids are very common and are facilitated by sedentary activities, inactivity, and constipation. They tend to bleed (pale red traces of blood on the toilet paper), burn, itch, and weep in the anal region. Anti-inflammatory ointments are applied in the initial stages. An attempt can also be made to widen the anal sphincter by inserting what is known as an anal expander. In advanced stages, the hemorrhoidal nodes can be removed through constriction, cauterization, or by surgery.

• *Anal fissure*

A tearing of the mucous membrane at the intestinal outlet. The symptoms are burning, sharp pains at the intestinal outlet, and light bleeding, and frequently occur in connection with other illnesses in the anal region (e.g., hemorrhoids, anal fistulas). The healing process is a protracted one as the fissures tear again easily. They can be treated surgically but are often treated conservatively with anti-inflammatory, wound-healing ointments or suppositories.

Digestive system

• ***Anal rhagade***

Superficial often radial damage to the mucocutaneous area around the anus, which often gives rise to burning pains and which occurs in connection with mechanical or infective trauma, hemorrhoids, and yeast or fungal infections.
Anal rhagades are treated with wound-healing, anti-inflammatory, and anti-fungal ointments.

• ***Anal fistula***

Small duct that develops between the anal canal and the exterior, opening in the skin around the anus. The illness is very painful because the fistula ducts usually become infected and inflamed. They need to be surgically removed.

• ***Anal cancer***

A rare form of cancer affecting the intestinal outlet. The tumor usually originates in the mucous membrane cells and is palpable. The lymph nodes in the inguinal region must always be examined for metastases and surgically removed if necessary. The cancerous growth itself is removed entirely due to the risk of metastases and this is followed by irradiation. Recovery chances are good if the tumor is detected and operated on in good time.

Liver

• ***Inflammation of the liver (hepatitis)***

Hepatitis is a viral infection. The hepatitis virus types are A, B, C, D, and E, with types A and B being the most common.
Hepatitis A virus is transmitted by bodily secretions such as urine, stools, or saliva. The virus spreads through foodstuffs and drinking water, meaning that epidemics may occur under unhygienic conditions. The transmission of the hepatitis B virus is via the bloodstream, e.g. through contaminated syringes, tattoo or acupuncture needles, or blood transfusions, as well as via sexual contact.
The incubation period for hepatitis A is ten to 50 days and for hepatitis B up to 180 days.
The first signs of the illness are lassitude, loss of appetite, nausea, painful joints, fever, and a dull feeling of pressure on the right side of the upper abdomen. The damage to the liver cells leads to the transfer of bile pigments into the blood and thus to the yellow coloring of the skin and the whites of the eyes.
A hepatitis infection takes a protracted course and recovery can take 12 weeks or more. In rare cases, especially with hepatitis B, the illness can be fatal. Treatment of the cause is not yet possible as there is no effective medicine against the virus. The body's own defense system can be supported by injection of antibody proteins (immunglobulins), and strict bed-rest; a low-fat diet must be adhered to. Alcohol must be avoided. If these rules are followed, acute hepatitis usually heals well. However, there is a risk of it becoming chronic and developing into cirrhosis of the liver.
Protective vaccinations against hepatitis A and B are especially recommended for people whose jobs entail them coming into contact with body fluids or the blood of others (e.g., hospital staff), or travelling to areas where the virus is endemic.

• ***Cancer of the liver***

A malignant tumor which usually only results in discomfort in later stages. There is a proven connection to hepatitis B. Weight loss, loss of appetite, listlessness, and pain in the right upper abdomen are warning signs. The surgical removal of the tumor is considered in some cases, and a liver transplant may be possible.

• ***Cirrhosis of the liver***

The most frequent consequence of excessive, long-term, alcohol consumption. Toxins, viruses, bacterial toxins, inadequate nutrition, or malnutrition can be further causes of the conversion of functioning liver tissue into non-functioning connective and/or scar tissue. Cirrhosis of the liver can exist for years without causing discomfort and it is not infrequent that it is detected by chance.
There is no cure. Cirrhosis causes restricted blood flow in the liver, leading to back pressure in the blood draining from the intestinal organs to the liver, and resulting in ascites, and esophageal varices (risk of hemorrhage). The gradual build-up of toxins in the blood ultimately leads to impaired brain function, and hepatic coma. A liver transplant is the only possible method of treatment.

• ***Hepatic coma***

Coma (profound loss of consciousness) as a consequence of advanced liver failure brought about by a viral liver infection, poisoning, e.g., alcoholic, or cirrhosis of the liver. This life-threatening situation is probably triggered by brain damage and derives from the toxic metabolic products produced by destruction of active liver tissue.

Gall bladder and bile ducts

• *Cholecystitis*

Inflammation of the gall bladder, usually related to gallstones, which commonly cause obstruction of the bile duct. Inflammation caused by bacteria entering the gall bladder from the intestine, liver, or blood system, is less common. The inflammation of the gall bladder causes severe colicky pain in the upper right abdomen, often extending to the right shoulder, and is accompanied by nausea, vomiting, and fever. Treatment consists of removal of the obstructions blocking the drainage of the bile, and combating the pathogens with antibiotics. There is an increased risk of cancer of the gall bladder in the case of chronic inflammation and the gall bladder is therefore often removed, together with gallstones.

• *Cholangitis*

Inflammation of the biliary duct that commonly occurs in conjunction with cholecystitis. The characteristic causes, symptoms, and treatments, are as for cholecystitis.

• *Gall bladder abscess*

Encapsulated abscess in the gall bladder resulting from an inflammation caused by gallstones obstructing the drainage of the bile, or through an infection caused by germs entering the gall bladder from the intestine and/or via the blood vessels. It leads to severe pain in the upper abdomen extending to the right shoulder. The underlying disease needs to be treated as a priority. Antibiotics are required if there is a risk of rupture of the gall bladder, or of the inflammation spreading to neighboring organs. The abscess may need to be surgically removed in some cases.

• *Ruptured gall bladder*

Rupture of the wall of the gall bladder due to an increase in internal pressure caused by inflammation or blocked drainage. There is a risk of gall bladder rupture as a complication of a gall bladder abscess or gallstones. As a result of the rupture, the contents can flow into the stomach, intestine, or, in the worst case, into the abdominal cavity. The related peritonitis and infection can be life-threatening.

• *Bile duct stenosis*

Disruption of the drainage of the bile into the intestine caused by gallstones, inflammation, or tumors in the bile ducts. This results in a build up of bile causing colic, pain, and jaundice. Acute biliary colic, which is accompanied by severe pain, can be triggered by a large gallstone. The stone will need to be removed surgically if it does not dissolve by itself.

• *Small gallstones (gravel)*

Small, gravel-like gallstones which often do not cause any discomfort and therefore usually go unnoticed, are formed when the normal composition of the bile has been disrupted, the solubility of the bile salts decreases, and crystal substances are produced. The precise cause is not known but there are a number of factors involved: excess weight, pregnancy, and metabolic disorders.

• *Gallstones*

Stones in the gall bladder and the bile ducts are common, especially with increasing age, and often remain unnoticed. The incidence is significantly higher among women than men. It is not entirely clear how gallstones arise but the risk factors include excess weight, repeated pregnancies, and metabolic disorders such as diabetes. The composition of gallstones, ranging from gravel- to walnut-size forms, varies considerably. There are mixed stones comprising bilirubin and calcium carbonate, as well as purely cholesterol stones. The symptoms range from a bloated feeling after eating, fat intolerance, flatulence, and a feeling of pressure in the upper right abdomen with slight discomfort, through to acute biliary colic caused by a stone becoming stuck in the bile duct. Gallstones can be detected by bile duct endoscopy as well as by an ultrasound or X-ray examination.

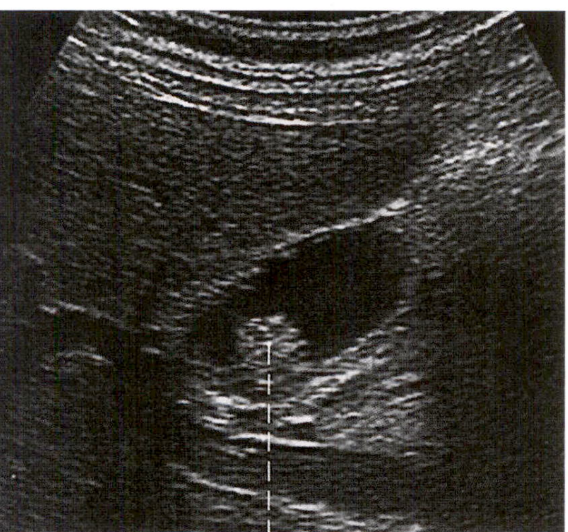

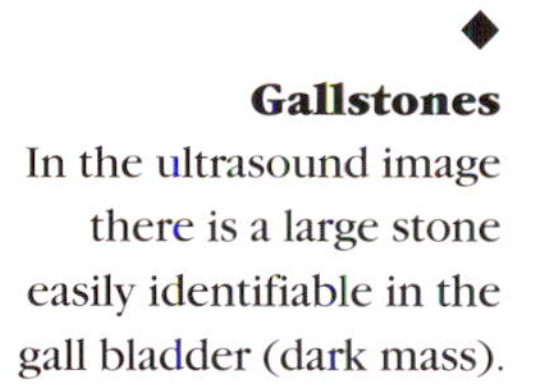

◆ **Gallstones**
In the ultrasound image there is a large stone easily identifiable in the gall bladder (dark mass).

Diagnostic methods

Gastroscopy

Examination of the interior of the stomach with an endoscope, an illuminated fiber-optic instrument that makes it possible to see inside the stomach. The diagnosis of inflammations, ulcers, cancer, or foreign bodies in the stomach and duodenum can be established in this manner. Tissue samples from a suspect growth can also be taken during a gastroscopy, bleeding vessels can be closed, and non-malignant growths or swallowed foreign bodies removed. A gastroscopy is especially important for the early detection of cancer.
With the development of ever-smaller medical instruments, a gastroscopy is by no means as unpleasant as many people fear. Medication can make the examination easier for the patients, suppressing both pain and the gagging reflex.

Gastrointestinal endoscopy

Examination of the intestine using an endoscope with a light source as well as a fiber-optic device attached to the end to enable detailed observation of the intestinal mucosa. A gastroscope is inserted through the mouth to examine the upper intestine (esophagus, stomach, and duodenum). For the large bowel, entry is via the anus. Rigid tubes are used to examine the lower sections (rectum and lower colon) while a flexible tube is used if the upper section of the large intestine is to be examined. A camera can be attached to the endoscope, if necessary, in order to record the examination process. Small instruments, such as forceps or scissors, can also be inserted into the endoscope in order to remove polyps or take tissue samples.

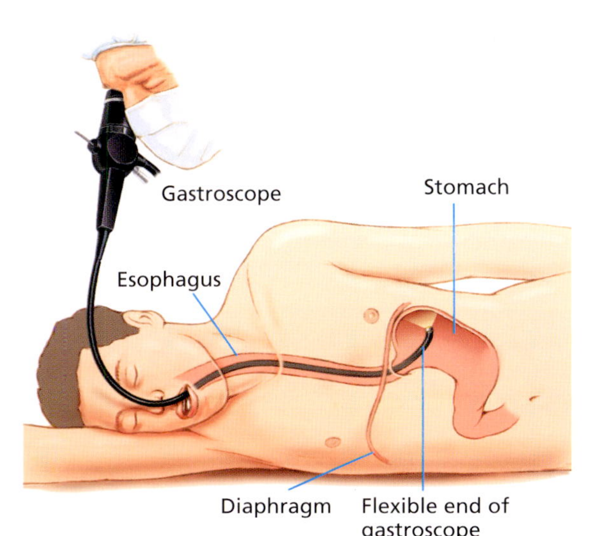

Gastroscopy
A gastroscope (a type of endoscope) is inserted into the stomach via the esophagus.

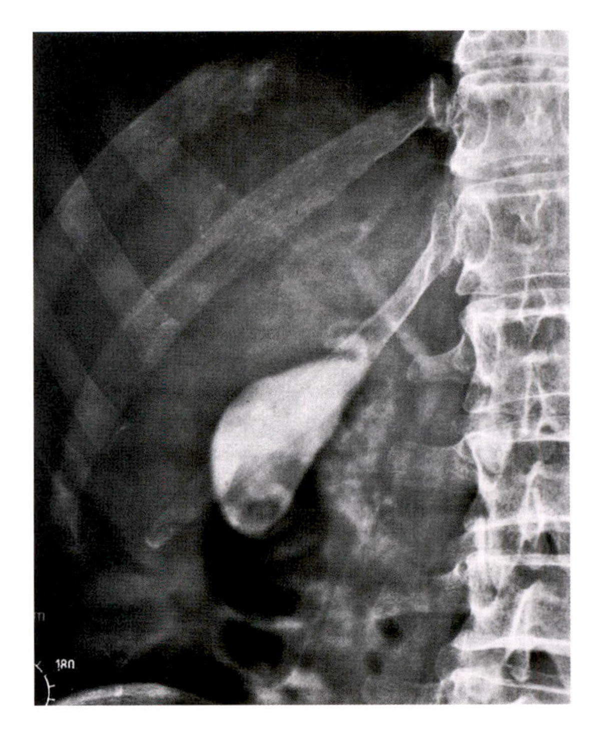

Gall bladder image
The gall bladder and bile ducts can be made visible in an X-ray image by using a contrast agent.

Liver biopsy

Removal of a tissue sample from the liver. A hollow needle is inserted into the liver via a small incision in the skin under local anesthetic and the tissue sucked out. The sample can then be examined under a microscope to determine the specific diagnosis.

Hepatic puncture

A procedure in which a hollow needle is inserted into the liver under local anesthetic. A puncture is carried out either for diagnostic purposes (microscopic examination of the liver tissue), or for the direct injection of medication.

Gall bladder image

A special X-ray method for imaging the gall bladder and the bile ducts. An X-ray-opaque contrast agent is either swallowed or injected directly into a vein or the bile ducts. It accumulates in the gall bladder and the bile ducts, making their contours visible on the X-ray image. Constrictions or obstructions, e.g., by gallstones, can be detected in this way.

Bile duct endoscopy

An endoscopic retrograde cholangio-pancreatography (ERCP) is a procedure in which an endoscope is passed through the mouth towards the stomach. The endoscope is used to inject a special dye into the bile and pancreatic ducts, which enables the identification of any pathological changes and also the removal of tissue or smaller stones at the same time.

Treatment methods

Vagotomy

The vagus nerve controls the production of gastric acid. Too much gastric acid can lead to gastric and duodenal ulcers. The production of gastric acid can be reduced if the nerve fibers are severed in part or entirely, causing the ulcers to recede. The reduced gastric acid production after the operation can sometimes lead to digestive problems, nausea, and a bloated feeling. With the appropriate medication, however, these side effects disappear within a matter of weeks. This operation has largely been made redundant by modern medications.

Gastric resection (gastrectomy)

An operation to remove (resect) either a diseased section of the stomach or the entire stomach. This can be a life-saving measure in the case of stomach cancer or perforation of the stomach.
A persistently recurrent stomach ulcer may also necessitate such an operation, especially if it is causing bleeding. Narrowing of the stomach exit can be corrected surgically. Digestion is temporarily restricted following a gastric resection, and diarrhea, vomiting, stomach pain, and a bloated feeling can sometimes be experienced initially. The small intestine is connected to what remains of the stomach thus forming a replacement stomach. See also Billroth's operation.

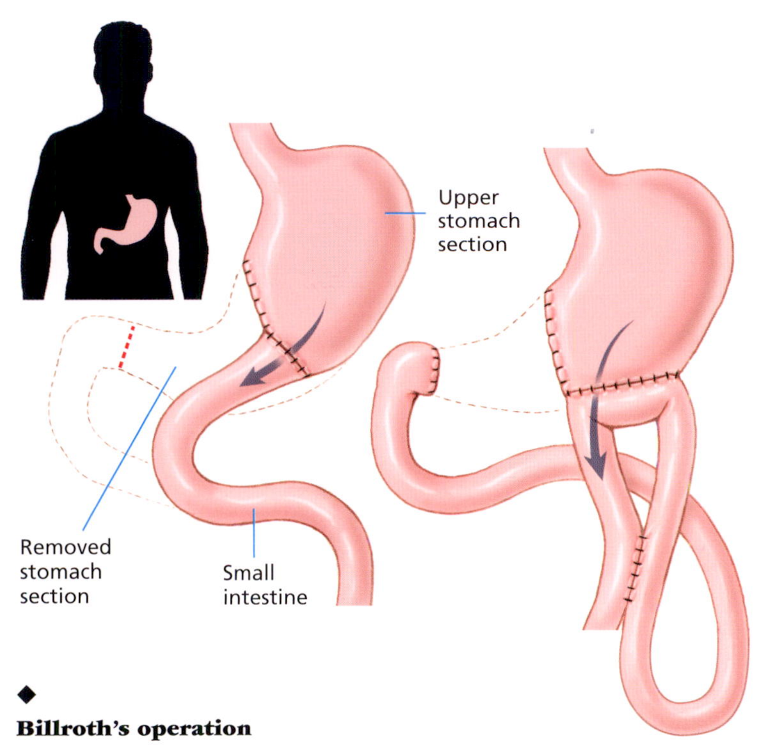

◆
Billroth's operation
If the lower section of the stomach needs to be removed, either the start of the small intestine is attached directly to the stomach remnant (left), or a lower intestinal loop is attached to the stomach remnant, and what was previously the beginning of the intestine is closed (right).

Billroth's operation

An Austrian surgeon, Theodor Billroth (1829–1894), developed two classical operations for gastric resection, Billroth I and II, which remain the basis of modern gastric surgery. Both are techniques for removing various amounts of the lower section of the stomach.

Gastric lavage

The washing of toxins or other harmful substances out of the stomach. To this end the stomach is filled with water several times via a tube inserted through the mouth (sometimes with the administration of medication as well) and emptied again. Gastric lavage is often carried out in cases of acute poisoning.

Pancreatic transplant

An organ transplant for a patient suffering from severe diabetes, particularly type I diabetes, whose pancreas is so severely damaged that it no longer produces insulin. A pancreatic transplant is usually accompanied by a kidney transplant, as diabetes also leads to severe kidney damage associated with the kidney's excretory function. It may enable the patient to become independent of insulin injections.

Artificial anus

An artificial intestinal opening (also known as an anus praeter or stoma), made during an operation in which the end of the intestine is sewn onto the abdominal wall; the contents of the intestine are drained off at this point, and an artificial intestinal opening is then fitted. An artificial anus may be made, for example, following surgery to remove intestinal tumors situated in direct proximity to the anal sphincter muscle. It is sometimes fitted on a temporary basis in order to relieve the intestine and allow healing in deeper sections of the intestine following surgery.

Enteroclysis (high enema)

The introduction of fluid into the colon via a tube inserted in the anus. This type of irrigation is generally carried out to clean the intestine prior to an operation.

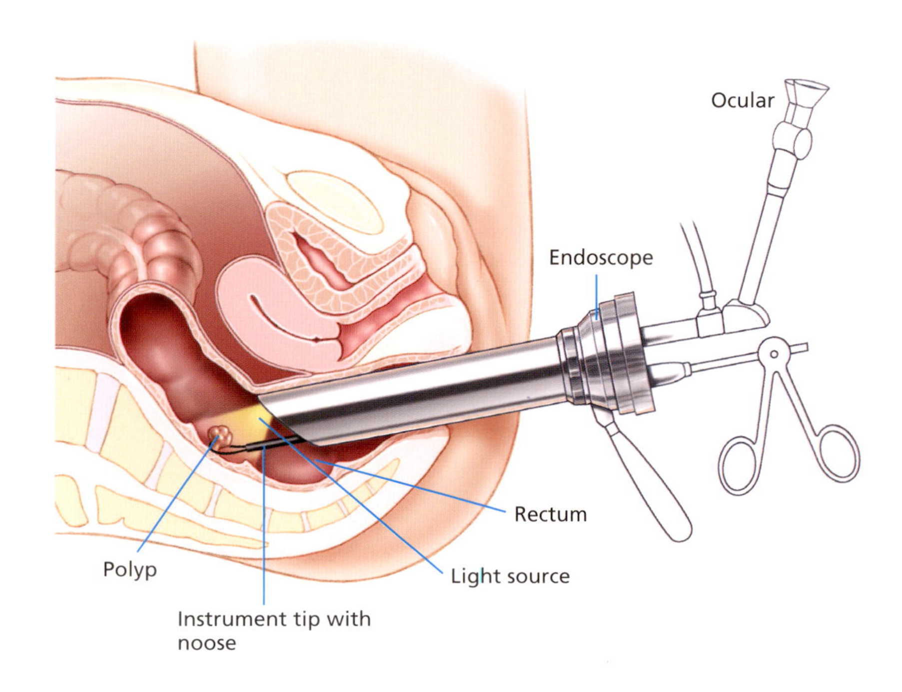

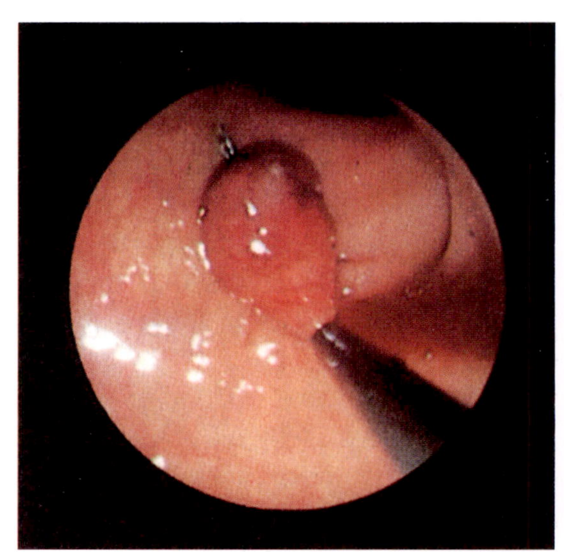

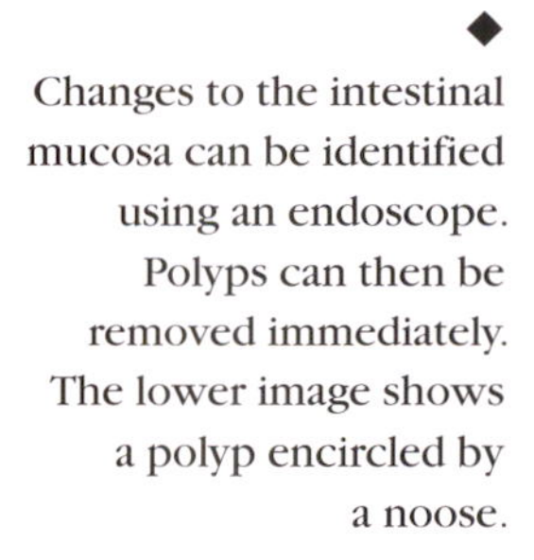

◆
Changes to the intestinal mucosa can be identified using an endoscope. Polyps can then be removed immediately. The lower image shows a polyp encircled by a noose.

Liver transplant

The transplant of an organ donor's liver into the body of a recipient. A liver transplant is mainly considered for patients with severe liver disease, such as cirrhosis, or with chronic liver inflammation. The number of possible donor organs available is insufficient to match the number of patients requiring an urgent liver transplant. However, the transplant of liver sections from living donors offers new hope for these patients. The liver is the only organ capable of regeneration. It is therefore able to regain full functionality even if up to two thirds of its tissue is removed.

Cholelitholysis

The dissolving of gallstones using medication (ursodeoxycholic acid), this is only likely to be successful in the case of small cholesterol stones free of calcium carbonate. The method of treatment is a protracted one because the medication needs to be taken daily over the course of a year. The method is used if the stones are not causing severe discomfort, or in patients whose general condition does not allow an operation to be performed.

Cholelithotripsy

Fragmentation of the gallstones without surgery using ultrasound waves administered from outside of the body. The gallstones are thus reduced in size and can be excreted via the intestine. The combination of cholelithotripsy and cholelitholysis is increasingly used as conservative therapy. This method has significantly fewer complications than surgery, but is only an option for stones with a specific composition and size, and when there are only a small number of stones.

Cholagogues

Agents that stimulate the production of bile and the emptying of the gall bladder. They primarily include fats, fatty acids, and bile acids. Medicinal plants such as celandine, artichokes, and wormwood have bile-stimulating effects. Specific hormones in the small intestine also stimulate the emptying of the gall bladder.

Endoscopic surgical methods

Endoscopic surgery (minimally invasive surgery) is now carried out in many specialist areas of medicine, usually under general anesthesia. Local anesthetics (e.g., low spinal anesthesia) may be selected for specific gynecological or orthopedic operations, or for surgery relating to the urinary tract or the sexual organs (urology).

For the removal of a diseased gall bladder, the endoscope is inserted through a trocar via a small incision at the navel. The operation is carried out following an initial diagnostic inspection and the gall bladder is removed from the abdomen via another trocar. Small incision wounds are all that remain. For intestinal operations, the endoscope is usually inserted via the intestinal exit. Polyps on the intestinal mucosa can be removed in this manner using a noose.

Kidneys and urinary tract

The human kidneys are vital organs: they filter toxins and substances the body is not able to utilize out of the blood and excrete these via the urine. They also regulate the mineral and water balance. The urine is stored in the renal pelvis, ureter, bladder, and urethra, until it is passed out of the body by micturition.

The two kidneys lie to the left and right of the lumbar spine, directly beneath the diaphragm. The bean-shaped organs are about 12 centimeters long, 6 centimeters wide and 4 centimeters thick, and each of them weighs around 150 grams. Each kidney is enclosed in a strong outer capsule and enveloped in connective and fatty tissue, both of which protect it from injury.

Structure

Two tissue layers are recognizable with the naked eye in a kidney that has been dissected lengthwise: the outer cortex and the horizontally striped renal medulla with its pyramid-shaped segments which open into the renal pelvis. This is where the urine is collected prior to being transported to the bladder via the ureter.
The kidneys receive an ample blood supply in order to fulfill their diverse tasks and also have a sophisticated vascular system. Each of the around one million renal corpuscules (glomeruli) in the renal cortex is supplied by a small artery which branches out to form a ball-like vascular network and exits again as an efferent artery. The blood from the efferent vessels of the renal corpuscules and the other kidney tissues collects in the renal vein and is channeled to the right side of the heart via the inferior vena cava.

Kidney function

The first phase of urine production takes place in the renal corpuscules. It is here that an aqueous filtrate is extracted from the blood flowing through the ball-like vascular network and transferred into the associated renal canal. The renal canals are surrounded by a network of very small blood vessels (capillary network) so as to enable the constant exchange of substances between the filtrate and the blood (second phase of urine production). The collecting tubules of the individual renal corpuscules form a canal system that conveys the urine into the renal pelvis, from which it passes, via the ureter, to the bladder.

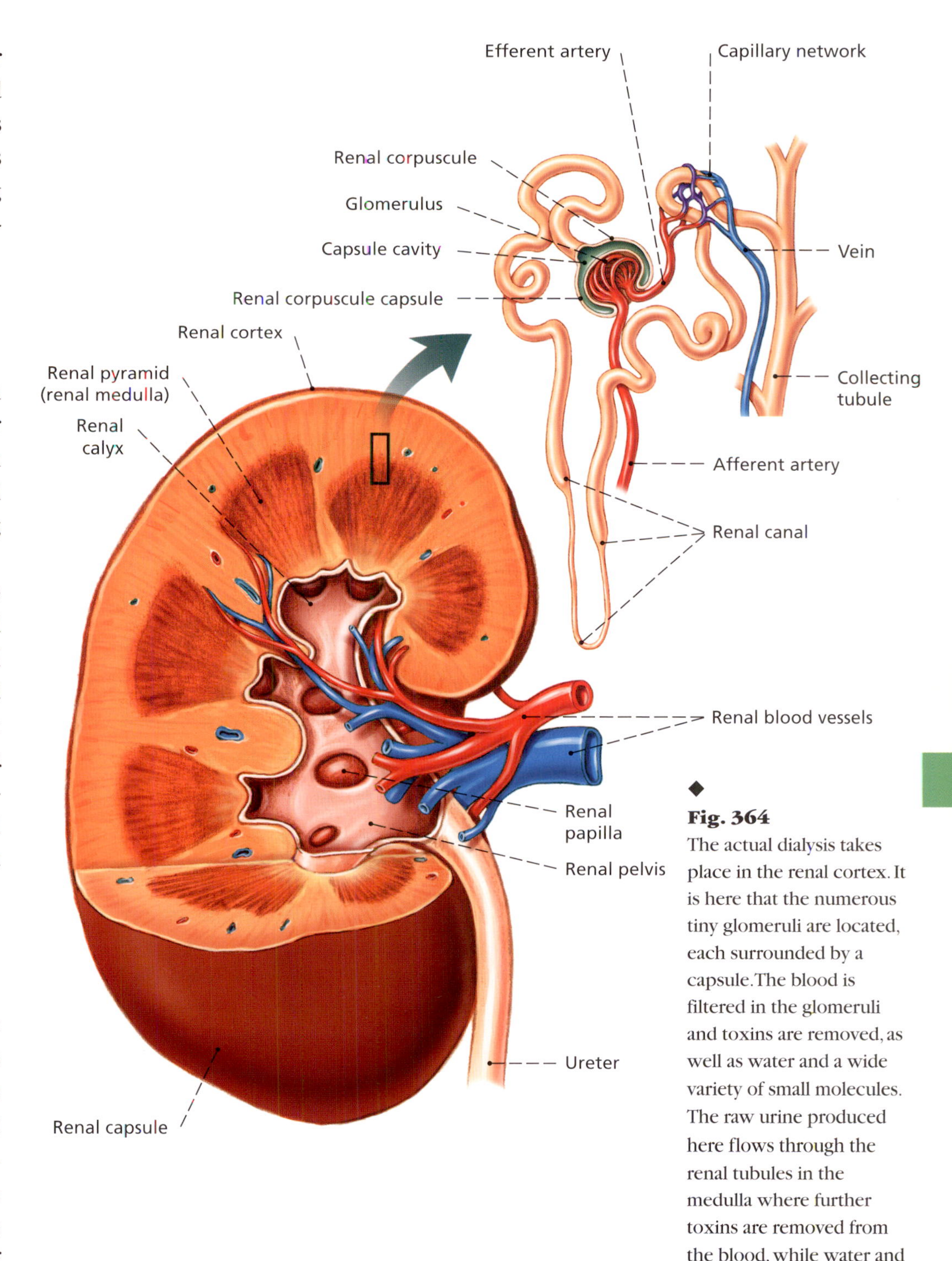

Fig. 364
The actual dialysis takes place in the renal cortex. It is here that the numerous tiny glomeruli are located, each surrounded by a capsule. The blood is filtered in the glomeruli and toxins are removed, as well as water and a wide variety of small molecules. The raw urine produced here flows through the renal tubules in the medulla where further toxins are removed from the blood, while water and some other materials are selectively returned to it, under the influence of renal hormones (see pages 365–366).

One of the kidneys' key tasks is to free the blood of metabolic waste products. These are primarily substances deriving from the breakdown of proteins, namely urea and ammonia. Toxins and medication are also excreted via the kidneys.

The renal corpuscules also filter substances the body can still utilize, such as protein and glucose. In order for these to be retained by the body they pass from the renal canals into the network of small blood vessels surrounding them. Urine therefore normally contains neither protein nor glucose.

The kidneys are also responsible for ensuring that the water content of the blood remains constant and that the blood is neither too concentrated nor too dilute. The kidneys therefore excrete more water in the urine in the event of too much water in the blood, making the urine pale yellow in color. Conversely, less water is excreted in the urine if the water content of the blood is too low; the urine then becomes very concentrated and dark yellow in color. This regulatory mechanism is controlled by a pituitary gland hormone, namely, antidiuretic hormone.

The kidneys regulate not only the water but also the mineral content of the body. They ensure that the concentration of the important minerals (sodium, potassium, calcium, and chloride) in the blood remains constant. If their concentrations drop, more minerals are returned to the blood vessels from the renal canals. If, however, their concentration increases, then the excess mineral salts are excreted with the urine.

Since every human being normally has two kidneys, the body is able to sustain the impairment or loss of one of these organs by the remaining kidney taking on the tasks of the diseased organ. The situation is far more serious, however, if both kidneys cease to function. In this case the blood needs to be cleansed regularly by an artificial kidney (kidney dialysis). A person suffering the loss of both kidneys can receive long-term help by the transplantation of a donor kidney, which enables them to live a normal life.

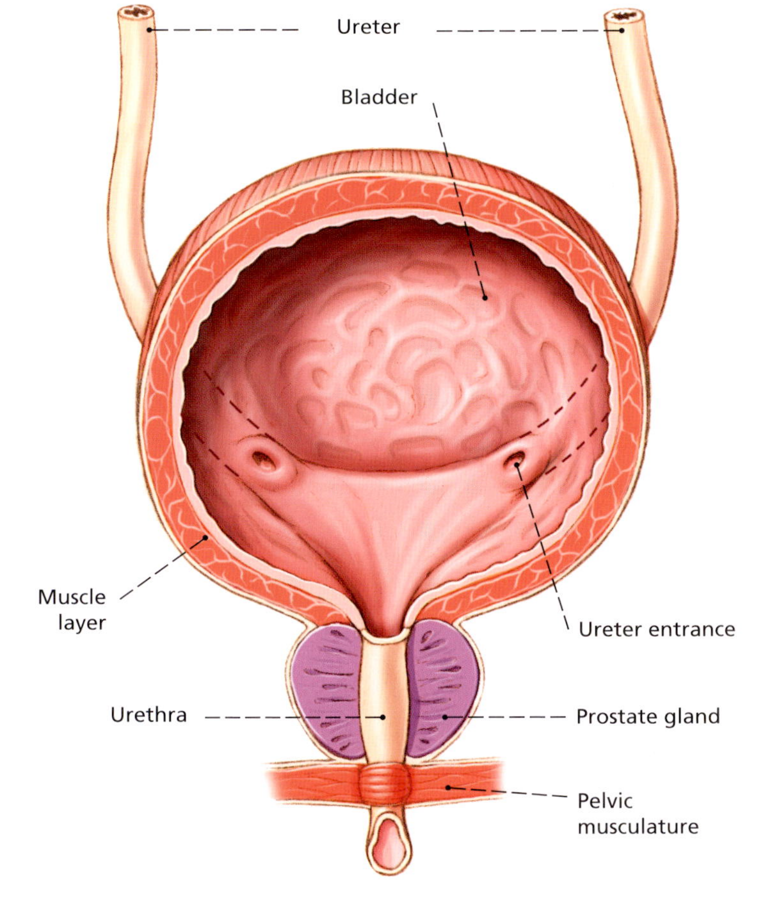

Fig. 365
The bladder wall contains a thick layer of muscle and the urine drains out when this muscle contracts. In men, the prostate gland is located at the lower pole of the bladder.

The urinary tract collection system

The urine is conveyed from the renal pelvis via the ureter to the bladder. The two ureters are tubes measuring about 30 centimeters long with a diameter of approximately 2–3 millimeters. Their walls are made up of muscle and they convey the urine in the direction of the bladder by means of rhythmic movements. A valvular mechanism at the entrance to the bladder ensures than the urine is able to flow only in one direction: from the ureter into the bladder.

The bladder is a hollow organ with a wall of muscle fibers, the inside of which is lined with a mucous membrane. It has a capacity of about 750 milliliters. When the bladder is full, the resultant pressure triggers the urge to empty the bladder, a process that can be consciously controlled.

The urine flows out of the bladder via the urethra. The latter is considerably shorter in women than in men, such that pathogens are more easily able to "ascend" into the female bladder, causing urinary tract infections.

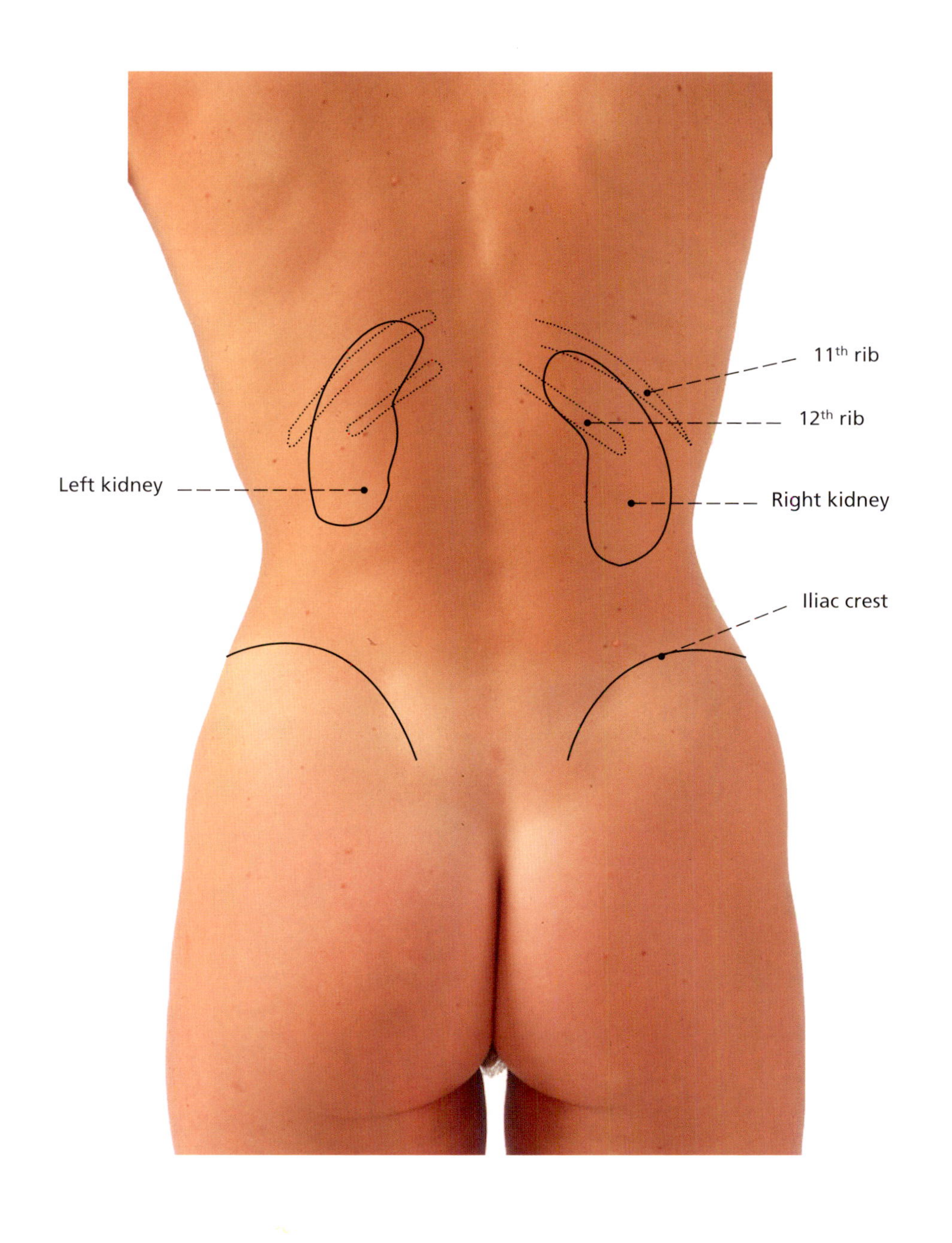

Fig. 366
The position of the kidneys. The outlines have been drawn on the skin. The right kidney is located about 2–3 centimeters lower than the left kidney.

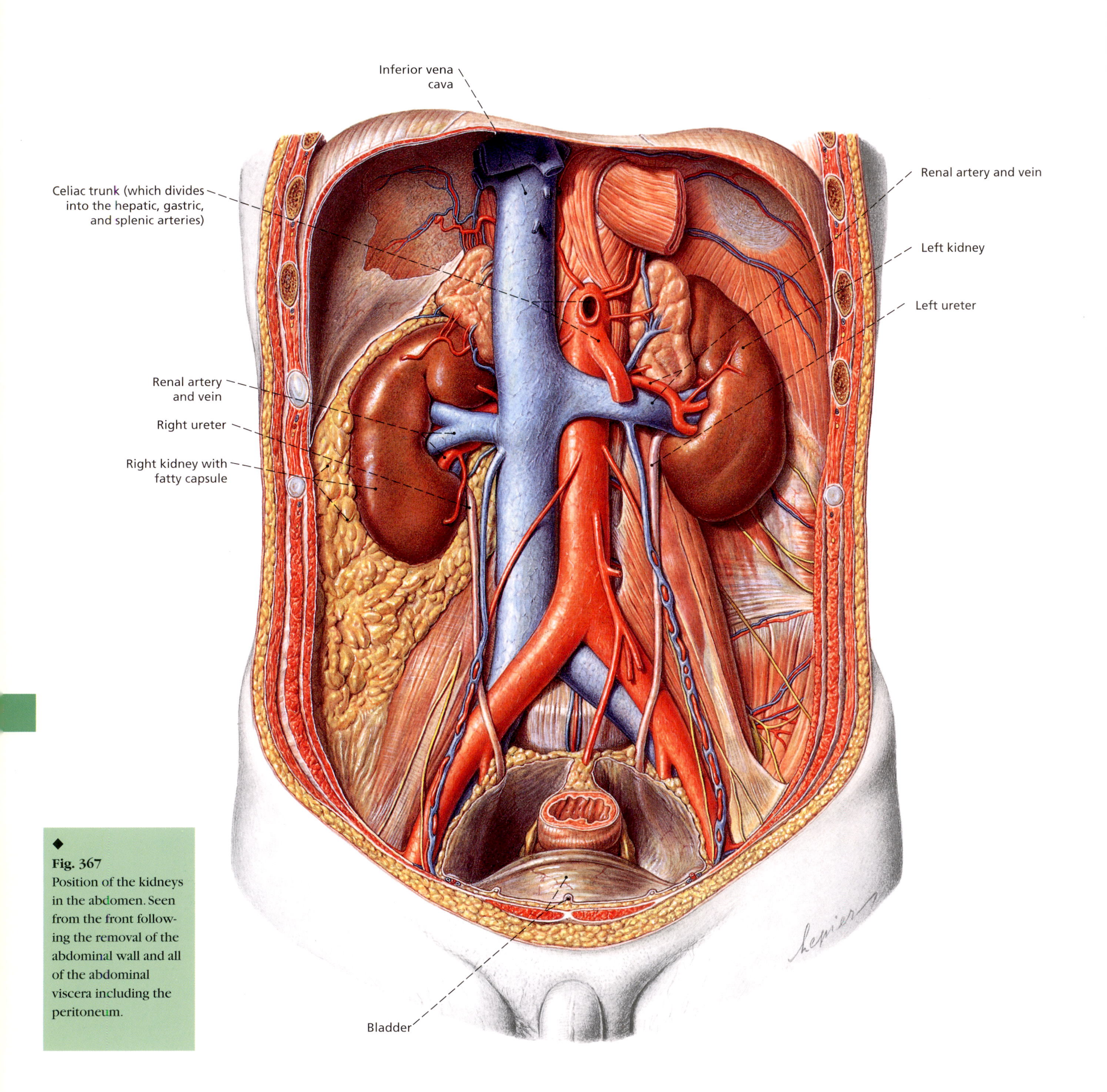

◆
Fig. 367
Position of the kidneys in the abdomen. Seen from the front following the removal of the abdominal wall and all of the abdominal viscera including the peritoneum.

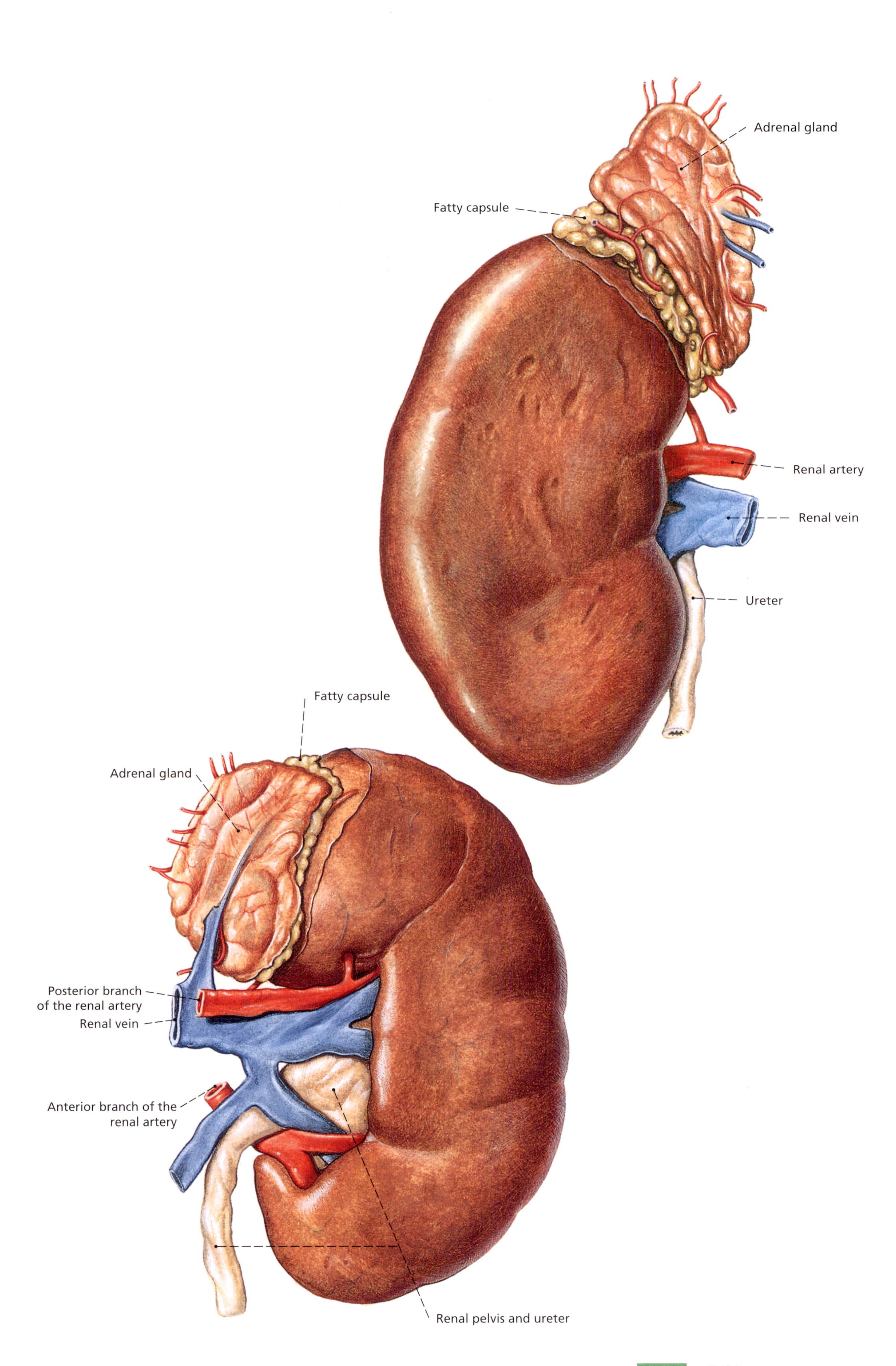

◆
Fig. 368
The right kidney, seen from the front. The adrenal gland is located on the upper renal pole like a capsule. The renal artery and vein, as well as the ureter, enter the kidney on the convex side.

◆
Fig. 369
The left kidney from the front. Part of the renal pelvis is visible underneath the arteries and veins.

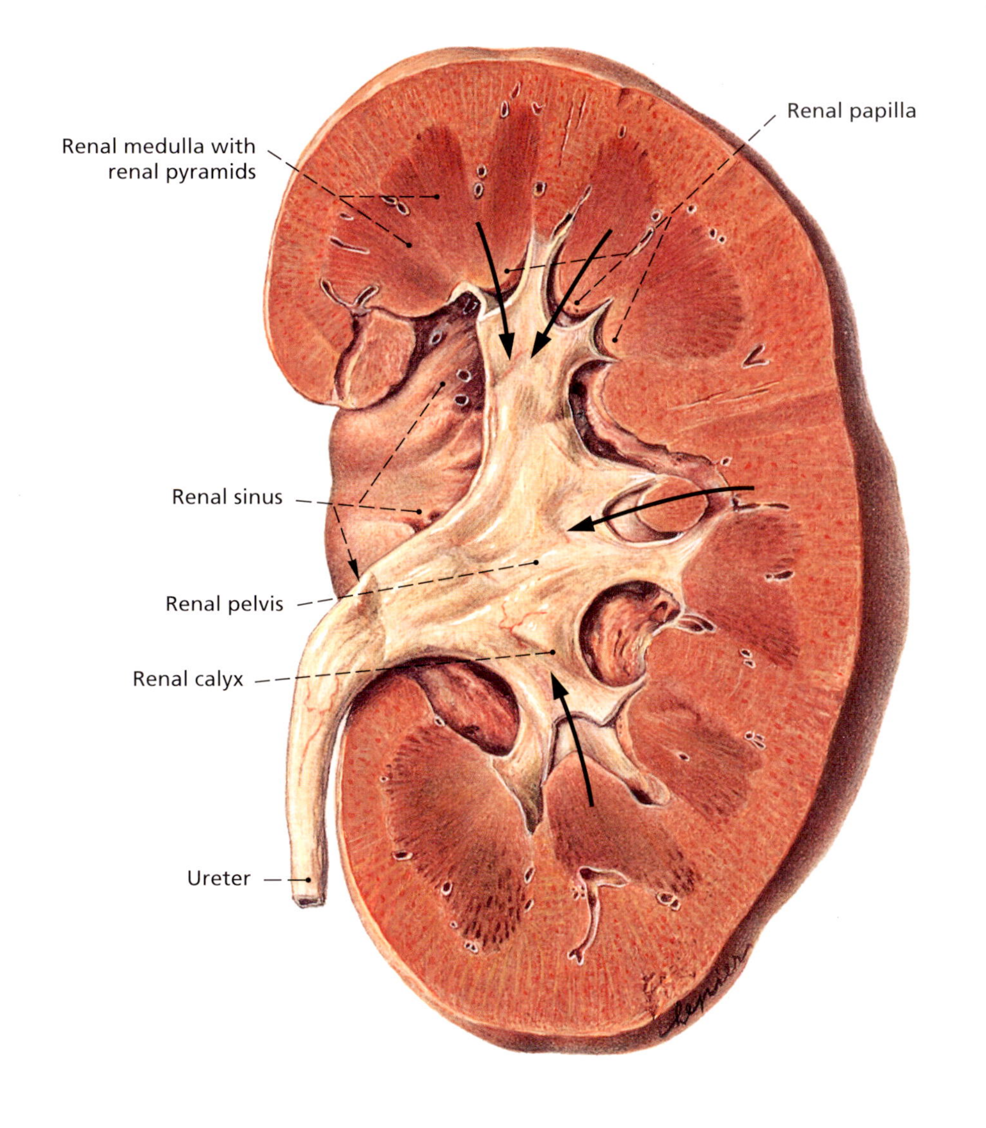

◆
Fig. 370
Longitudinal section of the left kidney with the renal sinus following the removal of blood vessels and fatty tissue, seen from the front. The pyramids extend into the renal calyces through which the urine passes into the renal pelvis (arrows).

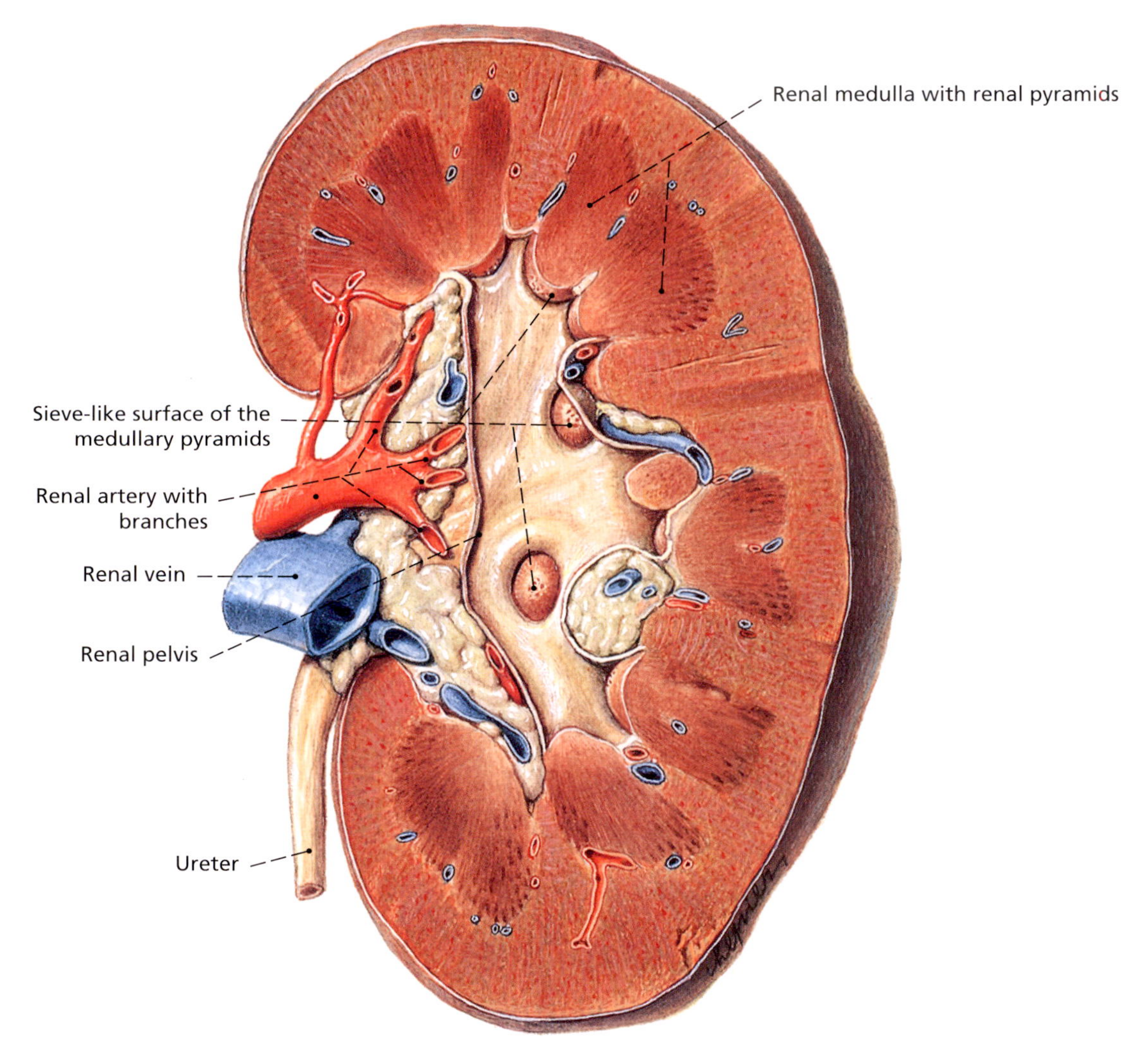

◆
Fig. 371
Longitudinal section of the left kidney, seen from the front. The renal pelvis has been opened. The renal artery and vein have been truncated. The renal pelvis is lined with a shiny mucous membrane into which the medullary pyramids extend.

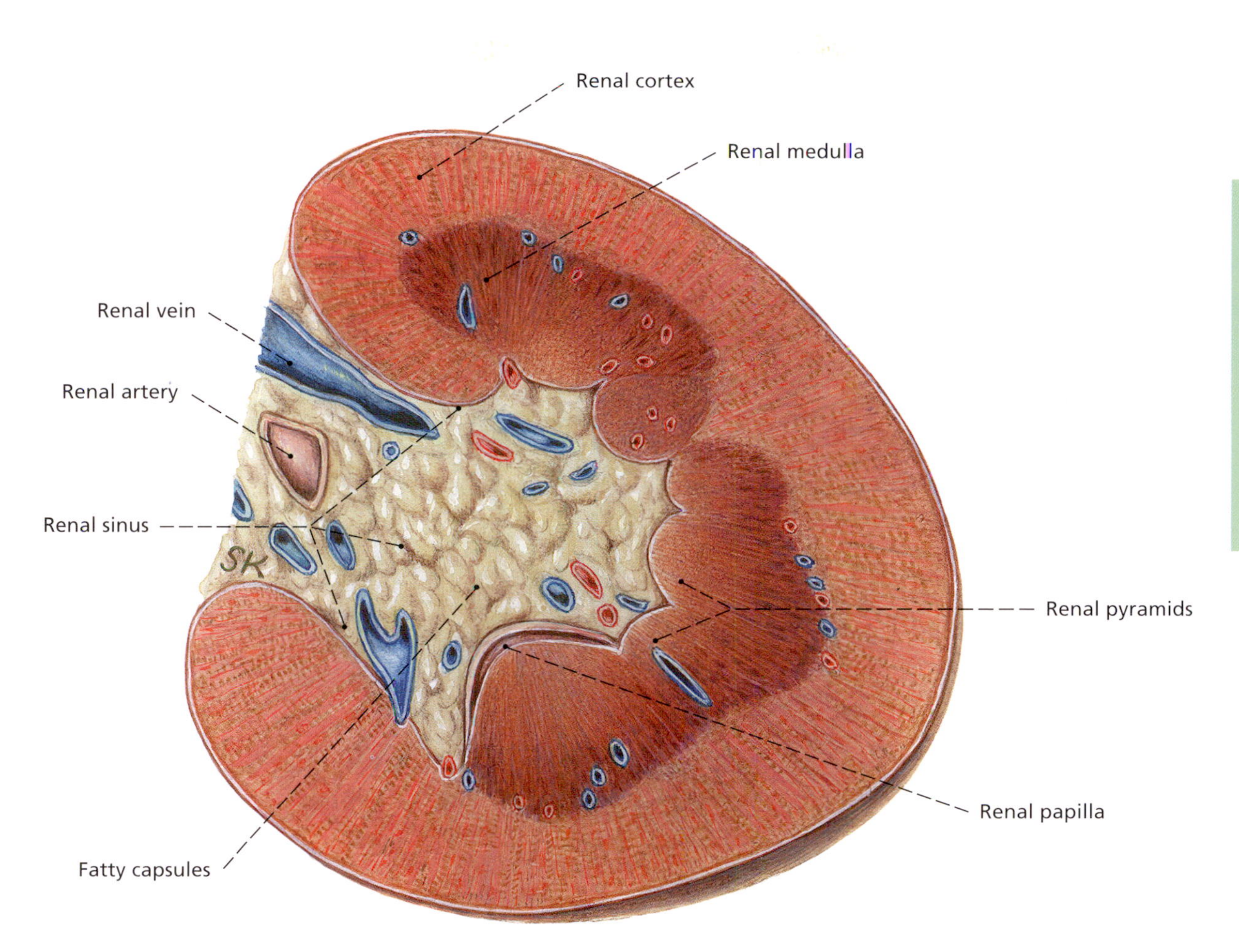

◆
Fig. 372
Cross-section of the left kidney. View from below to depict the fatty capsule of the renal sinus which overlies the renal pelvis.

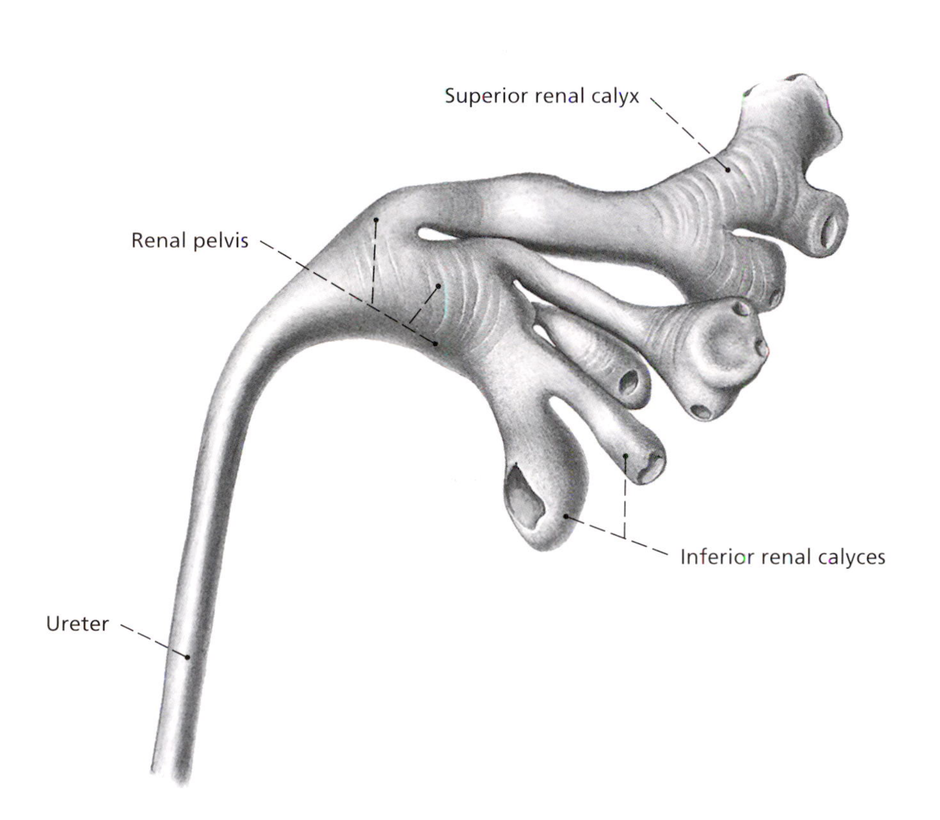

◆
Fig. 373
The renal pelvis with the renal calyces.
A plastic substance has been injected to better depict the structures.

◆
Fig. 374
An adult right kidney, seen from the front, in which the lobe structure of the early embryonic phase has been retained. A rare deviation from the norm.

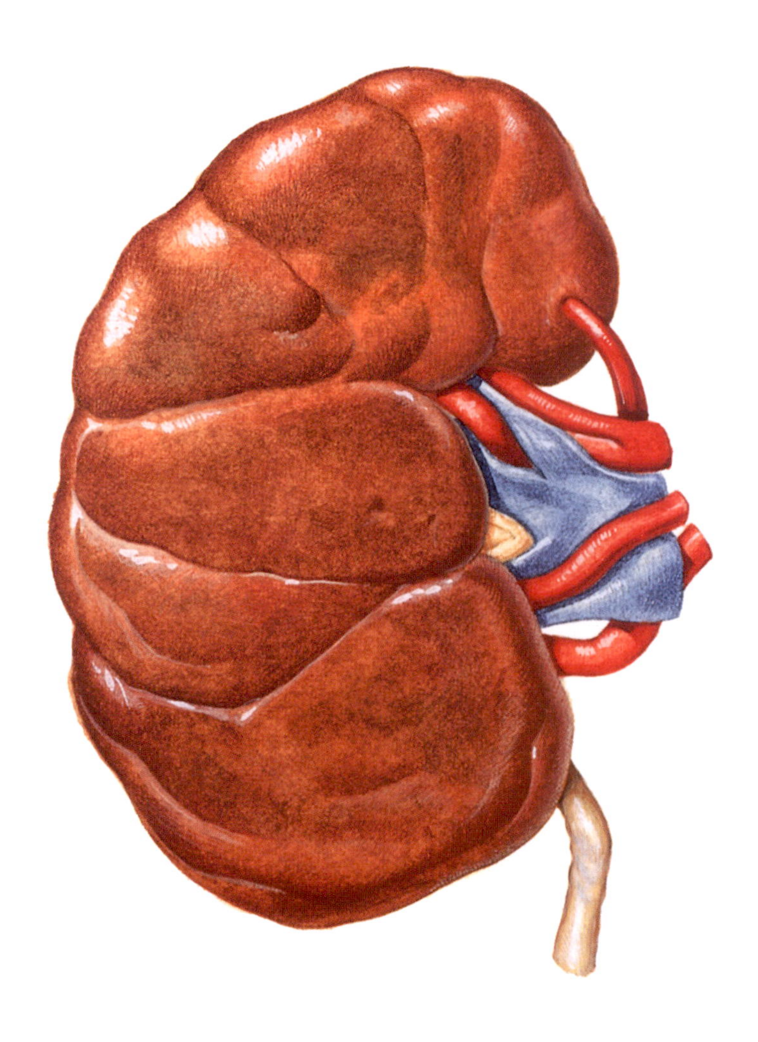

◆
Fig. 375
A so-called horseshoe kidney. A rare deformity in which both kidneys are joined at the lower pole.

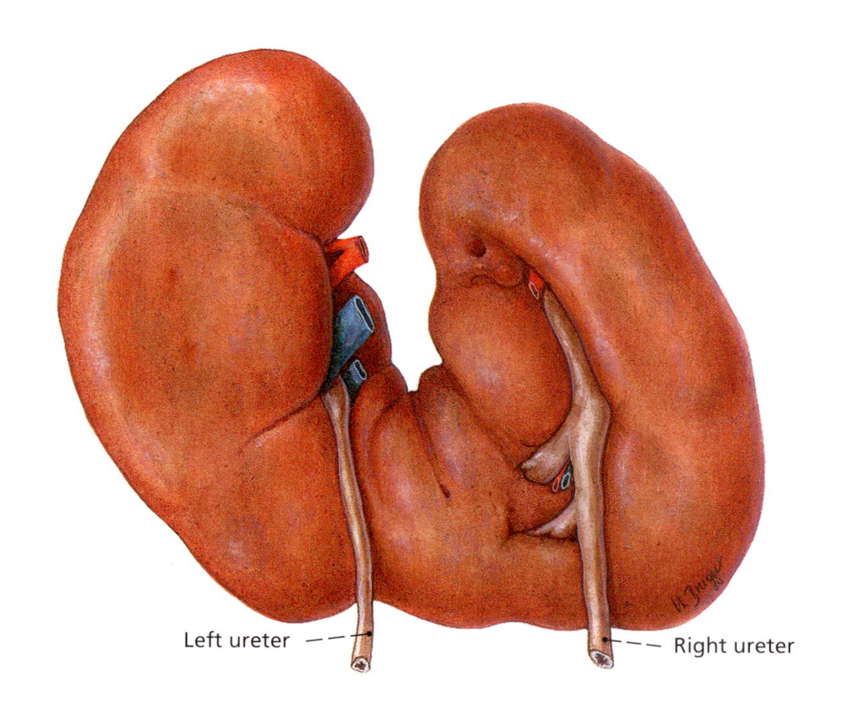

◆
Fig. 376
Depiction of the vascular tree comprising the renal arteries and veins as well as the renal pelvis. Different colored plastics have been injected into the renal vessels (arteries = red, veins = blue, renal pelvis = yellow). Right kidney, viewed from the front.

◆
Fig. 377
Depiction of the renal arteries and the renal pelvis. Different colored plastics have been injected into the renal artery (red) and renal pelvis (yellow). Right kidney, viewed from the front.

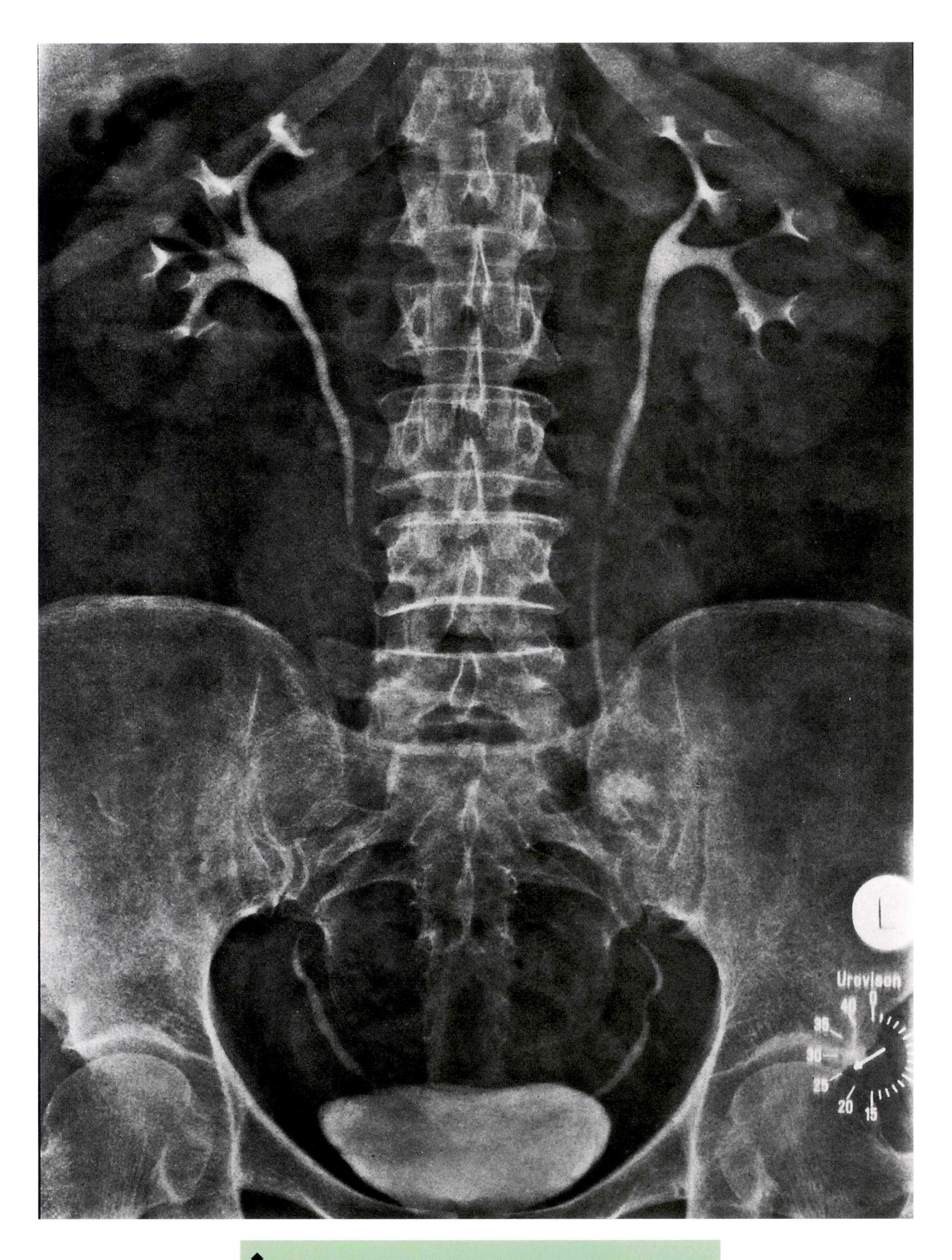

◆
Fig. 378
X-ray image of the urinary tract (renal pelvis and ureter) following the injection of a contrast agent, excreted via the kidneys, which causes the urinary tract to appear white (see detail accompanying the diagram on page 357).

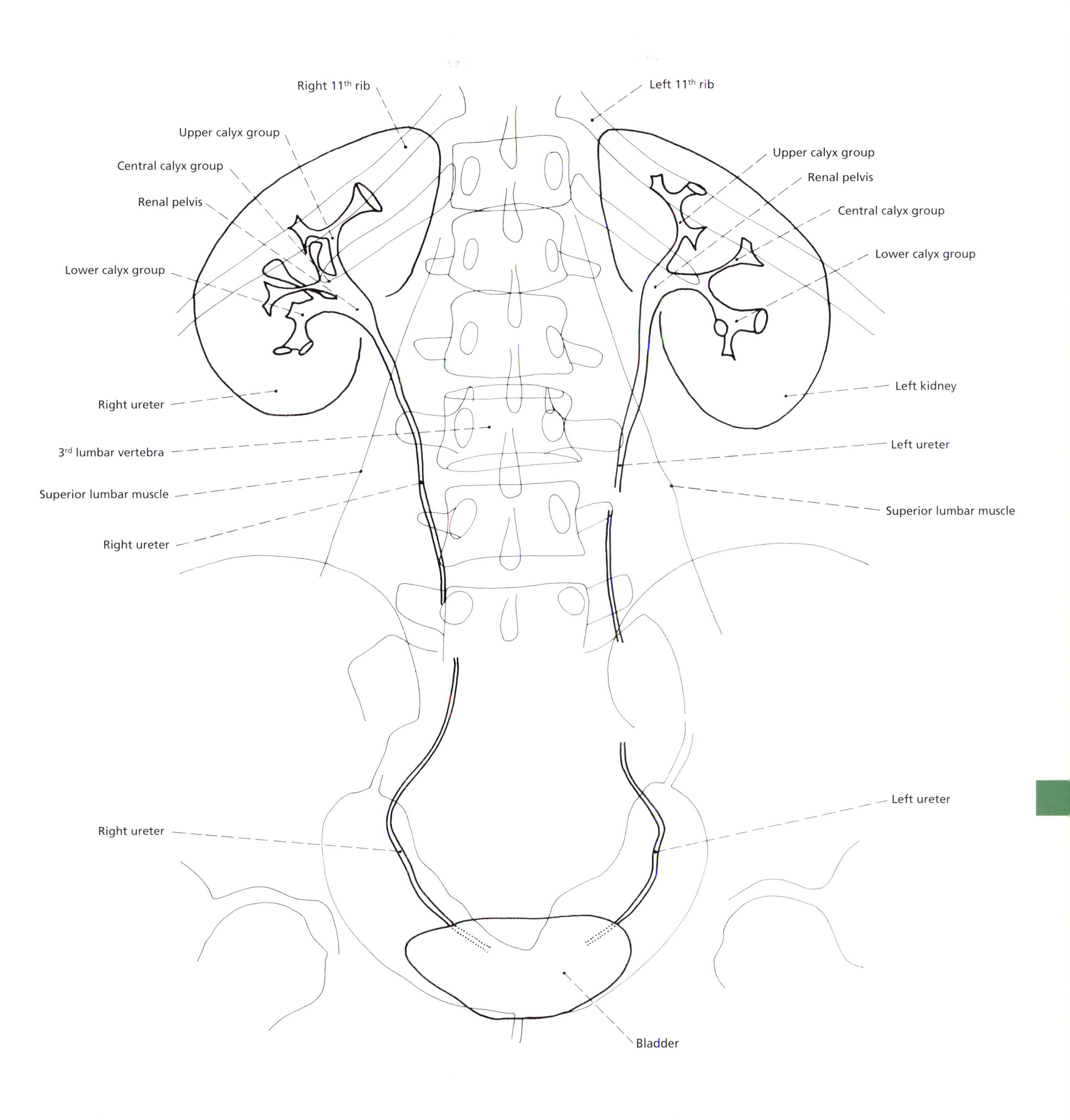

◆
Fig. 379
The urinary tract. Schematic representation of the image on page 356.

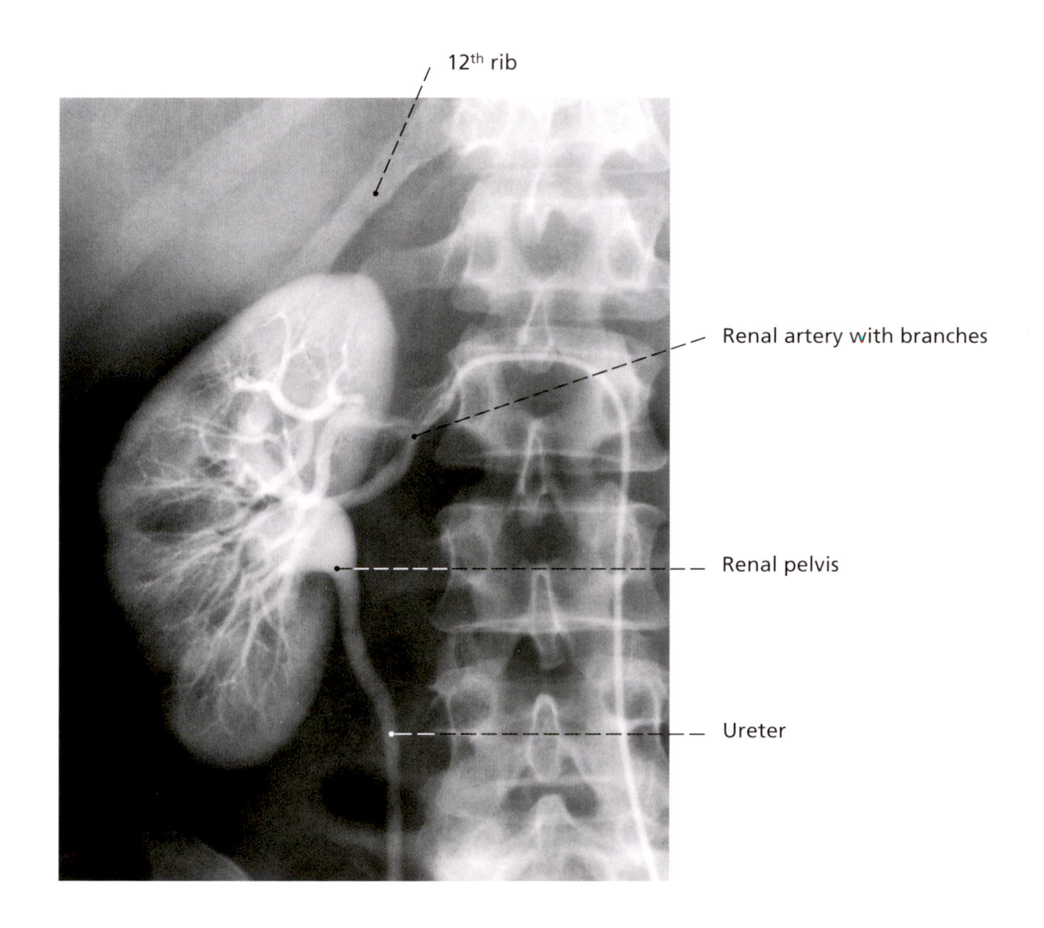

◆
Fig. 380
X-ray image of the right kidney showing the renal pelvis and tissue as well as the ureter. The depiction of the renal tissue is made possible by the injection of a contrast agent which accumulates in the urinary tract. Another contrast agent was simultaneously injected into the renal artery via a catheter in order to make the kidney's arterial vascular tree visible.

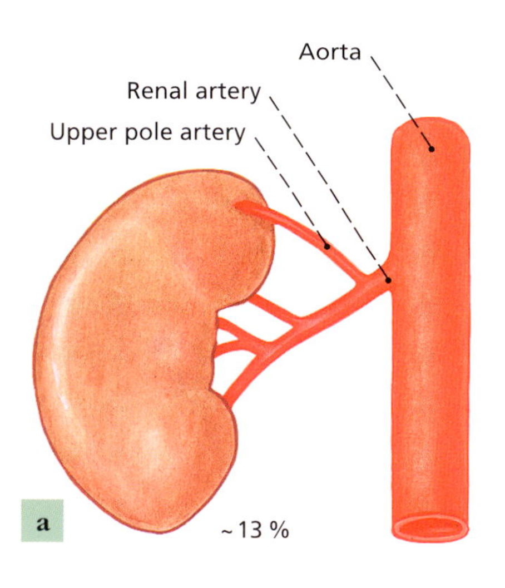

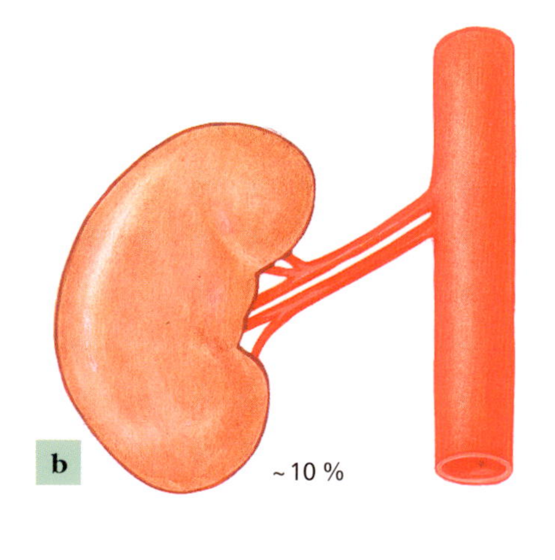

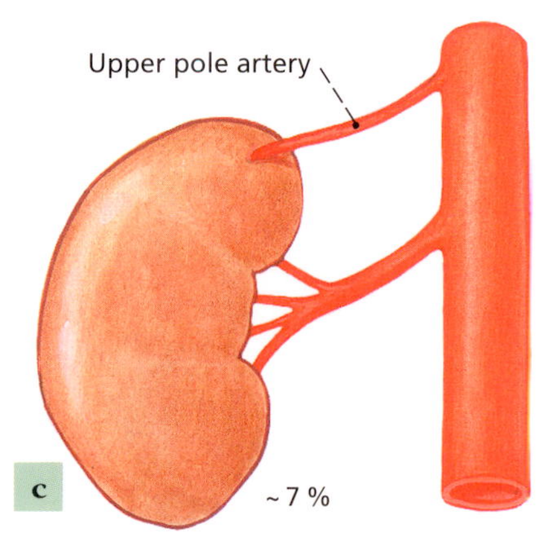

◆
Fig. 381
Deviations from the norm* in the arterial vessels supplying the kidney: (a) a renal artery with a so-called upper pole artery; (b) two renal arteries; (c) two renal arteries, one of which enters at the upper pole; (d) two renal arteries, one of which enters at the lower pole of the kidney.

* There is normally only one renal artery emanating from the aorta. Its branches enter the kidney on the convex side.

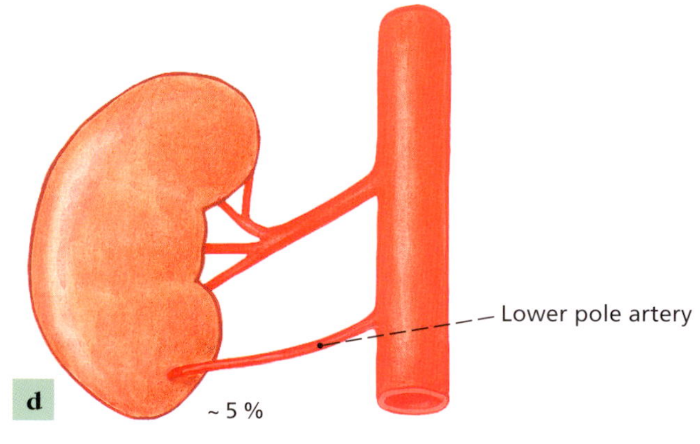

Symptoms and diseases

The kidneys

• ***Nephritis***

The collective name for acute or chronic kidney diseases, in which inflammatory changes occur in the smallest glomeruli in the kidney, namely the renal corpuscules, or in the connective tissue.
In the case of **acute glomerulonephritis**, inflammatory changes occur in the microfilters of the kidney glomeruli. These are usually a reaction to an earlier bacterial infection in another part of the body, and cause damage to the filter membrane in the glomeruli. Sufferers are usually children and teenagers under the age of 16. The disease begins suddenly with aching limbs as well as swelling of the joints and the face, especially the eyelids. There is usually an increase in blood pressure and impaired vision may occur as well as cramps, although these symptoms are less frequent. The quantity of urine is often significantly reduced and the urine is often cloudy due to the excretion of proteins. It is sometimes red in color due to the presence of red blood cells. Bed rest is required, and the intake and excretion of fluids must be closely monitored. Medicinal treatment leads to an improvement of the situation in the majority of cases. If the kidneys temporarily cease to function, the patient will need dialysis by means of an artificial kidney.
Chronic glomerulonephritis is often not detected at first. Eventually it manifests itself by increased protein in the urine, high blood pressure, the accumulation of tissue fluids, or a myocardial insufficiency. A low-sodium diet and a reduction of blood pressure are the treatment priorities; at an advanced stage regular dialysis is required or, in the event of increasing kidney failure, a kidney transplant may be required.
Pyelitis is an acute or chronic inflammation of the area of the kidney in which the urine collects before it is conveyed to the bladder via the ureter and ultimately emptied via the urethra. If this drainage is hindered at any point, e.g., due to a tumor, an enlarged prostate (men), or pregnancy (women), pathogens may enter the renal pelvis, where they can cause infections. Women are especially affected because they have a shorter urethra with a relatively wide opening, making it easier for germs to enter. The symptoms of acute pyelitis include fever, sometimes accompanied by chills, severe pains in the kidney region, and pain when urinating. The urine is usually cloudy, contains bacteria, and often also protein. The infection easily spreads from the renal pelvis to the renal tissue. Treatment comprises bed rest, plenty of fluids, and antibiotics.

• ***Renal gravel***

A large number of small fragments visible with the naked eye. They are usually a precursor to kidney stones.

• ***Kidney stones***

Urine salt deposits which agglomerate in the renal calyces and in the renal pelvis, initially as renal "sand" or "gravel", but ultimately forming large stones. A kidney stone can block the ureter once it has reached a specific size, causing a urinary obstruction with significant pain that can extend into the inguinal region. The urine is usually red in color due to the presence of blood. Kidney stones are especially likely to occur as a result of urinary tract infections and metabolic disorders such as gout.
Kidney stones are detected using X-ray images or ultrasound examination. The treatment comprises attempts to dissolve the stones with medication or to wash them out by drinking large quantities of fluids. If these measures are ineffective, fragmentation of the kidney stones can be carried out or, if this is unsuccessful, they can be removed surgically.

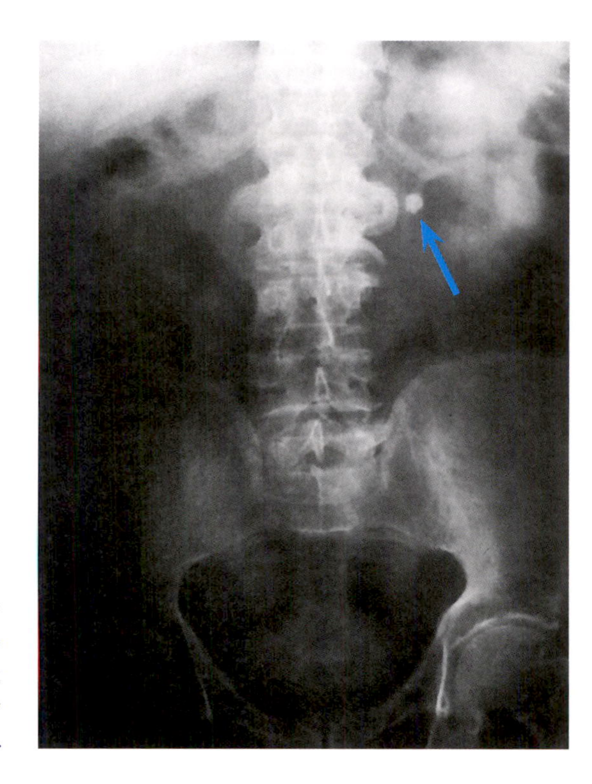

◆ **Kidney stones**
Visible as a white dot (arrow) on the X-ray image.

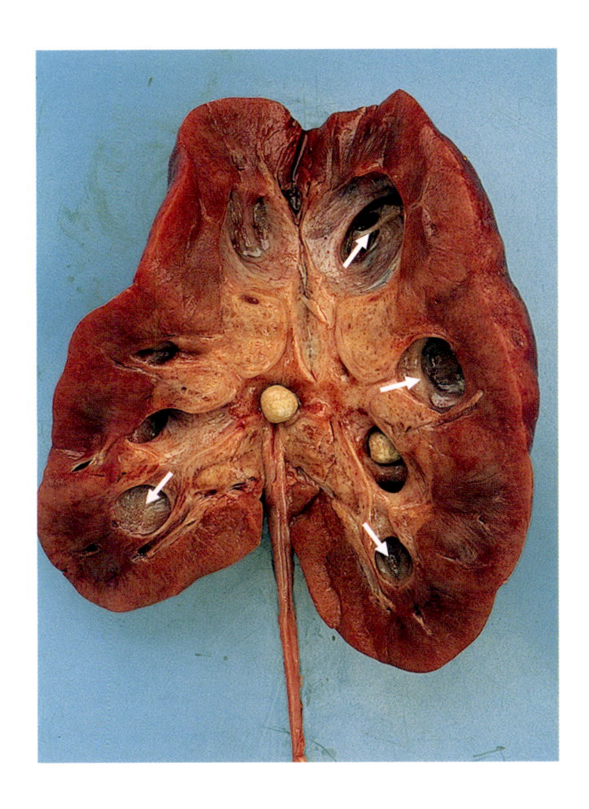

◆
Kidney stones
A dissected kidney in which kidney stones have formed. The reflux caused by the kidney stones leads to the expansion of the renal calyx system (arrows).

• ***Renal colic***

Spasmodic, usually severe, cramp-like pains occurring in the kidney region and which can extend to the bladder, sexual organs, and thigh. Almost always caused by kidney stones obstructing the drainage of urine and leading to a build-up of urine in the kidneys, with painful swelling. Renal colic can also be the result of circulatory disorders caused by the closure of a renal vessel.
It is treated with painkillers and anti-cramp medication. An attempt is made to flush out the kidney stones through the intake of large quantities of fluids. If this is unsuccessful, the stones are fragmented or removed surgically.

• ***Renal artery stenosis***

Constriction of a renal artery due to a blood clot, arteriosclerotic changes to the vessel wall, kinks in the renal artery, or tumors constricting or closing the blood vessel. Possible aftereffects include high blood pressure and restricted kidney function.
Renal artery stenosis can be detected by renal angiography. Depending on the cause, treatment can comprise administration of blood-thinning medication or surgery.

• ***Renal cancer***

Malignant tumor in the renal tissue, usually occurring in men over the age of 50, but which can also occur in children (nephroblastoma or Wilms' tumor). The risk of renal cancer is increased by the long-term intake of large quantities of painkillers and by smoking.
Blood in the urine is often the first sign of renal cancer, and may occur without the pain normally associated with kidney and bladder infections. Pains in the kidney region and high blood pressure can develop later.
The chances of recovery are good providing that the tumor is detected in good time. Since in most cases only one kidney is affected, the diseased organ can be removed completely without significant loss of function.

• ***Renal insufficiency***

A reduction in kidney performance through to complete kidney failure. Causes can include chronic kidney diseases, the calcification of the smallest renal vessels (resulting in severely restricted blood supply to the organ), or a blockage of the urinary tract collection system, e.g., by kidney stones or tumors. Reduced renal performance leads to a higher concentration of urea, uric acid, and creatinine in the blood, these metabolites normally being excreted via the kidneys. The quantity of urine is reduced. In the advanced stage, urine can no longer be produced at all and uremia is the result. The treatment consists initially of a diet adapted to the patient's individual metabolic state, taking into account the need to keep levels of proteins, electrolytes (potassium and sodium), and fluids within normal limits.
If kidney function ceases altogether then dialysis by means of an artificial kidney remains unavoidable. If the organ does not recover, the patient will need to undergo dialysis on a regular basis and a kidney transplant may need to be carried out.

• ***Acute kidney failure***

Sudden loss of kidney activity, usually as the result of circulatory shock in the case of severe blood loss, poisoning, or a pulmonary embolism. A disease of the kidneys themselves, such as acute nephritis, is seldom the cause.
The production of urine and therefore the excretion of metabolic toxins are reduced as a result of the loss of kidney function and fluid accumulates in the tissues. Dialysis is often required until such time as the kidney function has normalized again. If this does not occur, a kidney transplant will need to be considered.

The ureter

• ***Ureteritis***

Inflammation of the ureter caused by an infection, toxic urine components, or mechanical irritation. Mechanical irritation is usually caused by a kidney stone having entered the ureter. Infections usually derive from a bladder infection.

• ***Ureteral calculus***

A kidney stone that has entered the ureter. Large ureteral calculi often cause severe colic extending to the kidneys, testicles, or pudendal lips, as well as blood in the urine. In the long term there is a risk of infection and congestion damage in the kidneys due to the obstruction. About 80% of all ureteral calculi disappear on their own. The remaining 20% need to be removed with stone-dissolving and expelling medication or by surgery.

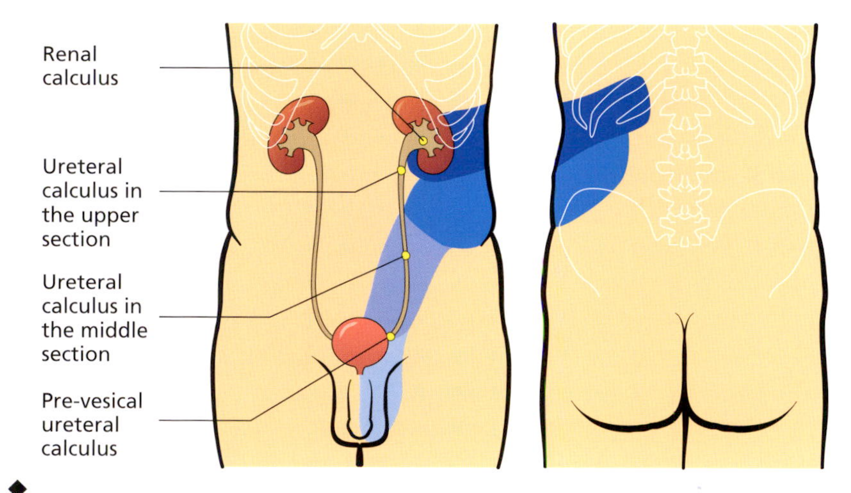

◆
Stone formation
Stones can become lodged in various parts of the body. The areas colored blue indicate the respective pain localizations.

• ***Ureteral constriction***

Congenital or acquired constriction of the ureter. Acquired constrictions can develop as a result of chronic inflammation of the ureter, or as a rare complication relating to the radiotherapy of genital tumors. A ureteral constriction encourages the build-up of urine as well as inflammation of the kidneys and should therefore be treated surgically.

The bladder

• ***Cystitis***

Inflammation of the bladder due to bacterial infection, ongoing incomplete micturation, or following surgery, especially in the case of micturation via a catheter.
Women suffer from bladder infections more frequently than men because their shorter and wider urethra facilitates the entry of germs. The symptoms of an acute bladder infection include burning and/or sharp pains when urinating; a frequent urge to urinate but at the same time reduced micturation; a general feeling of being unwell; and pains in the lower abdomen. These pains vary in severity, can be cramp-like, and are sometimes accompanied by fever. The same symptoms in a milder form occur in the case of a chronic bladder infection.

Treatment usually requires the intake of plenty of liquids (including special teas) in order to wash out the bladder. Local warmth (e.g., hot baths) may be effective. In persistent cases, treatment with antibiotics may be required.

• ***Vesicovaginal fistula***

A passage developing in women between the bladder and the vagina causing urinary incontinence with constant involuntary micturation. It can be caused by surgery or radiation therapy, undetected injury to the bladder, or by circulatory disorders. It is only rarely congenital and requires surgical correction.

• ***Urinary incontinence***

Weakness of the bladder sphincter, leading to an inability to retain urine. It can be improved by pelvic floor exercises in the initial stages, but bladder surgery may be required in advanced stages.

• ***Bladder cancer***

A malignant tumor on the inside wall of the bladder which rapidly breaks through the bladder wall and forms metastases, initially in the surrounding lymph nodes and later in the liver, lungs, bones, and peritoneum. The suspected causes are excessive nicotine intake and chronic bladder infections, as well as job-related contact with aniline.
Diagnosis is via cystoscopy. In the early stages the tumor can either be removed during the cystoscopy or destroyed by means of diathermy. Regular check-ups are required following removal. If the malignant tumor has spread, it will need to be irradiated and the bladder removed.

The urethra

• ***Urethritis***

Inflammation of the mucous membrane of the urethra, sometimes extending into the deeper tissue layers. The causes can include chemical or mechanical irritation, bacterial, fungal, viral, or parasitic infections, or allergic reactions. The symptoms are pain when urinating, disrupted urinary flow, and an abnormal discharge.

• ***Urethral stenosis***

Too narrow a urethra is usually congenital but can sometimes be acquired (e.g., through chronic inflammation). It favours the development of urinary tract infections and urinary obstruction. It must be widened using special distension instruments.

Frequent urge to urinate

A healthy human being empties their bladder an average of four to six times during the day, and at night only in exceptional cases. If somebody needs to urinate much more often than this, it is referred to as urinary urgency. This is normal when drinking large amounts, particularly when drinking generous quantities of beer, but can also be indicative of a bladder infection or of a so-called irritable bladder. It is occasionally symptomatic of kidney or prostate disease. Diabetes should also be considered if it occurs in combination with excessive thirst. The urge to urinate can also have mental causes.

Urinary incontinence

Involuntary urination (enuresis). The cause is often an impairment of the nerves or muscles responsible for the bladder-closing mechanism; incontinence is often associated with an excessive urge to urinate. Older people are generally more affected. Urinary incontinence is often triggered by strong, sudden physical effort (e.g., sneezing), and sometimes also occurs as a temporary accompaniment to urinary tract infections. In many women, urinary incontinence caused by a prolapse of the pelvic floor (following repeated pregnancies) can be alleviated by muscle-strengthening pelvic exercises. Urinary incontinence as a result of an irritable bladder can be improved with special medication (anticholinergics). Overflow incontinence, characterized by the constant dribbling of urine, usually requires surgery to remove a urinary calculus or enlarged prostate obstructing the flow.

Urinary obstruction

Build-up of urine in the urinary tract caused by an obstruction to the outflow of urine. This can be due to constrictions (congenital or acquired), or obstruction, e.g., urinary calculus, tumors, injury or chronic inflammation of the urinary tract. The ongoing build-up of urine leads to overexpansion of the urinary tract sections, encourages urinary tract infections, and damages the kidneys. Complete urinary obstruction leads to a total inability to urinate spontaneously. In this case an emergency procedure is necessary to drain off the urine via a bladder catheter.

Hydrophrenosis

Pathological change to the kidneys as a result of ongoing urinary obstruction. This can be determined by means of special X-ray examinations. Hydrophrenosis often has no symptoms for a long time. A dull feeling of pressure later develops in the kidney region and blood is occasionally present in the urine. At this stage, however, the damage is usually no longer completely curable which is why any suspicion of urinary obstruction should be followed by further investigation.

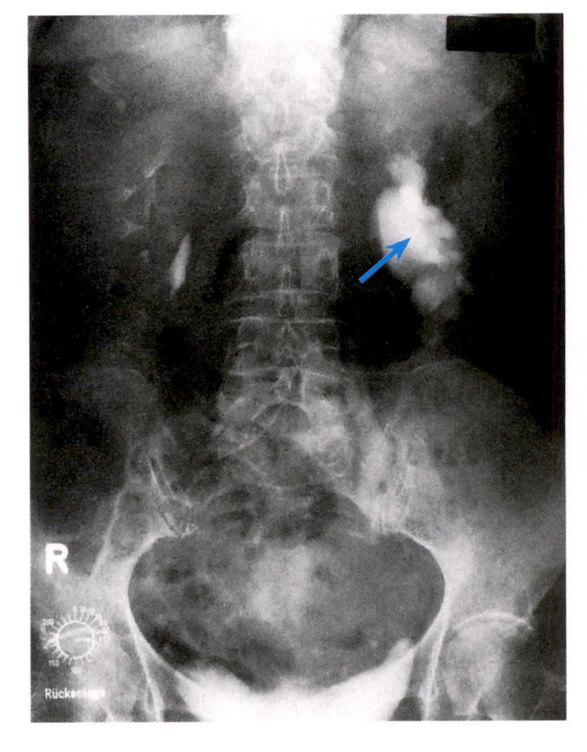

◆ **Hydrophrenosis**
In an X-ray examination, the enlarged renal pelvis is clearly visible in a kidney with hydrophrenosis (arrow) as the lighter section.

Uremia

An increase in urea and other toxins in the blood, which should normally have been excreted in the urine, caused by advanced renal insufficiency. Patients with uremia exhibit severe signs of being unwell with increased blood pressure, heart and breathing problems, spasmodic cramps and ultimately unconsciousness. Uremia can result in death if left untreated.
Depending on the kidneys' ability to recuperate, the treatment consists of dialysis on a temporary or ongoing basis. Irreversible kidney damage requires a kidney transplant.

Blood in the urine

The presence of blood in the urine, which is either visible or which can only be detected by chemical diagnostic procedures (e.g., using test strips). Causes can be tumors in the kidneys, renal pelvis, ureter, bladder, or prostate; stones in or inflammation of the bladder and kidneys; injury to the urinary organ and tract; or renal cysts.
Colorants ingested with food, such as beetroot, can also cause the urine to appear red.

Diagnostic methods

Urine test strips

Paper strips with a special layer for detecting specific substances in urine. They are dipped in the urine and change color, depending on the substance, to different degrees. The composition of the urine enables conclusions to be drawn about the kidney function. An increased sugar content, such as may occur in diabetes, can also be quickly detected by this means.

Renal angiography

Depiction of the renal vessels in an X-ray image following the injection of a contrast agent into the aorta in order to visualize the blood supply to the kidneys and to detect any changes such as vascular obstruction, vascular malformations, or kidney tumors.

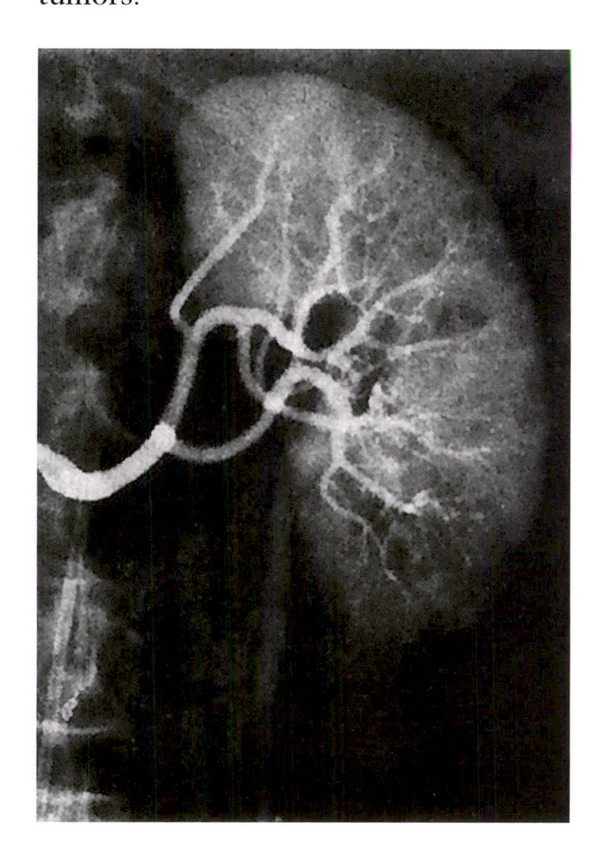

◆
Renal angiography
The blood vessels of the kidney are made visible on the X-ray image through the use of a contrast agent.

Urogramm

X-ray image of the kidneys and urinary tract. A special contrast agent is injected into the bloodstream and is excreted via the kidneys shortly thereafter. Successive X-ray images at short intervals can determine whether the discharge of urine from the kidneys is disrupted by constrictions or kidney stones.

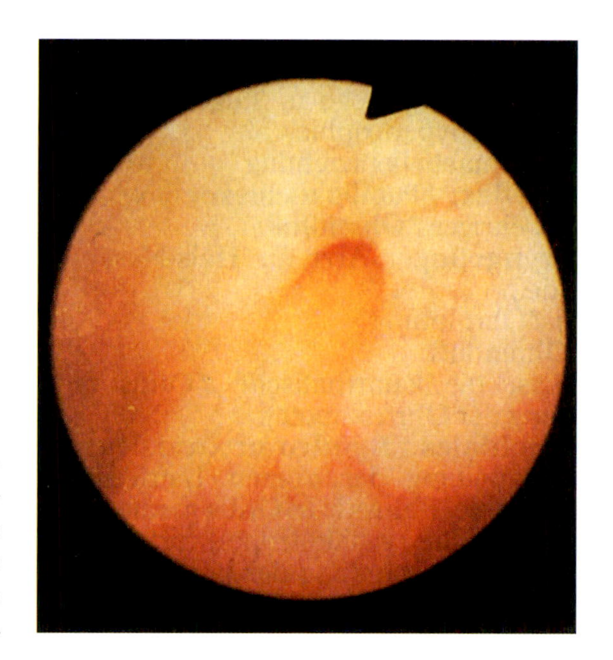

◆
Cystoscopy
The entry point of the ureter in the bladder is clearly visible through the endoscope.

Renal biopsy

Removal of renal tissue for detailed examination using a hollow needle inserted into the kidney via the skin, often under visual control, e.g., during an abdominal operation or under X-ray control. A renal biopsy serves to confirm diagnosis in the case of ambiguous kidney diseases and thus to enable targeted treatment.

Cystoscopy

Examination of the bladder filled with a sterile liquid using a rigid or flexible endoscope. Urinary calculus, bladder tumors, and inflammatory changes within the bladder, can be detected in this manner. If cancer of the bladder is suspected, tissue can be extracted during the cytoscopy for microscopic examination of the cells.

Therapeutic methods

Dialysis

A process for the separation of very small particles dissolved in a liquid from larger dissolved particles, using a semi-permeable membrane whose specific pore size makes it passable only for the smallest of particles and only in one direction. This principle is used in the medical field to carry out hemodialysis when kidney function is either extremely limited or has ceased completely. The blood of the kidney patient is fed through a membrane system surrounded by a flushing liquid (dialysate). The membrane system allows only the smallest particles to pass

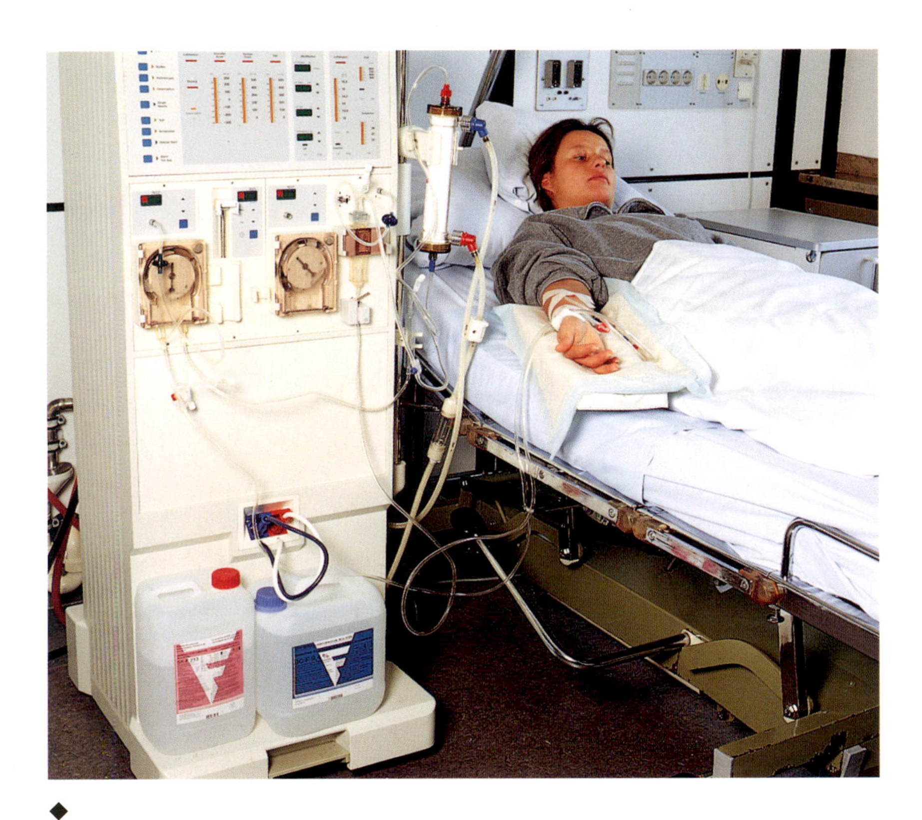

◆
Dialysis
A patient with kidney damage needs to be connected to an artificial kidney for several hours at a time for up to three times a week.

through to the dialysate. The toxins and waste products from the blood, which are generally small molecules normally excreted via the kidneys, now pass through the pores of the membrane systems into the flushing liquid and are flushed out. The cleansed blood is then pumped back into the body.
In peritoneal dialysis, the body's own peritoneum is used as the membrane. In this case the flushing liquid is fed into the abdominal cavity via a thin catheter and drained off again when dialysis is complete.

Kidney stone fragmentation

The disintegration of kidney stones using laser beams, ultrasound, or shock waves. Renal stones can be fragmented either with shock waves from an external source under X-ray control, or by ultrasound waves or laser beams introduced using an endoscope. The ultrasound or laser probes are brought into direct contact with the stones and fragmented without having any impact on the tissue.

Bladder irrigation

Flushing out of the bladder using a catheter and a syringe, usually carried out for the treatment of bladder infections. The syringe is filled with an aseptic flushing liquid at body temperature and this is slowly fed into the bladder; the liquid flowing back through the catheter is collected in a bowl. The process is repeated three to four times until the liquid is completely clear.

Artificial bladder

The introduction of a replacement bladder or enlargement of a bladder that is too small, using parts of the small or large intestine. An artificial bladder is required if the original organ has had to be removed due to a malignant tumor, if the closing mechanism of the bladder musculature no longer functions, or if uncontrollable urinary incontinence exists for other reasons. If there is sufficient residual bladder tissue, a piece of the small intestine is connected between the ureter and urethra and the urine passed through the urethra. Alternatively, the ureters may be connected to the lower end of the large intestine, and the urine drains through the anus.

Litholysis

Chemical, physical, or mechanical dissolution of urinary calculus in the kidney, ureter, or bladder region. Medication is either taken orally or administered directly into the region where the stone is located using a catheter (chemical dissolution). Alternatively, the stones are fragmented using shock waves (physical), or are broken up during an endoscopic operation using special forceps (mechanical). A combination of these procedures is sometimes required.

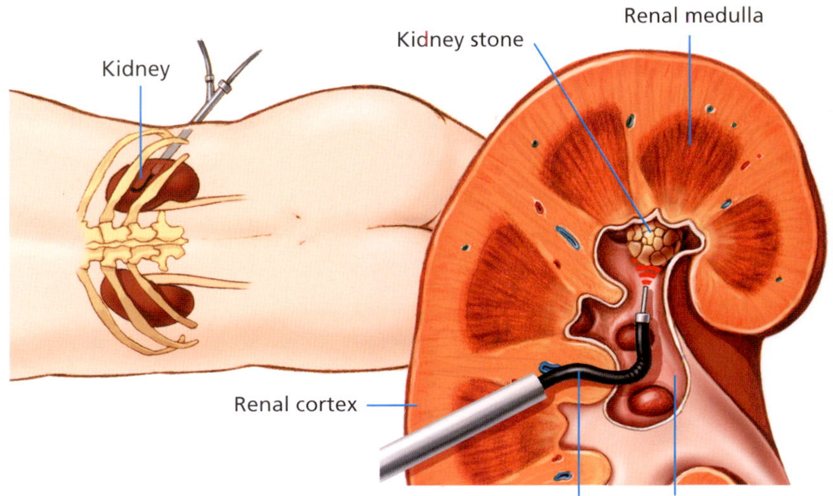

◆
Kidney stone fragmentation
An ultrasound probe is inserted directly into the renal pelvis using an endoscope in order to fragment a kidney stone.

Hormone system

Hormones are endogenous agents produced in specific cells or (glandular) tissues, and which have an influence on many crucial processes in the body. Hormones not only regulate metabolic processes, but also affect growth and development, sexuality, and behavior. The body's reaction to stress situations is also coordinated by hormones.

Almost all hormones are produced in special endocrine glands and are distributed within the body in the blood. Hormone production is controlled from specific centers in the brain: the hypothalamus (a part of the interbrain), the pituitary gland, and the epiphysis (pineal gland).

Hypothalamus

The hypothalamus is a brain structure lying close to the pituitary gland, which plays a commanding role in the control of essential organ functions. It registers the concentrations of most hormones in the blood and then, depending on the individual readings, may send impulses to the pituitary gland to stimulate hormone production.

Pituitary gland

The pituitary gland is divided into anterior and posterior lobes. It not only exercises a control function over other endocrine glands, but also synthesizes a number of hormones itself:

- antidiuretic hormone (ADH), which controls diuresis from the kidneys,
- follicle-stimulating hormone (FSH), which stimulates the ovaries or testes,
- luteinizing hormone (LH), which also acts on ovaries or testes,
- prolactin, which stimulates milk production in the mammary glands of pregnant and post-partum women,
- somatotropin, a growth hormone, which promotes growth and the production of proteins,
- adrenocorticotropin, which stimulates the adrenal glands,
- thyroid-stimulating hormone (TSH),
- melanocyte-stimulating hormone (MSH).

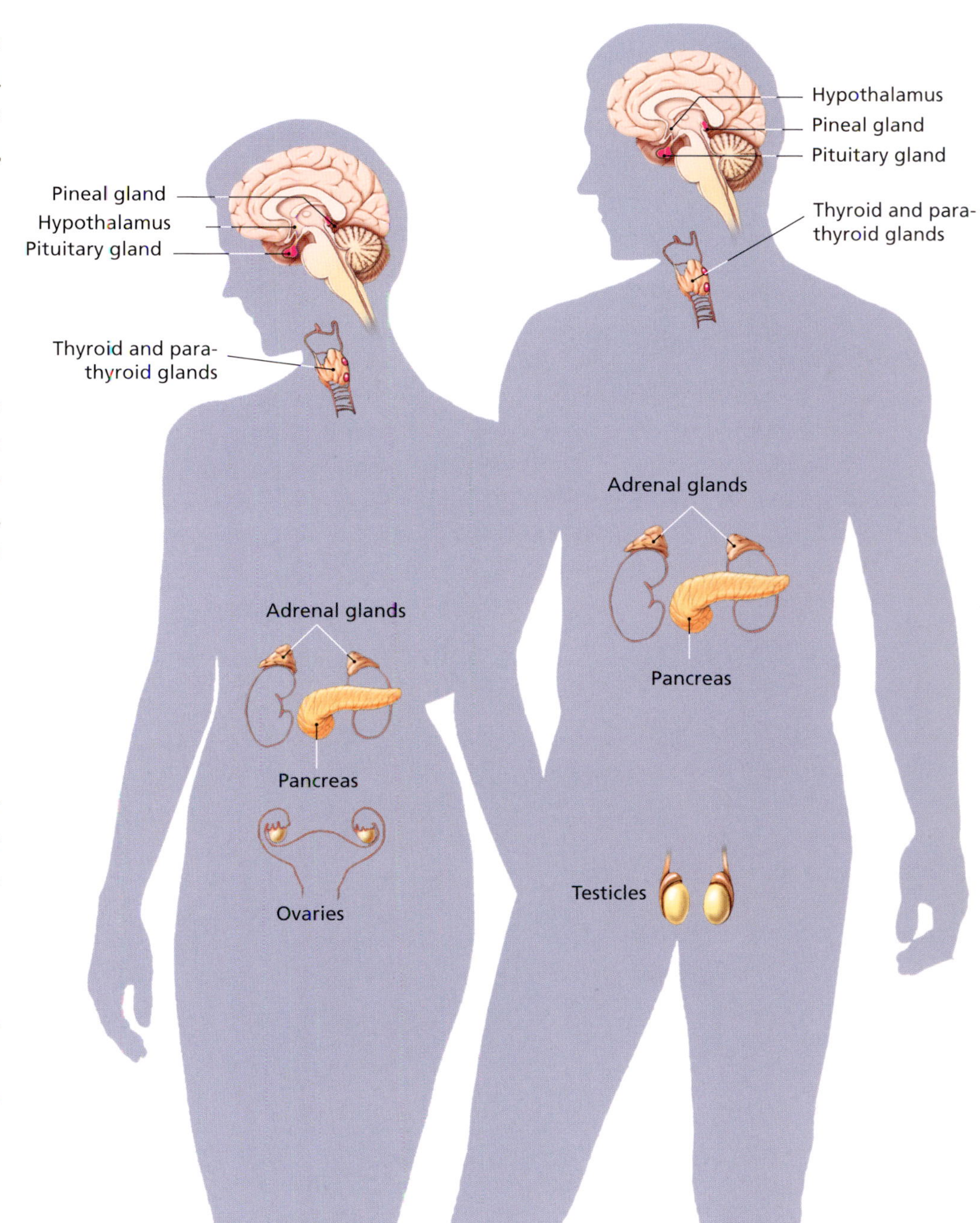

◆
Fig. 382
The higher centers for all the endocrine glands lie in the brain. The hormones reach their individual end organs through the circulatory system. The thymus gland, which is not shown on the general diagram, atrophies after puberty and is converted into fatty tissue.

Pineal gland (epiphysis)

The exact effect of the pineal hormone, melatonin, is not fully understood. It may influence circadian rhythm.

Thyroid gland

The gland, located under the larynx, produces the iodine-containing hormones thyroxine and triiodothyronine. They increase the basal metabolic rate, affect protein synthesis, and are essential to the proper development and differentiation of all cells in the human body.

Parathyroid gland

The four lentil-sized corpuscles of the parathyroid rest on the back of the thyroid gland. They produce parathormone, which releases calcium from the bones when there is an increased demand for it.

Adrenal glands

The two half-moon-shaped glands are placed like caps on the superior poles of the kidneys. They produce:

- Cortisone, which promotes protein degradation and lipolysis, and increases the blood-sugar level. It also has an anti-inflammatory effect, and suppresses allergies and immunoreactions.
- Aldosterone, which inhibits the elimination of salt and water from the kidneys, e.g., when the body's water content is reduced
- Androgens, which promote sexual development in men, as well as the growth of the (skeletal) musculature
- Epinephrine (adrenalin) and norepinephrine (noradrenalin), which are both catecholamines. They increase the blood pressure and heart rate, and are predominantly secreted in stress situations.

Pancreas

The pancreas produces two hormones which are involved in the metabolism of sugars. Insulin promotes the uptake of sugars from the blood into the cells of the body, and thus lowers blood sugar levels, while glucagon releases sugar from the cells and thus increases blood sugar levels.

Female sexual glands

The female sexual hormones (estrogen and progesterone) are produced in the ovaries. They promote the development of the female sexual characteristics and body form. In the sexually mature woman, estrogen stimulates the maturation of the egg until ovulation and builds up the endometrium. It also increases the sexual drive. Progesterone prepares the endometrium for the embedding of a fertilized ovum. Higher levels of progesterone induce fluid retention in the tissue.

Male sexual glands

The most important male sex hormone, testosterone, is mainly produced in the testicles. Testosterone promotes the formation of male secondary sex characteristics (body hair, growth of testicles and penis, etc.) and the sexual drive.

Thymus gland

This gland, which is located directly above the pericardium, starts to atrophy after puberty. Its hormone, thymosine, probably promotes the production and maturation of immune system defense cells.

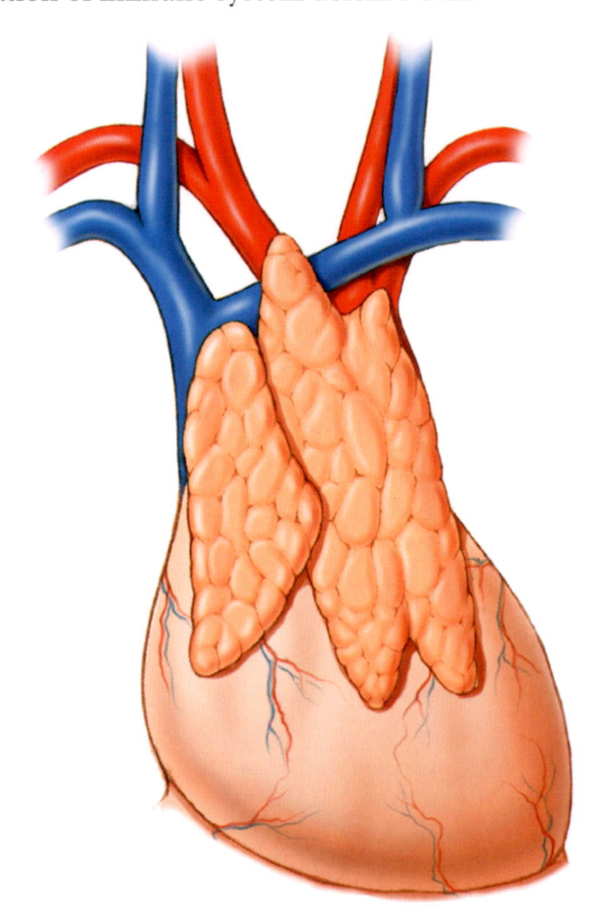

◆
Fig. 383
In children, the thymus gland is located directly in front of the heart.

Fig. 384
The hormone-producing structures of the brain, the hypothalamus and the pituitary gland, are control centers for the entire hormone system, and are located very close to one another. The pineal gland can also be detected, slightly above the cerebellum.

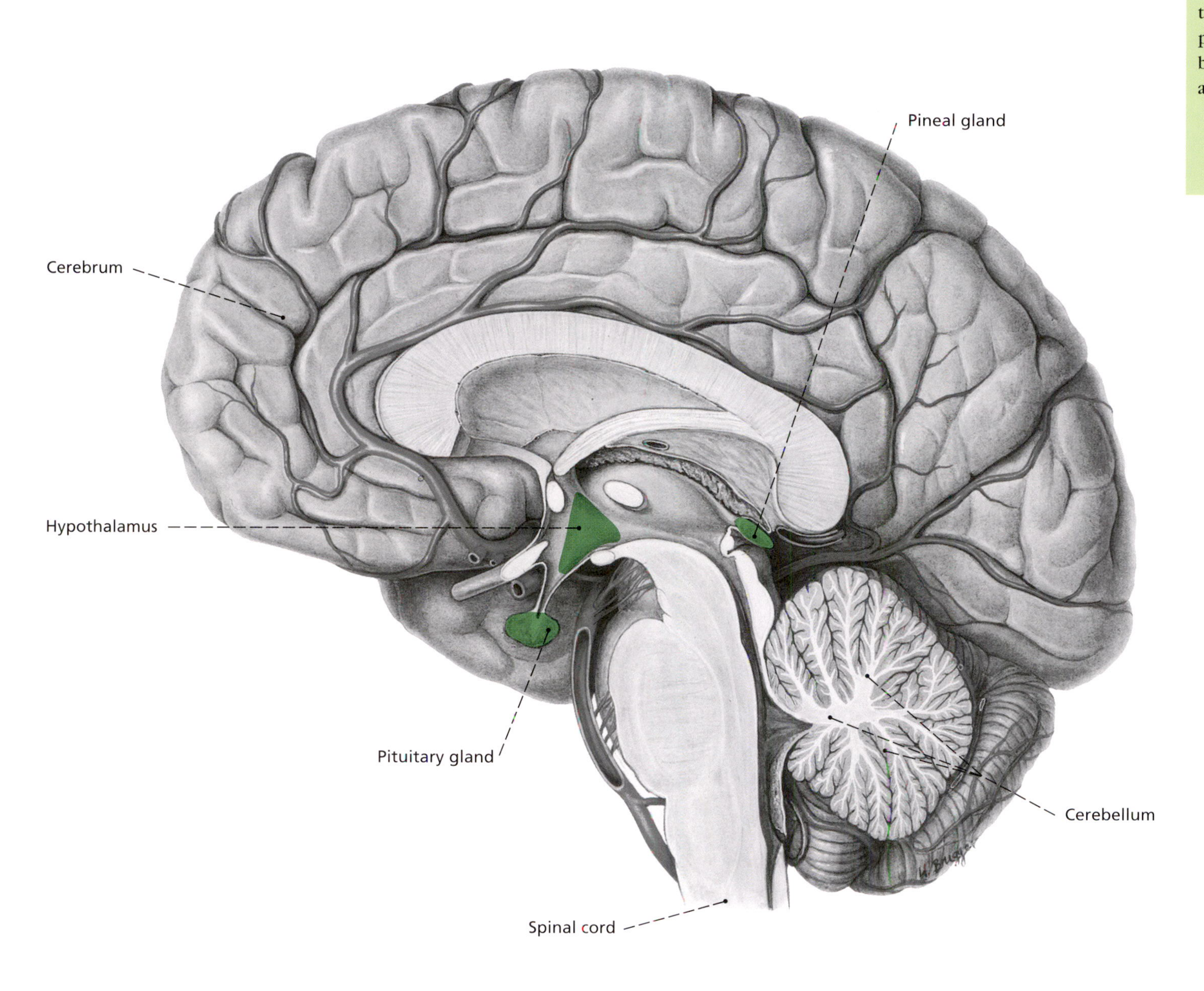

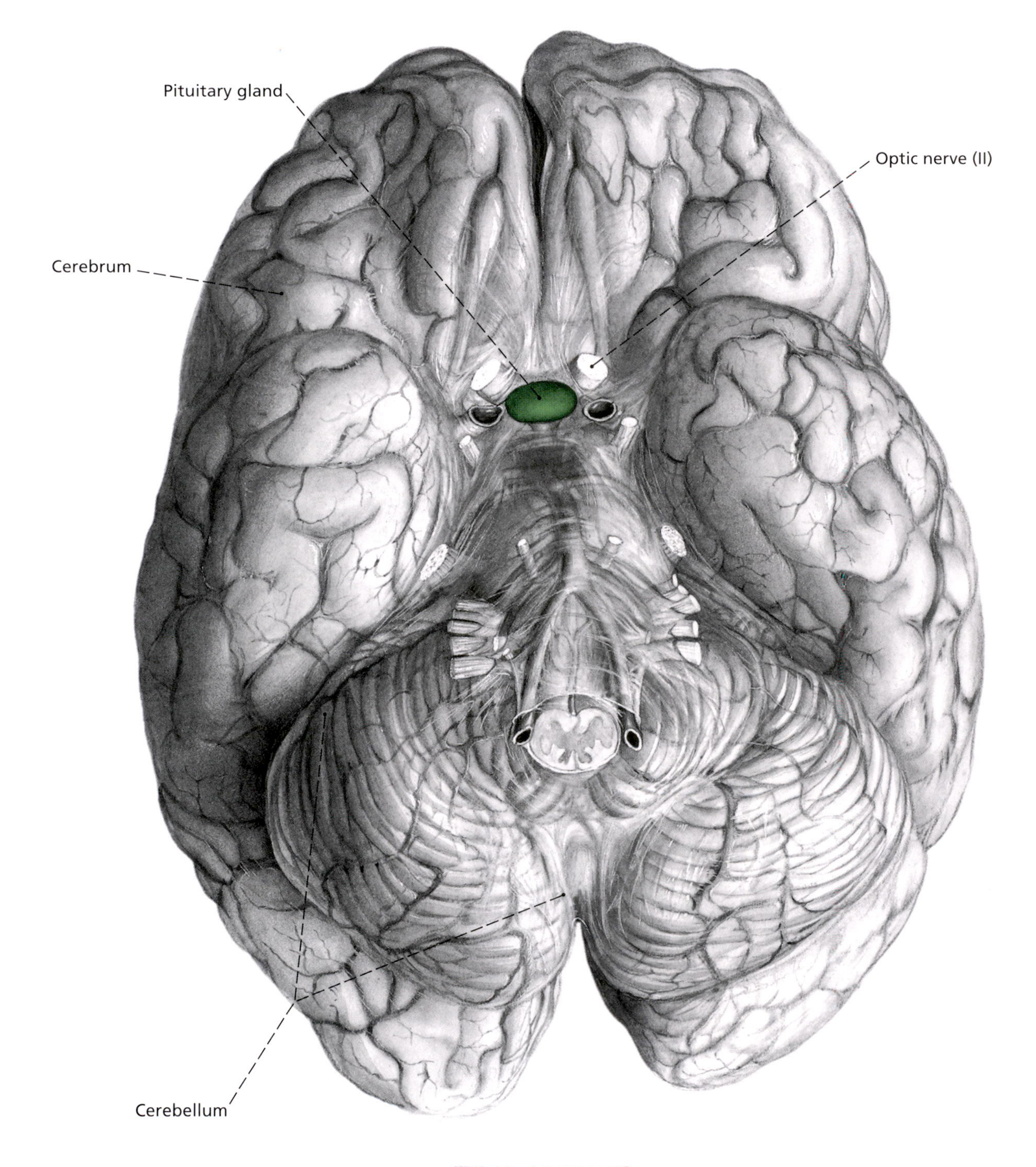

◆
Fig. 385
The pituitary gland, which is about the size of a cherry, and is shown here colored green, is located directly beneath the optic nerve intersection at the base of the brain (transected on this illustration), as seen from below.

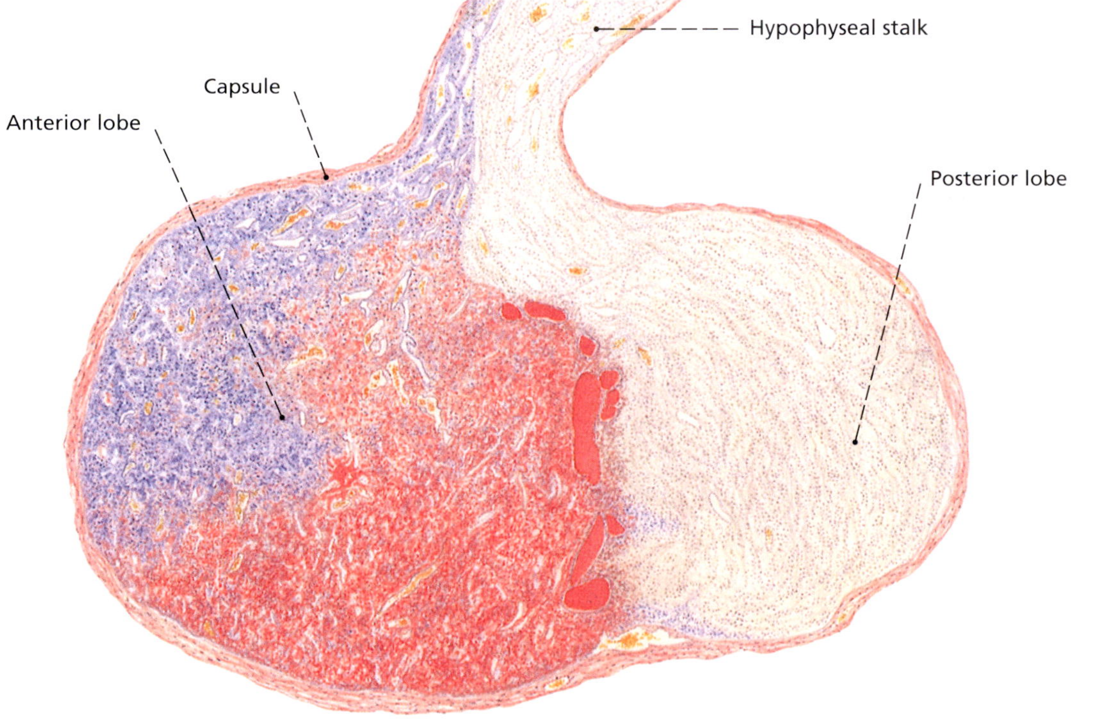

◆
Fig. 386
Pituitary gland in longitudinal section. Microscopic image (×10 magnification). The anterior lobe is colored blue/red and the posterior lobe is whitish.

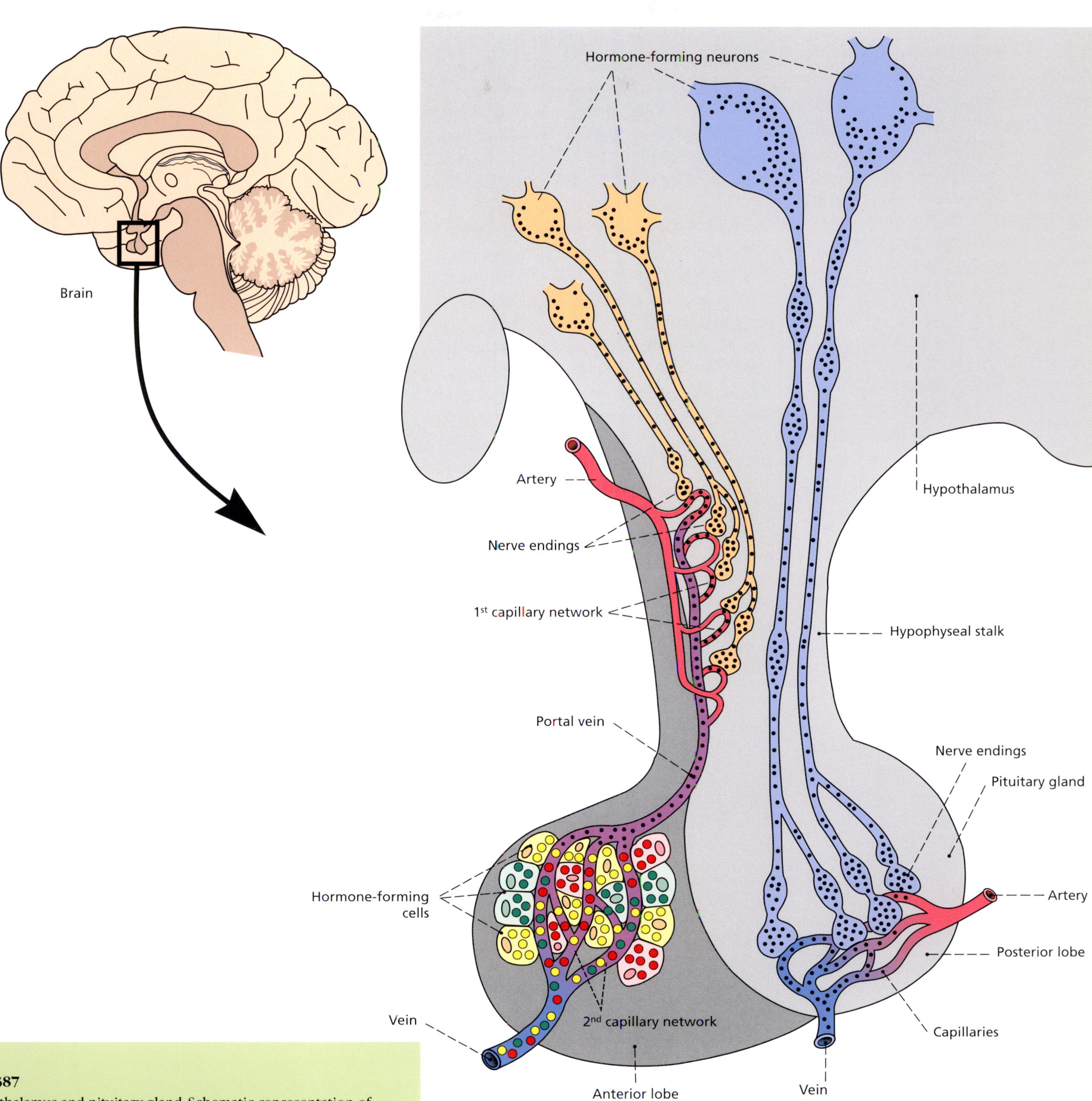

◆
Fig. 387
Hypothalamus and pituitary gland. Schematic representation of relation between the two control centers of the hormone system. The hormones produced in the neurons of the hypothalamus arrive at the pituitary gland by two routes: they reach the hormone-forming cells of the anterior lobe through a special blood vessel system, in which two capillary networks are connected to one another through the so-called portal vein; and they reach the posterior lobe directly via the nerve fibers. The endings of these hormone-forming neurons are directly joined to the capillary system of the posterior lobe. All hormones from the pituitary gland finally pass through veins into the systemic circulation.

◆
Fig. 388
Position of pineal gland (colored green) within brain, seen from above. The gland is about the size of a hazelnut, and is located above the quadrigeminal body.

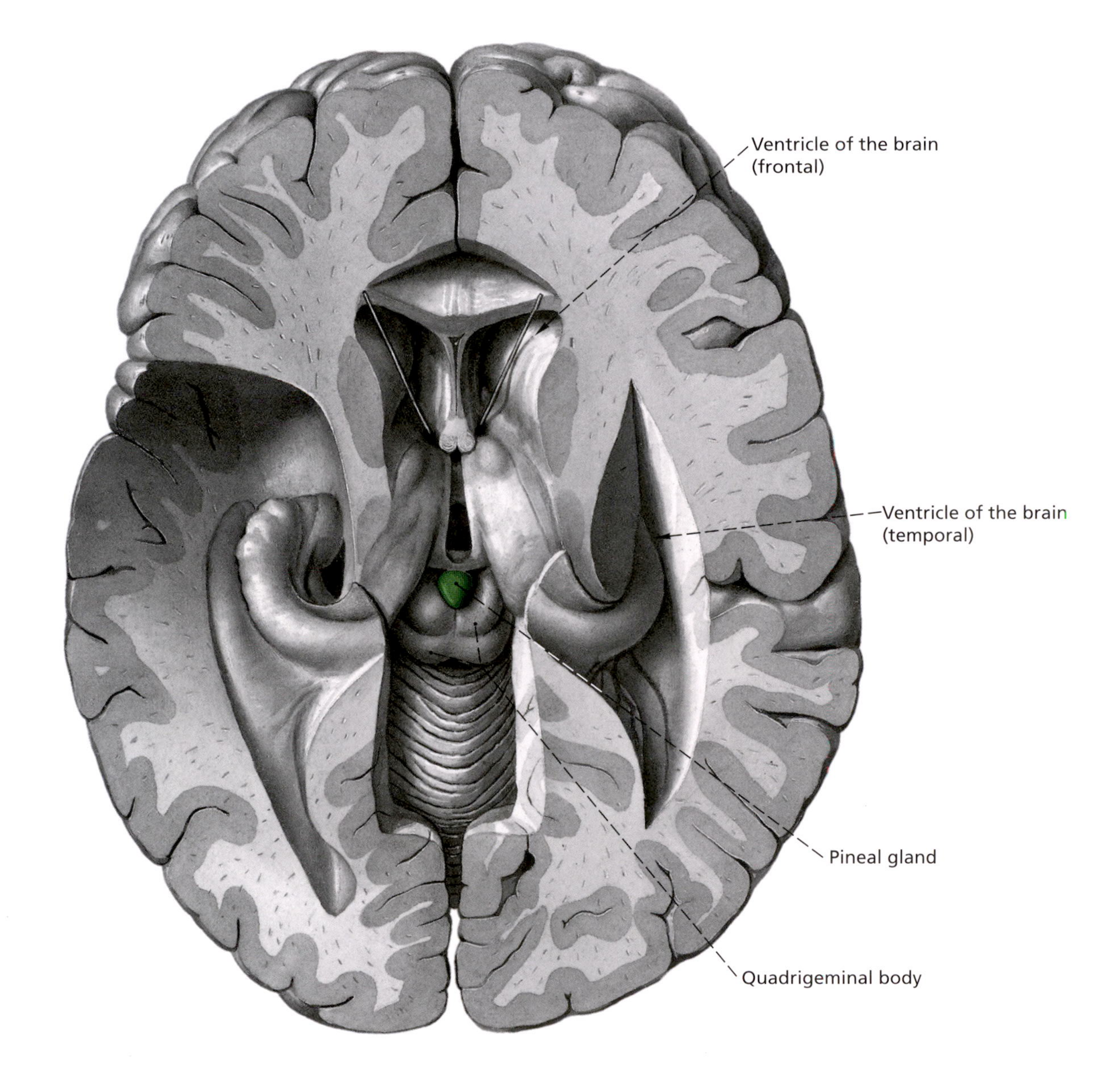

◆
Fig. 389
Pineal gland, after removal of a large part of the surrounding brain structure, seen at an angle, from behind and from above.

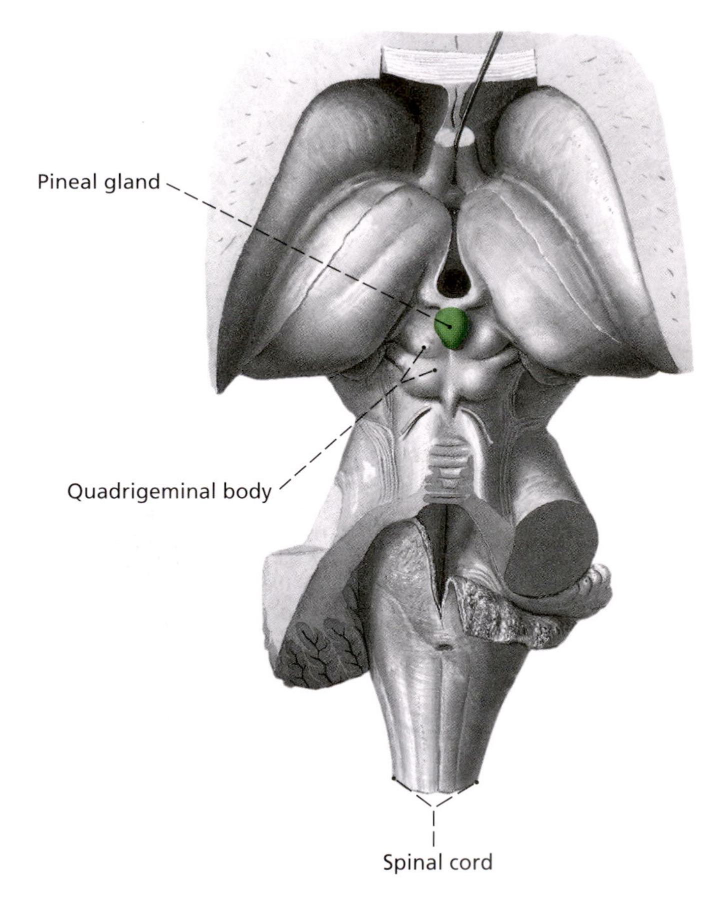

◆
Fig. 390
Thymus gland in the adolescent (colored green). The organ can be detected between the two upper sections of the lungs (chest wall removed). It later atrophies completely.

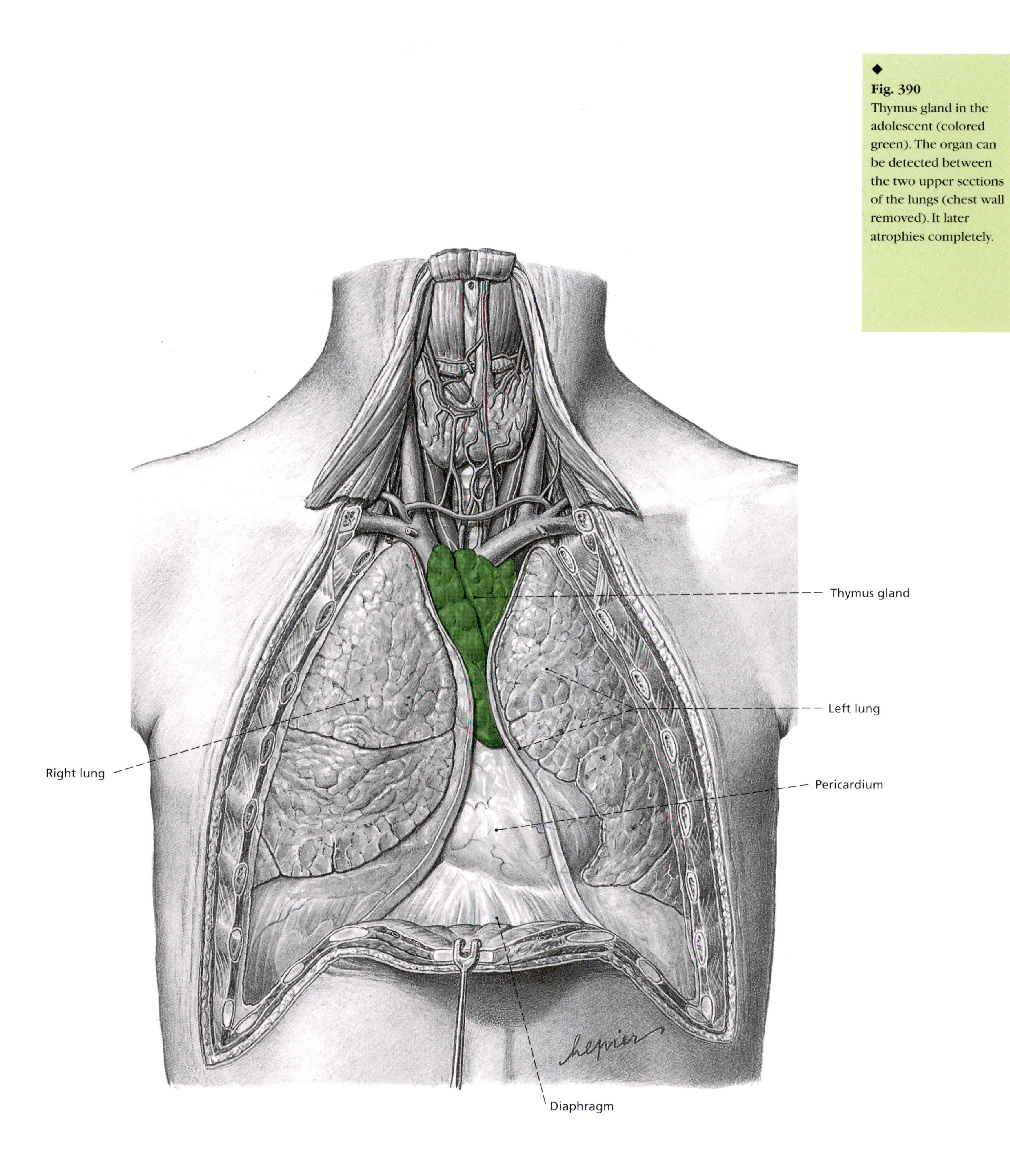

◆
Fig. 391
Thymus gland in the adolescent (colored green), seen after removal of the chest wall. The upper lobe of the left lung has been pulled slightly outwards. The thymus gland is lobed, and lies directly on the upper section of the pericardium.

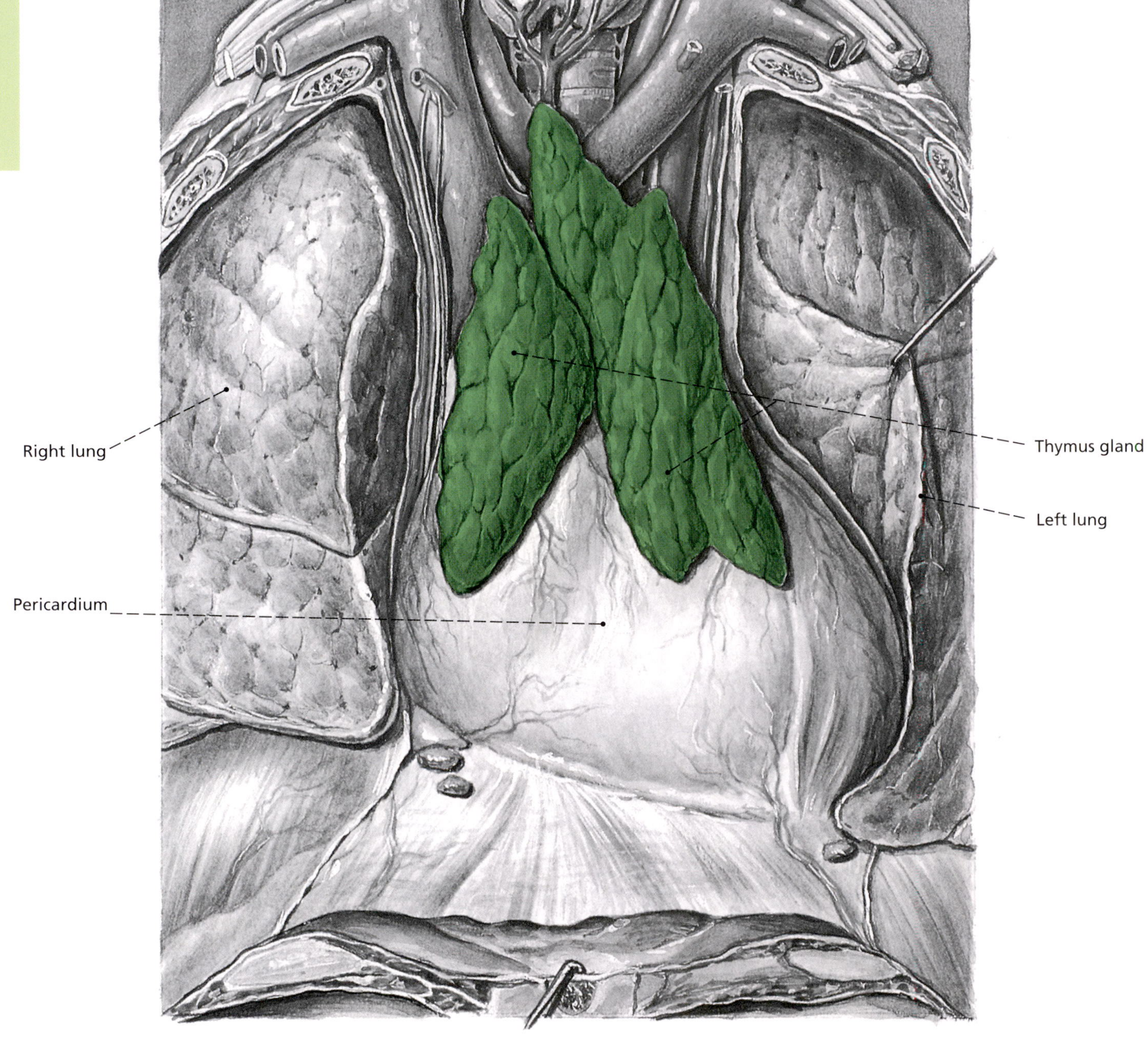

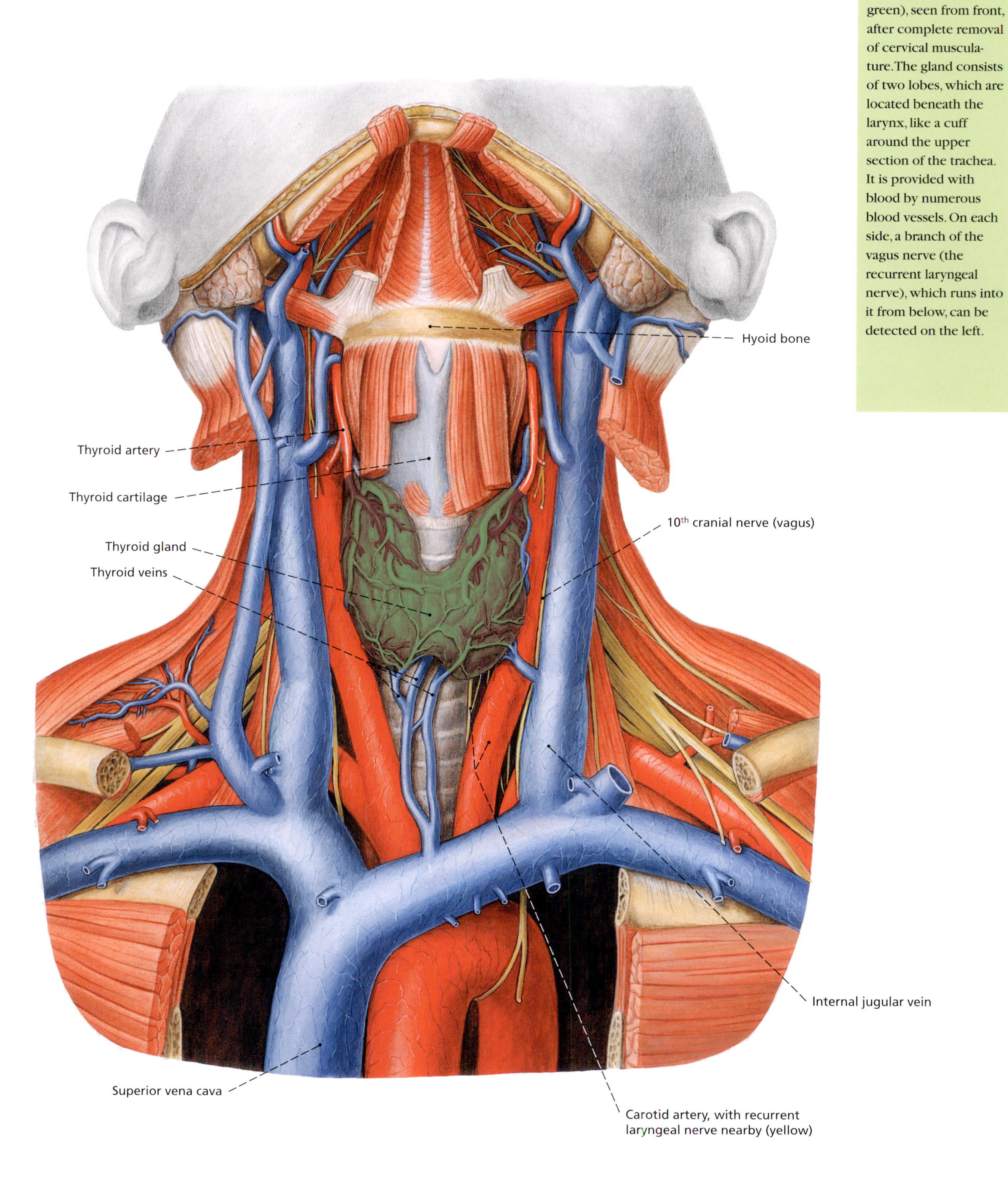

◆
Fig. 392
Pituitary gland (colored green), seen from front, after complete removal of cervical musculature. The gland consists of two lobes, which are located beneath the larynx, like a cuff around the upper section of the trachea. It is provided with blood by numerous blood vessels. On each side, a branch of the vagus nerve (the recurrent laryngeal nerve), which runs into it from below, can be detected on the left.

◆
Fig. 393
Lateral sections of thyroid gland (colored green, under blood vessels) with laryngeal inlet and trachea, viewed from rear.

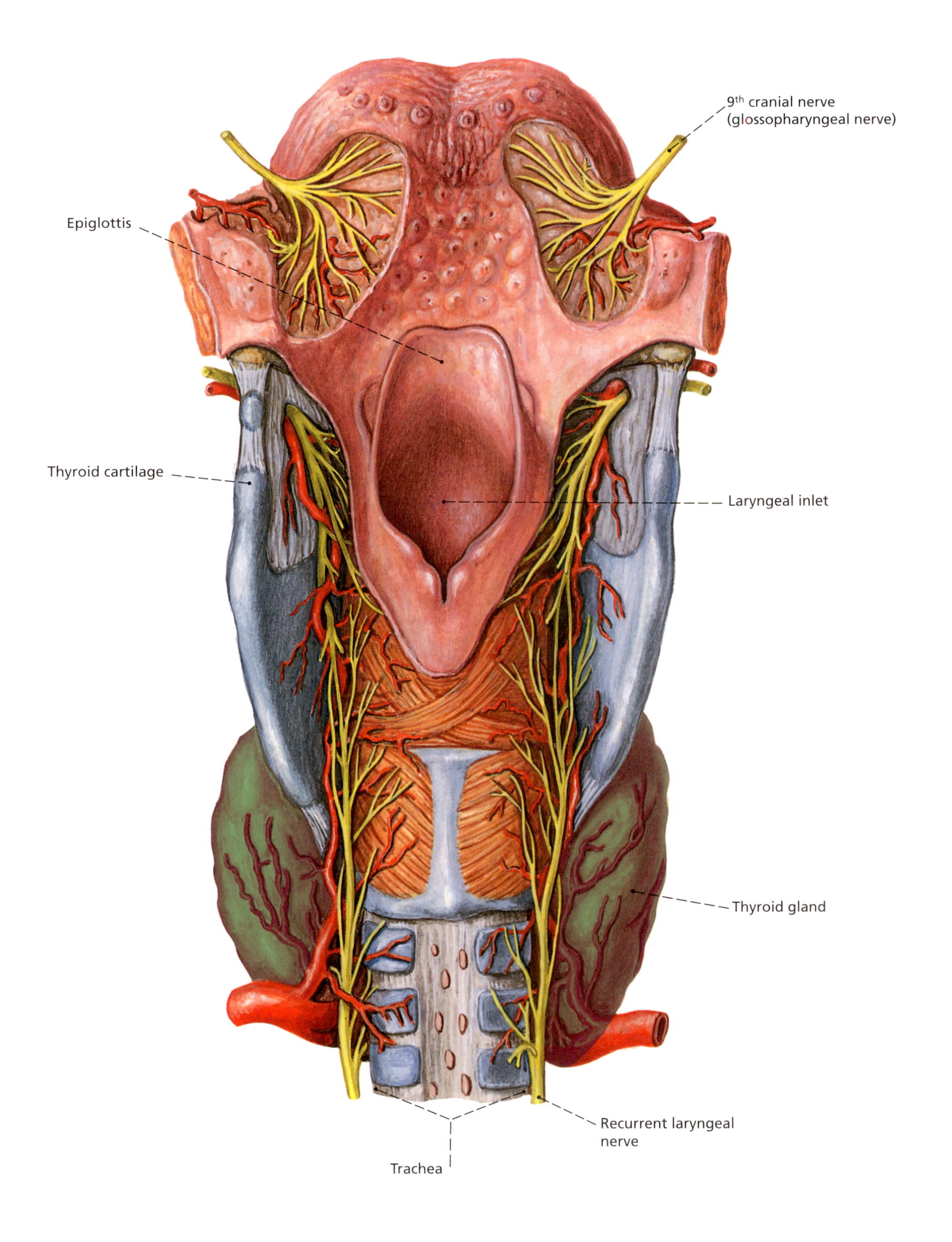

◆
Fig. 394
Parathyroid glands (colored green). View of the pharyngeal musculature, as seen from the rear. The paired organs are not even as big as peas and are located on the free rear edges of the two thyroid lobes.

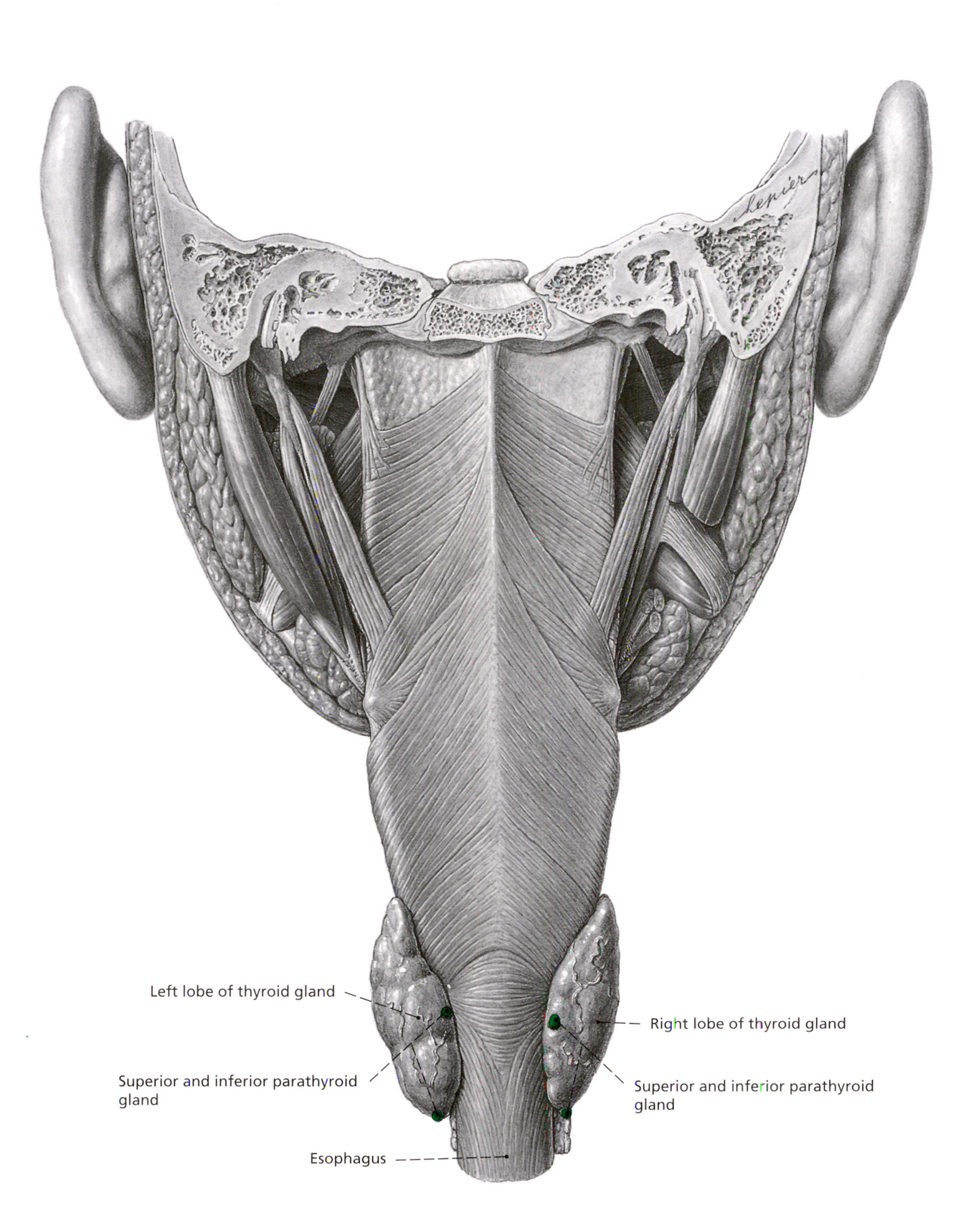

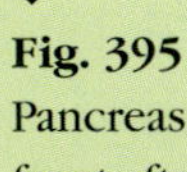

Fig. 395
Pancreas, seen from the front after removal of abdominal viscera, apart from the duodenum and kidneys. The gland is located in front of the inferior vena cava and the aorta. Its head presses into the C-shaped curve of the duodenum. The tail lies crosswise in the epigastrium and extends across the front surface of the left kidney.

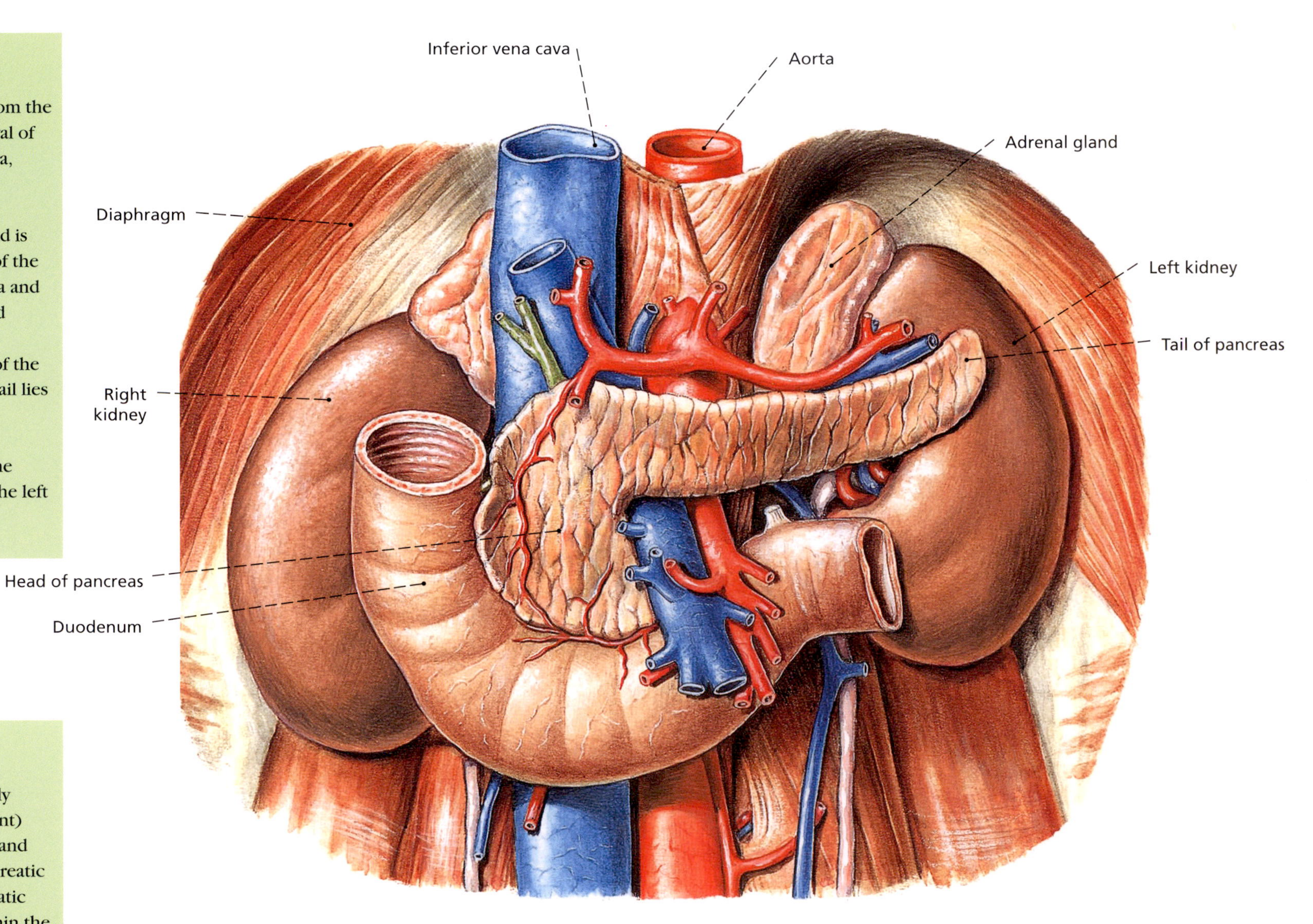

◆
Fig. 396
Pancreas (partially opened up at front) with duodenum and entrance to pancreatic duct. The pancreatic duct divides within the pancreas and enters the duodenum through two apertures. The lower glandular duct usually has a common outlet with the bile duct (papilla duodeni major). The section on the right shows the acini (green) and one of the Islets of Langerhans (orange), the production sites of digestive juices, and insulin and glucagon, respectively. While the digestive juices flow out of the acini into the bowel, insulin and glucagon are released directly into the blood from the Islets of Langerhans.

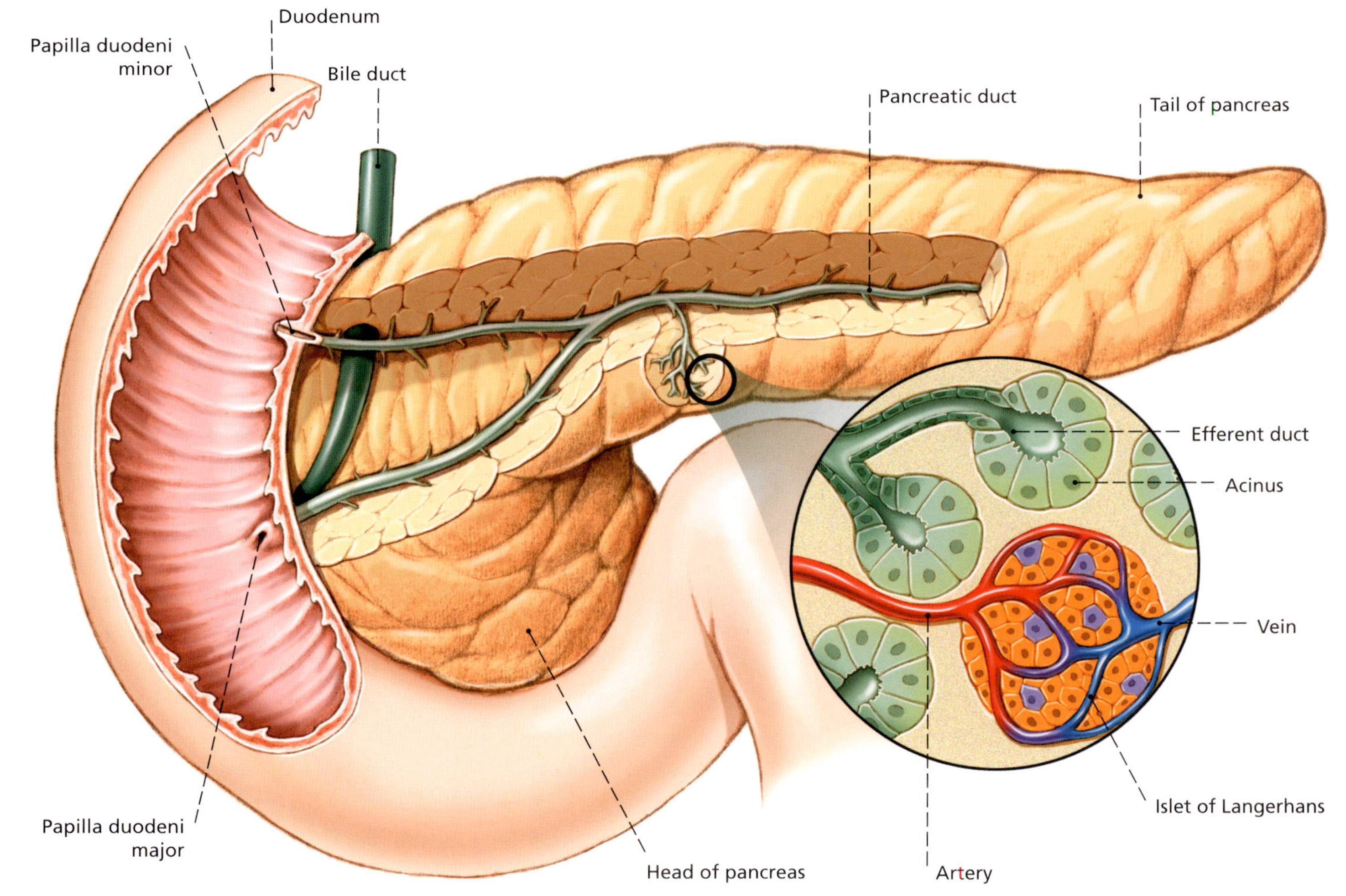

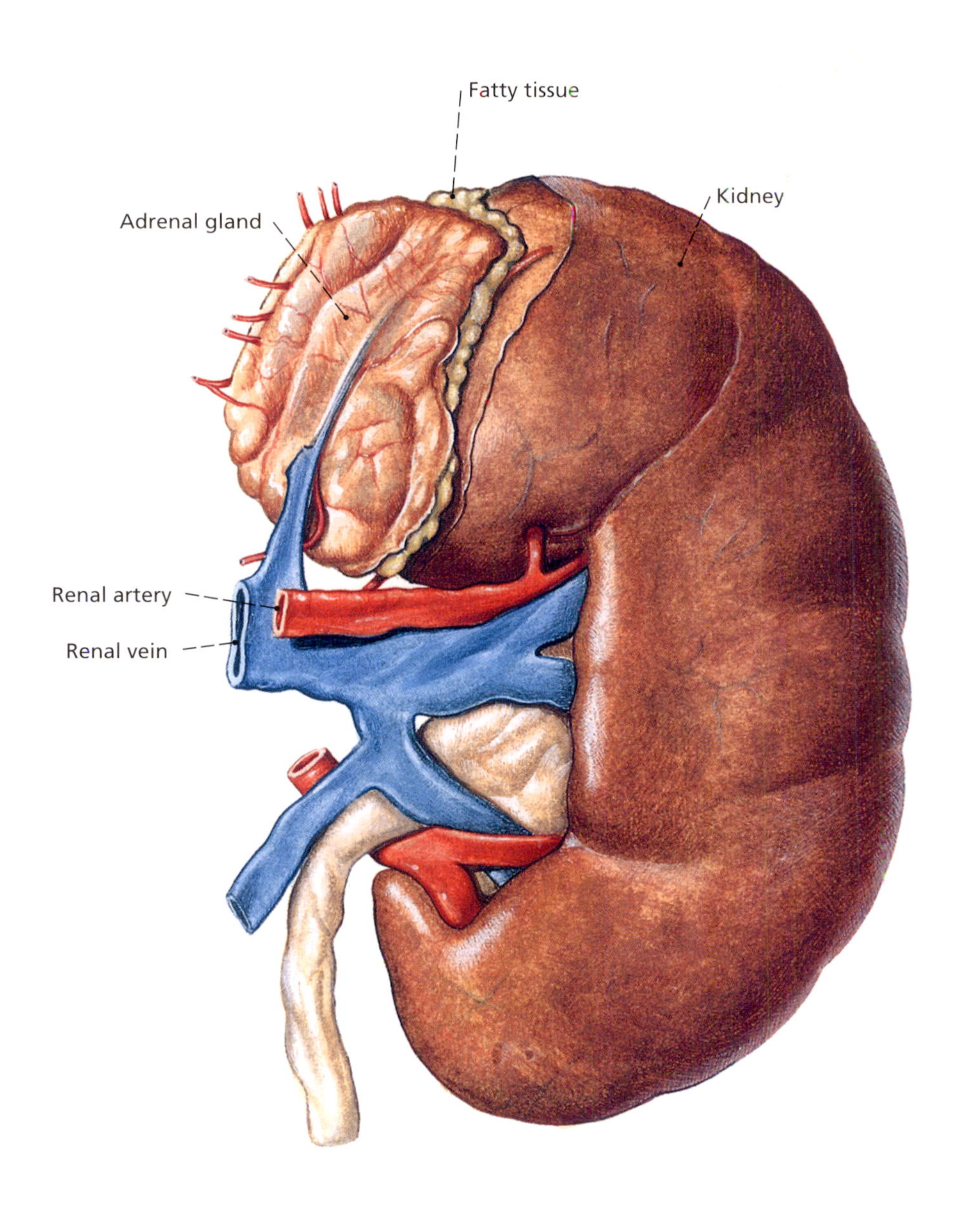

◆
Fig. 397
Left adrenal gland, seen from the front. It sits on the kidney like a cap. Between it and the kidney lies a thin layer of fatty tissue.

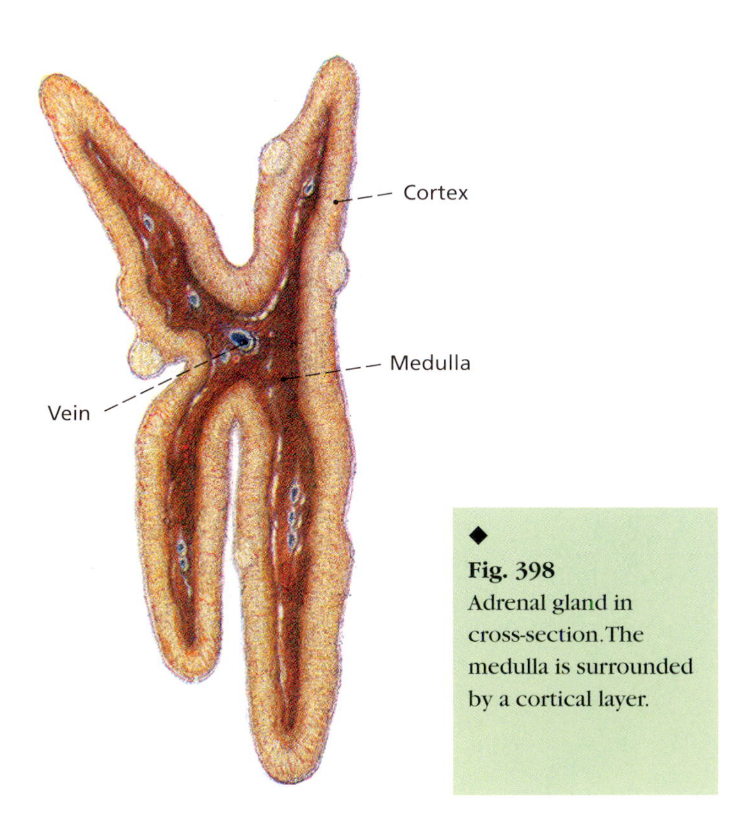

◆
Fig. 398
Adrenal gland in cross-section. The medulla is surrounded by a cortical layer.

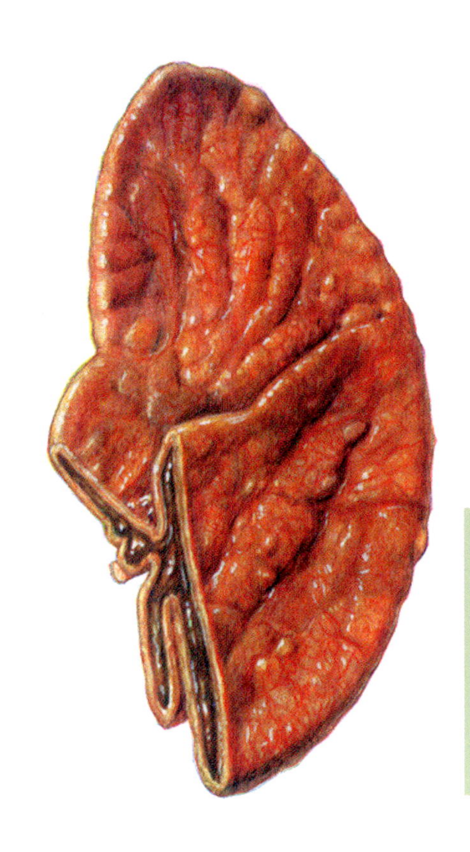

◆
Fig. 399
The same adrenal gland, seen from front. The lower section of the organ has been transected.

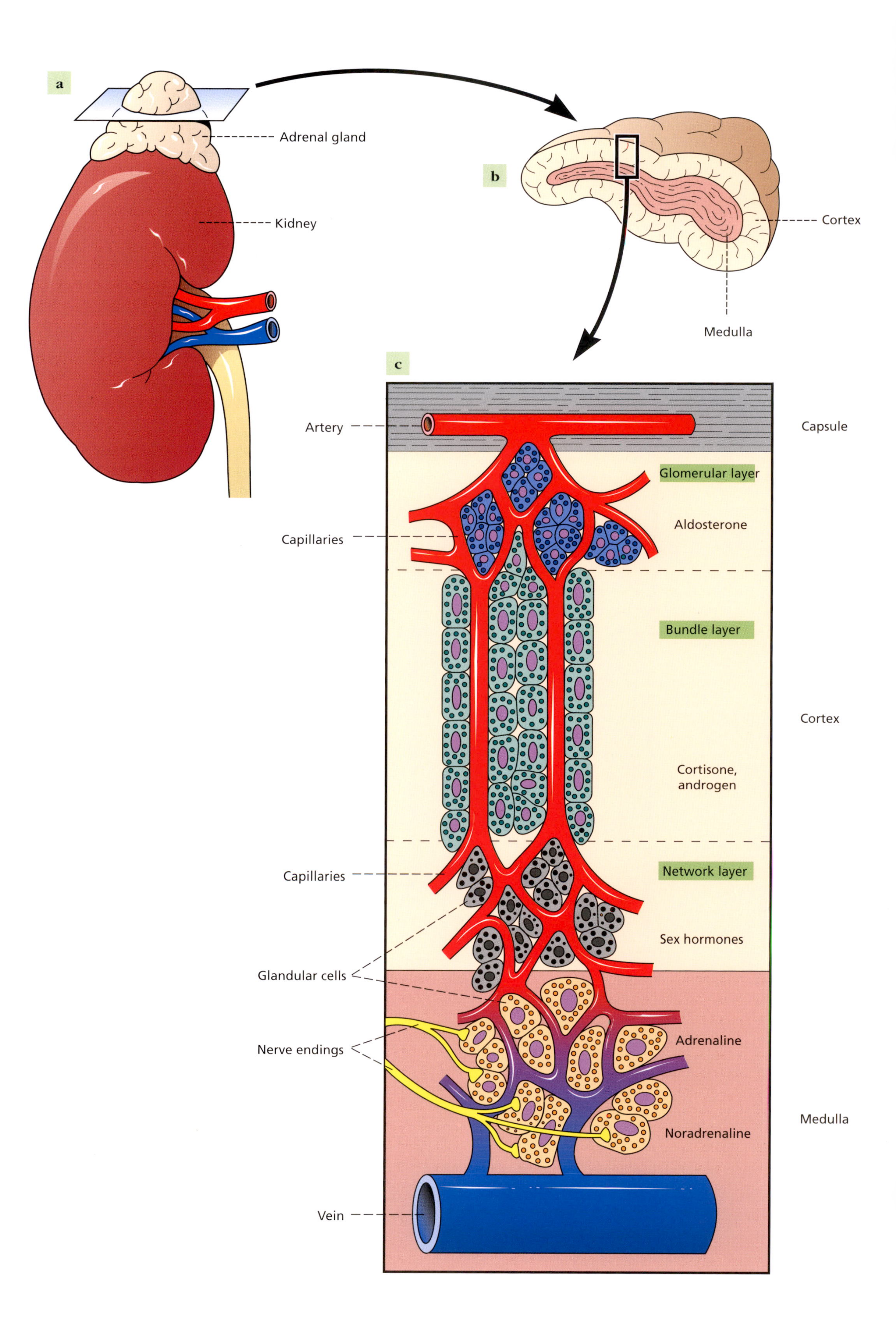

◆
Fig. 400
Adrenal glands. Location relative to kidney and fine structure of organ with production sites for various hormones (schematic representation):
(a) View from the front, with plane of incision.
(b) Cross-section to show spatial extent of cortex and medulla.
(c) Cortex and medulla with various layers.

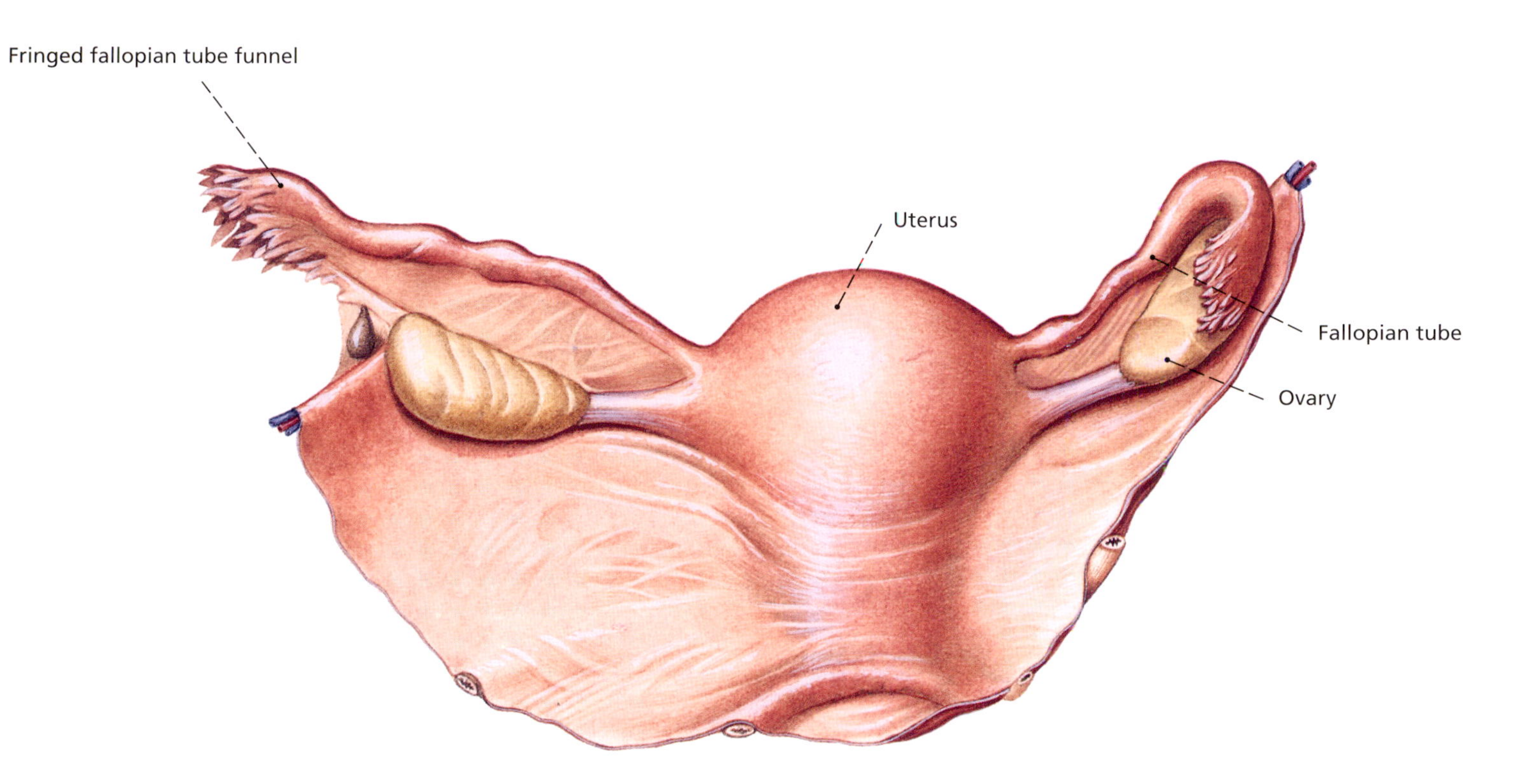

◆
Fig. 401
Representation of ovaries and uterus, seen from rear. The fringed fallopian tube funnel is detached from the ovary on the left-hand side.

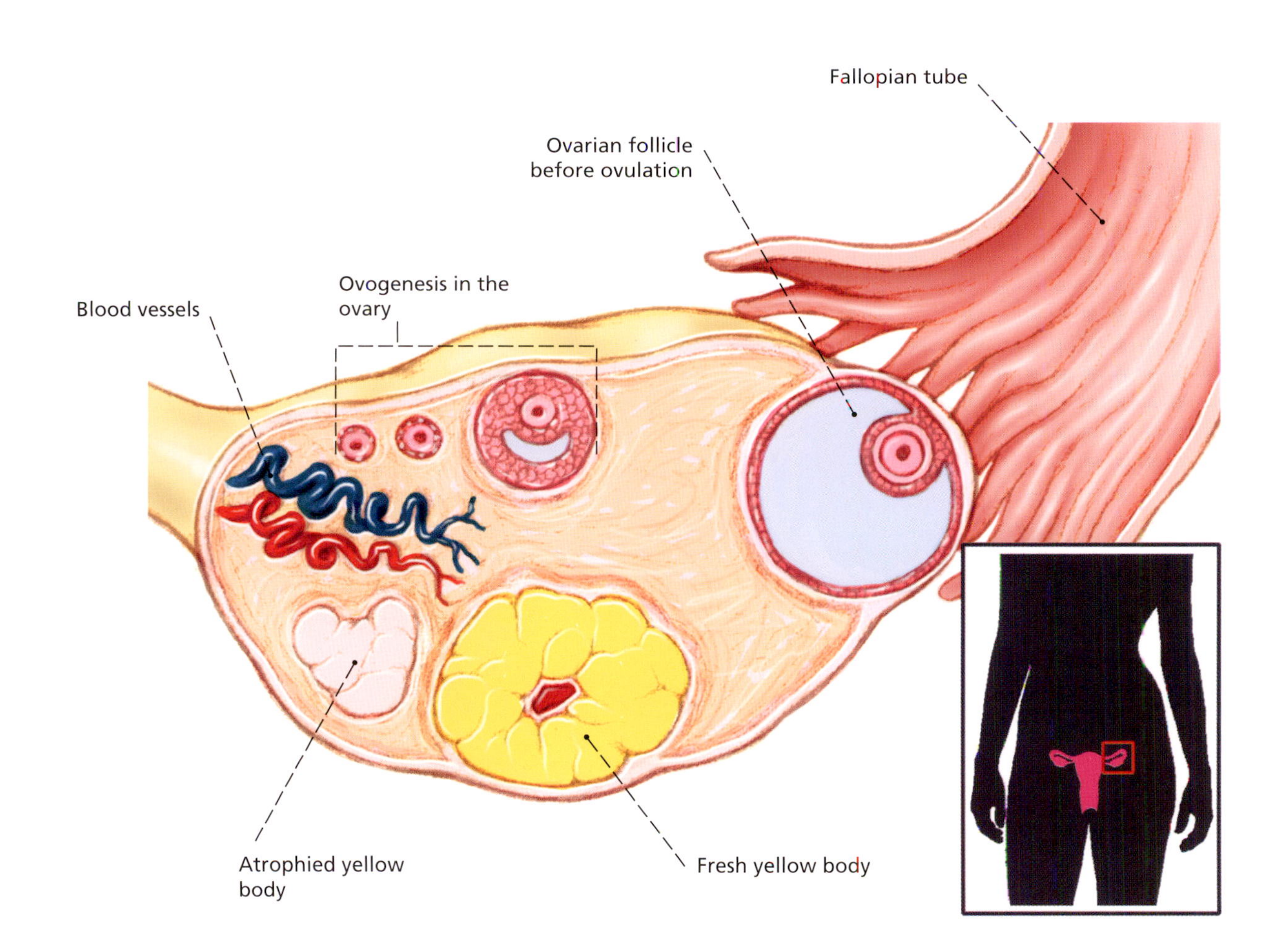

◆
Fig. 402
Ovary in longitudinal section. Schematic representation, with overall view for spatial orientation. The development of the yellow body is shown, which produces progesterone. Progesterone plays an essential role in the preparation of the endometrium for the implantation of the fertilized egg.

◆
Fig. 403
Testicles with epididymis after opening up of scrotum.

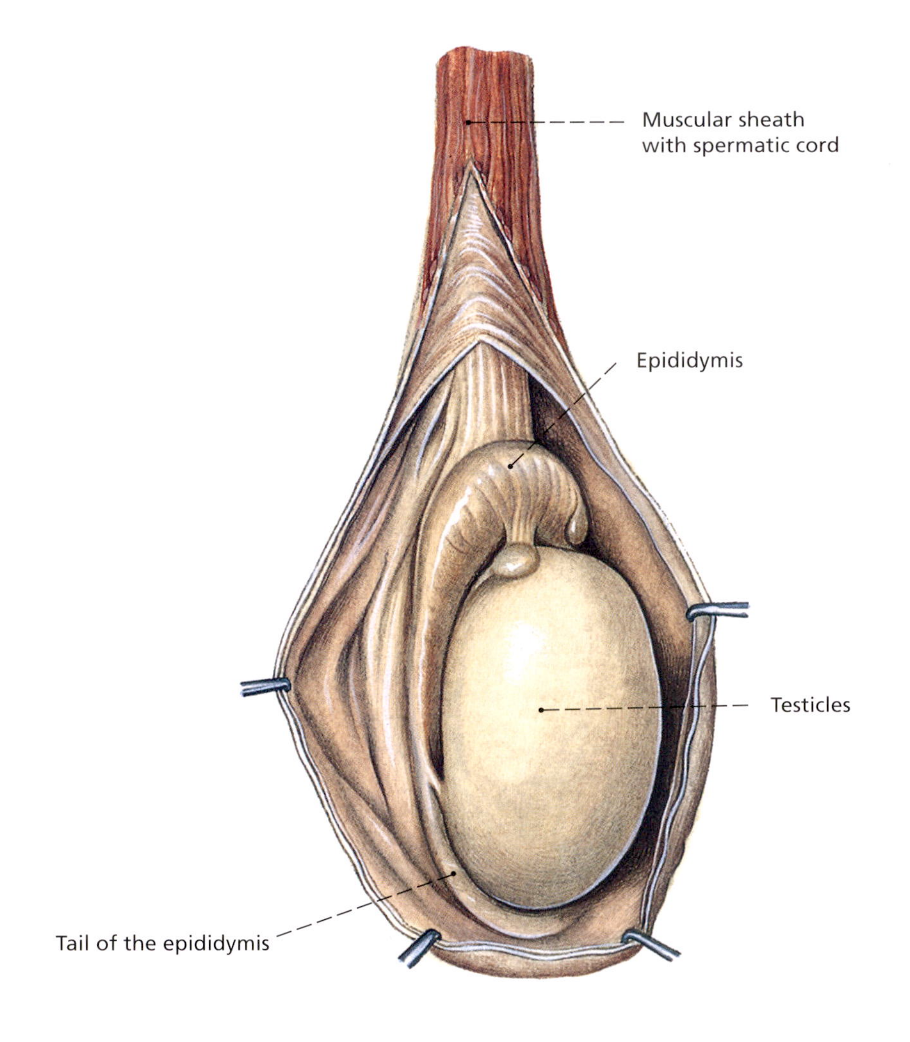

◆
Fig. 404
Testicles in cross-section. The testicular sheath can be detected, along with the cremaster muscle, the fibers of which surround the entire testicle, plus the lobular structure of the testicular tissue. The hormone testosterone is produced in the connective tissue. The epididymis can be seen in the lower part of the image.

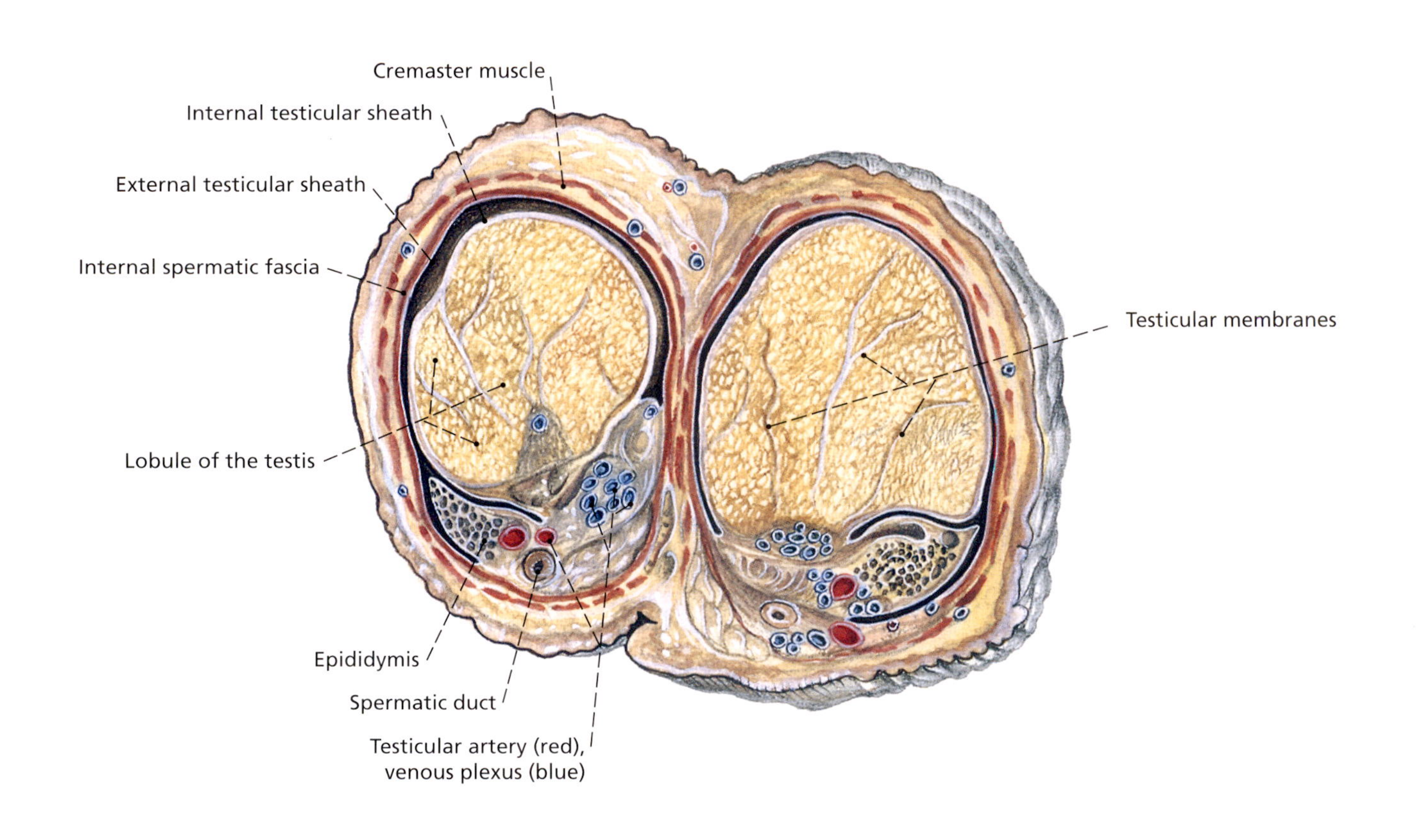

Symptoms and diseases

Pituitary gland (hypophysis)

• *Pituitary tumor*

Tumoral enlargement of pituitary gland, usually benign. Pituitary tumors can cause either an increase or a decrease in production of pituitary hormones, manifest by excess or deficient activity at the target tissue of the hormone. Larger pituitary tumors may exert pressure on the optic nerves, and thus lead to impaired vision, with visual field loss.
Pituitary tumors are detected through hormone level measurements and X-rays of the skull. In most cases, they have to be surgically removed. Radiation and hormone-regulating medication are rarely sufficient.

• *Growth disturbances*

Conspicuous departures from the normal height-for-age in children and young people. A pathological increase is described as gigantism, and a pathological decrease as stunted growth or dwarfism.
The causes of growth disturbances are numerous. **Gigantism** can be inherent, or the result of a growth in the pituitary gland, which then produces too much growth hormone (somatotropin). While affected children experience above-average overall growth, as they are still growing, in adults the excessive growth is usually restricted to the bony areas of the face, hands, and feet; the tongue may also expand. Later, diabetes, high blood pressure, hardening of the arteries, and muscular weakness may occur. Even children and adolescents can be affected by these symptoms. Depending on the position of the tumor, it may be surgically removed or treated with radiation.
On the other hand, **stunted growth** can be the result of inadequate functioning of the pituitary gland, or of malnutrition during fetal development and in early childhood.
As we can now manufacture growth hormone by genetic engineering, specific hormone therapy is possible. However, this is controversial as it is suspected that such treatment may be associated with the development of malignant tumors.

Thyroid gland

• *Inflammation of the thyroid gland (thyroiditis)*

Staphylococcal or streptococcal infections can trigger a **subacute** inflammation of the thyroid gland, with fever, neck pains, difficulty in swallowing, and hoarseness. It usually clears up on its own, but the doctor may prescribe pain-killers and/or cortisone to speed up the cure.
Chronic thyroiditis, is more often seen in adults. It is the result of an auto-immune disease in which the body's own defense system attacks the thyroid gland cells. Chronic inflammation of the thyroid gland is treated with thyroid hormones.

• *Hyperthyroidism (hyperthyreosis)*

Hyperthyroidism is caused by increased production of the thyroid hormone, thyroxine, in the thyroid gland. This leads to an imbalance between hormone requirements and hormone production. A frequent cause of hyperthyroidism is Graves' disease (see below).

• *Graves' disease (Basedow's disease)*

An over-active thyroid gland leads to increased hormone production, hyperexcitability, cardiac problems (increased heart rate, palpitations, and arrhythmia), weight loss, and problems with sleep and digestion. In over 70% of cases it is caused by an autoimmune condition, known as Graves' disease or Basedow's disease. Gleaming, protruding eyes are typical signs and are accompanied by an enlarged thyroid gland (goiter). Psychological stress, infections, and hormonal changes during puberty, pregnancy, or the menopause, can all trigger hyperactivity. The origins of the disease are unknown.
Treatment consists of medication to reduce hormone production by the thyroid gland. Alternatively, surgical reduction of the thyroid gland and treatment with radioactive iodine serve the same purpose.

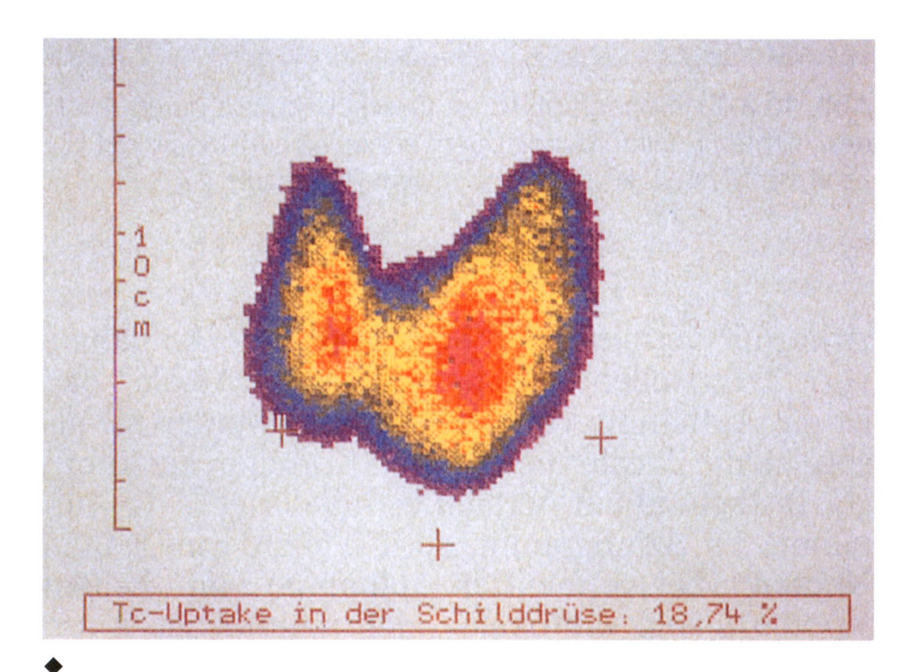

Graves' disease
Representation of a scintiscan of a Graves' disease patient with hyperthyroidism. The thyroid gland is clearly enlarged.

• *Hypothyroidism (hypothyreosis)*

Under-development of the glandular tissue in the thyroid due to reduced hormone production. The metabolism is slowed down with the result that affected persons are sluggish, sleepy, and overweight. They also have a tendency towards depression, and undergo personality changes. Their skin thickens and becomes rough, their eyelids become swollen, and their hair thins out and becomes tangled and brittle.
Congenital hypothyroidism occurs mainly in regions where there is a deficiency of iodine. It can be detected in good time as part of a medical check-up for newborns, and treated through the administration of thyroid hormones, so that the child's development is largely normal.
Acquired hypothyroidism may arise as a result of atrophy or inflammation of the thyroid tissues. This is also treated with thyroid hormones.
People refer to a cold nodule if no radioactive material has been stored in an area during a scintiscan, or if less is being stored than in the surrounding region. The majority of cold nodules are benign, but a small percentage may be cancerous.

• *Thyroid cancer*

Rare malignant thyroid tumors can usually be cured by the removal of the entire organ in good time. The causes are unknown. Radiation treatment can favor the occurrence of a tumor. Symptoms suggesting the presence of a thyroid tumor include continued growth of the thyroid gland, or an individual section of it, hardening of the gland, increasing pain, speech problems, or hoarseness.

• *Goiter (struma)*

Enlargement of thyroid gland due to an increase in thyroid-stimulating hormone from the pituitary in response to a defect in normal hormone synthesis within the thyroid gland, causing the thyroid to enlarge. It may also be caused by an inadequate supply of iodine in the food or drinking water. Goiter is not normally associated with either hyperthroidism or hypothyroidism. Goiter may not be very pronounced, but it can also take on decidedly noticeable dimensions. This looks unpleasant and can also lead to symptoms such as feelings of pressure and tightness in the neck area and feelings of numbness when wearing high-necked clothing, through to constriction of the trachea and esophagus with hoarseness, respiratory problems, and difficulties in swallowing. As the condition progresses, it can trigger impairments in the functioning of the thyroid gland. The risk of developing thyroid cancer is seven times greater where an untreated goiter is present.
Treatment usually requires only the administration of iodine (tablets) or thyroid hormones for a few months, with subsequent provision of adequate levels of iodine.

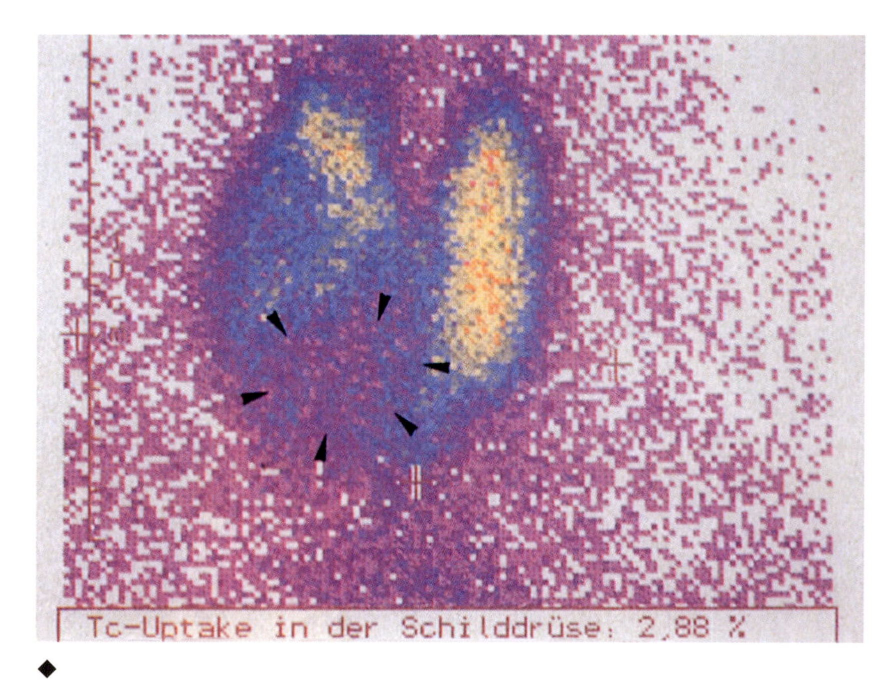

Cold nodule
The arrows point to an area in which little or no radioactive material has been stored during a scintiscan. There is no longer any activity here, and we speak of a "cold nodule" in the thyroid gland.

Adrenal glands

• *Cushing's syndrome*

Cushing's syndrome is a multisystem disorder resulting from chronic exposure to inappropriately elevated concentrations of free circulating glucocorticoids. The disturbance is caused by a problem in either the pituitary gland, which regulates hormone production (e.g., a tumor), or in the suprarenal cortex, which produces the hormone. The most frequent trigger is an enlargement of the suprarenal cortex due to a tumor.
Patients initially complain of being tired, which contrasts with their otherwise healthy appearance. Symptoms accompanying Cushing's syndrome are obesity of the trunk with slim arms and legs, a red, round face, acne, dark red/violet streaks on the sides of the stomach and the hips, high blood pressure, and diabetes. Back pains are frequently noticeable, which can be traced back to pronounced brittleness of the bone (osteoporosis). Women are about four times more commonly affected than men.
Treatment always involves reducing the increased hormone production. Surgical intervention in the vicinity of the pituitary gland or the adrenal glands is often unavoidable.

• *Addison's disease*

Hypofunctioning of the suprarenal cortex. Produced in the suprarenal cortex, hormones are needed for the metabolism of minerals and sugar in the body. In Addison's disease, an insufficiency of these hormones leads, above all, to a biochemical imbalance due to disturbances in electrolyte and fluid metabolism. The symptoms are digestive problems, loss of weight, vomiting, reduced body temperature, and increasing muscular weakness, leading to thickening of the blood, slower heartbeat, and weakening of the heart sound. The skin takes on a brown/yellow color, which is why this illness is also known as bronzed disease.
Treatment involves the administration of the deficient hormones, cortisone, and aldosterone.

Chromophile tumor

Rare, usually benign, tissue mass in the suprarenal medulla, which mostly occurs between the ages of 30 and 60. Typical symptoms are a rise in blood pressure, which is accompanied by headaches, impaired vision, and cardiac arrhythmia, together with increased metabolism, and accompanied by sweating, increased blood-sugar levels, and loss of weight. Anxiety states and trembling also occur.
The tissue mass is diagnosed through repeated urine testing, ultrasound, computed tomography, or nuclear magnetic resonance imaging, as well as scintiscans.
Treatment involves the surgical removal of the tissue mass. It is usually benign, but in rare cases turns out to be malignant, in which case surgery is followed by chemotherapy.

Masculinization (virilism)

The development of male bodily characteristics in women. A woman's adrenal glands and ovaries actually always produce small quantities of male sex hormones. Excessive production of male sex hormones can result from diseases affecting the adrenal glands or from specific types of ovarian tumor, so a thorough examination is necessary.
The symptoms are excessive body hair, male hair-line (frontal baldness), menstrual irregularities, clitoral enlargement, alterations in typical female fat distribution, prominent arm and shoulder musculature, and a deep voice. How pronounced these external changes are depends on the quantity of male sex hormone being produced.
Treatment involves either the removal of the tumor or the use of medication.

Hirsutism

The development of male hair patterns in women (e.g., chest hair, stomach hair, pudendal hair or hairy legs, with a more or less pronounced beard).
Hirsutism is usually constitutional or is the expression of an excess of male hormones. When pronounced hirsutism develops suddenly, this can be a sign of serious illness, such as a tumor of the suprarenal cortex. Other possible causes are the secondary effects of medication containing hormones, dope, or certain types of medication for the suppression of the immune system.
The hair is bleached or removed. The balance between male and female hormones is re-established with the help of medication.

The menopause

The development of the organism goes through several phases of life, from birth through to being a senior citizen. During this process, physical functions change in ways typical of the phase. When the female menopause starts, a woman's fertility begins to come to an end. This stage is accompanied by physical and mental changes, which are experienced more or less intensively by all women. Many women find this stage of their lives to be very stressful, while others see it as the beginning of a new and positive phase of their lives.

As recently as the beginning of the twentieth century, a woman in her fifties was considered to be an old lady, even in industrialized countries. Today, due mainly to an extended life expectancy, she is in the middle of her life, and many women at this stage of their lives reflect the image of a self-confident, independent, mature woman. On the other hand, many women feel they have been pushed out in the cold, having lived a life full of problems and suffered at the hands of fate, and think of themselves as old and worn out.

A break with former life

The menopause is characterized, not only by bodily changes, but, for many women, also changes in the habits and the circumstances of their lives. Many women in this phase of their lives have a feeling of loneliness. They have finished bringing up their children. Their husbands are busy with their work. If a woman has not previously had an independent career, she may not have any activities with which to fill her life. Her marriage may be in crisis, or already over. Women who are still, or once again, living alone, may believe they have not achieved what they wanted to in life, and that they are no longer capable of achieving it. And with many women, there is also the feeling that they are losing their attractiveness. Psychological problems may thus be an extra burden during the menopause.

Seeing changes as an opportunity

The change to a new phase of life need not necessarily bring problems with it, however. Aging can also present new opportunities. The pre-condition for this is that the women in question do not just look back on the years of their youth with wistful nostalgia, but also look forward, accepting the bodily changes which are occurring, many of which may not even be noticeable. Moreover, the menopause, frequently the object of so much dread, with its many levels of symptoms, has largely lost its terrors, since most of them can now be handled, thanks to medical progress.

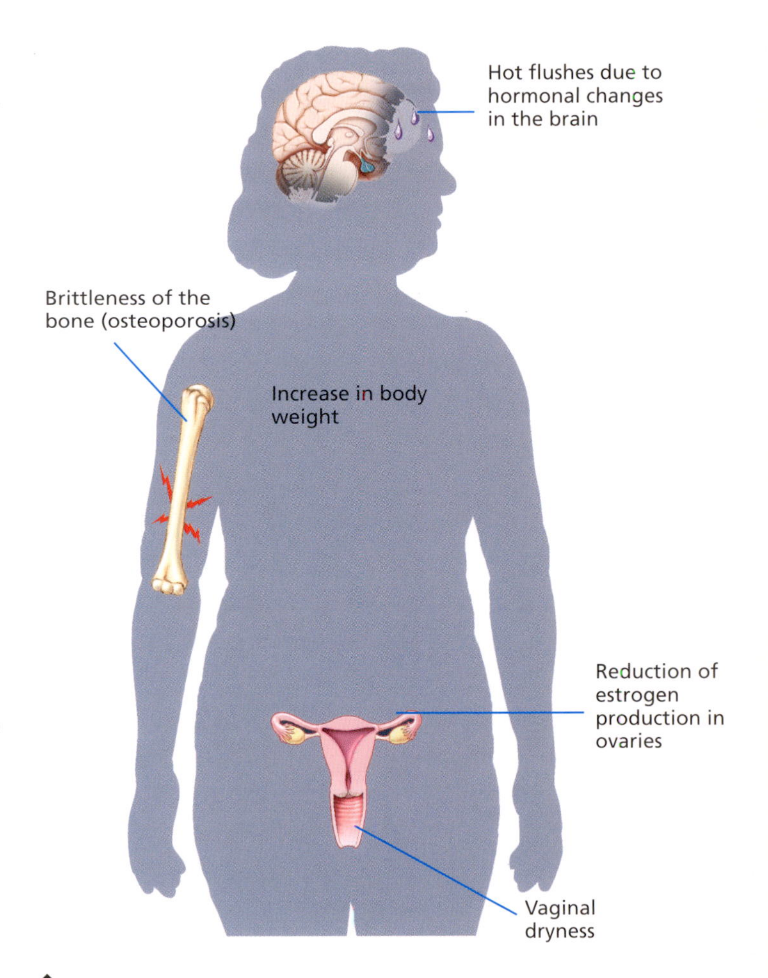

◆ Falling hormone levels are responsible for most of the organic changes which occur in women during the menopause.

You don't notice anything at first

Most women go through the menopause (climacteric) between the ages of about 48 and 52. The climacteric announces its arrival slowly, with various indications. The slight wrinkles around the eyes, chin and mouth deepen, and bad moods can become more frequent than previously. Headaches, sleep disturbances, circulation problems, vertigo, tachycardia, and hot flushes can be troublesome for many women. Finally, menstrual periods become irregular, and at some point stop completely. This is a sign to every woman of the end of the stage in life in which she can have children. Some women find this difficult to cope with and become depressed, because they now feel that their lives are pointless.

Hormones control the process

The origin of these processes lies in the hormonal balance. As has already happened once at puberty, the female organism experiences a hormonal change. The production of the sex hormone estrogen in the ovaries diminishes, and finally dies off completely. The first result of this is that an egg no longer matures in the ovary each month. Ovulation becomes rarer, and finally the ovarian function dies out, as a result of which periods become irregular and then cease forever.

Typical bodily indications

The numerous physical changes that go together with the hormonal change are not noticed at first, as they begin gradually. Thus, the uterus and the mucous membranes in the vagina, urethra and bladder, gradually atrophy. Because the vagina is too dry, sexual intercourse can be painful. The skin and the breasts start to become slacker. Since the estrogen needed for bone metabolism is no longer available in sufficient quantities, the risk of osteoporosis increases.

Remaining vital and attractive

The declining production of estrogen is predominantly responsible for many of the physical and psychological symptoms during the menopause, though treatment is not normally necessary. However, around a third of women suffer a lot from symptoms caused by lack of estrogen, and may decide to have this hormone medically replaced in order to alleviate their symptoms, and most of all to prevent osteoporosis. Estrogen (hormone replacement therapy) can be given in the form of tablets, plasters or ointments for a few years but chronic treatment is no longer recommended.
A healthy lifestyle, balanced nutrition, and sufficient sleep, plus playing an active part in life, e.g., by starting new activities, play a considerable part in remaining attractive (to yourself, in particular), even when you are older.

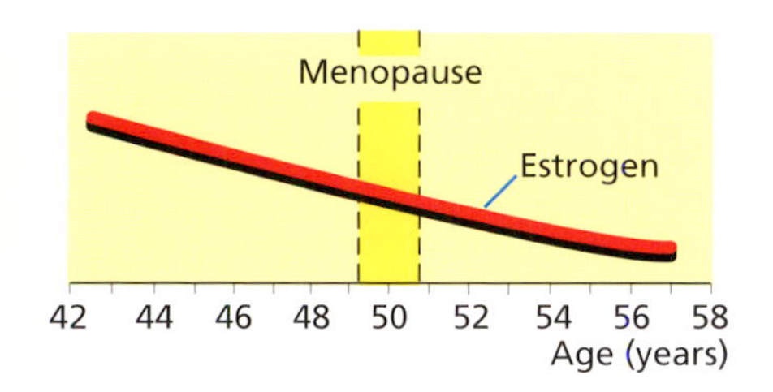

◆ From the age of about 40 onwards, estrogen levels in the blood fall continuously.

◆ What is especially important during the menopause is not to look back on your youth with wistful nostalgia, but to accept your current life situation and take a positive view.

Sexuality

Desire and love are not a privilege of youth. Pleasure can still be obtained from physical love, even when you are older, and the capacity for orgasm can be retained. There are many women who experience much stronger sexuality after the change of life. Many women also enjoy their sex life more because they no longer need to be afraid of an unwanted pregnancy (though contraception should still be used during sexual intercourse until there have been no more periods for a year).

When men reach middle age

Changes comparable to the menopause in women also take place to a lesser extent in men. Production of the male sex hormone testosterone is reduced, semen production and physical strength deteriorate, and the skin becomes less taut. However, since these changes occur very slowly, and the production of hormones does not cease altogether, the symptoms are not as marked as they are in women. Even if sexual desire is reduced in many men in their fifties, potency is retained until much later.

Diagnostic methods

Hormonal analysis

Suspicions of hormonal disturbance are usually aroused by a simple physical examination, i.e., an assessment of external appearance (skin, hair, increased neck girth), gait and posture. A description of symptoms may confirm these suspicions, causing the physician to take a blood sample so that hormone levels can be checked in the laboratory.
Hormonal function tests (hormonal stimulation or limitation tests) may also be necessary, depending on which hormonal disturbance is suspected. Such tests serve to monitor the hormonal control circuit. Under normal circumstances, the hypothalamus and the pituitary, as control centers, distribute the so-called "control" hormones, which stimulate the individual organs (e.g., the pituitary or the adrenal glands) to produce their own hormones. When sufficient hormone is produced, the organs communicate this to the relevant control center, which then limits the distribution of the control hormone. This feedback mechanism ensures that hormone levels are always adequate.
In order to carry out a hormonal function test, the physician injects a specific control hormone, and can then see from the hormone level whether the control hormone is indeed having the intended effect in the hormone-producing organ for which it is responsible. Thus, the type and location of the hormonal disturbance can be identified unambiguously.

Radioiodine test

Thyroid gland function test. Iodine is normally accumulated in the thyroid gland. By ingesting radioactive iodine (I-131), thyroid tissue activity can be displayed and any cold nodes or functional disturbances in the thyroid gland can be recognized. During the test, the patient drinks I-131, diluted with water. The amount of I-131 stored in the thyroid gland is measured at fixed intervals over a period of two days.

Thyroid scintigraphy scan

A procedure that gives an image representing the thyroid gland, with the help of a short-lived radioactive substance.
The substance is injected into the blood vessels and accumulates in the thyroid gland. Radiation from the thyroid gland tissue is picked up by a special camera and made visible.
In this way, information on the position, form, and size of the thyroid gland is obtained, together with information on the presence of tumors, cysts, and abscesses (indicated by the accumulation of radioactive materials in specific regions).

Treatment methods

Thyreostatics

Medications for treatment of hyperthyroidism. Thyreostatics limit the production and distribution of thyroid hormone in the thyroid and its release it into the bloodstream. The success of the therapy is monitored by determining the hormone level in the blood through laboratory testing.

Radioactive iodine therapy

Treatment with I-131 in the event of severe hyperthyroidism, a large goiter or thyroid tumor, which cannot be surgically removed. Radioactive treatment acts almost exclusively on the thyroid gland tissue, since this is where the iodine is most likely to be stored, and it is precipitated out again after a short time. Due to the dangers of increased radioactive contamination in susceptible periods, I-131 is not used in pregnant women and young children.

Hormone therapy

Therapeutic use of hormones or hormone precursors, which are converted to active hormones in the body. It is the most common form of treatment for hormonal disturbances.
Hormone therapy can be carried out to replace natural hormones that are no longer being produced in sufficient quantities in the body. This includes estrogen-replacement therapy for women undergoing the menopause, or who have had their ovaries removed.
Excessive or undesirable production of the body's endogenous hormones can also be combated using anti-hormones. Those most frequently used are anti-androgens, which are directed against male sex hormones. In women, they are given to combat excessive hair growth, while in men they combat prostatic cancer, both of which are promoted by male sex hormones. Hormone therapies can be swallowed, injected or administered through a skin plaster.

Sexual organs

The primary function of the male and female sexual organs is that of reproduction. They perform a variety of tasks in this regard: they produce gametes, namely egg cells in the woman and sperm cells in the man, and sexual hormones, as well as secretions to create the ideal conditions for the transportation and fusion of gametes.

The female sexual organs

The distinction is made between the external and the internal sexual organs. The woman's external sexual organs include the large and small pudendal lips, the clitoris, and the vulval vestibule. The large pudendal lips, comprising fatty and connective tissue, are covered with hair on the outside and contain sebaceous, sweat, and apocrine sweat glands. The small pudendal lips, on the other hand, situated further inside, are soft, hairless folds of skin which frame the vulval vestibule. The clitoris, which is only 1–2 centimeters in size, is surrounded by spongy bodies and swells as a result of sexual arousal. A membrane, the hymen, is located between the vulval vestibule and the vagina, and this membrane tears when sexual intercourse is experienced for the first time ("deflowering").
The internal sexual organs include the vagina, uterus, fallopian tubes, and ovaries. The vagina provides the link between the lower end of the uterus, the cervix, and the external genitals, and is about 10 centimeters long. Its mucous membrane secretes an acidic fluid and also contains bacteria (known as "vaginal flora") which protect it from germs entering from outside.

The uterus, a pear-shaped hollow muscle, is lined with a thick mucous membrane. Under the influence of the female sexual hormones, it undergoes significant changes during the course of a monthly cycle. Initially it thickens and prepares itself for the implantation of a fertilized egg. If fertilization does not take place, the thickened mucous membrane layer is disposed of and (menstrual) bleeding takes place.

The fallopian tubes, which are about 15 centimeters in length, enter the upper uterus on each side. They cover the surface of the ovaries with their extended, funnel-shaped, fringed ends. Their task is to collect the egg following ovulation and to transport it to the uterus. The ovaries, which are each about the size of a plum, produce the female sexual hormones. In women of childbearing age (between puberty and the menopause), an egg cell capable of being fertilized develops inside an ovarian follicle in one of the ovaries about once a month. The follicle bursts at the time of ovulation, enabling the egg to reach the fallopian tube through which it is transported to the uterus.

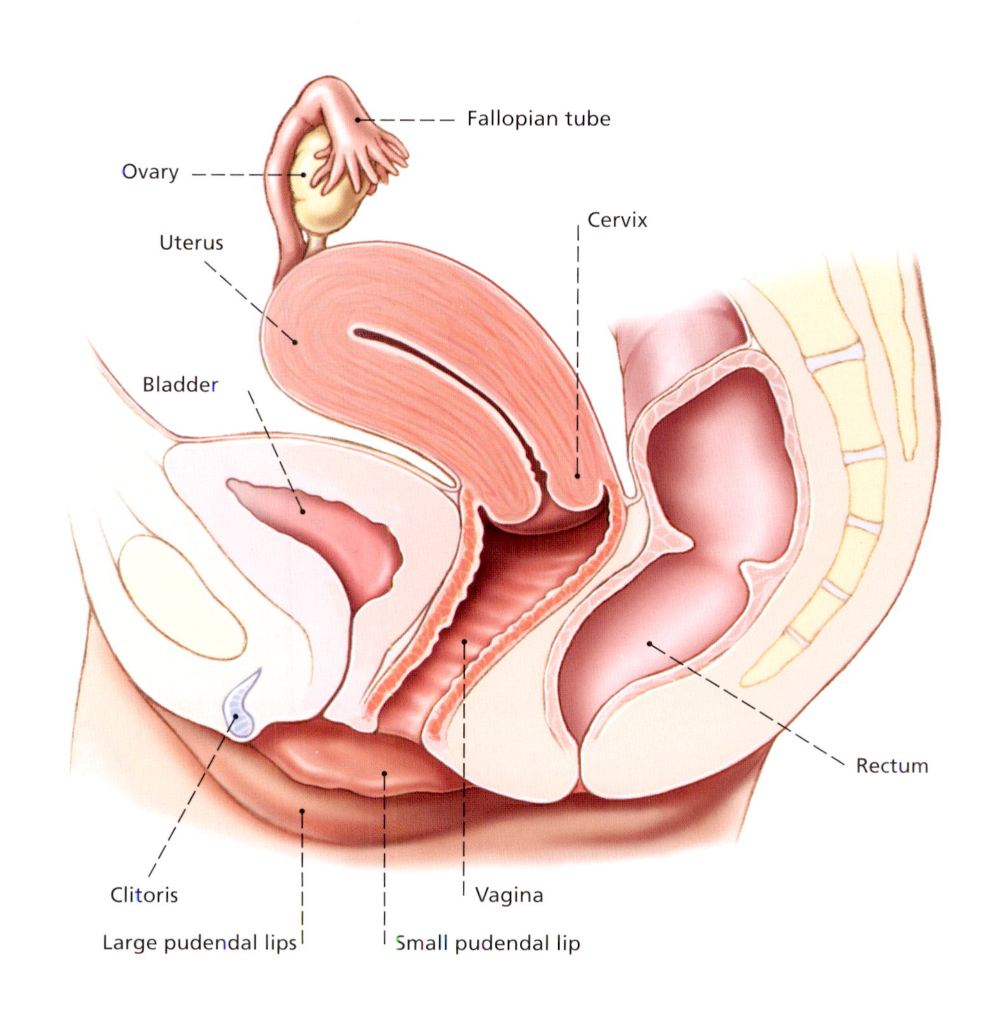

◆
Fig. 405
The female sexual organs (longitudinal section through the center of the pelvis).

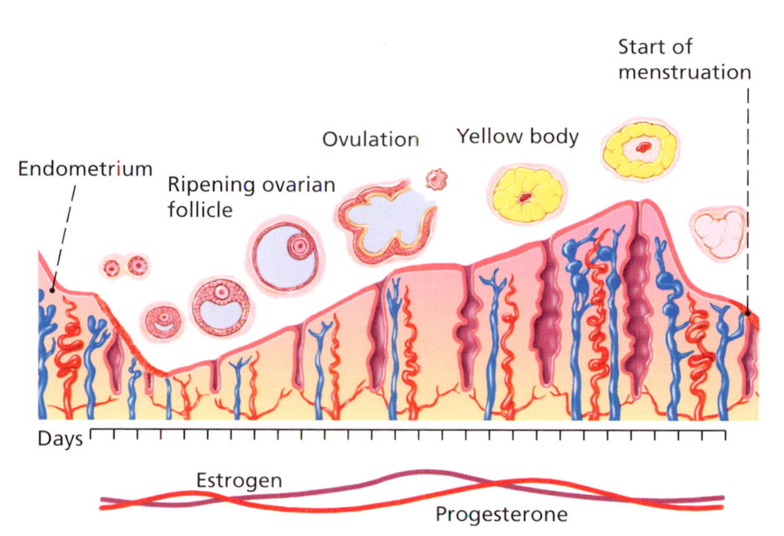

◆
Fig. 406
The menstrual cycle lasts about 28 days. During this time the endometrium thickens under the influence of the hormones estrogen and progesterone. Ovulation takes place in the middle of the cycle. The egg perishes if it is not fertilized and its follicle then transforms into a yellow body. The endometrium is shed at the end of the second half of the cycle (menstruation).

The male sexual organs

In the man, too, a distinction is made between the external and the internal sexual organs. The external sexual organs comprise the penis and the scrotum, while the testicles and the epididymi, spermatic duct, prostate gland, and seminal vesicles make up the internal sexual organs.

The penis is covered with flexible, moveable skin forming a double-layered fold over the tip of the penis (glans), known as the foreskin or prepuce. It is joined to the glans on the underside. The sebaceous glands of the foreskin produce a white secretion. Spongy bodies are located on the upper side of the penis. These comprise a sponge-like, connective tissue network, the hollows of which fill with blood during sexual arousal. The venal blood flow through the veins of the penis is restricted at this time, causing the penis to become stiff and erect (erection). The blood drains away again once arousal has subsided and the penis becomes flaccid.

The pair of testicles produces the male sexual hormone, testosterone, and it is also in the testicles that the male gametes (sperm) are produced. These move from the testicles to the epididymi where they are stored and where they mature. The testicles and epididymi are housed in the scrotum, together with part of the spermatic duct. The temperature in the scrotum is a number of degrees below normal body temperature, an essential condition for the formation of the gametes. This heat regulation is carried out by the skin covering the testicles, which is very temperature-sensitive. It contracts when cold to reduce its surface and to therefore give off less heat. If it is too warm, on the other hand, the skin becomes flaccid and gives off more heat.

The spermatic ducts, which are about 50 centimeters long, extend from the epididymi through the abdominal canal to the abdomen. They collect the secretions of the seminal vesicles, located behind the bladder, shortly before they enter the prostate gland. The spermatic ducts and excretory duct of the prostate gland combine to form one duct and enter the urethra together. Both the seminal vesicle and the prostate gland produce milky secretions together forming the seminal fluid. The prostate gland is located directly beneath the bladder and encircles the urethra. An enlargement of the prostate is therefore always accompanied by a disruption to bladder emptying.

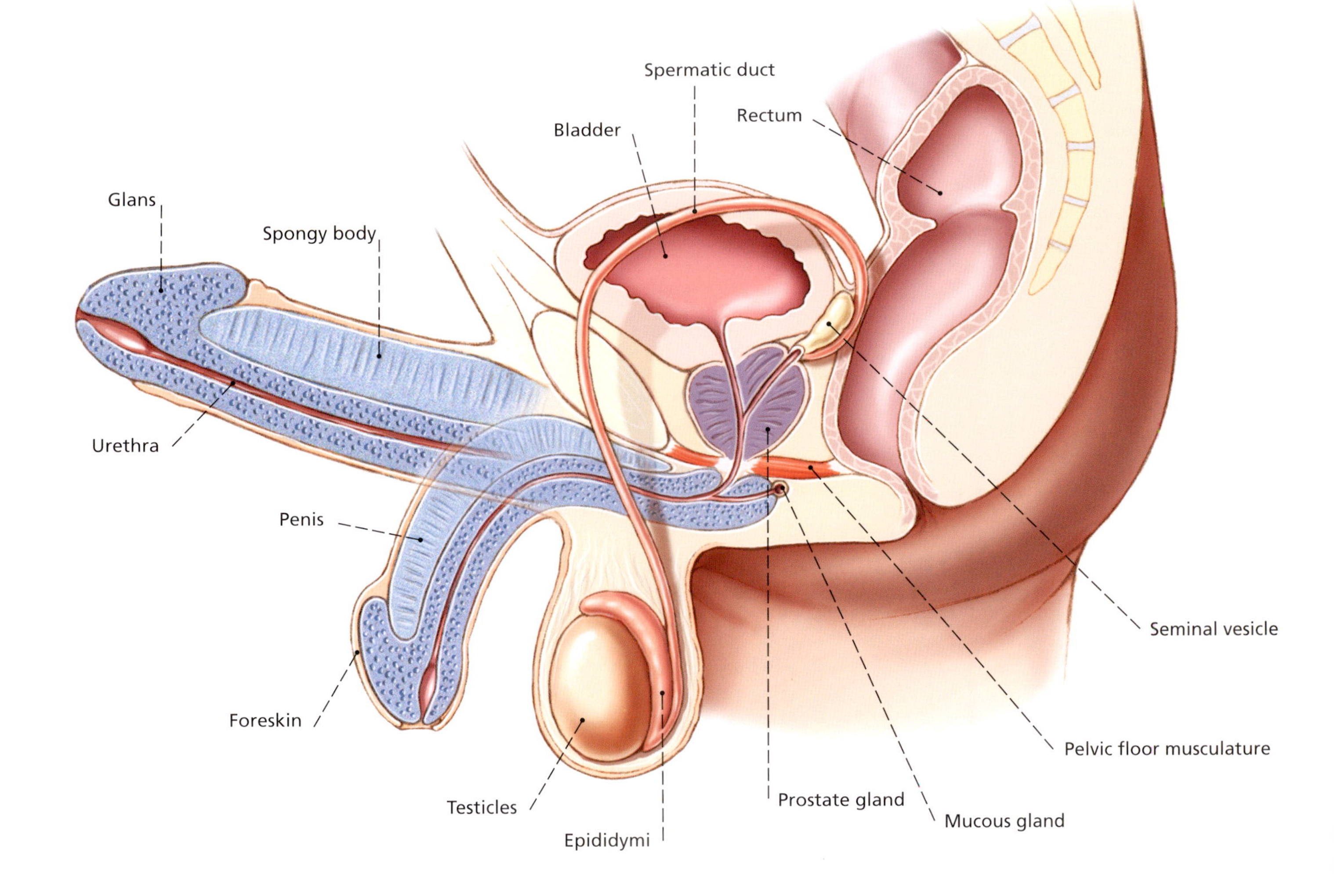

◆
Fig. 407
The male sexual organ (longitudinal section through the center of the pelvis). The penis can swell to three times its volume during an erection.

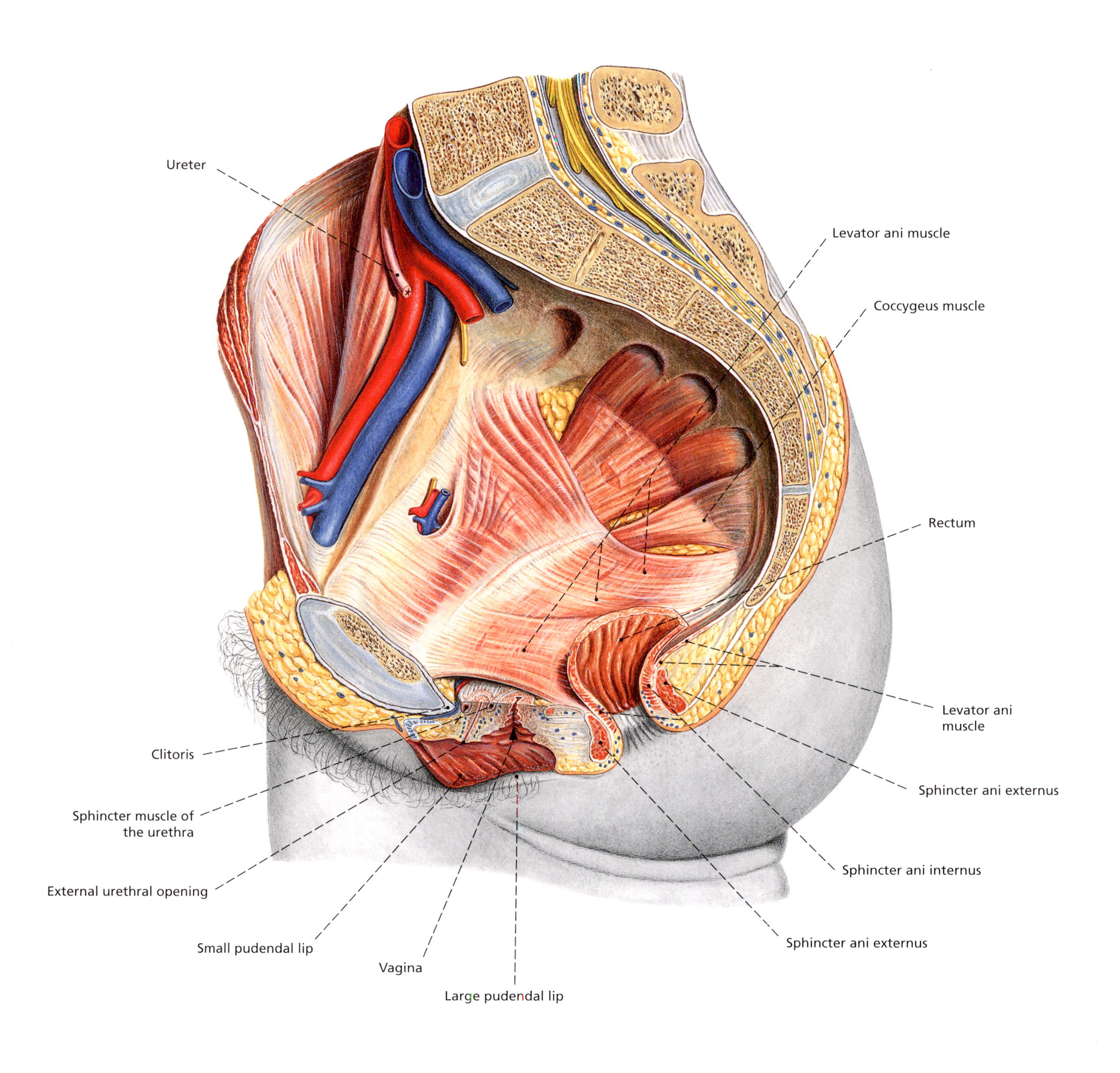

◆
Fig. 408
Musculature of the female pelvic floor.

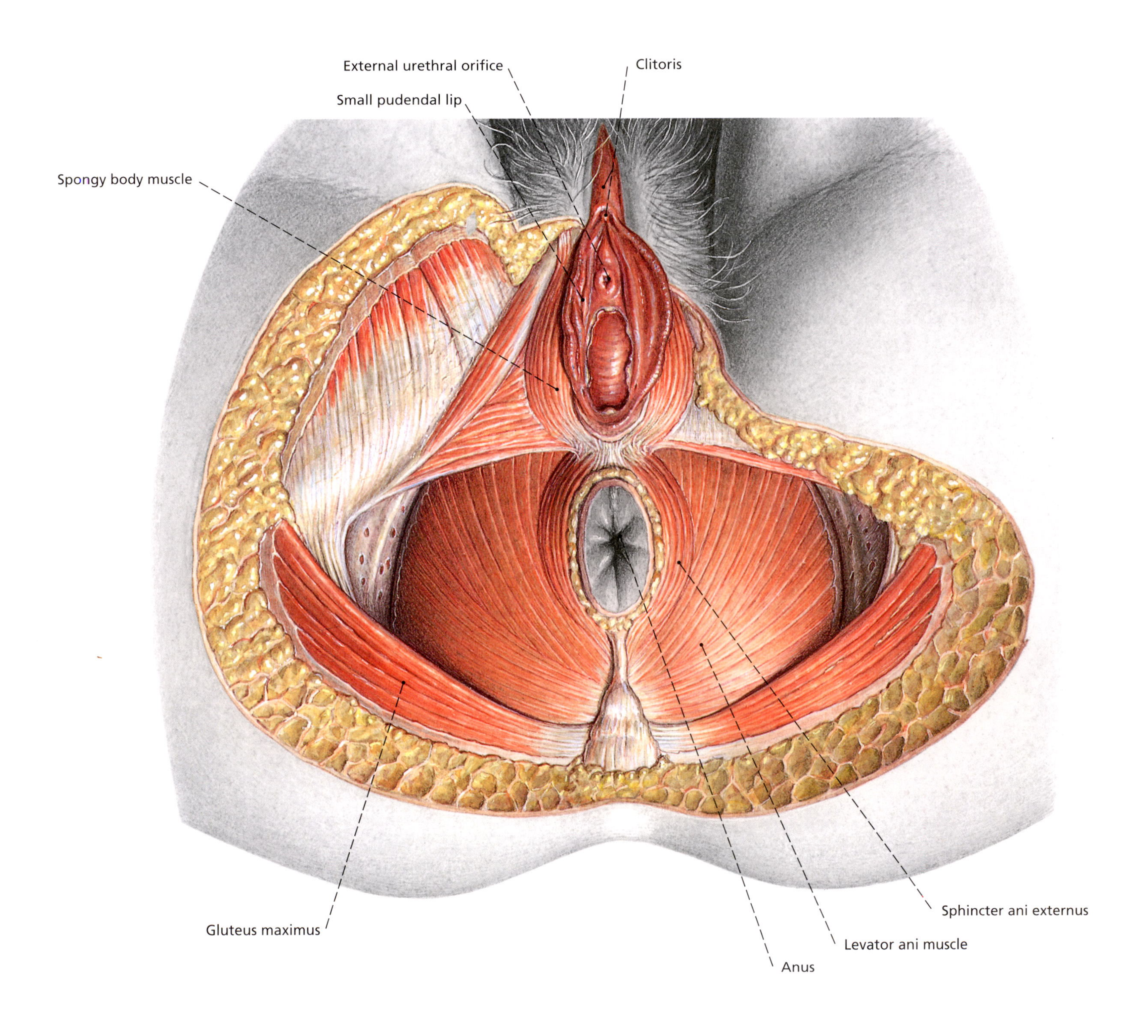

◆
Fig. 409
The perineal region in a woman following the removal of the skin and the fatty tissue. The outer layer of the musculature surrounding the vagina and anus is depicted.

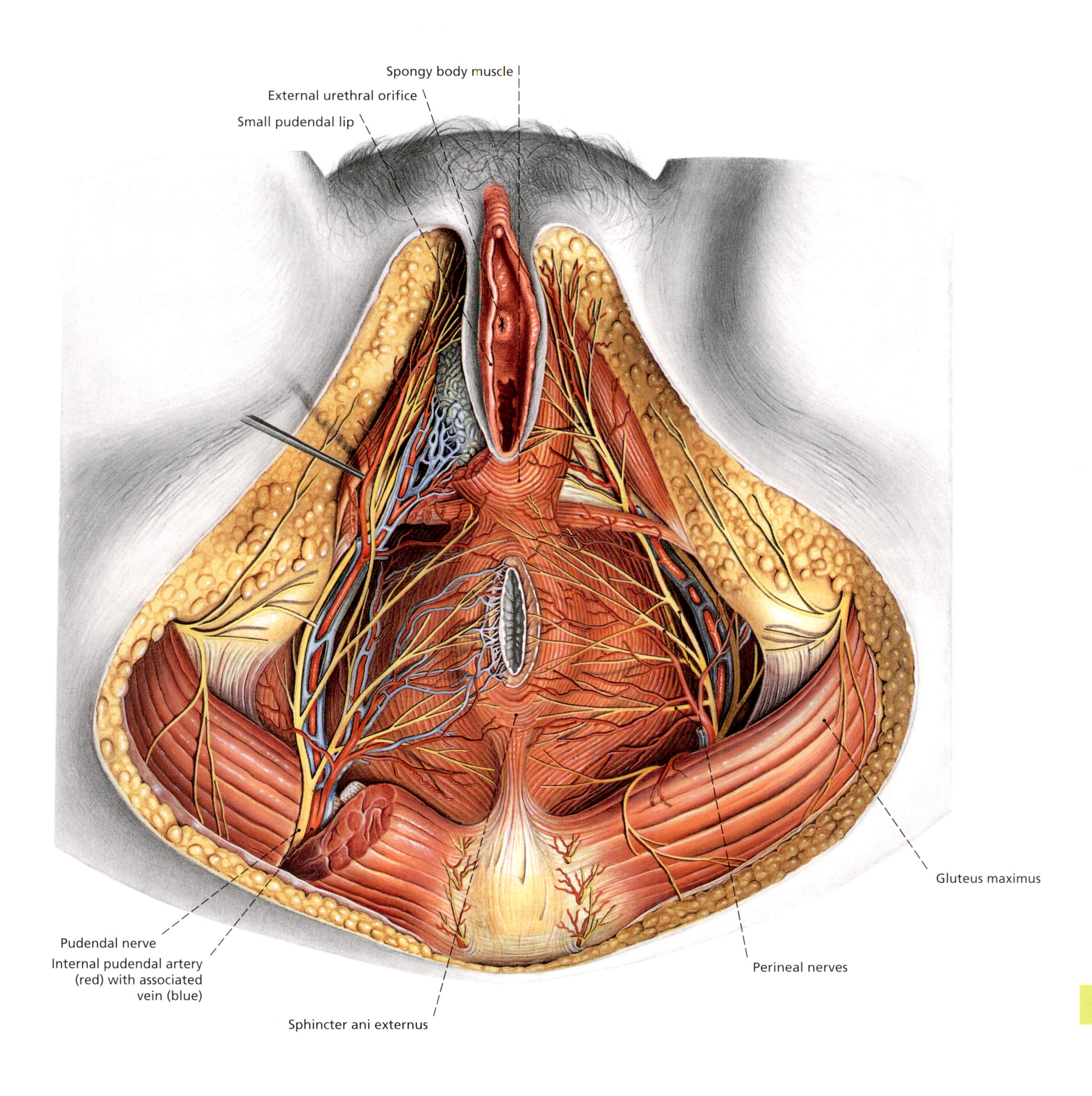

◆
Fig. 410
The perineal region in a woman, showing the vessels and nerves. The skin and the fatty tissue have been removed. The gluteus maximus to the right has been cut through to show the course of the nerves and vessels.

◆
Fig. 411
The female pelvic floor muscles from below.

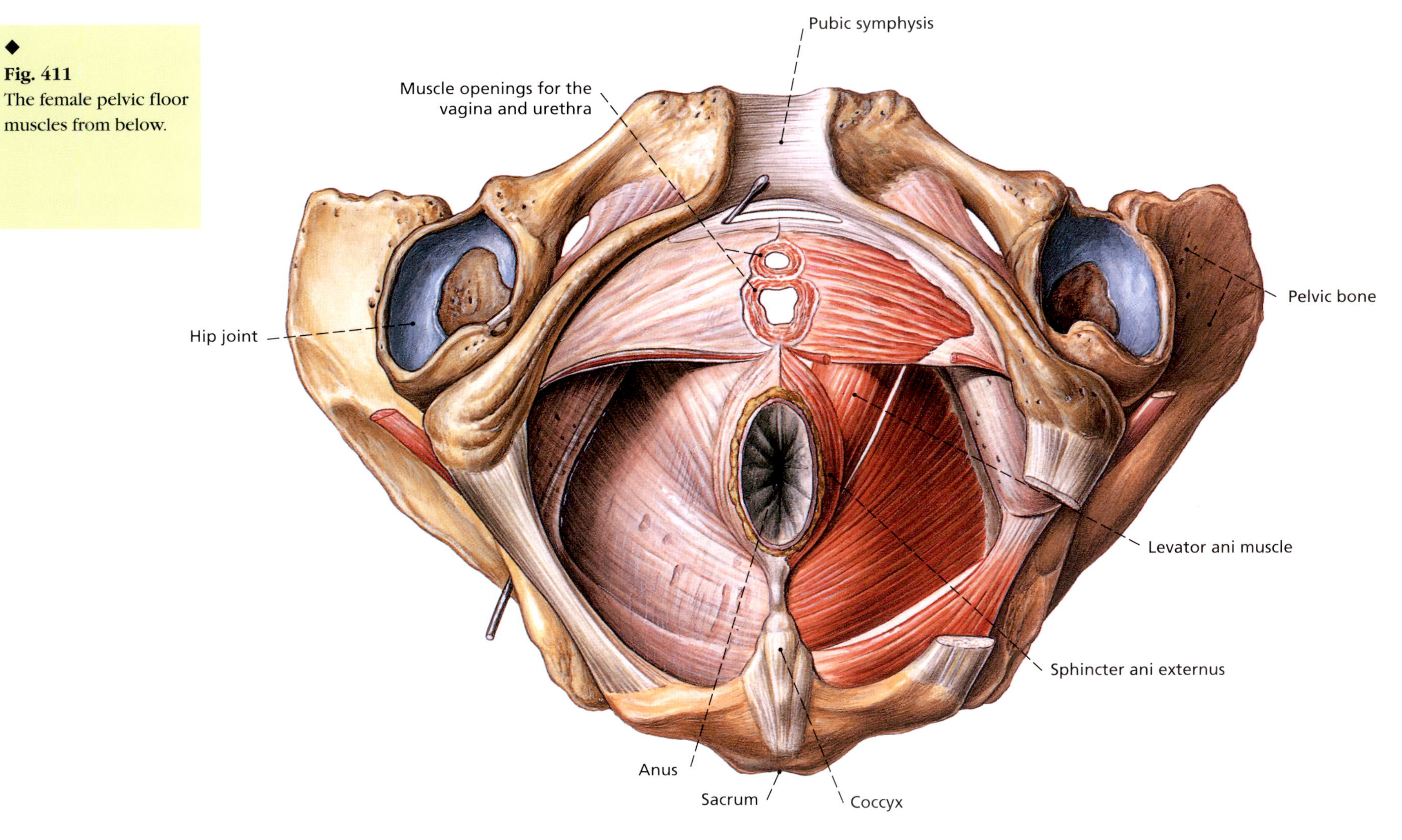

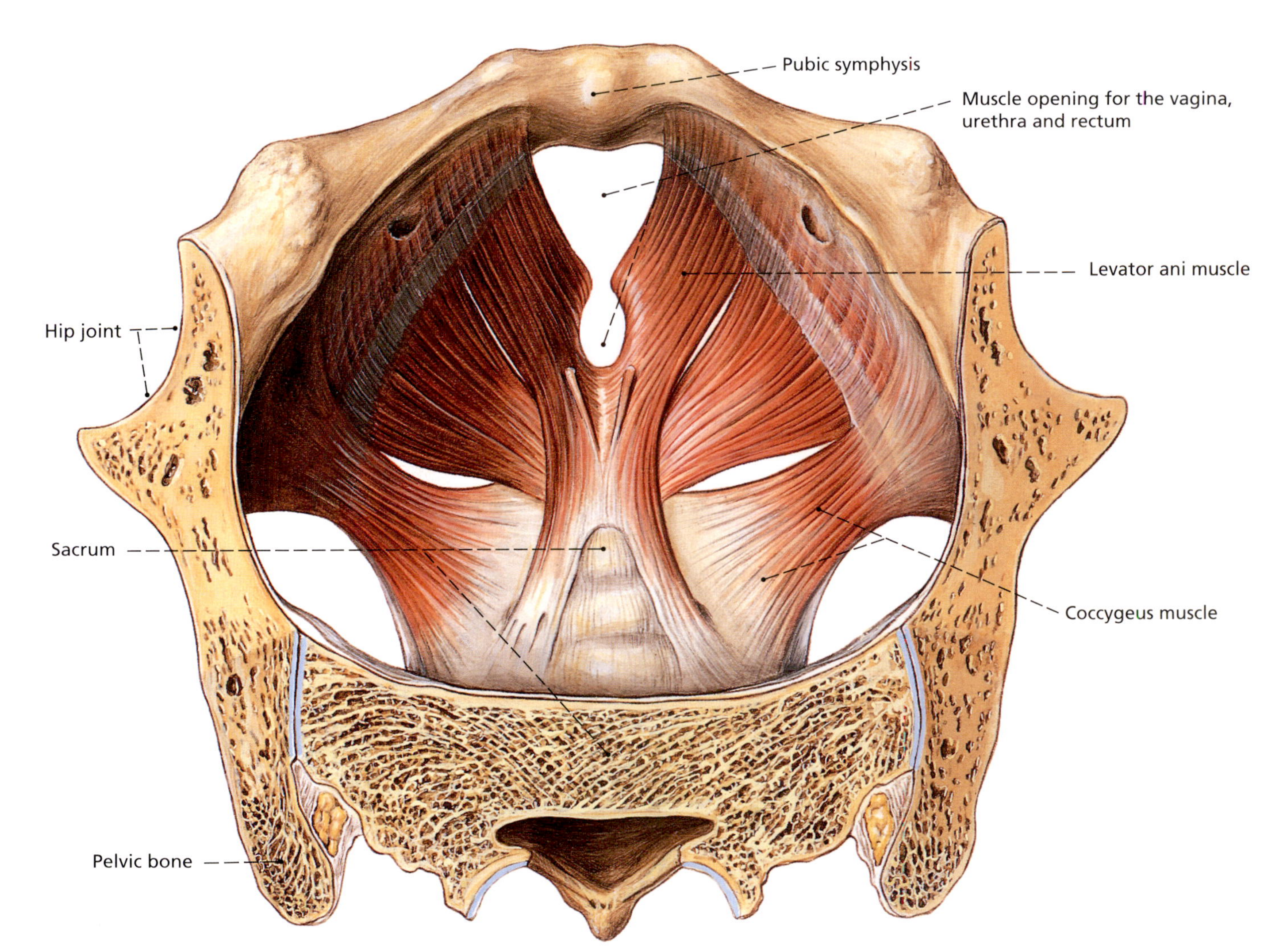

◆
Fig. 412
The female pelvic floor musculature. Viewed from above into the pelvic cavity. Both the upper section of the pelvic bone and the sacrum can be seen.

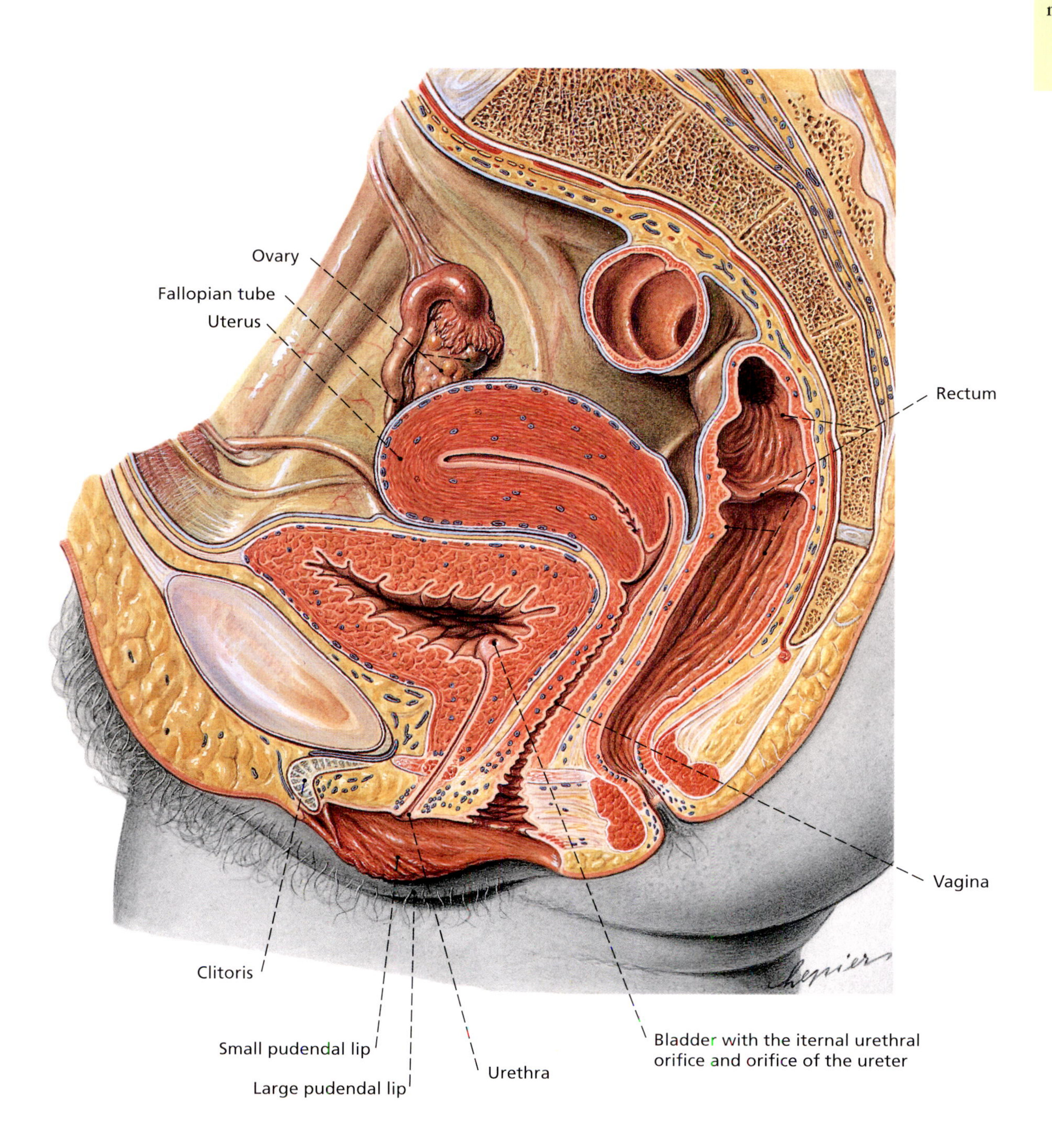

◆
Fig. 413
The female pelvis. Cross-section through the pubic symphysis. The large intestine has been removed as far as the rectum (longitudinal section).

◆
Fig. 414
The uterus with the fallopian tubes and ovaries. Seen from the front. The abdominal wall has been removed completely up to the pubic hairline. The uterus is held up slightly with a hook in order to depict the peritoneal fold between it and the bladder. The left fallopian tube is held up with a band in order to show the fringed funnel.

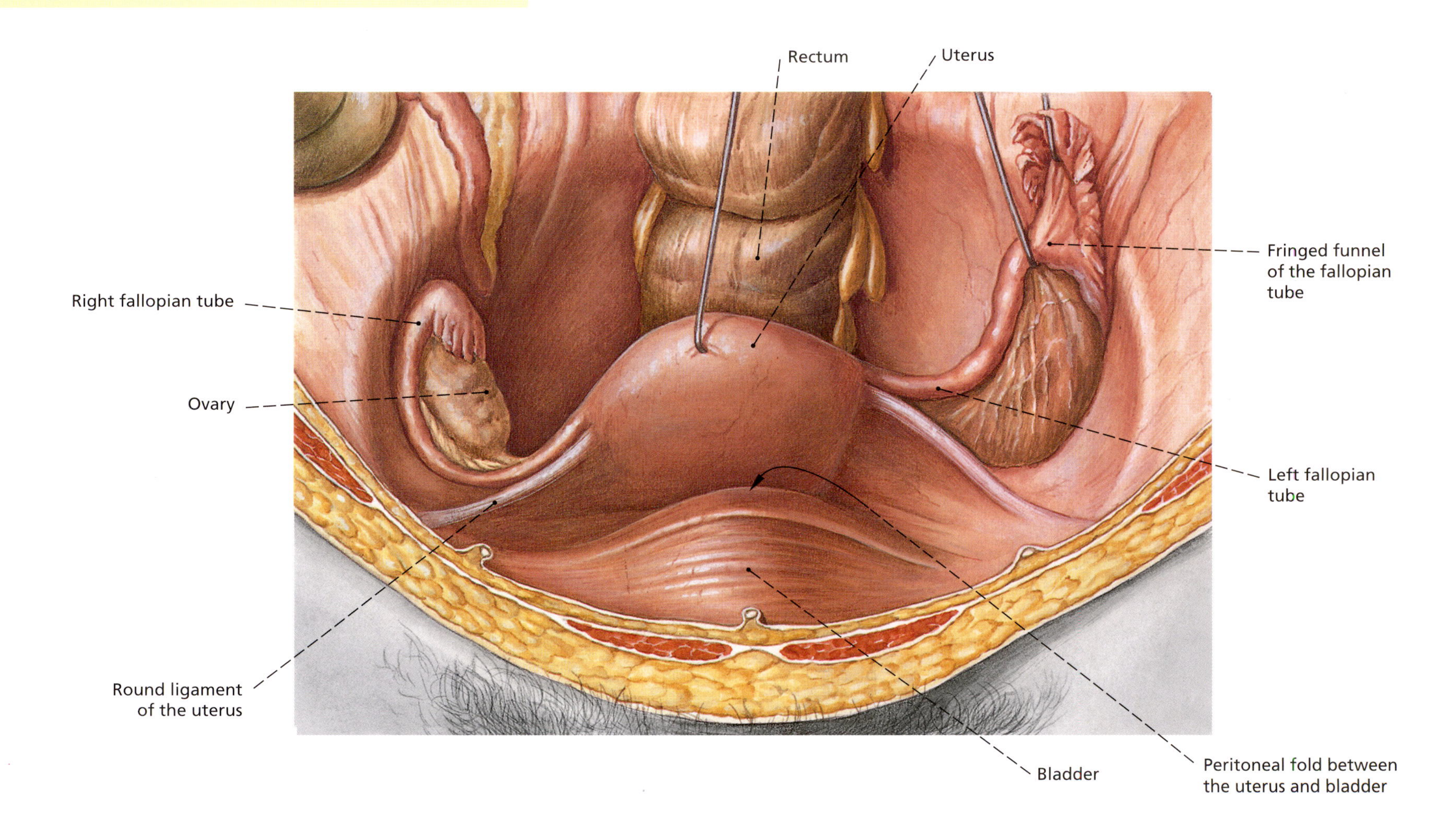

◆
Fig. 415
A longitudinal section of the uterus depicting the uterine cavity seen from the side. The fundus indicates the direction of the anterior abdominal wall.

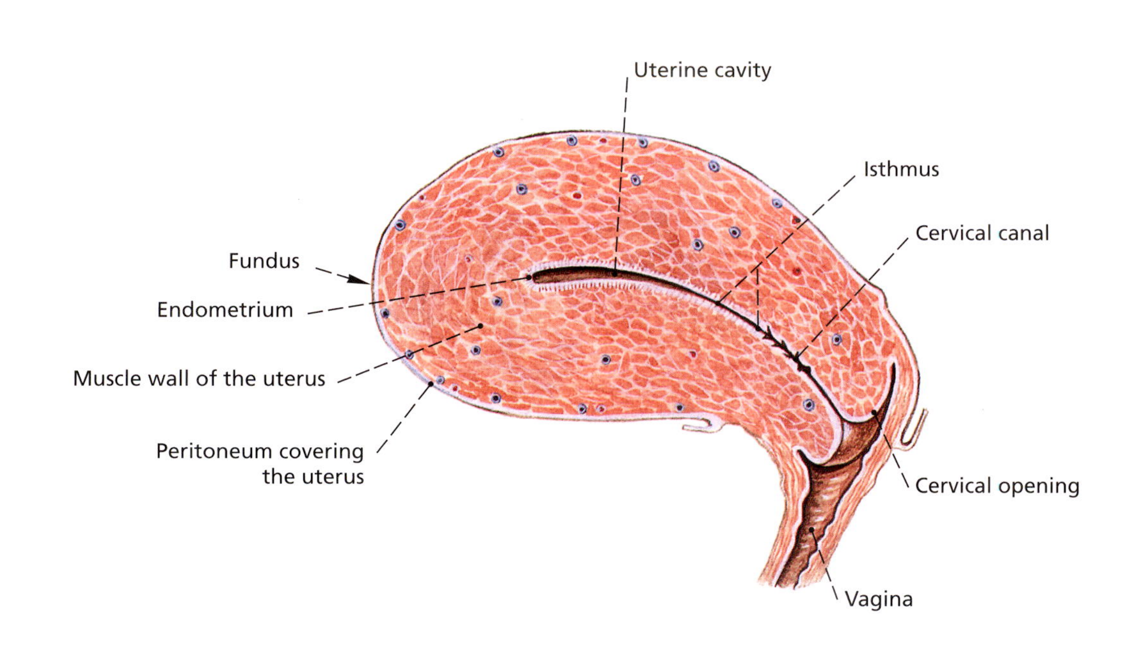

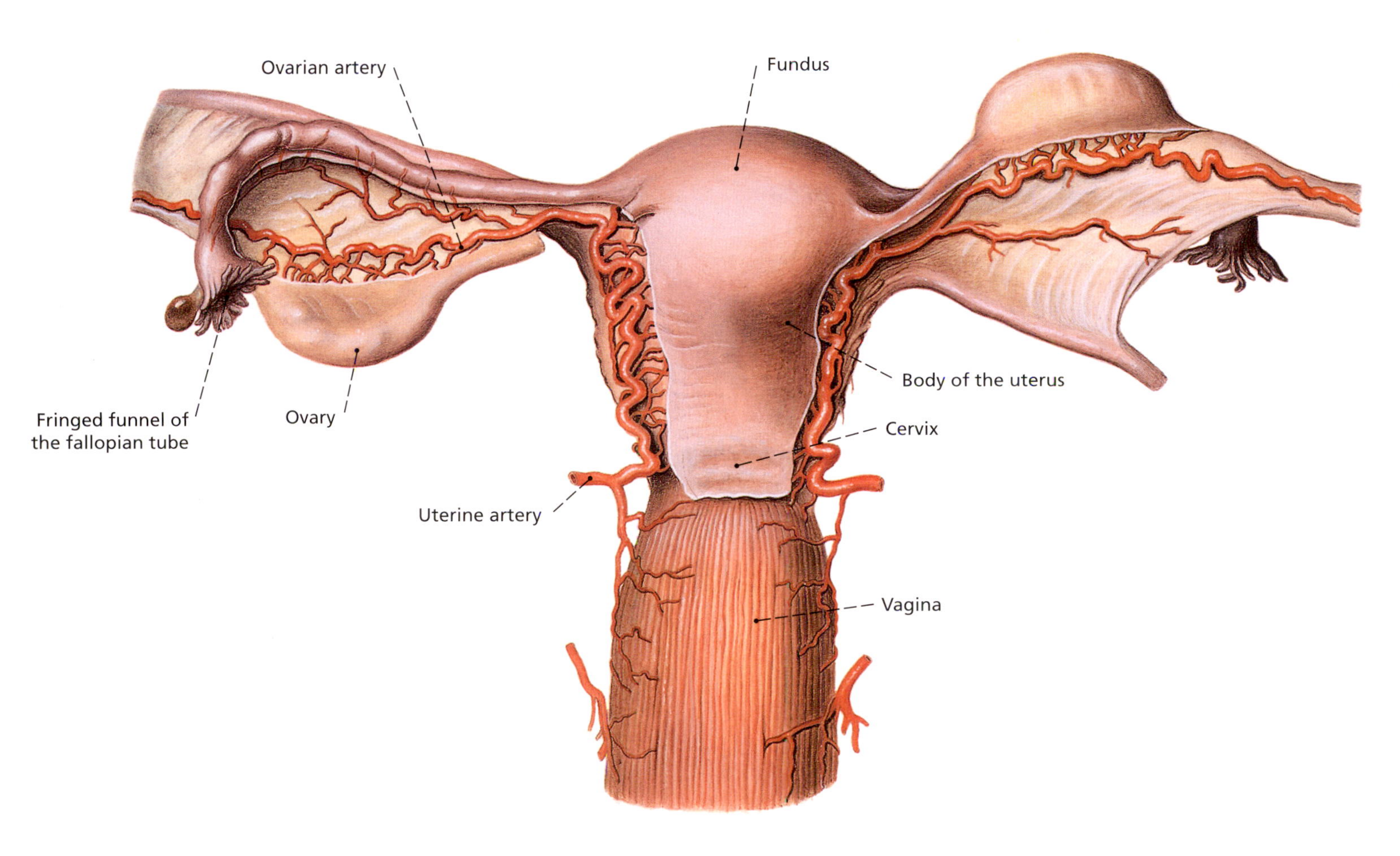

◆
Fig. 416
The arterial vessels of the female internal sexual organs. Part of the outer mucous membrane layer (the peritoneum) has been removed.

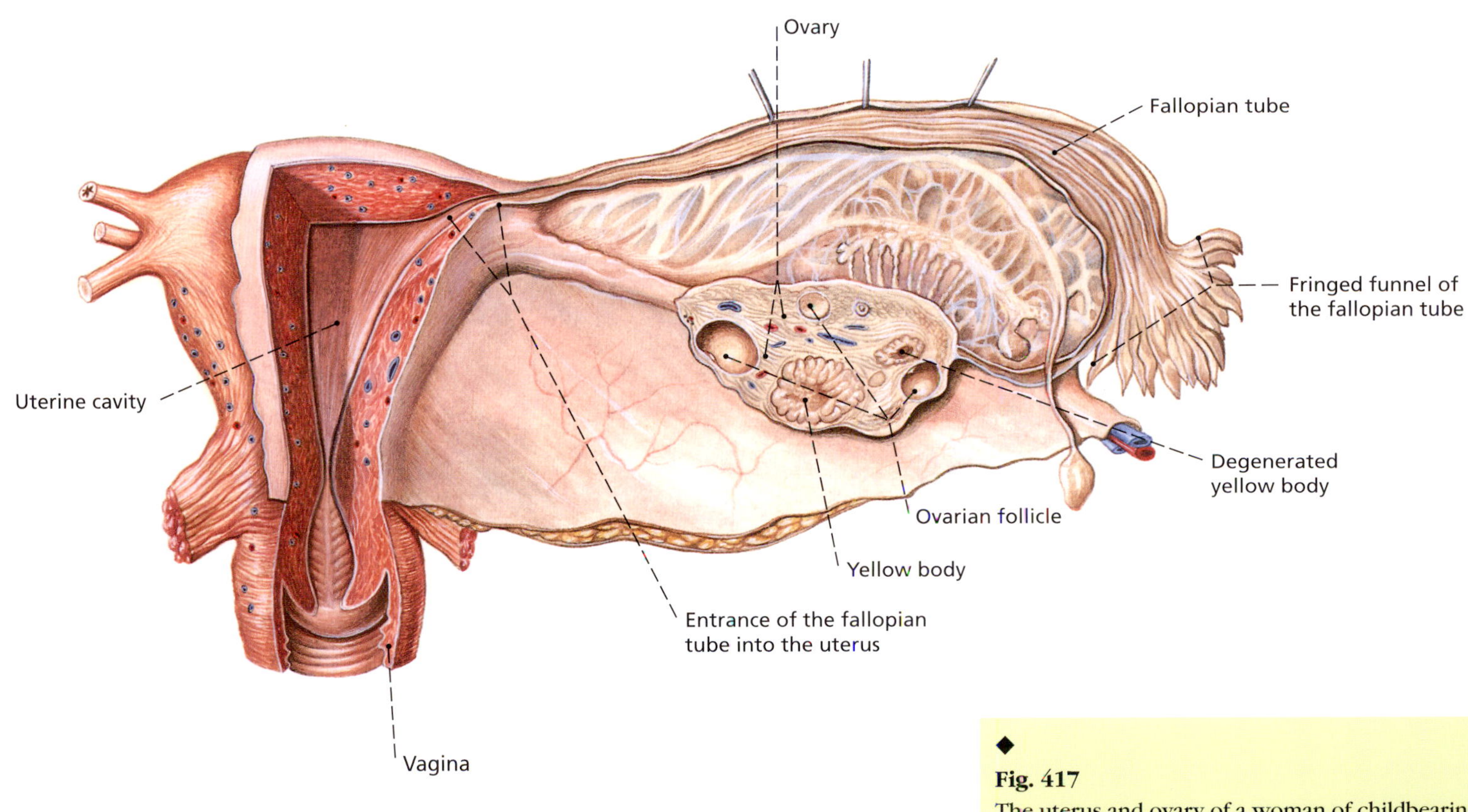

◆
Fig. 417
The uterus and ovary of a woman of childbearing age. A wedge-shaped section of the uterine wall has been removed in order to reveal the uterine cavity. The ovary and fallopian tube are shown in longitudinal section.

◆
Fig. 418
The position of a baby in the mother's abdomen shortly prior to birth. The baby's head is generally pointing downwards. The intestines of the pregnant woman are pushed upwards and to the side.

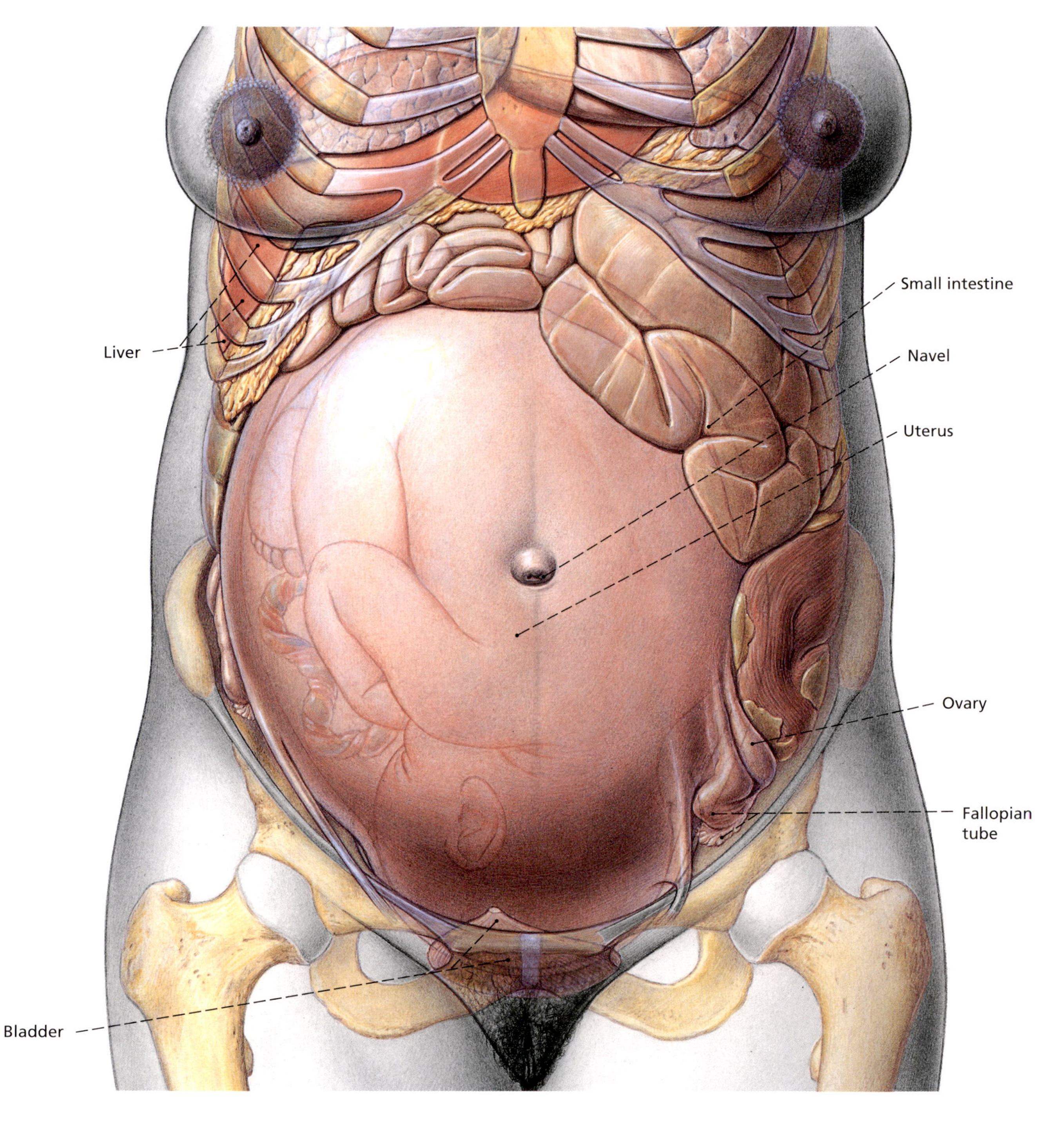

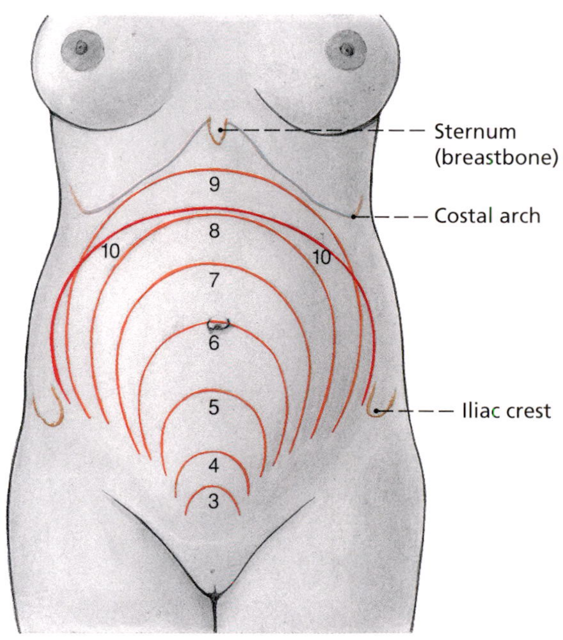

◆
Fig. 419
Changes to the height of the fundus during pregnancy. The numbers refer to the end of each corresponding pregnancy month, which are calculated at 28 days each. The uterus sinks again shortly before the birth, in the 10th month of pregnancy, because the baby's head then enters the lower pelvis.

Fig. 420
The position of a baby in the uterus, seen from the side. The fundus is tilted forwards, and the baby's head is pointing towards the cervical opening. The wall of the uterus becomes thinner towards the end of pregnancy.

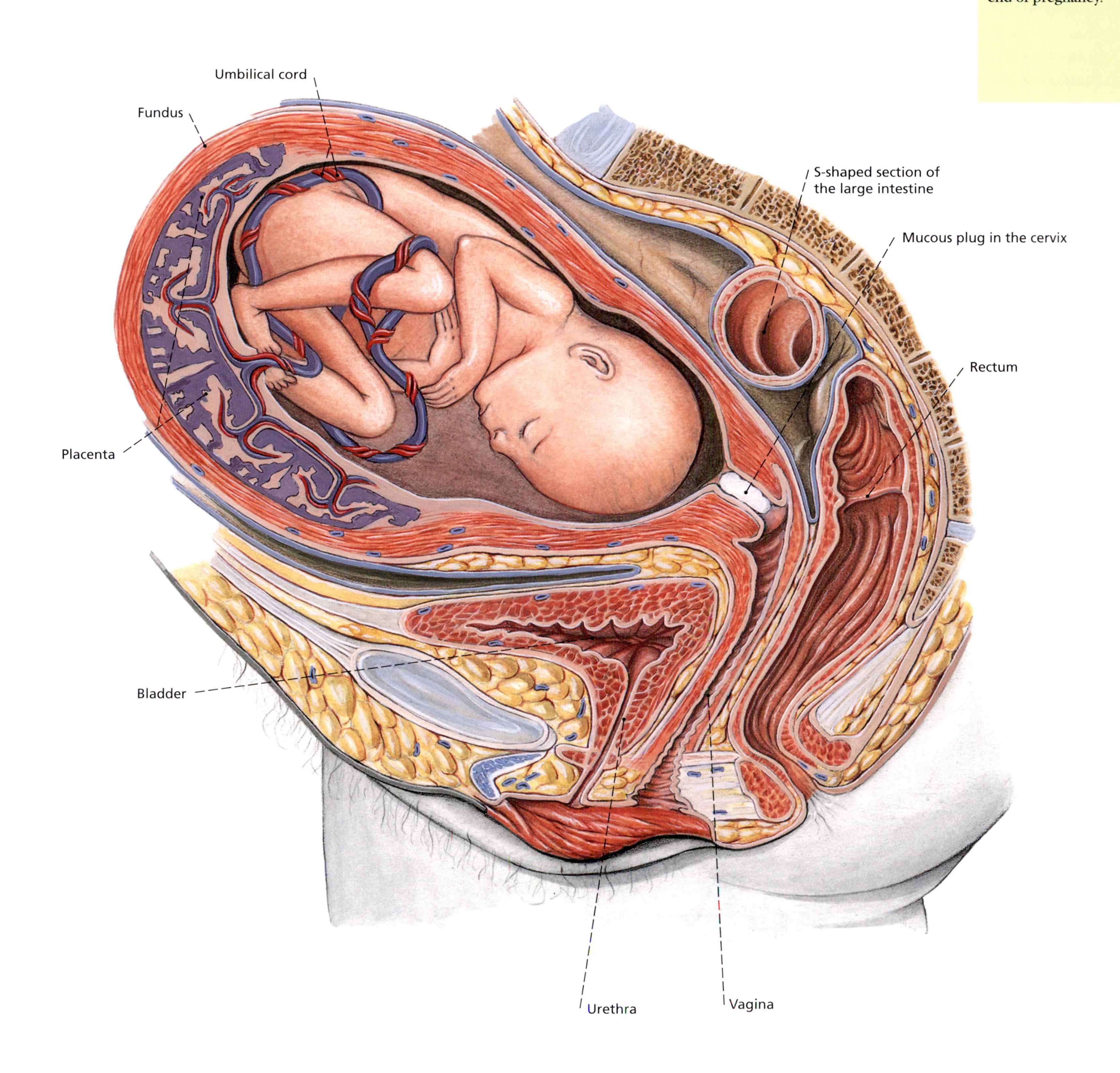

◆
Fig. 421
A schematic representation of the placenta. The organ is usually located on the rear wall of the uterus. From about the 4th month of pregnancy the unborn baby is supported entirely via the umbilical artery. The metabolic exchange between the baby's blood circulation system and that of the mother takes place in the villi of the placenta (*detail*).

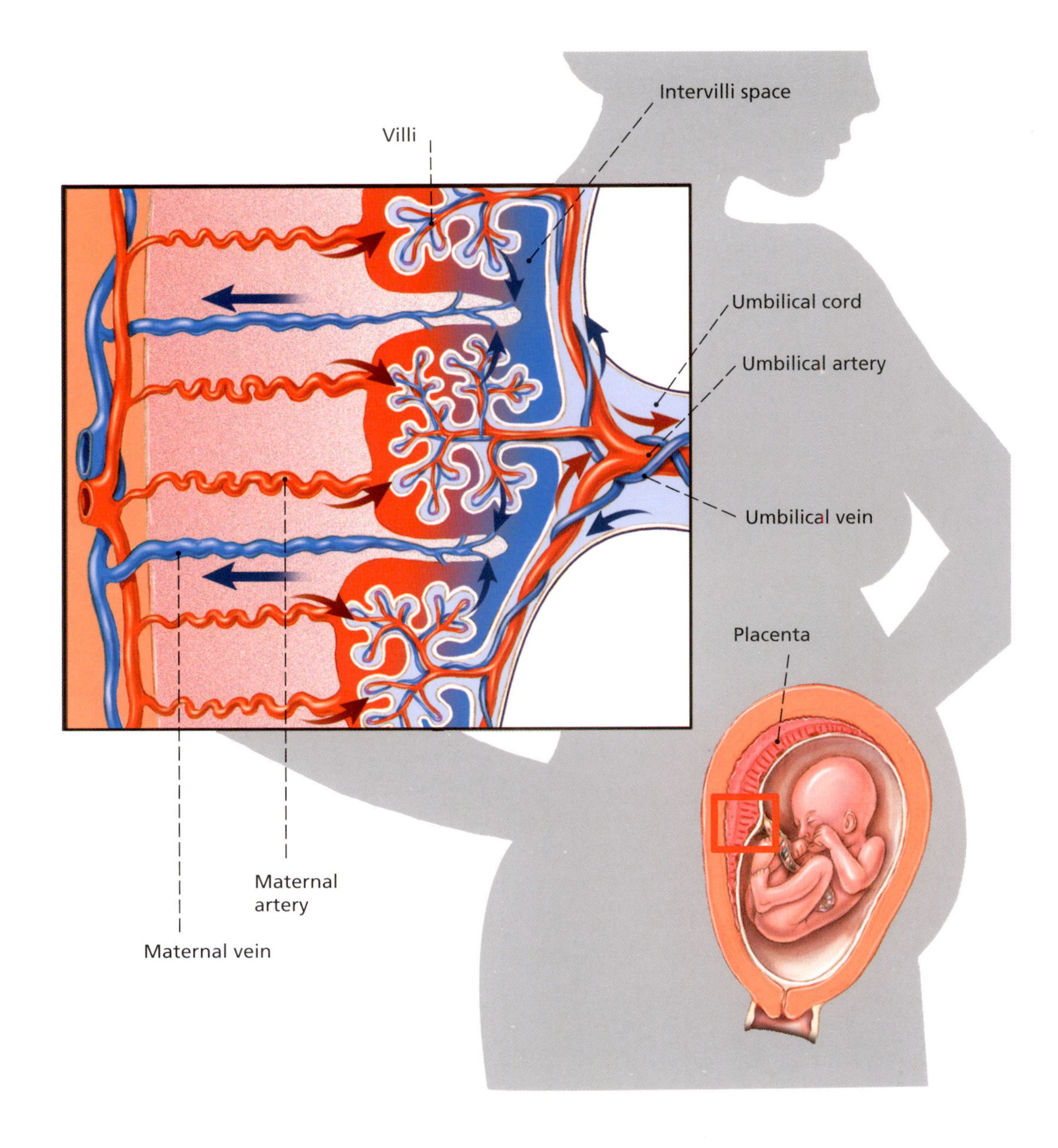

◆
Fig. 422
The placenta is covered with a delicate, white skin. The umbilical cord with the umbilical artery and vein (dissected here) enters in the middle of the organ.

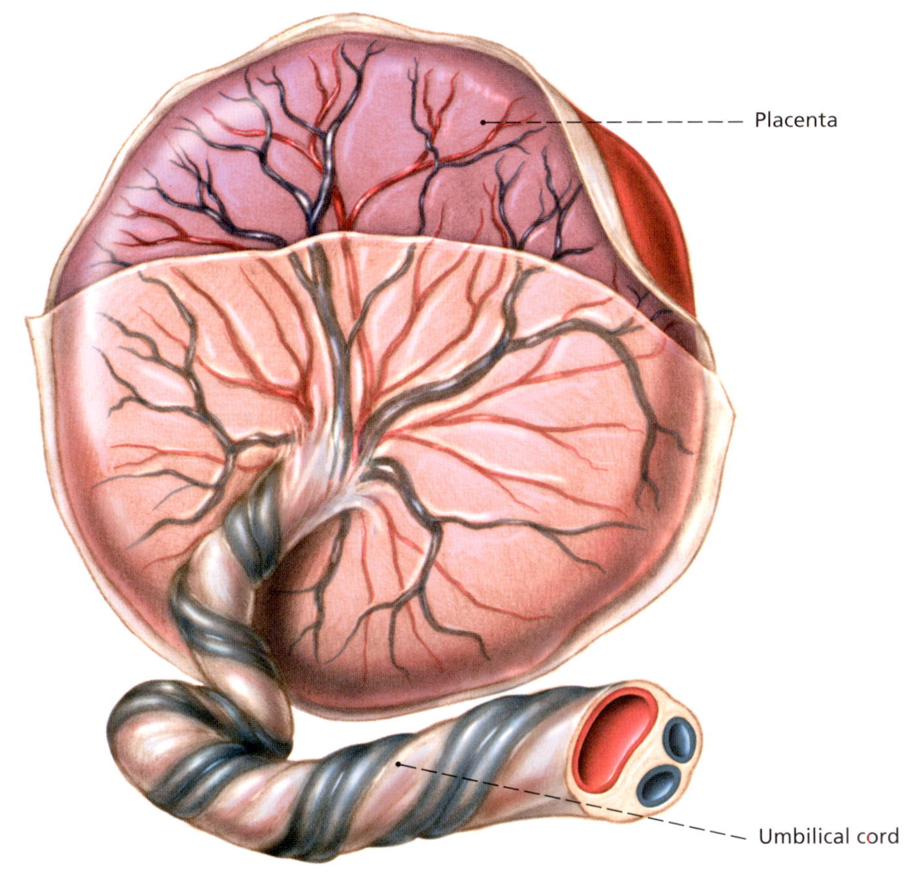

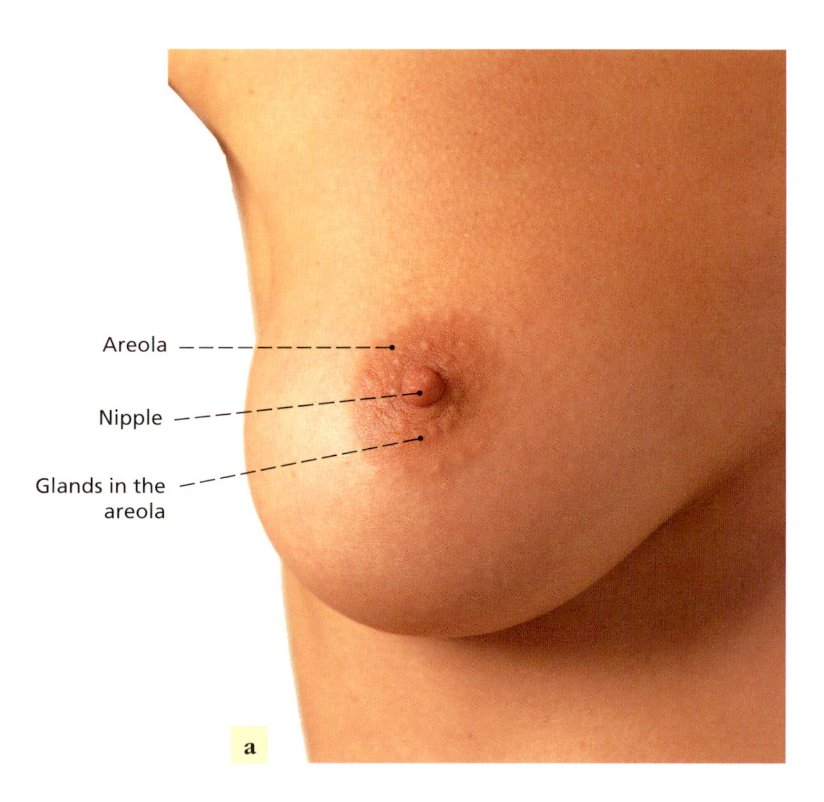

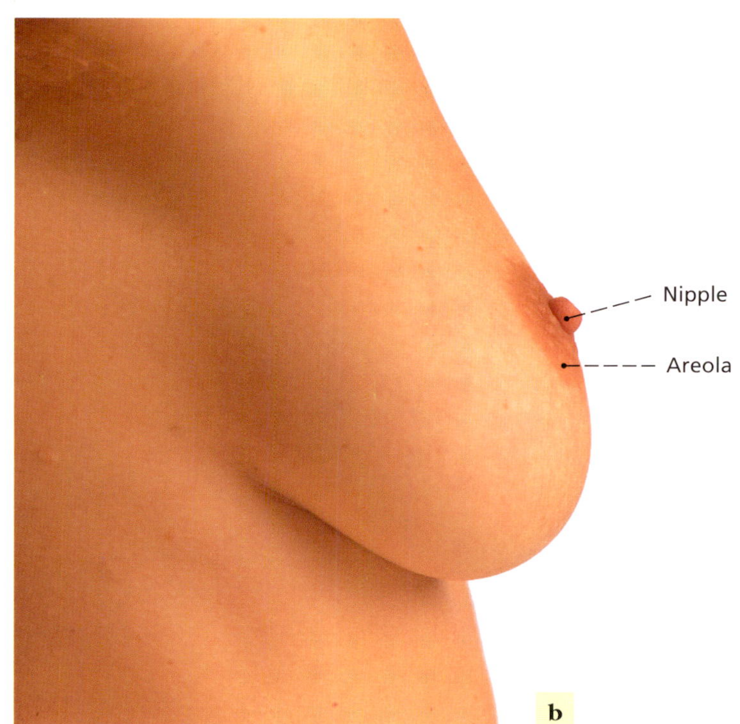

Fig. 423
The female breast from the front (a) and from the side (b).

Fig. 424
Longitudinal section of the female mammary gland.

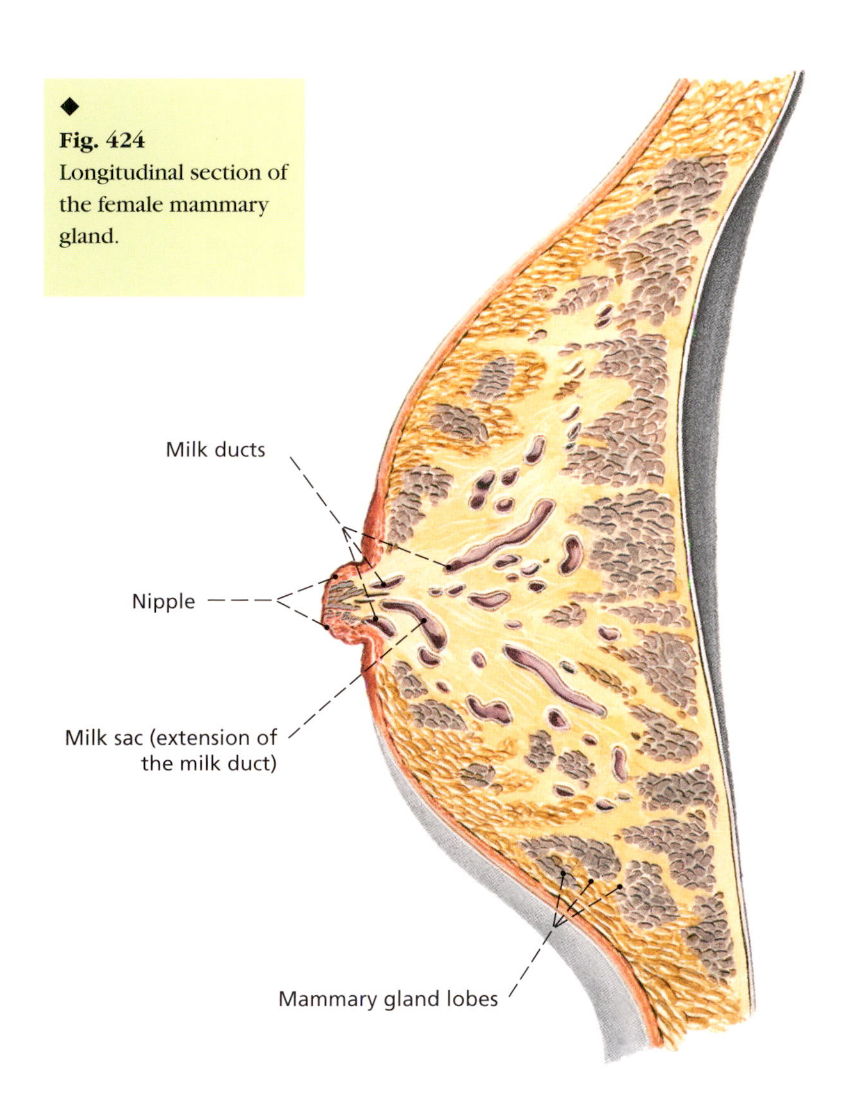

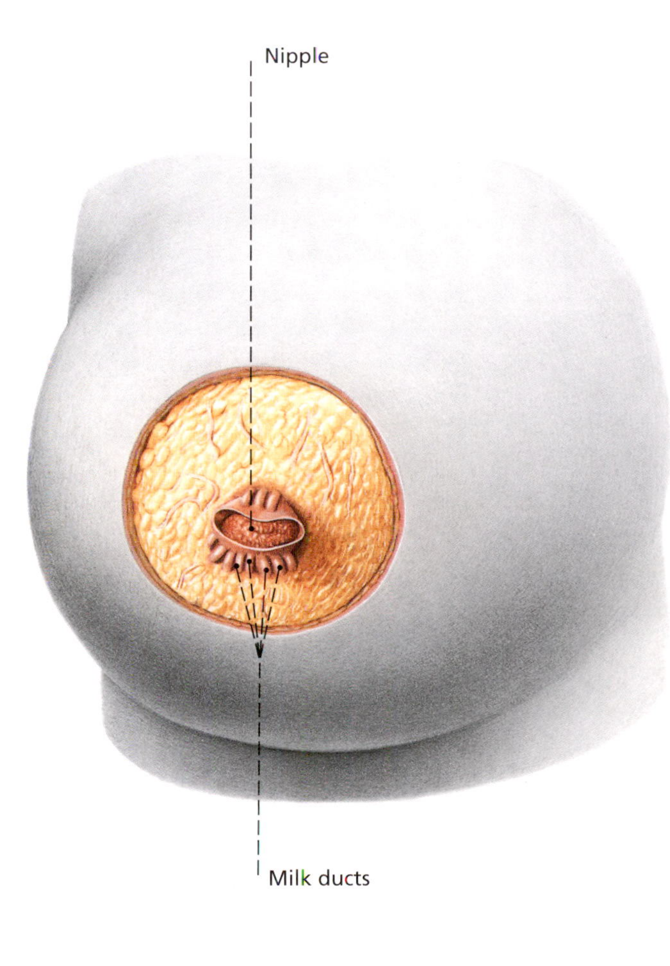

Fig. 425
The mammary gland of a pregnant woman with the skin around the nipple folded back to show the entry of the milk ducts into the nipple.

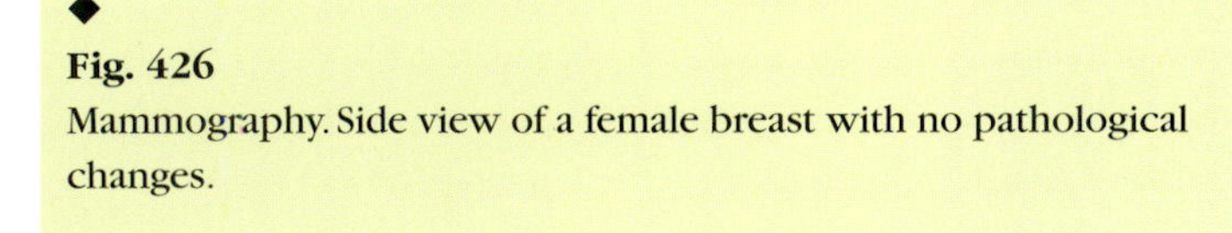

◆
Fig. 426
Mammography. Side view of a female breast with no pathological changes.

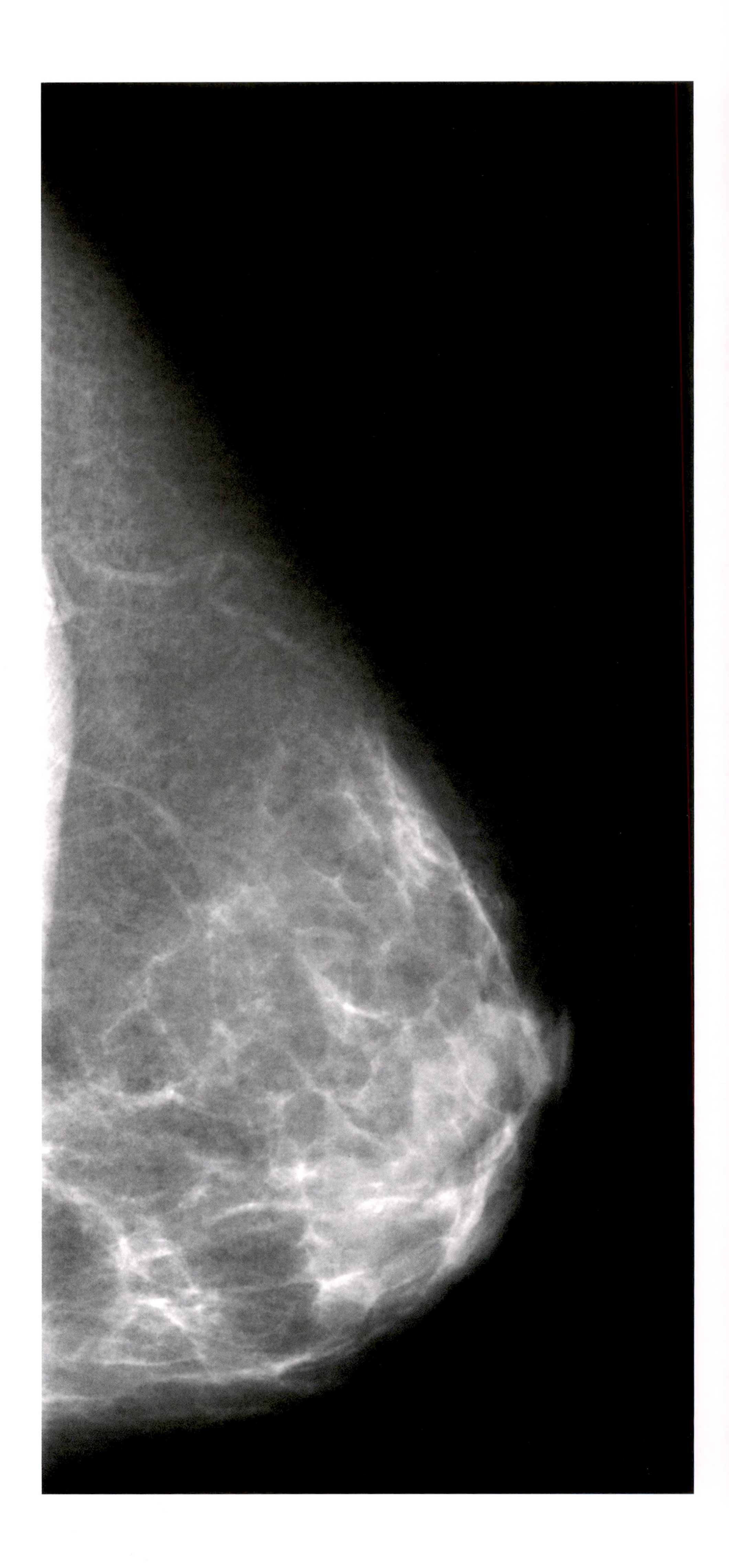

◆
Fig. 427
The external male sexual organs. The skin has been partly removed in order to show the course of the vessels and nerves.

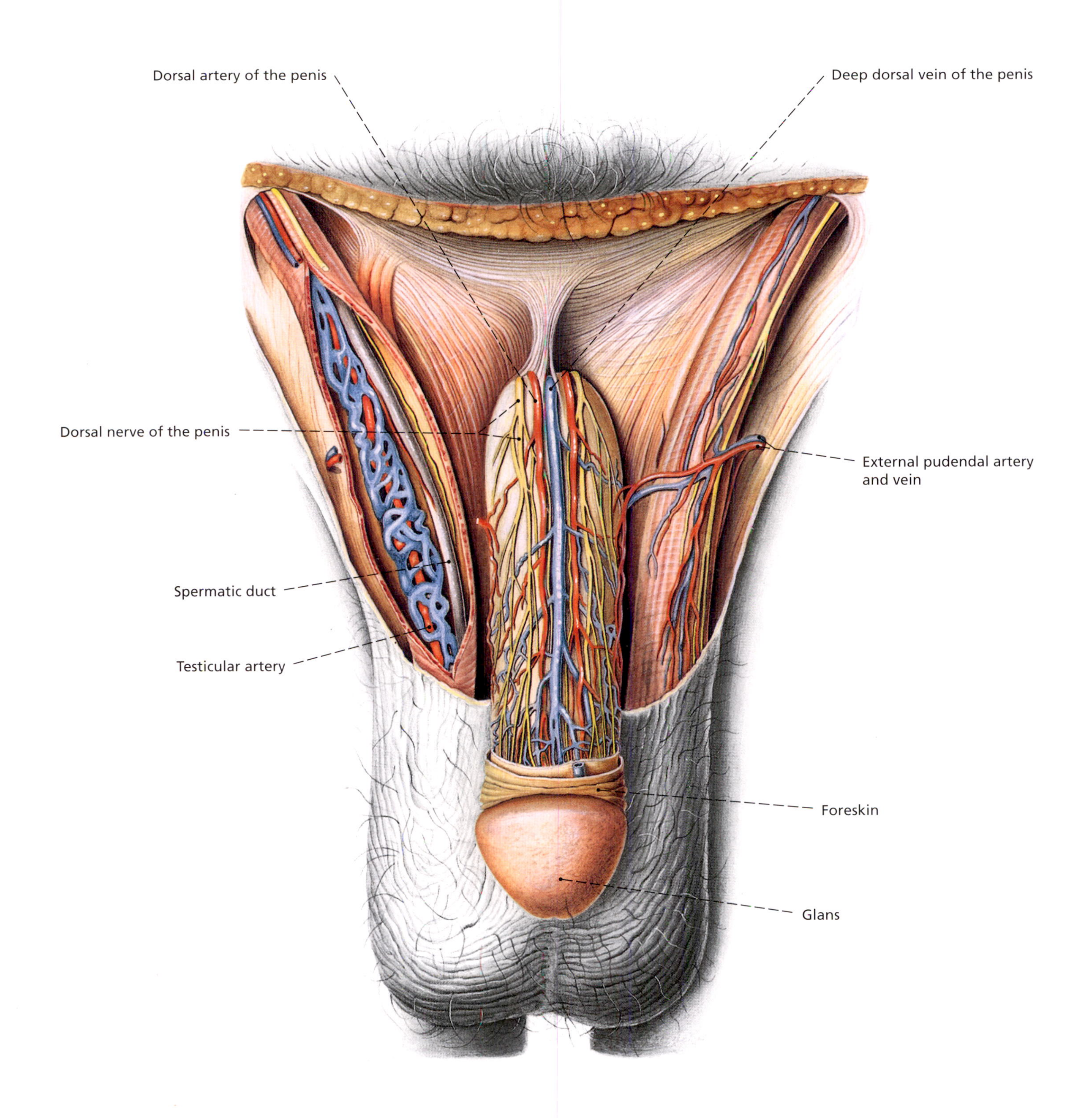

Fig. 428
The spongy body of the penis with the urethra, prostate gland, and bladder. The organ parts are shown in longitudinal section in order to see their structures better. The spongy bodies, which become filled with venous blood when the penis is erect, are clearly visible.

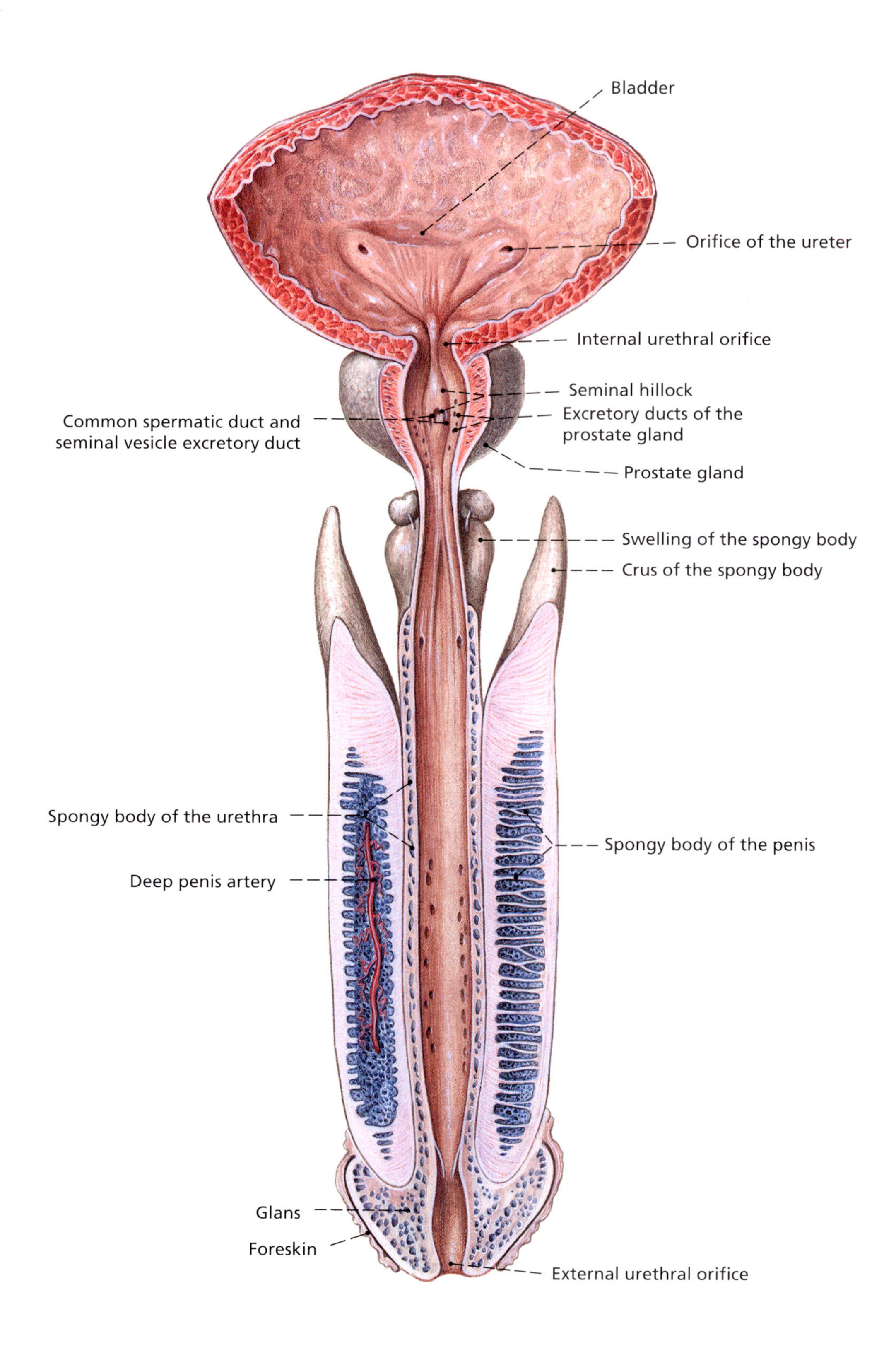

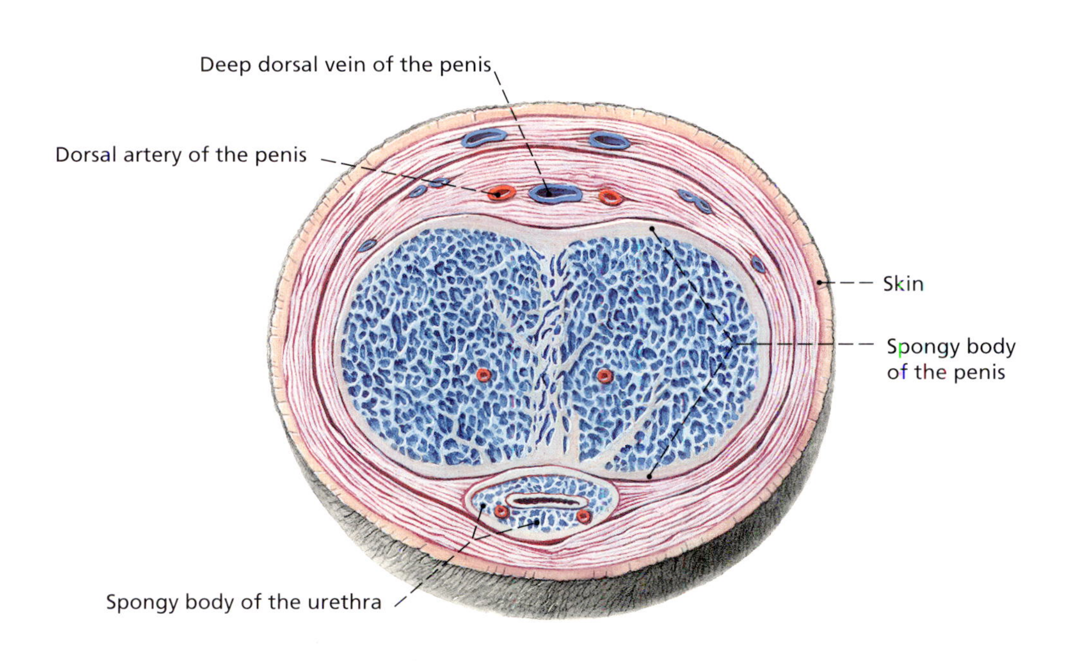

◆
Fig. 429
The penis (cross-section). The two spongy bodies are clearly visible.

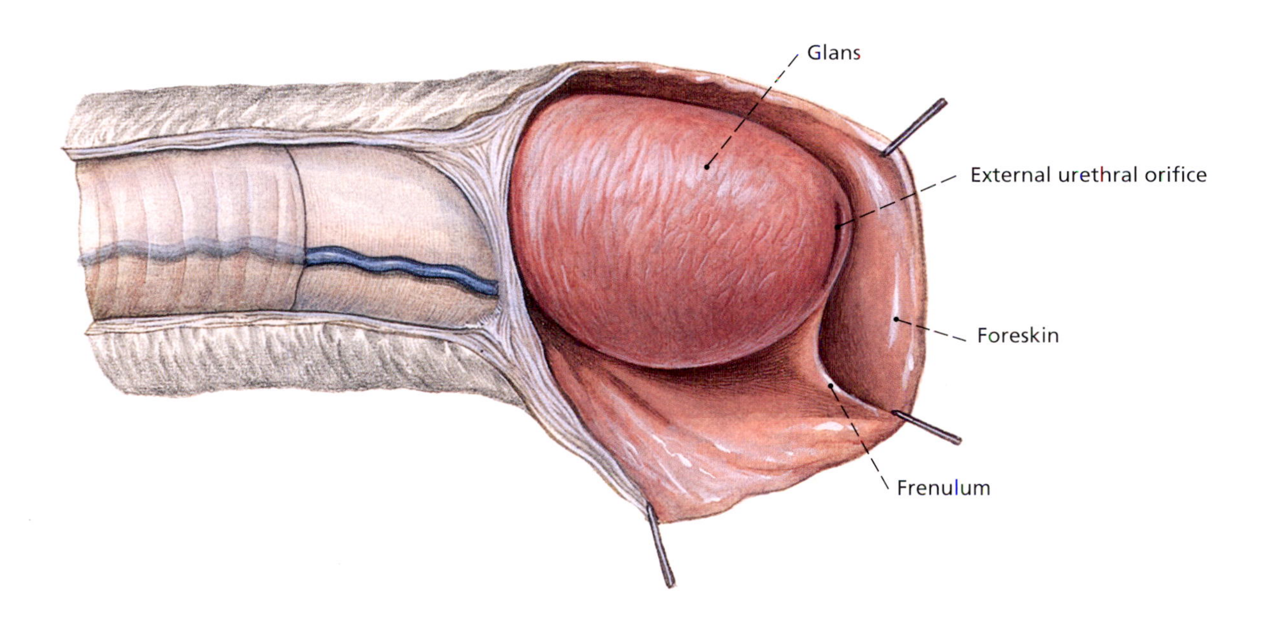

◆
Fig. 430
The penis with the glans and foreskin. The skin has been partly removed and the foreskin cut back.

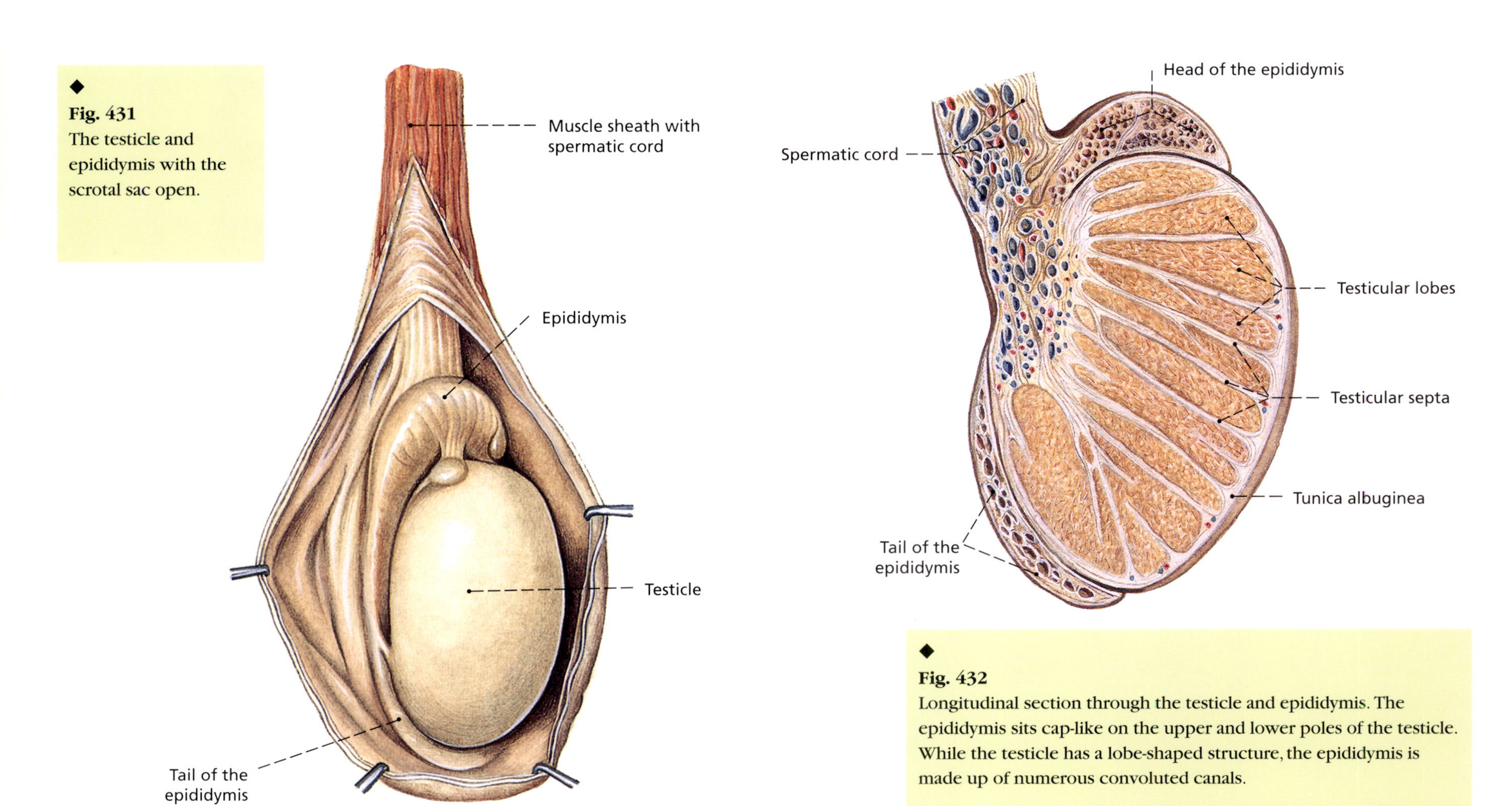

◆
Fig. 431
The testicle and epididymis with the scrotal sac open.

◆
Fig. 432
Longitudinal section through the testicle and epididymis. The epididymis sits cap-like on the upper and lower poles of the testicle. While the testicle has a lobe-shaped structure, the epididymis is made up of numerous convoluted canals.

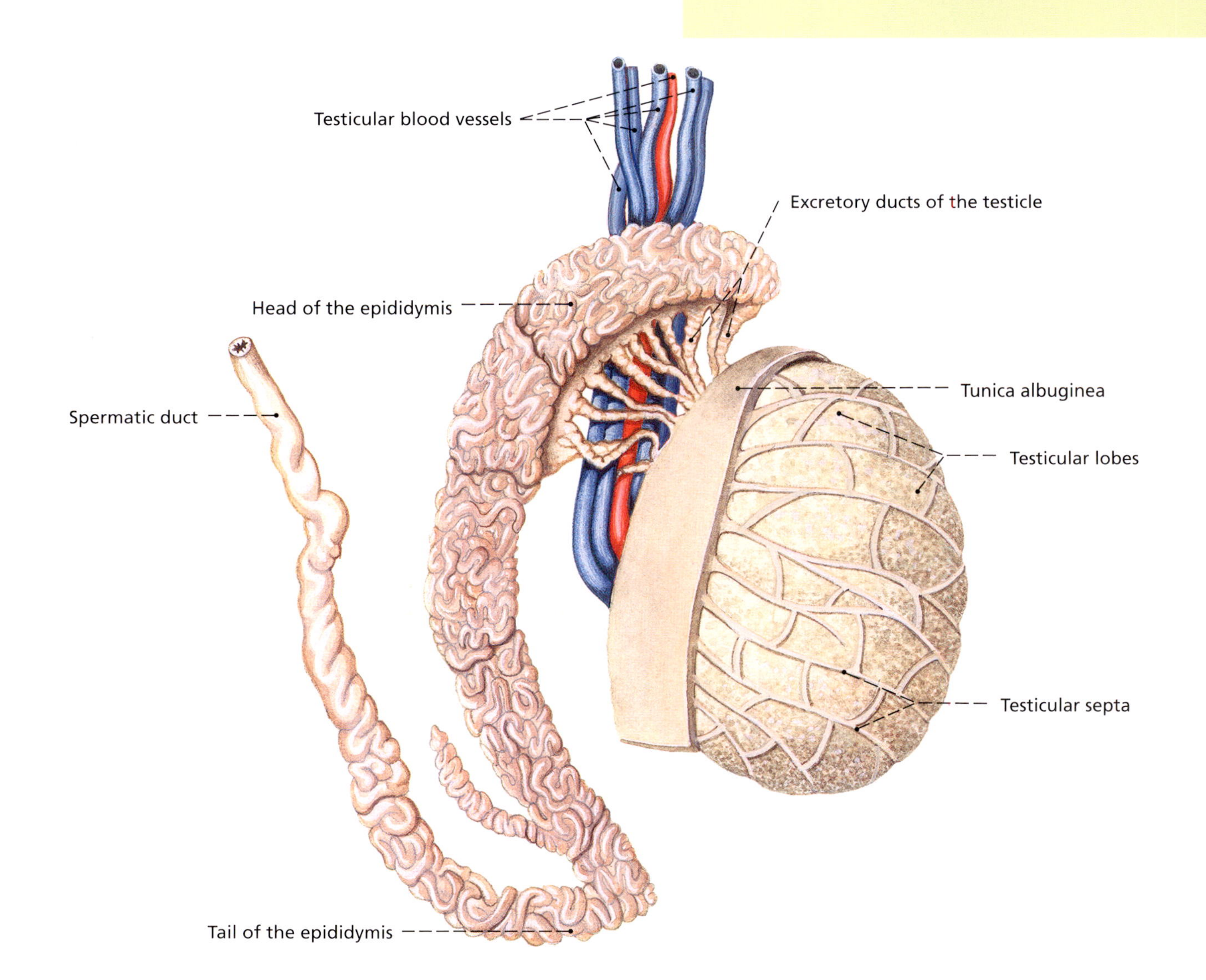

◆
Fig. 433
A testicle with epididymis, the scrotal sac having been partly removed. The testicular septa and the convoluted course of the epididymal canals, in which the sperm are stored, are clearly visible. These canals open into the spermatic duct, through which the sperm are transported to the seminal vesicles.

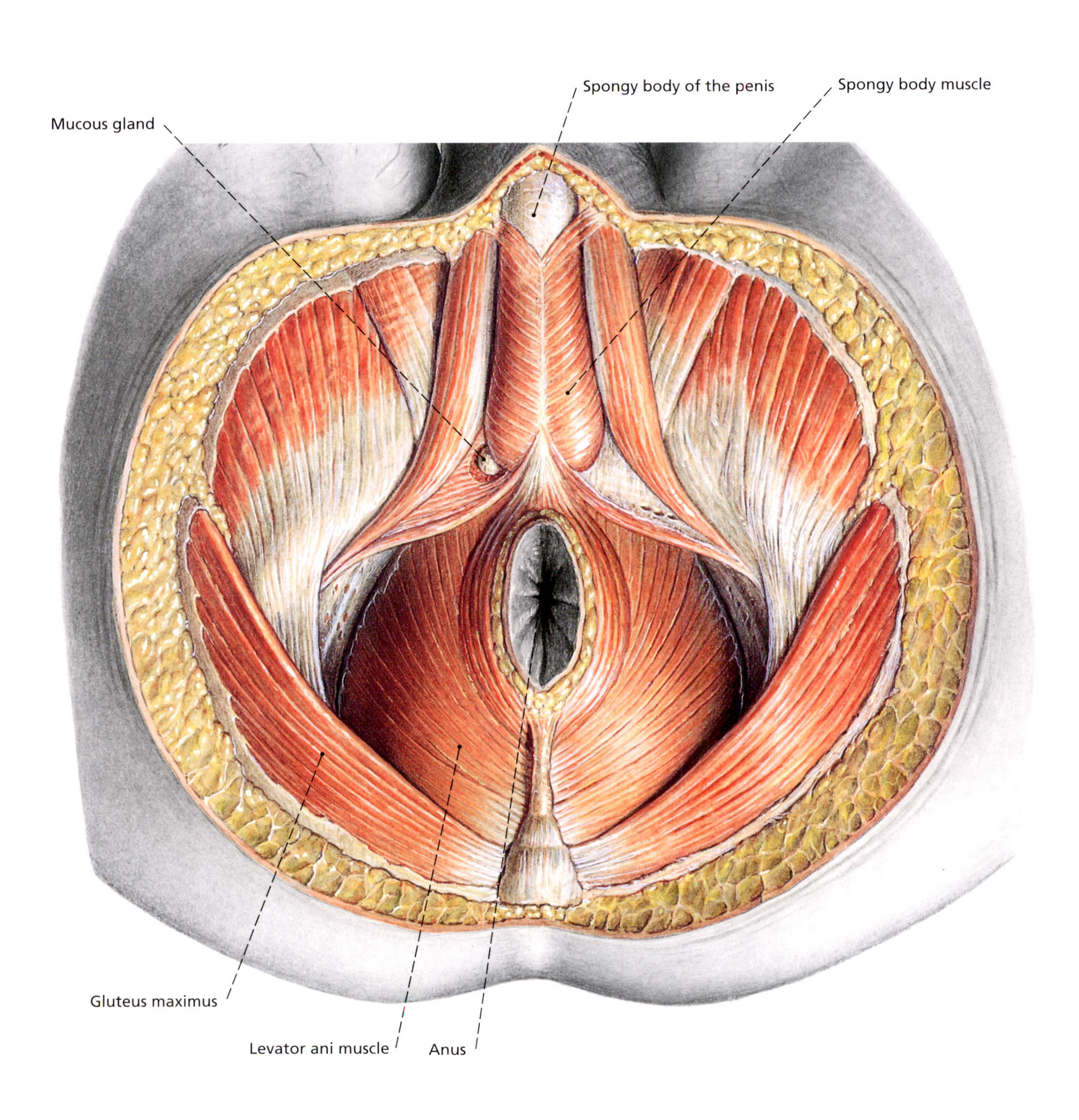

Fig. 434
The perineum region in a man following removal of the skin and fatty tissue, showing the outer layer of the musculature around the base of the penis and the anus.

◆
Fig. 435
The perineum region in a man following removal of the skin and fatty tissue, showing the vessels and nerves. The gluteus maximus to the right has been cut through in order to show the course of the nerves and vessels.

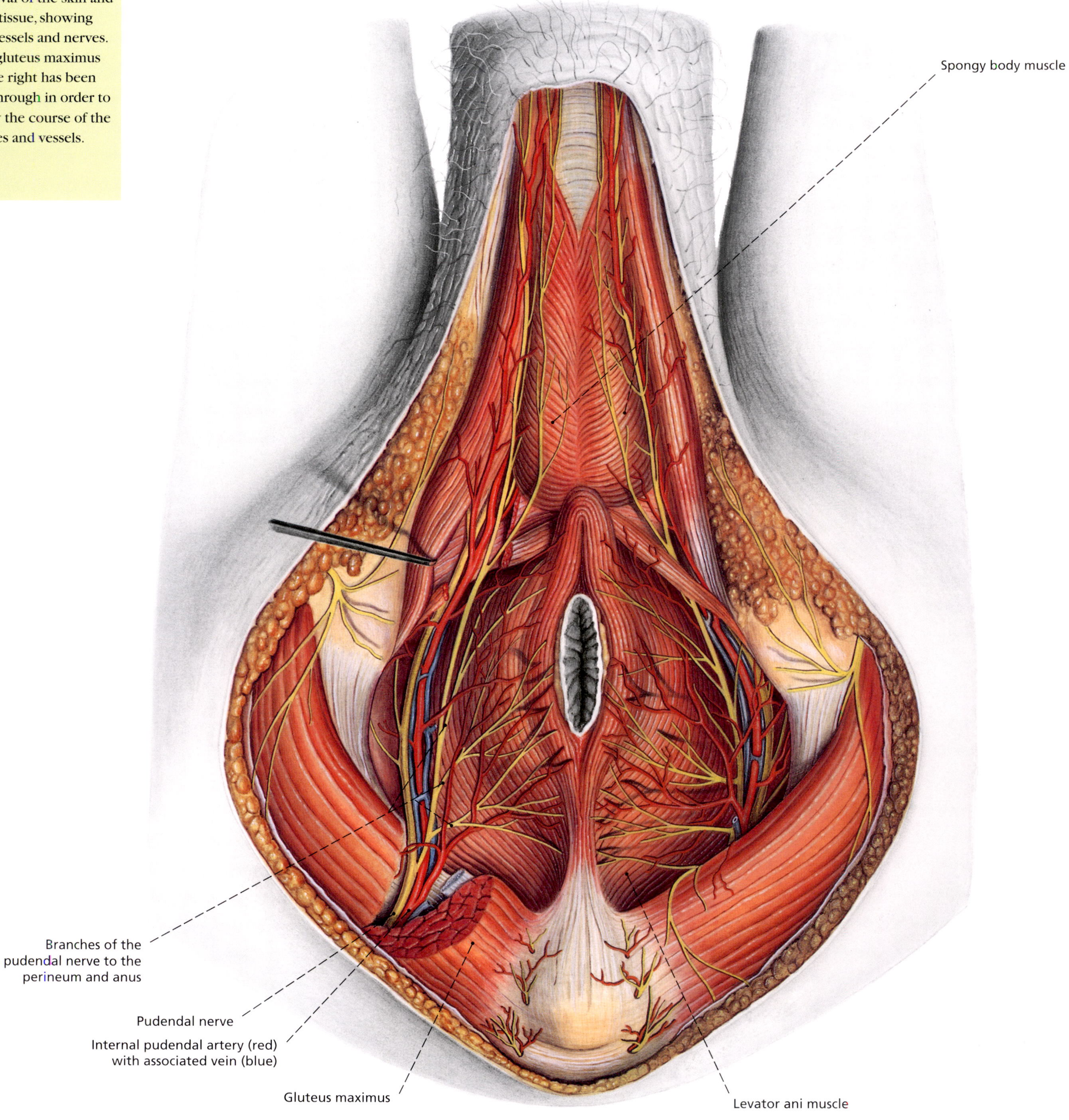

◆
Fig. 436
The male pelvis.
Cross-section through the pubic symphysis.

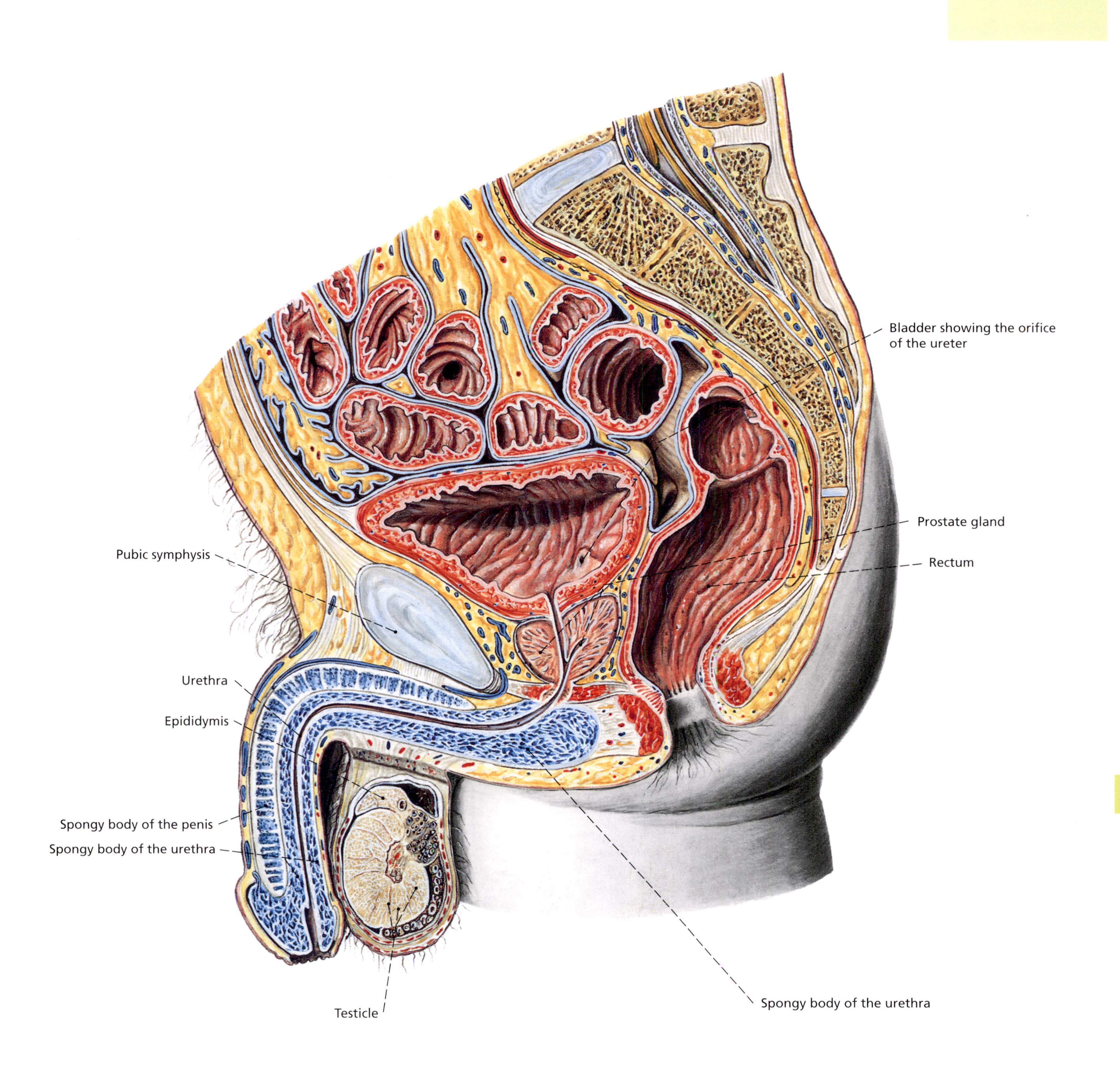

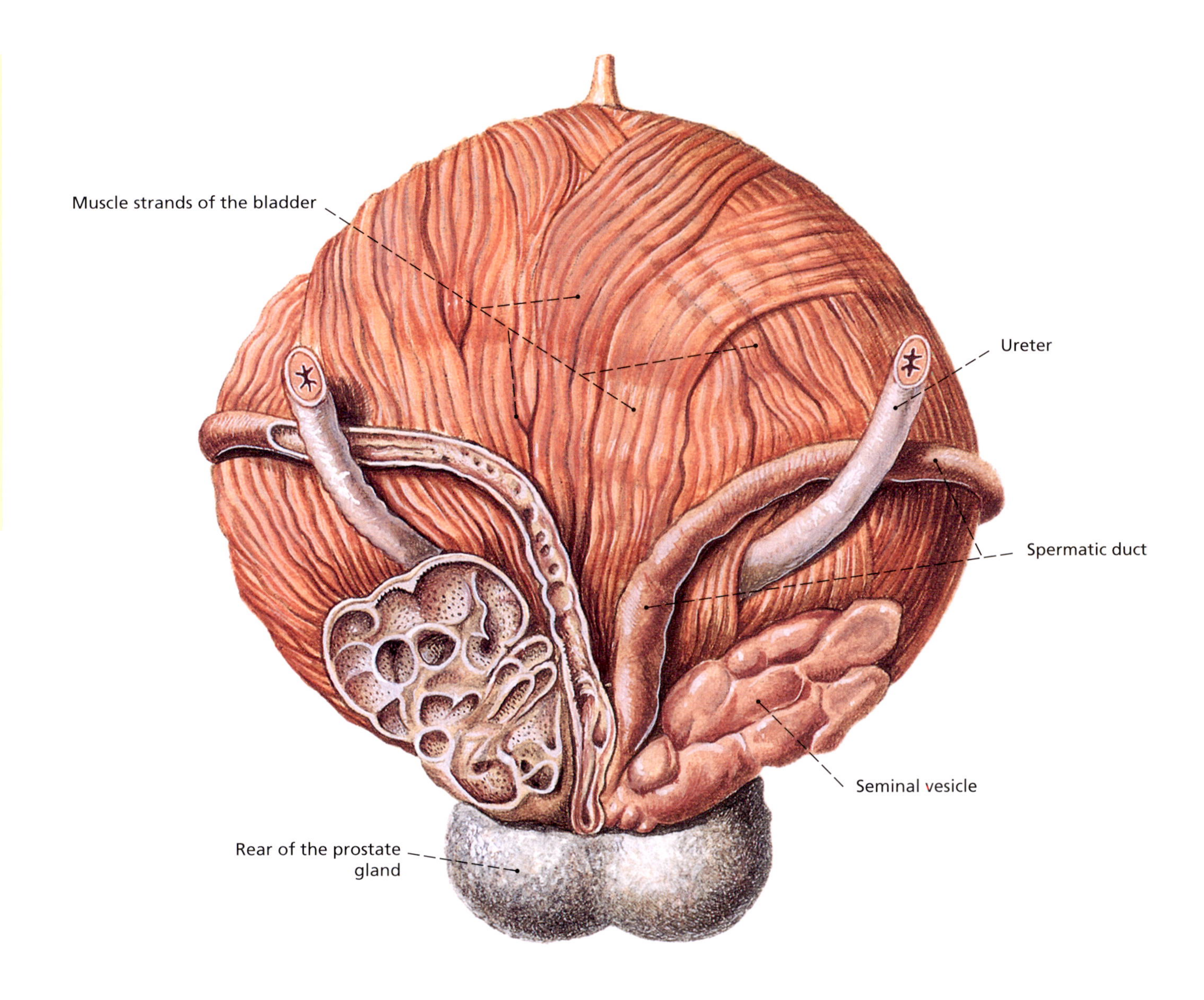

◆
Fig. 437
The male bladder from the rear. The mucous membrane covering the bladder has been cut through to the outer muscle layer. The left lobe of the seminal vesicle has been opened as has the left spermatic duct. The ureters have been cut through just before they enter the bladder.

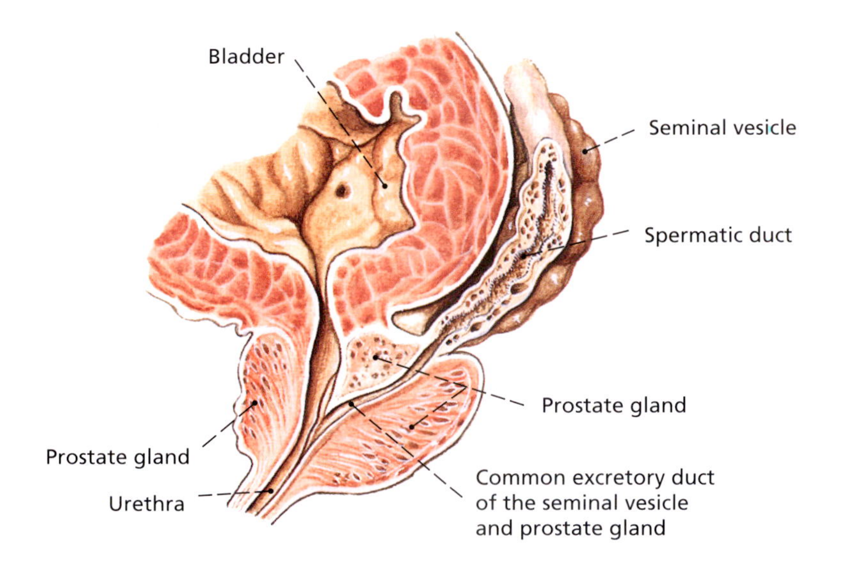

◆
Fig. 438
The base of the bladder with the prostate gland and seminal vesicle. Diagonal section to show the entry of the common seminal vesicle and epididymal duct in the urethra.

Symptoms and diseases

Diseases of the female sexual organs

• *Mastitis*

An inflammation of the milk ducts in the breast, usually extending to the armpit, in which the fluid usually secreted by the breast, congeals. Mastitis can, however, also be accompanied by the secretion of a stringy fluid from the breast or by bleeding from the milk ducts. It occurs almost solely during breastfeeding. A confined area of skin becomes red and painfully swollen if the inflammation forms an abscess, and the axillary lymph nodes are swollen. The inflammation usually heals quickly following removal of the abscess and treatment with antibiotics.

• *Mastopathy*

Benign change to the milk ducts and the connective tissue of the breast. Most frequently takes the form of fluid-filled cavities (cysts) in the glandular tissue. They can often be felt from the outside as lumps, and can sometimes be painful prior to menstrual bleeding. Usually caused by hormonal disorders. Therapy usually comprises the administration of hormonal medication.

• *Ovaritis*

Inflammation of the ovary, usually occurring in combination with the much more frequent salpingitis (inflammation of the fallopian tube). The symptoms are pains on one or both sides of the abdomen, as well as fever. The disease is triggered either by bacteria reaching the ovary via the bloodstream, or germs entering the fallopian tubes from the vagina. Adhesions and suppurative abscesses sometimes form as a result of the inflammation. The ovaries can suffer permanent damage if the inflammation is not treated with medication in good time.

• *Tumors of the ovary*

Malignant or benign swelling on the ovary, most commonly occurring as fluid-filled cavities known as cysts. They can occur if a ripe egg cell does not detach from the ovary and its follicle continues to grow, or can be caused by a hormonal disorder. The majority of cysts do not cause any discomfort, although larger cysts can sometimes cause pain in the lower abdomen. The ripening of the egg cells can be disrupted if a large number of cysts form on an ovary. This can be a cause of involuntary childlessness. Malignant growths on the ovary are far less common but they, too, usually only cause discomfort when relatively large. Regular check-ups with a gynecologist help to ensure that any changes to the ovaries are detected early on.

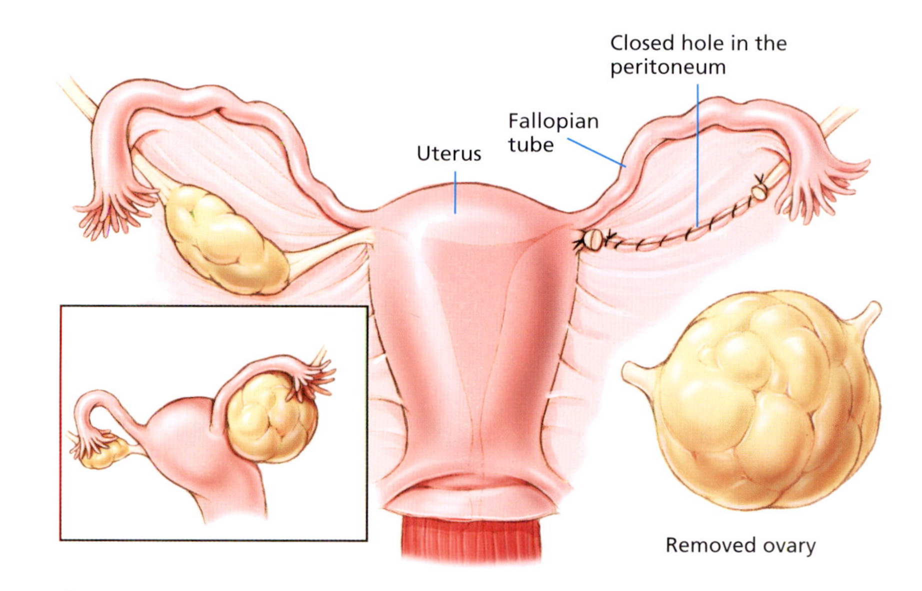

◆
Ovarian tumor
The removal of just the affected ovary is usually sufficient in the case of benign growths.

• *Salpingitis*

Usually caused by germs entering the fallopian tubes via the vagina and the uterus. The inflammation can result in severe pains in the lower abdomen, fever, and discharge. It is important that salpingitis be diagnosed and treated quickly, otherwise the fine hairs on the inside can be become stuck together, or the fallopian tubes can become blocked completely. If salpingitis is not treated in good time, therefore, it can be one of the most frequent causes of involuntary childlessness.

• *Ectopic pregnancy (tubal pregnancy)*

The path of the egg can be blocked if the fallopian tube has undergone changes due to a constriction following an inflammation, a kink, or bulges. The egg then implants in the mucous membrane of the fallopian tube where it develops into an embryo. This is known as an ectopic pregnancy, which cannot be carried to term. The fallopian tube bursts once the embryo reaches a critical size. Sudden, severe pain on one side of the lower abdomen in the early stage of pregnancy is an alarm signal and medical assistance must be called immediately. The implanted egg can be removed if an ectopic pregnancy is detected before the fallopian tube bursts. The fallopian tube can then be repaired and a normal pregnancy is again possible.

Breast cancer

The prospects of recovery from this disease are now very good if it is detected at an early stage. Treatment methods have improved significantly in recent years such that two out of every three patients are able to retain their breast. Even if the removal of the breast is unavoidable, it can be reconstructed cosmetically with good results.

Breast cancer occurs when cells in the milk ducts and/ or the glandular lobes of the breast degenerate, spread into the body of the gland, and form a tumor. The cells of this type of malignant tumor, also known as carcinoma of the breast, later spread into the surrounding tissue and destroy it. They then enter the lymphatic system (the vessels containing lymph), sometimes also the blood vessels, thus reaching more distant parts of the body where they can implant and form metastases. Metastases occur most frequently in the bones (spinal column, pelvis, ribs, thigh, and skull), followed by the lungs, liver, skin, ovaries, and brain. The prospects of recovery are good if a malignant tumor is detected and removed in time, before metastases form.

Early diagnosis

The majority of women detect a lump in their breast themselves. Self-examination of the breast should be carried out once a month following menstruation. It is the most important method for the early diagnosis of breast cancer. The following changes to the breast can be (but are not necessarily) a sign of breast cancer:

- palpable lumps in the breast which do not go away after the monthly period,
- inverted nipples,
- tangible hardening of the breast,
- secretion of fluid from the nipple,
- sudden changes to the size and extent of the breast,
- swelling of the lymph nodes under the armpit (axilla).

The standard medical examination is mammography, which is also able to detect lumps in the breast that are not yet palpable. An ultrasound scan (sonography) may also play a significant role in younger women with a suspected tumor. If neither of these methods produces sufficient clarity, then magnetic resonance imaging (MRI) of the breast is carried out. Absolute certainty as to whether a lump in the breast is malignant or benign can be achieved microscopically only from a biopsy sample or once the lump has been surgically removed.

Treatment

Surgery is the mainstay of treatment, followed by adjuvant hormonal therapy, irradiation, or chemotherapy. The tumor is removed, together with the surrounding tissue, so that there are no remaining tumor cells able to spread to the rest of the body. In many cases, the breast can be retained. A key factor in a breast-retaining operation is the relationship of the breast size to tumor size. The breast may have to be removed in the case of a large tumor and a small breast, or in the case of several tumor clusters. It can be reconstructed later by plastic surgery, however. If the breast cancer spreads, the lymph nodes under the armpit are infected first of all and can be felt as hard lumps. They are the first collection point in the body for migrating tumor cells and the removal of all axillary lymph nodes is therefore standard practice in breast cancer surgery today, irrespective of whether a breast-retaining operation or a full removal of the breast (mastectomy) is carried out.

Aftercare

Irradiation is standard practice following breast-retaining operations. This can sometimes lead to skin irritation, fatigue, and lassitude, and less frequently to headaches, loss of appetite, and depression.

Hormone treatment is used if the breast cancer has occurred as the result of the influence of the hormones estrogen and/or progesterone. A range of medications is used to suppress the production of these hormones in the ovaries. This often leads to side effects, similar to the symptoms that accompany the menopause.

Chemotherapy entails treatment with medication that has a toxic effect on the cancer cells. Today, the vomiting that used to be caused by chemotherapy can be suppressed by the simultaneous intake of appropriate medications. Hair loss may also occur. As chemotherapy weakens the immune system, it can also result in a temporary increase in the risk of infection.

• ***Endometritis***

Inflammation of the endometrium and the uterine musculature in the cervix region. Caused by germs from the external sexual organs, intestine, and/or bladder entering the uterus via the vagina. This may occur following delivery or a miscarriage, if the cervical opening is not yet completely closed, or through use of the contraceptive coil. The symptoms are discharge and a dragging pain in the lower abdomen that is aggravated by movement and contact. Bleeding disorders and fever can also occur.

• ***Cervical polyp***

Usually a benign tumor originating in the mucous membrane and forming within the cervix. Provided it is small it usually does not cause any discomfort. Larger and/or numerous polyps can lead to a mucous, bloody discharge, as well as to breakthrough bleeding, and their benign nature should be confirmed. Cervical polyps are removed surgically.

• ***Cervical myoma (fibroid)***

Benign tumor in the muscle of the uterus. Myomas are exceptionally common and, provided they are small, do not usually cause any discomfort. Larger myomas need to be removed surgically if they exert pressure in neighboring organs such as the bladder, or if their position compromises a pregnancy. Myomas often cause bleeding disorders.

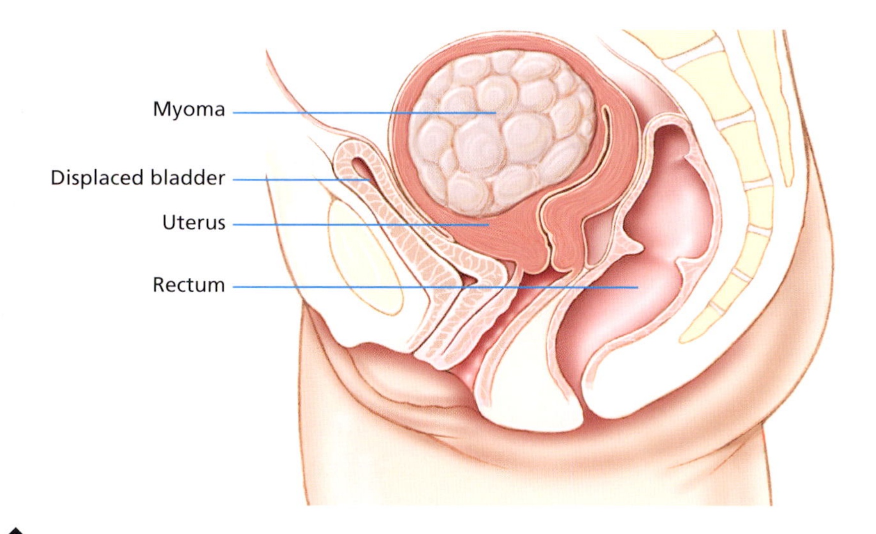

◆
Cervical myoma
One of the most common types of muscle tumor, also known as fibroids; larger or more numerous myomas may need to be removed surgically.

• ***Endometriosis***

Benign endometrial growth outside of the uterus. For reasons that are not fully understood, endometrial tissues can become dispersed in the abdominal cavity (usually in the ovaries, fallopian tubes, intestine, or bladder), where they implant and continue to grow. The dispersed clusters expand and recede cyclically, together with the endometrium, according to the menstrual cycle, but the dispersed tissue cannot be discharged via the vagina during menstruation. Blood-filled cysts therefore often develop and adhesions result in the affected regions.
The symptoms of endometriosis are usually severe pains one to two days prior to and during menstrual periods. They may be associated with the development of adhesions, which is a common cause of involuntary childlessness. It is therefore important that endometriosis be diagnosed and treated early on. Diagnosis is made through laparoscopy. Treatment is often achieved using medication to suppress the hormones responsible for endometrial growth. The endometriosis clusters can also be removed surgically, often by laparoscopy.

• ***Cervical cancer***

This form of cervical cancer develops in the outer cervical opening region and largely affects women aged between 40 and 60 years.
There are no characteristic symptoms as such. Consequently, any bleeding or bloody discharge that occurs outside menstruation (intermenstrual bleeding or spotting), or contact bleeding following sexual intercourse, can be a warning sign and a physician should be consulted.
Cervical cancer develops from what are initially harmless prestages. The first cell changes can be detected at an early stage by means of a smear test (known as a Pap test) of the surface of the cervical opening, which is carried out as part of any cancer check-up.
In the case of minor to medium cell changes, especially in women wanting to have children, the surface of the affected mucous membrane can be sloughed off (e.g., with laser beams). This destroys the affected cells.

A procedure known as conization, which also serves diagnostic purposes, is carried out in the case of cell changes that are not restricted solely to the surface and involves a wedge-shaped piece of tissue being cut out from the cervical opening. If the tumor has also penetrated the lower tissue layers, the entire uterus, including the adjoining lymph vessels and, if necessary, the ovaries will have to be removed. This operation is followed by irradiation if the tumor has already spread to one or more of the neighboring organs, namely the vagina, bladder, or rectum.

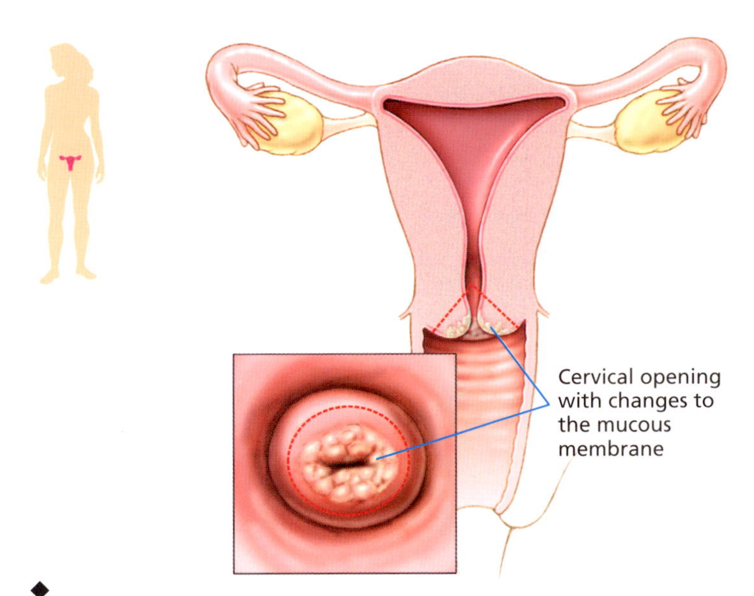

◆
Conization
In the case of cervical cancer that has not yet extended beyond the cervical opening, the removal of a small section of the cervical opening is usually sufficient

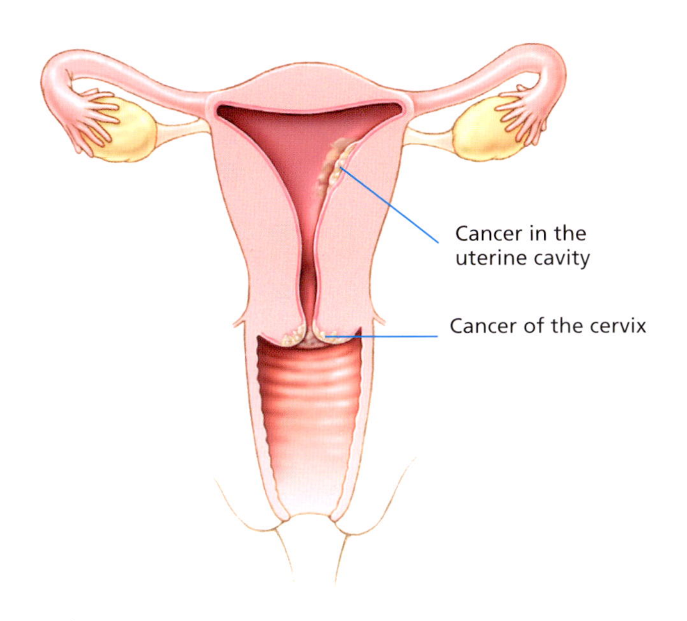

◆
Cancer of the uterus
Cancer of the cervix and cancer of the uterus differ according to their location and to the treatment options.

• *Cancer of the uterus*

Largely affects women over the age of 50 years. The recovery prospects are very good if the cancer is detected and treated early on. If left untreated, however, it can spread to the neighboring organs (vagina, fallopian tubes and ovaries, as well as the bladder and rectum) or even to more remote organs, such as the liver and the bones.

As there is no reliable screening test for this form of cancer, attention needs to be paid to any potential early symptoms. These include irregular bleeding and bleeding after the menopause which is often misdiagnosed as a "return of menstrual bleeding," as well as spotting and any discharge containing blood and pus. Such bleeding can, of course, be harmless but is considered to be a potential sign of carcinoma until this possibility has been eliminated by thorough gynecological examination.

The type of therapy (surgery, irradiation, hormone therapy, or chemotherapy) primarily depends on the extent to which the disease has spread and on the age of the patient. In the early stages the uterus may be removed, together with the fallopian tubes and ovaries (hysterectomy). If the tumor is already more advanced, surgery may be extended to include the surrounding lymph vessels, and also followed by radiation therapy. Regular check-ups are required for a period of at least 5 years in order to be able to detect and treat any late complications.

• *Uterine descensus*

The sinking of the uterus towards the vagina due to a weakness of the ligaments and the pelvic musculature, which are no longer able to hold the uterus in its original position in the lower pelvis. Uterine descensus often occurs as the result of repeated pregnancies and births, obesity, and heavy physical labor. The sinking causes a feeling of pressure on the vagina and leads to disruptions in urination to the extent of urinary incontinence. Involuntary urination when coughing or sneezing is also indicative of a weakness in the musculature, which can be strengthened by special pelvic floor exercises. The insertion of a support ring in the vagina can provide temporary relief. Severe sinking leads to a prolapse of the uterus.

• *Prolapse of the uterus*

The emergence of the uterus from the pelvis into the vagina, sometimes pulling the intestine and the bladder down with it. Caused by a relaxation of the ligaments and the pelvic musculature. A feeling of pressure, pain while sitting and walking, bladder complaints, back pain, discharge, and inflammation can occur. The uterus usually needs to be removed.

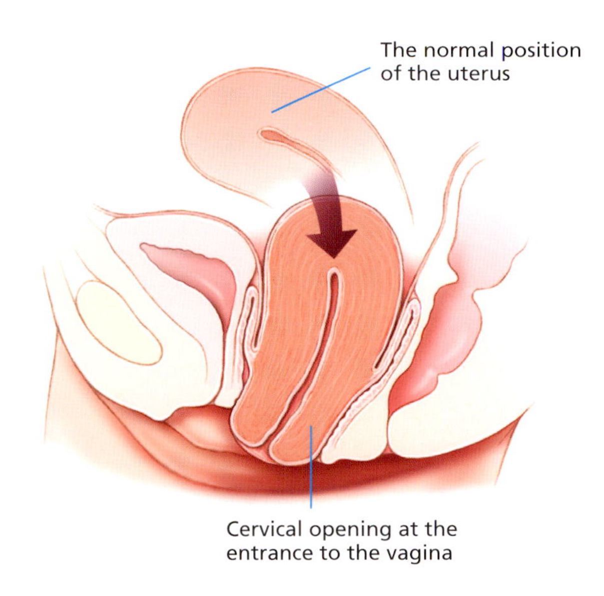

◆
Prolapse of the uterus
In prolapse of the uterus, the cervical opening may be visible at the entrance to the vagina.

• ***Colpitis***

Inflammation of the mucous membrane in the vagina accompanied by a greenish, unpleasantly smelling discharge, sometimes also with itching in the pudendal lip region, and a burning feeling in the area around the vaginal opening. Inadequate hygiene in association with sexual intercourse, or incorrect use of toilet paper allowing intestinal bacteria to enter the vagina, as well as frequent changes of sexual partner and hormonal changes, can alter the vaginal flora to such an extent that inflammation results. Treatment comprises vaginal pessaries containing antibiotics and lactic acid. Treatment of the partner is also always necessary.

Diseases of the male sexual organs

• ***Enlarged prostate***

Benign prostatic hyperplasia (BPH) is the gradual enlargement of the prostate gland that occurs as part of the normal aging process. As the prostate enlarges, the layer of tissue surrounding it stops it from expanding, causing it to press against the urethra. The bladder wall becomes thicker and the bladder begins to contract even when it contains only small amounts of urine, causing frequent urination. Eventually, the bladder weakens, becomes distended, and loses the ability to empty itself, with some urine remaining in the bladder. Urinary frequency, fever, bladder infections, and blood in the urine can also occur. The condition is diagnosed initially by digital rectal examination. A blood test for prostate-specific antigen (PSA) may be recommended to rule out cancer. Several medications can be used to relieve the symptoms, or the condition can be treated by a variety of techniques to reduce the size of the prostate. In complete urinary obstruction, surgical removal is required.

• ***Prostatitis***

Acute or chronic inflammation of the prostate gland. Pathogens are able to reach the prostate gland via the urethra or the bloodstream causing an infection characterized by frequent urination (small quantities), painful urination and bowel evacuation, fever, and chills. Pain also occurs during sexual intercourse. Prostatitis is diagnosed by digital rectal examination. The glandular secretion and urine are also examined for bacteria. Treatment is with antibiotics. The sufferer must also drink plenty of liquids in order to thoroughly rinse out the urinary tract.

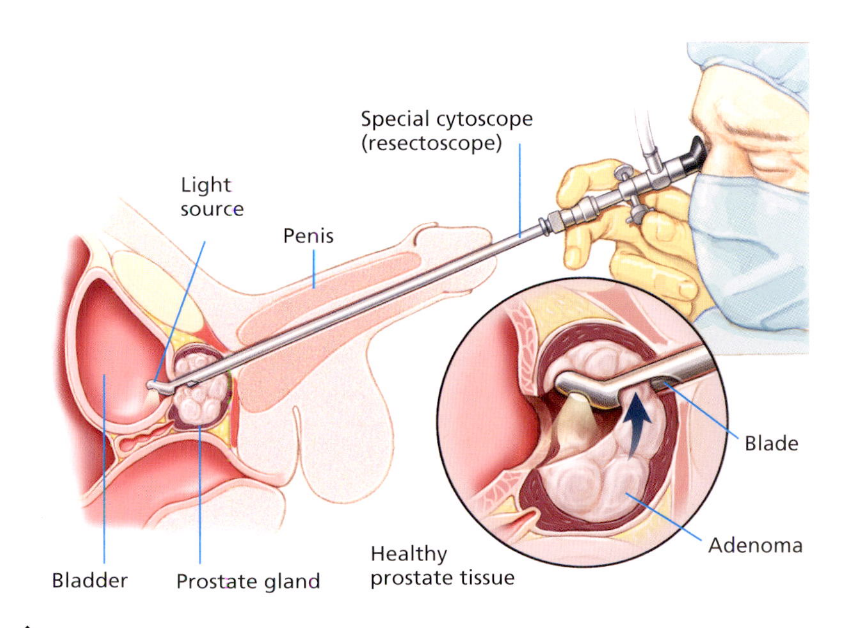

◆
Enlarged prostate
An enlarged prostate can be removed by means of a special instrument inserted via the urethra.

• ***Prostate cancer***

Malignant enlargement of the prostate gland. Symptoms characteristic of prostate enlargement, irrespective of whether benign or malignant, are the same (frequent urge to urinate, including at night, a thin urinary stream, urine dribbling, and continually reduced quantities of urine). Advanced prostate cancer can cause the presence of blood in the urine or seminal fluid.

A prostate tumor can almost always be operated on successfully if it is detected in time. The enlarged prostate can be felt through the anus. The size and shape of the gland is often also checked by means of an ultrasound examination. The level of PSA can be an indication of malignancy. Microscopic examination of the individual cells confirms the diagnosis.

Surgery offers the best prospects of recovery, especially in the case of smaller tumors. In this case the entire prostate gland, part of the ureter, the seminal vesicle situated behind the prostate, and the neighboring lymph nodes are removed. Hormone therapy also often produces good results, especially with larger tumors. This involves disabling the growth-stimulating influence of the male hormones on the prostate, either by administering female hormones (estrogen) or substances that suppress hormone production and effects. The removal of the testicles has the same effect.

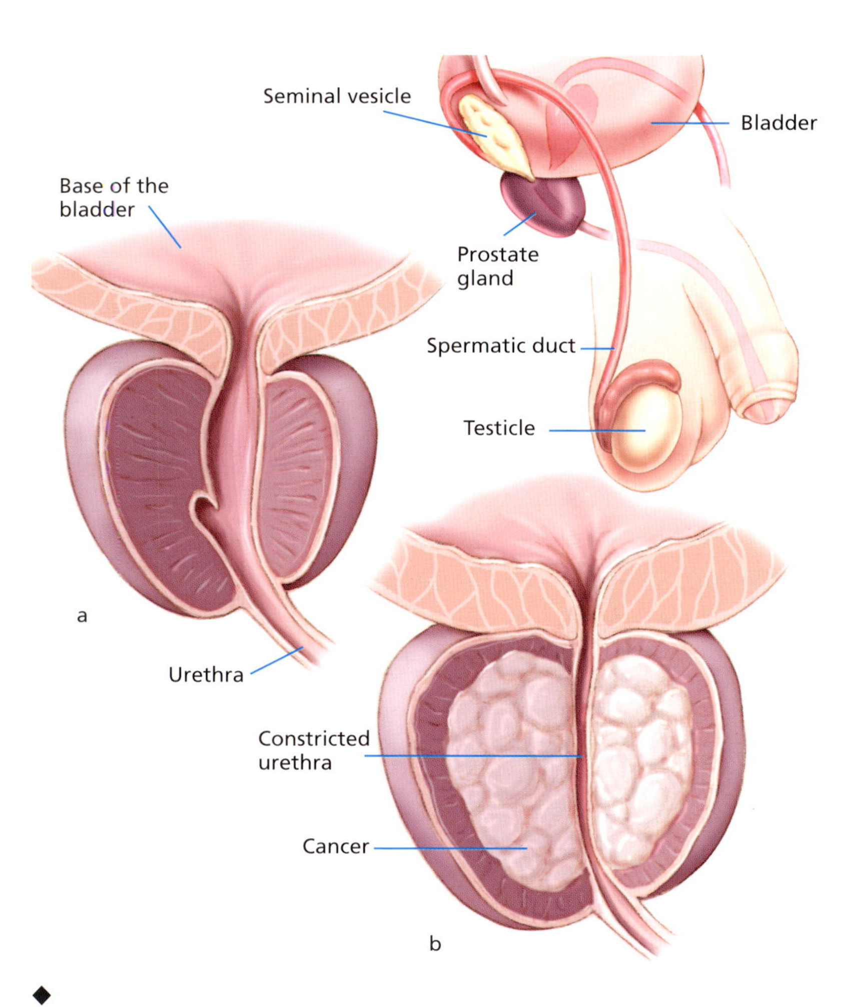

◆
Prostate cancer
The urethra runs unobstructed through the healthy prostate gland (a), into which the spermatic ducts also open. With a tumor the prostate becomes enlarged and strongly constricts the urethra, which can lead to problems when urinating (b).

• *Phimosis*

Congenital or acquired constriction of the foreskin of the penis. The distinction is made between full phimosis, where the foreskin cannot be retracted over the glans, and partial constriction of the foreskin. In the latter case the foreskin is only able to retract with difficulty when the penis is erect.

The acquired causes are inflammation or injury. If left untreated a constriction of the foreskin can lead to problems with urination and during sexual intercourse. Further complications can include constriction or inflammation of the glans. Constriction also means that adequate hygiene is impossible; residual secretions can promote the development of a penile carcinoma in the long term and the foreskin is therefore usually removed.

• *Orchitis*

Inflammation of a testicle with swelling and pain that can extend to the inguinal region and the lower back. The cause is usually an infection from other parts of the body spreading via the bloodstream. For example, around 25% of adult men who get mumps also develop testicular inflammation. It is usually treated with antibiotics. The most feared possible complication resulting from untreated orchitis is infertility.

• *Undescended testicle*

One, or sometimes both, testicles remaining in the abdominal or the inguinal cavity. The testicles usually migrate from the abdominal cavity through the inguinal cavity to the scrotum during embryonic development. In rare cases, the descent of the testicles fails to take place until after birth. Hormone therapy can induce the descent of the testicles in the majority of cases.
However, if the testicles are not fully descended by the age of three, this will need to be corrected surgically in order to avoid the risk of a loss of function or malignant change in the undescended testicle.

• *Testicular tumor*

Usually malignant, normally painless tumor on the testicle. Malignant testicular tumors are rare but do tend to occur in relatively young men. The recovery prospects are good if detected early on, so every man should have his testicles checked on a regular basis. A physician should be consulted in the event of the sudden enlargement of a testicle or changes to its surface structure, as well as a feeling or heaviness or dragging in the lower abdomen.

Diagnostic methods

Check-ups

The most important means for avoidance and/or early detection of serious diseases of either male or female sexual organs is attendance at regular check-ups. Check-ups for women usually comprise a physical examination of the lower abdomen and breasts, a smear test of the cervical opening and cervix, mammography of the breasts, and ultrasound examination of the lower abdomen if required. Women should also check their breasts and men their testicles for changes on a regular basis. Check-ups for men, are primarily focused on the prostate gland, which the doctor is able to feel rectally.

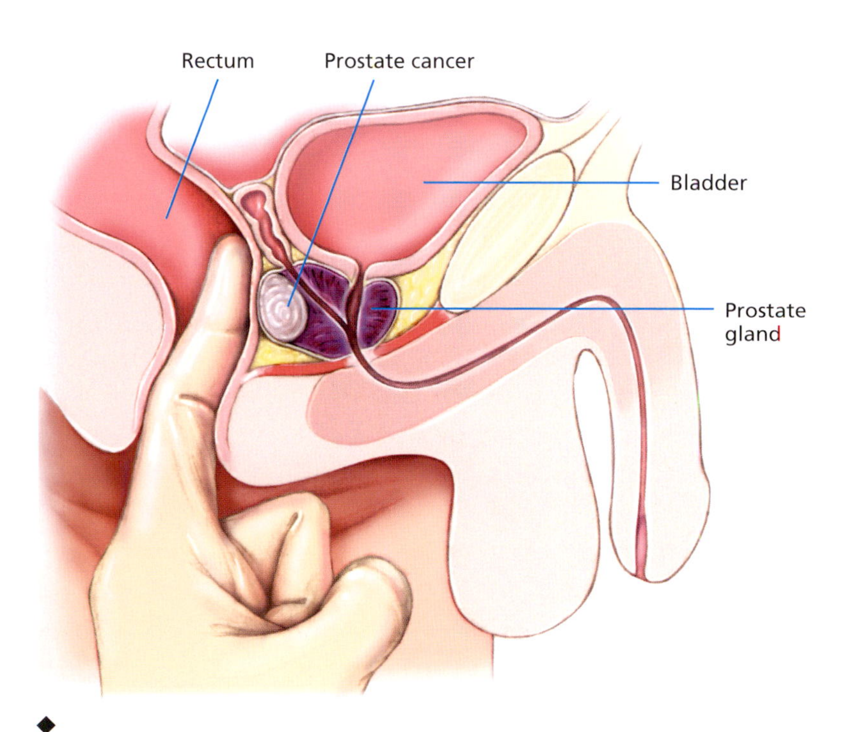

◆
Prostate check-up
An enlargement of the prostate can be felt relatively easily through the anus. This examination forms part of regular cancer screening.

Colposcopy

Examination of the vagina, its mucous membrane, and the uterus entrance with a magnifying glass known as a colposcope, a device with which the doctor is able to view the inside of the vagina at a magnification of 20–30 times. This enables early detection of changes to the vaginal mucous membrane and the entrance to the uterus. It may enable the pre- and early stages of cervical cancer to be detected and treated in good time.

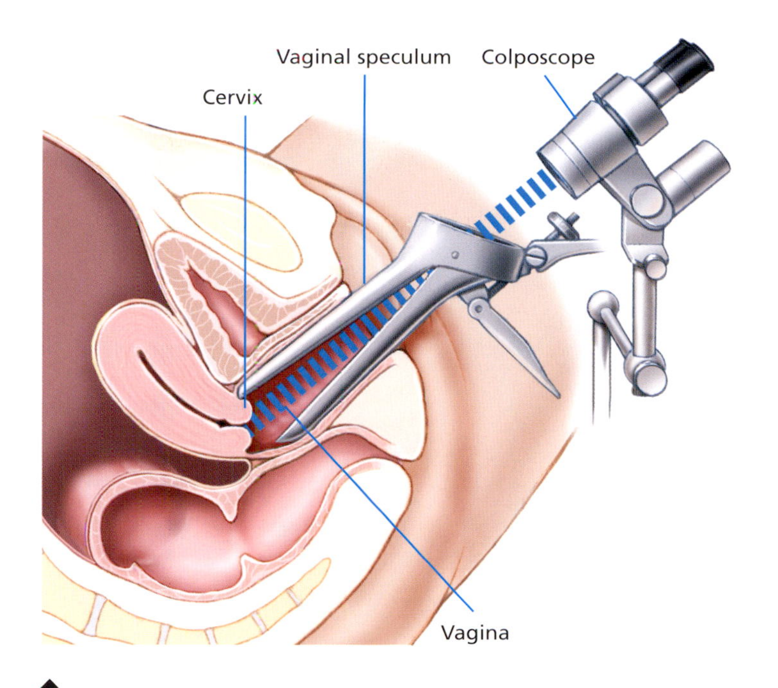

◆
Colposcopy
The examination of the mucous membrane of the vagina and the cervix with a colposcope is part of a cancer check-up.

Cervical smear

The removal of mucous membrane cells from the cervix for examination as part of a cancer check-up. This is particularly recommended for early detection of cervical cancer and also to detect pathogens. The cells are taken from the entrance to the uterus and from the cervix and examined under a microscope. Changes can be detected early on as cancer cells are distinct from healthy cells.

Pregnancy check-ups

Every mother-to-be should attend pregnancy check-ups. These check-ups are usually monthly to begin with, increasing to every two weeks in the last months prior to the birth. At the first visit, the physician or gynecologist will note all of the diseases affecting the mother-to-be and her family and determine whether there are any additional risks for the pregnant woman and the baby (e.g., diabetes, hereditary conditions, previous multiple births or miscarriages). Gynecological and general examinations are carried out and the anticipated date of birth is calculated. The urine is examined at regular intervals to check for diabetes, while blood pressure and weight are also monitored. A blood test will establish the blood group and rhesus factor status. Ultrasound is used to monitor fetal development. Amniocentesis, or other less invasive tests such as chorionic villus sampling, or maternal marker screening, may be carried out if growth disturbances, deformities, or other risks to the baby are suspected.

The cell

The cell is the smallest structural and functional unit of the organism. All animals are "metazoans," i.e., they consist of a huge number of individual cells, all of which are specialized to a greater or lesser extent to enable them to carry out a variety of different tasks and thereby enable the organism to survive. The cells may be permanently linked to one another (e.g., in organs), but can also be floating free in a fluid (e.g., blood cells). The form and size of individual cells can vary greatly, due to their different tasks, but their basic structure (except for the red blood cells) is always the same. They consist of a cell body with a ground substance (cell fluid or cytoplasm); a cell nucleus; various small cellular organs or organelles in the cytoplasm, such as mitochondria and endoplasmic reticulum; and a cell wall, a membrane that surrounds the cell.

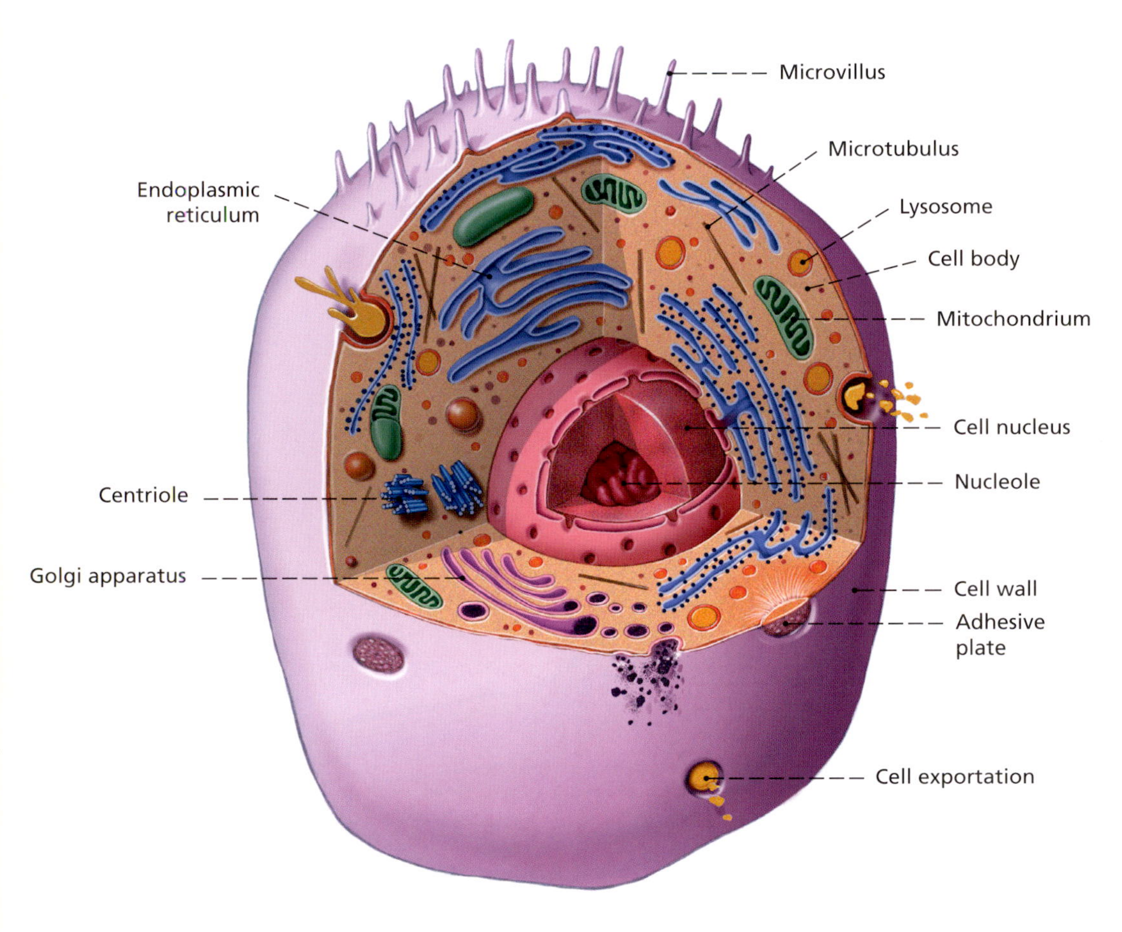

◆
Fig. 439
Cells all have a similar basic structure, with one exception, red blood cells (erythrocytes). Depending on their purpose, they may be of different sizes. The diameter of the smallest cells is no more than two thousandths of a millimeter.

Because of the numerous tasks each cell has to carry out, they are well organized, each component fulfilling a specific requirement:

- Microvilli, are small cell evaginations, which are responsible for absorbing nutrients from the surrounding environment.
- Adhesive plates link the cells to one another.
- The endoplasmic reticulum is a branched cavity system bounded by a membrane; they are responsible for the transportation of materials and fluids into the cell.
- The Golgi apparatus ejects waste matter from the cell.
- Lysosomes have the job of digesting any impurities absorbed.
- Microtubuli are the cell's messengers. They pass on information from the cell surface to the cell nucleus.
- Mytochondria generate the energy required for the cell metabolism and are therefore also known as the cell's "power house".
- Nucleoles and centrioles play an important part in cell division.
- The cell nucleus contains the cell's genetic information, organized into chromosomes, and also plays a decisive part in the control of the cell metabolism.

Glossary

A

Abdomen Anterior segment of trunk between thorax and pelvis, consisting of abdominal wall, abdominal cavity and abdominal organs.
Abdominal Relating to the abdomen, e.g., abdominal cavity.
Abdominal wall The wall of the abdominal cavity, partially covered by peritoneum. The front and side sections are formed by the diagonal abdominal muscles; the rear section is formed by the spinal column and muscles attached to it; the upper limit is formed by the diaphragm, and the lower limit by the pelvic floor and wings of the ilium with muscles attached to them.
Abducens 6th cranial nerve. Innervates lateral rectus muscle of the eye.
Abduction Lateral movement of part of body away from the vertical axis of the body front to back.
Abductor Muscle causing abduction, e.g., for raising the arm outwards.
Abscess Encapsulated gathering of pus, resulting from tissue necrosis due to toxins (e.g., bacterial).
Accessory nerve 11th cranial nerve. Controls the muscles of the leg.
Achilles tendon reflex Flexion of foot due to contracture of calf musculature following tapping of Achilles tendon previously passively tensed.
Adduction Bringing of part of body nearer to vertical axis of the body front to back.
Adductor Muscle causing adduction, e.g., for lowering the arm from horizontal to the side of the body.
Adenoid Pharnygeal tonsil. *See* Tonsil.
Adrenal glands A pair of endocrine glands located on the upper poles of the kidneys. They are each divided into an adrenal medulla and an outer adrenal cortex. Hormones are formed in both parts of the organ, epinephrin (adrenalin) and norepinephrin (noradrenalin) in the adrenal medulla, while hormones deriving from cholesterol, such as cortisone and a number of sexual hormones, are produced in the adrenal cortex.
Afferent Going towards, e.g., afferent nerve conducts impulses to the central nervous system.
Afterbirth The placenta, which is discharged together with the fetal membranes and the umbilical cord after the birth of a child.
Air vesicles *see* Alveoli.
Airways Any part of the respiratory tract through which air passes during breathing (pharynx, larynx, trachea, bronchi and bronchioles).
Alveolus A small cavity or sac-like dilatation, e.g., dental alveolus: tooth socket; pulmonary alveolus: thin-walled air sac.
Amnion Thin, avascular inner fetal membrane, which secretes amniotic fluid; part of the amniotic sac.
Amniotic fluid Fluid formed by amnion at beginning of pregnancy and subsequently by fetus, which fills up amniotic sac. By the end of pregnancy, about 1,100 milliliters.
Amniotic sac Envelope around embryo or fetus, containing amniotic fluid. It is formed from the amnion, the innermost of three birth membranes, and bursts during the birth process.
Amygdaloid body An almond-shaped conglomeration of nerve cells in the temporal lobe of the cerebrum, the amygdaloid body belongs to the basal ganglia and is functionally a part of the limbic system.
Anal Relating to the anus.
Anatomy Science of the structure of the body. Can be subdivided into functional, systematic, topographic and microscopic anatomy.
Aneurism Circumscribed, usually asymmetric, abnormal bulge in wall of arterial blood vessel or cardiac chamber, due to a weakening in the wall.
Ankle bone Talus, the first tarsal bone.
Ankle joint Hinged joint between the bones of the lower leg (tibia, fibula) and the ankle bone (talus). The talus articulates, in turn, with the heel bone (calcaneus) and the navicular bone of the foot.
Anterior Forward or front.
Anterior horn The anterior or frontal section of the gray matter in the spinal cord (as it appears in cross-section), containing the motor nerve cells (spinal motor neurones), the processes of which extend into the striated musculature.
Antrum Cavity, hollow space.
Anus Lowest segment of rectum, delineated by anal ring, i.e., end of lower bowel, opening on perineum.
Anvil *see* Incus
Aorta Main artery going from left ventricle of heart, delivers blood to all tissues except the lungs.
Apical At tip (apex) of an organ
Apparatus System of structures and/or organs with common function, e.g., vestibular apparatus
Appendix An appendage. The term is most commonly used to refer to the vermiform appendix, which extends from the blind end of the cecum. *See* Vermiform appendix.
Arterial Relating to arteries.
Arteriole Smallest arterial blood vessel preceding the capillary.
Arterio-venous Relating to both arteries and veins.
Artery A pulsating blood vessel with typical three-layer wall structure, which transports blood away from the heart to different parts of the body.
Articulation
1) A joint between bones.
2) The process of producing speech.
Atlas 1st cervical vertebra (which does not have a vertebral body).
Atrial Relating to atrium.
Atrio-ventricular node also referred to as AV node: Structure composed of special muscle fibres which is part of conduction system of heart. Situated near the orifice of the coronary sinus. Activated by the sinoatrial node. Transmits impulses to the ventricular muscles, causing contraction.
Atrium Vestibule or chamber of a hollow organ. The term is usually used as an abbreviation for the left and right atria of the heart.
Auditory nerve (audiovestibular nerve) 8th cranial nerve. Sensory nerve that conveys impulses deriving from the hearing process to the hearing center in the brain.
Auricle
1) Pinna of the ear, consisting of flexible cartilage and skin.
2) Pouch-like appendage projecting from the upper anterior portion of each atrium of the heart.
Autonomic nervous system The part of the nervous system that controls organ functions which are essential to life (vital functions such as breathing, blood circulion, metabolism, fluid balance). These are all regulated largely unconsciously and are almost impossible to influence voluntarily. The autonomic nervous system consists of two parts, the sympathetic and parasympathetic nervous systems, which differ in structure and often have opposing effects. While the sympathetic nervous system generally increases the performance of the body, the parasympathetic nervous system controls processes such as rest and digestion.
Axis 2nd cervical vertebra
Axon (neuraxon) Cylindrical, solitary cytoplasmic process of a nerve cell (neuron).bursa.

B

Back
1) The posterior aspect of a part of the body.
2) The posterior aspect of the trunk. The muscles of the back are divided into a deep layer along the spinal column (autochthonous back musculature) and a superficial layer. The autochthonous musculature of the back extends, stabilizes, and rotates the spinal column. The superficial muscles of the back are primarily responsible for movement of the shoulder blade and arm.
Bartholin's glands (greater vestibular glands) Paired mucous glands in vestibule of the vagina. Their secretions lubricate the vagina.
Basal ganglia Collections of neuron bodies, which lie deep within the white matter of each cerebral hemisphere. They serve as important links between various motor pathways of the central nervous system.
Belly button *see* Navel.
Biceps Abbreviation for musculus biceps brachii, the upper arm muscle, whose main function is to flex the forearm and arm.
Bile Fluid produced in hepatocytes containing water, electrolytes, and organic molecules, such as bile pigments and salts that help in the breakdown of fats.
Birth membranes Membranes which surround the unborn child and amniotic fluid. They have nourishment, excretory and protection functions for the embryo or fetus during pregnancy.
Bladder *see* Urinary bladder, Gall bladder.
Blood cells Collective name for erythrocytes (red blood cells), leucocytes (white blood cells), thrombocytes (blood platelets) and other cells in the blood.
Blood Fluid circulating in blood vessels, which is pumped through the organism by the heart. In adults, it makes up approximately 8% of body weight. Blood consist of a yellow fluid, plasma, in which are carried red blood cells (erythrocytes), white blood cells (leucocytes), and platelets (thrombocytes). Essential functions are transportation of matter to the tissues (e.g., oxygen, nutrients, messenger substances), transport of waste products to the excretory system, immune defense, heat regulation, wound closure through coagulation, and regulating the acid base balance of the body.
Blood pressure Pressure of blood flow through the arteries, specified in mmHg.
Blood–brain barrier Physiological restriction to substances in the blood entering nerve tissues in the brain.
Bone Dense, calcified, supporting structure forming the skeleton. Deposits of calcium salts make bones very robust. After their structural function, the central task of the bones is hematopoiesis, formation and development of blood cells. From the outside to the inside bones consist of periosteum, a hard bone cortex, and a soft framework of delicate trabecula, filled with bone marrow.
Bone marrow The substance between the spongy sections of a bone, also known as medulla ossium. Hematopoiesis takes place in the red bone marrow, which comprises reticulate connective tissue and stem cells. The yellow bone marrow largely consists of fat cells and connective tissue.
Boundary membrane Boundary layer between connecting tissue and non-connecting tissue structures, e.g., muscular fibers or epithelial tissues.
Bowels *see* Intestines.
Bowman's capsule Cup-like sac at the beginning of the tubular part of the renal nephron enclosing the glomerulus.
Brain Part of nervous system surrounded by bones of skull, which, together with spinal cord, forms the central nervous system. The entire brain is surrounded by the skull and cranial membranes. It has cavities (cerebral ventricles), which are filled with clear cerebrospinal fluid. Anatomically, brain is divided into the end brain (cerebrum, cerebral ventricles, basal ganglia, olfactory brain, and corpus callosum), the interbrain (pineal gland, thalamus, hypothalamus, and parts of pituitary), the mid-brain (part of brainstem) and the hind-brain (bridge, cerebellum, and extended cord). The gray matter consists mainly of neuron bodies and neuroglia cells, while white matter consists mainly of cord-rich nerve fibers. Nerve fibers connect the two cerebral hemispheres together, acting as commissures.
Brainstem Lowest part of the brain, beneath the cerebrum and cerebellum. In terms of development history the brainstem is the oldest part of the brain. It encompasses the midbrain, pons, and medulla oblongata. It is the source of most of the cranial nerves.
Breastbone *see* Sternum.
Bronchiole Small terminal branch of a bronchus.
Bronchus Part of the respiratory tract between the trachea and bronchioles. Bronchi are lined with a mucous membrane with ciliated epithelium, which fulfils important defensive functions (e.g., sweeping foreign matter back towards the throat so it can be coughed up); inhaled air is moistened and cleaned.
Bursa Pocket-shaped or bag-shaped body cavity, e.g., synovial bursa.

C

Café-au-lait spot Light brown spot on the skin, the color of milky coffee, which is irregularly but sharply delineated; it is usually harmless.
Calf (sura) The curved rear portion of the lower leg.
Callus
1) Fracture callus: initially connective tissue which later becomes compacted due to calcium deposits occurring in the case of secondary fracture healing and which bonds the broken ends.
2) Callosity: rough thickening of the corneous layer of the skin due to extensive mechanical wear.
Canalis Channel, groove, pipe. A small channel is a canaliculus. Example: C. vertebralis: Vertebral canal.
Cancer Malignant disease characterized by uncontrolled cell division. Cancer cells are able to destroy the surrounding tissue and can grow uncontrollably. They may form metastases in other parts of the body. In most cases, the cancer originates from a single abnormal cell.
Capillary The smallest blood vessels which connect arterioles to venules, and in which the exchange of gases, electrolytes, fluids, and nutrients takes place in tissues and organs.
Caput (head)
1) Part of the body which, among other things, contains the brain and sensory organs for vision, hearing, and smell.
2) Area of origin of a muscle or thickened end of a bone or organ. Example: C. mandibulae: head of lower jawbone (mandible).
Cardiac valves Flaps of specialized connective tissue lying between auricles and ventricles, as well as between ventricles and efferent vessels, which act to regulate blood flow, ensuring that it flows only in one direction.
Cardiovascular system Mass of all blood vessels and the heart.
Carpal joint A series of articulated joints between the carpal bones of the hand.
Carpus Wrist, *see* Hand.
Cartilage
1) A dense connective tissue composed of specialized cells called chondrocytes that produce an extracellular matrix composed of collagen fibers, ground substance, and elastin fibers. It has a low degree of metabolic activity, very few blood vessels and minimal regenerative capacity. A distinction is made between hyaline,

elastic, and fibrous cartilage (according to increasing fiber content), depending on the ratio between fibers and the cartilage matrix that surrounds the cells.
2) Term applied to cartilaginous organ and/or part of an organ, e.g., the cricoid cartilage on the larynx.
Caudal Towards the tail (coccyx) or posterior.
Cavernous bodies Spaces, which, depending on their function, may fill with blood so that the tissues become swollen.
Cavitas Cavity, pit, recess. Example: C. abdominalis: abdominal cavity.
Cavity A hollow occurring naturally in the body or as the result of disease, such as caries in the teeth.
Cecum Blind end of the large bowel at the junction with the small intestine.
Celom Embryonic cavity, which later forms the pleural, pericardial and peritoneal cavities.
Central nervous system (CNS) *see* Nervous system.
Cerebellar tracts Nerve fibers in the spinal cord that convey the impulses from the locomotor apparatus to the cerebellum.
Cerebellum Section of the brain at the back of the cranial fossa beneath the occipital lobe of the cerebrum. The cerebellum's main functions are control of balance and other complex motor functions. It comprises a vermicular middle section and two hemispheres with delicate furrows and twists.
Cerebrospinal fluid (CSF) A clear, colorless fluid, secreted in the cerebral ventricles of the brain, which surrounds the brain and the spinal cord. It both protects these sensitive tissues, and plays a part in their metabolism.
Cerebrum Largest section of the brain, the seat of reason, memory, consciousness and voluntary functional centers. Divided into two hemispheres.
Cervical opening The opening at the neck of the uterus where it joins the upper end of the vagina. The width of the cervical opening is a classification criterion for the phases of childbirth.
Cervix Commonly used as an abbreviation for the term Cervix uteri (neck of the uterus)
Chest cavity Thoracic cavity.
Chest Common term for Thorax.
Chiasma An area in which nerve fibers or tendons cross over. Most common example is the optic chiasma (C. opticum).
Choana Paired openings between the nasal cavity and the nasopharynx.
Chorionic sac Fetal membrane which completely surrounds the unborn child. The chorionic plate, which represents the infantile section of the after-birth (placenta), contains numerous extroversions (villi) for substance exchange between the maternal and infantile organisms.
Ciliated epithelium Epithelial cells with little hairs which have a transport function, e.g., to convey secretions in airways.
Cingulum
1) Collection of white matter fibers in the brain connecting components of the limbic system.
2) Lingual lobe of an anterior tooth.
3) Adjective meaning ring-shaped or girdle-shaped.
Circulation Movement of blood around the body, which conveys oxygen and nutrients to all somatic cells and removes metabolic products and carbon dioxide. Consist of two parts: pulmonary circulation (deoxygenated blood leaves the heart, collects oxygen in the lungs, and returns to the heart), and systemic circulation (oxygenated blood leaves the heart, runs through closed blood vessel system into tissues and organs of body, and returns to heart, by which time it is deoxygenated).
Clavicle Collar bone, *see also* Shoulder girdle.
Clitoris Female sexual organ at the front end of the small pudendal lip. Similarly to the male penis, it is able to become erect and has a wealth of sensitive nerve endings.
Coagulation system The mechanism whereby blood clots (thrombi) are formed, preventing blood loss following damage to a blood vessel.
Cochlea A small conical structure resembling a snail shell, the cochlea is the part of the inner ear that converts mechanical energy (vibrations) into nerve impulses that are sent to the brain. Also known as the organ of hearing.
Collateral vessels Blood vessels that branch off from a main vessel and then run alongside it to the same supply area. Should a closure of the main vessel take place, secondary collateral vessels can produce a functioning by-pass system (collateral circulation).
Colon *see also* Intestine. The large intestine, extending from the cecum to the rectum, and divided into the cecum, the ascending, transverse, descending and sigmoid sections of the colon, and the rectum.
Common carotid arteries Pair of arteries running along each side of the neck which supply the neck and head (including brain) with blood.
Compartment
1) Cell organelles, or parts thereof, in which specific biochemical reactions take place, separated from each other by membranes.
2) An enclosed space in the body, e.g., a canal for vessels or nerves.
Complement system A functional system of proteins found either in the blood fluid or on the surface of cells that serve the purposes of immune defense. Activities include the impairment of foreign cells or the attraction of defense and scavenger cells.
Concha
1) Part of external ear.
2) Alternative name for nasal turbinate.
Condyle A rounded prominence at the end of a bone, most often for articulation with another bone (knuckle).
Cone cell *see* Retina.
Connective tissue Basic type of body tissue, which consists of connective tissue cells and intercellular matter formed from the latter (albumen, carbon hydrate compounds, and various fibers).
Corn A thickened corneal layer of the skin.
Cornea A translucent, curved, vessel-free layer in the front section of the eyeball.
Coronary Relating to the coronary vessels.
Coronary vessels Blood vessels supplying the heart muscle with blood (coronary arteries) and removing oxygen-deficient blood from the tissues (coronary veins).
Corpus callosum System of transverse nerve fibers, which connect the two hemispheres of the brain cerebrum. The corpus callosum lies deep in the brain at the base of the longitudinal sulcus.
Corpus (body) Can be used to designate the human body, or the main part of an organ or structure.
Cortex Outer layer and/or structure of an organ; usually refers to the cerebral cortex (Cortex cerebri).
Corti, organ of Arrangement of sensory cells (hair cells) in the cochlea of the inner ear, which are used in hearing. The sensory cells are surrounded by supporting cells with fibers of the auditory nerve at their base.
Corticalis The compact outer layer of a bone.
Costa rib *see* Rib cage.
Costal pleura *see* Pleura.
Coxa *see* Hip.
Cranial Belonging to the head or lying towards the head. Opposite of caudal.
Cranial nerves Twelve pairs of nerve fiber bundles which innervate the head and neck area, as well as a large part of the internal and sensory organs. The cranial nerves are numbered according to the order in which they exit the brain and are cate-

gorized according to their function (sensory, motor, or mixed).
Cranium *see* Skull.
Crown Crown-shaped or coronal object. Example: C. dentis: tooth crown, visible part of tooth.
Crus Leg-like shape.
Cubitus *see* Elbow.

D

Dead space The part of the respiratory tract that plays no part in gaseous exchange, because it does not enter the alveoli.
Deltoid muscle Muscle of the pectoral girdle. It is wrapped around the shoulder joint like a scarf and raises the arm to the horizontal, turns it outwards or inwards, and pulls it backwards or forwards.
Dendrite The branched projections of a neuron that transfer electrical impulses to the cell body.
Dental alveolus (alveolus dentalis) Tooth socket. An indentation in the part of the bone of the upper and lower jaw in which the root of a tooth is firmly anchored. Teeth with more than one root have alveoli subdivided into the corresponding number of compartments.
Dermis *see* Skin.
Diaphragm (interseptum) Designation of a natural partition between parts of the body or organs. Usually refers to the thoracic diaphragm, a flat muscle forming a partition between the thoracic cavity and the abdominal cavity. In the middle, the diaphragm has a broad central tendon; it is penetrated at various points by blood vessels, the esophagus, and nerves. The diaphragm is the most important muscle involved in breathing.
Diaphysis Mid-section of a tubular bone. The diaphysis surrounds the medullary cavity.
Diencephalon Part of the brain functioning as a relay station between the cerebrum and the brainstem. The diencephalon consists of the thalamus, hypothalamus, pineal gland, and part of the hypophysis. It is closely linked to the limbic system and is also considered part of the brainstem.
Digestive organs Organs whose main task is to release nutrients from the food we eat, and to break down, absorb, and process these (e.g., excretion of waste products). They include the organs of the digestive tract and the various digestive glands.
Digit Individual finger or toe.
Diskus Disc-shaped structure. Example: D. articularis: joint disc, D. nervi optici: blind spot.
Distal Further away from the centre of the body. Opposite of proximal.
Diverticulum Congenital or acquired sac-shaped wall extroversion of a hollow organ. Diverticula occur most often in the large bowel, and less frequently in the esophagus, stomach, small bowel or urinary bladder.
Dorsal extension Backward movement of parts of body, also referred to as dorsal flexion (backward flexion).
Dorsal Relating to back and/or rear side (e.g., of an organ or part of body) or as a designation of direction. Opposite of ventral.
Dorsum *see* Back.
Duct Tubular passage; a small passage is referred to as a ductulus. Example: Ductulus biliferi: small bile duct.
Duodenum Part of the small intestine. *See* Intestine.
Dura Abbreviation for dura mater of the brain (D. mater encephali) or spinal cord (D. mater spinalis).

E

Ear One of a pair of hearing and equilibrium organs which register sound waves and changes to the position of the head, as well as transmitting signals to the brain for processing. They are divided into the outer ear (auricle) and outer auditory canal ending in the eardrum, middle ear (tympanic cavity with the auditory ossicles and the opening of the Eustachian tube, the connection to the pharynx) ending in the oval window, which connects to the inner ear (vestibule, semicircular canals, and cochlea).
Efferent To move away from, e.g., efferent nerve conveys stimuli away from central nervous system to periphery (muscles or internal organs). Opposite to afferent.
Elbow (cubitus) Region at transition between upper arm and forearm. Elbow comprises elbow bend and elbow joint.
Embryo Designation of the developing organism from conception until the end of the first two months of pregnancy.
Empyema Gathering of pus in a pre-formed cavity, e.g., pericardium, abdominal cavity, or paranasal sinuses.
Endocardium Smooth internal lining of heart cavities (including cardiac valves).
Endocrine gland Gland that releases hormones directly into the bloodstream. The most important are the pineal gland, the hypothalamus, the pituitary gland, the thyroid gland, the thymus, the adrenal gland, the Islets of Langerhans in the pancreas, the ovaries, and the testicles.
Endolymph Fluid in semicircular canals (labyrinth) of inner ear.
Endometrium Mucous membrane of uterus, consisting of simple ciliated epithelium.
Endothelium Single-layer assembly of flat cells, which line blood and lymph vessels, and body cavities.
Ependyma Fine cell layer of neuroglia cells, which lines subarachnoid spaces, i.e. cranial cavities, and central canal of spinal cord.
Epicardium Inner layer of pericardium which grows together with outer surface of cardiac muscle as outermost layer of cardiac wall.
Epidermis Outer layer of body's skin consisting, in upper layers, of horny flat epithelial cells (cutis).
Epididymis One of a pair of seminal ducts (epididymi) lying at the rear side of the testicle in which the sperm are stored. The epididymi gradually merge into the spermatic duct.
Epidural cavity Space between external dura mater of spinal cord and lining of spinal column, or cavity between dura mater of brain and periosteum of skull.
Epiglottis Plate made of flexible cartilage covering the laryngeal inlet during swallowing; open during respiration.
Epiphyseal plate An area of cartilage between epiphysis and metaphysis of tubular bone, from which lengthwise growth of bone commences; also referred to as growth plate. The bone can only grow in length if the epiphyseal plate is not closed, i.e., ossified. Ossification of the epiphysial plate is also termed closure of the epiphysis.
Epiphysis
1) *see* Pineal gland.
2) One of two ends of long tubular bone. The epiphysis is the head of the bone and is covered by cartilage.
Erythema Reddening of the skin as a result of increased skin blood flow, usually due to inflammation.
Esophagus (gullet) Muscular tube between the pharynx and the stomach through which food is transported from the mouth to the stomach. The esophageal wall contains a layer of muscle which helps push the food along, due to rhythmic contractions running from top to bottom (peristalsis).
Ethmoid bone (os ethmoidale) A lightweight bone in the cranium with a spongy, porous, structure, located between the orbital cavities of the eyes.
Exocrine gland Gland that excretes its secretion through a duct. Example: sweat glands.
Extensor Abbreviation for a muscle that causes extension. Antagonistic muscles of extensor are flexors.
Extracellular space Body space outside somatic cells occupied by fluid.

Extraperitoneal Outside the peritoneum, but located in the abdomen.
Extremity Limb branching off from trunk (arm or leg).
Eyelashes Short, curved hairs at the margin of the eyelid. Numbering 150-200, the eyelashes protect the eye from foreign bodies and the rays of the sun.

F

Facial nerve 7th cranial nerve. Mixed nerve with many ramifications, which predominantly controls facial musculature, as well as secretomotor innervation to salivary (except parotid) and lacrimal glands.
Fallopian tubes Paired tubes through which the unfertilized ovum is transported from the ovary to the uterus.
Fascia Soft tissue component of connective tissue surrounding muscles or internal organs.
Fascicle Bundle of nerve or muscle fibers.
Fatty tissue Tissue made up of fat cells, woven together to form fatty pads by reticular fibers. A special form of reticular connecting tissue.
Fetus Unborn child following completion of organ development (approximately 9th week of pregnancy to birth).
Fibula Bone in lower leg. *See* Lower leg.
Field of vision Area that can be seen without moving the eyes and head.
Fixing apparatus Mass of ligaments holding an organ in its position.
Follicle Vesicular structure. Example: F. pili: hair follicle.
Fontanelle Bone gaps on infant's skull, which are bridged over by soft connecting tissue. The fontanelles lie in areas in which three or more bone plates of skull meet; they close up within first three years of life.
Foot Lowest section of the leg. On standing, foot carries entire body load, and therefore has particularly compact bones, which are tensed by means of muscular traction and ligaments. The three main areas of the foot are: tarsus, metatarsus, and toes.
Foramen (pl foramina) Hole, aperture. Example: F. intervertebrale (intervertebral foramen): exit aperture for spinal nerves on side of cervical, thoracic or lumbar vertebra.
Forearm (antebrachium) The part of the arm that is connected to the upper arm by the elbow joint and extends as far as the wrist. It has two bones, the ulna and the radius. The muscles of the ulna may be divided into four groups: pronators and supinators for rotating the ulna, flexors for bending, and extensors for stretching the elbow joint.
Forehead Region of the face above the eyebrows. The bony base of the forehead is the frontal bone (os frontale) containing the frontal sinus.
Foreskin (prepuce) A fold of skin in the genital area. The foreskin covers the clitoris in the female and the head of the penis in the male.
Fossa Depression or hollow in bone, tissue or organs. Example: F. cranii (cranial fossa): bone pit of inner base of skull.
Fovea Round pit in bone, tissue or organs. Example: F. centralis: deepened point in yellow spot, a retinal area at rear pole of eyeball, is point at which vision is sharpest.
Frenulum (pl frenula) Small fold of tissue that secures or restricts the motion of a mobile organ. Example: F. linguae (labial frenulum): mucous membrane fold located in middle of underside of tongue, which connects tongue with floor of mouth.
Frontal brain Part of cerebrum formed by two frontal lobes.
Frontal sinus *see* Paranasal sinus.
Fundus Base or floor of hollow organ. Often used as an abbreviation for gastric fundus (F. ventriculi) or ocular fundus (F. oculi).
Funiculus Small, thin cord, usually nerve tract.

G

Gall bladder Thin-walled, pear-shaped muscular, membranous sac on intestinal surface of liver. Stores bile prior to secretion into the gastrointestinal system via the bile duct.
Ganglion Nerve node outside central nervous system, acting as a synapse between neurons emerging from central nervous system and neurons which run from ganglion to end organs.
Gastro-intestinal tract Collective designation for mouth, pharynx, esophagus, stomach, small bowel, large bowel, and rectum.
Genitalia Internal: *see* Gonads. External: in males, the penis; in females, the clitoris and vulva. The sex organs, in which egg cells (female) or sperm cells (male) are generated, along with sex hormones.
Gland Individual cell or collection of specialized epithelial cells that produce and secrete substances. Divided into endocrine (hormone glands, secrete directly into blood stream), exocrine (secrete through duct), serous (aqueous secretion), mucous (mucilaginous secretion), and mixed glands.
Glomerulus Small capillary bed area in renal cortex, surrounded by Bowman's capsule. Together, these form the renal corpuscle.
Glossopharyngeal nerve 9th cranial nerve. Motor innervation to parotid gland and stylopharyngeus muscle; receives taste sensation from posterior third of tongue.
Glottis A combination of the vocal folds and the area between the folds; structures in the larynx that contribute to voice formation.
Gonads A term for the ovaries and the testicles which produce the gametes (egg cells, sperm), sexual hormones and various secretions.
Gooseflesh Skin change in which smooth muscles of hair follicles draw together through reflex action, hairs become rather erect, and small papuloid protrusions of skin arise. Gooseflesh is triggered by cold, or by psychological influences.
Gray matter Areas of brain and spinal cord, which appear gray in section and predominantly contain nerve cell bodies and neuroglia (as distinct from White matter).
Groin Triangular-shaped region of the abdomen at the junction with the thigh, left and right of the pubic region.
Gullet *see* Esophagus.
Gyrus A ridge or convolution on the surface of the cerebral cortex.

H

Hair cell Sensory cell with fine little hairs at free end; important for hearing.
Hair Thread-like surface extension made from keratinized cells, consisting of hair shaft and hair root. Hair papilla, a cone of connecting tissue rich in blood vessels, provides growing hair with nutrients. Inner root sheaths (protrusions of epidermis) are surrounded by a root sheath made from connecting tissue. The whole structure is called the hair follicle.
Hammer *see* Malleus.
Hand Section of arm connected to forearm. Consists of wrist, metacarpus, and fingers.
Hearing center The part of the cerebral cortex responsible for the processing of acoustic signals.
Heart Hollow, muscular organ. Nerve pulses are converted into a rhythmic sequences of muscle contractions and relaxations, by means of which the blood is pumped round the body. A healthy heart is the size of a fist and is divided into four chambers, two atria and two ventricles, by the cardiac septum and connective tissue.
Heel bone (calcaneus) Rearmost tarsal bone. Heel bone has a protuberance to which Achilles tendon is attached.

Hematocrit Designation of percentage volume fraction of solid blood constituents (96% of which are erythrocytes) in the overall blood volume. Normal values are around 0.41–0.51 (for men) and 0.36–0.46 (for women).
Hemisphere Designation of half of the cerebellum or cerebrum. Hemispheres of cerebellum are separated by vermicular mid-section of cerebellum; hemispheres of cerebrum are connected by corpus callosum.
Hemoglobin Red blood pigment containing iron, mainly found in red corpuscles (erythrocytes), where its main tasks are to transport oxygen and carbon dioxide.
Henlé, loop of Section of tubular apparatus of nephrons in kidney, primarily responsible for re-absorption of water and electrolytes.
Hilum of lung Depression on the inner side of a lung facing the heart, point of entry of the pulmonary artery, pulmonary veins, main bronchi and the lymphatic vessels into the lung.
Hip joint The joint between the pelvis and the femur, comprising the iliac bone, the ischium and the pubic bone.
Hippocampus Anatomically part of the cerebrum, functionally part of the limbic system.
Horn cells Flat cells in the epidermis which are entirely filled with the protein keratin and therefore have a horn-like structure.
Humerus Upper arm bone.
Hymen Fold of mucous membrane largely closing the entrance to the vagina in young girls.
Hyoid bone (os hyoideum or lingual bone) Bone between the lower jaw (mandible) and the larynx. The lingual bone is the only bone in the body that does not directly join, or form part of a joint with, another bone.
Hypoglossal nerve 12th cranial nerve. Provides motor innervation to the tongue and other muscles.
Hypothalamus The lowest section of the diencephalon, the hypothalamus controls physical and emotional processes as well as acting as a link between the nervous and hormonal systems. The hypothalamus regulates body temperature, water content, and hormone levels, as well as coordinating the sleep—wake cycle and circulatory functions.

I

Ileum Part of the small intestine, *see* Intestine.
Immune system All of the tissues, cells, and cell products whose main task is to eliminate foreign substances (antigens) from the body. The immune system is divided into non-specific and specific defense. Non-specific defense includes natural barrier mechanisms, white blood cells, natural killer cells, enzymes (e.g., lysozyme in tear fluid), the complement system, and protein substances directed against pathogens (e.g., cytokines). Specific defense comprises lymphocytes (T and B cells) and the antibodies formed by the plasma cells, a subspecies of the B cells.
Impression An indentation in organs or other body parts.
in situ In place, in the natural position in the body. For example: a carcinoma in situ is a cancer tumor that is limited to its place of origin.
in vitro Taking place outside of the living organism.
in vivo Taking place in the living organism.
Incus (anvil) One of the three auditory ossicles, loosely connected to the other two (malleus and stapes) bones.
Inguinal canal (canalis inguinalis) Tube-shaped connection between the abdominal cavity and the outer pubic region. In the male the spermatic cord runs through the inguinal canal en route from the testicle to the prostate gland. In the female the inguinal canal contains a suspension ligament for the uterus.
Inner ear (labyrinth) The part of the ear located in the petrous bone of the cranium. The inner ear comprises the cochlea, the vestibule, and the three perpendicular semicircular canals. Anatomically, the inner ear comprises two parts: the bony labyrinth filled with perilymph and the membraneous labyrinth filled with endolymph situated between the canals of the bony labyrinth.
Innervation The nerve supply to an organ, region of the body, or tissue.
Intercostal Between the ribs.
Interstitium Space between the specific cells of an organ (parenchyma). The interstitium contains connective tissue and brings nerves and vessels to the organ's parenchymal cells. Hence, interstitial fluid and interstitial tissue.
Intervetebral cartilage (disk) Flexible disc between two vertebrae. It has a viscous gelatinous core in its center, which is surrounded by a ring made up of fibrous cartilage and connecting tissue. The inter-vertebral discs increase the mobility of the spine and cushion impacts.
Intestinal Relating to or deriving from the intestine.
Intestine Tubular digestive tract between pylorus and anus. Divided into small intestine (duodenum, jejunum, and ileum), large intestine (cecum and colon), and rectum.
Intima Abbreviated term for the Tunica intima, the innermost layer of an artery, vein or lymphatic vessel wall.
Intra-articular ligament Ligament within a joint, e.g., the knee joint or the ankle.
Intra-ocular fluid Clear fluid filling the anterior and posterior ocular chambers. The composition of the intra-ocular fluid is similar to that of the cerebrospinal fluid and serves to maintain the shape of the eyeball as well supplying the cornea and the lens.
Iris The colored part of the eye.
Islets of Langerhans Cells in the pancreas, which produce insulin and glucagon.

J

Jaws Bones in the cranium supporting the teeth, divided into upper and lower jaw.
Jejunum Part of the small intestine, *see* Intestine.
Joint Connecting point between bones. Synovial joints, which are usually more mobile, with joint cavities (diarthroses), are differentiated from fixed joints, without joint cavities (synarthroses). A synovial joint has a head, socket, joint cavity with synovial fluid, and a joint capsule. Stabilizing ligaments are often juxtaposed with the joint capsule, and there can be synovial bursae at mechanically stressed points. Types of joints include ball and socket joints (e.g., hip), saddle joints (e.g., thumb root), and hinge joints (e.g., elbow).

K

Kidneys A pair of bean-shaped urinary tract organs. The kidneys are the most important organs for the regulation of water content, electrolyte metabolism, and osmotic pressure. The main functional unit is the nephron, comprising renal corpuscules and the tubule apparatus, and it is here that urine is formed.
Knee joint Hinge joint between the lower end of the femur and the top of the shinbone (tibia). The key components of the knee joint are the menisci (crescent-shaped articulated discs made of fibrous cartilage) the ligaments, such as the cruciate ligaments, and the inner and outer collateral ligaments.
Kneecap *see* Patella.

L

Labia Lips.
Labial Relating or belonging to the lips, towards the lips.
Labrum Lip-shaped formation, e.g., the fibrous cartilage ring

forming the socket on an acetabulum such as that of the hip or shoulder joint.
Labyrinth *see* Inner ear.
Lacrimal apparatus Collective term for the lacrimal glands and tear ducts. The lacrimal apparatus is one of the protective mechanisms of the eye. Tears are secreted into the eye and collect in the inner corner, from where they are channeled via the lacrimal canaliculi, lacrimal sac, and the nasolacrimal duct into the nasal sinus.
Language center A general term for the regions of the cerebral cortex responsible for speech and understanding of written or spoken language. Broca's motor speech center is responsible for coordination of the muscles producing speech; the sensory speech center (Wernicke's speech area) for recall of the spoken word; and the optic speech center for the recognition and comprehension of the written word.
Lanugo Hair of unborn child, formed from fourth month of pregnancy onwards, and initially covering entire body. Shed and replaced by vellus hair at about 40 weeks gestation.
Large bowel *see* Intestines.
Larynx Organ containing the vocal cords located at the entrance to the trachea, largely responsible for sound production as well as providing a reversible closure for the lower air passages. It consists of a tube-shaped cartilage casing, which can be moved by the adjoining muscles, the inside of which is covered with mucous membrane.
Lateral Proceeding away from the middle, situated on the side. Opposite to medial.
Lateral funiculus
1) Nerve pathways in the lateral region of the white matter of the spinal cord.
2) Lymphatic tissue in the lateral wall of the pharynx.
Lateral incisor Second incisor next to the central incisor each side of the upper and lower jaws.
Lens Transparent, vessel-free body with two convex interfaces, situated in front of the vitreous body of the eye.
Ligament Usually fibrous connective tissue band connecting one bone to another, to provide greater stability, especially at the joints (e.g., cruciate ligaments of the knee). Peritoneal folds, attaching internal organs to the abdominal wall are also called ligaments.
Limbic system Functional unit comprising structures from different parts of the brain, thought to control emotional and behavioral patterns. In terms of development history, the limbic system is one of the older parts of the brain. It encompasses the cortex regions (limbic cortex) and the diencephalon and midbrain regions, and surrounds the core regions of the brainstem and the corpus callosum like a border (limbus).
Lingual bone *see* Hyoid bone.
Lips Structures forming the front boundary of the mouth. The lips are shaped by the orbicularis oris muscle, covered by a layer of skin on the outside and mucous membrane on the inside.
Liver Organ with diverse metabolic functions. The liver is the largest gland in the human body and weighs 1.5–2 kilograms. It is where bile, a number of coagulation factors, and glycogen (a reserve of glucose) are formed. The liver also performs important detoxification functions with regard to medicines, ammonia (from the break down of proteins), and alcohol.
Liver spot (lentigo) A brown patch derived from the proliferation of pigment-forming cells on the surface of the skin (melanocytes).
Locomotor system (musculoskeletal system) Bones and joints with ligaments, tendons, and skeletal muscles attached to them, responsible for active movement.
Lower jaw *see* Mandible.
Lower leg (crus) The part of the leg between the knee and the upper ankle. The lower leg has two bones, the shinbone (tibia) on the inside and the calf bone (fibula) on the outside. Its muscles can be divided into the following groups: the extensor group (for bending the dorsum of the foot up in the direction of the shin), the peroneal group (for lifting the outer edge of the foot), the superficial flexor group (for lifting the inner edge of the foot) and the deep flexor group (for plantar flexion of the foot and toes).
Lumbar vertebrae *see* Spinal column.
Lumbosacral joint Joint between the 5th lumbar vertebrae and the sacrum.
Lungs Paired respiratory organs located in the thoracic cavity and linked to the outside air by the respiratory tract. The lungs are covered by the inner layer of the pleura and adjoin the rib cage, diaphragm, and mediastinal space. The pulmonary circulation vessels branch out into the smallest blood capillaries which cover the pulmonary alveoli, where oxygen and carbon dioxide are exchanged.
Lymph A clear fluid that circulates around the body tissues, containing a high number of lymphocytes (white blood cells). Plasma leaks out of the capillaries to surround and bathe the body tissues. This then drains into the lymph vessels. The lymph transports lymphocytes, fats and proteins through the lymphatic system to the biggest lymph vessel, the thoracic duct. The thoracic duct empties back into the blood circulation.
Lymph nodes Lymphatic organs, varying in size, which absorb lymph from the small lymph vessels, cleanse it and pass it on to the larger vessels. Lymphocytes attack any bacteria or viruses they find in the lymph as it flows through the lymph nodes. The differentiation of lymphocytes, which are formed in the bone marrow, also takes place in the lymph nodes. A lymph node has a medulla and a cortex, which are surrounded by connective tissue.
Lymphatic organs Organs containing lymphatic or lymphoid tissue, i.e., reticular connective tissue infiltrated with lymphocytes, involved in immune defense. They include lymph nodes, thymus, spleen, and the tonsils. Lymphatic tissues are also found in the gastrointestinal and respiratory systems.
Lymphatic system All organs and structures involved with draining the lymph out of the tissue into the venous system. The small lymph vessels start off blind and combine to form ever-larger vessels which empty into the blood after passing the lymph nodes, which act as filter stations.

M

Macula (spot) A small area differing in appearance from the surrounding area, e.g.,
1) Blind spot in eye.
2) Spots on the skin due to change in pigment content, or venous filling.
Malleolus Bony protuberance on either side of the ankle.
Malleus (hammer) One of the three auditory ossicles, loosely connected to the other two (Incus and Stapes) bones.
Mammary glands Cutaneous glands in breasts of mammals, juxtaposed with pectoral muscle. In humans, develop further in females after puberty. In pregnancy, mammary gland becomes lactiferous gland.
Mandible (os mandibulare) A bone of the skull articulating by means of a bilateral joint with the rest of the skull. It consists of the U-shaped, curved jawbone (body of mandible) and two upturned extremities (ramus of mandible). At the upper end of the ramus and to the front is the coronoid process to which muscles are attached, primarily those enabling mastication, while to the rear is the articular process (head of the condyloid process of mandible), which forms the temporomandibular joint.
Meatus Passage or duct. Example: Meatus nasalis: one of the three nasal passages (superior, medius and inferior meatus nasalis), which open into the nasal cavity.
Medial Pertaining to the middle, in the middle of the body. Opposite of lateral.

Glossary

in the mouth is mixed, partly digested, and temporarily stored, before being passed on to the small intestine in portions.
Stratum Flat layer of cells, e.g., individual layers of the epidermis, the cornea, the retina, or the tendinous sheaths.
Stroma *see* Organ.
Subcortical Beneath the cortex.
Subcutaneous Lying or occurring beneath the skin.
Subdural Beneath the tough outer layer of the meninges (Dura mater) covering the brain.
Sulcus A groove formed as a result of anatomical factors such as the configuration of bones or organ segments, e.g., the grooves of the cerebrum such as the central sulcus (Sulcus centralis).
Supporting tissue Collective term for tissue that gives the body its shape.
Surfactant A substance produced in the lungs which is insoluble in water and which coats the inner surfaces of the alveoli with a film, thus reducing surface tension.
Suture
1) The term used by surgeons for a stitch.
2) A fixed union between two bones. During growth, the edges of the bones are connected by a thin layer of fibrous connective tissue. The most important sutures are those between the bones of the skull, the cranial sutures.
Sweat gland (glandula sudorifera) Skin glands, the secretions of which are known as perspiration or sweat. The tiny, coiled glands are distributed over practically all the body surface; the glands producing odoriferous secretions are found in areas covered with hair (e.g., the armpit, the pubic area) and around the nipple.
Sympathetic nervous system *see* Autonomic nervous system.
Symphysis Union between two bones consisting of fibrous cartilage and connective tissue, e.g., the pubic symphysis. *See* Synarthrosis.
Synapse The junction between the end of a neuraxon at which a stimulus is relayed to another nerve cell, a muscle cell (motor end-plate or myoceptor), or a gland cell.
Synarthrosis A fixed union between two bones consisting of cartilaginous tissue and/or taut connective tissue, which, in contrast with a genuine joint, does not possess an articular space. Synarthroses are also referred to as fibrous or cartilaginous joints and hold bones in place as rigidly as possible. Example: bone plates of the cranium.

T

Tactile sense The sense that receives mechanical stimuli (e.g., pressure, vibration).
Tarsus
1) The part of the foot with the seven tarsal bones of the instep.
2) Rigid, half-moon-shaped fibrous plates of tissue in the upper and lower eyelids.
Telencephalon Endbrain, part of the CNS.
Temporomandibular joint *see* Mandible
Tendon sheath *see* Tendon.
Tendon (sinew) A tough band of fibrous connective tissue that usually attaches muscle to bone and is capable of withstanding tension. The tendon is covered by a double-walled sheath which envelopes it at exposed points.
Tentorium (of the cerebellum) A wall of connective tissue formed from the tough Dura mater, covering the cerebellum and the posterior cranial fossa like a tent. The tentorium cerebelli has a circular opening, the tentorial notch, through which the brainstem passes.
Testicle (testis) One of paired male sexual organs, suspended elastically in the scrotum. The testicular canals in the testicles contain germinal epithelium, as well as clusters of Leydig cells, which produce the male sexual hormones (esp. testosterone).
Thalamus The largest group of cells or nuclei in the diencephalon of the brain. The thalamus gathers and filters internal or external information, e.g., sensory perceptions, feelings, and pain, and relays them to the cerebral cortex where they become part of conscious experience, as well as to the cerebellum, spinal cord and limbic system.
Thigh Upper part of the leg connecting the hip with the lower leg via the femur. The most important thigh muscles are the sartorial muscle (Musculus sartorius) and the quadriceps (Musculus quadriceps femoris) at the front, and the biceps muscle of the thigh (Musculus biceps femoris) at the back.
Thoracic Relating to the thorax (chest).
Thorax Part of trunk between neck and diaphragm. The thoracic cavity is surrounded by the thoracic wall and its protective layers and separated from the abdominal cavity by the diaphragm. The main organs it contains are the lungs, heart, thymus gland, and esophagus.
Throat *see* Pharynx.
Thumb Pollex. The first of five fingers. In contrast to the other four fingers, the thumb has only two sections. The thumb root joint is saddle-shaped, making it possible to oppose thumb to other fingers, of vital importance for gripping function of hand.
Thymus The primary lymphoid organ in children and adolescents, located in the upper part of the thoracic cavity, above the pericardium. The thymus is where T-lymphocytes and the hormones controlling the maturation of the cells in the lymph nodes are produced, and is thus responsible for immunological function. The thymus reduces in size and weight after puberty.
Thyroid (glandula thyroidea) A large, butterfly-shaped hormonal gland located directly beneath the thyroid cartilage and encircling the trachea. On its posterior wall are the parathyroid glands. The thyroid produces the hormones thyroxine and triiodothyronine, which are involved in many bodily functions, such as basic metabolism, body temperature, development of the skeletal musculature, and heart rate and pumping capacity.
Thyroid cartilage *see* Larynx.
Tibia Shinbone. *See also* Lower leg.
Tissue Group of cells and associated inter-cellular matter, which goes through similar development and fulfils a defined function. Four basic types can be differentiated: epithelial tissue, connecting and supporting tissue, muscle tissue, and nerve tissue.
Tongue (glossa) A muscle with a covering of mucosa in the oral cavity.
Tonsil Immune defense organs located at the back of the nasal and oral cavity. The tonsils are lymphatic tissues where, in the case of infections of the upper respiratory tract, lymphocytes come into contact with the components of invading pathogens. The most important tonsils are the palatine, pharyngeal (adenoid), and the lingual tonsils.
Tooth (dens) A bony structure designed to masticate food and forming part of the dentition. The crown of the tooth protrudes from the gum and is separated at the neck of the tooth from the root, which is buried in the bone of the jaw. From the center outwards, the tooth is divided into the dental pulp, dentine, cement (on roots), and enamel (on crowns).
Trachea A tubular air passage whose patency is maintained by rings of hyaline cartilage. The trachea is located in front of the esophagus; it begins below the larynx and ends with the branching of the two main bronchi.
Tragus Small projection of cartilage at the front of the entrance to the ear.
Triangularis Bone of wrist.
Trigeminal nerve The 5th cranial nerve, supplying most of the face.
Trochlear nerve The 4th cranial nerve innervates the superior oblique muscle of the eye.
Trunk The body without the head, neck and limbs.

Tubule A small tube, e.g., T. renalis: the smallest tubes (tubules) in the kidney.
Tunica A layer of tissue, e.g., T. intima: the innermost layer of the wall of an artery, vein or lymph vessel.
Tympanic cavity Air-filled bone cavity in the middle ear in which the three auditory ossicles are located.
Tympanic membrane (eardrum) A thin fibrous membrane forming the boundary between the outer and middle ear. The tympanic membrane relays acoustic vibrations to the bones of the middle ear (auditory ossicles).

U

Ulna (cubitus) *see* Forearm.
Ulnar nerve (nervus ulnaris) Large brachial nerve, originating from brachial plexus (plexus brachialis) innervates flexor muscles in forearm, as well as hand muscles and skin areas located near ulna.
Umbilical cord (funiculus umbilicalis) The cord connecting the abdominal wall of an unborn child with the mother's placenta. It is 50-60 centimeters long, almost always spiraled and contains two umbilical arteries from the child to the placenta and an umbilical vein which conveys blood gases and nutrients from the placenta to the child.
Unguis *see* Nail.
Upper abdomen Upper part of the abdominal cavity. It encompasses the liver, spleen, pancreas, stomach, the beginning of the small intestine, and the transverse section of the large intestine.
Upper arm (brachium) The bone in the upper arm (humerus) is connected to the shoulder girdle at the shoulder joint and to the lower arm at the elbow joint. The most important upper arm muscles are the lower arm flexors (biceps and brachials) at the front, and the triceps, the lower arm extensors, at the back.
Upper jaw (os maxillare) Maxilla. Pair of bones in the middle of the cerebral cranium forming part of the palate, the floor of the eye sockets, and the cheek bone, and carrying the teeth of the upper jaw. It encloses the two maxillary sinuses which form part of the paranasal sinuses.
Urachus A tubular canal in the embryo extending from the urinary bladder to the umbilicus, of which, all that normally remains at the time of birth is a fibrous cord.
Ureter Paired tubes, approx. 4 millimeters thick and 30 centimeters long, between the renal pelvis and urinary bladder.
Urethra Section of urinary tract that takes urine from the urinary bladder to the outside world. The male urethra is approx. 20-25 centimeters long and simultaneously acts as spermatic duct. By contrast, the female urethra is only 3-5 centimeters long.
Urinary bladder Distensible hollow muscular organ of the urinary system which acts as a reservoir for urine, with a capacity of approximately 500 milliliters.
Urinary duct *see* Ureter
Uterus (womb) Pear-shaped internal sex organ of female, into which Fallopian tubes emerge, and which opens into vagina. Site of implantation of fertilized egg and development of fetus.
Uvula Part of the soft palate.

V

Vagina
1) An elastic tube of muscle containing a large amount of connective tissue, which links the uterus with the external genitalia in women. Inside, the vagina is lined with many layers of tissue, the cells of which change with the menstrual cycle, and a few glands. The secretions of the vagina are produced by Bartholin's glands (vulvovaginal glands to the side of the vagina). Conditions inside the vagina are acidic (pH approx. 4.0), preventing pathogenic microbes ascending towards the uterus.
2) A sheath consisting largely of connective tissue and surrounding tendons, blood vessels or nerves, e.g., V. tendinis: tendon sheath.
Vagus Common term for the Nervus vagus, the 10th cranial nerve. Motor innervation to most laryngeal and pharyngeal muscles; parasympathetic innervation of most thoracic and abdominal viscera; transmits taste sensations from the epiglottis.
Valva A valve regulating the flow of liquid in the body, e.g., V. aortae: aortic valve.
Valvula Small, pocket-shaped valve or mucosal fold, e.g., V. venosa: venous valve.
Vas A small tubular duct conveying a body fluid such as blood, lymph, or sperm. Example: Vas deferens: the small duct that conveys sperm from the epididymis to the ejaculatory duct.
Vascular Relating to blood vessels.
Vasomotor function Causing dilatation and constriction of the blood vessels due to the action of nerve impulses on the musculature of the vascular wall.
Vegetative nervous system *see* Autonomic nervous system
Vein (vena) A thin-walled blood vessel carrying blood back to the heart. The venous system accommodates about two-thirds of the total blood volume.
Vellus Short fine hair covering most of the body.
Venous Relating to a vein; in relation to blood, low in oxygen. Opposite of arterial.
Venous system Collective term for the veins of the body.
Ventral Relating to the front or anterior of any structure. Opposite of dorsal.
Ventricle (ventriculus) A hollow cavity. Usually used to refer to the ventricles of the heart (V. cordis).
Ventricular septum Interventricular septum of the heart.
Vermiform appendix (commonly called the appendix) A slender, worm-shaped structure approximately 5-8 centimeters long, extending from the blind end of the cecum. Its mucosa are similar to those in the intestine, but the mucosal wall incorporates numerous lymph follicles, which serve to fight infection, particularly in childhood.
Vertebral body (vertebra) Individual building block from which the spinal column is built. The vertebral or spinal canal, through which the spinal cord passes, is bounded by the vertebral body, a thick, round bone, at the front, two vertebral pedicles at the sides, and the vertebral arch at the back. The vertebral arch has a number of bony processes: a spinous process pointing to the rear and downwards, a transverse process on the left and right, and an upper and lower articular process connecting the vertebra with the one above and below, respectively. The spinal nerves exit through lateral intervertebral foramina.
Vertebral column (columna vertebralis) Mobile skeletal structure (backbone) forming the vertical axis of the body and supporting it. The vertebral column surrounds and protects the spinal cord, supports the head and is the point of attachment for many muscles. It consists of seven cervical, 12 thoracic, five lumbar, five sacral, and four to five coccygeal vertebra (the latter are partially fused).
Vessel Flexible tube in which body fluid flows, i.e., lymph vessels or blood vessels.
Vestibular Relating to a vestibule.
Vestibule The space at the entrance of a canal, e.g., the oral vestibule or external oral cavity or the cavity of the osseous labyrinth of the inner ear.
Villi Protruberances on a superficial layer of cells, e.g., the intestinal mucosa (Villi intestinales). In the small intestine there are approximately four million villi, measuring about 1 mm in height; they are finger-shaped and serve to increase the inner surface area of the intestine.
Viscera Overall designation for organs lying within abdominal, chest, pelvic and cranial cavities.

Visceral cavity Body cavity, e.g., thoracic cavity or abdomen.
Visual center Region of the cerebrum, the destination of the optic nerve on its journey from the retina. Conscious perception of the visual information takes place in the primary visual center in the occipital lobe of the cerebral cortex. These images are processed further in the secondary visual center.
Visual system Parts of eye involved with vision: cornea, lens, iris, vitreous body, and aqueous humor.
Vitelline sac (umbilical vesicle) Sac containing fluid arising from fertilized egg, which guarantees initial provision of nutrients to fetus from the 9th embryonic day. This function is subsequently taken over by placenta.
Vocal chord (ligamentum vocale) Upper, free edge of the vocal fold.
Vulva A collective term for the various parts of the external female reproductive organs.

W

White matter (substantia alba) Bundles of myelinated nerve cell processes (neuraxons) connecting areas of gray matter (nerve cell bodies) in the brain to each other. White matter is responsible for the conduction of nerve impulses.
Windpipe *see* Trachea.
Wisdom tooth (dens sapientiae) The third molar, which is the last of the back teeth to erupt, usually after the age of 18 years.
Wrist Mobile connection between hand and forearm, consisting of condylar joint near forearm, and loose-jointed but almost immobile connections between individual carpal bones and between carpal and metacarpal bones.bursa.

Z

Zygomatic bone (os zygomaticum) Bone forming part of the cranium. The zygomatic bone forms part of the wall of the eye orbit. The processes of the zygomatic and temporal bones make up the zygomatic arch (cheek bone).

D

E

Picture credits

The pictures in this atlas are taken from the following books published by Urban & Schwarzenberg, Munich – Vienna – Baltimore:

A. Benninghoff: Anatomie, Ed. D. Drenckhahn/W. Zenker, Vol 1., 15th ed. 1994
A. Benninghoff/K. Goerttler: Makroskopische und mikroskopische Anatomie des Menschen, Vol. 2, 8th ed. 1967
R. Birkner: Das typische Röntgenbild des Skeletts, 3rd ed. 1996
H. Kremer/W. Dobrinski: Sonographische Diagnostik, 4th ed. 1994
G. Reiffenstuhl/W. Platzer/P. G. Knapstein: Die vaginalen Operationen, 2nd Ed. 1994
Roche Lexikon Medizin, 4th ed. 1999
J. Sobotta: Atlas der Anatomie des Menschen, Ed. R. Putz/R. Pabst, 2 Vols, 20th ed. 1993
J. Sobotta: Histologie, rev. by U. Welsch, 5th ed. 1997
J. Sobotta/H. Becher: Atlas der Anatomie des Menschen, Vol. 2, 16th ed. 1965
E.-J. Speckmann/W. Wittkowski: Bau und Funktionen des menschlichen Körpers, 18th ed. 1994
P. R. Wheater/H. G. Burkitt/V. G. Daniels: Funktionelle Histologie, 2nd ed. 1987
L. Wicke: Röntgen-Anatomie, 5th ed. 1995
and from:
Das Lexikon der Gesundheit, ADAC Verlag, Munich 1996